KINEMATIC DIFFERENTIAL GEOMETRY AND SADDLE SYNTHESIS OF LINKAGES

KINEMATIC DIFFERENTIAL GEOMETRY AND SADDLE SYNTHESIS OF LINKAGES

Delun Wang and Wei Wang

Dalian University of Technology, China

WILEY

This edition first published 2015

Registered office
John Wiley & Sons Singapore Pte. Ltd., I Fusionopolis Walk, #07-01 Solaris South Tower, Singapore 138628.

For details of our global editorial offices, for customer services and for information about how to apply for permission to reuse the copyright material in this book please see our website at www.wiley.com.

Library of Congress Cataloging-in-Publication Data applied for.

ISBN: 9781118255049

Typeset in 10/12pt TimesLTStd by Laserwords Private Limited, Chennai, India

1 2015

Contents

Preface

This book introduces the kinematic geometry of linkages in both analysis and synthesis, and builds up a theoretical system from planar, spherical, to spatial. The presentation differs from traditional ones in the approaches of both the differential geometry for kinematic geometry and the saddle point program for kinematic synthesis of linkages. Kinematic geometry provides the theoretical basis for the kinematic synthesis, both precise and approximated, of linkages by invariants.

The kinematic geometry of a rigid body, logically the combination of the kinematics of a rigid body and the geometry of graphs, tries to study the local geometrical properties of loci from the point of view of continuous motion along trajectories, while this continuous motion can certainly be visualized as the differential of the Frenet frame of the trajectories with respect to its arc length. Therefore, differential geometry, of course, may be the first choice in research on the kinematic geometry of a rigid body. However, the current research situation is unfortunately quite different, and this is one of the reasons for the authors writing this book.

There are currently many methods to study the kinematic geometry of a rigid body, such as geometry, algebra, screws, matrices, complex numbers, vectors, etc., each with their own merit in different application cases. In fact, this originates from the geometry by Burmester, which converts the displacement (or movement) of a lamina at several finite separated planar positions into a geometrical graph by means of corresponding poles of rotation. The algebraic equations are then built up to analyze the properties of geometrical graphs, which expand the object of research to all graphs of the lamina. For modern mathematics with expressions of vector algebra and invariants of geometrical graphs, it is difficult to identify them as belonging to the traditional geometry or algebra. For example, the differential geometry of curves and surfaces, both persists in geometrical significance and avoids the effects of external factors on geometrical graphs. In particular, a moving Frenet frame with three mutually orthogonal axes, or the natural trihedron of a curve or ruled surface, moving along the curve or surface is introduced to examine the intrinsic geometrical properties in differential geometry, whose derivatives can be viewed as the motion conversion for a rigid body at infinitesimally separated positions, just like the poles in finite separated positions, which is believed to be a powerful tool of kinematic geometry for a rigid body, in both planar and spatial motion. The kinematic geometry of a rigid body with multiple degrees of freedom is studied in multi-dimensional space. Of course, there is a natural extension from two or three dimensions to multiple dimensions while the classical differential geometry is developed into modern differential geometry, such as differential manifolds, Lie groups, and Lie algebras, although these are much more non-representational mathematical methods and the reader may have more difficulty understanding them.

The discrete kinematic geometry of a rigid body, naturally combining the discrete kinematics of a rigid body and the geometry of discrete graphs, studies the global geometrical properties of discrete trajectories, comprised of a series of discrete points or lines, which are globally compared with constraint curves or surfaces, while the differences between them or their errors have to be defined and estimated in terms of their invariants. Hence, the best uniform approximation in multi-dimensional space, or the saddle point programming approach, may be adopted first since it developed from one-dimensional space, or the interpolating approach of the Chebyshev polynomial originally, initially applied in the functional synthesis of linkages. The saddle point programming approach has been applied widely in geometrical error evaluations for manufacturing and measuring. However, the current objective function in the optimal synthesis of linkages for multiple positions, or the error evaluation method, is the least square structural error or the best square approximation, which intensively depends on the initial values and may be valid for the special cases but invalid for the general problems since the structural error is not uniformly defined and the design variables are redundant, other than the invariants in the approach of the saddle point program. This is another reasons for authors to write the book.

The book has seven chapters and two appendices in the order of planar, spherical, and spatial kinematic geometry of a rigid body and synthesis of linkages, so it is easy for readers to gain familiarity with the differential geometry and gradually build up the theoretical system. Also, for the reader's convenience, the required elemental knowledge of differential geometry is partly arranged in Chapter 1 for planar curves and Chapter 3 for space curves and surfaces. Chapters 1 and 2 describe the kinematic geometry and synthesis of planar linkages. Chapters 4 and 5 state the kinematic geometry and synthesis of spherical linkages, which is the bridge between the planar and spatial motion and a transition, even though it can be visualized as a special case of spatial motion. The kinematic geometry and synthesis of spatial linkages are respectively discussed in Chapters 6 and 7 in detail. In the appendices, the displacements of the spatial linkage RCCC are solved to provide the data for the numerical examples of kinematic geometry and synthesis of spatial linkages in the book.

Acknowledgments

Time goes so fast, and it has been over two decades since I first started research on the kinematic geometry of linkages as a PhD student under the guidance of Professor Dazhun Xiao and Professor Jian Liu (Dalian University of Technology). I can still remember the day when Professor Xiao told me honestly how arduous and challenging the study of the kinematic geometry of mechanisms was. I was deeply intrigued, and keen to discover the challenges and problems in the field of mechanisms. From that day on, I have been fully aware of the direction my research would take. Thanks to those unforgettable discussions with Professor Liu, I have gained many wonderful research ideas; he always supported and inspired me. I was also enlightened by many classical books, such as *Kinematic Geometry of Mechanisms* written by K.H. Hunt. This book would not have come into being if I had not been supported and encouraged by previous generations of scholars, both domestic and abroad, in the field of mechanisms, such as Professor Q.X. Zhang (Beihang University), Professor Y.L. Xiong (Huazhong University of Science and Technology), Professor H.M. Li (Harbin Institute of Technology), Professor S.X. Bai (Beijing University of Technology), Professor Z. Huang (Yanshan University), Professor T.L. Yang (Sinopec Jinling Petrochemical Co., Ltd), Professor H.J. Zou (Shanghai Jiao Tong University), Professor H.S. Yan (National Cheng Kung University), Professor C. Zhang (Tianjin University), Professor J.S. Dai (King's College London), Professor J.M. McCarthy (University of California, Irvine), Professor Kwun-Lon Ting (Tennessee Tech University), Professor J.Q. Ge (Stony Brook University), and so on. I should also mention the new generation of mechanism scholars in China, such as Professor T. Huang (Tianjin University), Professor F. Gao (Shanghai Jiao Tong University), Professor Z.Q. Deng (Harbin Institue of Technology), Professor Y.Q. Yu (Beijing University of Technology), Professor J. Xie (Southwest Jiaotong University), Professor X.L. Ding (Beihang University), Professor Y.H. Yang (Tianjin University), Professor S. Lin (Tongji University), Professor S.J. Li (Northeastern University), and Professor W.Z. Guo (Shanghai Jiao Tong University).

Many thanks go to my students and colleagues, because the book contains not just my own PhD dissertation but also those of three of my students – Mr. Wei Wang (co-author), Dr. Tao Li, and Dr. Shufen Wang – and eight Master's dissertations by Ms. Lihua Xiao, Mr. Jincang Zhou, Ms. Tianjian Li, Mr. Pengcheng Zheng, Mr. Baoying Zhang, Mr. Jianjun Zhang, Ms. Jie Chai, and Mr. Jinlei Li. Professor Huimin Dong, my classmate in graduate school, has been working with and helping me for 30 years. Professor Shudong Yu (Ryerson University), my classmate in college, Professor Yimin Tong, Professor Huili Wang, and Dr. Jin Qiu (Dalian University of Technology) have also given me enormous help in writing this book.

I want to express my gratitude to the National Natural Science Foundation of China (grants 59305033 and 59675003) for supporting my research into the kinematic geometry of mechanisms.

I would also like to thank the staff of John Wiley & Sons for their warm collaboration in presenting this book as accurately as possible in all its details.

1

Planar Kinematic Differential Geometry

Kinematics, a branch of dynamics, deals with displacements, velocities, accelerations, jerks, etc. of a system of bodies, without consideration of the forces that cause them, while kinematic geometry deals with displacements or changes in position of a particle, a lamina, or a rigid body without consideration of time and the way that the displacements are achieved. As a combination of kinematic geometry and differential geometry both in content and approach, kinematic differential geometry describes and studies the geometrical properties of displacements.

There are a number of articles and books on kinematic geometry. Pioneers such as Euler (1765), Savary (1830), Burmester (1876), Ball (1871), Bobillier (1880), and Müller (1892) established the theoretical foundation and developed the classical geometrical and algebraic approaches for studying kinematic geometry in two dimensions some hundred years ago. The classical geometric and algebraic approaches are still in use today. Differential geometry is favored by many researchers studying the geometrical properties of positions of a planar object, changes in its positions, and their relationships. Invariants, independent of coordinate systems, are introduced to describe the geometric properties concisely. Thanks to the moving Frenet frame for describing infinitesimally small variations of successive positions, the positional geometry can be naturally and conveniently connected to the time-independent differential movement of a planar object.

This chapter deals with the kinematic characteristics of a two-dimensional object (a point, a line) in a plane without consideration of time by means of differential geometry. Though abstract, the explanation is judiciously presented step by step for ease of understanding and will be a necessary foundation for studying the kinematic characteristics of a three-dimensional object by means of differential geometry in later chapters.

Kinematic Differential Geometry and Saddle Synthesis of Linkages, First Edition. Delun Wang and Wei Wang.

1.1 Plane Curves

1.1.1 Vector Curve

A plane curve Γ is represented in rectangular coordinates as

$$\begin{cases} x = x(t) \\ y = y(t) \end{cases} \tag{1.1}$$

where t is a parameter. The above equation can be rewritten in the following way by eliminating the parameter t:

$$y = F(x) \tag{1.2}$$

or in implicit form as

$$F(x, y) = 0 \tag{1.3}$$

In a fixed coordinate frame $\{\boldsymbol{O};\ \boldsymbol{i}, \boldsymbol{j}\}$, the vector equation of curve Γ can be written as

$$\Gamma : \boldsymbol{R} = x(t)\boldsymbol{i} + y(t)\boldsymbol{j} \tag{1.4}$$

or

$$\boldsymbol{R} = \boldsymbol{R}(t) \tag{1.5}$$

Obviously, both the magnitude and direction of $\boldsymbol{R}$ in equation (1.5) vary.

To describe a curve in the vector form, a real vector function, represented by a unit vector $\boldsymbol{e}_{\mathrm{I}(\varphi)}$ with an azimuthal angle φ with respect to axis $\boldsymbol{i}$, measured counterclockwise, is defined as a vector function of a unit circle (see Fig. 1.1). A plane curve Γ can be denoted by the following vector function:

$$\boldsymbol{R} = r(\varphi)\boldsymbol{e}_{\mathrm{I}(\varphi)} \tag{1.6}$$

In the above equation, the magnitude and direction of vector $\boldsymbol{R}$ depend on the scalar function $r(\varphi)$ and the vector function of a unit circle $\boldsymbol{e}_{\mathrm{I}(\varphi)}$.

Another vector function of a unit circle $\boldsymbol{e}_{\mathrm{II}(\varphi)} = \boldsymbol{e}_{\mathrm{I}(\varphi+\pi/2)}$ can be obtained by rotating $\boldsymbol{e}_{\mathrm{I}(\varphi)}$ counterclockwise about $\boldsymbol{k}$ by $\pi/2$ (in Chapters 1 and 2, $\boldsymbol{k}$ is the unit vector normal to the paper and directed toward the reader).

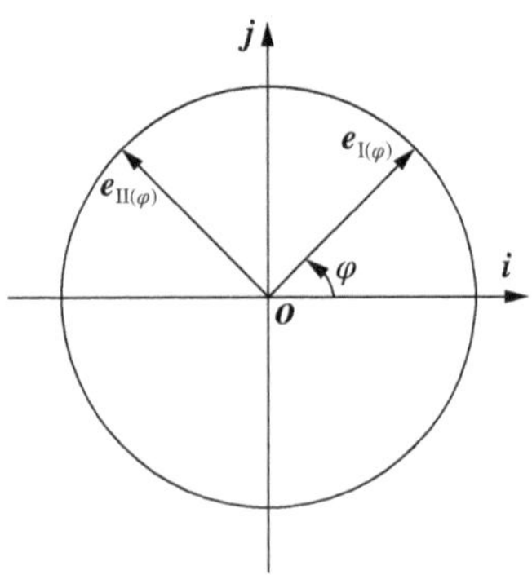

Figure 1.1 Vector function of a unit circle

The vector function of a unit circle has the following properties:

1. Expansion

$$\begin{cases} \boldsymbol{e}_{\mathrm{I}(\varphi)} = \cos\varphi \boldsymbol{i} + \sin\varphi \boldsymbol{j} \\ \boldsymbol{e}_{\mathrm{II}(\varphi)} = -\sin\varphi \boldsymbol{i} + \cos\varphi \boldsymbol{j} \end{cases} \tag{1.7}$$

2. Orthogonality

For a unit orthogonal right-handed coordinate system $\{\boldsymbol{O}; \boldsymbol{e}_{\mathrm{I}(\varphi)}, \boldsymbol{e}_{\mathrm{II}(\varphi)}, \boldsymbol{k}\}$ consisting of $\boldsymbol{e}_{\mathrm{I}(\varphi)}$, $\boldsymbol{e}_{\mathrm{II}(\varphi)}$, and $\boldsymbol{k}$, we have the following identities:

$$\boldsymbol{e}_{\mathrm{I}(\varphi)} \cdot \boldsymbol{e}_{\mathrm{II}(\varphi)} = 0, \quad \boldsymbol{e}_{\mathrm{I}(\varphi)} \times \boldsymbol{e}_{\mathrm{II}(\varphi)} = \boldsymbol{k} \tag{1.8}$$

3. Transformation

$$\begin{cases} \boldsymbol{e}_{\mathrm{I}(\theta+\varphi)} = \cos(\theta+\varphi)\boldsymbol{i} + \sin(\theta+\varphi)\boldsymbol{j} = \cos\theta \boldsymbol{e}_{\mathrm{I}(\varphi)} + \sin\theta \boldsymbol{e}_{\mathrm{II}(\varphi)} \\ \boldsymbol{e}_{\mathrm{II}(\theta+\varphi)} = -\sin(\theta+\varphi)\boldsymbol{i} + \cos(\theta+\varphi)\boldsymbol{j} = -\sin\theta \boldsymbol{e}_{\mathrm{I}(\varphi)} + \cos\theta \boldsymbol{e}_{\mathrm{II}(\varphi)} \end{cases} \tag{1.9}$$

4. Differentiation

$$\frac{\mathrm{d}\boldsymbol{e}_{\mathrm{I}(\varphi)}}{\mathrm{d}\varphi} = \boldsymbol{e}_{\mathrm{II}(\varphi)}, \quad \frac{\mathrm{d}\boldsymbol{e}_{\mathrm{II}(\varphi)}}{\mathrm{d}\varphi} = -\boldsymbol{e}_{\mathrm{I}(\varphi)} \tag{1.10}$$

The descriptive form of a curve depends on the chosen parameters and coordinates. A curve may have many descriptive forms, which differ in complexity if the parameters and reference coordinates are chosen differently. Below are three examples.

Example 1.1 A circle with radius r and center point C is shown in Fig. 1.2. Write its equation in both vector and parameter forms.

Solution

The parameter equation of a circle in rectangular coordinates $\{\boldsymbol{O}; \boldsymbol{i}, \boldsymbol{j}\}$ can be written as

$$\begin{cases} x = x_C + r\cos\varphi \\ y = y_C + r\sin\varphi \end{cases} \quad (0 \leq \varphi < 2\pi) \tag{E1-1.1}$$

where (x_C, y_C) are the coordinates of the center of the circle in the reference frame $\{\boldsymbol{O}; \boldsymbol{i}, \boldsymbol{j}\}$.

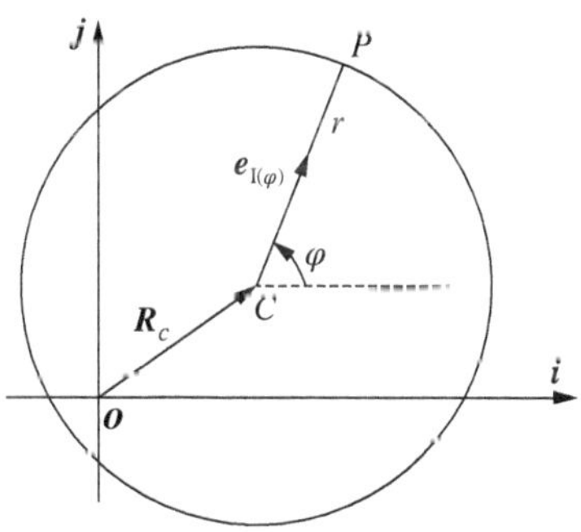

Figure 1.2 A circle

Alternatively, the same circle can be represented as a vector function of a unit circle:

$$\boldsymbol{R} = \boldsymbol{R}_C + r\boldsymbol{e}_{\mathrm{I}(\varphi)} \tag{E1-1.2}$$

Example 1.2 An involute is shown in Figs 1.3 and 1.4. Write its equation in both vector and parameter forms.

Solution

The equation of an involute can be written in three different forms using polar coordinates, rectangular coordinates, and a vector function of a unit circle, where r_b is the radius of the base circle.

1. Polar coordinates:

$$\begin{cases} r = \dfrac{r_b}{\cos\alpha} \\ \theta = \tan\alpha - \alpha \end{cases} \tag{E1-2.1}$$

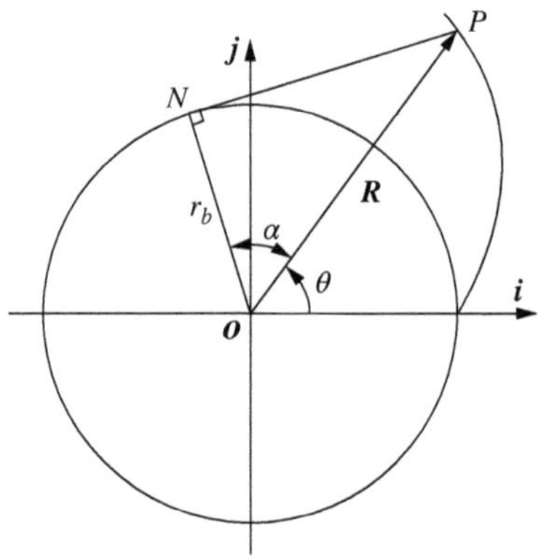

Figure 1.3 An involute

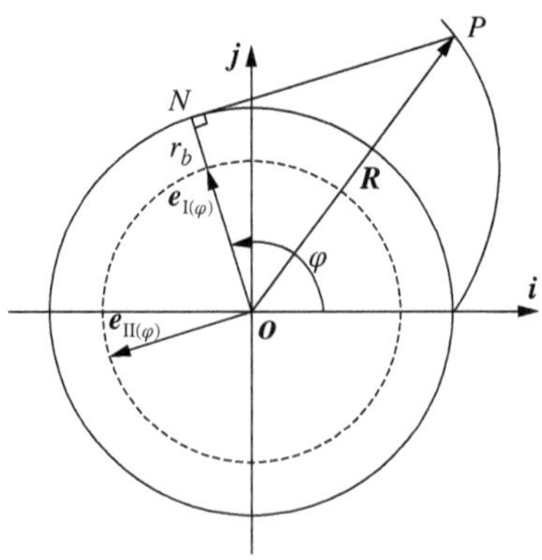

Figure 1.4 An involute with a unit circle vector function

2. Rectangular coordinates:

$$\begin{cases} x = r_b \cos\varphi + r_b\varphi \sin\varphi \\ y = r_b \sin\varphi - r_b\varphi \cos\varphi \end{cases} \tag{E1-2.2}$$

3. Vector function of a unit circle:

$$\boldsymbol{R} = r_b \boldsymbol{e}_{\mathrm{I}(\varphi)} - r_b\varphi \boldsymbol{e}_{\mathrm{II}(\varphi)} \tag{E1-2.3}$$

Example 1.3 A planar four-bar linkage is shown in Fig. 1.5. Write the equation of the coupler curve in both parameter and vector forms.

Solution

As shown in Fig. 1.5, links *BC* and *AB*, in a planar four-bar linkage *ABCD* with link lengths a_1, a_2, a_3, a_4, form an inclination angle γ and φ with respect to the fixed link. A moving rectangular coordinate system $\{\boldsymbol{B}; \boldsymbol{i}_m, \boldsymbol{j}_m\}$ attached to link *BC* and a fixed coordinate system $\{\boldsymbol{A}; \boldsymbol{i}_f, \boldsymbol{j}_f\}$ attached to the fixed link are established. Point *P* in the coupler with polar coordinates (r_P, θ_P) can be represented in the coordinate system $\{\boldsymbol{B}; \boldsymbol{i}_m, \boldsymbol{j}_m\}$ as

$$\begin{cases} x_m = r_P \cos\theta_P \\ y_m = r_P \sin\theta_P \end{cases} \tag{E1-3.1}$$

1. The parameter equation of coupler curves

A coupler curve traced by point *P* can also be expressed in the fixed frame $\{\boldsymbol{A}; \boldsymbol{i}_f, \boldsymbol{j}_f\}$ as

$$\begin{cases} x = r_P \cos(\theta_P + \gamma) + a_1 \cos\varphi \\ y = r_P \sin(\theta_P + \gamma) + a_1 \sin\varphi \end{cases} \tag{E1-3.2}$$

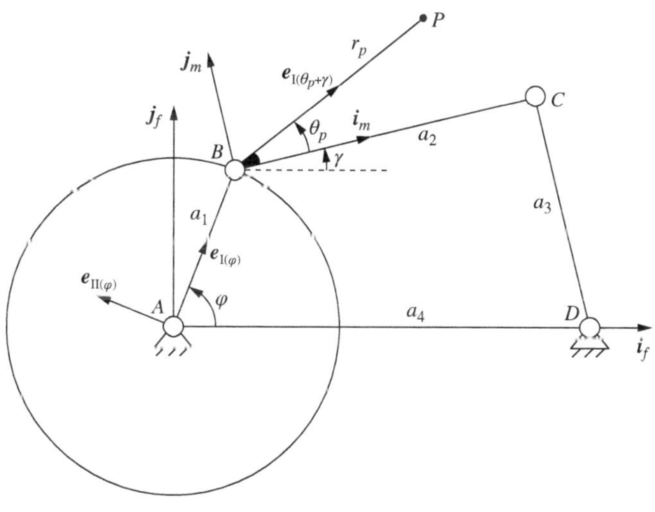

Figure 1.5 A planar four-bar linkage

A sextic algebraic equation can be deduced for a coupler curve if parameters γ and φ are replaced by function $\gamma = \gamma(\varphi)$ in the displacement solution of a four-bar linkage.

2. The vector equation of coupler curves

Link AB rotates about joint A of the fixed link AD, and link BC rotates about joint B of link AB. Since a circle can be expressed by a vector function of a unit circle, a coupler curve of a four-bar linkage can be written as

$$\boldsymbol{R}_P = a_1\boldsymbol{e}_{\mathrm{I}(\varphi)} + r_P\boldsymbol{e}_{\mathrm{I}(\theta_P+\gamma)} \tag{E1-3.3}$$

A point in link AB traces a circle vector $a_1\boldsymbol{e}_{\mathrm{I}(\varphi)}$ in the fixed frame $\{\boldsymbol{A};\boldsymbol{i}_f,\boldsymbol{j}_f\}$. A point in coupler link BC produces a circle vector $r_P\boldsymbol{e}_{\mathrm{I}(\theta_P+\gamma)}$ in the reference frame of link AB. The subscripts inside the brackets are independent variables. Here, we deal with the coupler point relative to the coordinate system by the vector function of a unit circle.

Based on the above three examples, we observe that the description of a plane curve in terms of a vector function of a unit circle is simpler than the traditional algebraic equation. Moreover, since a vector function of a unit circle has intrinsic properties, its successive derivatives with respect to the chosen parameters can be conveniently obtained.

Invariants of a curve, independent of the coordinate system used, can be used to simplify the equation of the curve, which is considered a general rule in differential geometry. The arc length of a curve, which is also termed a natural parameter, is an invariant. Other invariants will be introduced in the later of this chapter and other chapters of the book. For equation (1.4), t can be replaced by s. The differential relationship between s and t can be written as

$$s = \int_{t_a}^{t_b} \left|\frac{\mathrm{d}\boldsymbol{R}}{\mathrm{d}t}\right| \mathrm{d}t,\ \mathrm{d}s = |\mathrm{d}\boldsymbol{R}| = \sqrt{\left(\frac{\mathrm{d}x}{\mathrm{d}t}\right)^2 + \left(\frac{\mathrm{d}y}{\mathrm{d}t}\right)^2 + \left(\frac{\mathrm{d}z}{\mathrm{d}t}\right)^2}\,\mathrm{d}t \tag{1.11}$$

Then, the vector equation of curve Γ is expressed in terms of s as

$$\Gamma : \boldsymbol{R} = \boldsymbol{R}(s),\ s_a \leq s \leq s_b \tag{1.12}$$

It is recognized that $|\mathrm{d}\boldsymbol{R}/\mathrm{d}s| = 1$. Using the Taylor expansion, curve Γ can be expressed in the neighborhood Δs of point s by

$$\boldsymbol{R}(s+\Delta s) = \boldsymbol{R}(s) + \frac{\mathrm{d}\boldsymbol{R}(s)}{\mathrm{d}s}\Delta s + \frac{1}{2!}\frac{\mathrm{d}^2\boldsymbol{R}(s)}{\mathrm{d}s^2}(\Delta s)^2 + \cdots + \frac{1}{n!}\frac{\mathrm{d}^n\boldsymbol{R}(s)}{\mathrm{d}s^n}(\Delta s)^n + \varepsilon_n(s,\Delta s)(\Delta s)^n \tag{1.13}$$

where $\lim\limits_{\Delta s\to 0} \varepsilon_n(s,\Delta s) = 0$.

1.1.2 Frenet Frame

In a fixed frame, a curve is traced by a point of a moving body. There exists a connection between the point path and the moving body. A frame that moves along the curve can be employed to study the intrinsic geometrical properties of the curve.

Assume that the unit tangent vector of a plane curve $\boldsymbol{\alpha} = \mathrm{d}\boldsymbol{R}(s)/\mathrm{d}s$ is always in the direction of increasing arc length. Adopting the right-handed rule, as in the case of equation (1.8), the unit normal vector of a curve may be defined as $\boldsymbol{\beta} = \boldsymbol{k} \times \boldsymbol{\alpha}$. A unit orthogonal right-handed coordinate system $\{\boldsymbol{\alpha},\boldsymbol{\beta},\boldsymbol{k}\}$ may be uniquely established for each point s on the curve.

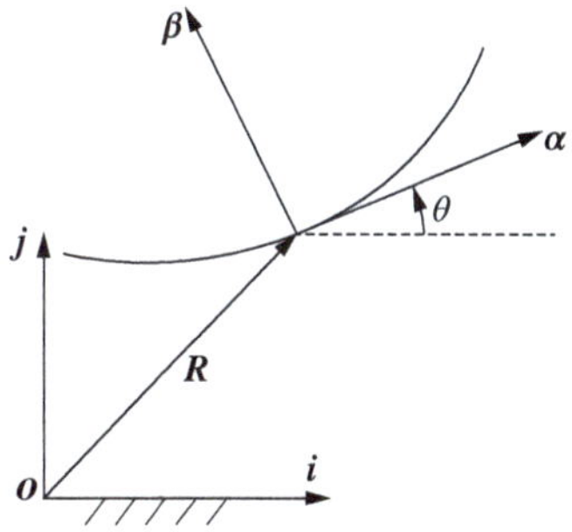

Figure 1.6 The Frenet frame of a plane curve

This moving Cartesian reference frame is called the Frenet frame, or the moving frame of a plane curve (see Fig. 1.6). The Frenet frame for a plane curve may be defined as

$$\begin{cases} \dfrac{\mathrm{d}\boldsymbol{R}}{\mathrm{d}s} = \boldsymbol{\alpha} \\ \dfrac{\mathrm{d}\boldsymbol{\alpha}}{\mathrm{d}s} = k\boldsymbol{\beta} \\ \dfrac{\mathrm{d}\boldsymbol{\beta}}{\mathrm{d}s} = -k\boldsymbol{\alpha} \end{cases} \tag{1.14}$$

where k, an invariant of the curve, is the curvature. Performing a dot product of both sides of the second equation in (1.14) with vector $\boldsymbol{\beta}$, we obtain

$$k = \frac{\mathrm{d}\boldsymbol{\alpha}}{\mathrm{d}s}\cdot\boldsymbol{\beta} = \left(\frac{\mathrm{d}\boldsymbol{\alpha}}{\mathrm{d}s}, \boldsymbol{k}\times\boldsymbol{\alpha}\right) = \left(\boldsymbol{k}, \frac{\mathrm{d}\boldsymbol{R}}{\mathrm{d}s}, \frac{\mathrm{d}^2\boldsymbol{R}}{\mathrm{d}s^2}\right) \tag{1.15}$$

If a vector equation with a general parameter t is given, as in equation (1.4) for a plane curve Γ, the unit tangent vector $\boldsymbol{\alpha}$ can be expressed as

$$\boldsymbol{\alpha} = \frac{\mathrm{d}\boldsymbol{R}}{\mathrm{d}s} = \frac{\mathrm{d}\boldsymbol{R}}{\mathrm{d}t}\cdot\frac{\mathrm{d}t}{\mathrm{d}s} = \frac{\mathrm{d}t}{\mathrm{d}s}\left(\frac{\mathrm{d}x}{\mathrm{d}t}\boldsymbol{i} + \frac{\mathrm{d}y}{\mathrm{d}t}\boldsymbol{j}\right) \tag{1.16}$$

Utilizing the identity equation $\boldsymbol{\beta} = \boldsymbol{k}\times\boldsymbol{\alpha}$, the unit normal vector $\boldsymbol{\beta}$ is obtained as

$$\boldsymbol{\beta} = \frac{\mathrm{d}t}{\mathrm{d}s}\left(-\frac{\mathrm{d}y}{\mathrm{d}t}\boldsymbol{i} + \frac{\mathrm{d}x}{\mathrm{d}t}\boldsymbol{j}\right) \tag{1.17}$$

According to equation (1.11), the relationship between s and t is $\frac{\mathrm{d}t}{\mathrm{d}s} = 1/\sqrt{\left(\frac{\mathrm{d}x}{\mathrm{d}t}\right)^2 + \left(\frac{\mathrm{d}y}{\mathrm{d}t}\right)^2}$. Hence, the curvature k of a plane curve Γ can be written in terms of t as

$$k = \frac{\mathrm{d}\boldsymbol{\alpha}}{\mathrm{d}s}\cdot\boldsymbol{\beta} = \frac{\mathrm{d}t}{\mathrm{d}s}\frac{\mathrm{d}\boldsymbol{\alpha}}{\mathrm{d}t}\cdot\boldsymbol{\beta} = \frac{\dfrac{\mathrm{d}x}{\mathrm{d}t}\cdot\dfrac{\mathrm{d}^2y}{\mathrm{d}t^2} - \dfrac{\mathrm{d}y}{\mathrm{d}t}\cdot\dfrac{\mathrm{d}^2x}{\mathrm{d}t^2}}{\left[\left(\dfrac{\mathrm{d}x}{\mathrm{d}t}\right)^2 + \left(\dfrac{\mathrm{d}y}{\mathrm{d}t}\right)^2\right]^{3/2}} \tag{1.18}$$

In order to explain the geometrical meaning of k, $\boldsymbol{\alpha}$ and $\boldsymbol{\beta}$ are projected onto each axis of the fixed coordinate system $\{\boldsymbol{O}; \boldsymbol{i}, \boldsymbol{j}\}$. We have $\boldsymbol{\alpha} = \{\cos\theta, \sin\theta\}$ and $\boldsymbol{\beta} = \boldsymbol{k} \times \boldsymbol{\alpha} = \{-\sin\theta, \cos\theta\}$, where θ is the angle between $\boldsymbol{\alpha}$ and axis $\boldsymbol{i}$. The differentiation of $\boldsymbol{\alpha}$ with respect to s can be written as

$$\frac{\mathrm{d}\boldsymbol{\alpha}}{\mathrm{d}s} = \frac{\mathrm{d}\boldsymbol{\alpha}}{\mathrm{d}\theta}\frac{\mathrm{d}\theta}{\mathrm{d}s} = \frac{\mathrm{d}\theta}{\mathrm{d}s}\{-\sin\theta, \cos\theta\} = \frac{\mathrm{d}\theta}{\mathrm{d}s}\boldsymbol{\beta} \tag{1.19}$$

From the second equation in (1.14), another expression for the curvature, $k = \frac{\mathrm{d}\theta}{\mathrm{d}s}$, can be obtained. The geometrical meaning of k is the rate of change in angle θ with respect to s.

The curvature here may be positive or negative, depending on its direction around the curve, while the current curvature is always positive (used in the book for readers to identify the difference between them). As shown in Fig. 1.7, if $\boldsymbol{\alpha}$ points in the direction of increasing arc length, then $\boldsymbol{\beta}$ always points toward the left-hand side of the curve. Hence, if $\boldsymbol{\alpha}$ and $\boldsymbol{\beta}$ lie on opposite sides of the curve at a point, k is positive, otherwise, k is negative. In a special case, if k at a point is zero, $\boldsymbol{\alpha}$ is coincidental with the curve at that point. The point is then called an inflection point.

With the definition of the curvature, both local and global properties of a plane curve can be determined. We have the following theorem:

Theorem 1.1 *Given a continuously differentiable function $k(s)$ in interval (s_a, s_b) and an initial point $\boldsymbol{R}_a$ along with a unit tangent vector $\boldsymbol{\alpha}_a$, there exists only one regular plane curve having the given curvature $k(s)$.*

Although a plane curve may be expressed in different forms for different coordinate systems and parameters, the curvature determines the curve uniquely according to Theorem 1.1 because it is an invariant of the curve and independent of the coordinate system used. The curvature function is usually termed the natural equation of a curve in two dimensions.

As special cases, a circle has constant curvature while a straight line has zero curvature. The local shape of a plane curve is close to a circle in the neighborhood of point s if the curvature of the curve is constant at point s or in its vicinity. To define closeness, a contact order between two curves is introduced. Two plane curves have two common points at two infinitesimally separated positions if they are tangent to each other. We define this case as "first-order contact." Order-n contact is defined as the contact between two curves having n+1 common points at

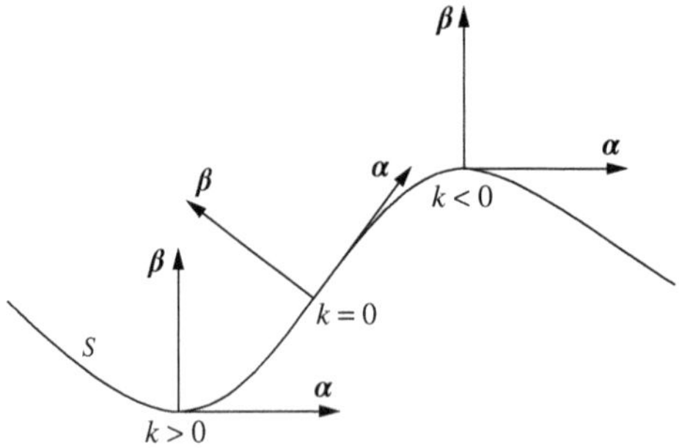

Figure 1.7 The curvature of a plane curve

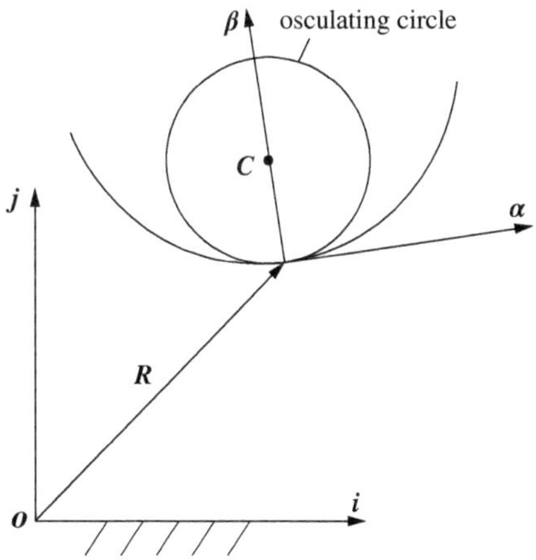

Figure 1.8 The osculating circle of a plane curve

infinitesimally separated positions. According to this definition, a plane curve and a circle have first-order contact if the curve is tangent to the circle. Second-order contact implies that a curve and a circle have three common points at infinitesimally separated positions; the circle is then called the osculation circle (see Fig. 1.8), located by three infinitesimally separated points and whose radius is the radius of the curvature circle at the contact point. Third-order contact between a curve and a circle indicates that there are four common points at infinitesimally separated positions. In this case, the differential curvature of the curve with respect to the natural parameter at point s must be zero, or $\mathrm{d}k/\mathrm{d}s = 0$. Similarly, if the contact order between a curve and a circle is n, the successive derivatives up to order $n-2$ of the curvature with respect to the natural parameter at position s must be zero. The vector of center of the osculating circle, or the curvature center of a curve Γ at the contact point s, can be written as

$$\boldsymbol{R}_C = \boldsymbol{R} + \frac{1}{k}\boldsymbol{\beta} \tag{1.20}$$

Each point on curve Γ has a corresponding curvature center. The loci of all centers of the curvature circles of a curve is another curve, called the evolute of curve Γ. In a special case, if curve Γ is a circle, $k =$ const., the evolute degenerates to a fixed point.

We can also define contact orders between a curve and a straight line. A plane curve and a straight line have first-order contact, which means that the line is tangent to the curve at the contact point. Second-order contact implies that a curve and a straight line have three common points at infinitesimally separated positions; the curvature of the curve at the contact point, or the inflection point of the curve, must be zero. Third-order contact requires that a curve and a line have four common points at infinitesimally separated positions, in which the curvature of the curve at the contact point is zero and whose differential with respect to the natural parameter at position s is zero, or $\mathrm{d}k/\mathrm{d}s = 0$.

For convenience, in discussing the global geometrical properties of a curve in two dimensions, we introduce the following definitions.

- ***Closed plane curve.*** A plane curve Γ is a closed curve during interval $[s_a, s_b]$ if $\boldsymbol{R}(s_a) = \boldsymbol{R}(s_b)$ is satisfied.

- ***Simple closed plane curve.*** A closed plane curve is a simple closed plane curve if it has no intersections with itself, or is without self-intersection points.
- ***Simple convex closed plane curve.*** A simple closed plane curve is a simple convex closed plane curve if the tangent vector of the curve at every point is always located on one side of the curve.

As examples, the curve in Fig. 1.9(a) is a simple closed and non-convex plane curve. The curve in Fig. 1.9(b) is a simple closed and convex plane curve, often called an ***oval curve*** because of its similarity to a goose egg shape. The curve in Fig. 1.9(c) is a non-simple closed plane curve.

From the definition of a simple convex closed plane curve and the geometrical meaning of curvature, we have:

Theorem 1.2 *A simple closed plane curve is a simple convex closed plane curve if and only if the curvature remains greater than or equal to zero ($k \geq 0$) at every point for a properly chosen positive direction of increasing arc length.*

In particular, a simple convex closed plane curve is an oval curve if the curvature is not equal to zero at any point of the curve. From Theorem 1.2, we have the following corollary:

Corollary 1.1 *A simple closed plane curve must be an oval curve if the curvature remains sign-invariant at every point of the curve for a properly chosen positive direction of increasing arc length.*

1.1.3 Adjoint Approach

A point P moves along a plane curve Γ_P in the fixed coordinate system $\{\boldsymbol{O};\ \boldsymbol{i}, \boldsymbol{j}\}$. Another point P^*, which does not belong to curve Γ_P, traces a different curve Γ_P^* in the same coordinate system $\{\boldsymbol{O};\ \boldsymbol{i}, \boldsymbol{j}\}$. If each position of point P^* at Γ_P^* always corresponds to a position of point P at Γ_P, point P^* is said to be adjoint to point P, and curve Γ_P^* is said to be adjoint to curve Γ_P. Γ_P is defined as the ***original curve***, and Γ_P^* is called the ***adjoint curve*** of Γ_P (see Fig. 1.10).

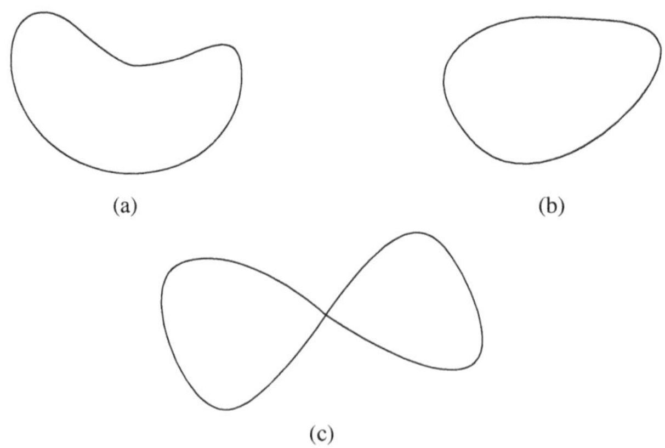

Figure 1.9 (a) A simple closed and non-convex plane curve (b) An oval curve (c) A non-simple closed plane curve

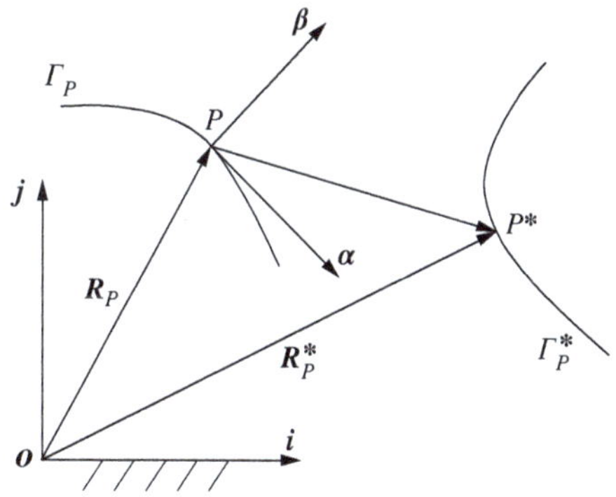

Figure 1.10 A plane curve adjoint to another plane curve

The Frenet frame $\{\boldsymbol{R}_P; \boldsymbol{\alpha}, \boldsymbol{\beta}\}$ is set up at the original curve Γ_P, and the vector equation of the adjoint curve Γ_P^* can be written as

$$\Gamma_P^* : \boldsymbol{R}_P^* = \boldsymbol{R}_P + u_1\boldsymbol{\alpha} + u_2\boldsymbol{\beta} \tag{1.21}$$

where (u_1, u_2) are the coordinates of point P^* in the Frenet frame $\{\boldsymbol{R}_P; \boldsymbol{\alpha}, \boldsymbol{\beta}\}$ of the original curve Γ_P. Based on the Frenet formulas (1.14), the first derivative of the above equation with respect to the arc length s of the original curve Γ_P (not the arc length s^* of the adjoint curve Γ_P^*) is given by

$$\begin{cases} \dfrac{\mathrm{d}\boldsymbol{R}_P{}^*}{\mathrm{d}s} = A_1\boldsymbol{\alpha} + A_2\boldsymbol{\beta} \\ A_1 = 1 + \dfrac{\mathrm{d}u_1}{\mathrm{d}s} - ku_2 \\ A_2 = ku_1 + \dfrac{\mathrm{d}u_2}{\mathrm{d}s} \end{cases} \tag{1.22}$$

where $\mathrm{d}\boldsymbol{R}_P^*/\mathrm{d}s$ is the tangent vector of the plane curve Γ_P^*. The absolute motion of point P^* is examined in $\{\boldsymbol{O}; \boldsymbol{i}, \boldsymbol{j}\}$ and expressed by the moving Frenet frame $\{\boldsymbol{R}_P; \boldsymbol{\alpha}, \boldsymbol{\beta}\}$ of the original curve Γ_P. Hence, $\left(\frac{\mathrm{d}u_1}{\mathrm{d}s}, \frac{\mathrm{d}u_2}{\mathrm{d}s}\right)$ are the rates of change of the coordinates in the Frenet frame $\{\boldsymbol{R}_P; \boldsymbol{\alpha}, \boldsymbol{\beta}\}$; here, (A_1, A_2) are the rates of change of the absolute motion of point P^* in the fixed frame and expressed by the Frenet frame $\{\boldsymbol{R}_P; \boldsymbol{\alpha}, \boldsymbol{\beta}\}$. In particular, P^* is a fixed point in $\{\boldsymbol{O}; \boldsymbol{i}, \boldsymbol{j}\}$; the absolute coordinates of point P^* do not change with s of the original curve Γ_P, or (A_1, A_2) are zero (i.e., $\mathrm{d}\boldsymbol{R}_P^*/\mathrm{d}s = 0$). The last two expressions of (1.22) can be written as

$$\begin{cases} A_1 = 1 + \dfrac{\mathrm{d}u_1}{\mathrm{d}s} - ku_2 = 0 \\ A_2 = ku_1 + \dfrac{\mathrm{d}u_2}{\mathrm{d}s} = 0 \end{cases} \tag{1.23}$$

The above equations are called the ***fixed point conditions*** of an adjoint plane curve, or ***Cesaro's fixed point conditions***. That is, point P^* in the Frenet frame $\{\boldsymbol{R}_P; \boldsymbol{\alpha}, \boldsymbol{\beta}\}$ at that instant has to meet the same conditions as if it remained absolutely still in the fixed coordinate system

$\{\boldsymbol{O};\, \boldsymbol{i}, \boldsymbol{j}\}$. The differential equation (1.23), in fact, implies the relationship between the motion of the Frenet frame $\{\boldsymbol{R}_P; \boldsymbol{\alpha}, \boldsymbol{\beta}\}$ with respect to the original curve $\varGamma_P$ and the motion of point P^* relative to the fixed frame $\{\boldsymbol{O};\, \boldsymbol{i}, \boldsymbol{j}\}$.

Cesaro once used a metaphor for this: the original curve $\varGamma_p$ is like a winding river, whereas the Frenet frame $\{\boldsymbol{R}_P; \boldsymbol{\alpha}, \boldsymbol{\beta}\}$ is like a boat flowing and going with the water in the river (see Fig. 1.11). The tangent vector $\boldsymbol{\alpha}$ goes ahead downstream with the second axis, or the normal vector $\boldsymbol{\beta}$, perpendicular to the boat and toward the left. The boatman can see the breathtaking scenery of banks and mountains all from the viewpoint of the boat as the coordinate system, and hence he knows everything about the curves of the river and detailed information about the banks.

A point P moves along a plane curve $\varGamma_P$ in $\{\boldsymbol{O};\, \boldsymbol{i}, \boldsymbol{j}\}$. A straight line L passing through point P^*, which does not belong to curve $\varGamma_P$, traces a set of lines $\varGamma_l^*$ in $\{\boldsymbol{O};\, \boldsymbol{i}, \boldsymbol{j}\}$, but each position of line L with point P^* always corresponds to a position of point P at $\varGamma_P$, or line L with point P^* is adjoint to point P. Hence, $\varGamma_l^*$ is adjoint to curve $\varGamma_P$. We designate $\varGamma_P$ as an original curve and $\varGamma_l^*$ as the set of adjoint lines of $\varGamma_P$ (see Fig. 1.12).

The Frenet frame $\{\boldsymbol{R}_P; \boldsymbol{\alpha}, \boldsymbol{\beta}\}$ is set up at the original curve $\varGamma_P$, and the vector equation of the set of adjoint lines $\varGamma_l^*$ can be written as

$$\varGamma_l^* : \boldsymbol{R}_l^* = \boldsymbol{R}_P^* + \lambda \boldsymbol{l} = \boldsymbol{R}_P + u_1\boldsymbol{\alpha} + u_2\boldsymbol{\beta} + \lambda(l_1\boldsymbol{\alpha} + l_2\boldsymbol{\beta}), l_1^2 + l_2^2 = 1 \tag{1.24}$$

where λ is a parameter for the straight line, $\boldsymbol{l}$ is a unit direction vector of the straight line described in the Frenet frame $\{\boldsymbol{R}_P; \boldsymbol{\alpha}, \boldsymbol{\beta}\}$, which is a function of the arc length s of the original

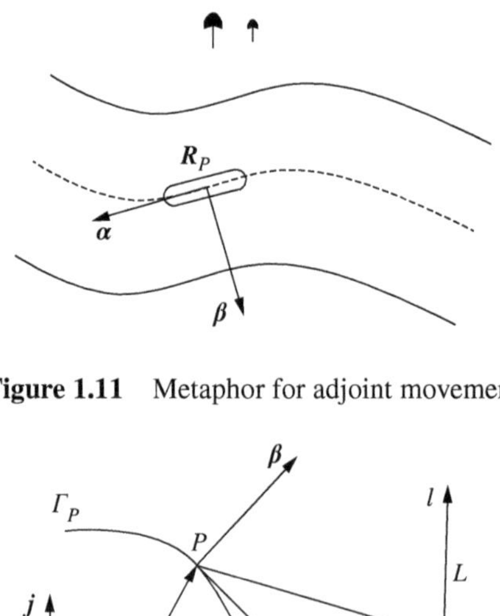

Figure 1.11 Metaphor for adjoint movement

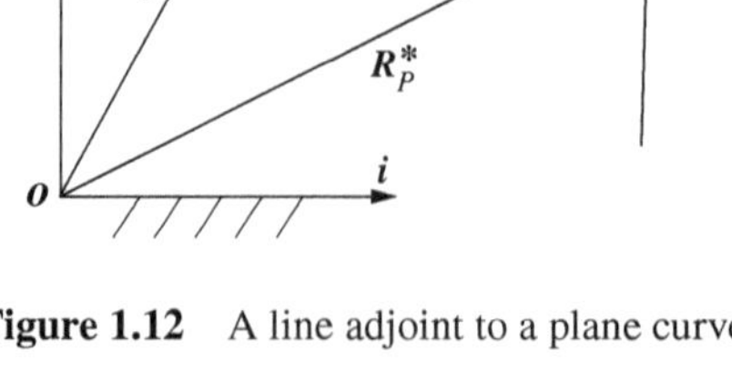

Figure 1.12 A line adjoint to a plane curve

curve Γ_P. Based on the Frenet formulas in equation (1.14), the first derivative of the above equation with respect to s of Γ_P is given by

$$\begin{cases} \dfrac{\mathrm{d}\boldsymbol{R}_l^*}{\mathrm{d}s} = A_1\boldsymbol{\alpha} + A_2\boldsymbol{\beta} + \lambda\left(B_1\boldsymbol{\alpha} + B_2\boldsymbol{\beta}\right) \\ A_1 = 1 + \dfrac{\mathrm{d}u_1}{\mathrm{d}s} - ku_2,\ A_2 = ku_1 + \dfrac{\mathrm{d}u_2}{\mathrm{d}s} \\ B_1 = \dfrac{\mathrm{d}l_1}{\mathrm{d}s} - kl_2,\ B_2 = kl_1 + \dfrac{\mathrm{d}l_2}{\mathrm{d}s} \end{cases} \tag{1.25}$$

If all points on the straight line L are fixed points in $\{\boldsymbol{O};\ \boldsymbol{i}, \boldsymbol{j}\}$ and do not change with s of Γ_P, the line is a fixed line. We define it as an ***absolute fixed line***. Hence, the last four expressions of equation (1.25) are equal to zero in this case, which leads to the conditions of an absolute fixed line in the fixed frame $\{\boldsymbol{O};\ \boldsymbol{i}, \boldsymbol{j}\}$:

$$\begin{cases} A_1 = 1 + \dfrac{\mathrm{d}u_1}{\mathrm{d}s} - ku_2 = 0,\ A_2 = ku_1 + \dfrac{\mathrm{d}u_2}{\mathrm{d}s} = 0 \\ B_1 = \dfrac{\mathrm{d}l_1}{\mathrm{d}s} - kl_2 = 0,\ B_2 = kl_1 + \dfrac{\mathrm{d}l_2}{\mathrm{d}s} = 0 \end{cases} \tag{1.26}$$

If line L is always collinear with a fixed line in $\{\boldsymbol{O};\ \boldsymbol{i}, \boldsymbol{j}\}$, but slides along the fixed line, the line L satisfies the following conditions:

$$\frac{\mathrm{d}\boldsymbol{l}}{\mathrm{d}s} = 0,\ \frac{\mathrm{d}\boldsymbol{R}_P^*}{\mathrm{d}s} \times \boldsymbol{l} = 0 \tag{1.27}$$

Substituting equation (1.24) into the above equations, we introduce the ***quasi-fixed line conditions***

$$B_1 = 0,\ B_2 = 0,\ l_2A_1 = l_1A_2 \tag{1.28}$$

Example 1.4 Represent a coupler curve by an adjoint curve for a planar linkage.

Solution

As shown in Example 1.3, a coupler curve of a four-bar linkage is concisely expressed in terms of a vector function of a unit circle. In fact, the coupler point is always adjoint to joint B when it traces a coupler curve. Path Γ_B of joint B in the fixed frame $\{\boldsymbol{O}; \boldsymbol{i}_f, \boldsymbol{j}_f\}$ is a circle, and is consequently taken as the original curve, while a coupler curve is viewed as an adjoint curve of Γ_B. Hence, the Frenet frame $\{\boldsymbol{R}_B;\ \boldsymbol{\alpha}, \boldsymbol{\beta}\}$ of the original curve Γ_B is established as in Fig. 1.13. A coupler curve Γ_P, or the adjoint curve of Γ_B, can be expressed by the Frenet frame $\{\boldsymbol{R}_B;\ \boldsymbol{\alpha}, \boldsymbol{\beta}\}$ as

$$\boldsymbol{R}_P = \boldsymbol{R}_B + u_1\boldsymbol{\alpha} + u_2\boldsymbol{\beta} = u_1\boldsymbol{\alpha} + (u_2 - a_1)\boldsymbol{\beta} \tag{E1-4.1}$$

where (u_1, u_2), the coordinates of a coupler point P in the Frenet frame $\{\boldsymbol{R}_B;\ \boldsymbol{\alpha}, \boldsymbol{\beta}\}$, are

$$\begin{cases} u_1 = r_P \sin\left(\theta_P - \varphi + \gamma\right) \\ u_2 = -r_P \cos\left(\theta_P - \varphi + \gamma\right) \end{cases} \tag{E1-4.2}$$

Here, angles φ, γ are identical to those in Example 1.3. Both angles are measured in the counterclockwise sense.

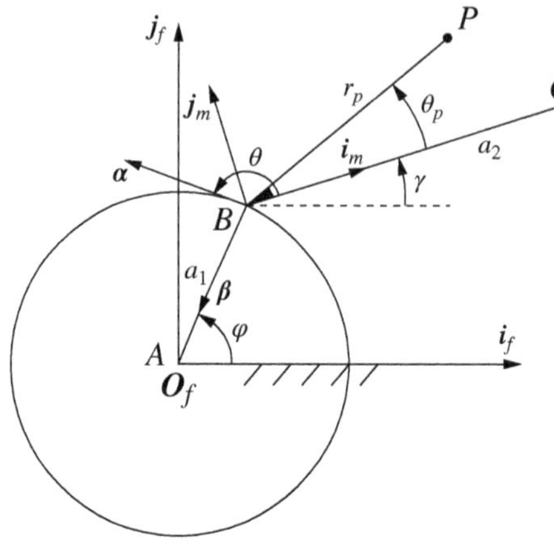

Figure 1.13 A coupler curve adjoint to a circle

1.2 Planar Differential Kinematics

1.2.1 Displacement

1.2.1.1 A General Description of Plane Displacement

To describe the displacement of a moving body $\sum^*$ relative to a fixed body $\sum$ from one position to another, different reference frames are established. A moving Cartesian reference frame $\{\boldsymbol{O}_m; \boldsymbol{i}_m, \boldsymbol{j}_m\}$ is set up and attached to the moving body $\sum^*$ and a fixed Cartesian reference frame $\{\boldsymbol{O}_f; \boldsymbol{i}_f, \boldsymbol{j}_f\}$ is established in the fixed body $\sum$ (see Fig. 1.14). For planar motion, a body has three-degrees-of-freedom motion, moving along both $\boldsymbol{i}_f$ and $\boldsymbol{j}_f$ and rotating about $\boldsymbol{k}$, or two linear displacements and one angular displacement. A point $O_m(x_{Om}, y_{Om})$ of $\sum^*$ is taken as a reference point, whose linear displacements (x_{Omf}, y_{Omf}) present a given movement of $\sum^*$ in $\{\boldsymbol{O}_f; \boldsymbol{i}_f, \boldsymbol{j}_f\}$, and the angular γ of $\sum^*$ denotes the angular displacement around $\boldsymbol{k}$, completely describing a planar motion of $\sum^*$ relative to $\sum$.

An arbitrary point $P_m(x_{Pm}, y_{Pm})$ in $\sum^*$ corresponds to a position $P(x_{Pf}, y_{Pf})$ or a displacement in $\{\boldsymbol{O}_f; \boldsymbol{i}_f, \boldsymbol{j}_f\}$, which are related by

$$\begin{bmatrix} x_{Pf} \\ y_{Pf} \\ 1 \end{bmatrix} = \begin{bmatrix} \cos\gamma & -\sin\gamma & x_{Omf} \\ \sin\gamma & \cos\gamma & y_{Omf} \\ 0 & 0 & 1 \end{bmatrix} \begin{bmatrix} x_{Pm} \\ y_{Pm} \\ 1 \end{bmatrix} \tag{1.29}$$

We can analyze and determine the displacements of a point on $\sum^*$ if any two linear displacements (x_{Omf}, y_{Omf}) and an angular displacement γ of $\sum^*$ relative to $\sum$ are given. In equation (1.29), the given movement of $\sum^*$ may be continuous, or discrete. For the former, its content belongs to the continuous kinematic geometry, or the kinematic geometry at infinitesimally separated positions, but the authors prefer the kinematic differential geometry since the continuous movement is achieved by the differential of the moving frame. The latter belongs to the discrete kinematic geometry (the authors appreciate this term, although its contents are not mature so far) or the kinematic geometry at finite separated positions corresponding to the classical curvature theory or the classical Burmester theory. All of these form a theoretical basis for kinematic synthesis of linkage. The kinematic geometry, either with continuous or discrete movement, has similar geometrical properties.

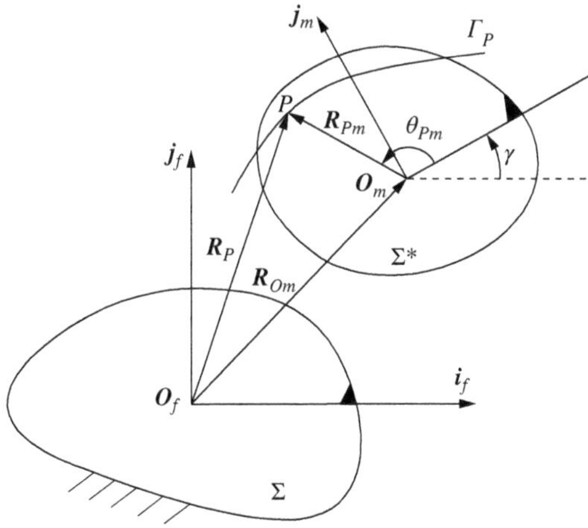

Figure 1.14 A moving body $\sum^*$ relative to a fixed body $\sum$

1.2.1.2 Descriptions of Plane Displacement by an Adjoint Approach

The point P of $\sum^*$ is located by the Cartesian coordinates (x_{Pm}, y_{Pm}) or the polar coordinates (r_{Pm}, θ_{Pm}) in $\{\boldsymbol{O}_m; \boldsymbol{i}_m, \boldsymbol{j}_m\}$ of $\sum^*$, whose vector equation is

$$\boldsymbol{R}_{Pm} = x_{Pm}\boldsymbol{i}_m + y_{Pm}\boldsymbol{j}_m = r_{Pm}\boldsymbol{e}_{\mathrm{I}(\theta_{Pm})} \tag{1.30}$$

The plane displacements of a moving body $\sum^*$ are examined in a fixed Cartesian reference frame $\{\boldsymbol{O}_f; \boldsymbol{i}_f, \boldsymbol{j}_f\}$ of $\sum$. The relationship between the coordinate axes of $\{\boldsymbol{O}_m; \boldsymbol{i}_m, \boldsymbol{j}_m\}$ and those of $\{\boldsymbol{O}_f; \boldsymbol{i}_f, \boldsymbol{j}_f\}$ is

$$\begin{cases} \boldsymbol{i}_m = \cos\gamma \boldsymbol{i}_f + \sin\gamma \boldsymbol{j}_f \\ \boldsymbol{j}_m = -\sin\gamma \boldsymbol{i}_f + \cos\gamma \boldsymbol{j}_f \end{cases} \tag{1.31}$$

where γ is the angle between $\boldsymbol{i}_m$ and $\boldsymbol{i}_f$. Hence, the plane displacements of $\sum^*$ can be represented by two linear displacements (x_{Omf}, y_{Omf}) of point O_m and the rotational displacement γ about $\boldsymbol{k}$. Point O_m moves along curve Γ_{Om}, which is given as a regular plane curve with higher-order continuity.

A given curve Γ_{Om} in the fixed frame $\{\boldsymbol{O}_f; \boldsymbol{i}_f, \boldsymbol{j}_f\}$ is designated as the curve traced by the origin point O_m of the moving frame $\{\boldsymbol{O}_m; \boldsymbol{i}_m, \boldsymbol{j}_m\}$ of $\sum^*$ (see Fig. 1.15); its vector equation can be written as

$$\Gamma_{Om} : \boldsymbol{R}_{Om} = x_{Omf}\boldsymbol{i}_f + y_{Omf}\boldsymbol{j}_f \tag{1.32}$$

Point $P(x_{Pm}, y_{Pm})$ of $\sum^*$ traces a path Γ_P in the fixed frame $\{\boldsymbol{O}_f; \boldsymbol{i}_f, \boldsymbol{j}_f\}$; its vector equation is

$$\Gamma_P : \boldsymbol{R}_P = \boldsymbol{R}_{Om} + \boldsymbol{R}_{Pm} = \boldsymbol{R}_{Om} + x_{Pm}\boldsymbol{i}_m + y_{Pm}\boldsymbol{j}_m \tag{1.33}$$

Taking the derivative of equation (1.33) with respect to time t, the absolute velocity of point $P(x_{Pm}, y_{Pm})$ can be written as

$$\boldsymbol{V}_P = \frac{\mathrm{d}\boldsymbol{R}_P}{\mathrm{d}t} = \frac{\mathrm{d}\boldsymbol{R}_{Om}}{\mathrm{d}t} + x_{Pm}\frac{\mathrm{d}\boldsymbol{i}_m}{\mathrm{d}t} + y_{Pm}\frac{\mathrm{d}\boldsymbol{j}_m}{\mathrm{d}t} + \frac{\mathrm{d}x_{Pm}}{\mathrm{d}t}\boldsymbol{i}_m + \frac{\mathrm{d}y_{Pm}}{\mathrm{d}t}\boldsymbol{j}_m \tag{1.34}$$

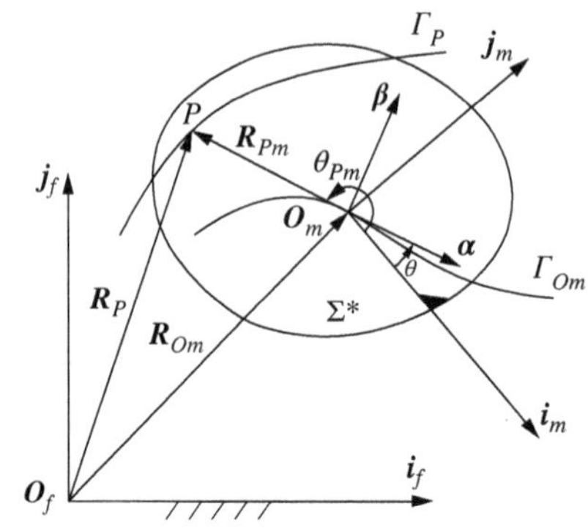

Figure 1.15 Adjoint descriptions of plane displacements

The absolute velocity of the point P can be viewed as the superposition of two velocities, the following velocity of the moving frame $\{\boldsymbol{O}_m; \boldsymbol{i}_m, \boldsymbol{j}_m\}$ and the relative velocity of point P with reference to the moving frame $\{\boldsymbol{O}_m; \boldsymbol{i}_m, \boldsymbol{j}_m\}$. The position of the origin O_m and the orientation of $\{\boldsymbol{O}_m; \boldsymbol{i}_m, \boldsymbol{j}_m\}$ in $\sum^*$ not only affect the property and complexity of the following velocity in equation (1.34), $\frac{d\boldsymbol{R}_{Om}}{dt} + x_{Pm}\frac{d\boldsymbol{i}_m}{dt} + y_{Pm}\frac{d\boldsymbol{j}_m}{dt}$, but also change the magnitude of the relative velocity with reference to the moving frame.

In general, point P is a fixed point in the moving body $\sum^*$, and the moving frame $\{\boldsymbol{O}_m; \boldsymbol{i}_m, \boldsymbol{j}_m\}$ is also fixed in $\sum^*$. There is no relative motion between P and the moving frame $\{\boldsymbol{O}_m; \boldsymbol{i}_m, \boldsymbol{j}_m\}$, or the components of the relative velocity in equation (1.34), dx_{Pm}/dt and dy_{Pm}/dt, are zero. The following velocity of $\{\boldsymbol{O}_m; \boldsymbol{i}_m, \boldsymbol{j}_m\}$ can be determined by equations (1.31) and (1.32).

As mentioned above, both the given curve Γ_{Om} and the relative angular displacement γ of $\sum^*$ completely define the kinematic properties of a moving body $\sum^*$. A point P of $\sum^*$ traces a path Γ_P in $\{\boldsymbol{O}_f; \boldsymbol{i}_f, \boldsymbol{j}_f\}$, while the origin point O_m of $\sum^*$ moves along the given curve Γ_{Om} in $\{\boldsymbol{O}_f; \boldsymbol{i}_f, \boldsymbol{j}_f\}$ simultaneously. Each position of P at Γ_P always corresponds to a position of O_m at Γ_{Om}. In other words, P is adjoint to O_m. Hence, any point of $\sum^*$ traces a path, which can be viewed as an adjoint curve of the given curve Γ_{Om}, or the displacements of $\sum^*$ in plane motion can be expressed by the adjoint approach of curve Γ_{Om} if the given curve Γ_{Om} is a regular curve.

Utilizing the analytic properties of a given curve Γ_{Om}, the natural arc length s is chosen as the parameter of the given curve Γ_{Om}. We can set up the Frenet frame $\{\boldsymbol{R}_{Om}; \boldsymbol{\alpha}, \boldsymbol{\beta}\}$ of the given curve Γ_{Om} as

$$\boldsymbol{\alpha} = d\boldsymbol{R}_{Om}/ds,\ \boldsymbol{\beta} = \boldsymbol{k} \times \boldsymbol{\alpha} \tag{1.35a}$$

or

$$\begin{cases} \boldsymbol{\alpha} = \left(\dfrac{dx_{Omf}}{ds}\boldsymbol{i}_f + \dfrac{dy_{Omf}}{ds}\boldsymbol{j}_f\right) \Big/ \left[\left(\dfrac{dx_{Omf}}{ds}\right)^2 + \left(\dfrac{dy_{Omf}}{ds}\right)^2\right]^{1/2} \\ \boldsymbol{\beta} = \left(-\dfrac{dy_{Omf}}{ds}\boldsymbol{i}_f + \dfrac{dx_{Omf}}{ds}\boldsymbol{j}_f\right) \Big/ \left[\left(\dfrac{dx_{Omf}}{ds}\right)^2 + \left(\dfrac{dy_{Omf}}{ds}\right)^2\right]^{1/2} \end{cases} \tag{1.35b}$$

The Frenet formulas are

$$\frac{\mathrm{d}\boldsymbol{\alpha}}{\mathrm{d}s} = k_{Om}\boldsymbol{\beta},\ \frac{\mathrm{d}\boldsymbol{\beta}}{\mathrm{d}s} = -k_{Om}\boldsymbol{\alpha} \tag{1.36}$$

where invariant k_{Om} is the curvature of the given curve Γ_{Om}.

The given curve Γ_{Om} is taken as the original curve. Path Γ_P traced by point P of $\sum^*$ is an adjoint curve of Γ_{Om}. Hence, the adjoint curve Γ_P in $\{\boldsymbol{O}_f; \boldsymbol{i}_f, \boldsymbol{j}_f\}$ can be expressed in the Frenet frame $\{\boldsymbol{R}_{Om}; \boldsymbol{\alpha}, \boldsymbol{\beta}\}$ of Γ_{Om}, or

$$\Gamma_P : \boldsymbol{R}_P = \boldsymbol{R}_{Om} + u_1\boldsymbol{\alpha} + u_2\boldsymbol{\beta} \tag{1.37}$$

where (u_1, u_2) are the coordinates of point P in $\{\boldsymbol{R}_{Om}; \boldsymbol{\alpha}, \boldsymbol{\beta}\}$ and the function of s. Unit vectors $\{\boldsymbol{\alpha}, \boldsymbol{\beta}\}$ can be derived from equation (1.35). We do not forget that point P of $\sum^*$ has the Cartesian coordinates (x_{Pm}, y_{Pm}) or the polar coordinates (r_{Pm}, θ_{Pm}) in $\{\boldsymbol{O}_m; \boldsymbol{i}_m, \boldsymbol{j}_m\}$.

Let θ be the angle between the unit vector $\boldsymbol{\alpha}$ and the axis $\boldsymbol{i}_m$ of $\{\boldsymbol{O}_m; \boldsymbol{i}_m, \boldsymbol{j}_m\}$, which can be obtained from equations (1.31) and (1.35) (see Fig. 1.15). We obtain

$$\begin{cases} \boldsymbol{\alpha} = \cos\theta\boldsymbol{i}_m + \sin\theta\boldsymbol{j}_m \\ \boldsymbol{\beta} = -\sin\theta\boldsymbol{i}_m + \cos\theta\boldsymbol{j}_m \end{cases} \tag{1.38}$$

Substituting equations (1.33) and (1.38) into (1.37), we have

$$\begin{cases} u_1 = r_{Pm}\cos\left(\theta_{Pm} - \theta\right) \\ u_2 = r_{Pm}\sin\left(\theta_{Pm} - \theta\right) \end{cases} \tag{1.39}$$

Note that the path Γ_P is examined in $\{\boldsymbol{O}_f; \boldsymbol{i}_f, \boldsymbol{j}_f\}$ but expressed by equations (1.29), (1.33), and (1.37) in the different frames respectively, such as the moving frame $\{\boldsymbol{O}_m; \boldsymbol{i}_m, \boldsymbol{j}_m\}$ of $\sum^*$ and the Frenet frame $\{\boldsymbol{R}_{Om}; \boldsymbol{\alpha}, \boldsymbol{\beta}\}$ of Γ_{Om}. Equation (1.37) takes advantage of the analytic properties of the given curve by the adjoint approach.

Differentiating equation (1.37) with respect to s of the original curve Γ_{Om}, and making use of the Frenet formulas in equation (1.14), we obtain

$$\begin{cases} \dot{\boldsymbol{R}}_P = A_1\boldsymbol{\alpha} + A_2\boldsymbol{\beta} \\ A_1 = 1 + \dot{u}_1 - k_{Om}u_2 = 1 - \left(k_{Om} - \dot{\theta}\right)u_2 \\ A_2 = k_{Om}u_1 + \dot{u}_2 = (k_{Om} - \dot{\theta})u_1 \end{cases} \tag{1.40}$$

where an overdot represents the derivative with respect to s and k_{Om} is the curvature of the original curve Γ_{Om} and a function of s.

Comparing equation (1.40) with equation (1.23), we apply Cesaro's fixed point conditions and have a point of $\sum^*$ with coordinates (u_1, u_2) at an instant, which is

$$\begin{cases} A_1 = 1 - \left(k_{Om} - \dot{\theta}\right)u_2 = 0 \\ A_2 = (k_{Om} - \dot{\theta})u_1 = 0 \end{cases} \tag{1.41}$$

Using the chain rule of differentiation, we have $\frac{\mathrm{d}\boldsymbol{R}_P}{\mathrm{d}s} = \frac{\mathrm{d}\boldsymbol{R}_P}{\mathrm{d}t}\frac{\mathrm{d}t}{\mathrm{d}s} = \boldsymbol{V}_P\frac{\mathrm{d}t}{\mathrm{d}s} = 0$. This condition is the kinematic explanation of the Cesaro fixed point conditions, which is an instantaneous center of velocity for $\sum^*$ relative to $\sum$, or the velocity center for short. From equation (1.41), the coordinates (u_1, u_2) in the Frenet frame $\{\boldsymbol{R}_{Om}; \boldsymbol{\alpha}, \boldsymbol{\beta}\}$ of Γ_{Om} for the velocity center at an instant of $\sum^*$ can be represented by

$$\begin{cases} u_1 = 0 \\ u_2 = 1/\left(k_{Om} - \dot{\theta}\right) \end{cases} \tag{1.42}$$

From the above equations and the Frenet frame $\{\boldsymbol{R}_{Om}; \boldsymbol{\alpha}, \boldsymbol{\beta}\}$, the velocity center is located at the normal of the original curve Γ_{Om} with a distance $1/(k_{Om} - \dot{\theta})$. If the coordinates (u_1, u_2) of the velocity center are transformed into the moving frame $\{\boldsymbol{O}_m; \boldsymbol{i}_m, \boldsymbol{j}_m\}$ from the Frenet frame $\{\boldsymbol{R}_{Om}; \boldsymbol{\alpha}, \boldsymbol{\beta}\}$ of the original curve Γ_{Om}, the moving centrode π_m of the moving body $\sum^*$ can be obtained. If the coordinates (u_1, u_2) are transformed into the fixed frame $\{\boldsymbol{O}_f; \boldsymbol{i}_f, \boldsymbol{j}_f\}$, the fixed centrode π_f of the fixed body $\sum$ is formed. Detailed discussions are given in later sections.

1.2.2 Centrodes

1.2.2.1 Moving Centrode π_m and its Curvature k_m

The velocity center of a moving body at an instant is a point in or at the extension part of $\sum^*$, whose velocity is zero relative to the fixed body. For all instants, the velocity centers trace a curve π_m in $\sum^*$, called the ***moving centrode*** π_m. Of course, the moving centrode π_m is a curve of $\sum^*$ and described in $\{\boldsymbol{O}_m; \boldsymbol{i}_m, \boldsymbol{j}_m\}$ of $\sum^*$. By means of equations (1.37) and (1.42), we have the following vector expression of π_m:

$$\pi_m : \boldsymbol{R}_m = \frac{1}{k_{Om} - \dot{\theta}}\boldsymbol{\beta} \tag{1.43}$$

where vector $\boldsymbol{\beta}$ belongs to the Frenet frame $\{\boldsymbol{R}_{Om}; \boldsymbol{\alpha}, \boldsymbol{\beta}\}$ of Γ_{Om}, whose direction changes with the arc length s of Γ_{Om}. For $\sum^*$, $\{\boldsymbol{\alpha}, \boldsymbol{\beta}\}$ are similar to a set of orthogonal vectors, whose origin $\boldsymbol{R}_{Om}$ is coincident with the origin $\boldsymbol{O}_m$ of $\{\boldsymbol{O}_m; \boldsymbol{i}_m, \boldsymbol{j}_m\}$, shown in Fig. 1.16. Vector $\boldsymbol{\alpha}$ makes angle θ with respect to $\boldsymbol{i}_m$. Vectors $\boldsymbol{\alpha}$, $\boldsymbol{\beta}$ can be viewed as vector functions of a unit circle $\boldsymbol{\alpha} = \boldsymbol{e}_{\mathrm{I}(\theta)}$, $\boldsymbol{\beta} = \boldsymbol{e}_{\mathrm{II}(\theta)}$, with θ as the parameter at the origin point $\boldsymbol{O}_m$. Here, the Frenet frame $\{\boldsymbol{R}_{Om}; \boldsymbol{\alpha}, \boldsymbol{\beta}\}$ of Γ_{Om} has no significance in $\sum^*$. Based on the properties of a vector function of a unit circle, the derivatives of $\{\boldsymbol{\alpha}, \boldsymbol{\beta}\}$ with respect to θ are obtained as $\mathrm{d}\boldsymbol{\alpha}/\mathrm{d}\theta = \boldsymbol{\beta}$, $\mathrm{d}\boldsymbol{\beta}/\mathrm{d}\theta = -\boldsymbol{\alpha}$, and the tangent vector of π_m can be written as

$$\frac{\mathrm{d}\boldsymbol{R}_m}{\mathrm{d}s} = -\frac{\dot{\theta}}{k_{Om} - \dot{\theta}}\boldsymbol{\alpha} - \frac{\dot{k}_{Om} - \ddot{\theta}}{(k_{Om} - \dot{\theta})^2}\boldsymbol{\beta} \tag{1.44}$$

where s is a general parameter representing the arc length of Γ_{Om}, not the natural parameter of π_m. If the arc length of π_m is denoted as σ_m, from the above equation, we may obtain the following differential relationship between σ_m and s:

$$\mathrm{d}\sigma_m = |\mathrm{d}\boldsymbol{R}_m| = \left[\frac{\dot{\theta}^2}{\left(k_{Om} - \dot{\theta}\right)^2} + \frac{(\dot{k}_{Om} - \ddot{\theta})^2}{(k_{Om} - \dot{\theta})^4}\right]^{1/2} \mathrm{d}s \tag{1.45}$$

The Frenet frame $\{\boldsymbol{R}_m; \boldsymbol{E}_{1m}, \boldsymbol{E}_{2m}\}$ of π_m can be established as

$$\begin{cases} \boldsymbol{E}_{1m} = \dfrac{\mathrm{d}\boldsymbol{R}_m}{\mathrm{d}\sigma_m} = \dfrac{\mathrm{d}\boldsymbol{R}_m}{\mathrm{d}s}\dfrac{\mathrm{d}s}{\mathrm{d}\sigma_m} = \left[-\dfrac{\dot{\theta}}{k_{Om} - \dot{\theta}}\boldsymbol{\alpha} - \dfrac{\dot{k}_{Om} - \ddot{\theta}}{\left(k_{Om} - \dot{\theta}\right)^2}\boldsymbol{\beta} \right] \cdot \dfrac{\mathrm{d}s}{\mathrm{d}\sigma_m} \\ \boldsymbol{E}_{2m} = \boldsymbol{k} \times \boldsymbol{E}_{1m} = \left[\dfrac{\dot{k}_{Om} - \ddot{\theta}}{\left(k_{Om} - \dot{\theta}\right)^2}\boldsymbol{\alpha} - \dfrac{\dot{\theta}}{k_{Om} - \dot{\theta}}\boldsymbol{\beta} \right] \cdot \dfrac{\mathrm{d}s}{\mathrm{d}\sigma_m} \end{cases} \tag{1.46}$$

This can be expressed in the moving frame $\{\boldsymbol{O}_m; \boldsymbol{i}_m, \boldsymbol{j}_m\}$ of $\sum^*$ as

$$\begin{cases} \boldsymbol{R}_m = \dfrac{1}{k_{Om} - \dot{\theta}}\left(-\sin\theta \boldsymbol{i}_m + \cos\theta \boldsymbol{j}_m\right) \\ \boldsymbol{E}_{1m} = \left[\dfrac{-\dot{\theta}\left(k_{Om} - \dot{\theta}\right)\cos\theta + (\dot{k}_{Om} - \ddot{\theta})\sin\theta}{(k_{Om} - \dot{\theta})^2}\boldsymbol{i}_m - \dfrac{\dot{\theta}(k_{Om} - \dot{\theta})\sin\theta + (\dot{k}_{Om} - \ddot{\theta})\cos\theta}{(k_{Om} - \dot{\theta})^2}\boldsymbol{j}_m \right] \cdot \dfrac{\mathrm{d}s}{\mathrm{d}\sigma_m} \\ \boldsymbol{E}_{2m} = \left[\dfrac{\dot{\theta}\left(k_{Om} - \dot{\theta}\right)\sin\theta + (\dot{k}_{Om} - \ddot{\theta})\cos\theta}{(k_{Om} - \dot{\theta})^2}\boldsymbol{i}_m + \dfrac{-\dot{\theta}(k_{Om} - \dot{\theta})\cos\theta + (\dot{k}_{Om} - \ddot{\theta})\sin\theta}{(k_{Om} - \dot{\theta})^2}\boldsymbol{j}_m \right] \cdot \dfrac{\mathrm{d}s}{\mathrm{d}\sigma_m} \end{cases} \tag{1.47}$$

The curvature of π_m, deduced from the Frenet formulas (1.14), is written as

$$\begin{aligned} k_m &= \frac{\mathrm{d}\boldsymbol{E}_{1m}}{\mathrm{d}\sigma_m} \cdot \boldsymbol{E}_{2m} \\ &= \frac{[\dot{\theta}(\ddot{k}_{Om} - \dddot{\theta}) - \ddot{\theta}(\dot{k}_{Om} - \ddot{\theta}) + \dot{\theta}^3(k_{Om} - \dot{\theta})](k_{Om} - \dot{\theta})^3}{[\dot{\theta}^2(k_{Om} - \dot{\theta})^2 + (\dot{k}_{Om} - \ddot{\theta})^2]^{3/2}} \end{aligned} \tag{1.48}$$

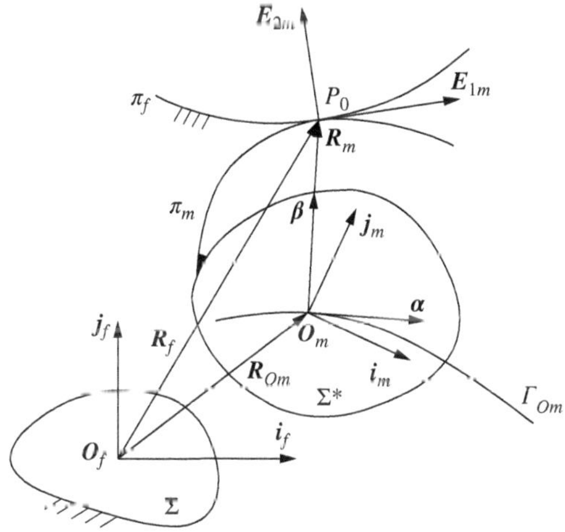

Figure 1.16 The moving centrode π_m adjoint to the given curve

1.2.2.2 Fixed Centrode π_f and its Curvature k_f

The velocity centers trace a curve π_f in the fixed body $\sum$ for all instants, called the ***fixed centrode*** π_f, which is attached to the fixed body $\sum$ and described in $\{\boldsymbol{O}_f;\boldsymbol{i}_f,\boldsymbol{j}_f\}$ of $\sum$. Substituting equation (1.42) into equation (1.37), we have the vector expression of π_f:

$$\pi_f : \boldsymbol{R}_f = \boldsymbol{R}_{Om} + \frac{1}{k_{Om} - \dot{\theta}}\boldsymbol{\beta} \tag{1.49}$$

From the above equation, the fixed centrode π_f is expressed as an adjoint curve of Γ_{Om}. The tangent vector of π_f may be obtained by differentiating equation (1.49) with respect to s:

$$\frac{\mathrm{d}\boldsymbol{R}_f}{\mathrm{d}s} = -\frac{\dot{\theta}}{k_{Om} - \dot{\theta}}\boldsymbol{\alpha} - \frac{\dot{k}_{Om} - \ddot{\theta}}{(k_{Om} - \dot{\theta})^2}\boldsymbol{\beta} \tag{1.50}$$

Based on the above equation, the differential relationship between s and the arc length σ_f of π_f can be derived as

$$\mathrm{d}\sigma_f = |\mathrm{d}\boldsymbol{R}_f| = \left[\frac{\dot{\theta}^2}{\left(k_{Om} - \dot{\theta}\right)^2} + \frac{(\dot{k}_{Om} - \ddot{\theta})^2}{(k_{Om} - \dot{\theta})^4}\right]^{1/2} \mathrm{d}s \tag{1.51}$$

Comparing equation (1.45) with equation (1.51), we find that the differential arc length of π_f is equal to that of π_m, $\mathrm{d}\sigma_f = \mathrm{d}\sigma_m$, or $\mathrm{d}\sigma$ for short. That is, the moving centrode π_m rolls on the fixed centrode π_f without sliding.

The Frenet frame $\{\boldsymbol{R}_f;\boldsymbol{E}_{1f},\boldsymbol{E}_{2f}\}$ of π_f is established as

$$\begin{cases} \boldsymbol{E}_{1f} = \dfrac{\mathrm{d}\boldsymbol{R}_f}{\mathrm{d}\sigma_f} = \dfrac{\mathrm{d}\boldsymbol{R}_f}{\mathrm{d}s}\dfrac{\mathrm{d}s}{\mathrm{d}\sigma_f} = \left[-\dfrac{\dot{\theta}}{k_{Om} - \dot{\theta}}\boldsymbol{\alpha} - \dfrac{\dot{k}_{Om} - \ddot{\theta}}{\left(k_{Om} - \dot{\theta}\right)^2}\boldsymbol{\beta}\right] \cdot \dfrac{\mathrm{d}s}{\mathrm{d}\sigma} \\ \boldsymbol{E}_{2f} = \boldsymbol{k} \times \boldsymbol{E}_{1f} = \left[\dfrac{\dot{k}_{Om} - \ddot{\theta}}{\left(k_{Om} - \dot{\theta}\right)^2}\boldsymbol{\alpha} - \dfrac{\dot{\theta}}{k_{Om} - \dot{\theta}}\boldsymbol{\beta}\right] \cdot \dfrac{\mathrm{d}s}{\mathrm{d}\sigma} \end{cases} \tag{1.52}$$

The relationship between the Frenet frame $\{\boldsymbol{R}_f;\boldsymbol{E}_{1f},\boldsymbol{E}_{2f}\}$ of π_f and $\{\boldsymbol{O}_f;\boldsymbol{i}_f,\boldsymbol{j}_f\}$ of $\sum$ is

$$\begin{cases} \boldsymbol{R}_f = \left[x_{Omf} - \dfrac{\sin(\theta+\gamma)}{k_{Om} - \dot{\theta}}\right]\boldsymbol{i}_f + \left[y_{Omf} + \dfrac{\cos(\theta+\gamma)}{k_{Om} - \dot{\theta}}\right]\boldsymbol{j}_f \\ \boldsymbol{E}_{1f} = \left[\dfrac{-\dot{\theta}\left(k_{Om} - \dot{\theta}\right)\cos(\theta+\gamma) + (\dot{k}_{Om} - \ddot{\theta})\sin(\theta+\gamma)}{(k_{Om} - \dot{\theta})^2}\boldsymbol{i}_f \right. \\ \qquad\qquad \left. + \dfrac{\dot{\theta}\left(k_{Om} - \dot{\theta}\right)\sin(\theta+\gamma) + (\dot{k}_{Om} - \ddot{\theta})\cos(\theta+\gamma)}{(k_{Om} - \dot{\theta})^2}\boldsymbol{j}_f\right] \cdot \dfrac{\mathrm{d}s}{\mathrm{d}\sigma} \\ \boldsymbol{E}_{2f} = \left[\dfrac{\dot{\theta}\left(k_{Om} - \dot{\theta}\right)\sin(\theta+\gamma) + (\dot{k}_{Om} - \ddot{\theta})\cos(\theta+\gamma)}{(k_{Om} - \dot{\theta})^2}\boldsymbol{i}_f \right. \\ \qquad\qquad \left. + \dfrac{-\dot{\theta}\left(k_{Om} - \dot{\theta}\right)\cos(\theta+\gamma) + (\dot{k}_{Om} - \ddot{\theta})\sin(\theta+\gamma)}{(k_{Om} - \dot{\theta})^2}\boldsymbol{j}_f\right] \cdot \dfrac{\mathrm{d}s}{\mathrm{d}\sigma} \end{cases} \tag{1.53}$$

Differentiating $\boldsymbol{E}_{1f}$ of the above equation with respect to σ, we have the following curvature expression for π_f:

$$\begin{aligned} k_f &= \frac{\mathrm{d}\boldsymbol{E}_{1f}}{\mathrm{d}\sigma} \cdot \boldsymbol{E}_{2f} \\ &= \frac{[\dot{\theta}(\ddot{k}_{Om} - \dddot{\theta}) - \ddot{\theta}(\dot{k}_{Om} - \ddot{\theta}) + (\dot{k}_{Om} - \ddot{\theta})^2 + \dot{\theta}^2 k_{Om}(k_{Om} - \dot{\theta})](k_{Om} - \dot{\theta})^3}{[\dot{\theta}^2(k_{Om} - \dot{\theta})^2 + (\dot{k}_{Om} - \ddot{\theta})^2]^{3/2}} \end{aligned} \tag{1.54}$$

For the fixed centrode π_f and the moving centrode π_m at an instant, comparing equation (1.46) with equation (1.52), we find that their Frenet frames coincide at the velocity center. From both equations (1.45) and (1.51), the differentials of their arc lengths are equal to each other, $\mathrm{d}\sigma_f = \mathrm{d}\sigma_m$. We have the following conclusion for plane motion of a rigid body:

For plane motion of a body $\sum^*$ relative to a fixed body $\sum$, there exist a moving centrode π_m on the moving body and a fixed centrode π_f on the fixed body for all instants, π_m rolls on π_f without sliding, and the two Frenet frames of π_m and π_f are coincident at that instant.

The above conclusion may be expressed as

$$\mathrm{d}\sigma_f = \mathrm{d}\sigma_m,\ \boldsymbol{R}_f = \boldsymbol{R}_{Om} + \boldsymbol{R}_m,\ \boldsymbol{E}_{1f}//\boldsymbol{E}_{1m},\ \boldsymbol{E}_{2f}//\boldsymbol{E}_{2m} \tag{1.55}$$

As a result, the moving centrode π_m contacts tangentially the fixed centrode π_f at the velocity center. Consequently, the induced curvature of centrodes, $k^* = k_f - k_m$, appears in the curvature analysis of point paths, implies the geometrical and kinematical properties of the moving body at the instant, and is also an important invariant for the plane motion of a body, which can be derived by subtracting equation (1.48) from equation (1.54), or

$$k^* = k_f - k_m = \frac{(k_{Om} - \dot{\theta})^3}{[\dot{\theta}^2(k_{Om} - \dot{\theta})^2 + (\dot{k}_{Om} - \ddot{\theta})^2]^{1/2}} \tag{1.56}$$

1.2.2.3 Centrodes of a Planar Four-bar Linkage

For a general plane motion of the coupler for a planar four-bar linkage, both the moving and fixed centrodes can be derived by means of the adjoint approach. There exist special points on the coupler, or joints between the coupler and the other binary links, which constrain the motion of the coupler. As in Example 1.4, point B is the joint of link AB and coupler BC. It traces a circle in the fixed frame. At an instant, the coupler can only move along the path of point B and rotate about joint B. Thus, the path, a circle, traced by joint B is taken as the original curve Γ_{Om}; its arc length and differential are $s = a_1\varphi$ and $\mathrm{d}s = a_1\mathrm{d}\varphi$. The Frenet frame of the original curve Γ_{Om} is comprised of a unit tangent vector and a unit normal vector of the circle. The curvature of the original curve Γ_{Om} is a constant, which is the reciprocal of the radius of the circle or the length of link AB, $k_a = 1/a_1$. The rate of change of k_a with respect to the arc length s is zero, or $\dot{k}_a = 0$.

As defined in equation (E1-4.2) and shown in Fig. 1.13, the inclination angle θ between axis $\boldsymbol{i}_m$ of $\{\boldsymbol{O}_m; \boldsymbol{i}_m, \boldsymbol{j}_m\}$ and $\boldsymbol{\alpha}$ of $\{\boldsymbol{R}_B; \boldsymbol{\alpha}, \boldsymbol{\beta}\}$ is $\theta = \pi/2 + \varphi - \gamma$, which leads to $k_a - \dot{\theta} = \dot{\gamma}$.

Substituting $k_{Om} = k_a$ into equation (1.42), we obtain the coordinates $(0, 1/\dot{\gamma})$ of the velocity center in $\{\boldsymbol{R}_B; \boldsymbol{\alpha}, \boldsymbol{\beta}\}$ at an instant, and the polar coordinates in $\{\boldsymbol{O}_m; \boldsymbol{i}_m, \boldsymbol{j}_m\}$ of the coupler plane are

$$\begin{cases} \theta_{P0} = n\pi + (\varphi - \gamma),\ n = 0,\ 1 \\ r_{P0} = \mp 1/\dot{\gamma} \end{cases} \tag{1.57}$$

Then we have $\dot{\gamma} = \frac{d\gamma}{ds} = \frac{d\gamma}{dt}\frac{dt}{ds} = \frac{d\gamma}{dt}\frac{dt}{a_1 d\varphi} = \frac{\omega_2}{a_1\omega_1}$, where ω_1, ω_2 are the angular velocities of link *AB* and coupler *BC*, respectively. Substituting the curvature of the original curve $k_{Om} = k_a$, the rotation angle θ, and equation (1.57) into the vector equations of (1.43) and (1.49), we can locate the moving centrode π_m in $\{\boldsymbol{B}; \boldsymbol{i}_m, \boldsymbol{j}_m\}$ of the coupler and the fixed centrode π_f in $\{\boldsymbol{A}; \boldsymbol{i}_f, \boldsymbol{j}_f\}$ of the fixed link:

$$\begin{cases} \pi_m : \boldsymbol{R}_m = \dfrac{1}{\dot{\gamma}}\boldsymbol{\beta} = -\dfrac{\cos(\varphi - \gamma)}{\dot{\gamma}}\boldsymbol{i}_m - \dfrac{\sin(\varphi - \gamma)}{\dot{\gamma}}\boldsymbol{j}_m \\ \pi_f : \boldsymbol{R}_f = \left(\frac{1}{\dot{\gamma}} - a_1\right)\boldsymbol{\beta} = \left(a_1 - \frac{1}{\dot{\gamma}}\right)(\cos\varphi \boldsymbol{i}_f + \sin\varphi \boldsymbol{j}_f) \end{cases} \tag{1.58}$$

From equation (E1-4.2), the relationship between $\{\boldsymbol{B}; \boldsymbol{i}_m, \boldsymbol{j}_m\}$ of the coupler and $\{\boldsymbol{A}; \boldsymbol{i}_f, \boldsymbol{j}_f\}$ of the fixed link can be written as

$$\begin{cases} \boldsymbol{R}_B = a_1 \cos\varphi \boldsymbol{i}_f + a_1 \sin\varphi \boldsymbol{j}_f \\ \boldsymbol{i}_m = \cos\gamma \boldsymbol{i}_f + \sin\gamma \boldsymbol{j}_f \\ \boldsymbol{j}_m = -\sin\gamma \boldsymbol{i}_f + \cos\gamma \boldsymbol{j}_f \end{cases} \tag{1.59}$$

Substituting $k_{Om} = k_a$ and $\dot{k}_a = 0$ into equations (1.48) and (1.54), we obtain the two curvatures k_m of the moving centrode π_m and k_f of the fixed centrode π_f for a planar four-bar linkage, respectively:

$$\begin{cases} k_m = \dfrac{\left[\ddot{\theta}^2 - \dot{\theta}\dddot{\theta} + \dot{\theta}^3\left(k_a - \dot{\theta}\right)\right](k_a - \dot{\theta})^3}{[\dot{\theta}^2(k_a - \dot{\theta})^2 + \ddot{\theta}^2]^{3/2}} \\ k_f = \dfrac{[2\ddot{\theta}^2 - \dot{\theta}\dddot{\theta} + \dot{\theta}^2 k_a(k_a - \dot{\theta})](k_a - \dot{\theta})^3}{[\dot{\theta}^2(k_a - \dot{\theta})^2 + \ddot{\theta}^2]^{3/2}} \end{cases} \tag{1.60}$$

In addition, we have the induced curvature k^* of π_m and π_f for a planar linkage from equation (1.56):

$$k^* = k_f - k_m = \frac{(k_a - \dot{\theta})^3}{[\dot{\theta}^2(k_a - \dot{\theta})^2 + \ddot{\theta}^2]^{1/2}} \tag{1.61}$$

Example 1.5 Find the moving centrode π_m and the fixed centrode π_f for a planar four-bar linkage.

Solution

A planar four-bar linkage *ABCD* with the moving Cartesian reference frame $\{\boldsymbol{B}; \boldsymbol{i}_m, \boldsymbol{j}_m\}$ and a fixed link with the fixed Cartesian frame $\{\boldsymbol{A}; \boldsymbol{i}_f, \boldsymbol{j}_f\}$ are shown in Fig. 1.17. The displacements of the coupler, both the path of point *B* and the angle γ, have to be solved to describe the

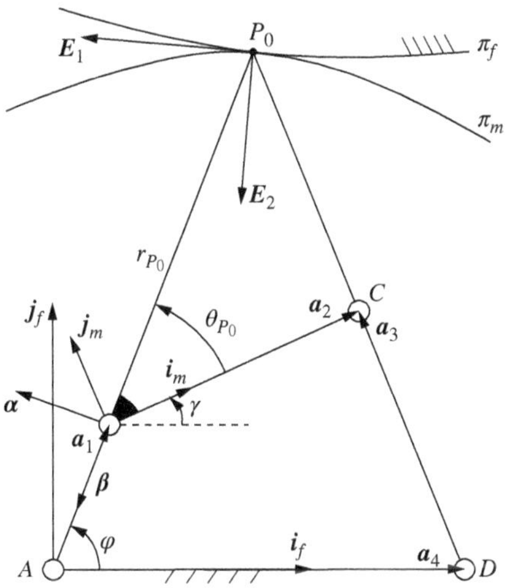

Figure 1.17 The velocity center of the coupler of a planar four-bar linkage

movement of the coupler, or the displacement analysis of a four-bar linkage. In general, we can use the following vector form of the loop-closure equation:

$$\boldsymbol{a}_1 + \boldsymbol{a}_2 = \boldsymbol{a}_3 + \boldsymbol{a}_4 \tag{E1-5.1}$$

Projecting the loop-closure equation into the axes $\boldsymbol{i}_f$ and $\boldsymbol{j}_f$, and simplifying, we have

$$a_1{}^2 + a_2{}^2 + a_4{}^2 - a_3{}^2 + 2a_1a_2\cos(\varphi - \gamma) - 2a_1a_4\cos\varphi - 2a_2a_4\cos\gamma = 0 \tag{E1-5.2}$$

The above nonlinear equation defines a relationship between γ and φ. Singularity may appear as the binary link AB takes on different positions for four-bar linkages with different dimensions. It is a convenient way to solve the displacement equations of a planar four-bar linkage [1], which is converted into one of the two basic types of linkage by changing its fixed frame: a crank rocker for a Grashof kinematic chain and a double rocker for a non-Grashof one.

Differentiating the above equation with respect to the arc length of the original circle, we obtain

$$\dot{\gamma} = \frac{d\gamma}{ds} = [a_2\sin(\varphi - \gamma) - a_4\sin\varphi]/(a_2a_4\sin\gamma + a_1a_2\sin(\varphi - \gamma)) \tag{E1-5.3}$$

1. Crank-rocker linkage

   A crank-rocker linkage has the following dimensions: $a_1 = 1$, $a_2 = 3$, $a_3 = 3.5$, $a_4 = 5$. By substituting $\gamma = \gamma(\varphi)$ into equation (E1-5.3) to get $\dot{\gamma}(\varphi)$, the coordinates of the velocity center can be located in both $\{\boldsymbol{B};\boldsymbol{i}_m,\boldsymbol{j}_m\}$ and $\{\boldsymbol{A};\boldsymbol{i}_f,\boldsymbol{j}_f\}$ by equation (1.58). For all instants $\varphi \in (0, 2\pi)$, the moving centrode π_m and the fixed centrode π_f are calculated, which have first-order contact, or contact in tangency at instant $\varphi = 90°$ (as shown in Fig. 1.18). The velocity center $P_0(r_{P0}, \theta_{P0})$ locates at $(4.2279, 59.3498°)$ on the coupler plane. The curvatures k_m of π_m and k_f of π_f are calculated as -0.1271 and -0.0874, respectively, from equation (1.60), and the induced curvature k^* is 0.0398.

2. Double-crank linkage

A double-crank linkage, shown in Fig. 1.19, is obtained by fixing link AB of the four-bar linkage in Fig. 1.17. The path of joint C, a circle in the fixed frame AB, is taken as an original curve, and the rotation angle of coupler CD around joint C is $\gamma = \gamma(\varphi)$. For $\varphi \in (0, 2\pi)$, the rotation angle of BC relative to AB, the moving centrode π_m, and the fixed centrode π_f can be calculated, as shown in Fig. 1.20.

At $\varphi = 90°$, the velocity center, in polar coordinates, locates at (5.0910, 244.4286°) of the coupler. The curvature k_m of π_m is −0.7219, k_f of π_f is −1.0015, and the induced curvature k^* is −0.2796.

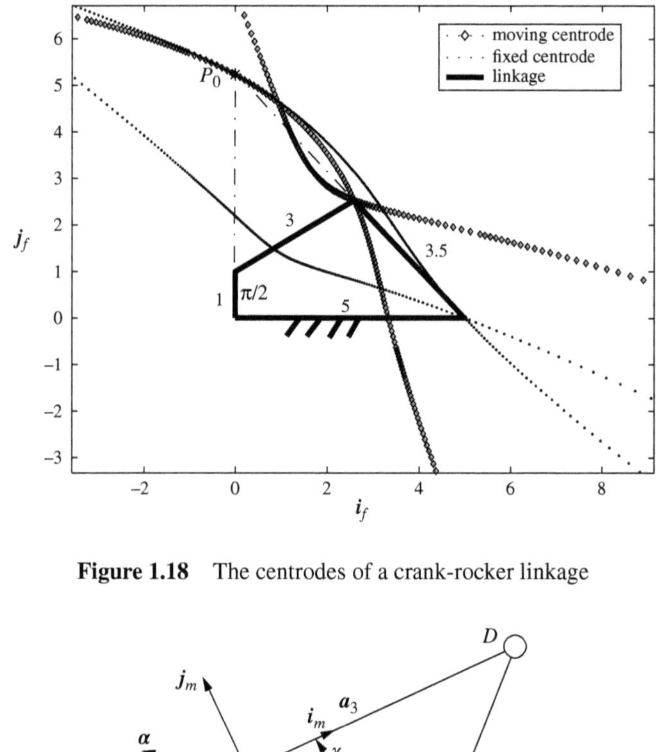

Figure 1.18 The centrodes of a crank-rocker linkage

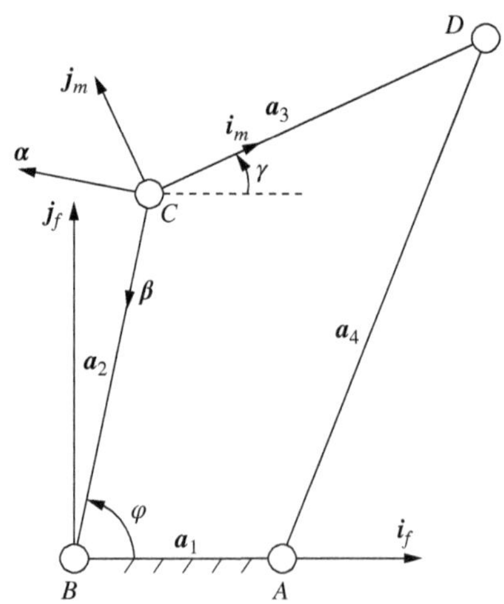

Figure 1.19 A double-crank linkage

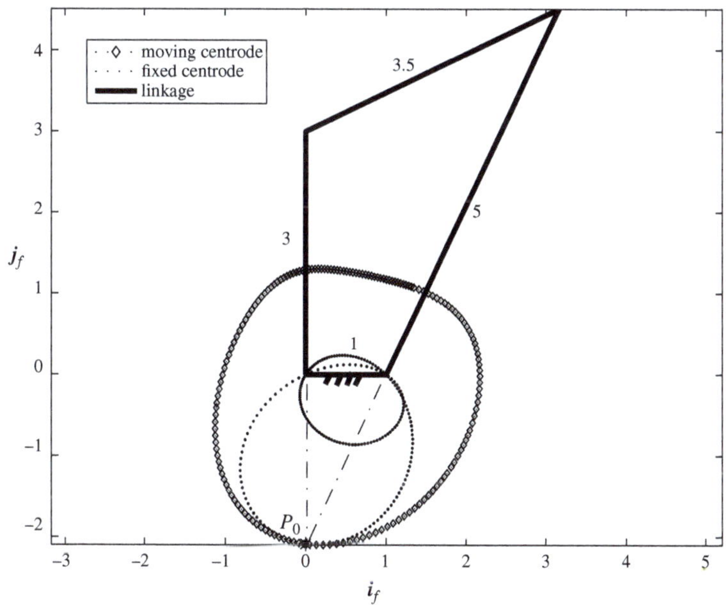

Figure 1.20 The centrodes of a double-crank linkage

Example 1.6 A slider-crank linkage is shown in Fig. 1.21. Write the vector equation of both the moving centrode π_m and the fixed centrode π_f.

Solution

For a slider-crank linkage, the loop-closure equation may be written as

$$\boldsymbol{a}_1 + \boldsymbol{a}_2 = \boldsymbol{S} + \boldsymbol{E} \tag{E1-6.1}$$

Projecting the above equation onto axis $\boldsymbol{j}_f$ of the fixed frame, we obtain

$$a_1 \sin\varphi + a_2 \sin\gamma - e = 0 \tag{E1-6.2}$$

The above equation determines the relationship between the rotation angle φ of link AB and the inclination angle $\gamma = \gamma(\varphi)$. Differentiating the above equation with respect to the arc length of the original curve, a circle in the fixed frame, we have

$$\dot{\gamma} = \frac{\mathrm{d}\gamma}{\mathrm{d}s} = -\frac{\cos\varphi}{a_2\cos\gamma} \tag{E1-6.3}$$

The velocity center point, both in $\{\boldsymbol{B}; \boldsymbol{i}_m, \boldsymbol{j}_m\}$ of the coupler and $\{\boldsymbol{A}; \boldsymbol{i}_f, \boldsymbol{j}_f\}$ of the base link, can be readily located by means of equation (1.58) with a_1, φ, $\gamma(\varphi)$, and $\dot{\gamma}(\varphi)$. For the linkage with $a_1 = 1$, $a_2 = 2.5$, $e = 1.2$, the moving centrode π_m and the fixed centrode π_f are calculated, and at the instant $\varphi = 45°$, the velocity center $P_0(r_{P0}, \theta_{P0})$ locates at $(3.4661, 33.6292°)$ and $k_f = -0.0044$, $k_m = -0.0543$, $k^* = k_f - k_m = 0.0499$, as shown in Fig. 1.22.

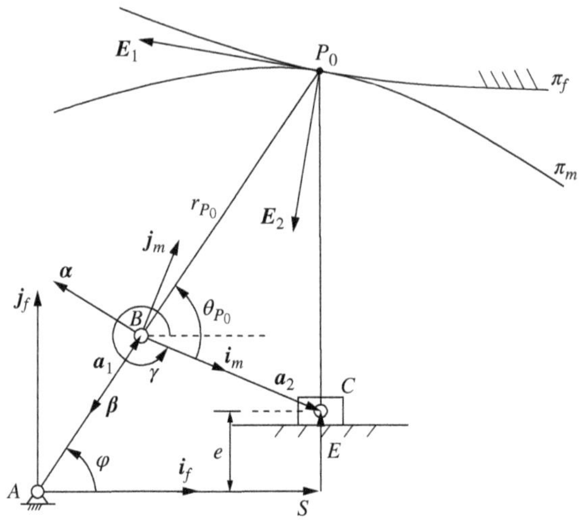

Figure 1.21 A slider-crank linkage

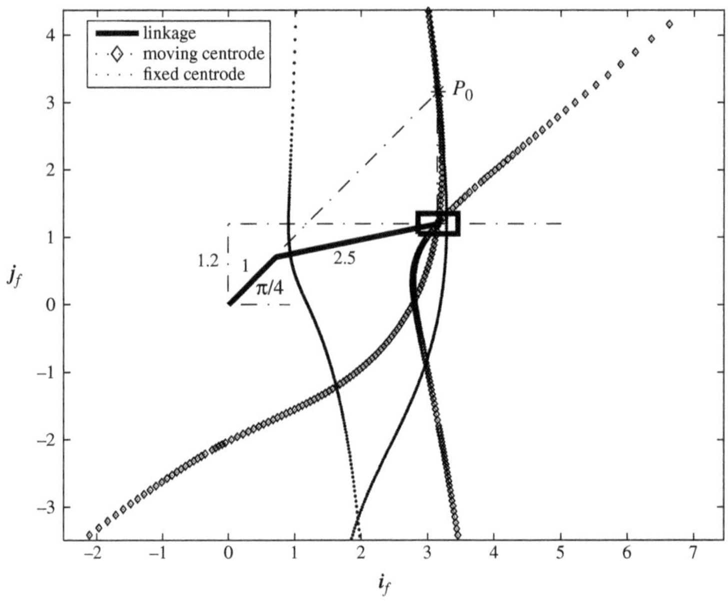

Figure 1.22 The centrodes of a slider-crank linkage

1.2.3 *Euler–Savary Equation*

The planar motions of a moving body $\sum^*$ relative to the fixed body $\sum$ with three degrees of freedom, two translational displacements and a rotational one, are expressed as three equations (1.29), (1.33), and (1.37). Unfortunately, the intrinsic connection between the kinematics and the geometry of the movement of $\sum^*$ is not intuitively revealed in the three equations. In fact, the movement of $\sum^*$ is equivalent to that of the moving centrode π_m, or the moving centrode π_m rolls on the fixed centrode π_f of the fixed body $\sum$ without sliding, which can be viewed as the differential movement of the Frenet frame $\{\boldsymbol{R}_f; \boldsymbol{E}_{1f}, \boldsymbol{E}_{2f}\}$ along the fixed centrode π_f. For both the moving and fixed centrodes, the curvature theory in instantaneous plane kinematics (in particular, the Euler–Savary equation) can be revealed in a natural, intuitive, and analytical way.

In Section 1.2.2, a velocity center locates at both the moving and the fixed centrodes at an instant, or generates a fixed centrode in the fixed frame and a moving centrode in the moving frame for all instants. Hence, point $P(x_{Pm}, y_{Pm})$ or $P(r_{Pm}, \theta_{Pm})$ in $\{\boldsymbol{O}_m; \boldsymbol{i}_m, \boldsymbol{j}_m\}$ of $\sum^*$ traces a point path Γ_P in $\{\boldsymbol{O}_f; \boldsymbol{i}_f, \boldsymbol{j}_f\}$ of $\sum$, which can be designated as an adjoint curve of the fixed centrode π_f, or the fixed centrode π_f is taken as an original curve. Referring to Fig. 1.23, we have

$$\Gamma_P : \boldsymbol{R}_P = \boldsymbol{R}_f + v_1 \boldsymbol{E}_{1f} + v_2 \boldsymbol{E}_{2f} \tag{1.62}$$

where (v_1, v_2), the coordinates of point P of $\sum^*$ in the Frenet frame $\{\boldsymbol{R}_f; \boldsymbol{E}_{1f}, \boldsymbol{E}_{2f}\}$ of π_f, are functions of the arc length σ, represented by the polar coordinates (r, α) in the Frenet frame $\{\boldsymbol{R}_f; \boldsymbol{E}_{1f}, \boldsymbol{E}_{2f}\}$ as follows:

$$v_1 = r\cos\alpha,\ v_2 = r\sin\alpha \tag{1.63}$$

Differentiating equation (1.62) with respect to the arc length σ, and simplifying the so-obtained equation with the Frenet formulas in equation (1.14), we have

$$\frac{\mathrm{d}\boldsymbol{R}_P}{\mathrm{d}\sigma} = \left(1 + \frac{\mathrm{d}v_1}{\mathrm{d}\sigma} - k_f v_2\right)\boldsymbol{E}_{1f} + \left(\frac{\mathrm{d}v_2}{\mathrm{d}\sigma} + k_f v_1\right)\boldsymbol{E}_{2f} \tag{1.64}$$

The point P, a fixed point attached to the moving body $\sum^*$, can be represented in the Frenet frame $\{\boldsymbol{R}_m; \boldsymbol{E}_{1m}, \boldsymbol{E}_{2m}\}$ of π_m; its position does not change in the moving frame $\{\boldsymbol{O}_m; \boldsymbol{i}_m, \boldsymbol{j}_m\}$ as

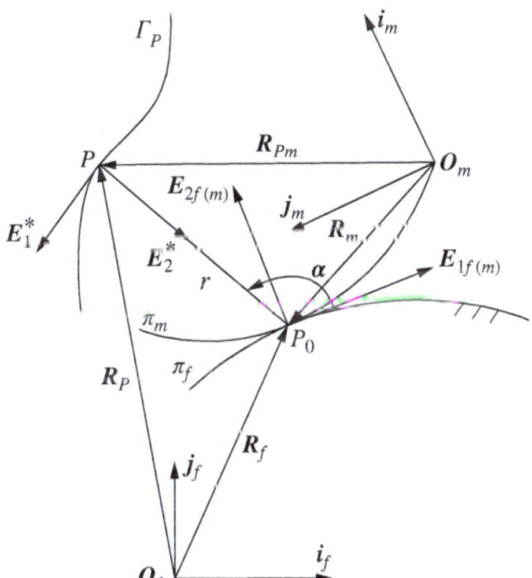

Figure 1.23 The point path adjoint to the centrodes of a moving body

$\sum^*$ moves, which can be expressed as

$$\boldsymbol{R}_{Pm} = \boldsymbol{R}_m + u_1\boldsymbol{E}_{1m} + u_2\boldsymbol{E}_{2m} \tag{1.65}$$

where (u_1, u_2) are the coordinates of point P in $\{\boldsymbol{R}_m; \boldsymbol{E}_{1m}, \boldsymbol{E}_{2m}\}$ of π_m, and functions of the arc length σ. In particular, vector $\boldsymbol{R}_{Pm}$ in equation (1.65) points from the origin O_m of $\{\boldsymbol{O}_m; \boldsymbol{i}_m, \boldsymbol{j}_m\}$ to the point P via the velocity center P_0 in $\sum^*$ at the instant. While P_0 moves along π_m, the coordinates (u_1, u_2) change their values, although the positions of the starting point O_m and the end point P do not change. Hence, the point P of $\sum^*$ is examined in the moving frame $\{\boldsymbol{R}_m; \boldsymbol{E}_{1m}, \boldsymbol{E}_{2m}\}$ to be a path, which can be viewed as an adjoint curve of π_m, or π_m is an original curve. Differentiating equation (1.65) with respect to σ_m, and utilizing the Frenet formulas, we have

$$\frac{\mathrm{d}\boldsymbol{R}_{Pm}}{\mathrm{d}\sigma_m} = \left(1 + \frac{\mathrm{d}u_1}{\mathrm{d}\sigma_m} - k_m u_2\right)\boldsymbol{E}_{1m} + \left(\frac{\mathrm{d}u_2}{\mathrm{d}\sigma_m} + k_m u_1\right)\boldsymbol{E}_{2m} \tag{1.66}$$

The position of point P in $\{\boldsymbol{O}_m; \boldsymbol{i}_m, \boldsymbol{j}_m\}$ does not change since it is a fixed point in $\sum^*$, or $\mathrm{d}\boldsymbol{R}_{Pm}/\mathrm{d}\sigma_m = 0$, which is Cesaro's fixed point conditions:

$$\begin{cases} 1 + \dfrac{\mathrm{d}u_1}{\mathrm{d}\sigma_m} - k_m u_2 = 0 \\ \dfrac{\mathrm{d}u_2}{\mathrm{d}\sigma_m} + k_m u_1 = 0 \end{cases} \tag{1.67}$$

or

$$\frac{\mathrm{d}u_1}{\mathrm{d}\sigma_m} = k_m u_2 - 1, \ \frac{\mathrm{d}u_2}{\mathrm{d}\sigma_m} = -k_m u_1 \tag{1.68}$$

Note that $\{\boldsymbol{R}_m; \boldsymbol{E}_{1m}, \boldsymbol{E}_{2m}\}$ moves along π_m, and the position change of point P is described in the Frenet frame $\{\boldsymbol{R}_m; \boldsymbol{E}_{1m}, \boldsymbol{E}_{2m}\}$ as $\left(\frac{\mathrm{d}u_1}{\mathrm{d}\sigma_m}, \frac{\mathrm{d}u_2}{\mathrm{d}\sigma_m}\right)$ (just like the breath-taking scenery of the banks examined by the boatman, which implies the geometrical properties of the river). Hence, $\left(\frac{\mathrm{d}u_1}{\mathrm{d}\sigma_m}, \frac{\mathrm{d}u_2}{\mathrm{d}\sigma_m}\right)$ express the movement of point P of $\sum^*$ relative to the Frenet frame $\{\boldsymbol{R}_m; \boldsymbol{E}_{1m}, \boldsymbol{E}_{2m}\}$ of π_m, and imply the geometrical properties of the moving centrode π_m.

As mentioned in Section 1.2.2, the Frenet frame $\{\boldsymbol{R}_f; \boldsymbol{E}_{1f}, \boldsymbol{E}_{2f}\}$ of the fixed centrode π_f is coincident with the Frenet frame $\{\boldsymbol{R}_m; \boldsymbol{E}_{1m}, \boldsymbol{E}_{2m}\}$ of the moving centrode π_m at an instant. A point P of $\sum^*$ has the same coordinates and changing rates in these two Frenet frame:

$$u_1 = v_1, \ u_2 = v_2, \ \frac{\mathrm{d}u_1}{\mathrm{d}\sigma_m} = \frac{\mathrm{d}v_1}{\mathrm{d}\sigma_f}, \ \frac{\mathrm{d}u_2}{\mathrm{d}\sigma_m} = \frac{\mathrm{d}v_2}{\mathrm{d}\sigma_f} \tag{1.69}$$

Substituting equation (1.68) into (1.64) and utilizing equation (1.63), we have the first derivative of a curve Γ_P:

$$\frac{\mathrm{d}\boldsymbol{R}_P}{\mathrm{d}\sigma} = (k_f - k_m)(-v_2\boldsymbol{E}_{1f} + v_1\boldsymbol{E}_{2f}) = rk^*(-\sin\alpha\boldsymbol{E}_{1f} + \cos\alpha\boldsymbol{E}_{2f}) \tag{1.70}$$

The identical magnitudes can be taken from either side of the above equation. The differential relationship between the arc length σ_P of Γ_P and the arc length σ of the fixed centrode π_f is

$$\mathrm{d}\sigma_P = \left|\frac{\mathrm{d}\boldsymbol{R}_P}{\mathrm{d}\sigma}\right|\mathrm{d}\sigma = rk^*\mathrm{d}\sigma \tag{1.71}$$

Differentiating equation (1.63) with respect to σ and utilizing the conditions of the fixed point in equation (1.68), we obtain

$$\frac{\mathrm{d}r}{\mathrm{d}\sigma} = -\cos\alpha, \quad \frac{\mathrm{d}\alpha}{\mathrm{d}\sigma} = -k_m + \frac{\sin\alpha}{r} \tag{1.72}$$

Now the Frenet frame $\{\boldsymbol{R}_P; \boldsymbol{E}_1^*, \boldsymbol{E}_2^*\}$ of the point path Γ_P can be established by equations (1.70) and (1.71) as

$$\begin{cases} \boldsymbol{E}_1^* = \dfrac{\mathrm{d}\boldsymbol{R}_P}{\mathrm{d}\sigma_P} = \dfrac{\mathrm{d}\boldsymbol{R}_P}{\mathrm{d}\sigma} \Big/ \left|\dfrac{\mathrm{d}\boldsymbol{R}_P}{\mathrm{d}\sigma}\right| = -\sin\alpha \boldsymbol{E}_{1f} + \cos\alpha \boldsymbol{E}_{2f} \\ \boldsymbol{E}_2^* = \boldsymbol{k} \times \boldsymbol{E}_1^* = -\cos\alpha \boldsymbol{E}_{1f} - \sin\alpha \boldsymbol{E}_{2f} \end{cases} \tag{1.73}$$

Differentiating $\boldsymbol{E}_1^*$ in the above equation with respect to the arc length σ_P, and dotting $\boldsymbol{E}_2^*$, we can write the curvature k_Γ of Γ_P as

$$k_\Gamma = \frac{\mathrm{d}\boldsymbol{E}_1^*}{\mathrm{d}\sigma_P} \cdot \boldsymbol{E}_2^* = \frac{1}{r} + \frac{\sin\alpha}{r^2 k^*} = \frac{r + \sin\alpha / k^*}{r^2} \tag{1.74a}$$

This is the famous Euler Savary equation in the Frenet frame of the centrodes. It indicates the relationship among the position (r, α) of the point P on the moving body, the kinematic invariant k^* of the movement of the rigid body, and the curvature k_Γ of the point path Γ_P at an instant. The parameters in equation (1.74a), such as r, α, k^*, are functions of the arc length σ of the centrodes. If the curvature k_Γ is replaced by the reciprocal of the radius of curvature ρ_Γ, another expression of the Euler–Savary equation is

$$\rho_\Gamma \left(r + \frac{\sin\alpha}{k^*}\right) = r^2 \tag{1.74b}$$

This form of the Euler–Savary equation in the Frenet frame of centrodes reveals the radius ρ_Γ of curvature, the polar coordinates (r, α) of the tracing point, and the induced curvature k^* of the centrodes.

The radius of curvature ρ_Γ is replaced by vector $\overrightarrow{PO_P}$ from the point P to the curvature center O_P. The polar coordinates (r, α) of point P are represented by vector $\overrightarrow{PP_0}$ from P to P_0, the instant velocity center. The induced curvature k^* is the reciprocal of vector $\overrightarrow{PP_I}$ from the point P to the ***inflection point*** P_I of the line $\overrightarrow{PO_P}$. The geometrical meaning of the Euler–Savary equation (1.74) can be written as

$$\overrightarrow{PP_I} \cdot \overrightarrow{PO_P} = (\overrightarrow{PP_0})^2 \tag{1.74c}$$

As shown in Fig. 1.24, the inflection point P_I and the curvature center O_P always lie on one side of the point path since the right-hand side of the above equation is always positive. In particular, the curvature center O_P of the point path Γ_P is at infinity when the point P coincides with the inflection point P_I, or $\overrightarrow{PP_I} = 0$.

The point P has polar coordinates (r, α) in the Frenet frame $\{\boldsymbol{R}_m; \boldsymbol{E}_{1m}, \boldsymbol{E}_{2m}\}$, and the curvature k_Γ is derived from the point path Γ_P in the Frenet frame $\{\boldsymbol{R}_f; \boldsymbol{E}_{1f}, \boldsymbol{E}_{2f}\}$, such as equations (1.74a). From equations (1.63), (1.65), and (1.69), the polar coordinates (r, α) of the point P are replaced by their Cartesian coordinates (x_{Pm}, y_{Pm}) or polar coordinates (r_{Pm}, θ_{Pm})

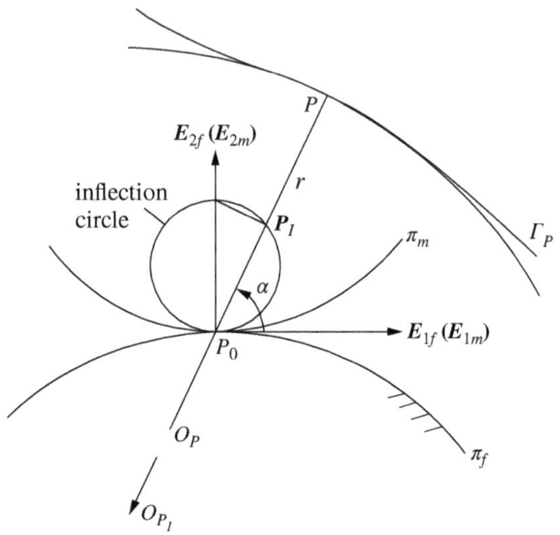

Figure 1.24 Euler–Savary equation of a point path

in the moving frame $\{\boldsymbol{O}_m;\boldsymbol{i}_m,\boldsymbol{j}_m\}$ of $\sum^*$. We have

$$\begin{cases} k_\Gamma = F/G \\ F = \left(x_{Pm} - a\right)^2 + (y_{Pm} - b)^2 - D^2/4 \\ G = \{[x_{Pm} + \sin\theta/(k_{Om} - \dot{\theta})]^2 + [y_{Pm} - \cos\theta/(k_{Om} - \dot{\theta})]^2\}^{3/2} \\ a = -\dfrac{(\dot{k_{Om}} - \ddot{\theta})\cos\theta + (2k_{Om} - \dot{\theta})(k_{Om} - \dot{\theta})\sin\theta}{2(k_{Om} - \dot{\theta})^3} \\ b = \dfrac{-(\dot{k_{Om}} - \ddot{\theta})\sin\theta + (2k_{Om} - \dot{\theta})(k_{Om} - \dot{\theta})\cos\theta}{2(k_{Om} - \dot{\theta})^3} \end{cases} \tag{1.74d}$$

As mentioned in Section 1.1.2, a straight line contacts a curve in second order, which results in the contact point being the inflection point of the curve. By letting the curvature k_Γ of Γ_P in equation (1.74) be zero, or the point path in the fixed frame $\{\boldsymbol{O}_f;\boldsymbol{i}_f,\boldsymbol{j}_f\}$ be an inflection point at an instant, the positions of all points on $\sum^*$ can be located by

$$r + D\sin\alpha = 0 \tag{1.75}$$

where $D = 1/k^*$. The above equation is the equation of a circle in the Frenet frame $\{\boldsymbol{R}_f;\boldsymbol{E}_{1f},\boldsymbol{E}_{2f}\}$, called an ***inflection circle*** of $\sum^*$. We can visualize (r,α) as two independent variables, since they are obtained by converting the Cartesian coordinates (x_{Pm}, y_{Pm}) or the polar coordinates (r_{Pm}, θ_{Pm}) of the point P in the moving frame $\{\boldsymbol{O}_m;\boldsymbol{i}_m,\boldsymbol{j}_m\}$ into those in $\{\boldsymbol{R}_f;\boldsymbol{E}_{1f},\boldsymbol{E}_{2f}\}$. Hence, the inflection circle exists on $\sum^*$ and has the same equation in the Frenet frame $\{\boldsymbol{R}_m;\boldsymbol{E}_{1m},\boldsymbol{E}_{2m}\}$, since the two Frenet frames are coincidental at the instant. Any point of this circle is an inflection point at that instant; its path has $k_\Gamma = 0$, except for the instant center (see Fig. 1.24). The inflection circle with diameter $D = |1/k^*|$ is tangent to π_m and the inflection circle center lies on the normal of π_m. Both the diameter and the inflection circle center vary as $\sum^*$ moves. From the viewpoint of different instants, or different positions of $\sum^*$ corresponding to different values of the centrode arc length σ, equation (1.75) may be regarded as the equation of a family of inflection circles with parameters (r, α, σ), or (x_{Pm}, y_{Pm}, σ). In particular, the inflection circle passes through the velocity center point P_0 at that instant, but the velocity center point P_0 is an singular point of the inflection circle, belongs to the moving centrode π_m, and traces a path with an infinite curvature $k = \infty$, a cusp, at this instant by equation (1.74).

Example 1.7 A crank-rocker linkage with moving centrode π_m and fixed centrode π_f is shown in Example 1.5. Determine the Euler–Savary equation.

Solution

The Frenet frame $\{\boldsymbol{R}_m; \boldsymbol{E}_{1m}, \boldsymbol{E}_{2m}\}$ of the moving centrode π_m can be established by equation (1.47), as shown in Fig. 1.17:

$$\begin{cases} \boldsymbol{R}_m = \dfrac{1}{k_a - \dot{\theta}} \left(-\sin\theta \boldsymbol{i}_m + \cos\theta \boldsymbol{j}_m\right) \\ \boldsymbol{E}_{1m} = \dfrac{-[\dot{\theta}(k_a - \dot{\theta})\cos\theta + \ddot{\theta}\sin\theta]\boldsymbol{i}_m + [-\dot{\theta}(k_a - \dot{\theta})\sin\theta + \ddot{\theta}\cos\theta]\boldsymbol{j}_m}{[\dot{\theta}^2(k_a - \dot{\theta})^2 + \ddot{\theta}^2]^{1/2}} \\ \boldsymbol{E}_{2m} = \dfrac{[\dot{\theta}(k_a - \dot{\theta})\sin\theta - \ddot{\theta}\cos\theta]\boldsymbol{i}_m - [\dot{\theta}(k_a - \dot{\theta})\cos\theta + \ddot{\theta}\sin\theta]\boldsymbol{j}_m}{[\dot{\theta}^2(k_a - \dot{\theta})^2 + \ddot{\theta}^2]^{1/2}} \end{cases} \tag{E1-7.1}$$

The inflection circle equation (1.75) in the Frenet frame $\{\boldsymbol{R}_m; \boldsymbol{E}_{1m}, \boldsymbol{E}_{2m}\}$ is rewritten in the moving frame $\{\boldsymbol{B}; \boldsymbol{i}_m, \boldsymbol{j}_m\}$ of the coupler link as

$$(x_m - a)^2 + (y_m - b)^2 = D^2/4 \tag{E1-7.2a}$$

where the coordinates (a, b) of the inflection circle center are

$$\begin{cases} a = -\dfrac{-\ddot{\theta}\cos\theta + \left(2k_a - \dot{\theta}\right)(k_a - \dot{\theta})\sin\theta}{2(k_a - \dot{\theta})^3} \\ b = \dfrac{\ddot{\theta}\sin\theta + (2k_a - \dot{\theta})(k_a - \dot{\theta})\cos\theta}{2(k_a - \dot{\theta})^3} \end{cases} \tag{E1-7.2b}$$

At $\varphi = 238°$, the polar coordinates (r_{P0}, θ_{P0}) of the velocity center P_0 in the moving frame $\{\boldsymbol{B}; \boldsymbol{i}_m, \boldsymbol{j}_m\}$ of the coupler are $(2.6995, 15.9604°)$, the curvatures of centrodes are $k_f = 0.3712$, $k_m = 0.5892$, and the induced curvature is $k^* = -0.2180$. Based on equation (1.74b), the Euler–Savary equation in the Frenet frame $\{\boldsymbol{R}_m; \boldsymbol{E}_{1m}, \boldsymbol{E}_{2m}\}$ for the crank-rocker linkage at this instant is

$$\rho_\Gamma \left(r - \frac{\sin\alpha}{0.2180}\right) = r^2 \tag{E1-7.3}$$

In terms of coordinates, the inflection circle at an instant can be expressed in the moving frame $\{\boldsymbol{B}; \boldsymbol{i}_m, \boldsymbol{j}_m\}$ of the coupler as

$$(x_m - 4.8075)^2 + (y_m - 1.3502)^2 = 5.2627 \tag{E1-7.4}$$

From equation (E1-7.3), a coupler point P, with polar coordinates $(5, 250°)$ in the Frenet frame $\{\boldsymbol{R}_m; \boldsymbol{E}_{1m}, \boldsymbol{E}_{2m}\}$ and Cartesian coordinates $(-2.3882, 1.1463)$ in the moving frame $\{\boldsymbol{B}; \boldsymbol{i}_m, \boldsymbol{j}_m\}$, traces the path with a radius of curvature $\rho_\Gamma = 2.6849$ at this instant. The curvature center O_P locates at $(2.3151, 250°)$ in the Frenet frame $\{\boldsymbol{R}_f; \boldsymbol{E}_{1f}, \boldsymbol{E}_{2f}\}$ and at $(-0.9384, 0.0349)$ in the fixed frame $\{\boldsymbol{O}_f; \boldsymbol{i}_f, \boldsymbol{j}_f\}$, while point P_I has polar coordinates $(4.3114, 70°)$ and Cartesian coordinates $(6.8928, 0.3940)$ in the moving frame $\{\boldsymbol{B}; \boldsymbol{i}_m, \boldsymbol{j}_m\}$.

For convenience of plotting, all points are expressed in the fixed frame $\{\boldsymbol{O}_f; \boldsymbol{i}_f, \boldsymbol{j}_f\}$, as shown in Fig. 1.25.

Example 1.8 A slider-crank linkage with moving centrode π_m and fixed centrode π_f is shown in Example 1.6. Determine the Euler–Savary equation.

Solution

The inflection circle is located by equations (E1-7.2) and (E1-7.3) in the moving frame $\{\boldsymbol{B}; \boldsymbol{i}_m, \boldsymbol{j}_m\}$ of the coupler.

At $\varphi = -30°$, the instant center $P_0(r_{P0}, \theta_{P0})$ has polar coordinates $(2.1166, -72.8463°)$ in $\{\boldsymbol{B}; \boldsymbol{i}_m, \boldsymbol{j}_m\}$, and the curvatures are $k_f = -0.1076$, $k_m = -0.2585$, $k^* = 0.1509$. Based on equation (1.74b), the Euler–Savary equation in the Frenet frame $\{\boldsymbol{R}_m; \boldsymbol{E}_{1m}, \boldsymbol{E}_{2m}\}$ for the slider-crank linkage is

$$\rho_\Gamma \left(r + \frac{\sin\alpha}{0.1509} \right) = r^2 \tag{E1-8.1}$$

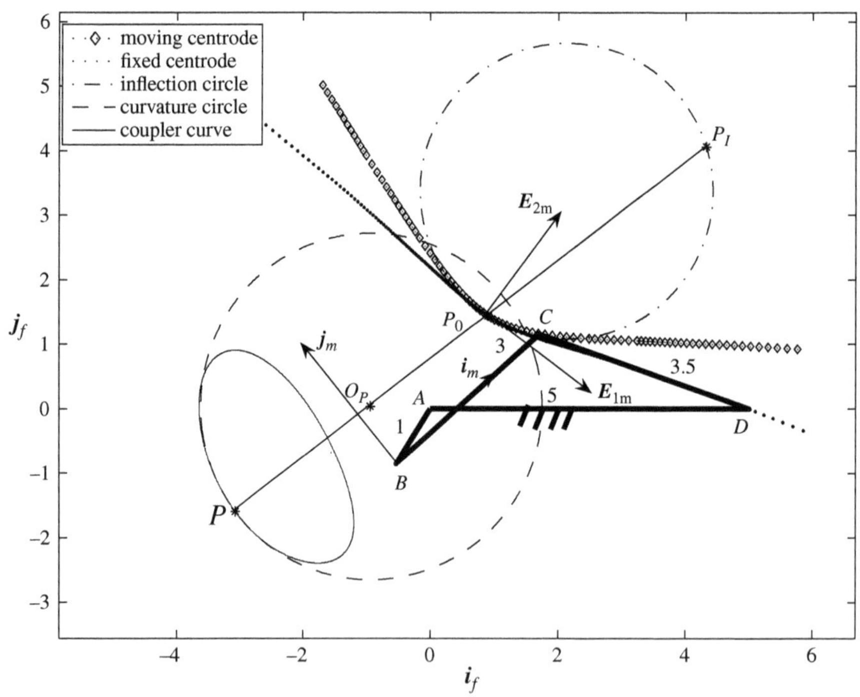

Figure 1.25 Euler–Savary equation of the point path for a crank-rocker linkage

The inflection circle can be located instantaneously by its center coordinates (a, b) and expressed in the moving frame $\{\boldsymbol{B}; \boldsymbol{i}_m, \boldsymbol{j}_m\}$ as

$$(x_m + 0.6465)^2 + (y_m - 1.0372)^2 = 10.9760 \qquad \text{(E1-8.2)}$$

A coupler point P with polar coordinates $(5.7, 45°)$ in $\{\boldsymbol{R}_m; \boldsymbol{E}_{1m}, \boldsymbol{E}_{2m}\}$ and Cartesian coordinates $(-1.5518, -7.2907)$ in $\{\boldsymbol{B}; \boldsymbol{i}_m, \boldsymbol{j}_m\}$ traces a path with radius of curvature $\rho_\Gamma = 3.1285$ by equation (E1-8.1) at this instant. The curvature center O_P has polar coordinates $(2.5715, 45°)$ in the Frenet frame $\{\boldsymbol{R}_f; \boldsymbol{E}_{1f}, \boldsymbol{E}_{2f}\}$, or rectangular coordinates $(3.5954, -3.9686)$ in the fixed frame $\{\boldsymbol{A}; \boldsymbol{i}_f, \boldsymbol{j}_f\}$, while the point P_I has polar coordinates $(4.6853, 225°)$ in the Frenet frame $\{\boldsymbol{R}_m; \boldsymbol{E}_{1m}, \boldsymbol{E}_{2m}\}$ or rectangular coordinates $(2.4131, 2.3080)$ in the moving frame $\{\boldsymbol{B}; \boldsymbol{i}_m, \boldsymbol{j}_m\}$, as shown in Fig. 1.26.

1.2.4 Curvatures in Higher Order

The curvature of a point path only reveals the geometrical property at three infinitesimally separated positions surrounding a nominal point. Sometimes, higher-order curvature is needed in the neighborhood of a point on a point path. In particular, the derivative of the curvature has a special value, such as zero, which indicates something important for the kinematics and geometry of linkages.

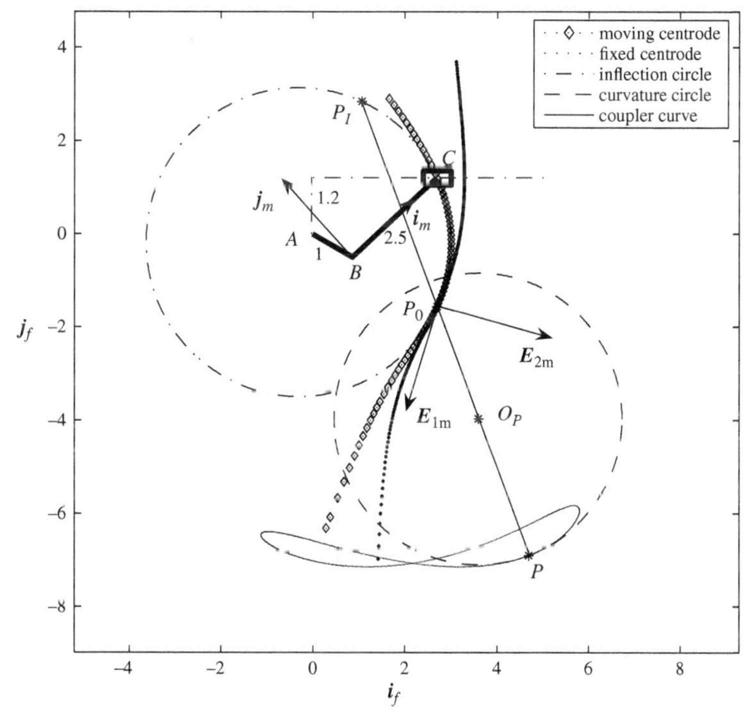

Figure 1.26 Euler–Savary equation of the point path for a slider-crank linkage

Differentiating equation (1.74a) once and twice with respect to the arc length of the centrode, we have

$$\frac{\mathrm{d}k_{\Gamma}}{\mathrm{d}\sigma} = \frac{1}{H}\left(\frac{1}{M\sin\alpha} + \frac{1}{N\cos\alpha} - \frac{1}{r}\right) \tag{1.76}$$

$$\frac{\mathrm{d}^2k_{\Gamma}}{\mathrm{d}\sigma^2} = \left(T - \frac{1}{H}\frac{\mathrm{d}H}{\mathrm{d}\sigma}\right)\frac{\mathrm{d}k^*}{\mathrm{d}\sigma} - H\cos\alpha\left(\frac{\mathrm{d}k^*}{\mathrm{d}\sigma}\right)^2 + \frac{1}{H}RQ \tag{1.77}$$

where

$$\begin{cases} \dfrac{1}{H} = -\dfrac{3\sin\alpha\cos\alpha}{r^2k^*}, \ \dfrac{1}{M} = \dfrac{k_m - k^*}{3}, \ \dfrac{1}{N} = \dfrac{\mathrm{d}k^*/\mathrm{d}\sigma}{3k^*}, \ T = \dfrac{3}{M}\cot\alpha + \dfrac{1}{N}\left(2 - \tan^2\alpha\right) \\ R = \dfrac{1}{N^2}\dfrac{\cos\alpha}{\sin^2\alpha}, \ Q = \tan^4\alpha + c_1\tan^3\alpha + c_2\tan^2\alpha + c_3\tan\alpha + c_4 \\ c_1 = -k_mN + \dfrac{N}{M}, \ c_2 = -\left(1 + \dfrac{\mathrm{d}N}{\mathrm{d}\sigma}\right), \ c_3 = -\dfrac{\left(\dfrac{\mathrm{d}M}{\mathrm{d}\sigma}N + 3M\right)N}{M^2}, \ c_4 = \dfrac{(Mk_m - 2)N^2}{M^2} \end{cases} \tag{1.78}$$

All of these are described in the Frenet frame $\{\boldsymbol{R}_f; \boldsymbol{E}_{1f}, \boldsymbol{E}_{2f}\}$, and the same as in the Frenet frame $\{\boldsymbol{R}_m; \boldsymbol{E}_{1m}, \boldsymbol{E}_{2m}\}$, while the parameters (r, α, σ), or (x_{Pm}, y_{Pm}, σ), are independent variables.

1.2.4.1 Cubic Stationary Curvature

Point P of $\sum^*$ traces a path Γ_P in the fixed frame $\{\boldsymbol{O}_f; \boldsymbol{i}_f, \boldsymbol{j}_f\}$. For a certain instant, the point P happens to be at a point of path Γ_P with $\frac{\mathrm{d}k_{\Gamma}}{\mathrm{d}\sigma} = 0$ in equation (1.76); we obtain

$$\frac{1}{r} = \frac{1}{M\sin\alpha} + \frac{1}{N\cos\alpha} \tag{1.79}$$

The above relation is described in the Frenet frame $\{\boldsymbol{R}_f; \boldsymbol{E}_{1f}, \boldsymbol{E}_{2f}\}$, and may be simplified as a cubic equation in the coordinates of the moving frame $\{\boldsymbol{O}_m; \boldsymbol{i}_m, \boldsymbol{j}_m\}$ if (r, α) and (x_{Pm}, y_{Pm}) are taken as independent variables for instant σ. The curve defined by the above equation is called the ***cubic of stationary curvature*** at that instant. All points on the cubic curve of $\sum^*$ trace point paths with $\mathrm{d}k_{\Gamma}/\mathrm{d}\sigma = 0$ at instant σ. The point path Γ_P and a circle have third-order contact at this instant. We have a family of cubic curves on $\sum^*$ with respect to the arc length σ if (x_{Pm}, y_{Pm}, σ) are all viewed as independent variables.

1.2.4.2 Ball Point

Letting $k_{\Gamma} = 0$ and $\mathrm{d}k_{\Gamma}/\mathrm{d}\sigma = 0$ simultaneously, and solving equations (1.75) and (1.79), we obtain one point on $\sum^*$ in the Frenet frame $\{\boldsymbol{R}_m; \boldsymbol{E}_{1m}, \boldsymbol{E}_{2m}\}$. Its polar coordinates are

$$\begin{cases} \tan\alpha = -\dfrac{k^*\left(k_m + 2k^*\right)}{\mathrm{d}k^*/\mathrm{d}\sigma} \\ r = -\dfrac{\sin\alpha}{k^*} \end{cases} \tag{1.80}$$

The point, the intersection of the inflection circle with the cubic of stationary curvature, is called the ***Ball point***. As mentioned in Section 1.1.2, the point path Γ_P in the fixed frame $\{\boldsymbol{O}_f;\boldsymbol{i}_f,\boldsymbol{j}_f\}$, traced by the Ball point of $\sum^*$, contacts a line in third order at this instant.

1.2.4.3 Burmester Point

Furthermore, from equations (1.76) and (1.77), the point path Γ_P with $\mathrm{d}k_\Gamma/\mathrm{d}\sigma = 0$, and $\mathrm{d}^2k_\Gamma/\mathrm{d}\sigma^2 = 0$ at instant σ, we have the following expressions in the Frenet frame $\{\boldsymbol{R}_m;\boldsymbol{E}_{1m},\boldsymbol{E}_{2m}\}$:

$$\begin{cases} \dfrac{1}{r} = \dfrac{1}{M\sin\alpha} + \dfrac{1}{N\cos\alpha} \\ \tan^4\alpha + c_1\tan^3\alpha + c_2\tan^2\alpha + c_3\tan\alpha + c_4 = 0 \end{cases} \tag{1.81}$$

According to equation (1.79), $\mathrm{d}k_\Gamma/\mathrm{d}\sigma = 0$ is a cubic equation with two independent variables, while $\mathrm{d}^2k_\Gamma/\mathrm{d}\sigma^2 = 0$ in (1.81) is a quartic equation in $\tan\alpha$; it may have four real roots, two real roots, or no real roots. In order for equations to be true at a particular instant, there may exist four real solutions, two real solutions, or even no real solutions. The solutions correspond to the points on the moving body, called the ***Burmester points*** of $\sum^*$, which trace the path with $\mathrm{d}k_\Gamma/\mathrm{d}\sigma = 0$ and $\mathrm{d}^2k_\Gamma/\mathrm{d}\sigma^2 = 0$ in the fixed frame $\{\boldsymbol{O}_f;\boldsymbol{i}_f,\boldsymbol{j}_f\}$ at that instant. The point path Γ_P and a circle have five common points at infinitesimally separated positions in the neighborhood of the Burmester point, or they have fourth-order contact at this instant.

1.2.4.4 Ball Curve and Singularity

The arc length σ of the centrode is designated as an independent variable while either (r,α) or (x_{Pm}, y_{Pm}) hold equation (1.80) true, or equation (1.80) yields a single Ball point on the moving body at an instant. All Ball points form a curve on the moving body for all instants, called ***Ball curve***.

For a point $P(x_{Pm}, y_{Pm})$ of $\sum^*$, the curvature equation (1.74d) can be rewritten in the moving frame $\{\boldsymbol{O}_m;\boldsymbol{i}_m,\boldsymbol{j}_m\}$ as

$$k_\Gamma = F(x_{Pm}, y_{Pm}, s)/G(x_{Pm}, y_{Pm}, s) \tag{1.82}$$

where F and G are given in equation (1.74d) as functions of (x_{Pm}, y_{Pm}, s).

Differentiating equation (1.82) with respect to s of the original curve Γ_{Om}, and letting $\mathrm{d}k_\Gamma/\mathrm{d}s = 0$, we have

$$\frac{\mathrm{d}k_\Gamma}{\mathrm{d}s} = \left(\frac{\mathrm{d}F}{\mathrm{d}s}G - \frac{\mathrm{d}G}{\mathrm{d}s}F\right)/G^2 = 0 \tag{1.83}$$

The above equation can be simplified as a cubic equation in terms of the coordinates of points of $\sum^*$, which is identical to equation (1.79).

Letting equation (1.82) be zero and solving it and equation (1.83) simultaneously, we have the conditions for a Ball point:

$$F(x_{Pm}, y_{Pm}, s) = 0,\ \frac{\partial F}{\partial s} = 0 \tag{1.84}$$

In a time interval, $F(x_{Pm}, y_{Pm}, s) = 0$ in equation (1.84) represents a family of inflection circles; $\frac{\partial F}{\partial s} = 0$ is the partial derivative of an inflection circle with respect to s. From the viewpoint of continuity, equation (1.84) stands for the envelope of the inflection circles in the moving body.

Substituting equation (1.74d) into (1.84), we obtain two sets of solutions:

$$\begin{cases} x_m = u_1^{(1)} \cos\theta - u_2^{(1)} \sin\theta \\ y_m = u_1^{(1)} \sin\theta + u_2^{(1)} \cos\theta \end{cases} \tag{1.85a}$$

where

$$\begin{cases} u_1^{(1)} = 0 \\ u_2^{(1)} = 1/(k_{Om} - \dot{\theta}) \end{cases} \tag{1.85b}$$

and

$$\begin{cases} x_{bm} = u_1^{(2)} \cos\theta - u_2^{(2)} \sin\theta \\ y_{bm} = u_1^{(2)} \sin\theta + u_2^{(2)} \cos\theta \end{cases} \tag{1.86}$$

$$\begin{cases} u_1^{(2)} = \dfrac{C\dot{\theta}(k_{Om} - \dot{\theta}) - C^2(\dot{k}_{Om} - \ddot{\theta})}{(1 + C^2)(k_{Om} - \dot{\theta})^3} \\ u_2^{(2)} = \dfrac{1}{k_{Om} - \dot{\theta}} + \dfrac{C^2\dot{\theta}(k_{Om} - \dot{\theta}) + C(\dot{k}_{Om} - \ddot{\theta})}{(1 + C^2)(k_{Om} - \dot{\theta})^3} \\ C = \dfrac{\dot{\theta}(k_{Om} - \dot{\theta})^2(\ddot{k}_{Om} - \dddot{\theta}) + \dot{\theta}^2(k_{Om} - \dot{\theta})^3(2k_{Om} - \dot{\theta}) + (k_{Om} - \dot{\theta})^2(\dot{k}_{Om} - \ddot{\theta})(2\dot{k}_{Om} - 3\ddot{\theta})}{\dot{\theta}\ddot{\theta}(k_{Om} - \dot{\theta})^3 + (k_{Om} - \dot{\theta})(\dot{k}_{Om} - \ddot{\theta})(\ddot{k}_{Om} - \dddot{\theta}) - 2\dot{\theta}^2(\dot{k}_{Om} - \ddot{\theta})(k_{Om} - \dot{\theta})^2 - 3(\dot{k}_{Om} - \ddot{\theta})^3} \end{cases} \tag{1.86b}$$

Equations (1.85a) and (1.86) represent two enveloping curves of the inflection circles. Equation (1.85a), identical to (1.42) in Section 1.2.2, describes the moving centrode of $\sum^*$; equation (1.86) describes another form of the Ball curve. The two enveloping curves are not equidistant curves because the radii of the inflection circles vary with position.

Furthermore, we are interested in the geometrical properties of the Ball curve of $\sum^*$. The conditions of the Ball points are represented in equation (1.84) and their solutions are given in equation (1.86). Based on these equations, the Ball curve can be rewritten as

$$x_{bm} = x_{bm}(s),\ y_{bm} = y_{bm}(s) \tag{1.87}$$

To determine the instant and location at which a singularity of the Ball curve takes place, we use the following conditions:

$$\dot{x}_{bm} = \frac{\mathrm{d}x_{bm}}{\mathrm{d}s} = 0,\ \dot{y}_{bm} = \frac{\mathrm{d}y_{bm}}{\mathrm{d}s} = 0 \tag{1.88}$$

We rewrite equation (1.84) as

$$\begin{cases} F_{x_{bm}}\dot{x}_{bm} + F_{y_{bm}}\dot{y}_{bm} + F_s = 0 \\ F_{sx_{bm}}\dot{x}_{bm} + F_{sy_{bm}}\dot{y}_{bm} + F_{ss} = 0 \end{cases} \tag{1.89}$$

where a subscript on F represents the partial derivative with respect to that subscript (e.g., $F_s = \partial F/\partial s$). Solving the above equations simultaneously, we have

$$\begin{cases} \dot{x_{bm}} = F_{ss}F_{y_{bm}} / \left(F_{x_{bm}}F_{sy_{bm}} - F_{y_{bm}}F_{sx_{bm}}\right) \\ \dot{y_{bm}} = F_{ss}F_{x_{bm}} / \left(F_{x_{bm}}F_{sy_{bm}} - F_{y_{bm}}F_{sx_{bm}}\right) \end{cases} \tag{1.90}$$

Substituting equation (1.90) into (1.88), the singularity of the Ball curve is

$$F_{ss} = 0,\ F_{x_{bm}}F_{sy_{bm}} - F_{y_{bm}}F_{sx_{bm}} \neq 0 \tag{1.91}$$

Combining equations (1.84) and (1.91), we have the following singularity conditions of the Ball curve:

$$F = 0,\ F_s = 0,\ F_{ss} = 0,\ F_{x_{bm}}F_{sy_{bm}} - F_{y_{bm}}F_{sx_{bm}} \neq 0 \tag{1.92}$$

The above indicates that the singularity of the Ball curve occurs for the point path Γ_P with $k_\Gamma = 0$, $\frac{dk_\Gamma}{d\sigma} = 0$, and $\frac{d^2k_\Gamma}{d\sigma^2} = 0$ in the fixed frame $\{\boldsymbol{O}_f; \boldsymbol{i}_f, \boldsymbol{j}_f\}$ at that instant. Obviously, $k_\Gamma = 0$ and $\frac{dk_\Gamma}{d\sigma} = 0$ are the conditions for the Ball point while $\frac{dk_\Gamma}{d\sigma} = 0$ and $\frac{d^2k_\Gamma}{d\sigma^2} = 0$ are the conditions for the Burmester point. The Ball point of $\sum^*$ happens to be a Burmester point at this instant, called the ***Ball–Burmester point***.

The point path Γ_P and a circle have five common points at infinitesimally separated positions in the neighborhood of a Burmester point, or fourth-order contact at this instant. The point path Γ_P and a line have five common points at infinitesimally separated positions in the neighborhood of the Ball–Burmester point, or fourth-order contact at this instant; while the Ball–Burmester points appear in pairs, their positions can be located by the roots of a quartic equation.

1.2.4.5 Burmester Curve and its Singularity

According to equation (1.81), there may exist Burmester points in $\sum^*$ at an instant. In general, from the viewpoint of continuity, equation (1.81) may have continuous solutions only for a time interval throughout a cycle of motion. Real solutions to equation (1.81) appear in pairs for a quartic equation. All Burmester points form a few curves for a cycle of motion, called ***Burmester curves***. More details about Burmester curves will be given later in examples of four-bar linkages.

Similarly, the curvature of a point path Γ_P in the fixed frame $\{\boldsymbol{O}_f; \boldsymbol{i}_f, \boldsymbol{j}_f\}$ can also be written as equation (1.82). By means of $\frac{dk_\Gamma}{d\sigma} = 0$ and $\frac{d^2k_\Gamma}{d\sigma^2} = 0$, the conditions for a point to be a Burmester point (x_{Bm}, y_{Bm}) of $\sum^*$ are

$$\begin{cases} \left(F_sG - G_sF\right)/G^2 = 0 \\ (F_{ss}G - G_{ss}F)/G^4 = 0 \end{cases} \tag{1.93a}$$

or

$$\begin{cases} H\left(x_{Bm}, y_{Bm}, s\right) = F_sG - G_sF = 0 \\ H_s(x_{Bm}, y_{Bm}, s) = F_{ss}G - G_{ss}F = 0 \end{cases} \tag{1.93b}$$

In a time interval, the equations in (1.93b) represent a family of cubic curves of stationary curvature and their partial derivative with respect to s. Therefore, from the continuity point of view, equation (1.93b) stands for the enveloping curves of a family of cubic curves of stationary curvature in the moving body, which are Burmester curves.

The singularity of the Burmester curve is identified by the following conditions:

$$H_{ss}(x_{Bm}, y_{Bm}, s) = 0,\ H_{x_{Bm}} H_{sy_{Bm}} - H_{y_{Bm}} H_{sx_{Bm}} \neq 0 \tag{1.94}$$

Combining the Burmester point equation (1.93b) and the above equations, we have the singularity conditions of the Burmester curve:

$$\begin{cases} H\left(x_{Bm}, y_{Bm}, s\right) = F_s G - G_s F = 0 \\ H_s(x_{Bm}, y_{Bm}, s) = F_{ss} G - G_{ss} F = 0 \\ H_{ss}(x_{Bm}, y_{Bm}, s) = F_{sss} G - G_{sss} F + F_{ss} G_s - G_{ss} F_s = 0 \\ H_{x_{Bm}} H_{sy_{Bm}} - H_{y_{Bm}} H_{sx_{Bm}} \neq 0 \end{cases} \tag{1.95}$$

It can be seen that the singularity of a Burmester curve occurs when the curvature of curve Γ_P in the fixed frame $\{\boldsymbol{O}_f; \boldsymbol{i}_f, \boldsymbol{j}_f\}$ has $\frac{\partial k_\Gamma}{\partial s} = 0$, $\frac{\partial^2 k_\Gamma}{\partial s^2} = 0$, and $\frac{\partial^3 k_\Gamma}{\partial s^3} = 0$ at the special point. Hence, curve Γ_P contacts a circle in fifth order at this special point, or has six common points with a circle at infinitesimally separated positions in the neighborhood of the contact point.

1.2.4.6 Ball Curve and Burmester Curves for Planar Four-bar Linkages

For a double-crank linkage as in Example 1.5, the inflection circle can be calculated by using equations (E1-7.2) and (E1-7.3) at an instant. A family of inflection circles is computed in a step of 0.1 and shown in Fig. 1.27 for $\varphi \in [0, 2\pi]$.

A family of inflection circles envelops two curves – the Ball curve and moving centrode on the coupler of the double-crank linkage, as shown in Fig. 1.27. Substituting $k_{Om} = k_a$ into equation (1.86), we have the vector equation of the Ball curve. The coupler curves traced by three Ball points $P_1(-5.0876, -0.3534)$, $P_2(0.1186, -1.0982)$, and $P_3(2.1283, -4.9436)$ are shown in Fig. 1.27(b).

Example 1.9 A four-bar linkage in Fig. 1.5 has link lengths $a_1 = 1$, $a_2 = 1$, $a_3 = 1.3$, and $a_4 = 0.3$. Calculate the Burmester points at an instant and the Burmester curves for all instants.

Solution

For a planar four-bar linkage, the two joints on the coupler, or the Burmester points, whose polar coordinates are designated as (r_1, α_1) and (r_2, α_2) in the Frenet frame $\{\boldsymbol{R}_m; \boldsymbol{E}_{1m}, \boldsymbol{E}_{2m}\}$, move along two different circles. In this case, equation (1.81b), a quartic algebraic equation in $\tan\alpha$, is degenerated to a quadratic equation in $\tan\alpha$:

$$\tan^2\alpha + (c_1 + p)\tan\alpha + c_4/w = 0 \tag{E1-9.1}$$

where

$$p = \tan\alpha_1 + \tan\alpha_2, w = \tan\alpha_1 \cdot \tan\alpha_2 \tag{E1-9.2}$$

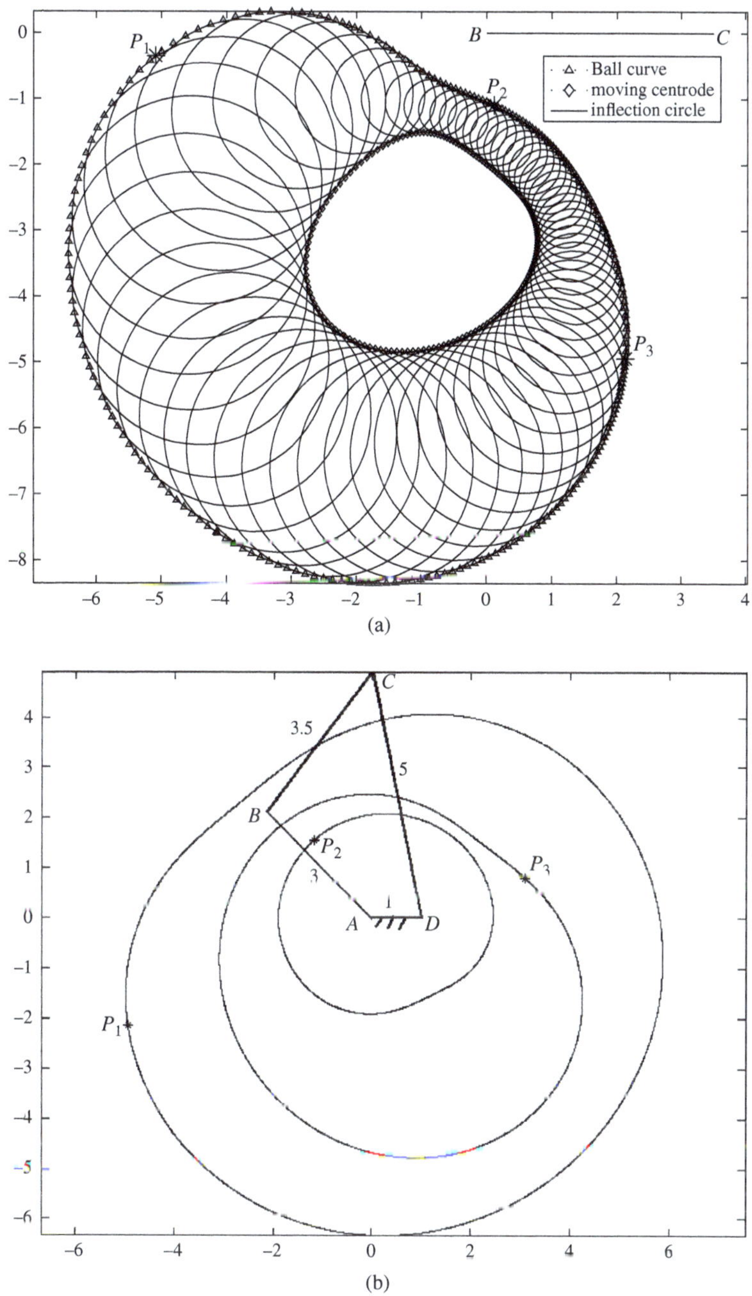

Figure 1.27 (a) Ball curve for a double-crank linkage (b) Coupler curves traced by three Ball points

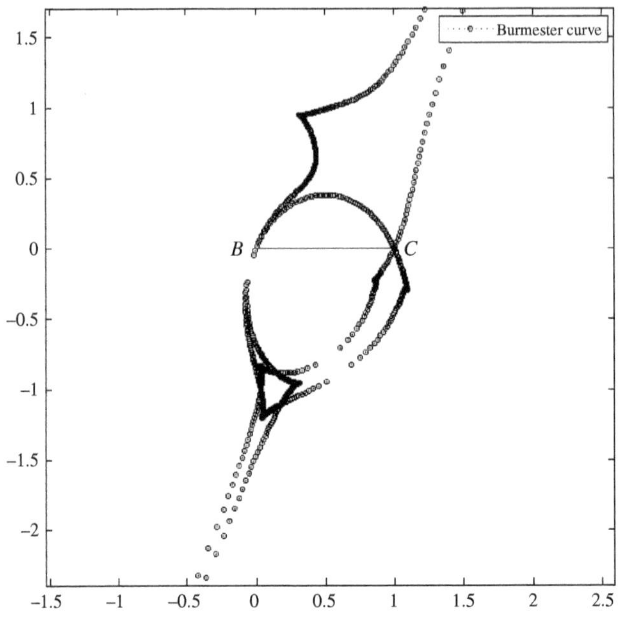

(a)

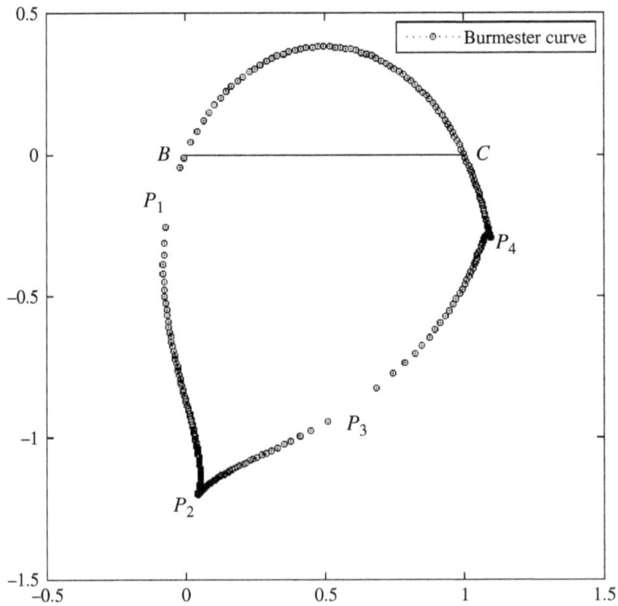

(b)

Figure 1.28 (a) Burmester curves for a double-crank linkage (b) A Burmester curve

The two roots α_3 and α_4 of equation (E1-9.1) are

$$\tan\alpha_3 = \frac{1}{2}\left[-(c_1+p)+\sqrt{(c_1+p)^2-4c_4/w}\right] \tag{E1-9.3a}$$

$$\tan\alpha_4 = \frac{1}{2}\left[-(c_1+p)-\sqrt{(c_1+p)^2-4c_4/w}\right] \tag{E1-9.3b}$$

Substituting α_3 and α_4 into equation (1.81a), we obtain the corresponding values for r_3 and r_4. The coordinates of the four Burmester points in the coupler fixed frame $\{\boldsymbol{B};\boldsymbol{i}_m,\boldsymbol{j}_m\}$ may be determined by equation (E1-7.1). Depending on the values of factor $(c_1+p)^2-4c_4/w$, equations (E1-9.3a) and (E1-9.3b) may have (i) two different real roots, (ii) two identical roots, or (iii) no real root. From the continuity point of view, (r_3,α_3) and (r_4,α_4) change continuously as the linkage moves. Each root corresponds to a Burmester point, and produces a branch of the Burmester curve on the coupler. The two branches together form a complete Burmester curve. When factor $(c_1+p)^2-4c_4/w$ becomes zero, the two branches coincide at a point, which corresponds to a double root of equation (E1-9.2), or a ***double Burmester point***. The Burmester curve disappears when equation (E1-9.2) has no root as the linkage continues to move. Another Burmester curve appears with two branches starting at this point and ending at the next double Burmester point. A double Burmester point, the beginning or the end of a Burmester curve, has fourth-order contact with a circle at this instant. Referring to Fig. 1.28, the position of a double Burmester point is very sensitive to the coupler position.

If a Ball point of the coupler happens to be a Burmester point at an instant, this special point is called a Ball–Burmester point. The polar coordinates of a Ball–Burmester point in the Frenet frame $\{\boldsymbol{R}_m;\boldsymbol{E}_{1m},\boldsymbol{E}_{2m}\}$ satisfy the conditions in equations (1.80) and (E1-9.2) simultaneously. Substituting equation (1.80) into equation (E1-9.2), we have

$$\tan\alpha = -\frac{k^*(k_m+2k^*)}{\mathrm{d}k^*/\mathrm{d}\sigma} = \frac{1}{2}[-(c_1+p)\pm\sqrt{(c_1+p)^2-4c_4/w}] \tag{E1-9.4}$$

Substituting equation (1.78) into the above equation, we obtain, after some simplification, the condition for the singularity of the Ball curve:

$$\left(\frac{1}{w}-3\right)-\left(3+\frac{1}{w}\right)\frac{k_m}{k^*}+p\frac{\mathrm{d}k^*/\mathrm{d}\sigma}{k^{*2}}=0 \tag{E1-9.5}$$

The Ball–Burmester point appears when parameters k_m, k^*, $\mathrm{d}k^*/\mathrm{d}\sigma$ and the polar coordinates of the two moving joints, (r_1,α_1) and (r_2,α_2), satisfy equation (E1-9.5).

The Ball–Burmester point is not only a Burmester point, but also a double Burmester point since equation (E1-9.2) has a double root, called the ***Ball–double Burmester point***. By combining equations (1.80) and (E1-9.2), we have

$$\tan\alpha = -\frac{k^*(k_m+2k^*)}{\mathrm{d}k^*/\mathrm{d}\sigma} = -\frac{1}{2}(c_1+p) \tag{E1-9.6}$$

The above equation can be simplified by equation (E1-9.1):

$$p=\frac{4k_mk^*+5k^{*2}}{\mathrm{d}k^*/\mathrm{d}\sigma}=\tan\alpha_1+\tan\alpha_2,\ w=\frac{k_m-k^*}{k_m+2k^*}=\tan\alpha_1\tan\alpha_2 \tag{E1-9.7}$$

These are the conditions for a coupler point to be a Ball–double Burmester point for a planar hinged four-bar linkage.

As we know, the Ball–Burmester point or the Ball–double Burmester point is a Burmester point, whose path contacts a line in fourth order at the instant.

For the double-crank linkage with $a_1 = 1$, $a_2 = 1$, $a_3 = 1.3$, and $a_4 = 0.3$, two joints $B(0,0)$ and $C(1,0)$ in the moving frame $\{\boldsymbol{O}_m; \boldsymbol{i}_m, \boldsymbol{j}_m\}$ of the coupler are represented by the polar coordinates (r_B, α_B) and (r_C, α_C) at instant $\varphi = 90°$ in $\{\boldsymbol{R}_m; \boldsymbol{E}_{1m}, \boldsymbol{E}_{2m}\}$ of π_m by equation (E1-7.1), or $r_B = 1.4741$, $\alpha_B = -60.3611°$, $r_C = 1.8610$, $\alpha_C = 87.3130°$. Substituting the above values into equations (E1-9.1) and (E1-9.2), we have $p = 19.5433$, $w = -37.4403$, and a quadratic equation:

$$\tan^2\alpha + 1072.4\tan\alpha - 5438.2 = 0 \tag{E1-9.8}$$

The two roots of the above equation are $\tan\alpha_{B1} = 5.0475$, $\tan\alpha_{B2} = -1077.4$, and two Burmester points on the coupler are located by equation (1.81):

$$(r_{B1} = 1.7089, \alpha_{B1} = 258.7936°),\ (r_{B2} = 0.3271, \alpha_{B2} = -89.9486°)$$

The Burmester curves on the coupler plane of the double-crank linkage for $\varphi \in (0, 2\pi)$ are shown in Fig. 1.28(a). A Burmester curve is amplified in Fig. 1.28(b), in which the Burmester curve begins at point P_1 and develops into two branches, then ends at point P_3, or the Burmester point disappears, becoming a closed curve. Two double Burmester points locate at $P_3(0.5965, -0.8890)$ and $P_1(-0.0452, -0.1318)$ on the coupler at $\varphi = 60.4761°$ and $151.4286°$. The two singularities of two branches of the Burmester curve appear simultaneously, at $P_2(0.0417, -1.2026)$ and $P_4(1.0963, -0.2953)$ at $\varphi = 93.62°$.

1.2.5 Line Path

As mentioned in Section 1.2.3, point $P(x_{Pm}, y_{Pm})$ in $\{\boldsymbol{O}_m; \boldsymbol{i}_m, \boldsymbol{j}_m\}$ of $\sum^*$ traces a point path Γ_P in the fixed frame $\{\boldsymbol{O}_f; \boldsymbol{i}_f, \boldsymbol{j}_f\}$, which can be expressed as an adjoint curve of the fixed centrode π_f. Meanwhile, a straight line L belongs to $\sum^*$, passes through point P, and traces a line path (a set of lines Γ_l) in $\{\boldsymbol{O}_f; \boldsymbol{i}_f, \boldsymbol{j}_f\}$. The line path Γ_l can be viewed as the adjoint lines of π_f, while π_f is taken as an original curve, shown in Fig. 1.29, and expressed in the Frenet frame $\{\boldsymbol{R}_f; \boldsymbol{E}_{1f}, \boldsymbol{E}_{2f}\}$ of $\pi_f$as

$$\Gamma_l : \boldsymbol{R}_l = \boldsymbol{R}_f + \boldsymbol{r}_P + \lambda\boldsymbol{l} = \boldsymbol{R}_f + u_1\boldsymbol{E}_{1f} + u_2\boldsymbol{E}_{2f} + \lambda(l_1\boldsymbol{E}_{1f} + l_2\boldsymbol{E}_{2f}) \tag{1.96}$$

where $\boldsymbol{l}$ is the unit vector of the straight line L with components (l_1, l_2) in the Frenet frame $\{\boldsymbol{R}_f; \boldsymbol{E}_{1f}, \boldsymbol{E}_{2f}\}$ and $l_1^2 + l_2^2 = 1$. Here λ is the parameter of line L and (u_1, u_2) are the coordinates of point P in the Frenet frame $\{\boldsymbol{R}_f; \boldsymbol{E}_{1f}, \boldsymbol{E}_{2f}\}$. Differentiating equation (1.96) with respect to σ, the arc length of the centrode, we have

$$\begin{cases} \mathrm{d}\boldsymbol{R}_l = \left(\dfrac{\mathrm{d}\boldsymbol{R}_f}{\mathrm{d}\sigma} + \dfrac{\mathrm{d}\boldsymbol{r}_P}{\mathrm{d}\sigma} + \lambda\dfrac{\mathrm{d}\boldsymbol{l}}{\mathrm{d}\sigma}\right)\mathrm{d}\sigma + \boldsymbol{l}\,\mathrm{d}\lambda \\ \dfrac{\mathrm{d}\boldsymbol{R}_f}{\mathrm{d}\sigma} + \dfrac{\mathrm{d}\boldsymbol{r}_P}{\mathrm{d}\sigma} = \left(1 + \dfrac{\mathrm{d}u_1}{\mathrm{d}\sigma} - k_f u_2\right)\boldsymbol{E}_{1f} + \left(\dfrac{\mathrm{d}u_2}{\mathrm{d}\sigma} + k_f u_1\right)\boldsymbol{E}_{2f} \\ \dfrac{\mathrm{d}\boldsymbol{l}}{\mathrm{d}\sigma} = \left(\dfrac{\mathrm{d}l_1}{\mathrm{d}\sigma} - k_f l_2\right)\boldsymbol{E}_{1f} + \left(\dfrac{\mathrm{d}l_2}{\mathrm{d}\sigma} + k_f l_1\right)\boldsymbol{E}_{2f} \end{cases} \tag{1.97}$$

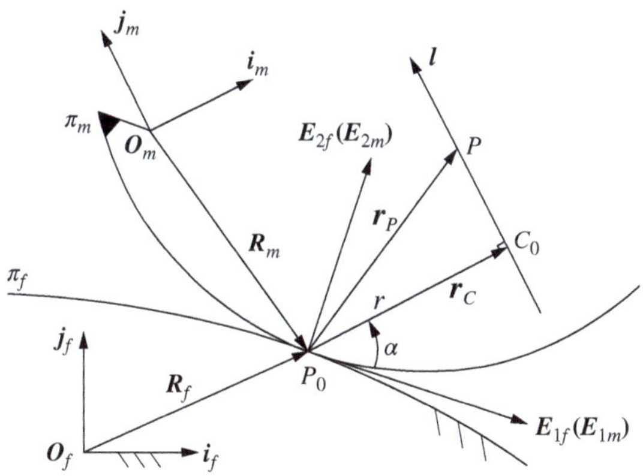

Figure 1.29 The line path adjoint to the centrodes

Line L in $\sum^*$ can also be viewed as adjoint to the moving centrode π_m and represented in the Frenet frame $\{R_m; E_{1m}, E_{2m}\}$ as

$$R_{lm} = R_m + r_{Pm} + \lambda l_m = R_m + u_1 E_{1m} + u_2 E_{2m} + \lambda(l_1 E_{1m} + l_2 E_{2m}) \tag{1.98}$$

where the parameters of line L in the Frenet frame $\{R_m; E_{1m}, E_{2m}\}$ are the same as in the Frenet frame $\{R_f; E_{1f}, E_{2f}\}$ since they are coincident at the instant velocity center. Differentiating equation (1.98) with respect to σ, we obtain

$$\begin{cases} \mathrm{d}R_{lm} = \left(\dfrac{\mathrm{d}R_m}{\mathrm{d}\sigma} + \dfrac{\mathrm{d}r_{Pm}}{\mathrm{d}\sigma} + \lambda\dfrac{\mathrm{d}l_m}{\mathrm{d}\sigma}\right)\mathrm{d}\sigma + l_m \mathrm{d}\lambda \\ \dfrac{\mathrm{d}R_m}{\mathrm{d}\sigma} + \dfrac{\mathrm{d}r_{Pm}}{\mathrm{d}\sigma} = \left(1 + \dfrac{\mathrm{d}u_1}{\mathrm{d}\sigma} - k_m u_2\right) E_{1m} + \left(\dfrac{\mathrm{d}u_2}{\mathrm{d}\sigma} + k_m u_1\right) E_{2m} \\ \dfrac{\mathrm{d}l_m}{\mathrm{d}\sigma} = \left(\dfrac{\mathrm{d}l_1}{\mathrm{d}\sigma} - k_m l_2\right) E_{1m} + \left(\dfrac{\mathrm{d}l_2}{\mathrm{d}\sigma} + k_m l_1\right) E_{2m} \end{cases} \tag{1.99}$$

The conditions for a fixed line may be obtained in the Frenet frame $\{R_m; E_{1m}, E_{2m}\}$ by equation (1.26) since line L is a fixed line in $\sum^*$ while point P and unit vector l are fixed to $\sum^*$:

$$\begin{cases} \dfrac{\mathrm{d}u_1}{\mathrm{d}\sigma} = k_m u_2 - 1, \ \dfrac{\mathrm{d}u_2}{\mathrm{d}\sigma} = -k_m u_1 \\ \dfrac{\mathrm{d}l_1}{\mathrm{d}\sigma} = k_m l_2, \ \dfrac{\mathrm{d}l_2}{\mathrm{d}\sigma} = -k_m l_1 \end{cases} \tag{1.100}$$

Substituting equation (1.100) into (1.97), we have

$$\begin{cases} \mathrm{d}R_l = \left(\dfrac{\mathrm{d}R_f}{\mathrm{d}\sigma} + \dfrac{\mathrm{d}r_P}{\mathrm{d}\sigma} + \lambda\dfrac{\mathrm{d}l}{\mathrm{d}\sigma}\right)\mathrm{d}\sigma + l\mathrm{d}\lambda \\ \dfrac{\mathrm{d}R_f}{\mathrm{d}\sigma} + \dfrac{\mathrm{d}r_P}{\mathrm{d}\sigma} = k^*(-u_2 E_{1f} + u_1 E_{2f}) \\ \dfrac{\mathrm{d}l}{\mathrm{d}\sigma} = k^*(-l_2 E_{1f} + l_1 E_{2f}) \end{cases} \tag{1.101}$$

A set of lines Γ_l envelop a curve Γ_C in the fixed frame $\{\boldsymbol{O}_f; \boldsymbol{i}_f, \boldsymbol{j}_f\}$ as $\sum^*$ moves, with vector equation defined by

$$\begin{cases} \boldsymbol{R}_l = \boldsymbol{R}_f + \boldsymbol{R}_P + \lambda \boldsymbol{l} = \boldsymbol{R}_f + u_1 \boldsymbol{E}_{1f} + u_2 \boldsymbol{E}_{2f} + \lambda \left(l_1 \boldsymbol{E}_{1f} + l_2 \boldsymbol{E}_{2f} \right) \\ \partial \boldsymbol{R}_l / \partial \sigma \times \partial \boldsymbol{R}_l / \partial \lambda = 0 \end{cases} \tag{1.102}$$

The second of the above equations represents the tangent contact condition between a straight line and the enveloping curve Γ_C, which can be simplified to $\lambda = -(l_1 u_1 + l_2 u_2)$ by equation (1.101) if $k^* = k_f - k_m \neq 0$. The vector equation of the enveloping curve Γ_C can be written as

$$\boldsymbol{\rho} = \boldsymbol{R}_f + \boldsymbol{r}_P + \lambda \boldsymbol{l} = \boldsymbol{R}_f + (l_2 u_1 - l_1 u_2)(l_2 \boldsymbol{E}_{1f} - l_1 \boldsymbol{E}_{2f}) \tag{1.103}$$

For given σ, the above equation represents a point C_0 on one of the set of lines contacting the enveloping curve Γ_C, whose position vector $\boldsymbol{R}_C$ and direction $\boldsymbol{l}$ of the line meet the following condition:

$$(\boldsymbol{R}_C - \boldsymbol{R}_f) \cdot \boldsymbol{l} = 0 \tag{1.104}$$

The above condition reveals that point C_0 is the foot of a perpendicular from P_0 to line L, or Γ_C is traced by the foot point of line L on $\sum^*$ for plane motion. To avoid confusion, we define hereinafter that the instant contact point C_0 is described in the fixed frame while the foot point is a fixed point of $\sum^*$, although they are coincident with each other at an instant.

Point C_0 is designated with polar coordinates (r, α) in the Frenet frame $\{\boldsymbol{R}_f; \boldsymbol{E}_{1f}, \boldsymbol{E}_{2f}\}$ of the fixed centrode as

$$\begin{cases} \boldsymbol{r}_C = \left(c_1, c_2 \right)^T = (r \cos \alpha, r \sin \alpha)^T \\ \boldsymbol{l} = (l_1, l_2)^T = (-\sin \alpha, \cos \alpha)^T \end{cases} \tag{1.105}$$

while point C_0 is not a fixed point of $\sum^*$ and can slide along the line. However, line L is examined in the moving centrode as a quasi-fixed line, by equation (1.28), whose conditions are

$$\begin{cases} l_2 \left(1 + \dfrac{\mathrm{d}c_1}{\mathrm{d}\sigma} - k_m c_2 \right) - l_1 \left(\dfrac{\mathrm{d}c_2}{\mathrm{d}\sigma} + k_m c_1 \right) = 0 \\ \dfrac{\mathrm{d}l_1}{\mathrm{d}\sigma} = k_m l_2, \ \dfrac{\mathrm{d}l_2}{\mathrm{d}\sigma} = -k_m l_1 \end{cases} \tag{1.106}$$

Substituting equation (1.105) into the above equation, we have

$$\mathrm{d}\alpha / \mathrm{d}\sigma = -k_m, \ \mathrm{d}r / \mathrm{d}\sigma = -l_2 \tag{1.107}$$

The first equation in (1.106) can be simplified as

$$\begin{cases} \dfrac{\mathrm{d}c_1}{\mathrm{d}\sigma} = \dfrac{\mathrm{d}r}{\mathrm{d}\sigma} \cos \alpha - r \sin \alpha \dfrac{\mathrm{d}\alpha}{\mathrm{d}\sigma} = -l_2^2 - r l_1 k_m \\ \dfrac{\mathrm{d}c_2}{\mathrm{d}\sigma} = \dfrac{\mathrm{d}r}{\mathrm{d}\sigma} \sin \alpha + r \cos \alpha \dfrac{\mathrm{d}\alpha}{\mathrm{d}\sigma} = l_1 l_2 - r l_2 k_m \end{cases} \tag{1.108}$$

The vector equation of the enveloping curve Γ_C can be rewritten in the Frenet frame $\{\boldsymbol{R}_f; \boldsymbol{E}_{1f}, \boldsymbol{E}_{2f}\}$ as

$$\boldsymbol{\rho} = \boldsymbol{R}_f + \boldsymbol{r}_C = \boldsymbol{R}_f + c_1 \boldsymbol{E}_{1f} + c_2 \boldsymbol{E}_{2f} \tag{1.109}$$

Differentiating the above equation with respect to the arc length σ and combining with equation (1.108), we obtain

$$\frac{\mathrm{d}\boldsymbol{\rho}}{\mathrm{d}\sigma} = (l_1 + rk^*)(l_1\boldsymbol{E}_{1f} + l_2\boldsymbol{E}_{2f}) \tag{1.110}$$

The differential arc length $\mathrm{d}s_C$ of Γ_C is related to the differential arc length $\mathrm{d}\sigma$ as follows:

$$\mathrm{d}s_C = \left|\frac{\mathrm{d}\boldsymbol{\rho}}{\mathrm{d}\sigma}\right| \mathrm{d}\sigma = (l_1 + rk^*)\mathrm{d}\sigma \tag{1.111}$$

To determine the curvature of Γ_C, the Frenet frame $\{\boldsymbol{\rho}; \boldsymbol{E}_1^*, \boldsymbol{E}_2^*\}$ is established:

$$\begin{cases} \boldsymbol{E}_1^* = \dfrac{\mathrm{d}\boldsymbol{\rho}}{\mathrm{d}\sigma} \Big/ \left|\dfrac{\mathrm{d}\boldsymbol{\rho}}{\mathrm{d}\sigma}\right| = l_1\boldsymbol{E}_{1f} + l_2\boldsymbol{E}_{2f} \\ \boldsymbol{E}_2^* = \boldsymbol{k} \times \boldsymbol{E}_1^* = -l_2\boldsymbol{E}_{1f} + l_1\boldsymbol{E}_{2f} \end{cases} \tag{1.112}$$

The curvature k_C of Γ_C may be obtained from equation (1.14), and written as

$$k_C = \frac{\mathrm{d}\boldsymbol{E}_1^*}{\mathrm{d}s_C} \cdot \boldsymbol{E}_2^* = \frac{1}{r - \sin\alpha/k^*} \tag{1.113}$$

Based on the above equation, the curvature of Γ_C approaches infinity if $r - \frac{\sin\alpha}{k^*} = 0$. The point on Γ_C becomes a cusp. A circle, called a ***cusp circle***, or a ***return circle***, exists in $\sum^*$ at this instant. The cusp circle locates the lines of $\sum^*$ by their perpendicular foot points and their distance to the instant velocity center P_0. These lines envelop the curves with a cusp. By comparing the inflection circle in Section 1.2.3, the cusp circle and the inflection circle are symmetric about the tangent of the moving centrode at the instant velocity center, and their diameters are the reciprocal of the induced curvature of centrodes $|1/k^*|$.

Replacing k_C in equation (1.113) with the reciprocal of ρ_C, we have

$$r - \rho_C = \frac{\sin\alpha}{k^*} \tag{1.114}$$

The above equation is called the Euler–Savary equation for a line in plane movement.

For plane movement, a line of $\sum^*$ can be located by the foot point coordinates r and α. The position vector of the curvature center B_0 of the envelope curve may be expressed in the Frenet frame $\{\boldsymbol{R}_f; \boldsymbol{E}_{1f}, \boldsymbol{E}_{2f}\}$ as

$$\boldsymbol{r}_B = \boldsymbol{r}_C + \frac{\boldsymbol{E}_2^*}{k_C} = \frac{1}{k^*}\{\sin\alpha\cos\alpha, \sin^2\alpha\} \tag{1.115}$$

From the above equation, $\boldsymbol{r}_B$ does not depend on the distance r from the instant velocity center P_0 to the line. So, all parallel lines of $\sum^*$ envelop curves with the same curvature center in the fixed frame $\{\boldsymbol{O}_f; \boldsymbol{i}_f, \boldsymbol{j}_f\}$. These enveloping curves happen to be equidistant curves. The curvature centers actually lie on the cusp circle in the fixed frame $\{\boldsymbol{O}_f; \boldsymbol{i}_f, \boldsymbol{j}_f\}$ of the fixed body.

In particular, one of the parallel lines passes through the curvature center B_0 by chance, as shown in Fig. 1.30, and produces an enveloping curve with a zero radius of curvature, which means that this line passes a point at three infinitesimally separated positions in $\{\boldsymbol{O}_f; \boldsymbol{i}_f, \boldsymbol{j}_f\}$. Hence, the position of the line in $\sum^*$ is located by its foot point coordinates in the moving

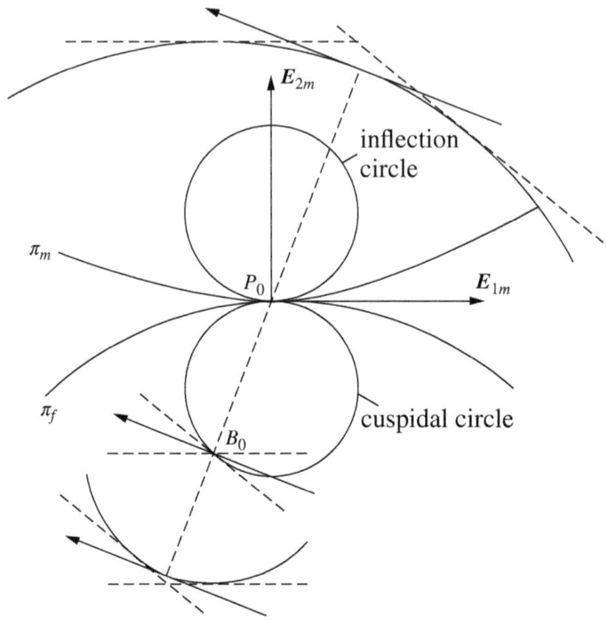

Figure 1.30 Euler–Savary equation of a line path

frame $\{O_m; i_m, j_m\}$. The foot point, also an ***inverted sliding point*** of $\sum^*$, meets the cusp circle equation $r - \frac{\sin\alpha}{k^*} = 0$. The cusp circle can be viewed as a collection of all the inverted sliding points at an instant.

Differentiating equation (1.113) with respect to σ once and twice, we obtain

$$\frac{dk_C}{d\sigma} = \frac{M\sin\alpha + N\cos\alpha}{H} \tag{1.116}$$

$$\frac{d^2k_C}{d\sigma^2} = \frac{1}{H}\left(R\sin\alpha + Q\cos\alpha - \frac{dH}{d\sigma}\frac{dk_C}{d\sigma}\right) \tag{1.117}$$

where M, N, H, R, and Q are given below:

$$\begin{cases} M = \dfrac{d(1/k^*)}{d\sigma},\ N = 1 - k_m/k^*,\ H = \left(r - \dfrac{1}{k^*}\sin\alpha\right)^2 \\ R = \dfrac{dM}{d\sigma} + k_m N,\ Q = \dfrac{dN}{d\sigma} - k_m M \end{cases} \tag{1.118}$$

Letting $dk_C/d\sigma$ in equation (1.116) be zero, we have

$$\tan\alpha = -N/M \tag{1.119}$$

The enveloping curve contacts a circle in the third order. There are a group of parallel lines determined by equation (1.119) on $\sum^*$. Each of them is tangent to a circle with radius

$r = \frac{\sin\alpha}{k^*}$ at four infinitesimally separated positions, since equation (1.119) depends only on the parameter of the direction angle α.

Particularly, a straight line in $\sum^*$, located by the foot point in the moving frame $\{\boldsymbol{O}_m; \boldsymbol{i}_m, \boldsymbol{j}_m\}$, passes through a fixed point in the fixed frame $\{\boldsymbol{O}_f; \boldsymbol{i}_f, \boldsymbol{j}_f\}$ at four infinitesimally separated positions at an instant, and the foot point of $\sum^*$ is designated as a ***higher-order inverted sliding point***, or an ***inverted Ball point***, when ρ_C, the radius of curvature of the enveloping curve, is zero. We then have $r = \frac{\sin\alpha}{k^*}$. The line is located in the Frenet frame $\{\boldsymbol{R}_m; \boldsymbol{E}_{1m}, \boldsymbol{E}_{2m}\}$ by

$$\begin{cases} \tan\alpha = -N/M \\ r = \dfrac{\sin\alpha}{k^*} \end{cases} \tag{1.120}$$

The curvature center of an enveloping curve is at the cusp circle. Equation (1.120) is rewritten in the Frenet frame $\{\boldsymbol{R}_m; \boldsymbol{E}_{1m}, \boldsymbol{E}_{2m}\}$ as

$$\begin{cases} r = \dfrac{\sin\alpha}{k^*} \\ \dfrac{\mathrm{d}\left(r - \dfrac{\sin\alpha}{k^*}\right)}{\mathrm{d}\sigma} = 0 \end{cases} \tag{1.121}$$

where the first equation represents a set of cusp circles and the second equation represents the partial derivative of the set of cusp circles with respect to the arc length of π_m. From the viewpoint of continuity, equation (1.121) stands for an enveloping curve of the set of cusp circles in the moving frame $\{\boldsymbol{O}_m; \boldsymbol{i}_m, \boldsymbol{j}_m\}$. It is not surprising that equation (1.121) is similar to equation (1.84), since the cusp circle is symmetrical to the inflection circle about the tangent line of the centrode at an instant.

The kinematic inversion means that the roles of the fixed body and the moving body are interchanged, such that the moving centrode π_m is viewed as a fixed centrode π_f while the fixed now becomes a moving centrode, which converts the inflection circle into a cusp circle. Hence, there is an inverted Ball curve in the fixed coordinate system $\{\boldsymbol{O}_f; \boldsymbol{i}_f, \boldsymbol{j}_f\}$, which is determined by equation (1.120). The higher-order curvature for a line in plane movement has properties similar to those for a point, such that the inflection circle corresponds to the cusp circle and the Ball curve to the inverted Ball curve.

The singularity of the enveloping curve of the cusp circles may be determined by letting $\frac{\mathrm{d}k_C}{\mathrm{d}\sigma} = 0$ and $\frac{\mathrm{d}^2 k_C}{\mathrm{d}\sigma^2} = 0$, which leads to

$$\begin{cases} M\sin\alpha + N\cos\alpha = 0 \\ R\sin\alpha + Q\cos\alpha = 0 \end{cases} \tag{1.122}$$

There is only one independent variable α in the Frenet frame $\{\boldsymbol{R}_m; \boldsymbol{E}_{1m}, \boldsymbol{E}_{2m}\}$ in the two equations above. By means of $MQ - NR = 0$, equation (1.122) has a solution when the following equation is satisfied:

$$M\frac{\mathrm{d}N}{\mathrm{d}\sigma} - N\frac{\mathrm{d}M}{\mathrm{d}\sigma} = k_m(M^2 + N^2) \tag{1.123}$$

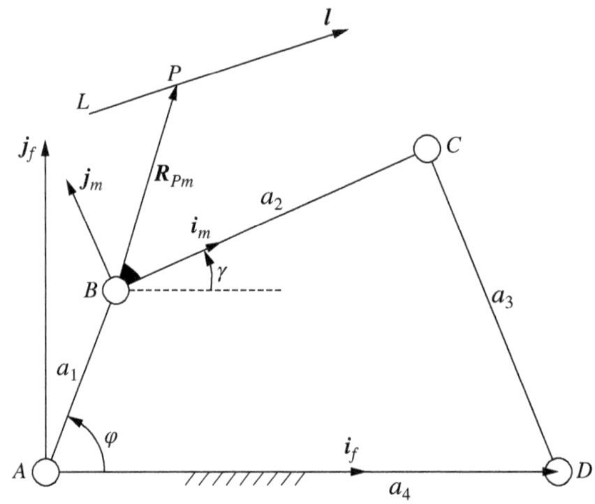

Figure 1.31 A line L of the coupler link of a planar four-bar linkage

which is determined by the properties of the plane motion. The singularity of the enveloping curve of the cusp circles in $\sum^*$ can be expressed in the Frenet frame $\{\boldsymbol{R}_m;\boldsymbol{E}_{1m},\boldsymbol{E}_{2m}\}$ by the solution of equation (1.122). The line of $\sum^*$, located by the foot point corresponding to the singularity of the enveloping curve in the moving frame $\{\boldsymbol{O}_m;\boldsymbol{i}_m,\boldsymbol{j}_m\}$, generates an enveloping curve at this instant. This enveloping curve contacts a circle in the fourth order, or has five common points at infinitesimally separated positions with a circle in the fixed coordinate system $\{\boldsymbol{O}_f;\boldsymbol{i}_f,\boldsymbol{j}_f\}$.

Example 1.10 A crank-rocker linkage, as shown in Fig. 1.31, has dimensions $a_1 = 1$, $a_2 = 3$, $a_3 = 3.5$, $a_4 = 5$. Determine the Euler–Savary equation for a line of a planar four-bar linkage at an instant.

Solution

A line L of the coupler is defined by a point $P(30, 30)$ and a unit vector $\boldsymbol{l}(0.4721, 0.8816)$ in the moving frame $\{\boldsymbol{B};\boldsymbol{i}_m,\boldsymbol{j}_m\}$. The moving centrode π_m and the fixed centrode π_f, the enveloping curve of the set of lines for $\varphi \in \left(0, \frac{\pi}{2}\right)$, are all shown in Fig. 1.32.

At $\varphi = 60°$, the instant velocity center P_0 locates at (3.0138, 1.1630) in the moving frame $\{\boldsymbol{B};\boldsymbol{i}_m,\boldsymbol{j}_m\}$. The curvatures of centrodes are $k_f = -0.1215$ and $k_m = -0.1920$, respectively; the induced curvature is $k^* = 0.0705$. The parameters (r, α) of line L in the Frenet frame $\{\boldsymbol{R}_m;\boldsymbol{E}_{1m},\boldsymbol{E}_{2m}\}$ are (10.1762, 56.2415°). The higher-order inverted sliding point can be obtained by equation (1.120) and the polar coordinates in the Frenet frame $\{\boldsymbol{R}_m;\boldsymbol{E}_{1m},\boldsymbol{E}_{2m}\}$ are (11.7912, 56.2415°), or the Cartesian coordinates in the fixed frame $\{\boldsymbol{A};\boldsymbol{i}_f,\boldsymbol{j}_f\}$ are (13.7003, 5.8588). The radius of curvature of the enveloping curve is 1.6150.

On the coupler, another line L_0, parallel to line L, is determined by a point (13.4084, −4.4034) and the same unit vector as line L in the moving frame $\{\boldsymbol{B};\boldsymbol{i}_m,\boldsymbol{j}_m\}$. At $\varphi = 60°$, the line L_0 just passes through the higher-order inverted sliding point at four infinitesimally separated positions.

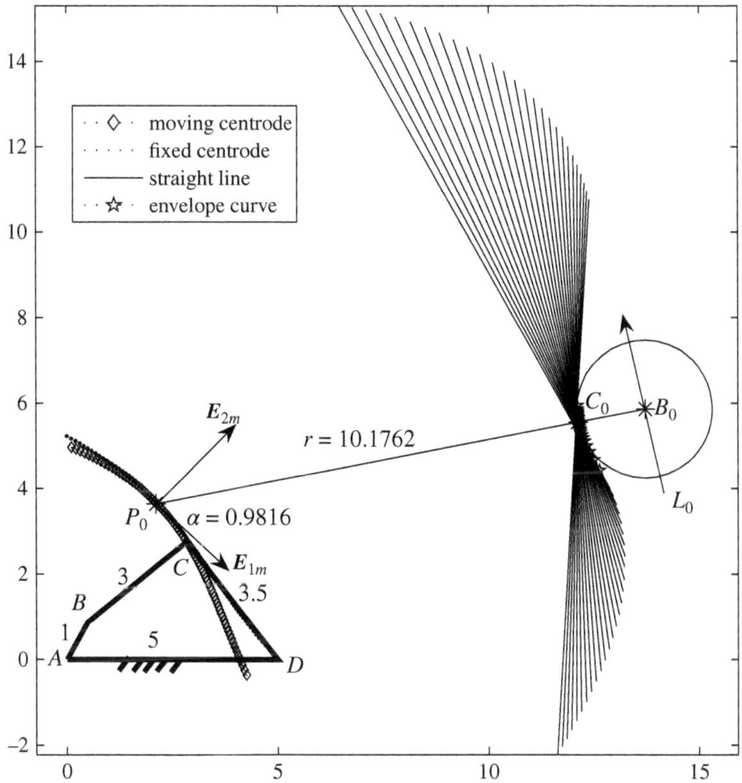

Figure 1.32 Euler–Savary equation of a line path for a crank-rocker linkage

1.3 Plane Coupler Curves

A plane coupler curve is generated by a coupler point of a planar linkage in the base link as the coupler moves over a cycle. A variety of coupler curves may be generated to meet various requirements of widely used planar linkages. The geometrical properties of the coupler curves and the distribution indicate the kinematics and geometry of motion of the coupler, and provide a theoretical basis for analysis and synthesis of linkages.

Based on the kinematic differential geometry in plane motion, the following local and global geometrical characteristics of the coupler curves are of interest: cusp, inflection point, double point (tacnode and crunode), ovality (goose egg shape), and symmetry. The plane motion here implies a general motion generated by planar linkages, not limited to any specific type or dimension. However, the results and drawings provided in this chapter are obtained for specific linkages such as a planar four-bar linkage.

1.3.1 Local Characteristics

For a planar linkage, some coupler points or joints of the coupler are constrained by the binary links connected to the base link. While the coupler moves, different points on the coupler trace different plane curves in the fixed frame. As mentioned in Section 1.2, the geometrical properties of a plane curve are determined by its curvature.

1.3.1.1 Cusp

The curvature of the coupler curve approaches infinity at a cusp point. This implies that the denominator of equation (1.74a) is zero at a cusp, that is $r = 0$, in the moving Frenet frame $\{\boldsymbol{R}_m; \boldsymbol{E}_{1m}, \boldsymbol{E}_{2m}\}$. The necessary and sufficient condition for a coupler curve to have a cusp is that the tracing point coincides with the instant velocity center between the coupler and the base link as the tracing point is at the cusp; that is, the tracing point lies on the centrode of the coupler.

1.3.1.2 Inflection Point and its Area

An inflection point of a coupler curve is a point at which the curvature k_Γ is zero, that is $r + D \sin \alpha = 0$, in the moving Frenet frame $\{\boldsymbol{R}_m; \boldsymbol{E}_{1m}, \boldsymbol{E}_{2m}\}$. As mentioned in Section 1.2, there is a family of inflection circles in the moving body that can be represented as equation (E1-7.2) in rectangular coordinates (x_{Pm}, y_{Pm}) or polar coordinates (r_{Pm}, θ_{Pm}) in the moving frame $\{\boldsymbol{O}_m; \boldsymbol{i}_m, \boldsymbol{j}_m\}$. Both the radius and the position of the inflection circle center change with the coupler position. From the viewpoint of a series of positions, the inflection circles change their position on the moving body in a continuous manner. The positions of all inflection circles form special areas on the coupler, called ***inflection point areas***. In an inflection point area, all points are touched by the inflection circle twice or more. A point in an inflection area traces a curve with inflection points. There are a few areas on the coupler where no inflection circles exist. Each of these areas is called a ***non-inflection point area***. A point in a non-inflection point area traces a curve with no inflection point. The entire coupler plane can be divided into an inflection point area and a non-inflection point area separated by boundary curves. The Ball curve and centrodes are the border curves, since they are the enveloped curves of the inflection circles.

1.3.1.3 Ball Curve

At a certain instant, a point on the coupler traces a curve with $k_\Gamma = 0$ and $\mathrm{d}k_\Gamma/\mathrm{d}\sigma = 0$. This point is known to be a Ball point. There is one Ball point at an instant and one Ball curve π_b on the coupler plane. As shown in Fig. 1.33, the coordinates of a Ball point are expressed by equation (1.80) in the Frenet frame $\{\boldsymbol{R}_m; \boldsymbol{E}_{1m}, \boldsymbol{E}_{2m}\}$ or in the moving frame $\{\boldsymbol{O}_m; \boldsymbol{i}_m, \boldsymbol{j}_m\}$. If the tracing point is the singularity point of the Ball curve, or a Ball–Burmester point, this point of the coupler traces a coupler curve with a special point, which has fourth-order contact with a line at this point. The Ball curve is enveloped by a family of inflection circles, and may be a boundary separating the inflection point area and the non-inflection point area.

1.3.1.4 Burmester Curve

A Burmester point of a moving body traces a path with $\mathrm{d}k_\Gamma/\mathrm{d}\sigma = 0$ and $\mathrm{d}^2k_\Gamma/\mathrm{d}\sigma^2 = 0$ in the fixed frame $\{\boldsymbol{O}_f; \boldsymbol{i}_f, \boldsymbol{j}_f\}$ at an instant. The curve Γ_P and a circle have five common points at infinitesimally separated positions in the neighborhood of a contacting point. For an instant,

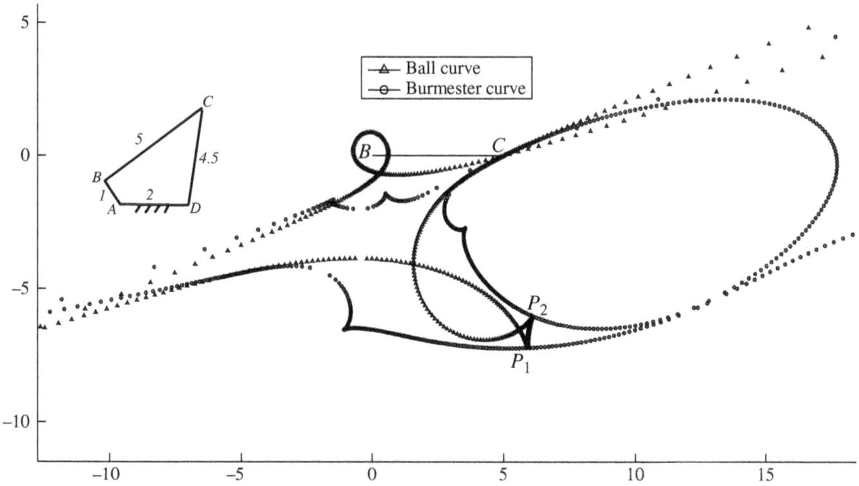

Figure 1.33 Ball curve and Burmester curve of a plane linkage

there exist fewer than four Burmester points, and even none at all. Therefore, there are a few Burmester curves on the coupler plane for different intervals of movement. Each Burmester curve is a closed curve. In the general case, a Burmester curve begins at a double Burmester point and ends with another double Burmester point. The singularity of the Burmester curve appears when $\frac{\partial k_\Gamma}{\partial s} = 0$, $\frac{\partial^2 k_\Gamma}{\partial s^2} = 0$ and $\frac{\partial^3 k_\Gamma}{\partial s^3} = 0$ in the fixed frame $\{\boldsymbol{O}_f; \boldsymbol{i}_f, \boldsymbol{j}_f\}$. A Burmester curve has singularities, which generate a coupler curve with the special point, and the coupler curve has fifth-order contact with a circle at this point. According to the examples, the position of a double Burmester point on the coupler plane is sensitive to the position parameters of the coupler plane just for $(c_1 + p)^2 - 4c_4/w = 0$.

1.3.2 Double Points

A ***double point*** (DP) refers to a point on a coupler curve, which is passed through by a corresponding tracing point twice at two different instants. There are two types of double point: a crunode at which the coupler curve crosses itself and a tacnode at which the coupler curve touches itself or is self-tangent. The tacnode, in fact, is a coalescence of two crunodes.

A coupler curve is a closed curve. As mentioned in Section 1.1.2, a coupler curve may be a simple curve or a non-simple curve, depending on the existence of DPs, which is a key indicator of the global behavior of coupler curves and also a theoretical basis for kinematic analysis and synthesis of linkages, especially for six-bar linkages.

For a planar linkage, link BC's coordinate system $\{\boldsymbol{B}; \boldsymbol{i}_m, \boldsymbol{j}_m\}$ and fixed coordinate system $\{\boldsymbol{A}; \boldsymbol{i}_f, \boldsymbol{j}_f\}$ are established in Fig. 1.34. Joint B traces a circle in the fixed frame $\{\boldsymbol{A}; \boldsymbol{i}_f, \boldsymbol{j}_f\}$. A coupler point with polar coordinates (r_P, θ_P) in the coordinate system $\{\boldsymbol{B}; \boldsymbol{i}_m, \boldsymbol{j}_m\}$ is always adjoint to joint B when it traces a coupler curve. The path Γ_B of joint B is consequently taken as an original curve, and a coupler curve Γ_p traced by a coupler point $P(r_P, \theta_P)$ is viewed as an adjoint curve of Γ_B. The Frenet frame $\{\boldsymbol{R}_B; \boldsymbol{\alpha}, \boldsymbol{\beta}\}$ of the original curve Γ_B is established

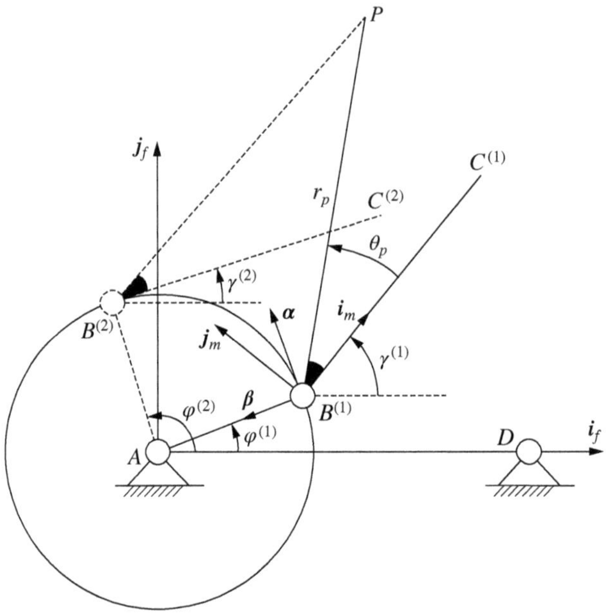

Figure 1.34 Double point of a coupler curve for plane linkages

in Fig. 1.34. Referring to Example 1.4, a coupler curve Γ_P as the adjoint curve of Γ_B can be expressed in the Frenet frame $\{\boldsymbol{R}_B; \boldsymbol{\alpha}, \boldsymbol{\beta}\}$ as

$$\boldsymbol{R}_P = \boldsymbol{R}_B + u_1\boldsymbol{\alpha} + u_2\boldsymbol{\beta} = u_1\boldsymbol{\alpha} + (u_2 - a_1)\boldsymbol{\beta} \tag{1.124a}$$

where (u_1, u_2) are the coordinates of a coupler point P in the Frenet frame $\{\boldsymbol{R}_B; \boldsymbol{\alpha}, \boldsymbol{\beta}\}$, which are given by

$$\begin{cases} u_1 = r_P \sin\left(\theta_P - \varphi + \gamma\right) \\ u_2 = -r_P \cos(\theta_P - \varphi + \gamma) \end{cases} \tag{1.124b}$$

In order for a point on a coupler curve to be a DP, we must have

$$\boldsymbol{R}_P^{(1)} = \boldsymbol{R}_P^{(2)} \tag{1.125}$$

where $\boldsymbol{R}_P^{(1)}$ and $\boldsymbol{R}_P^{(2)}$ represent, respectively, the position vectors of the tracing point $P(r_P, \theta_P)$ at two different instants, the superscripts 1 and 2 represent the corresponding instants $\varphi^{(1)}$ and $\varphi^{(2)}$. The above equation may be replaced by the following equations:

$$\begin{cases} \left|\boldsymbol{R}_P^{(1)}\right| = |\boldsymbol{R}_P^{(2)}| \\ \boldsymbol{R}_P^{(1)} \times \boldsymbol{R}_P^{(2)} = 0(\boldsymbol{R}_P^{(1)} \neq -\boldsymbol{R}_P^{(2)}, \boldsymbol{R}_P^{(1)} \neq 0, \boldsymbol{R}_P^{(2)} \neq 0) \end{cases} \tag{1.126}$$

Substituting equation (1.124a) into the first equation in (1.126), we have

$$(u_1^{(1)})^2 + (u_2^{(1)} - a_1)^2 = (u_1^{(2)})^2 + (u_2^{(2)} - a_1)^2 \tag{1.127}$$

where $(u_1^{(1)}, u_2^{(1)})$ and $(u_1^{(2)}, u_2^{(2)})$ are, respectively, the coordinates of the point P in the Frenet frame $\{\boldsymbol{R}_B; \boldsymbol{\alpha}, \boldsymbol{\beta}\}$ corresponding to $\varphi^{(1)}$ and $\varphi^{(2)}$, when the coupler point $P(r_P, \theta_P)$ moves adjoint to joint B. Based on the fact that the tracing point $P(r_P, \theta_P)$ belongs to a fixed point of the coupler plane and equation (1.124a), the identical length requirement of vectors $\boldsymbol{R}_P^{(1)}$ and $\boldsymbol{R}_P^{(2)}$ can be simplified as

$$r_P^2 + a_1^2 - 2a_1 r_P \sin(\theta_P - \theta^{(1)}) = r_P^2 + a_1^2 - 2a_1 r_P \sin(\theta_P - \theta^{(2)}) \tag{1.128}$$

which leads to

$$\sin(\theta_P - \theta^{(1)}) = \sin(\theta_P - \theta^{(2)}) \tag{1.129}$$

and

$$\theta_P - \theta^{(1)} = \theta_P - \theta^{(2)}, \text{ or } \theta_P - \theta^{(1)} = n\pi - (\theta_P - \theta^{(2)}) \tag{1.130}$$

where $\theta^{(1)} = \pi/2 + \varphi^{(1)} - \gamma^{(1)}$, $\theta^{(2)} = \pi/2 + \varphi^{(2)} - \gamma^{(2)}$ is the same as in Section 1.2, and n is an odd number.

Substituting equation (1.124a) into the second equation of (1.126), after some simplifications, we have the following expression for the identical azimuthal angle requirement for vectors $\boldsymbol{R}_P^{(1)}$ and $\boldsymbol{R}_P^{(2)}$:

$$\begin{aligned} &u_1^{(1)} u_1^{(2)} \sin\Delta\varphi - u_1^{(2)}(u_2^{(1)} - a_1)\cos\Delta\varphi \\ &\quad + u_1^{(1)}(u_2^{(2)} - a_1)\cos\Delta\varphi + (u_2^{(1)} - a_1)(u_2^{(2)} - a_1)\sin\Delta\varphi = 0 \end{aligned} \tag{1.131}$$

where $\Delta\varphi = \varphi^{(2)} - \varphi^{(1)}$. Substituting equation (1.124b) into the above equation, we have

$$\begin{aligned} r_P = a_1 \{ &\cos(\theta_P - \theta^{(1)} - \Delta\varphi) - \cos(\theta_P - \theta^{(2)} + \Delta\varphi) \\ &\pm \sqrt{[\cos(\theta_P - \theta^{(1)} - \Delta\varphi) - \cos(\theta_P - \theta^{(2)} + \Delta\varphi)]^2 - 4\sin\Delta\varphi \sin(\theta^{(1)} - \theta^{(2)} + \Delta\varphi)} \} \\ &/[2\sin(\theta^{(1)} - \theta^{(2)} + \Delta\varphi)] \end{aligned} \tag{1.132}$$

Equations (1.130) and (1.132) have three solutions. Two of them are $\boldsymbol{R}_P^{(1)} = 0$ and $\boldsymbol{R}_P^{(1)} = -\boldsymbol{R}_P^{(2)}$. The third solution is

$$\begin{cases} \theta_P - \theta^{(1)} = n\pi - (\theta_P - \theta^{(2)}) \\ r_P = a_1 \dfrac{\sin(\Delta\varphi/2)}{\cos(\theta_P - \theta^{(1)} - \Delta\varphi/2)} \end{cases} \tag{1.133}$$

Equation (1.133) indicates the relationship between the tracing point $P(r_P, \theta_P)$ of the coupler plane and two positions $B^{(1)}$ and $B^{(2)}$ of the joint B, or two positions of the Frenet frame $\{\boldsymbol{R}_B; \boldsymbol{\alpha}, \boldsymbol{\beta}\}$, in which a DP of the coupler curve appears and coincides with the tracing point $P(r_P, \theta_P)$ at the two instants, called the ***DP condition***. Based on this equation, we may find the position of the coupler point, which traces a coupler curve with a DP at two given instants. Similarly, for a given point $P(r_P, \theta_P)$ on the coupler, we may determine whether the point tracing a curve has a DP or not.

The first equation of (1.133) can be rewritten as

$$(\theta_P - \theta^{(1)}) + (\theta_P - \theta^{(2)}) = n\pi \tag{1.134}$$

Utilizing the above equation, the denominator of the second equation of (1.133) is simplified as

$$\theta_P - \theta^{(1)} - \Delta\varphi/2 = \theta_P - \theta^{(1)} - \frac{\theta^{(2)} + \gamma^{(2)} - (\theta^{(1)} + \gamma^{(1)})}{2}$$
$$= \theta_P - \frac{\theta^{(1)} + \theta^{(2)}}{2} - \frac{\gamma^{(2)} - \gamma^{(1)}}{2} = \frac{n}{2}\pi - \frac{\Delta\gamma}{2} \tag{1.135}$$

where $\Delta\gamma = \gamma^{(2)} - \gamma^{(1)}$ and $\cos(\theta_P - \theta^{(1)} - \Delta\varphi/2) = \pm\sin(\Delta\gamma/2)$. The second equation of (1.133) can be rewritten as

$$r_P = \pm a_1 \sin(\Delta\varphi/2)/\sin(\Delta\gamma/2) \tag{1.136}$$

The tracing point is just located on the $\boldsymbol{\beta}$-axis of the Frenet frame $\{\boldsymbol{R}_B; \boldsymbol{\alpha}, \boldsymbol{\beta}\}$ and the value of the polar radius r_P can be obtained by taking the limit of equation (1.136) as $\Delta\varphi$ approaches zero. We have

$$r_P = \pm \lim_{\Delta\varphi\to 0} \frac{a_1 \sin(\Delta\varphi/2)}{\sin(\Delta\gamma/2)} \tag{1.137a}$$

Using the rules of calculus, we find that

$$r_p = \pm \lim_{\Delta\varphi\to 0} \frac{a_1 \sin(\Delta\varphi/2)}{\sin(\Delta\gamma/2)} = \pm \lim_{\Delta\varphi\to 0} \frac{a_1 \cos(\Delta\varphi/2)\cdot \mathrm{d}\varphi}{\cos(\Delta\gamma/2)\cdot \mathrm{d}\gamma} = \pm\frac{1}{\dot{\gamma}} \tag{1.137b}$$

where the plus sign applies if r_P points in the positive $\boldsymbol{\beta}$-axis, the minus sign applies if r_P points in the negative $\boldsymbol{\beta}$-axis according to equation (1.135). For a planar linkage, γ is a function of φ; we have $\gamma^{(2)} \to \gamma^{(1)}$ and $\theta^{(2)} \to \theta^{(1)}$ if $\Delta\varphi \to 0$ or $\varphi^{(2)} \to \varphi^{(1)}$. Equation (1.134) can be rewritten as

$$\theta_P - \theta^{(1)} = \frac{n}{2}\pi \tag{1.138}$$

Comparing equation (1.57) with the above equation, we find that the tracing point on the coupler at a double point on its path is the instant velocity center at that position, which implies the cusp is a special case of the double point.

For a given type of planar linkage, the movement of the coupler is determined by the linkage dimensions, or γ is a function of φ. The coordinates (r_P, θ_P) of the double point's tracer are functions of φ, while $\Delta\gamma$ is also a function of φ by equation (1.134). For convenience in discussing the distribution region of the double point's tracers on the coupler plane, a θ_P-ray is defined as a line from the origin of the coordinate system $\{\boldsymbol{B}; \boldsymbol{i}_m, \boldsymbol{j}_m\}$, joint B, to infinity and makes an angle θ_P with the $\boldsymbol{i}_m$-axis. Based on equation (1.133), r_P varies in accordance with the point on the θ_P-ray of the coupler plane. If the variation of r_P is not monotonic, an extreme value must exist. Taking the first derivative of equation (1.133) with respect to the arc length of Γ_B, we have

$$\begin{cases} \dfrac{2}{a_1} + \dfrac{\mathrm{d}\Delta\varphi}{\mathrm{d}s} - \dot{\gamma^{(1)}} - \dot{\gamma^{(2)}}\left(1 + \dfrac{\mathrm{d}\Delta\varphi}{\mathrm{d}s}a_1\right) = 0 \\[2ex] \dfrac{\mathrm{d}r_P}{\mathrm{d}s} = \pm a_1 \dfrac{\cos(\Delta\varphi/2)\sin(\Delta\gamma/2)\dfrac{\mathrm{d}\Delta\varphi}{\mathrm{d}s} - \sin(\Delta\varphi/2)\cos(\Delta\gamma/2)\dfrac{\mathrm{d}\Delta\gamma}{\mathrm{d}s}}{2\sin^2(\Delta\gamma/2)} \end{cases} \tag{1.139}$$

Letting $\mathrm{d}r_P/\mathrm{d}s = 0$ in the second of the above equation and combining the first equation, we obtain

$$r_P = \pm\frac{(a_1\dot{\gamma}^{(1)} + a_1\dot{\gamma}^{(2)} - 2)\cos(\Delta\varphi/2)}{(2a_1\dot{\gamma}^{(1)}\dot{\gamma}^{(2)} - \dot{\gamma}^{(1)} - \dot{\gamma}^{(2)})\cos(\Delta\gamma/2)} \tag{1.140}$$

From the above equation and equation (1.133), the extreme value of the polar radius of the DP tracer on the θ_P-ray can be found. The condition equations are rewritten as follows:

$$\begin{cases} 2\theta_P - \varphi^{(1)} - \varphi^{(2)} + \gamma^{(1)} + \gamma^{(2)} = (n+1)\pi \\ r_P = a_1\dfrac{\sin(\Delta\varphi/2)}{\cos(\theta_P - \theta^{(1)} - \Delta\varphi/2)} \\ r_P = \pm\dfrac{(a_1\dot{\gamma}^{(1)} + a_1\dot{\gamma}^{(2)} - 2)\cos(\Delta\varphi/2)}{(2a_1\dot{\gamma}^{(1)}\dot{\gamma}^{(2)} - \dot{\gamma}^{(1)} - \dot{\gamma}^{(2)})\cos(\Delta\gamma/2)} \end{cases} \tag{1.141}$$

Equation (1.141) is a transcendental equation. Obviously, equation (1.137) satisfies equation (1.141). The polar radius of the instant center is an extreme value. To examine the geometrical meaning of the extreme value of the polar radius r_P on the θ_P-ray, it is necessary to distinguish the two types of DP. The condition for a tacnode is

$$\boldsymbol{R}_P^{(1)} = \boldsymbol{R}_P^{(2)}, \dot{\boldsymbol{R}}_P^{(1)} = \dot{\boldsymbol{R}}_P^{(2)} \tag{1.142}$$

which is actually a parallel condition for the tangent vectors in addition to the condition in (1.126) for the DP. Substituting equations (1.124) and (1.133) into equation (1.142), we have an equation identical to equation (1.141). This indicates that the self-tangent is the limit of the self-crossing, while the cusp is a special case of the tacnode.

1.3.3 Four-bar Linkage I

Equation (1.133) cannot easily be solved to locate the position of a DP tracer since ranges of variations of φ and γ depend on the dimension of a four-bar linkage. Recall that planar four-bar linkages can be divided into two groups: (i) linkages with a single crank and (ii) linkages without a single crank. Examples of the first group include crank-rocker, slider-crank, and inverted crank-slider linkages. Examples of the second group include double crank and double rocker in a Grashof kinematic chain, and double rocker in a non-Grashof kinematic chain. Linkages in the first group are discussed in this subsection, linkages in the other group are studied in the next.

For a crank-rocker linkage, there exists an exact characteristic that the crank rotates about the fixed pivot in $\varphi \in [0, 2\pi]$ while the coupler rotates with $\gamma(\varphi) = \gamma(2\pi + \varphi)$ and $\gamma \in [-\pi, \pi]$ relative to the base link. As an example, we choose the following dimensions: $a_1 = 10.5$, $a_2 = 19.5$, $a_3 = 19$, and $a_4 = 25.5$. By solving the displacement equations, we have the following function of the parameters:

$$f = \theta_P - \varphi + \gamma - \pi/2 \tag{1.143}$$

The graphs of $\gamma(\varphi)$ and $f = f(\varphi)$ can be calculated for $\varphi \in [0, 2\pi]$, as shown in Fig. 1.35.

Since θ_P is an invariant with φ, different values of θ_P only shift the φ-axis up or down, without affecting the shape of the graphs. Take a position $\varphi^{(1)}$ and $f^{(1)} = \theta_P - \varphi^{(1)} + \gamma^{(1)} -$

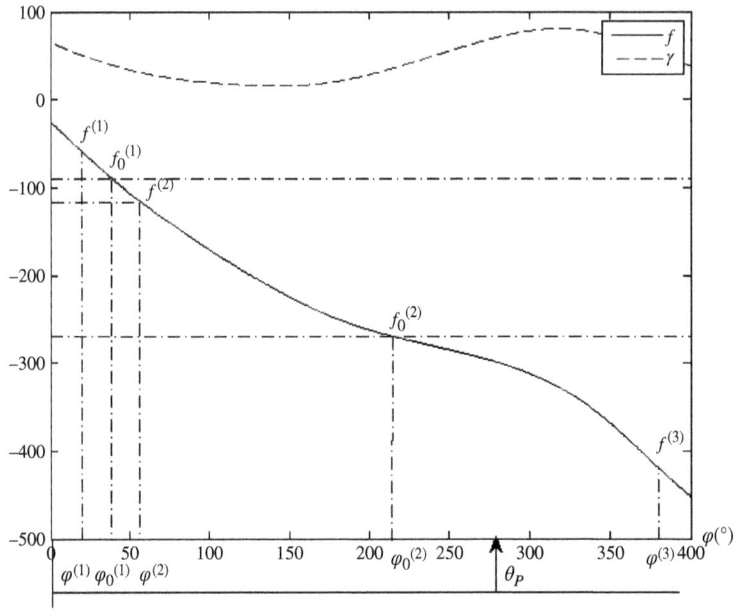

Figure 1.35 $\gamma - \varphi$ and $f = f(\varphi)$ of a crank-rocker linkage

$\pi/2$ as an example and let $\varphi^{(3)} = \varphi^{(1)} + 2\pi$, then $f^{(3)} = \theta_P - \varphi^{(3)} + \gamma^{(3)} - \pi/2$ and $f^{(3)} = f^{(1)} - 2\pi$, because $\gamma^{(3)} = \gamma^{(1)}$ by the periodicity of motion of the crank-rocker linkage. In view of continuity of $f - \varphi$ in the interval $[f^{(3)}, f^{(1)}]$, there must be $f(\varphi^{(2)}) = f^{(2)}$ and $f^{(3)} \le f^{(2)} \le f^{(1)}$ such that

$$\begin{cases} f^{(1)} + f^{(2)} = n\pi \\ f^{(2)} + f^{(3)} = (n-2)\pi \end{cases} \tag{1.144}$$

where n is an odd integer and $2f^{(1)}/\pi - 2 \le n \le 2f^{(1)}/\pi$. For an arbitrary θ_P-ray, equation (1.144) indicates that $\varphi^{(1)}$ corresponding to $\varphi^{(2)}$ and $\varphi^{(2)}$ corresponding to $\varphi^{(3)}$ satisfy the DP condition in equation (1.133). The interval of the crank rotation angle $\varphi \in [0, 2\pi]$ may be divided into two sub-intervals $[\varphi^{(1)}, \varphi^{(2)}]$ and $[\varphi^{(2)}, \varphi^{(3)}]$. Letting $\varphi^{(i)} = \varphi^{(1)} + \Delta\varphi^{(1)}$ and $\varphi^{(j)} = \varphi^{(2)} - \Delta\varphi^{(2)}$, we have $f^{(i)} = f(\varphi^{(1)} + \Delta\varphi^{(1)})$ and $f^{(j)} = f(\varphi^{(2)} - \Delta\varphi^{(2)})$, in which $\Delta\varphi$ is an increment of φ. For an arbitrary position $\varphi^{(i)}$ in $[\varphi^{(1)}, \varphi^{(2)}]$ or $[\varphi^{(2)}, \varphi^{(3)}]$, there must exist another position (instant) $\varphi^{(j)}$ in the same subinterval such that

$$f(\varphi^{(i)}) + f(\varphi^{(j)}) = n\pi \tag{1.145a}$$

or

$$f(\varphi^{(i)}) + f(\varphi^{(j)}) = (n-2)\pi \tag{1.145b}$$

Substituting $\varphi^{(i)}$ and $\varphi^{(j)}$ into the second equation in (1.133), we can find the position of the point on the θ_P-ray which will trace a curve with a DP and will be exactly at the DP in both $\varphi^{(i)}$ and $\varphi^{(j)}$ positions.

Therefore, it may be concluded that:

At a position between 0 and 2π of the crank for a four-bar linkage I defined by input angle φ, there exists a point on the θ_P-ray in the coupler which coincides with a DP on its coupler curve. Hence, there also exists a curve on the coupler for $\theta_P \in [0, 2\pi]$, an instant double-point tracer's curve, in which each point is a DP tracer and traces a coupler curve with a DP at this instant φ.

For a crank-rocker linkage ($a_1 = 10.5$, $a_2 = 19.5$, $a_3 = 19$, and $a_4 = 25.5$) at $\varphi = \pi/3$ and $\theta_P \in [0, 2\pi]$, the polar radius r_P of the double point can be calculated through equation (1.133). It can be verified that each of the following points, $P_1(68.1028, 85.9437°)$, $P_2(36.6060, 48.7014°)$, $P_3(31.2283, 297.9381°)$, and $P_4(64.7278, 289.9166°)$, is on a double-point tracer's curve of the coupler, shown in Fig. 1.36(a), and the coupler curves for P_1, P_2, P_3, and P_4 are shown in Fig. 1.36(b).

To locate the varying range for positions of tracing points on the coupler, whose coupler curve has a DP, further analysis follows. Figure 1.35 shows that there exist $f_0^{(1)} = -\pi/2$ in $[\varphi^{(1)}, \varphi^{(2)}]$ and $f_0^{(2)} = -3\pi/2$ in $[\varphi^{(2)}, \varphi^{(3)}]$. $\varphi^{(i)}$ and $\varphi^{(j)}$ in equation (1.145) lie on two sides of $\varphi_0^{(1)}$ (or $\varphi_0^{(2)}$). $\varphi^{(i)}$ and $\varphi^{(j)}$ approach $\varphi_0^{(1)}$ (or $\varphi_0^{(2)}$) from both sides as $\Delta\varphi = \varphi^{(j)} - \varphi^{(i)}$ approaches zero.

Since $\varphi^{(i)}$ and $\varphi^{(j)}$ correspond to each other, it is sufficient to limit φ to $(\varphi_0^{(1)}, \varphi_0^{(2)})$ rather than to $(0, 2\pi)$ in root searching. As for $\varphi_0^{(1)}$ and $\varphi_0^{(2)}$, they are two limited positions; each corresponds to itself according to equation (1.145). Referring to Fig. 1.35, when $f_0^{(1)} = -\pi/2$ and $f_0^{(2)} = -3\pi/2$, the tracing point lies on the $\boldsymbol{\beta}$-axis. The value of the polar radius can be obtained from the following equation:

$$r_P = \lim_{\Delta\varphi \to 0} \frac{a_1 \sin(\Delta\varphi/2)}{\sin(\theta_P - \varphi + \gamma - \Delta\varphi/2)} \tag{1.146a}$$

It can easily be shown that the polar radius approaches

$$r_P = \pm\frac{1}{\dot{\gamma}} = r_0 \tag{1.146b}$$

Equations (1.146) define a DP tracing point at the instant center for $\varphi_0^{(1)}$ and $\varphi_0^{(2)}$. Since the polar radius r_P corresponding to the positions (or instants) $\varphi_0^{(1)}$ and $\varphi_0^{(2)}$ is identical to that of the instant velocity center, $r_0^{(1)}$ and $r_0^{(2)}$, this means that r_P will vary from $r_0^{(1)}$ to $r_0^{(2)}$ in $(\varphi_0^{(1)}, \varphi_0^{(2)})$. If the variation is monotonic, the DP tracers on the θ_P-ray of the coupler plane will distribute between the two instant velocity centers, and the paths of those tracing points have only one DP – as shown in Fig. 1.37(b) for a coupler plane with points $P_1 - P_6$ on the θ_{P1}-ray in Fig. 1.37(a). Each point, between the two branches π_m^1 and π_m^2 of the moving centrode, $P_2 - P_4$, traces a closed curve with one double point. For non-monotonic variations of r_P, an extreme value exists. Taking the first derivative of equation (1.133) with respect to φ and letting $\mathrm{d}r_P/\mathrm{d}s = 0$, we obtain

$$r_P = [(a_1 r_0^{(1)} + a_1 r_0^{(2)} \mp 2r_0^{(1)} r_0^{(2)}) \cos(\Delta\varphi/2)]/[(\pm 2a_1 - r_0^{(1)} - r_0^{(2)}) \cos(\Delta\gamma/2)] \tag{1.147}$$

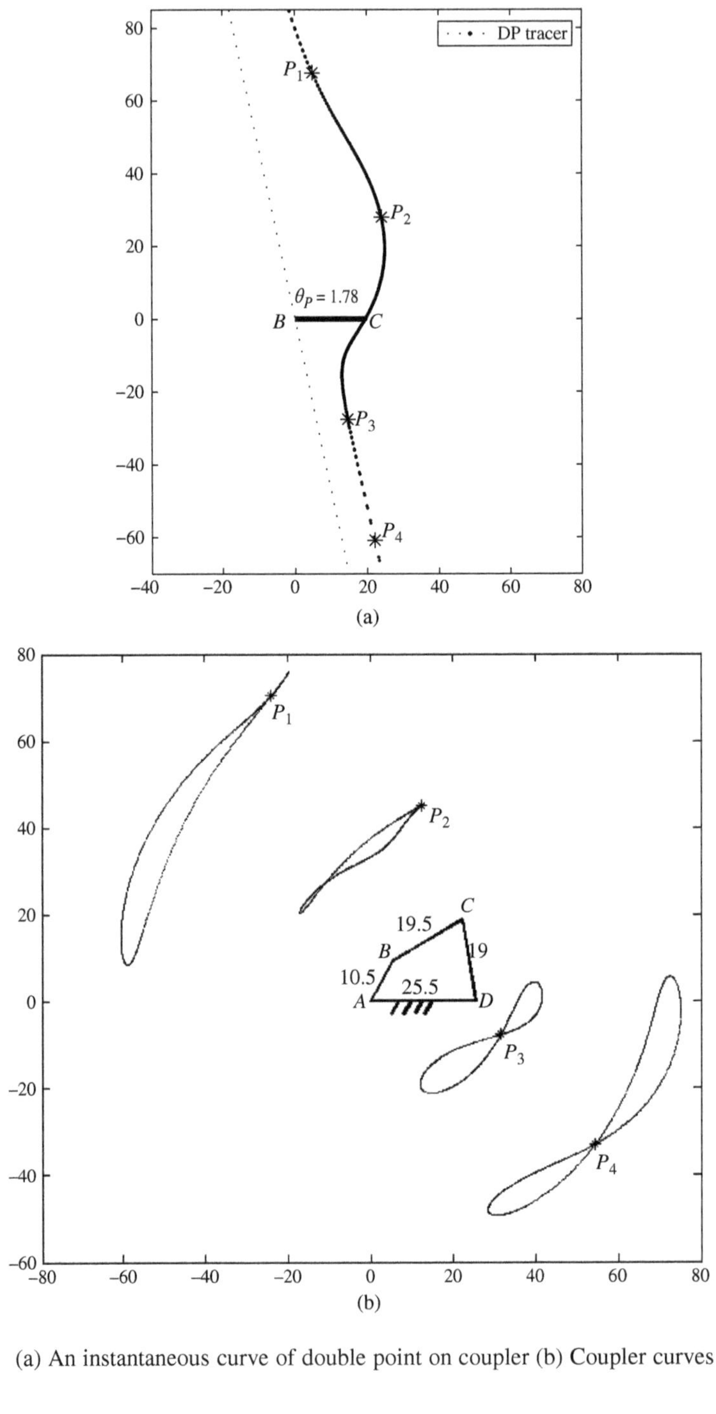

Figure 1.36 (a) An instantaneous curve of double point on coupler (b) Coupler curves with a double point

From equations (1.133) and (1.147), the extreme value of the polar radius r_P of the points on the θ_P-ray can be found. The equations are rewritten as follows:

$$\begin{cases} 2\theta_p - \varphi^{(1)} + \gamma^{(1)} - \varphi^{(2)} + \gamma^{(2)} = \pi \\ r_P = [a_1 \sin(\Delta\varphi/2)]/[\sin(\theta_p - \varphi^{(1)} + \gamma^{(1)} - \Delta\varphi/2)] \\ r_P = [(a_1 r_0^{(1)} + a_1 r_0^{(2)} \mp 2r_0^{(1)} r_0^{(2)})\cos(\Delta\varphi/2)]/[(\pm 2a_1 - r_0^{(1)} - r_0^{(2)})\cos(\Delta\gamma/2)] \end{cases} \tag{1.148}$$

Equation (1.148) is a transcendental equation. Obviously, equation (1.146) satisfies equation (1.148). The polar radius for the instant center is an extreme value, which is the same as that of equation (1.141) for a crank-rocker linkage.

For a given polar angle θ_{P2} in the coupler, if there exists an extreme value other than those for the instant centers $P_0^{(1)}$ and $P_0^{(2)}$, such as P_9 (r_9, θ_{P2}) on the θ_{P2}-ray in Fig. 1.37 where $r_9 < r_0^{(1)} < r_0^{(2)}$, the polar radius r_P of the point position on the θ_{P2}-ray varies from $P_0^{(1)}$ to $P_0^{(2)}$ via P_9. The range of r_P covers the interval $(r_9, r_0^{(1)})$ twice and the interval $(r_0^{(1)}, r_0^{(2)})$ only once, which indicates that the points in $(r_9, r_0^{(1)})$ on the θ_{P2}-ray trace curves with two DPs, as shown in Fig. 1.37(b) for P_9–$P_0^{(1)}$, while the points in $(r_0^{(1)}, r_0^{(2)})$ on the θ_{P2}-ray will trace curves with only one double point, as shown in Fig. 1.37(b) for $P_0^{(1)}$–$P_0^{(2)}$.

For $\theta_p \in [0, 2\pi]$, the coupler points, whose polar radius r_P have the extreme values, will constitute a curve on the coupler, called a ***self-tangent tracer curve*** or a transition curve (a tens-order curve [2, 3]), as shown by the dotted line in Fig. 1.37(a). Both ends of the self-tangent tracer curve coalesce with two branches π_m^1 and π_m^2 of the moving centrode, respectively. The coupler points on the self-tangent tracer curve will trace coupler curves with a tacnode in the fixed frame. The coupler points between the self-tangent tracer curve and the moving centrode, known as the ***area of two double points (2DP) tracers***, will trace coupler curves with two crunodes. The coupler points between the two branches π_m^1 and π_m^2 of the moving centrode, known as the ***area of a single double-point (1DP) tracer***, will trace coupler curves with single crunodes. The coupler points in other areas, known as the ***area of a simple curve (0DP) tracer***, will trace simple coupler curves.

A region is shown in figure 6 of Ref. [4] between the two branches of the moving centrode (identical to the fixed centrode for the specific dimensions of a crank-rocker linkage) where points are claimed to have three DPs in their paths. However, our verification indicates that this is not the case. Any point in that region traces a curve with one crunode. Besides, the area of 2DP tracers is not shown in that figure, but we find that an area of 2DP tracers similar to that in figure 4 of Ref. [4] does exist.

From the above discussion, it is derived that, for a crank-rocker linkage, if the areas of DP and 2DP tracers on some ray in the coupler plane shrink to one single point, such as joint C of the coupler, shown in Fig. 1.37(a), this point traces a path on which every point is a double point. As a matter of fact, the path of joint C comprises two circular arcs coincident with each other.

If there are a few extreme values of the polar radius r_P on the θ_P-ray other than those for the instant velocity centers by equation (1.141), there exist several self-tangent tracer curves on the coupler plane. The moving centrode with two branches and several self-tangent tracer curves are designated as boundaries; they divide the coupler plane into several areas with simple tracers, 1DP tracers, 2DP tracers, and 3DP tracers. Points in these areas are DP tracers whose paths in the fixed frame have one, two, or three DPs, respectively. The difference in number of

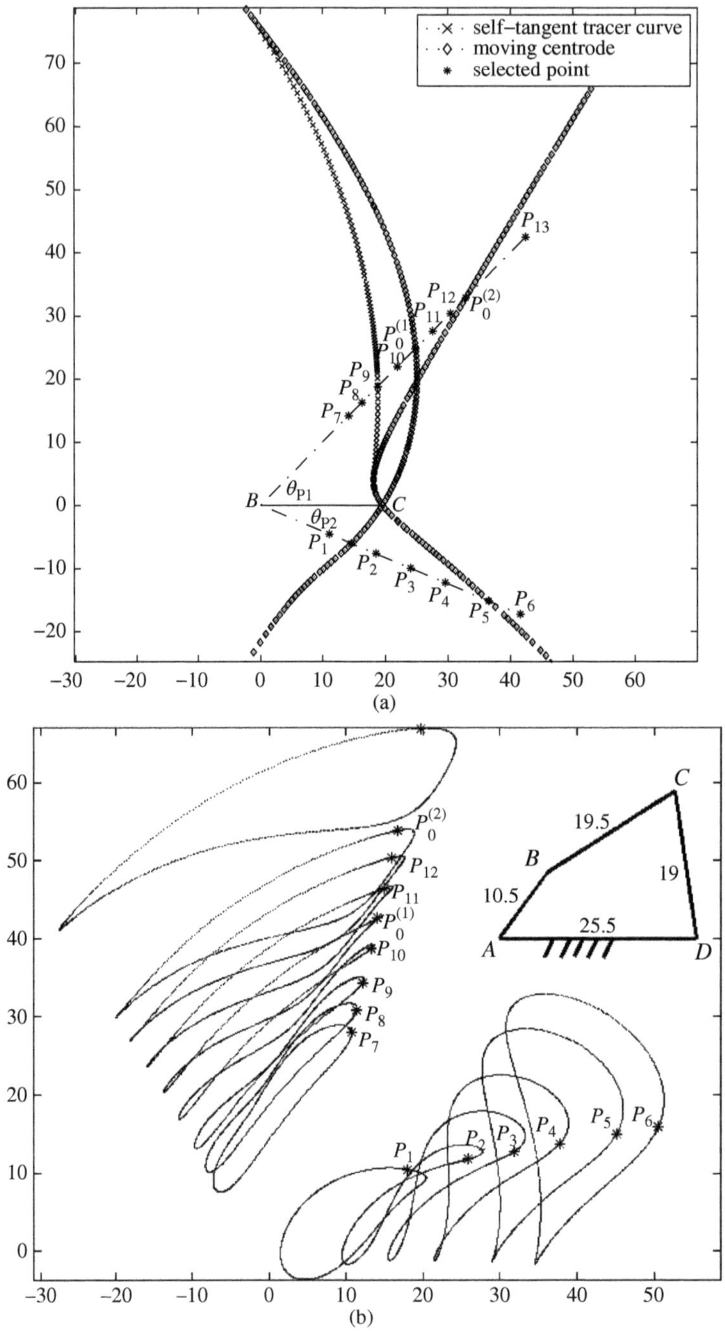

Figure 1.37 (a) Two lines of the coupler (b) The coupler curves of coupler points on two lines

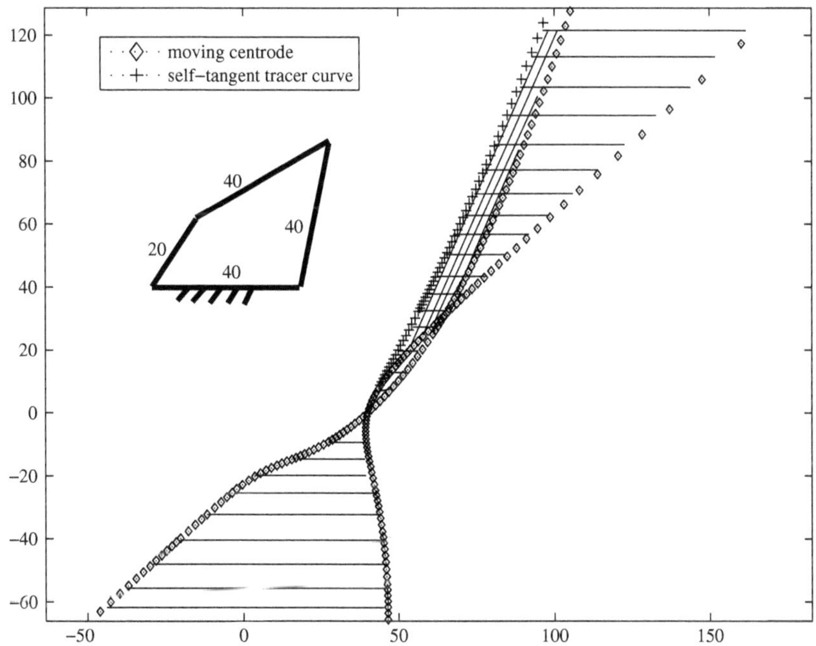

Figure 1.38 Distribution area of DP tracers for a crank-rocker linkage

DPs on the coupler curves is two if the coupler points are on either side of a self-tangent tracer curve. The difference is one if they are on either side of a moving centrode. For crank-rocker linkages, the area of 2DP tracers usually appears (Fig. 1.38). However, we have not found an area of 3DP tracers, although there theoretically exist three DPs in a coupler curve for a four-bar linkage [3].

For a single-crank linkage with one sliding joint and three revolute joints, substituting the $\gamma(\varphi)$ and $f = f(\varphi)$ relationships into equations (1.133) and (1.141), we have the distribution area of DP tracers in the coupler plane, as shown in Fig. 1.39.

In the same way as mentioned above, the distribution area of DP tracers in the coupler plane can be located for the linkage with given dimensions, as shown in Fig. 1.40.

1.3.4 Four-bar Linkage II

The conditions for a DP of a coupler curve for a four-bar linkage are discussed in Section 1.3.2. The distributions of DPs for a single-crank four-bar linkage are studied in Section 1.3.3. It is noted here that the results in these two sections are not applicable to all four-bar linkages. The second group of four-bar linkage includes those having double cranks, double rockers of Grashof kinematic chain, and double rockers of non-Grashof kinematic chain. The difference between the two groups of four-bar linkages is just the range of the rotation angle φ of the driving link (or reference binary link) and the inclination angle γ with respect to the base link. The graph $f(\varphi)$ is not a monotonic function: $\varphi^{(3)} = 2\pi + \varphi^{(1)}$ for a double-crank linkage; $\gamma^{(3)} = 2\pi + \gamma^{(1)}$ and $\varphi^{(1)} = \varphi^{(3)}$ for a double-rocker linkage of a Grashof kinematic chain; $\varphi^{(1)} = \varphi^{(3)}$

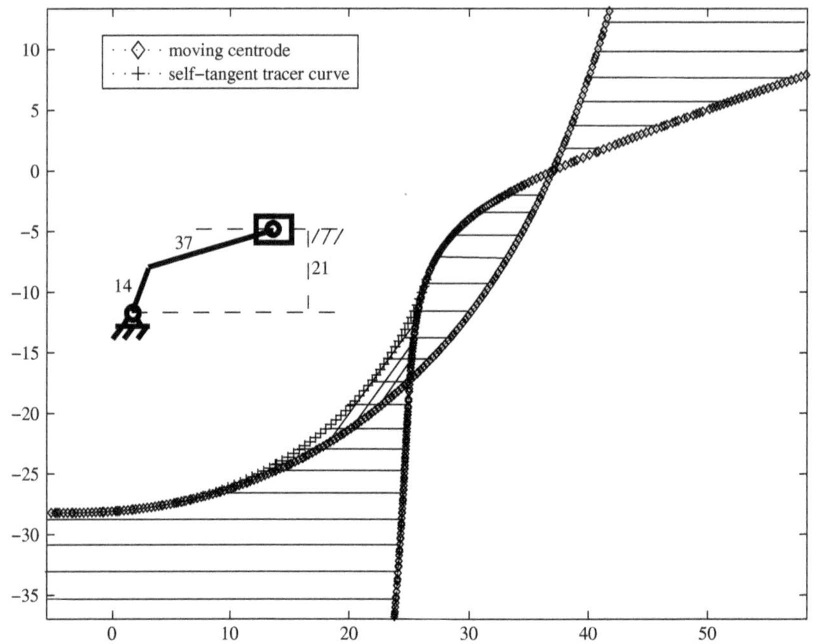

Figure 1.39 Distribution area of DP tracers for a slider-crank linkage

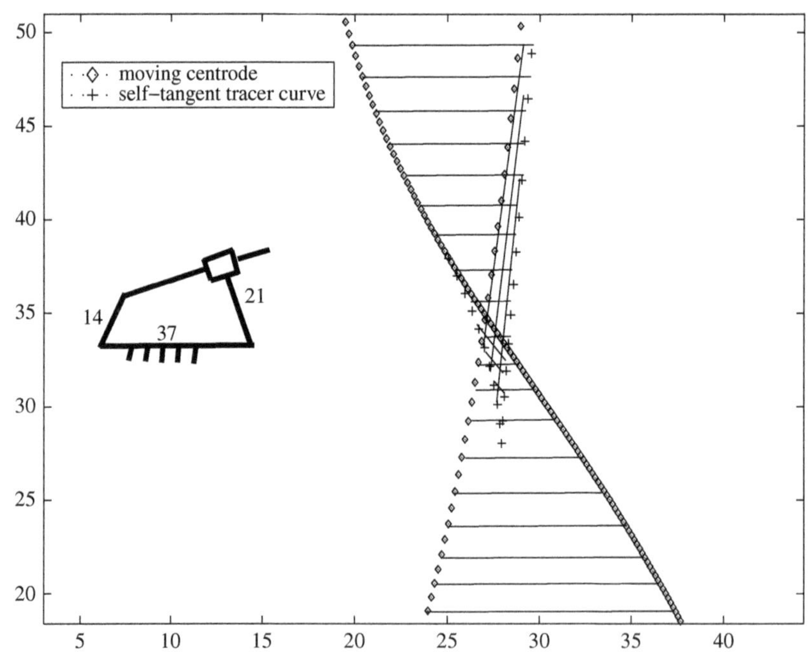

Figure 1.40 Distribution area of DP tracers for an inverted slider-crank linkage

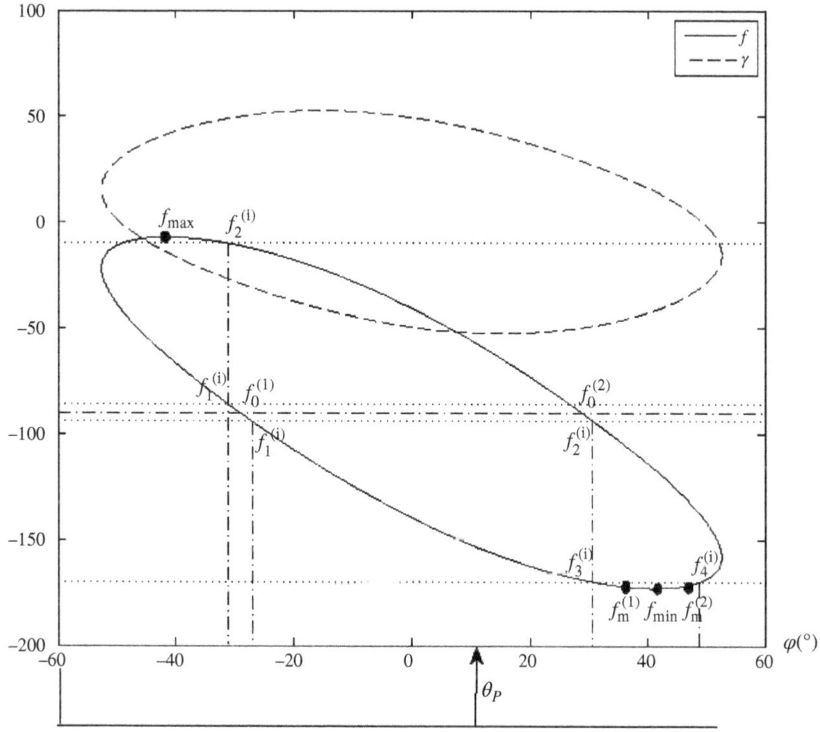

Figure 1.41 The graph of $\gamma(\varphi)$ and $f(\varphi)$ for a non-Grashof four-bar linkage

for all double-rocker linkages of non-Grashof kinematic chains. Special attention is needed to deal with those linkages without a single crank.

Two four-bar linkages of a non-Grashof kinematic chain are taken as examples. Their link lengths are $a_1 = 20$, $a_2 = 20$, $a_3 = 40$, $a_4 = 70$ and $a_1 = 20$, $a_2 = 30$, $a_3 = 40$, $a_4 = 60$. The graphs of $\gamma(\varphi)$ and $f(\varphi)$ are drawn in Fig. 1.41 ($\theta_P = 0°$) and Fig. 1.42 ($\theta_P = -100°$) for these two linkages. The θ_P-ray has been defined in Section 1.3.3. For a group II four-bar linkage, there may exist several positions, from two to four, in the graph of $f(\varphi)$ for $f = n\pi/2$ or $f = (n-2)\pi/2$. The maximum value f_{max} and the minimum value f_{min} of the graph $f(\varphi)$ can now be determined and two special cases will be discussed for two positions if $\Delta f_{max} = f_{max} - f_{min} < \pi$ and four positions if $\Delta f_{max} \geq \pi$.

1. $\Delta f_{max} < \pi$

   For $\Delta f_{max} < \pi$ in the graph of $f(\varphi)$, shown in Fig. 1.41, the two θ_P-rays are determined by $\theta_{P1} = n\pi/2 - f_{min} (n = -1, -3)$ and $\theta_{P2} = n\pi/2 - f_{max}$. This indicates that any θ_P-ray has two intersection points with the moving centrode π_m if $\theta_P \in [\theta_{P1}, \theta_{P2}]$ and no intersection point at all if $\theta_P \notin [\theta_{P1}, \theta_{P2}]$. The polar radius r_P of the intersection point can be located at $r_0^{(1)}$ and $r_0^{(2)}$ by $f_0^{(1)} = f_0^{(2)} = n\pi/2$ and equation (1.137). Meanwhile, the graph $f(\varphi)$ is divided into four regions: $[f_{min}, f_0^{(1)}]$, $[f_{min}, f_0^{(2)}]$, $[f_0^{(1)}, f_{max}]$, and $[f_0^{(2)}, f_{max}]$. Here, $f^{(i)} \in [f_{min}, f_0^{(1)}]$ (or $f^{(i)} \in [f_{min}, f_0^{(2)}]$) is given at first. Finally, $f^{(j)} \in [f_0^{(1)}, f_{max}]$ (or $f^{(j)} \in$

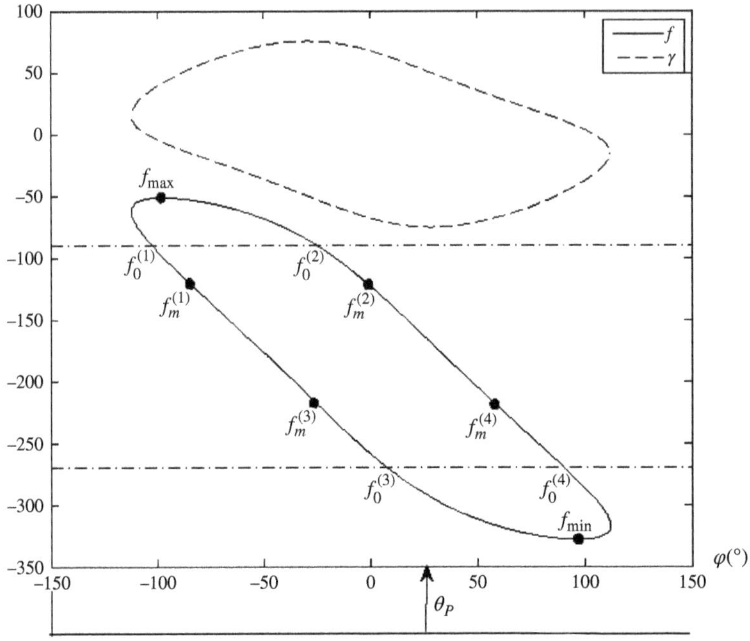

Figure 1.42 The graph of $\gamma(\varphi)$ and $f(\varphi)$ for a non-Grashof four-bar linkage

$[f_0^{(2)}, f_{max}]$) can be searched out by equation (1.145), or $f^{(i)}$ and $f^{(j)}$ satisfy equation (1.145), which correspond to two positions $\varphi^{(i)}$ and $\varphi^{(j)}$. The polar radius r_P of the DP tracer can be located by equation (1.133).

In case A, $f_{max} + f_{min} \le n\pi$, where n is an odd integer; there exist $f_m^{(1)} \in [f_{min}, f_0^{(1)}]$ and $f_m^{(2)} \in [f_{min}, f_0^{(2)}]$, which make $f_m^{(1)} + f_{max} = f_m^{(2)} + f_{max} = n\pi$ true. On the contrary, any $f^{(i)}$ belonging to $[f_{min}, f_m^{(1)}]$ (or $[f_{min}, f_m^{(2)}]$) cannot match another $f^{(j)}$ by $f^{(i)} + f^{(j)} = n\pi$ since $f_{max} + f_{min} \le n\pi$, which leads to the two regions $[f_{min}, f_m^{(1)}]$ and $[f_{min}, f_m^{(2)}]$ being invalid. The valid regions for calculating the DP tracer are four regions on the two branches of the graph $f(\varphi)$, such that $[f_m^{(1)}, f_0^{(1)}]$, $[f_0^{(1)}, f_{max}]$ and $[f_m^{(2)}, f_0^{(2)}]$, $[f_0^{(2)}, f_{max}]$. For a given θ_P-ray in the coupler plane at a given instant or position φ, $f^{(i)} \in [f_0^{(1)}, f_{max}]$ (or $f^{(i)} \in [f_0^{(2)}, f_{max}]$) can be calculated by the graph $f(\varphi)$ in Fig. 1.41; $f^{(j)} \in [f_m^{(1)}, f_0^{(1)}]$ and $f^{(j)} \in [f_m^{(2)}, f_0^{(2)}]$ can be determined by equation (1.145), corresponding to the parameters $\varphi^{(i)}$, $\gamma^{(i)}$, $\varphi^{(j)}$, and $\gamma^{(j)}$. As a result, the DP tracer can be located by equation (1.133). During the calculation of regions of DP tracer on the θ_P-ray of a coupler plane, the valid ranges of the position angle φ are $[f_m^{(1)}, f_0^{(1)}]$, $[f_0^{(1)}, f_{max}]$ and $[f_m^{(2)}, f_0^{(2)}]$, $[f_0^{(2)}, f_{max}]$. In fact, it will suffice to take $[f_0^{(1)}, f_{max}]$ and $[f_0^{(2)}, f_{max}]$, since $\varphi^{(i)}$ and $\varphi^{(j)}$ correspond to each other.

As shown in the $f(\varphi)$ graph in Fig. 1.41, for $f_{max} = -7.18°$, we have $f_m^{(1)} = f_m^{(2)} = -172.82°$ and $f_0^{(1)} = f_0^{(2)} = -90°$ according to $f_m^{(1)} + f_{max} = f_m^{(2)} + f_{max} = n\pi$. There are four regions: $[f_m^{(1)}, f_0^{(1)}]$, $[f_0^{(1)}, f_{max}]$ and $[f_m^{(2)}, f_0^{(2)}]$, $[f_0^{(2)}, f_{max}]$. For a given position $\varphi = -31°$, we obtain two points $f_1^{(i)} \in [f_0^{(1)}, f_{max}]$ with value $-85.85°$ and $f_2^{(i)} \in [f_0^{(2)}, f_{max}]$ with value $-10.07°$. $f_1^{(i)}$ has two corresponding positions $f_1^{(j)}$, $f_2^{(j)}$, and $f_2^{(i)}$ has corresponding $f_3^{(j)}$, $f_4^{(j)}$

by equation (1.145). $f_1^{(j)} \in [f_m^{(1)}, f_0^{(1)}]$ has value $-94.15°$ at $\varphi = -26.855°$, $f_2^{(j)} \in [f_m^{(2)}, f_0^{(2)}]$ has value $-94.15°$ at $\varphi = 31.005°$, $f_3^{(j)} \in [f_m^{(1)}, f_0^{(1)}]$ is $-169.93°$ at $\varphi = 31.02°$ and $f_4^{(j)} \in [f_m^{(2)}, f_0^{(2)}]$ is $-169.93°$ at $\varphi = 48.908°$.

The procedure for calculating the DP is shown in Fig. 1.43. For a group-II four-bar linkage, the $f(\varphi)$ graph is drawn based on equation $f = \theta_P - \varphi + \gamma - \pi/2$. For any $\theta_P \in [0, 2\pi]$, all of f_{max}, f_{min}, $f_0^{(1)}$, $f_0^{(2)}$, $f_m^{(1)}$, $f_m^{(2)}$ can be obtained through the $f(\varphi)$ graph, then the polar radius r_P of the DP on the θ_P-ray can be calculated through equation (1.133). For the viewpoint of continuity of the calculation results, the polar radius r_P of the DP tracer on the θ_P-ray must vary continuously. The calculation process of the position $f^{(i)}$ naturally starts from $f_0^{(1)}$ in the region $[f_0^{(1)}, f_{max}]$, extends to f_{max} with searching out the corresponding $f^{(j)} \in [f_m^{(1)}, f_0^{(1)}]$, via f_{max} through $[f_{max}, f_0^{(2)}]$ to $f_0^{(2)}$ with calculating the corresponding $f^{(j)} \in [f_m^{(1)}, f_0^{(1)}]$ again. Then, $f^{(i)}$ goes back from $f_0^{(2)}$ in the region $[f_0^{(2)}, f_{max}]$ with corresponding $f^{(j)} \in [f_m^{(2)}, f_0^{(2)}]$, via f_{max} again along $[f_0^{(1)}, f_{max}]$ to $f_0^{(1)}$ with corresponding $f^{(j)} \in [f_m^{(2)}, f_0^{(2)}]$ also stopping at $f_0^{(1)}$.

Since $f_0^{(1)}$ and $f_0^{(2)}$ correspond respectively to the instant centers $r_0^{(1)}$ and $r_0^{(2)}$, the polar radius r_P of a DP tracer on the θ_P-ray is calculated and varies from one instant center $r_0^{(1)}$ to another $r_0^{(2)}$ when $f^{(i)}$ or $f^{(j)}$ passes through four regions during the calculation process. However, there is no extreme value of the polar radius r_P if it is monotonic during its variation. Otherwise, an extreme value of the polar radius r_P, corresponding to a tacnode, can be determined by equation (1.141).

In case B, $f_{max} + f_{min} > n\pi$, there must exist $f_m^{(1)} \in [f_0^{(1)}, f_{max}]$ and $f_m^{(2)} \in [f_0^{(2)}, f_{max}]$, which make $f_m^{(1)} + f_{min} = f_m^{(2)} + f_{min} = n\pi$ true. In other words, any $f^{(i)}$ belonging to $[f_m^{(1)}, f_{max}]$ (or $[f_m^{(2)}, f_{max}]$) cannot match another $f^{(j)} \in [f_{min}, f_{max}]$ by $f^{(i)} + f^{(j)} = n\pi$. The two regions $[f_m^{(1)}, f_{max}]$ and $[f_m^{(2)}, f_{max}]$ are invalid. There exist also four regions $[f_{min}, f_0^{(1)}]$, $[f_0^{(1)}, f_m^{(1)}]$ and $[f_{min}, f_0^{(2)}]$, $[f_0^{(2)}, f_m^{(2)}]$ to calculate the regions of a DP tracer on the θ_P-ray, and the calculation process is the same as that mentioned above, the only difference is the range of the two regions, in which f_{min} is replaced by f_{max}.

2. $\Delta f_{max} \geq \pi$

In the case of $\Delta f_{max} \geq \pi$ for a four-bar linkage without a single crank, as shown in Fig. 1.42, there exist four positions corresponding to the intersection points on the coupler plane between the θ_P-ray and the moving centrode. The four positions (instant velocity centers) in the $f(\varphi)$ graph, $f_0^{(1)}$, $f_0^{(2)}$, $f_0^{(3)}$, and $f_0^{(4)}$, are

$$f_0^{(1)} = f_0^{(2)} = \frac{n\pi}{2}, f_0^{(3)} = f_0^{(4)} = \frac{(n-2)\pi}{2} \tag{1.149}$$

In the same way as mentioned above, the f_{max} and f_{min} of the graph $f(\varphi)$ can be determined, then we have

$$f_m^{(1)} = f_m^{(2)} = n\pi - f_{max}, f_m^{(3)} = f_m^{(4)} - (n - 2)\pi - f_{min} \tag{1.150}$$

The $f(\varphi)$ graph can be divided into two parts, which are the upper part and the lower part, respectively. The upper part includes the regions of $[f_m^{(1)}, f_{max}]$ and $[f_m^{(2)}, f_{max}]$ with $f_0^{(1)}$ and $f_0^{(2)}$, while the lower part has the regions of $[f_{min}, f_m^{(3)}]$ and $[f_{min}, f_m^{(4)}]$ with $f_0^{(3)}$ and $f_0^{(4)}$. As we know, $f_0^{(1)} - f_0^{(4)}$ correspond to the polar radius $r_0^{(1)} - r_0^{(4)}$ of the instant velocity

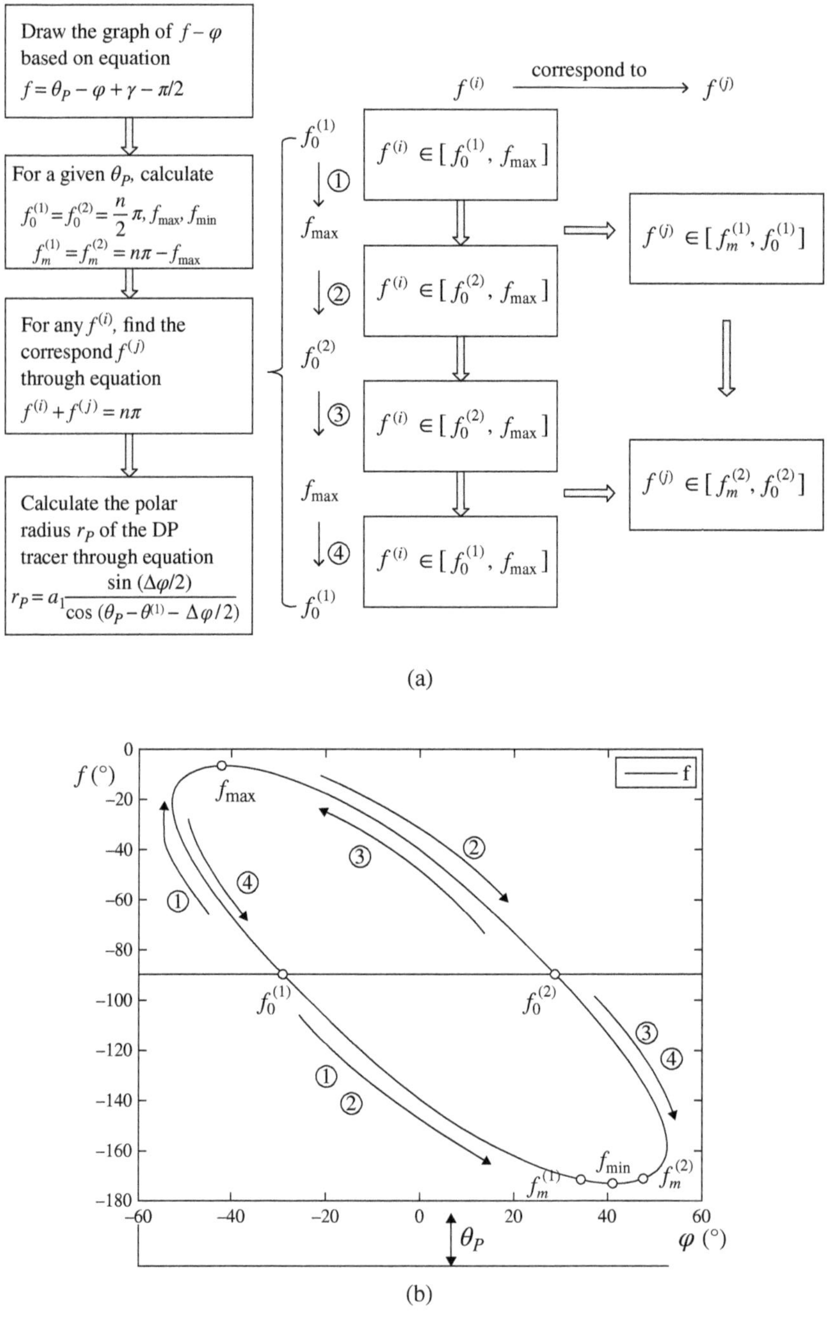

Figure 1.43 The procedure for calculating the double point

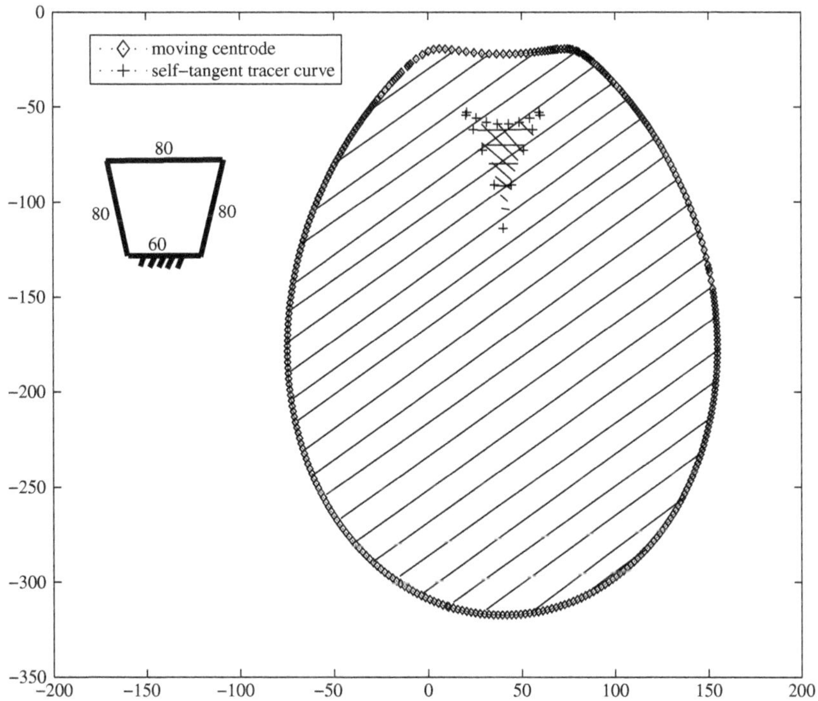

Figure 1.44 Distribution area of DP tracers for a double-crank linkage

centers or the θ_P-ray intersects the moving centrode π_m at the beginning (or end) points of the polar radius r_P of the DP tracer. The polar radius r_P of the DP tracers can respectively be located in two regions of $[r_0^{(1)}, r_0^{(2)}]$ and $[r_0^{(3)}, r_0^{(4)}]$ by equation (1.136). The two regions may be separated without intersection, or covered partly by each other on the θ_P-ray of the coupler plane. The polar radius r_P varies in different ways and the distribution of DP tracers is complex. In such a case, the analyzing and calculating approaches of the preceding section with $\Delta f_{\max} < \pi$, or case A with $f_{\max} + f_{\min} > n\pi$ and case B with $f_{\max} + f_{\min} \leq n\pi$, can also be applied in the case of $\Delta f_{\max} \geq \pi$.

Finally, six examples of group-II four-bar linkages are investigated: a double-crank linkage in Fig. 1.44, a double rocker of a Grashof kinematic chain in Fig. 1.45, and four double-rocker linkages of non-Grashof kinematic chains in Figs 1.46–1.49.

1.3.5 Oval Coupler Curves

A coupler curve without a double point (including the cusp) is a simple closed curve. A simple closed curve is referred to as the oval curve in differential geometry if the curvature remains sign-invariant at every point of the curve for a properly chosen positive direction of increasing arc length.

The curvature of a coupler curve is discussed in Section 1.2. The expression for the curvature of a coupler curve can be written as in equation (1.74). If $k_a - \dot{\theta} > 0$ at some position φ, the curvatures are (i) negative at points within the inflection circle, (ii) zero at points on

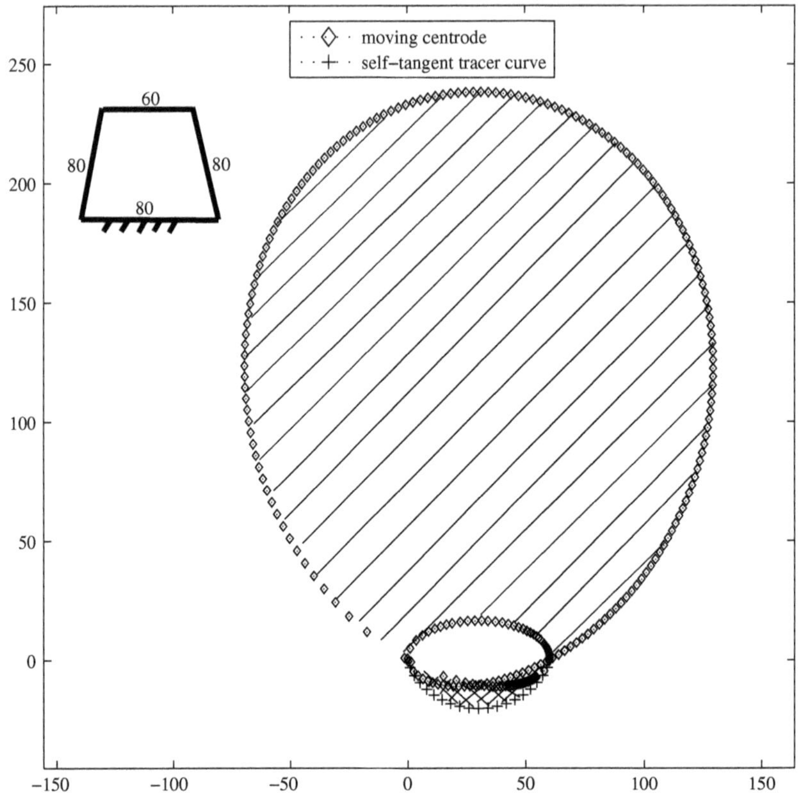

Figure 1.45 Distribution area of DP tracers for a Grashof double-rocker linkage

the inflection circle, and (iii) positive at points outside the inflection circle. Conversely, if $k_a - \dot{\theta} < 0$, items (i) and (iii) are opposite. For the linkage moves with $\varphi \in [0, 2\pi]$, some coupler points keep inside the inflection circle on the coupler if $k_a - \dot{\theta}$ remains sign-invariant, which means that the curvature of its coupler curve is sign-invariant for all instants, and the coupler curve would be an oval curve; the point is known as an oval point except that its position locates on the area of DP tracer on the coupler (including 1DP and 2DP). In contrast, some coupler points keep outside the inflection circle, which is the same case as that of inside, and leads to the curvature of the coupler curve with opposite sign but remaining sign-invariant.

Take a double-crank linkage as in Fig. 1.50 as an example. It holds that the sign of $k_a - \dot{\theta}$ remains unchanged. The coupler curves for those points within the moving centrode π_m, such as points P_4 and P_5, have no inflection point. They do have a crunode as non-simple closed curves. The tracer points locate outside the inflection circle, also outside the Ball curve π_b in the coupler plane for $\varphi \in [0, 2\pi]$, such as points P_1 and P_2, the curvature of whose coupler curves remains positive. Those coupler curves are oval curves. The circle traced by the moving joint of the crank is a special oval curve.

According to equation (1.74), when the moving centrode has two branches, the sign of $(k_a - \dot{\theta})$ will not remain unchanged for $\varphi \in [0, 2\pi]$. As shown in Fig. 1.51 for a crank-rocker linkage, one branch π_m^1 of the moving centrode π_m corresponds to $k_a - \dot{\theta} > 0$ and the other

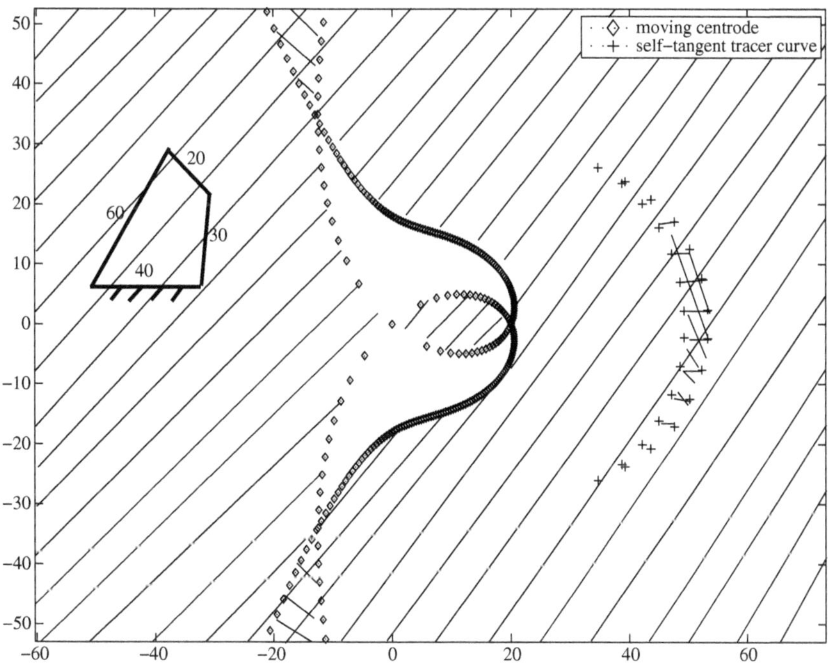

Figure 1.46 Distribution area of DP tracers for a non-Grashof double-rocker linkage

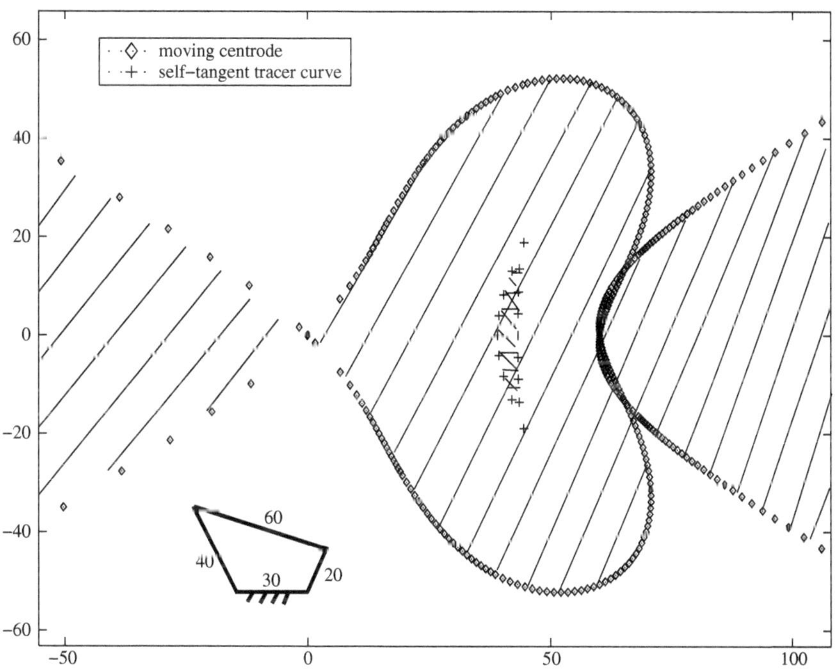

Figure 1.47 Distribution area of DP tracers for a non-Grashof double-rocker linkage

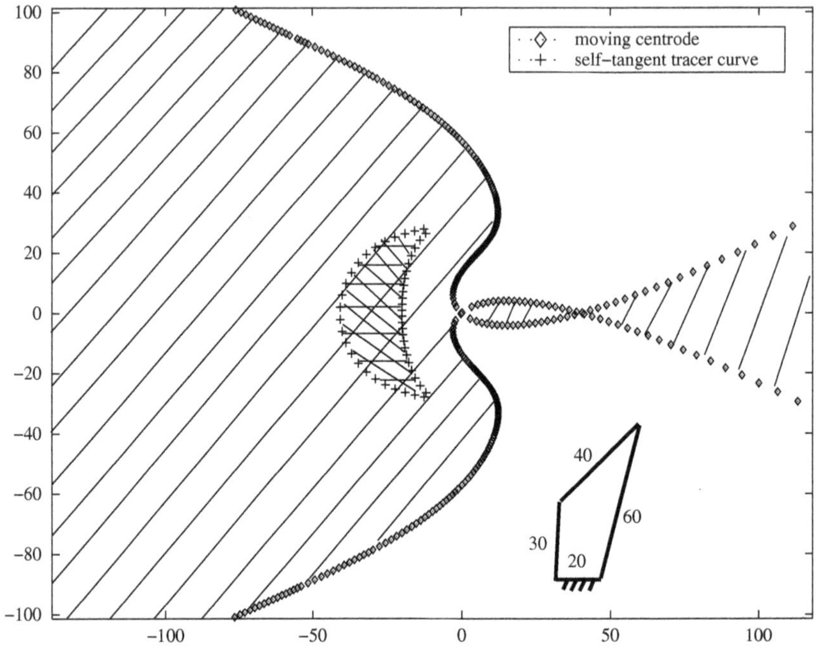

Figure 1.48 Distribution area of DP tracers for a non-Grashof double-rocker linkage

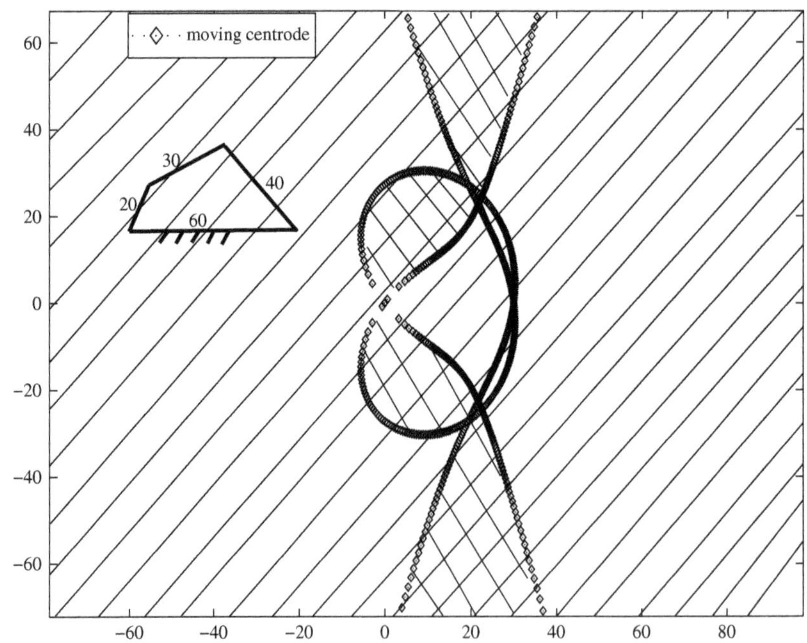

Figure 1.49 Distribution area of DP tracers for a non-Grashof double-rocker linkage

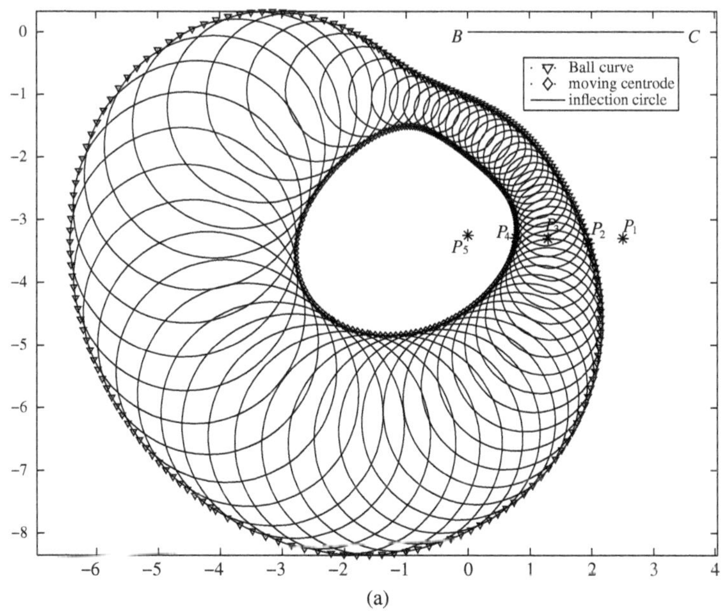

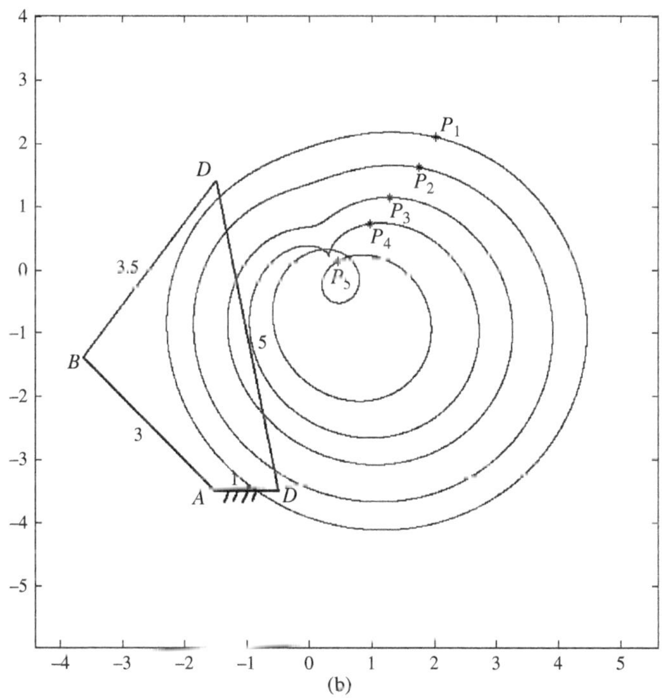

Figure 1.50 (a) Distribution of oval curves for a double-crank linkage (b) Coupler curves of a double-crank linkage

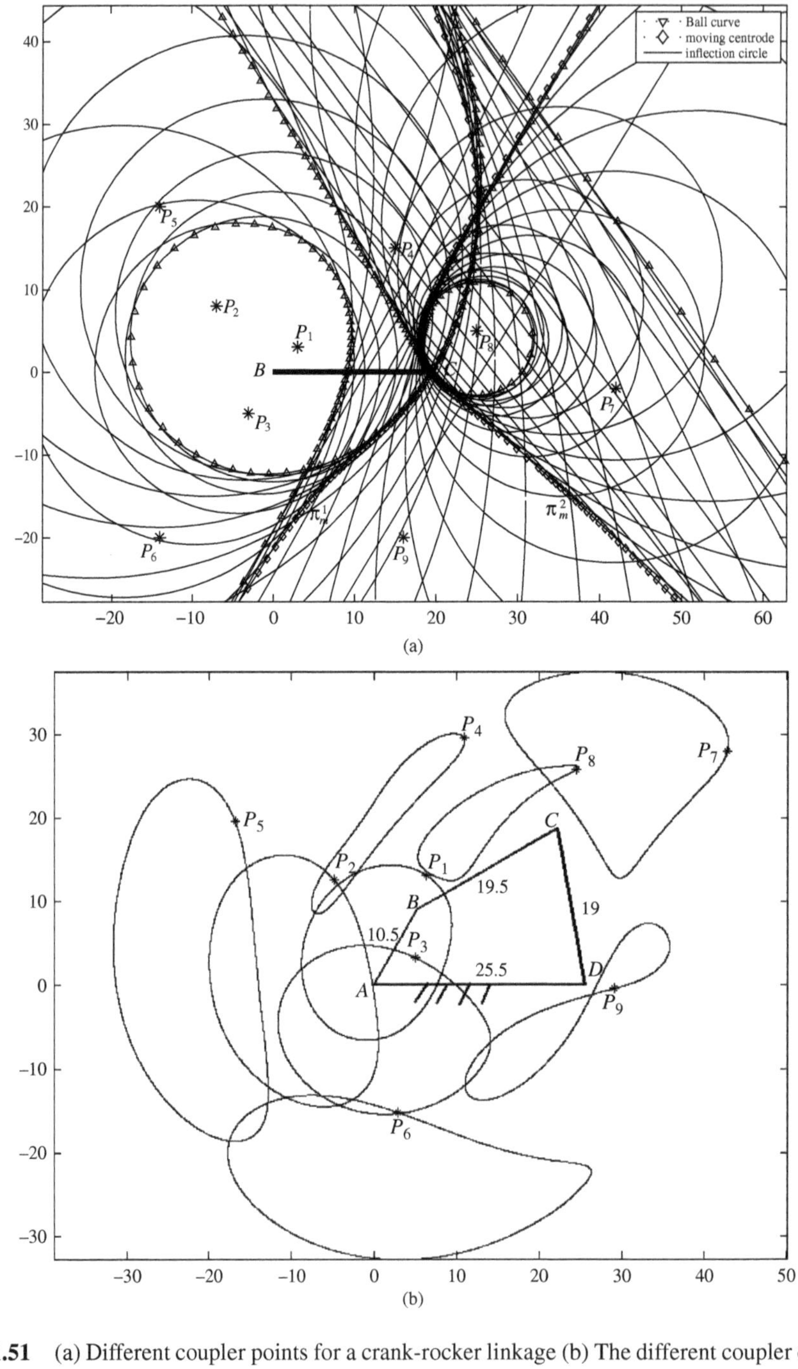

Figure 1.51 (a) Different coupler points for a crank-rocker linkage (b) The different coupler curves for a crank-rocker linkage

branch π_m^2 corresponds to $k_a - \dot{\theta} < 0$. As $k_a - \dot{\theta} = 0$, r_{P0} approaches infinity. The curvature sign of the coupler curves of point $P(r_P, \theta_P)$ is not only related to its location within or outside the inflection circle, but also to the sign of $(k_a - \dot{\theta})$. When one branch π_m^1 of the moving centrode rolls on the corresponding branch of the fixed centrode π_f, some points on one side of π_m^1 are always located within the inflection circle. Hence, they must be within the minimum inflection circle corresponding to π_m^1. The curvature sign of their coupler curves remains the same when the other branch π_m^2 of the moving centrode rolls on the corresponding branch of the fixed centrode π_f and those points remain outside the inflection circle corresponding to π_m^2. In such a case, the curvature sign of their coupler curves remains unchanged, the coupler curves are oval curves. The area in which those points are located is referred to as the ***oval point area***, whose boundary is the Ball curve π_b within the minimum inflection circle corresponding to π_m^1. Similarly, the oval point area may also exist on one side of another branch π_m^2 of the moving centrode. The basis for identifying the existence of the oval point area is whether the minimum inflection circle corresponding to π_m^1 (or π_m^2) is within the maximum inflection circle corresponding to π_m^2 (or π_m^1). If it is, there is no oval point area at all. Otherwise, an oval point area exists within the minimum inflection circle corresponding to π_m^1 (or π_m^2), a ***single-side minimum inflection circle***.

In the coupler as shown in Fig. 1.51, the minimum inflection circle corresponding to π_m^1 is a single-side inflection circle, and the area inside the circle is bounded by the Ball curve π_b, which is the oval point area. For instance, points $P_1 - P_3$ locate on the area; see Fig. 1.51(b) for their coupler curve. On the contrary, the minimum inflection circle corresponding to π_m^2 is contained in the maximum inflection circle corresponding to π_m^1, the area inside the circle is not an oval point area. Point P_8, for example, is not an oval point; see Fig. 1.51(b) for its coupler curve.

The oval point area depends on both the sign of $(k_a - \dot{\theta})$ and the location of a tracing point within or outside the inflection circle (not the DP tracer area for a planar linkage). The oval point area is readily located by boundaries with moving centrode, the Ball curve, and the single side minimum inflection circle. The numerical examples have revealed different results in two types of four-bar linkage. There usually exists an oval point area in the coupler plane for a Grashof four-bar linkage. However, for a non-Grashof linkage, no oval point area exist in its coupler plane because of the disappearance of a one-sided minimal inflection circle. None of the theory and approaches of the oval point area deals with any assumption about linkage, which can be applied to analyzing and calculating the oval point area of a coupler plane for planar four-bar and six-bar linkages

1.3.6 Symmetrical Coupler Curves

The global properties of coupler curves for four-bar linkages – such as a coupler curve with double points, an oval coupler curve – which display the dramatic change from one type of shape to another, are key characteristics in the kinematic synthesis of linkages. However, we have to remind readers about another important global characteristic, symmetry, of coupler curves. Symmetrically proportional four-bar linkages produce symmetrical coupler curves, which comes from the cognate linkages. The same coupler curve could be traced by three different four-bar linkages, if they have a common tracing point and similar coupler triangles generated by the tracing point and two joints of the coupler plane whose fixed pivots are two

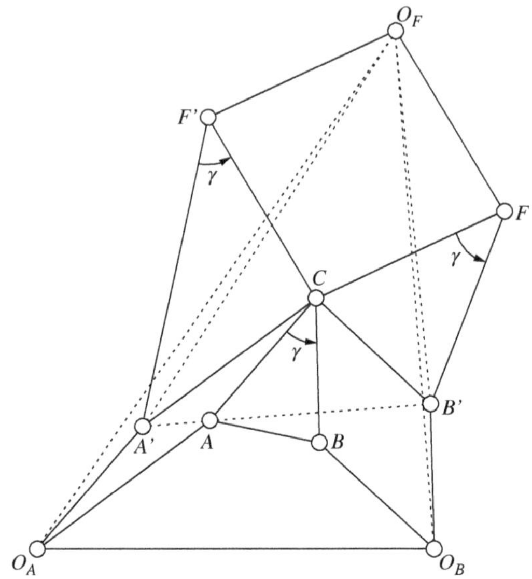

Figure 1.52 Cognate linkages

of three fixed pivots, which form a fixed triangle similar to the coupler triangle of the coupler plane (see Fig. 1.52). This is the ***cognate linkage***, also called the Roberts–Chebyshev theorem. Hence, the three cognate linkages may be different types of linkage: a Grashof four-bar linkage; a non-Grashof four-bar linkage; a crank rocker, double crank or double rocker for a Grashof kinematic chain; or a double rocker for a non-Grashof kinematic chain. In particular, if one of the three cognate linkages is a symmetrical linkage in which the lengths of two binary links are equal, such as the Roberts linkage in Fig. 1.53, then the coupler points on the axis of the coupler plane will produce symmetrical coupler curves. On the contrary, the two cognate linkages of the symmetrical linkage also trace the same symmetrical coupler curve although they are not a symmetrical linkage, while the length of the coupler link is equal to that of the follower, called ***symmetrically proportional four-bar linkages***. The axis of symmetry of the coupler curves passes through the follower pivot and forms a half angle with the based link, where the half angle is just the angle of the coupler triangle at the tracing point on the coupler. As a result, there exists a circle with radius of the coupler link on the coupler plane for the symmetrically proportional four-bar linkages, called the ***symmetrical circle*** (shown in Fig. 1.54), and all points on this circle will trace symmetrical coupler curves.

Furthermore, we would like to consider why symmetrically proportional four-bar linkages produce symmetrical coupler curves. There is nothing except the length of the coupler link equal to that of the follower. We believe that this is a special phenomenon of symmetrical coupler curves, resulting from cognate linkages. In a general four-bar linkage, there exist symmetrical coupler curves, such as a circle of the arc traced by the joints of the coupler. In a way similar to deriving the analytical conditions for the existence of double-points of coupler curves, we can use the approach of differential geometry to establish the necessary and sufficient conditions for the existence of symmetry of coupler curves for a general planar linkage, which are not restricted to the equal length of the coupler link and the follower.

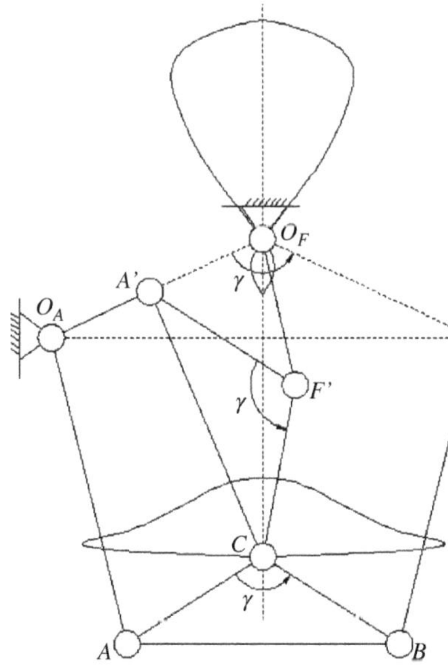

Figure 1.53 Roberts linkage

1.3.7 Distribution of Coupler Curves

A planar four-bar linkage has frequently been the final choice in practical designs due to its structural simplicity and kinematical complexity. A coupler curve generated by a planar four-bar linkage is a sextic algebraic curve. In general, coupler curves have the following shapes: eggs, pears, raindrops, bananas, figures-of-eight, double figures-of-eight, etc. However, what shape will a special point in the coupler plane trace? Is there any point tracing a curve of a given shape? And if any, where are they located? A given crank-rocker linkage is taken as an example to show the distribution of coupler curves with local and global properties. In particular, the boundaries are stressed, while other four-bar linkages have similar boundary curves although these boundary curves appear in different shapes and positions.

For a special dimension of crank-rocker linkage, the moving centrode, the self-tangent tracer curve, the Ball curve, the Burmester curve, and the minimum inflection circle can be drawn based on the formulations in Section 1.2. As a result, the distribution area of various types of coupler curve of the linkage can be determined.

In the coupler plane shown in Fig. 1.55(a), within a single-side minimum inflection circle, the area bounded by the Ball curve is the oval point area, within which a point such as P_1 traces an oval curve (see Fig. 1.55(b) for the coupler curve). A point on the Ball curve, such as P_2, traces a curve with the Ball point, which is widely adopted to generate an approximately straight line. A point on the Burmester curve, such as P_9, traces a curve with the Burmester point, which generates approximately a circle. A point on the self-tangent tracer curve, such as P_4, traces a curve with a tacnode. A point on the moving centrode, such as P_6, traces a curve with a cusp. The points between the self-tangent tracer curve and the moving centrode such as P_5, trace curves with two crunodes shaped like a double figure-of-eight, while the points between the two branches of the moving centrode, such as P_8, trace curves with a single crunode shaped

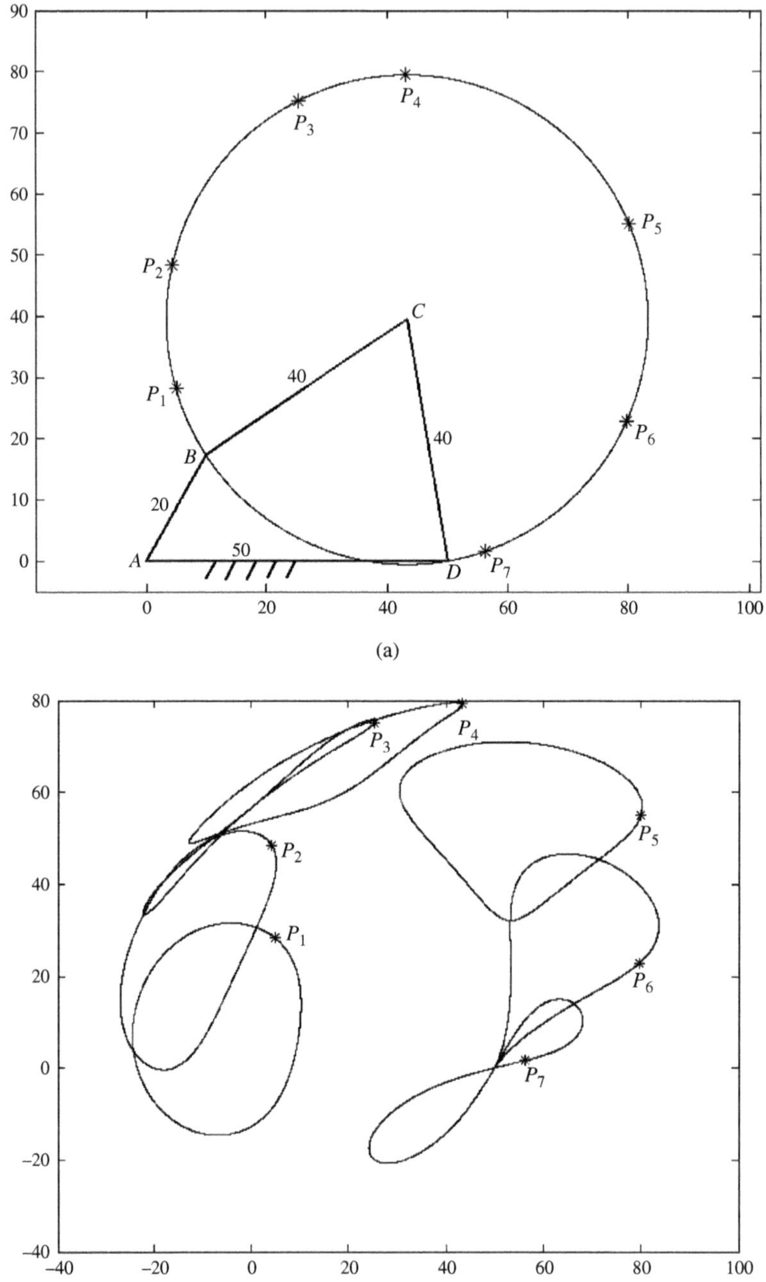

Figure 1.54 (a) Symmetrical circle of the crank-rocker linkage (b) Symmetrical coupler curves of the crank-rocker linkage

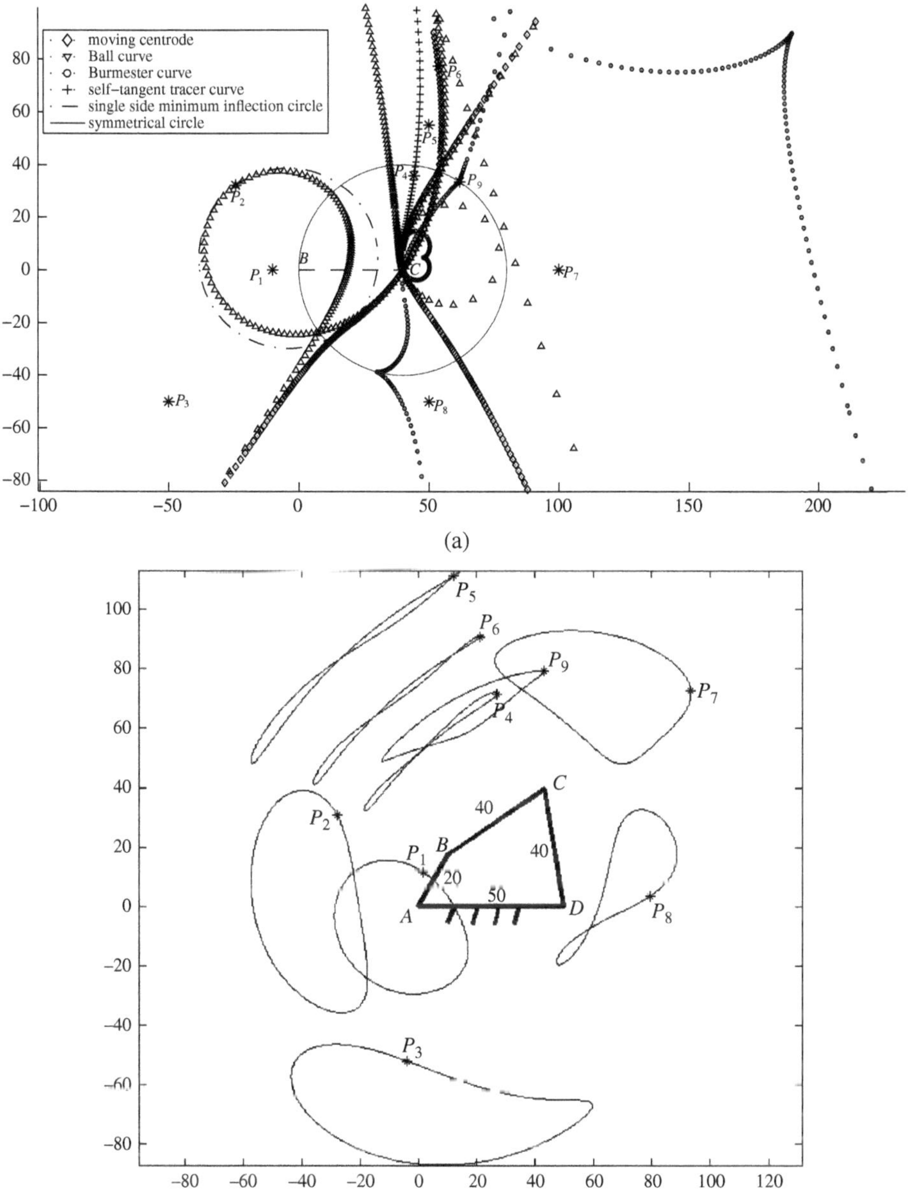

Figure 1.55 (a) Distribution of typical coupler points for four-bar linkages (b) The typical coupler curves of a crank-rocker linkage

like a single figure-of-eight. The points beyond the above-mentioned area, such as P_3, trace a single closed curve with inflection points.

For symmetrically proportional crank-rocker linkages, the length of the coupler is equal to that of the follower. All distribution areas mentioned above also exist in the coupler plane.

In addition, there is a symmetrical circle on the coupler, on which each point traces a symmetrical curve. In particular, the symmetrical circle almost passes through the different distribution areas on the coupler, and the points trace all kinds of symmetrical curve with characteristics, respectively, such as the Ball point, the Burmester point, cusps, crunodes, tacnodes, oval points, and so on. All of these provide a solid basis for the synthesis of a four-bar linkage. Using the symmetrical curves may lay the ground work for the synthesis of a six-bar dwell linkage.

Since the coupler curves vary gradually and continuously from one type of shape to another, there exists a transition region between the above-mentioned different areas. The coupler curve traced by a point in that region has a shape similar to those traced by the points on both sides. For example, the coupler curve with two crunodes is transmitted to that with a single crunode via a coupler curve with a cusp. The point close to the moving centrode traces a curve without a cusp but seeming to have a cusp. A point close to the oval point area traces a curve similar to but not like an oval curve, while the point near the self-tangent curve traces a curve that does not have a tacnode. Only if three curves and one or two circles are drawn in the coupler plane can the distribution area and variation of various types of coupler curve be readily determined.

1.4 Discussion

The kinematic geometry of a rigid body at infinitesimally separated positions, a theory studying the local geometrical properties of a point or a line on the moving body, is very classic and the start of the research can date back to the work of Bernoulli [5] about the instantaneous velocity center, and of Euler [6, 7] about the curvature formula in the eighteenth century. The nineteenth century was the developing period for the theory of the kinematic geometry of a rigid body, with important contributions to the theory including the Euler–Savary formula [8] in curvature theory, Cauchy's finding [9] about the moving centrode rolling on the fixed centrode without sliding for planar movement of a rigid body, the Bobillier theorem [10], the Ball point [11], the Burmester point [12], etc. The twentieth century saw a systematic development in the classical theory of the kinematic geometry of a rigid body in planar movement by Müller [2], and the great enrichment of the theory particularly with the rapid and widespread application of computing technology in the last 50 years, when Walford [13] applied Burmester theory to analyze the coupler curves' curvature for planar four-bar linkages. Freudenstein, Sandor, Veldkamp *et al.* [14–24] worked out the analytic expressions of the Burmester theory for planar movement and developed it for higher-order curvature, which provides the theoretical base for the synthesis of linkages, including what linkage can trace the coupler curves to reproduce what algebraic curves, and the solution existence of the kinematic synthesis in planar movement with higher-order as well as cognate linkages.

Hrones and Nelson [25] gave the atlas of coupler curves of planar four-bar linkages. Viodosic and Tesar [26] drew an atlas of coupler curves with straight lines for the synthesis of linkages, in which one or two Burmester points at infinitesimally separated positions are taken to coincide with the Ball point at the instant and become the Ball–Burmester point and the Ball–double Burmester point. The first author [27] pointed out that the singularity of the Ball curve is just a Burmester point, or the Ball point of the moving body happens to be a Burmester point at this instant, and that the singularity of the Burmester curve is just a higher Burmester point, or has six common points with a circle at infinitesimally separated positions in the neighborhood of the contact point while it is not meaningful for a double Burmester point in geometry. Sandor and Erdman [28] derived the Euler–Savary formula in complex number form, and applied it to the curvature analysis of a mechanism with higher pairs. Hunt [29, 30] discussed and classified the curvature of the enveloped curves in the fixed frame by lines on the coupler of planar four-bar linkages, expanding the kinematic geometry in the planar movement of a rigid body. Bottema, Roth, Koetsier, and Hunt [31–33] worked out the invariants of planar movement and described the curvature theory systemically; these are the classical books to read in the literature. Pennock [34, 35] presented the approach to determine the curvature radius of coupler curves for a geared seven-bar mechanism and double-flier eight-bar linkages.

The kinematic geometry in planar movement usually discusses the local geometrical properties of paths. It is essential for kinematic synthesis of linkages to expose the geometrical properties of paths, in particular the global ones such as cusps, crunodes, tacnodes, etc. Hartenberg [36] located the positions of a double point on a coupler plane by the circle of singular foci. Antuma [37] discussed the symmetrical coupler curves and triangular nomograms. Torfason [4] tried to locate the distribution of coupler curves with double points by means of the circles of singular foci, and it is believed that the results in this paper need to be verified. Hunt [38] and Wunderlich [39] presented the positions of the tacnode and the cranode of a coupler curve based on the circles of singular foci, though it is difficult to build up the distributions of the tacnodes and the cranodes on the coupler plane. Davis [40, 41] mapped all shapes of coupler curves into the coupler circle, whose center locates at the joint of the coupler and whose radius is the length of the link by means of cognate linkage. The first author [42, 43] worked out the global geometrical properties of coupler curves based on the approaches of differential geometry. The different characteristics of coupler curves – such as ovality, cranode, and tacnode – are discussed with vector analyzed equations, which leads to the distributions of coupler curves with different shapes for four-bar linkages being revealed.

In the last 200 years the kinematic geometry of a rigid body in planar instantaneous movement has been developed into a theoretical system based on moving and fixed centrodes, which reveals the curvature theory and its higher-order characteristics; the Euler–Savary formula, for instance, whose representation is so graceful that it is incomparable with any other expressions of the instantaneous planar kinematics. Even so, both the contents and the approaches, including invariants, may not be exposed completely since a similar representation has not yet been expanded to that of the kinematic geometry in spatial instantaneous kinematics.

In content, it is a challenging topic to discuss both the local and the global geometrical properties of centrodes in order to study further the kinematic geometry in planar movement. The moving centrode rolls on the fixed one without sliding, which can reveal completely the kinematics of the lamina. Therefore, there would be an intrinsic relationship between the geometrical properties of the centrodes and the movement of the lamina. Then, the global properties of the centrodes may depend on the dimensions of linkages, or vice versa. The approach for kinematic synthesis of mechanisms based on the centrodes is applicable not only to the synthesis of mechanisms with higher kinematic pairs, which is currently used in practice, but also to that with lower kinematic pairs, linkages, which does not appear in the literature. In particular, the centrodes would be the theoretical basis for higher-order kinematic synthesis of linkages. For example, the appearing and disappearing of Burmester points on the coupler plane of a four-bar linkage, or the existence of Burmester curves, depend on the positions of the coupler [27], which correspond to the geometrical shape of centrodes at the instant. Probably, the boundary, periodicity, and change of a coupler movement may present some clues to study centrodes. It is considerably difficult, of course, and more light needs to be shed.

On the contrary, the local geometrical properties of coupler curves are taken in at a glance by the theory of kinematic geometry in planar movement. However, it is not enough to study the global properties, and the relationship between the shapes of coupler curves and the position of tracing points on the coupler plane is not found clearly. For example, two joints of a link for a crank-rocker linkage respectively trace a full circle and part of a circle in the base link frame, while some coupler points on the line of two joints may produce coupler curves, respectively, in the shape of a goose egg, a pear, a drop of rain, a banana, a figure-of-eight, a double figure-of-eight, etc. even though they are all sextic algebraic curves. The first author just presented the boundary curves for a DP area on the coupler plane [27], but the geometrical properties are hardly known as yet.

The current methods to study the kinematic geometry of a rigid body in planar motion – such as geometry, algebra, screw, matrix, complex number, vector, etc. – have merit in different application cases. In fact, this originates from the geometry of Burmester, who converted the displacements (or movements) of a lamina in several finite separated positions into a geometrical graph by means of corresponding poles, and then built up the algebraic equations to analyze the properties of geometrical graphs, expanding the object of research to all the lamina. It is difficult to identify which method is the traditional geometry or algebra for modern mathematics with expressions of vector algebra and invariants of geometrical graphs, because the differential geometry of both curves and surfaces persists in geometric significance, avoiding the effects of external factors of geometrical graphs. In particular, a moving Frenet frame with three mutually orthogonal axes, or the natural trihedron of a curve or a ruled surface, is introduced into the examination of the geometrical graph in differential geometry by moving along the curve or surface whose derivative can be viewed as the motion conversion for a lamina at infinitesimally separated positions, just like the poles in finite separated positions, which is believed to be a powerful tool in the kinematic geometry of a rigid body, both in planar and spatial motion. The kinematic geometry of a rigid body with multiple degrees of freedom is studied in a multiple-dimensional space, of course, and it is naturally extended from two or three dimensions to multiple dimensions while the classical differential geometry develops into modern differential geometry, such as differential manifolds, Lie groups, and Lie algebra, although these are much more non-representational mathematical methods and more difficult for the reader to understand.

References

[1] D.Z. Xiao, Mathematical model for planar four-bar linkages. *Journal of Dalian Institute of Technology*, **28**(2), 1987 (in Chinese).

[2] R. Müller, *Einführung in die theoretische Kinematik*. Springer, Berlin, 1932.

[3] K.H. Hunt, *Kinematic Geometry of Mechanism*. Oxford University Press, Oxford, 1978.

[4] L.E. Torfason and A. Ahmed, Double points of a 4-bar linkage. *Mechanism and Machine Theory*, **13**(6), 593–601, 1978.

[5] J. Bernoulli, *Opera Omnia IV*. Hildesheim, 1968 (Nachdruck der Ausgabe Lausanne und Genf 1742).

[6] L. Euler, Supplementum de figura dentium rotarum. *Novi Commentarii Academiae Scientiarum Imperialis Petropolitanae* (pro Anno 1765), Tome XI, Petropoli, pp. 207–231, tables V–VIII, 1767.

[7] L. Euler, *Opera Omnia*, Series 2, Volume 17, Commentationes mechanicae ad theoriam machinarum pertinentes, volumen tertium (edited by C. Blanc and P. de Haller), Zurich, 1982.

[8] F. Savary, Lecons et cours autographies. Notes sur les machines. Ecole Polytechnique 1835–36. Unpublished lecture notes, available at the Bibliothèque Nationale, Paris.

[9] A.L. Cauchy, Sur les mouvements que peut prendre un systeme invariable, libre, ou assujetti à certaines conditions. Oeuvres IIe Série, pp. 94–120, 1899.
[10] É.E. Bobillier. *Cours de Géométrie*, 15th edn. Gauthier-Villars, Paris, 1880.
[11] R.S. Ball, Notes on applied mechanics. 1. Parallel motion. *Proceedings of the Irish Academy* (Series 2), 1, 1871.
[12] L. Burmester, *Lehrbuch der Kinematik*. Felix, Leipzig, 1888.
[13] J.C. Wolford, An analytical method for locating the Burmester points for five infinitesimally separated positions of the coupler plane of a four-bar mechanism. *ASME Journal of Applied Mechanics*, **27**(1), 182–186, 1960.
[14] F. Freudenstein and G.N. Sandor, On the Burmester points of a plane. *ASME Journal of Applied Mechanics*, **28**(1), 41–49, 1961.
[15] G.R. Veldkamp, *Curvature Theory in Plane Kinematics*. J.B. Wolters, Groningen, 1963.
[16] F. Freudenstein, On the variety of motions generated by mechanisms. *ASME Journal of Engineering for Industry*, **84**(1), 156–159, 1962.
[17] F. Freudenstein and E.J.F. Primrose, Geared five-bar motion. *ASME Journal of Applied Mechanics*, **30**(2), 161–175, 1963.
[18] G.N. Sandor, On the existence of a cycloidal Burmester theory in planar kinematics. *ASME Journal of Applied Mechanics*, **31**(4), 694–699, 1964.
[19] F. Freudenstein, Higher path-curvature analysis in plane kinematics. *ASME Journal of Engineering for Industry*, **87**(2), 184–190, 1965.
[20] G.R. Veldkamp, Some remarks on higher curvature theory. *Journal of Engineering for Industry*, **89**(1), 84–86, 1967.
[21] G.N. Sandor and F. Freudenstein, Higher-order plane motion theories in kinematic synthesis. *ASME Journal of Engineering for Industry*, **89**(2), 223–230, 1967.
[22] E.A. Dijksman, Calculation and construction of the Burmester points for five positions of a moving plane. *ASME Journal of Engineering for Industry*, **91B**, 66–74, 1969.
[23] B. Roth, On the multiple generation of coupler-curves. *ASME Journal of Engineering for Industry*, **87**(2), 177–183, 1965.
[24] S.X. Bai, *Advanced Mechanism*. Shanghai Science and Technology Press, Shanghai, 1988 (in Chinese).
[25] J.A. Hrones and G.L. Nelson, *Analysis of the Four-bar Linkage*. The Technology Press of MIT and Wiley, New York, 1951.
[26] J.P. Vidosic and D. Tesar, Selection of four-bar mechanisms having required approximate straight-line outputs. *Mechanism and Machine Theory*, **2**, 23–76, 1967.
[27] D.L. Wang, D.Z. Xiao, and J. Liu, Ball's curve and Burmester's curve for four-bar linkage. *Journal of Dalian University of Technology*, **34**(4), 411–417, 1994 (in Chinese).
[28] G.N. Sandor, A.G. Erdman, L. Hunt and E. Raghavacharyulu, New complex-number forms of the Euler–Savary equation in a computer-oriented treatment of planar path-curvature theory for higher-pair rolling contact. *ASME Journal of Mechanical Design*, **104**(1), 227–232, 1982.
[29] K.H. Hunt and E.F. Fichter, Equations for four-bar line-envelopes. *ASME Journal of Mechanical Design*, **103**(4), 743–749, 1981.
[30] J.E. Kimbrell and K.H. Hunt, A classification of coupler-line envelopes from hinged four-bar linkages. *ASME Journal of Mechanical Design*, **103**(4), 750–757, 1981.
[31] O. Bottema, Some remarks on theoretical kinematics. I. Instantaneous invariants. II. On the application of elliptic functions in kinematics. Proceedings of the International Conference for Teachers of Mechanisms, Yale University, New Haven, CT, pp. 156–167, 1961.
[32] O. Bottema and B. Roth, *Theoretical Kinematics*, North-Holland, New York, 1979.
[33] T. Koetsier, From kinematically generated curves to instantaneous invariants: Episodes in the history of instantaneous planar kinematics. *Mechanism and Machine Theory*, **21**(6), 489–498, 1986.
[34] G.R. Pennock and H. Sankaranarayanan, Path curvature of a geared seven-bar mechanism. *Mechanism and Machine Theory*, **38**(12), 1345–1361, 2003.
[35] G.R. Pennock and N.N. Raje, Curvature theory for the double flier eight-bar linkage. *Mechanism and Machine Theory*, **39**(7), 665–679, 2004.
[36] R.S. Hartenberg and J. Denavit, *Kinematic Synthesis of Linkages*. McGraw-Hill, New York, 1964.
[37] H.J. Antuma, Triangular nomograms for symmetrical coupler curves. *Mechanism and Machine Theory*, **13**(3), 251–268, 1978.
[38] K.H. Hunt and J.E. Kimbrell, A note on symmetrical self-osculating coupler curves. *Mechanism and Machine Theory*, **17**(3), 229–232, 1982.

[39] W. Wunderlich, Self-osculating coupler curves. *Mechanism and Machine Theory*, **18**(3), 207–212, 1983.
[40] T.H. Davis, Proposals for a finite 5-dimensional atlas of crank-rocker linkage coupler curves. *Mechanism and Machine Theory*, **10**(5), 517–530, 1981.
[41] T.H. Davis, Proposals for finite 5-dimensional atlas of all planar 4-bar linkage coupler curves. *Mechanism and Machine Theory*, **19**(2), 211–221, 1984.
[42] D.L. Wang and D.Z. Xiao, Distribution of coupler curves for crank-rocker linkages. *Mechanism and Machine Theory*, **28**(5), 671–684, 1993.
[43] D.L. Wang, *Kinematic Differential Geometry of Mechanisms*, Dalian University of Technology, 1995 (in Chinese).

2

Discrete Kinematic Geometry and Saddle Synthesis of Planar Linkages

In a planar four-bar linkage, the coupler link is constrained by two binary links with the kinematic pairs or joints, such as R–R, P–R, or R–P. Here, the first letters represent the kinematic pair connected to the base link, or the fixed joint, and the second ones denote the kinematic pair connected to the coupler, or the moving joint. The center point of a moving joint R is called a ***characteristic point***, or a circle point for a binary link R–R, and a sliding point for P–R, while the guidance line of the prismatic pair P is referred to as a ***characteristic line*** for a binary R–P, or an inverted sliding point on the base link. The special curves are traced (or enveloped) by the characteristic point (or line) of the binary links for plane linkages, called ***constraint curves***. The constraint curves are usually special simple plane curves, for instance, a circle, a straight line, or a line-envelope circle, whose curvatures are constant, and sometimes even zero. In fact, the kinematic synthesis of plane four-bar linkages is to locate the characteristic points or the characteristic lines on the coupler plane according to the given application cases.

The continuous or discrete movements of a rigid body, or lamina, the characteristic points on the moving body, and their paths in the fixed body, as well as the relationships among them, are studied in the plane kinematic geometry. These form the theoretical basis of the kinematic synthesis of plane linkages. As the ***continuous kinematic geometry*** was introduced in Chapter 1, the ***discrete kinematic geometry*** will be discussed in this chapter, although it is still to be completed for more than five separated positions. We use the word "discrete" for kinematics, movements, positions, or paths (trajectories in space) in this book, which means that a lamina occupies separated positions or the displacements or paths of a point of the lamina are not continuous, while the number of separated positions is not limited, so we avoid using the words "finite separated" since it is easily misunderstood that the case is limited to a few separated positions. The classical kinematic geometry of linkage deals with the circle points (or sliding points) on the lamina with less than six separated positions by means of rotation poles. The approach of ***saddle point programming*** is applied to estimate the global geometrical properties of the discrete curve by comparing it with constraint curves, and the relationships

---

Kinematic Differential Geometry and Saddle Synthesis of Linkages, First Edition. Delun Wang and Wei Wang.

among the discrete movements of a lamina, points on the lamina, and their discrete paths, as well as the normal errors between the discrete path and a constraint curve, are discussed, which set up the theoretical basis for the discrete kinematic geometry of linkages. The topics in the classical kinematic geometry of linkages are just the cases of the normal errors being zero at the finite separated positions.

Based on the discrete kinematic geometry, the characteristic points on a lamina are located by the approach of saddle point programming, which is called the ***discrete kinematic synthesis of planar linkages***, discussed in this chapter. Although the plane constraint curves are just the simple circle and the straight line, the term "constraint curve" is still adopted for the corresponding descriptions in later chapters. This chapter is just one part of the approach system for the discrete kinematic synthesis of linkages, including planar, spherical, and spatial linkages.

2.1 Matrix Representation

A point P of a moving body $\sum^*$ has coordinates (x_{Pm}, y_{Pm}) in the moving Cartesian coordinate system $\{\boldsymbol{O}_m; \boldsymbol{i}_m, \boldsymbol{j}_m\}$ and the vector can be represented by

$$\boldsymbol{R}_{Pm} = x_{Pm}\boldsymbol{i}_m + y_{Pm}\boldsymbol{j}_m \tag{2.1}$$

In the fixed Cartesian coordinate system $\{\boldsymbol{O}_f; \boldsymbol{i}_f, \boldsymbol{j}_f\}$ of the base body $\sum$, the displacement vector of P can be written as

$$\boldsymbol{R}_P = \boldsymbol{R}_{Om} + \boldsymbol{R}_{Pm} = \boldsymbol{R}_{Om} + x_{Pm}\boldsymbol{i}_m + y_{Pm}\boldsymbol{j}_m \tag{2.2}$$

The above equation can be represented by the coordinate transformation as $\boldsymbol{R}_P = [\mathbf{M}] \cdot \boldsymbol{R}_{Pm}$, it is

$$\begin{bmatrix} x_P \\ y_P \\ 1 \end{bmatrix} = [\mathbf{M}] \cdot \begin{bmatrix} x_{Pm} \\ y_{Pm} \\ 1 \end{bmatrix}, \quad [\mathbf{M}] = \begin{bmatrix} \cos\gamma & -\sin\gamma & x_{Omf} \\ \sin\gamma & \cos\gamma & y_{Omf} \\ 0 & 0 & 1 \end{bmatrix} \tag{2.3}$$

where $[\mathbf{M}]$ is the ***planar displacement matrix*** of the moving body $\sum^*$, the origin point O_m of $\{\boldsymbol{O}_m; \boldsymbol{i}_m, \boldsymbol{j}_m\}$ has the coordinates (x_{Omf}, y_{Omf}) in $\{\boldsymbol{O}_f; \boldsymbol{i}_f, \boldsymbol{j}_f\}$, and γ is the rotation angle of $\sum^*$ around the point O_m. The movement of $\sum^*$ relative to $\sum$ is represented as the displacements (x_{Omf}, y_{Omf}) of the reference point O_m of $\sum^*$ and the rotation angle γ. The movement of $\sum^*$ is continuous if the three parameters $(x_{Omf}, y_{Omf}, \gamma)$ are continuous, and $\sum^*$ occupies the discrete positions if $(x_{Omf}, y_{Omf}, \gamma)$ are designated values at different positions, represented by $(x^{(i)}_{Omf}, y^{(i)}_{Omf}, \gamma^{(i)})$, where i in the superscript blanket denotes the ith position of $\sum^*$. Hence, a point $P(x_{Pm}, y_{Pm})$ of $\sum^*$ has discrete displacements $\boldsymbol{R}^{(i)}_P = (x^{(i)}_P, y^{(i)}_P)^T$ in $\{\boldsymbol{O}_f; \boldsymbol{i}_f, \boldsymbol{j}_f\}$ and can be represented as $\boldsymbol{R}^{(i)}_P = [\mathbf{M}^{(i)}] \cdot \boldsymbol{R}_{Pm}$, or

$$\begin{bmatrix} x^{(i)}_P \\ y^{(i)}_P \\ 1 \end{bmatrix} = [\mathbf{M}^{(i)}] \cdot \begin{bmatrix} x_{Pm} \\ y_{Pm} \\ 1 \end{bmatrix}, \quad [\mathbf{M}^{(i)}] = \begin{bmatrix} \cos\gamma^{(i)} & -\sin\gamma^{(i)} & x^{(i)}_{Omf} \\ \sin\gamma^{(i)} & \cos\gamma^{(i)} & y^{(i)}_{Omf} \\ 0 & 0 & 1 \end{bmatrix} \tag{2.4}$$

where $[\mathbf{M}^{(i)}]$ is the ***planar discrete displacement matrix*** of $\sum^*$. The discrete paths $(x^{(i)}_P, y^{(i)}_P)$ of P in $\{\boldsymbol{O}_f; \boldsymbol{i}_f, \boldsymbol{j}_f\}$, represented by $\{\boldsymbol{R}^{(i)}_P\}$ for short, can be calculated by equation (2.4) if a series of discrete parameters $(x^{(i)}_{Omf}, y^{(i)}_{Omf}, \gamma^{(i)})$ are given first.

The other expression forms of the discrete movements of $\sum^*$ may be prescribed; for instance, the displacements of two reference points A and B of $\sum^*$ are given as (x_A, y_A) and (x_B, y_B) in $\{\boldsymbol{O}_f; \boldsymbol{i}_f, \boldsymbol{j}_f\}$. According to Ref. [1], the displacement (x_C, y_C) of a third point C can be obtained as

$$\begin{cases} x_C = x_A - (y_B - y_A) \\ y_C = y_A + (x_B - x_A) \end{cases} \tag{2.5}$$

The displacements of three points A, B, and C can be represented by the column matrices (A_{I}), (B_{I}), and (C_{I}) for position I, and (A_{II}), (B_{II}), and (C_{II}) for position II, and then the planar displacement matrix $[\mathbf{M}_{\mathrm{I\,II}}]$ of $\sum^*$ can be obtained as

$$[\mathbf{M}_{\mathrm{I\,II}}] = \begin{bmatrix} x_{A\mathrm{II}} & x_{B\mathrm{II}} & x_{C\mathrm{II}} \\ y_{A\mathrm{II}} & y_{B\mathrm{II}} & y_{C\mathrm{II}} \\ 1 & 1 & 1 \end{bmatrix} \cdot \begin{bmatrix} x_{A\mathrm{I}} & x_{B\mathrm{I}} & x_{C\mathrm{I}} \\ y_{A\mathrm{I}} & y_{B\mathrm{I}} & y_{C\mathrm{I}} \\ 1 & 1 & 1 \end{bmatrix}^{-1} \tag{2.6}$$

At the original position, $\{\boldsymbol{O}_m; \boldsymbol{i}_m, \boldsymbol{j}_m\}$ coincides with $\{\boldsymbol{O}_f; \boldsymbol{i}_f, \boldsymbol{j}_f\}$ and $[\mathbf{M}]$ is a unit matrix. The displacement matrix, from $\{\boldsymbol{O}_m; \boldsymbol{i}_m, \boldsymbol{j}_m\}$ to $\{\boldsymbol{O}_f; \boldsymbol{i}_f, \boldsymbol{j}_f\}$, can be obtained by the above equation for $\sum^*$ at any position.

So far, the discrete movements of $\sum^*$ can be represented by either the discrete parameters $(x^{(i)}_{Omf}, y^{(i)}_{Omf}, \gamma^{(i)})$ or the planar discrete displacement matrix $[\mathbf{M}^{(i)}]$, and the ***discrete path*** $\{\boldsymbol{R}^{(i)}_P\}$, comprised of the ***discrete point set***, traced by a point P of $\sum^*$, can be calculated by equation (2.4).

2.2 Saddle Point Programming

A four-bar linkage to be synthesized has an output function, which is expected to match the required function. The designer may minimize the deviation of the output function and the required function between precision points, or accuracy points, which is called the ***structural error*** of the linkage [2]. Generally, there are two methods adopted by the designers to minimize the structural error, of which one is the interpolation method and the other is the optimization method.

In the ***interpolation method***, the function $f_b(y, \boldsymbol{X})$ is taken to match the required function $f_a(y)$, where $\boldsymbol{X} = (x_1, x_2, \ldots, x_n)^T$ are the structural parameters of the linkage to be synthesized, such as the links' lengths, and $\boldsymbol{y}$ is the one-dimensional design variable, such as the different position of the linkage. There are k interpolating points $(y_1, y_2, \ldots, y_k)$ and the structural error function $F(\boldsymbol{X})$ will be zero at the interpolating points, or precision points, that is

$$F(\boldsymbol{X}) = f_b(y_i, \boldsymbol{X}) - f_a(y_i) = 0,\ i = 1, 2, \ldots, k \tag{2.7}$$

Equation (2.7) is the simultaneous system of equations of the structural parameters $x_1, x_2, \ldots, x_n$. We designate the values of $n-k$ structural parameters of the linkage freely if $k \le n$, and the others have to be solved by equation (2.7).

Obviously, the structural errors will only be zero at the precision points. At other points, the structural errors are unknown, even unpredictable. In the case $k \le n$, it is an effective way to minimize the maximum structural errors by arranging suitable interpolating points and letting the maxima and the minima of the structural error function $F(\boldsymbol{X})$ be identical. For instance, the interpolating points are chosen according to the zeros of ***the Chebyshev polynomials***, ***the best uniform approximation in one-dimensional space***.

In the case $k > n$, the optimization method is adopted to solve equation (2.7) and determine the structural parameters $x_1, x_2, \ldots, x_n$ of a linkage whose objective function is usually the sum of squares of the structural errors, or equation (2.7) is rewritten as

$$F(\boldsymbol{X}) = \sum_{i=1}^{k} [f_b(y_i, \boldsymbol{X}) - f_a(y_i)]^2 \tag{2.8}$$

It is the ***best-square approximation*** or the ***least-square method*** by letting equation (2.8) be least. This method is effective for the linear problem, but in the optimum synthesis of plane linkages it is not always valid since the structural error is not uniformly defined. The unified evaluation criterion of the deviation between the output function and the required function, such as the normal error, needs to be defined. The design variables, the coordinates of both fixed hinge points and moving hinge points, are redundant, leading to the gradient descent of the objective function values not exactly indicating the influence of the optimal variables, or non-convergence of the optimization algorithm. Hence, the current approach of kinematic synthesis of linkages usually depends on the initial values and may be valid for special cases but not for general problems. In this book, we define the unified evaluation criterion of the difference, or the approximation between the points' (lines') discrete paths (trajectories) in the fixed frame and the constraint curves (surfaces), which have been the key topics in discrete kinematic geometry.

The function, such as the equation of a point (or line) trajectory of a linkage, is assumed to be $f_a(\boldsymbol{y})$, and the function of the constraint curve (or surface) of a binary link is $f_b(\boldsymbol{y}, \boldsymbol{X})$. The error function $F(\boldsymbol{X}, \boldsymbol{y})$ between $f_b(\boldsymbol{y}, \boldsymbol{X})$ and $f_a(\boldsymbol{y})$ is first defined along the normal of the constraint curve (or surface). Thus, for given values of the structural parameters $\boldsymbol{X} \in \mathbf{R}^n$, the constraint function $f_b(\boldsymbol{y}, \boldsymbol{X})$ may have different values with variation of the variable $\boldsymbol{y} \in \mathbf{R}$, and the error function $F(\boldsymbol{X}, \boldsymbol{y})$ also has corresponding values. Hence, the proximity of $f_b(\boldsymbol{y}, \boldsymbol{X})$ and $f_a(\boldsymbol{y})$ depends on the value of the error function $F(\boldsymbol{X}, \boldsymbol{y})$, especially the maximum value at the position (or instant) $\boldsymbol{y} = \boldsymbol{y}^*$, or $F(\boldsymbol{X}, \boldsymbol{y}^*) = \|f_b(\boldsymbol{y}^*, \boldsymbol{X}) - f_a(\boldsymbol{y}^*)\|_\infty$. Then, for different structural parameters $\boldsymbol{X} \in \mathbf{R}^n$, the constraint function $f_b(\boldsymbol{y}, \boldsymbol{X})$ and the error function $F(\boldsymbol{X}, \boldsymbol{y})$ may have different values. The discrete kinematic synthesis of linkages involves finding the structural parameters $\boldsymbol{X} = \boldsymbol{X}^*$ that make the value of the error function $F(\boldsymbol{X}, \boldsymbol{y}^*)$ minimal, or

$$F(\boldsymbol{X}^*, \boldsymbol{y}) \le F(\boldsymbol{X}^*, \boldsymbol{y}^*) \le F(\boldsymbol{X}, \boldsymbol{y}^*) \tag{2.9}$$

The above equation can be described by a general description of ***mathematical programming***: the optimum variables are $\boldsymbol{X} = (x_1, x_2, \ldots, x_n)^T$, $\boldsymbol{y} = (y_1, y_2, \ldots, y_k)^T$, $\boldsymbol{X} \in \mathbf{R}^n$, $\boldsymbol{y} \in \mathbf{R}^k$, $F(\boldsymbol{X}, \boldsymbol{y}) \in \mathbf{R}^{k+n}$. The function $F(\boldsymbol{X}, \boldsymbol{y})$ achieves optimal value at $(\boldsymbol{X}^*, \boldsymbol{y}^*)$, or equation (2.9) holds true, which is expressed as

$$\begin{cases} \min\limits_{\boldsymbol{X}} \max\limits_{\boldsymbol{y}} F(\boldsymbol{X}, \boldsymbol{y}) \\ \text{s.t.} \quad g_i(\boldsymbol{X}) \le 0, \; i = 1, 2, \ldots, p \\ \qquad h_j(\boldsymbol{y}) \le 0, \; j = p+1, \ldots, m \end{cases} \tag{2.10}$$

where $F(\boldsymbol{X},\boldsymbol{y})$ is the objective function, $g_i(\boldsymbol{X})$ and $h_j(\boldsymbol{y})$ are the inequality constraint functions. $\boldsymbol{X} = (x_1, \ldots, x_n)^T$ and $\boldsymbol{y} = (y_1, \ldots, y_k)^T$ are the optimal variables. The set $G = \{(\boldsymbol{X},\boldsymbol{y})|g_i(\boldsymbol{X}) \leq 0, i = 1,2, \ldots, p; h_j(\boldsymbol{y}) \leq 0, j = p+1, \ldots, m\}$ defines the feasible region. The geometrical meaning of equation (2.10) at the optimal values $(\boldsymbol{X}^*,\boldsymbol{y}^*)$ is: the objective function $F(\boldsymbol{X},\boldsymbol{y})$ is maximum for the variables $\boldsymbol{y}$ and minimum for the variables $\boldsymbol{X}$, so the adjacent geometrical shape of point $(\boldsymbol{X}^*,\boldsymbol{y}^*)$ is saddle-like, which is defined as the ***hyperbolic point*** in Chapter 3. This optimization model is then referred to as ***saddle point programming***, which is a branch of mathematical programming, sometimes called the minimax problem. In fact, it is the ***best uniform approximation in multi-dimensional space***, the extension of the best uniform approximation in one-dimensional space. The solving method has been provided by the reliable software Matlab®. The theory of saddle point programming and its applications in mechanical engineering, such as the necessary and sufficient conditions for the existence of solutions, the geometric tolerance evaluation, and so on, have been introduced in detail in Ref. [3].

Based on saddle point programming, we study the discrete kinematic geometry of a rigid body in planar, spherical, and spatial movement, and set up a unified mathematical model of the kinematic synthesis of linkages in this chapter and Chapters 5 and 7.

Example 2.1 For a given ellipse, whose equation is $x = 5\cos\theta$, $y = 4\sin\theta$, $\theta \in [0, 2\pi)$, try to determine the best-fitting circle by the mathematical model of saddle point programming.

Solution

Based on equation (2.10), the function $f_a(\theta)$ of the given ellipse is $x = 5\cos\theta$, $y = 4\sin\theta$, and $f_b(\boldsymbol{X},\theta)$ is the function of the fitting circle. The design variables are $\boldsymbol{X} = (\boldsymbol{R}_C, r)^T = (x_C, y_C, r)^T$, or the coordinates (x_C, y_C) of the center point and the radius r of the fitting circle. As shown in Fig. 2.1, the normal error along the radial direction of the fitting circle is defined as the objective function, that is $F(\boldsymbol{X},\theta) = \left\{\left|\sqrt{(5\cos\theta - x_C)^2 + (4\sin\theta - y_C)^2} - r\right|\right\}$, $\theta \in [0, 2\pi)$.

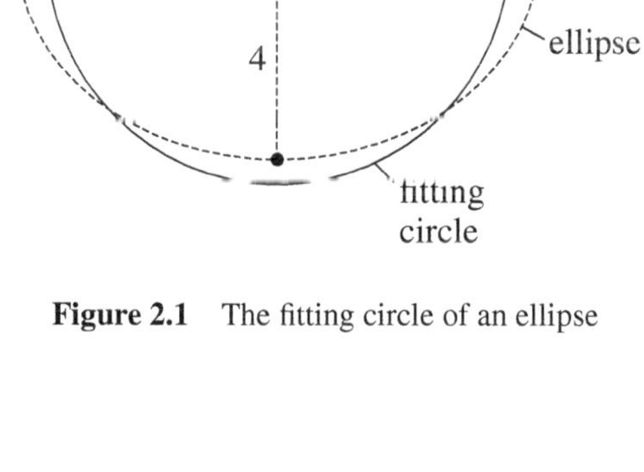

Figure 2.1 The fitting circle of an ellipse

The mathematical model of saddle point programming for the fitting error evaluation is

$$\begin{cases} \min\limits_{X}\max\limits_{\theta} F(X,\theta) \\ = \min\limits_{X}\max\limits_{\theta}\left\{\left|\sqrt{(5\cos\theta - x_C)^2 + (4\sin\theta - y_C)^2} - r\right|\right\} \\ \text{s.t. } \theta \in [0, 2\pi),\ r \in (0, +\infty) \\ X = (x_C, y_C, r)^T \end{cases} \tag{E2-1.1}$$

When the variable θ of the function $f_a(\theta)$ has values $\theta^* = 0, \frac{\pi}{2}, \pi, \frac{3\pi}{2}$, the objective function has the extreme values $F(X, \theta^*)$, they are $F(X, \theta^*) = \left\{\left|\sqrt{(5 - x_C)^2 + y_C{}^2} - r\right|,\right.$ $\left.\left|\sqrt{x_C{}^2 + (4 - y_C)^2} - r\right|, \left|\sqrt{(-5 - x_C)^2 + y_C{}^2} - r\right|, \left|\sqrt{x_C{}^2 + (-4 - y_C)^2} - r\right|\right\}$. Only when these four extreme values are equal to each other for optimization variables $X = (x_C, y_C, r)^T$ will the objective function have a minimum value, or the maximum fitting error get the minimum. That is

$$\left|\sqrt{(5 - x_C)^2 + y_C{}^2} - r\right| = \left|\sqrt{x_C{}^2 + (4 - y_C)^2} - r\right| = \left|\sqrt{(-5 - x_C)^2 + y_C{}^2} - r\right| = \left|\sqrt{x_C{}^2 + (-4 - y_C)^2} - r\right|$$

The solutions $x_C = 0$, $y_C = 0$, $r = 4.5$ can be solved by the above equation.

It is not easy to obtain the analytical solution of the mathematical model of saddle point programming. Generally, we only have the numerical solution from the effective algorithm and programming in Matlab.

When the numeral method is taken, the given ellipse is discretized into the point set $\{(x_P^{(i)}, y_P^{(i)})^T\}$, $i = 1, 2, \ldots, n$, where n is the number of discrete points. Then, equation (E2-1.1) of the mathematical model can be rewritten as

$$\begin{cases} \Delta = \min\limits_{X}\max\limits_{1\le i\le N}\left\{\left|\sqrt{(x_P^{(i)} - x_C)^2 + (y_P^{(i)} - y_C)^2} - r\right|\right\} \\ \text{s.t. } \theta \in [0, 2\pi],\ r \in (0, +\infty) \\ X = (x_C, y_C, r)^T \end{cases} \tag{E2-1.2}$$

To solve the above equation, it is necessary to provide the initial values for the optimization variables $X = (x_C, y_C, r)^T$. In this example, the accuracy of the circle fitting, or the fitting error, is independent of the number of discrete points of the ellipse and is equal to the analytical solution, if the discrete points of the given curve include the four vertexes of the ellipse, the ***fitting characteristic points***. It can be noticed that the mathematical model of saddle point programming deals only with the fitting characteristic points of a fitted curve, not all the discrete points, which may be one of the differences identifying it from the least-squares method.

2.3 Saddle Circle Point

The open kinematic chain R–R is currently a binary link of plane linkages, whose constraint curve is a circle. It is naturally one of the basic elements in the discrete kinematic geometry of planar linkages. The comparison between a circle and a discrete point path, traced by a point of the moving body in planar movement, is discussed in detail in this section.

2.3.1 Saddle Circle Fitting

For a set of discrete positions of a moving body $\sum^*$ defined by the parameter $(x^{(i)}_{Om}, y^{(i)}_{Om}, \gamma^{(i)})$ in $\{\boldsymbol{O}_f; \boldsymbol{i}_f, \boldsymbol{j}_f\}$, a point P of $\sum^*$ with Cartesian coordinates (x_{Pm}, y_{Pm}) in $\{\boldsymbol{O}_m; \boldsymbol{i}_m, \boldsymbol{j}_m\}$, called a ***tracing point***, traces a discrete path Γ_P or a discrete point set $\{\boldsymbol{R}^{(i)}_P\}$ in $\{\boldsymbol{O}_f; \boldsymbol{i}_f, \boldsymbol{j}_f\}$. Two points of $\sum^*$ have to be designated as the hinge points of the moving revolute joints for synthesizing a four-bar linkage. The constraint curve, corresponding to the binary link R–R, is a circle. Therefore, it is a key issue how to evaluate the degree of approximation between the discrete point path $\{\boldsymbol{R}^{(i)}_P\}$ and a circle in the kinematic synthesis. According to the properties of plane curves in Chapter 1, a plane curve with constant curvature is a circle. A circle with radius r and center point coordinates (x_C, y_C) is taken to fit the discrete point set $\{\boldsymbol{R}^{(i)}_P\}$, called ***fixed circle fitting*** since the radius and center point of the fitting circle are prescribed. The distance from the center point of the circle to a point of the discrete point set varies with the different points of $\{\boldsymbol{R}^{(i)}_P\}$, since they may not lie on the fitting circle. The deviations between the distances and the radius of the fitting circle are defined as normal fitting errors. How to estimate these errors is the key point for us in identifying different approaches. We focus on the maximum normal fitting error. For fixed circle fitting, the maximum normal fitting error depends on both the discrete point set $\{\boldsymbol{R}^{(i)}_P\}$ and the given values of the radius r and the center point coordinates. If we assign first just the radius r, the center point coordinates (x_C, y_C) can be determined adaptively by letting the maximum normal fitting error be minimum for all discrete points $\{\boldsymbol{R}^{(i)}_P\}$, called ***floating circle fitting***, as shown in Fig. 2.2. The maximum fitting error varies with the different radii of the floating circle. For all floating circles there must be a best one, whose maximum fitting error is the minima.

Definition 2.1 Based on the discrete point set $\{\boldsymbol{R}^{(i)}_P\}$, a circle is determined adaptively by letting the maximum fitting error be the minima; this is defined as a ***saddle circle***, and the corresponding error is referred to as a ***saddle circle error***.

Obviously, the properties of the discrete point set $\{\boldsymbol{R}^{(i)}_P\}$ determine the radius and the position of the saddle circle by letting the maximum fitting error be the minima. Based on $\{\boldsymbol{R}^{(i)}_P\}$,

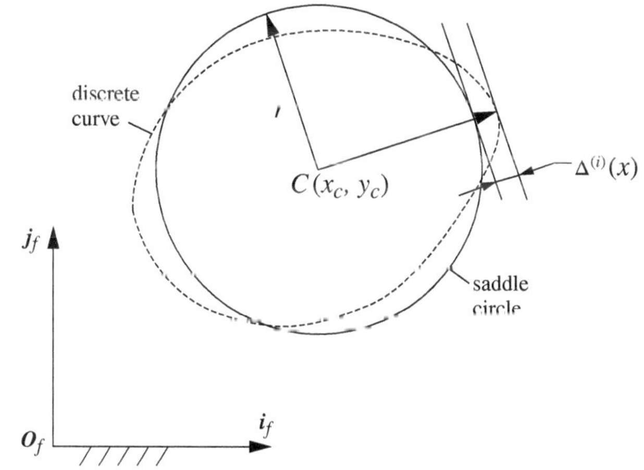

Figure 2.2 The saddle circle of a plane discrete curve

a mathematical model of saddle point programming for saddle circle fitting can be set up as

$$\begin{cases} \Delta_{rr} = \min\limits_{x} \max\limits_{1\le i\le n} \{\Delta^{(i)}(\boldsymbol{x})\} \\ \quad = \min\limits_{x} \max\limits_{1\le i\le n} \left\{ \left| \sqrt{(x_P^{(i)} - x_C)^2 + (y_P^{(i)} - y_C)^2} - r \right| \right\} \\ \text{s.t.} \quad r \in (0, +\infty) \\ \boldsymbol{x} = (x_C, y_C, r)^T \end{cases} \tag{2.11}$$

where Δ_{rr} is the saddle circle error; $\{\Delta^{(i)}(\boldsymbol{x})\}$ is the object function, the set of fitting errors for all points in $\{\boldsymbol{R}_P^{(i)}\}$ and the fitting circle, $\boldsymbol{x} = (x_C, y_C, r)^T$ are the optimization variables, in which r is the radius and (x_C, y_C) are the position coordinates of the center point of the fitting circle, n is the number of discrete points. The saddle circle has the following properties:

1. *Adaptivity*. The parameters of the saddle circle, the radius r, and the center point coordinates (x_C, y_C) are determined by minimizing the maximum fitting error according to the properties of the discrete point set $\{\boldsymbol{R}_P^{(i)}\}$.
2. *Uniqueness*. The saddle circle has a minimum fitting error among all fitting circles for a given discrete point set $\{\boldsymbol{R}_P^{(i)}\}$, which means that the maximum normal fitting error is the minima, or the first saddle point program of the circle fitting. There may be a few circles to fit the given set of discrete points $\{\boldsymbol{R}_P^{(i)}\}$, whose maximum normal fitting errors will give the extreme minimum values in the program, which depend on the geometrical shape of the given curve. It is easy to identify which one is the least. The saddle circle is unique since there is no longer any fitting circle with less fitting error.
3. *Comparability*. The fitting error of the fitting circle is measured in the normal direction, which reveals the approximation between a fitting circle and the discrete point set $\{\boldsymbol{R}_P^{(i)}\}$. Compared with other fitting circles, the saddle circle has the minimal maximum fitting error.

The initial values of the saddle circle fitting are needed to solve equation (2.11). A fitting circle, close to the saddle circle, can be determined by the least-squares method. We set up the initial values by the least-squares method according to the given discrete point set. The normal deviation e_i between the discrete points $\{\boldsymbol{R}_P^{(i)}\}$ and a fitting circle can be represented as

$$e_i = (x_P^{(i)} - x_C)^2 + (y_P^{(i)} - y_C)^2 - r^2, \; i = 1, \ldots, n \tag{2.12}$$

The sum of squares of the normal deviations e_i can be derived by

$$F = \sum_{i=1}^{n} e_i^2 = \sum_{i=1}^{n} [(x_P^{(i)} - x_C)^2 + (y_P^{(i)} - y_C)^2 - r^2]^2$$

$$= \sum_{i=1}^{n} [(x_P^{(i)})^2 + (y_P^{(i)})^2 + A \cdot x_P^{(i)} + B \cdot y_P^{(i)} + C]^2 \tag{2.13}$$

where $A = -2x_C$, $B = -2y_C$, $C = x_C^2 + y_C^2 - r^2$. Differentiating the above equation with respect to A, B, and C, then letting the three derivatives be zero, we have

$$\begin{cases} \dfrac{\partial F}{\partial A} = 2\sum_{i=1}^{n}[(x_P^{(i)})^2 + (y_P^{(i)})^2 + A \cdot x_P^{(i)} + B \cdot y_P^{(i)} + C] \cdot x_P^{(i)} = 0 \\ \dfrac{\partial F}{\partial B} = 2\sum_{i=1}^{n}[(x_P^{(i)})^2 + (y_P^{(i)})^2 + A \cdot x_P^{(i)} + B \cdot y_P^{(i)} + C] \cdot y_P^{(i)} = 0 \\ \dfrac{\partial F}{\partial C} = 2\sum_{i=1}^{n}[(x_P^{(i)})^2 + (y_P^{(i)})^2 + A \cdot x_P^{(i)} + B \cdot y_P^{(i)} + C] = 0 \end{cases} \tag{2.14}$$

The rectangular coordinates (x_C, y_C) of the center point and the radius r of the fitting circle, whose corresponding quadratic sum reaches an extreme value in the least sum of squares, can be determined by

$$\begin{cases} x_C = \dfrac{c_1b_2 - c_2b_1}{a_1b_2 - a_2b_1}, y_C = \dfrac{c_1a_2 - c_2a_1}{a_1b_2 - a_2b_1} \\ r = \sqrt{x_C^2 + y_C^2 - 2\dfrac{x_C}{n}\sum_{i=1}^{n} x_P^{(i)} - 2\dfrac{y_C}{n}\sum_{i=1}^{n} y_P^{(i)} + \dfrac{1}{n}\left[\sum_{i=1}^{n}(x_P^{(i)})^2 + \sum_{i=1}^{n}(y_P^{(i)})^2\right]} \end{cases} \tag{2.15a}$$

where

$$a_1 = 2\sum_{i=1}^{n}(x_P^{(i)})^2 - \frac{2}{n}\left(\sum_{i=1}^{n} x_P^{(i)}\right)^2, b_2 = 2\sum_{i=1}^{n}(y_P^{(i)})^2 - \frac{2}{n}\left(\sum_{i=1}^{n} y_P^{(i)}\right)^2$$

$$b_1 = a_2 = 2\sum_{i=1}^{n} x_P^{(i)}y_P^{(i)} - \frac{2}{n}\sum_{i=1}^{n} x_P^{(i)}\sum_{i=1}^{n} y_P^{(i)}$$

$$c_1 = \sum_{i=1}^{n}(x_P^{(i)})^3 + \sum_{i=1}^{n} x_P^{(i)}(y_P^{(i)})^2 - \frac{1}{n}\sum_{i=1}^{n} x_P^{(i)}\left[\sum_{i=1}^{n}(x_P^{(i)})^2 + \sum_{i=1}^{n}(y_P^{(i)})^2\right]$$

$$c_2 = \sum_{i=1}^{n}(y_P^{(i)})^3 + \sum_{i=1}^{n} y_P^{(i)}(x_P^{(i)})^2 - \frac{1}{n}\sum_{i=1}^{n} y_P^{(i)}\left[\sum_{i=1}^{n}(x_P^{(i)})^2 + \sum_{i=1}^{n}(y_P^{(i)})^2\right] \tag{2.15b}$$

The circle, determined by the least-squares method, is close to the saddle circle. Its parameters are considered to be good initial values for the saddle circle fitting.

Based on the above mathematical model of equation (2.11) and the initial values, a given discrete point set $\{\boldsymbol{R}_P^{(i)}\}$, $i = 1, \ldots, n$ in $\{\boldsymbol{O}_f; \boldsymbol{i}_f, \boldsymbol{j}_f\}$ corresponds to a saddle circle, which is located by the function **fminimax** in the software Matlab. The parameters are set as

[x, fval] = fminimax(fun,x0,A,b,Aeq,beq,lb,ub,nonlcon,options)

The options are defined as a default setting in function **fminimax**, and the variables on the left are respectively

x: the optimal solution $\boldsymbol{x} = (x_C, y_C, r)^T$, the parameters of the saddle circle corresponding to the set of discrete points $\{\boldsymbol{R}_P^{(i)}\}$.
fval: the value of the objective function $\Delta_{\rm rr}$, or the set of normal fitting errors of the discrete point set $\{\boldsymbol{R}_P^{(i)}\}$ and the saddle circle. The saddle circle error is max(fval).

The input variables for function **fminimax** are

fun: the objective function $\{\Delta^{(i)}(\boldsymbol{x})\}$, a set of normal fitting errors between the points of $\{\boldsymbol{R}_P^{(i)}\}$ and the floating fitting circle, calculated by equation (2.11).
x0: the initial values of the optimal parameters, the center point coordinates of the saddle circle and the radius, which are assigned by equation (2.15).
A,b,Aeq,beq,lb,ub,nonlcon: a subject function, including the constraint equation and the boundary of the optimization variables, in which the linear constraints **A**, **b**, **Aeq**, **beq** and the nonlinear constraints **nonlcon** are set to null. The lower bound of the optimization variables is lb = [−Inf, −Inf, 0] and the upper bound is ub = [Inf, Inf, Inf].

A ***sub-program ArrF*** is constructed for a saddle circle fitting to a discrete point set $\{\boldsymbol{R}_P^{(i)}\}$, whose input parameters are the coordinates of the discrete points and whose output parameters are the center point coordinates and the radius of the saddle circle, and the saddle circle error $\Delta_{\rm rr}$.

2.3.2 Saddle Circle

Based on Definition 2.1, a saddle circle is located by a given discrete path Γ_P in $\{\boldsymbol{O}_f; \boldsymbol{i}_f, \boldsymbol{j}_f\}$, whose saddle circle error $\Delta_{\rm rr}$ is maximal for all the points on Γ_P, but with maximal error $\Delta_{\rm rr}$ smallest for all the fitting circles.

For convenience in discussing the properties of the fitting error for saddle circle fitting, the given path Γ_P is assumed to be continuous. The center point vector of the saddle circle is $\boldsymbol{R}_C$ and the radius is r. As shown in Fig. 2.3, for a point P on Γ_P, where the normal fitting error to the saddle circle is Δ, we write the position vector $\boldsymbol{R}_P$ of point P as

$$\boldsymbol{R}_P = \boldsymbol{R}_C + (r + \Delta)\boldsymbol{n} \tag{2.16}$$

where $\boldsymbol{n}$ is the unit normal vector of the saddle circle, and it passes through the point P. The sign of the normal fitting error Δ is positive if point P locates outside the saddle circle and negative if P is inside. Then, the expression of Δ is

$$\Delta = (\boldsymbol{R}_P - \boldsymbol{R}_C) \cdot \boldsymbol{n} - r \tag{2.17}$$

For a saddle circle of Γ_P, the position vector $\boldsymbol{R}_C$ is a constant vector and the radius r is a constant. Differentiating equation (2.17) with respect to the arc length s of Γ_P, we have

$$\frac{\mathrm{d}\Delta}{\mathrm{d}s} = \frac{\mathrm{d}\boldsymbol{R}_P}{\mathrm{d}s} \cdot \boldsymbol{n} + (\boldsymbol{R}_P - \boldsymbol{R}_C) \cdot \frac{\mathrm{d}\boldsymbol{n}}{\mathrm{d}s} \tag{2.18}$$

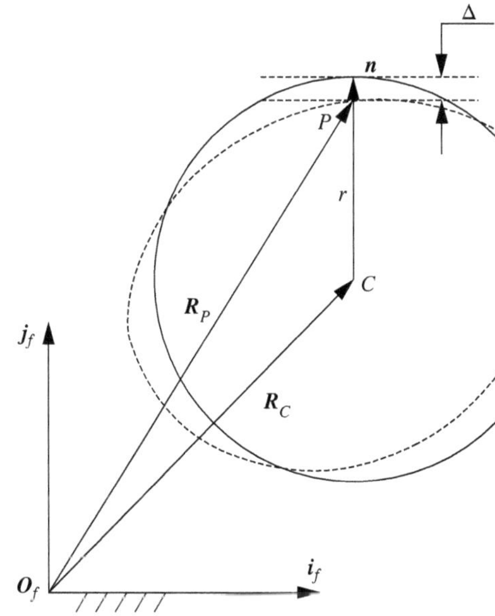

Figure 2.3 The normal fitting error of a curve and its saddle circle

From equation (2.16), we have $\boldsymbol{R}_P - \boldsymbol{R}_C = (r + \Delta)\boldsymbol{n}$ and $(\boldsymbol{R}_P - \boldsymbol{R}_C) \cdot \frac{\mathrm{d}\boldsymbol{n}}{\mathrm{d}s} = (r + \Delta)\boldsymbol{n} \cdot \frac{\mathrm{d}\boldsymbol{n}}{\mathrm{d}s} = 0$. By equation (2.18), the condition for the fitting error Δ to have an extreme value is $\frac{\mathrm{d}\Delta}{\mathrm{d}s} = 0$ at a point P of Γ_P, or

$$\frac{\mathrm{d}\boldsymbol{R}_P}{\mathrm{d}s} \cdot \boldsymbol{n} = 0 \tag{2.19}$$

When the tangent vector $\frac{\mathrm{d}\boldsymbol{R}_P}{\mathrm{d}s}$ of Γ_P at P is orthogonal to the normal vector $\boldsymbol{n}$ of the saddle circle, or the normal of Γ_P is also identical to that of the saddle circle, the normal fitting error Δ gets an extreme value, but may not be the maxima, which may appear at multiple points on Γ_P, such as the points $P^{(k+1)}$, $P^{(k+2)}$, and $P^{(k+3)}$ in Fig. 2.4. The saddle circle error is only the absolute maxima of all points.

Based on the convention of the right-handed system, such as equation (1.8), the normal vector of Γ_P at any point can be represented as $\frac{\mathrm{d}\boldsymbol{R}_P}{\mathrm{d}s} \times \boldsymbol{k}$. The position vector of the saddle circle's center can be expressed by

$$\boldsymbol{R}_C = \boldsymbol{R}_P - (r + \Delta)\frac{\mathrm{d}\boldsymbol{R}_P}{\mathrm{d}s} \times \boldsymbol{k} \tag{2.20}$$

In order to determine the three parameters (x_C, y_C, r) of the saddle circle and the saddle circle error Δ_{rr} for the given curve Γ_P, there are at least four points on Γ_P where equation (2.19) holds true, and the position vectors are substituted into equation (2.20) at the four points; (x_C, y_C, r) and Δ_{rr} can be solved.

Based on the theory of saddle point programming [3], saddle circle fitting by letting the maximum fitting error be the minima is equivalent to the given plane curve being contained by two concentric circles, or the given plane curve tangentially contacting the two concentric circles at four or more points, which leads to the maximum normal fitting errors being the

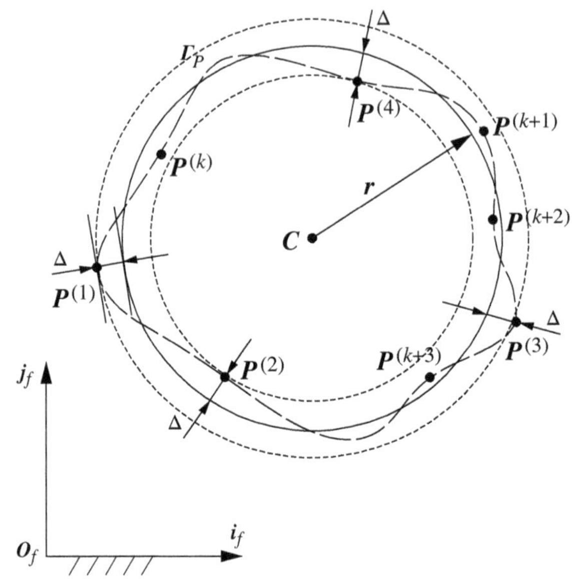

Figure 2.4 Two concentric circles and the saddle circle errors

minima with the same value as shown in Fig. 2.4. The four tangent points may distribute on both sides of the saddle circle in this manner, two inside the saddle circle and another two outside.

The four tangent points on Γ_P locate geometrically the saddle circle of Γ_P and are defined as the ***saddle circle fitting points*** on Γ_P. For the ellipse in Example 2.1, the fitting points are just the four vertexes. The corresponding discrete positions of the moving body are referred to as the ***saddle circle fitting positions***. The saddle circle error, the maxima of the normal fitting error, appears at the four saddle circle fitting points. If the saddle circle error is zero, the fitted plane curve must be a circle.

If the fitted plane curve Γ_P is discretized into a discrete trajectory point set $\{\boldsymbol{R}_P^{(i)}\}$, four saddle circle fitting points should be selected from $\{\boldsymbol{R}_P^{(i)}\}$ to determine the saddle circle. By equation (2.17), the normal fitting error $\Delta^{(i)}$ between a point in $\{\boldsymbol{R}_P^{(i)}\}$ and the saddle circle can be written as

$$\Delta^{(i)} = \left|\boldsymbol{R}_P^{(i)} - \boldsymbol{R}_C\right| - r = \sqrt{(x_P^{(i)} - x_C)^2 + (y_P^{(i)} - y_C)^2} - r, \ \boldsymbol{R}_P^{(i)} = [\mathbf{M}^{(i)}] \cdot \boldsymbol{R}_{Pm} \tag{2.21}$$

The saddle circle error Δ_{rr} is the absolute maximum value of all, or

$$\Delta_{\mathrm{rr}} = \max\{|\Delta^{(i)}|\} = \max\left\{\left|\sqrt{(x_P^{(i)} - x_C)^2 + (y_P^{(i)} - y_C)^2} - r\right|\right\} \tag{2.22}$$

The saddle circle error of a discrete point set $\{\boldsymbol{R}_P^{(i)}\}$ can be obtained by numerical calculation and comparison among all the normal errors $\Delta^{(i)}$ of $\{\boldsymbol{R}_P^{(i)}\}$ by equation (2.22).

Thus, the relationships among the separate positions of the moving body, the discrete point set, and the saddle circle can be established by the mathematical model of saddle circle fitting.

2.3.3 Four Positions

In order to discuss the global geometrical property of the plane discrete point set compared with a circle, we start at the four separated positions of the moving body $\sum^*$ in planar movement, the four saddle circle fitting positions, prescribed in $\{\boldsymbol{O}_f; \boldsymbol{i}_f, \boldsymbol{j}_f\}$ and numbered 1, 2, 3, 4. A point $P(x_{Pm}, y_{Pm})$ of $\sum^*$ traces a plane discrete point set $\{\boldsymbol{R}_P^{(i)}\}$, $i = 1, 2, 3, 4$, comprised of four points $P^{(1)}$, $P^{(2)}$, $P^{(3)}$, and $P^{(4)}$ in $\{\boldsymbol{O}_f; \boldsymbol{i}_f, \boldsymbol{j}_f\}$. Based on equation (2.11) of the saddle circle fitting model, we call the sub-program ArrF and obtain the saddle circle of $\{\boldsymbol{R}_P^{(i)}\}$, $i =$ 1, 2, 3, 4, called the ***four-point saddle circle***. In this case, the four points $P^{(1)}$, $P^{(2)}$, $P^{(3)}$, and $P^{(4)}$ are the saddle circle fitting points and the given four positions of $\sum^*$ are all the saddle circle fitting positions.

A plane discrete point path, comprised of four points, corresponds to a four-point saddle circle. According to the saddle point programming, the maximum fitting errors at the four points are minimum if and only if the fitting errors are the same as an extreme value when the four fitted points $P^{(1)}$, $P^{(2)}$, $P^{(3)}$, and $P^{(4)}$ distribute on both sides of the four-point saddle circle. There are three such distribution cases in total, $P^{(1)}P^{(2)}-P^{(3)}P^{(4)}$, $P^{(1)}P^{(3)}-P^{(2)}P^{(4)}$, and $P^{(1)}P^{(4)}-P^{(2)}P^{(3)}$, which correspond to three fitting circles, called ***distributing circles*** of the four points, as shown in Fig. 2.5.

The distribution case $P^{(1)}P^{(2)}-P^{(3)}P^{(4)}$ is taken as an example, two points $P^{(1)}$ and $P^{(2)}$ are inside the distributing circle and two points $P^{(3)}$ and $P^{(4)}$ are outside. By equation (2.21), the

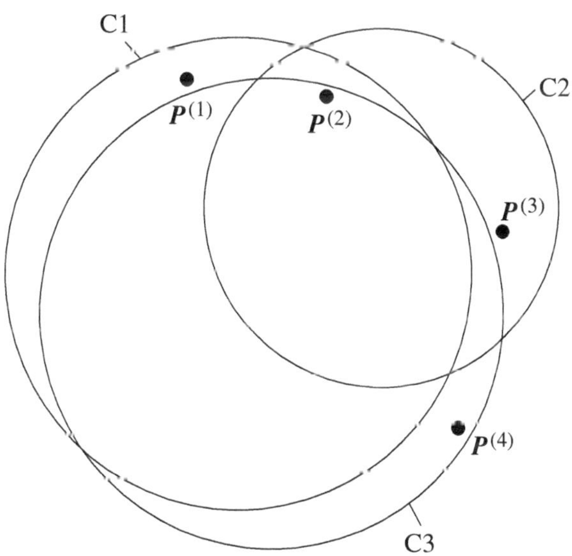

Figure 2.5 Three distributing circles of four points

equation of the fitting error Δ is

$$\begin{cases} (r-\Delta)^2 = (\boldsymbol{R}_P^{(i)} - \boldsymbol{R}_C)^2,\ i=1,2 \\ (r+\Delta)^2 = (\boldsymbol{R}_P^{(i)} - \boldsymbol{R}_C)^2,\ i=3,4 \end{cases}, \boldsymbol{R}_P^{(i)} = [\mathbf{M}^{(i)}]\cdot \boldsymbol{R}_{Pm} \tag{2.23}$$

The center C of the distributing circle is just the intersection point of two straight lines, the mid-perpendicular of the segment $P^{(1)}P^{(2)}$ and that of the segment $P^{(3)}P^{(4)}$. The position coordinates of C in $\{\boldsymbol{O}_f; \boldsymbol{i}_f, \boldsymbol{j}_f\}$ are

$$\begin{cases} y_C = \dfrac{\left((x_P^{(4)})^2 + (y_P^{(4)})^2 - (x_P^{(3)})^2 - (y_P^{(3)})^2\right)(x_P^{(2)} - x_P^{(1)}) - \left((x_P^{(2)})^2 + (y_P^{(2)})^2 - (x_P^{(1)})^2 - (y_P^{(1)})^2\right)(x_P^{(4)} - x_P^{(3)})}{2(x_P^{(2)} - x_P^{(1)})(y_P^{(4)} - y_P^{(3)}) - 2(x_P^{(4)} - x_P^{(3)})(y_P^{(2)} - y_P^{(1)})} \\ x_C = \dfrac{\left((x_P^{(4)})^2 + (y_P^{(4)})^2 - (x_P^{(3)})^2 - (y_P^{(3)})^2\right)(y_P^{(2)} - y_P^{(1)}) - \left((x_P^{(2)})^2 + (y_P^{(2)})^2 - (x_P^{(1)})^2 - (y_P^{(1)})^2\right)(y_P^{(4)} - y_P^{(3)})}{2(y_P^{(2)} - y_P^{(1)})(x_P^{(4)} - x_P^{(3)}) - 2(y_P^{(4)} - y_P^{(3)})(x_P^{(2)} - x_P^{(1)})} \end{cases} \tag{2.24}$$

The distributing circle error $\Delta_{12,34}$ for the distribution case $P^{(1)}P^{(2)}$–$P^{(3)}P^{(4)}$ can be obtained by the coordinates of the three points, C, $P^{(1)}$, and $P^{(3)}$. It is

$$\Delta_{12,34} = \frac{\left|\sqrt{(x_P^{(1)} - x_C)^2 + (y_P^{(1)} - y_C)^2} - \sqrt{(x_P^{(3)} - x_C)^2 + (y_P^{(3)} - y_C)^2}\right|}{2} \tag{2.25a}$$

Combining equation (2.24) and the above equation, we have

$$16\Delta_{12,34}{}^4 - 8\Delta_{12,34}{}^2\left((x_P^{(1)} - x_C)^2 + (y_P^{(1)} - y_C)^2 + (x_P^{(3)} - x_C)^2 + (y_P^{(3)} - y_C)^2\right)$$
$$+\left((x_P^{(1)} - x_C)^2 + (y_P^{(1)} - y_C)^2 - (x_P^{(3)} - x_C)^2 - (y_P^{(3)} - y_C)^2\right)^2 = 0 \tag{2.25b}$$

Substituting equation (2.4) into the above equation, we get the 12th-order algebraic equation of $(x_{Pm}, y_{Pm}, \Delta_{12,34})$. In a similar way, the distributing circle errors $\Delta_{13,24}$ for the case $P^{(1)}P^{(3)}$–$P^{(2)}P^{(4)}$ and $\Delta_{14,23}$ for the case $P^{(1)}P^{(4)}$–$P^{(2)}P^{(3)}$ can be obtained. According to Definition 2.1, the four-point saddle circle error Δ_{1234} of $\{\boldsymbol{R}_P^{(i)}\}$, $i = 1, 2, 3, 4$ is the minimum of the three distributing circle errors, or

$$\Delta_{1234} = \min(\Delta_{12,34}, \Delta_{13,24}, \Delta_{14,23}) \tag{2.26a}$$

Another form is

$$\Delta_{1234} = \frac{1}{2}\left|\left|\frac{1}{2}\left|\Delta_{12,34} + \Delta_{13,24}\right| - \frac{1}{2}\left|\Delta_{12,34} - \Delta_{13,24}\right|\right| + \Delta_{14,23}\right|$$
$$-\frac{1}{2}\left|\left|\frac{1}{2}\left|\Delta_{12,34} + \Delta_{13,24}\right| - \frac{1}{2}\left|\Delta_{12,34} - \Delta_{13,24}\right|\right| - \Delta_{14,23}\right| \tag{2.26b}$$

For the continuity of the coordinates of the points on $\sum^*$, the four-point saddle circle error surface is a 12th-order piecewise algebraic surface.

For the moving body $\sum^*$ with four separated positions, described by the motion parameters of the first four lines listed in Table 2.1, the coordinates (x_{Pm}, y_{Pm}) of the point P of $\sum^*$

Table 2.1 Ten positions of the rigid body

	$x_{Omf}^{(i)}$	$y_{Omf}^{(i)}$	$\gamma^{(i)}$
1	14.73	25.96	4.19
2	9.24	24.99	4.24
3	5.49	20.44	4.47
4	5.16	14.13	4.80
5	8.52	8.53	5.16
6	14.18	5.74	5.41
7	19.43	6.63	5.32
8	22.53	10.95	4.99
9	22.68	17.19	4.64
10	19.79	22.96	4.34

are viewed as the independent variables and the four-point saddle circle error Δ_{1234} as the function value. The 12th-order piecewise algebraic surface can be mapped onto a plane, in which all points with the same saddle circle errors are connected with contour lines, as shown in Fig. 2.6. In particular, if the saddle circle error is zero, we obtain an intersection curve between the error surface and the plane defined by $\Delta_{1234} = 0$, the circle point curve, shown in Fig. 2.6 as the heavy line.

2.3.4 *Five Positions*

Five planar separated positions of $\sum^*$ are prescribed in $\{\boldsymbol{O}_f; \boldsymbol{i}_f, \boldsymbol{j}_f\}$ and numbered 1, 2, 3, 4, 5. A point $P(x_{Pm}, y_{Pm})$ of $\sum^*$ traces a discrete point set $\{\boldsymbol{R}_P^{(i)}\}$, $i = 1, 2, 3, 4, 5$, comprised of five points $P^{(1)}$, $P^{(2)}$, $P^{(3)}$, $P^{(4)}$, and $P^{(5)}$ in $\{\boldsymbol{O}_f; \boldsymbol{i}_f, \boldsymbol{j}_f\}$. Based on equation (2.11) of the saddle circle fitting, we call the sub-program ArrF and have the saddle circle of $\{\boldsymbol{R}_P^{(i)}\}$, $i =$ 1, 2, 3, 4, 5, called the ***five-point saddle circle***. Similar to the four-position, taking the coordinates (x_{Pm}, y_{Pm}) as the independent variables and the five-point saddle circle error Δ_{12345} as the function value, we map the error surface on a plane. All points with the same values are connected by contour lines, shown in Fig. 2.7, with data from the first five lines listed in Table 2.1.

It is the same saddle circle fitting model for a discrete path with either four or five points. But from the viewpoint of discrete geometry, it is necessary to discuss both the difference and the relationship of the saddle circle fitting for four position and five-position cases.

The saddle circle of the five plane discrete points $P^{(1)}$–$P^{(5)}$ is geometrically determined by the four saddle circle fitting points, which correspond to the four saddle circle fitting positions. In order to reveal the relationship between the four-point saddle circle and the five-point saddle circle, we have to identify four points of the five saddle circle fitting points and four fitting positions of the five positions. The four-point saddle circle of $P^{(1)}$–$P^{(4)}$ has been discussed in the previous section. If another point $P^{(5)}$ is added, whether it is the saddle circle fitting point of the five points or not can be discussed in the following two cases.

(1) The fifth point $P^{(5)}$ is not the saddle circle fitting point, and position 5 is not the fitting position. The fitting error between point $P^{(5)}$ and the four-point saddle circle is less than the saddle circle error Δ_{1234}, or the saddle circle error Δ_{12345} of the five points $P^{(1)}$–$P^{(5)}$ is determined by the saddle circle fitting points $P^{(1)}$–$P^{(4)}$, or $\Delta_{12345} = \Delta_{1234}$, as shown in Fig. 2.8. For $\sum^*$ with five given separated positions, the points whose first four discrete points

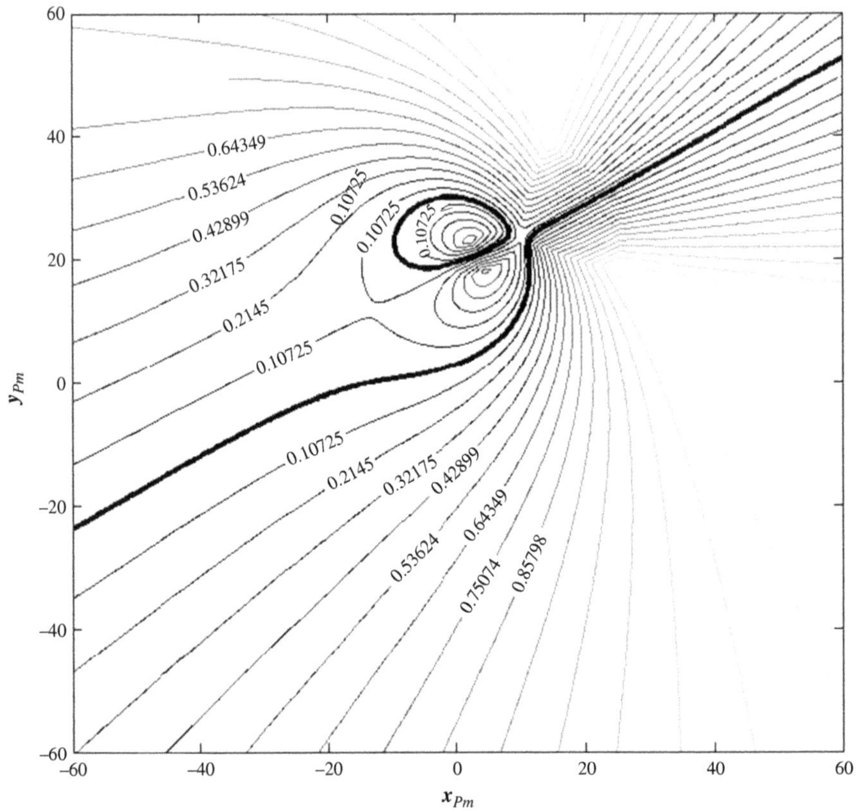

Figure 2.6 The four-position saddle circle errors

($P^{(1)}-P^{(4)}$) are the saddle circle fitting points distribute on the specific area of $\sum^*$, called the ***four-position area*** (1234), such as area I in Fig. 2.7. There is also another situation for the points on the four-position area (1234). Although the four points $P^{(1)}-P^{(4)}$ are the saddle circle fitting points of $P^{(1)}-P^{(5)}$, the saddle circle of $P^{(1)}-P^{(5)}$ is not the saddle circle of $P^{(1)}-P^{(4)}$. As shown in Fig. 2.9, the saddle circle of $P^{(1)}-P^{(4)}$ is the circle C1 with the distribution case $P^{(1)}P^{(3)}-P^{(2)}P^{(4)}$, or $\Delta_{1234} = \Delta_{13,24}$. The five-point saddle circle of $P^{(1)}-P^{(5)}$ is the circle C2, which is also determined by the four points $P^{(1)}-P^{(4)}$ in the distribution case $P^{(1)}P^{(2)}-P^{(3)}P^{(4)}$, or just a distributing circle.

(2) The fifth point $P^{(5)}$ is the saddle circle fitting point, and position 5 is the fitting position. In this case, position 5 and three of the four positions 1234 may become the saddle circle fitting positions, and there exist four possibilities: 1235, 1245, 1345, and 2345. Each possible four-position combination has three distributing circles; there are 12 distributing circles for the four possible combinations. The five-point saddle circle is one of the 12 distributing circles of $P^{(1)}-P^{(5)}$.

A point on $\sum^*$ with five given separated positions corresponds to a five-point saddle circle error, and all points on $\sum^*$ correspond to an error surface comprised of several four-point error surfaces, a 12th-order algebraic surface. The coordinate plane on $\sum^*$ can be divided into several four-position areas, such as area I corresponding to the positions 1234, area II to 1235,

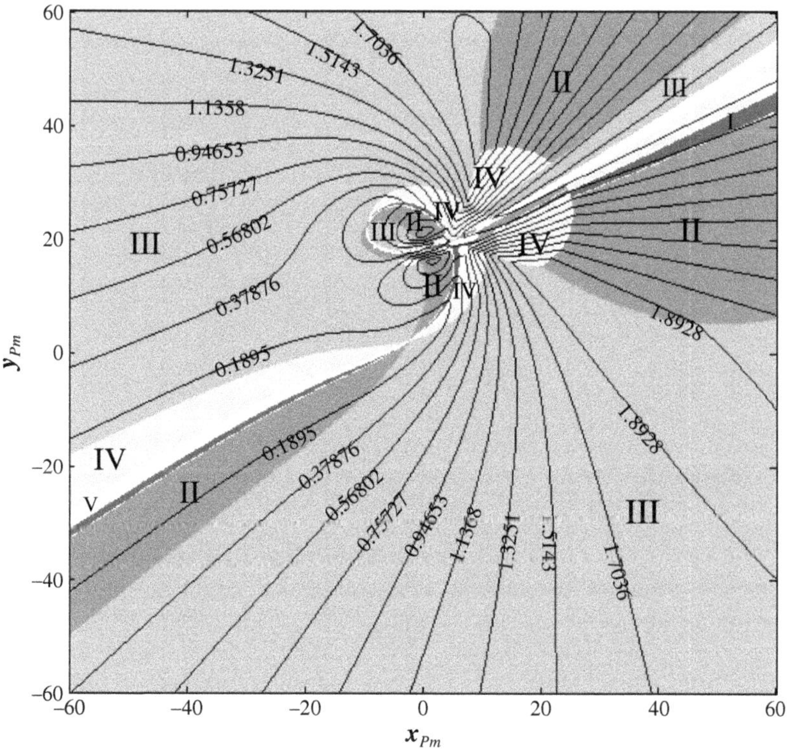

Figure 2.7 The five-position saddle circle errors

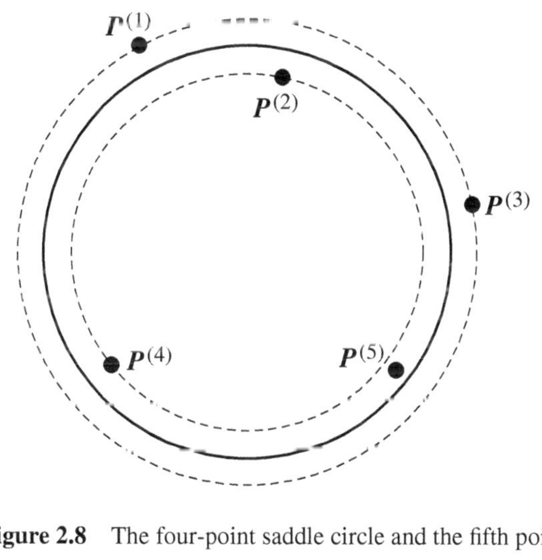

Figure 2.8 The four-point saddle circle and the fifth point

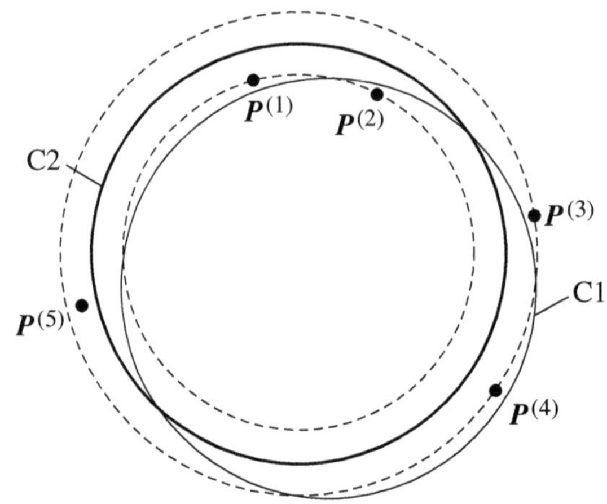

Figure 2.9 The five-point saddle circle is the four-point distributing circle

area III to 1245, area IV to 1345, and area V to 2345 in Fig. 2.7. The algebraic equations of the boundaries of the different four-position areas can be derived by equation (2.25). The point locating on the boundaries has the characteristics of the adjacent areas, belonging to both areas, and the five-point saddle circle of its path, corresponding to two four-point distributing circles with the same distributing circle error. If the two four-point distributing circles are coincident with each other, a new characteristic on the boundary of the error surface would appear, such as the Burmester points on the rigid body.

As previously mentioned, the saddle circle error surface of $\sum^*$ at five positions theoretically comprised of several four-point distributing circle error surfaces, a 12th-order algebraic surface, has many peak valleys. If the saddle circle error surface at four positions is compared with that at five positions, we can find the bottom of a trench (circle point curve at four positions) evolves into several valley bottoms (circle points at five positions). Globally, the saddle circle error at five positions is bigger than that at four positions. The geometrical properties of the saddle circle errors at both four and five positions need to be studied further by modern differential geometry.

2.3.5 Multiple Positions

For multiple planar separated positions of the moving body $\sum^*$, based on the approach of saddle point programming, the saddle circle of the discrete point path is still located by the four saddle circle fitting points, no matter how many points the discrete path has. For all points on $\sum^*$, the saddle circles can be determined in the same way as in the five-position case. Therefore, no matter how many separated positions of $\sum^*$ are prescribed, it can also be viewed as a combination of four positions. For n given positions, there are C_n^4 combinations of four positions, and there exist more four-position areas on $\sum^*$. For the 10 separated positions listed in Table 2.1, the error surface of the saddle circle is mapped onto contour plot with different four-position areas, as shown in Fig. 2.10.

The more separated positions of $\sum^*$ are prescribed, the smaller are the four-position areas of the saddle circle. When the multiply separated positions of $\sum^*$ tend to infinity, a discrete

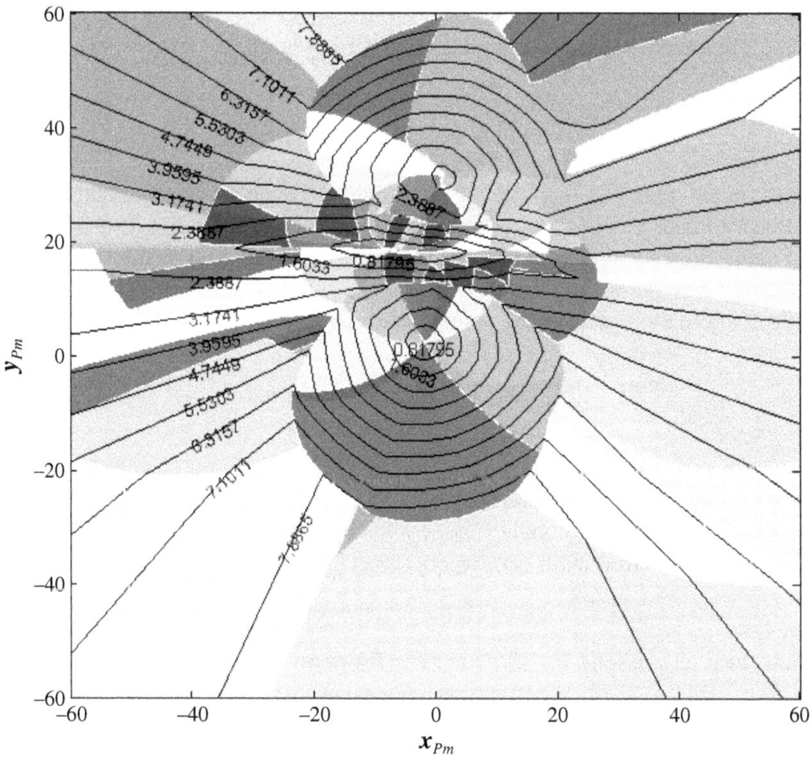

Figure 2.10 The 10-position saddle circle errors

path traced by a point of $\sum^*$ would approximate a continuous curve, whose saddle circle is still determined by the four fitting points. Those fitting points may change continuously as the tracing point's coordinates vary continuously on $\sum^*$.

2.3.6 Saddle Circle Point

A point of $\sum^*$ in planar movement traces a plane path, which corresponds to a saddle circle and its error. Different points on $\sum^*$ trace different paths and correspond to different saddle circle errors. In other words, the saddle circle error is the nonlinear function (a 12th-order algebraic equation) of the point's coordinates, and may have maximum or minimum value at some points.

Definition 2.2 A point of a moving body in planar movement, whose path in the fixed frame corresponds to the saddle circle, having minimum saddle circle error with respect to other tracing points in its neighborhood, is defined as a ***saddle circle point***, and the corresponding saddle circle error is called the ***saddle circle point error***.

If the saddle circle point error is zero, a saddle circle point is a ***circle point***. The saddle point programming needs to be carried out twice to locate the saddle circle point. The first one is for the saddle circle fitting. The second one is to locate the special point on $\sum^*$ to be a saddle circle point. The mathematic model of the saddle circle point can be written as

$$\begin{cases} \delta_{\rm rr} = \min \Delta_{\rm rr}(z) \\ z = (x_{Pm}, y_{Pm})^T \end{cases} \tag{2.27}$$

where $\Delta_{\rm rr}(z)$ is the objective function, or the saddle circle error for any point on $\sum^*$, and the optimization variables $z = (x_{Pm}, y_{Pm})^T$ are the coordinates of the point. It is necessary to give the initial values to solve equation (2.27). We do not know yet how many saddle circle points exist and where they lie on $\sum^*$; we have to assign a number of initial values randomly in the concave area where the saddle circle points locate. The search area is

$$a = \max\left(\max_{1\le i\le n} x_{Omf}^{(i)} - \min_{1\le i\le n} x_{Omf}^{(i)}, \max_{1\le i\le n} y_{Omf}^{(i)} - \min_{1\le i\le n} y_{Omf}^{(i)}\right)$$

$$-c_0 a \le x_{Pm}, y_{Pm} \le c_0 a \tag{2.28}$$

where c_0 is the coefficient of the search area.

Based on the mathematical model of the saddle circle point, equation (2.27) is solved by the function **fmincon** in Matlab with parameters set as follows:

[x,fval,exitflag] = fmincon (fun,x0,A,b,Aeq,beq,lb,ub,nonlcon,options)

The optimization conditions for the function **fmincon** are specified through the function **options**. The medium-scale algorithm is chosen as the optimization scheme. The stopping conditions are set through the maximum number of function evaluations allowed, or MaxFunEvals = 500. The maximum allowed number of iterations is set to 200. The termination tolerances are set to TolFun = 0.000001 and TolX = 0.000001.

The outputs of the function **fmincon** are described below.

x: the optimal solution, or final values of the optimization variables; the coordinates of the saddle circle points.
fval: the value of the objective function, or the saddle circle point errors.
exitflag: the exit condition, which describes the reason for the termination of the function **fmincon**.

The inputs of the function **fmincon** are respectively

fun: the objective function, or saddle circle error corresponding to a point's path, obtained by calling the sub-program ArrF.
x0: the initial values of the optimization variables, randomly generated by the function **rand** in the area $x_m \in [x_{m\min}, x_{m\max}]$, $y_m \in [y_{m\min}, y_{m\max}]$.
A,b,Aeq,beq,lb,ub,nonlcon: the subject function, including the constraint equation and the bounds of the optimization variables, in which the linear constraints **A**, **b**, **Aeq**, **beq**, and the nonlinear constraints **nonlcon** are all set to null. The lower bound of the optimization variable array is lb = $[x_{m\min}, y_{m\min}]$ and the upper bound is ub = $[x_{m\max}, y_{m\max}]$.

A sub-program ArrP is constructed for the location of saddle circle points on $\sum^*$ with multiple separated positions. The inputs are the motion parameters of the prescribed separated positions and the outputs are the coordinates and errors of the saddle circle points.

Even a coupler curve of the simplest four-bar linkage is a sextic algebraic curve. If the given planar motion is reproduced by a four-bar linkage, the path of a point on $\sum^*$ will correspond to a saddle circle, whose saddle circle error is a nonlinear function of the point's coordinates, a 12th-order piecewise algebraic equation. Therefore, there must be some extreme values of the saddle circle errors with respect to the points of $\sum^*$. Based on Definitions 2.1 and 2.2:

Theorem 2.1 *There exist several saddle circle points on the moving body in non-degenerate planar motion.*

As mentioned above, the properties of saddle circle and saddle circle point, such as adaptivity, minimum, uniqueness, and comparability, are described by Definitions 2.1 and 2.2. According to Theorem 2.1, the optimal solutions, or the saddle circle points, certainly exist on $\sum^*$. All of these provide a theoretical basis for the existence of optimum solutions and the convergence of the effective algorithm in saddle point programming.

As mentioned in Section 2.1, a coupler curve of a planar hinged four-bar linkage is a sextic algebraic curve. This means that a planar sextic algebraic curve can be reproduced by a planar RRRR four-bar linkage [4]. Different points on $\sum^*$ trace different paths. These paths change their shape gradually and continuously from one to another. The saddle circle errors corresponding to these paths are nonlinear functions of the locations of the points. There may be a few saddle circle points on $\sum^*$. Two saddle circle points have to be found for a planar hinged four-bar linkage. Of course, the values of the saddle circle point errors might not be zero, depending on the properties of planar movement, or the given movement may not be theoretically reproduced by a planar hinged four-bar linkage. We expect to prove Theorem 2.1 in terms of the nonlinear function of saddle circle error, but this can be completed by the reader.

The saddle circle point error is usually not zero for multiple given separated positions of $\sum^*$. There exist circle points when the saddle circle point errors happen to be zero in the case of five given separated positions, which are the special cases of the saddle circle points and can be located by the above mathematical model of saddle point programming. The circle points usually appear for four given separated positions of $\sum^*$, or on a curve of $\sum^*$ (the well-known Burmester curve). Of course, this can also be calculated easily by the above mathematical model of saddle point programming. Some numeral examples are given in the following. We need do no more than focus on the approach of saddle circle fitting, although this is a classical topic for the circle points of $\sum^*$ with four and five given separated positions. The saddle circle point errors, are shown in Figs 2.6 and 2.7. From four, five to the multiple given separated positions of $\sum^*$, the circle point curve, the circle points, and the saddle circle points on $\sum^*$ are respectively revealed on their corresponding saddle circle error surfaces, which may be globally examined for their variations with different numbers of given positions. In particular, both the kinematic differential geometry and the discrete kinematic geometry, from planar, spherical, and spatial movement, are connected by the approach of saddle point programming, which may be studied from a new angle.

Example 2.2 A planar crank-rocker linkage *ABCD* with dimensions $l_1 = 15$, $l_2 = 21$, $l_3 = 25$, $l_4 = 30$ is shown in Fig. 2.11. Determine the saddle circle error surface and the saddle circle points.

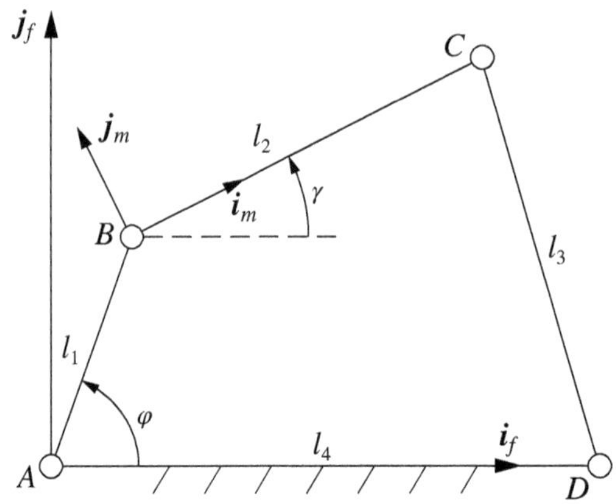

Figure 2.11 A planar crank-rocker linkage

Solution

The moving Cartesian coordinate system $\{\boldsymbol{B}; \boldsymbol{i}_m, \boldsymbol{j}_m\}$ is set up on the coupler link BC and the fixed system $\{\boldsymbol{A}; \boldsymbol{i}_f, \boldsymbol{j}_f\}$ is built on the base link AD, respectively. The rotation angle $\varphi \in [0, 2\pi]$ of the driving link AB is set in discrete positions with 5° step, which leads to the discrete positions $(x_B^{(i)}, y_B^{(i)}; \gamma^{(i)})$, $i = 1, \ldots, 72$ of the coupler BC by solving the displacement equations of the crank-rocker linkage.

For convenience, to reveal the distribution of the saddle circle error corresponding to the coupler points, 14 647 coupler points are taken with 0.5 step in the direction of both the $\boldsymbol{i}_m$ and the $\boldsymbol{j}_m$ axis in area $x_m \in [-20, 55]$, $y_m \in [-18, 30]$ on the coupler plane. Each coupler point has its discrete path $\{\boldsymbol{R}_P^{(i)}\}$ and corresponds to its saddle circle error. Then, we build up a saddle circle error surface in three-dimensional space, as shown in Fig. 2.12.

To represent this clearly, the error surface is mapped onto the coupler plane as several contour lines with the same value of the saddle circle error, as shown in Fig. 2.13. There are a few valleys on the surface or the contour – the special points at the bottom of the valleys, whose saddle circle errors have an extreme value (or minimum) at the corresponding coupler points, such as $P_1(0,0)$, $P_2(21,0)$, $P_3(21.2014,16.4933)$, and $P_4(41.2527, 1.7485)$. Two circle points P_1 and P_2 are the hinge points on the coupler plane, whose saddle circle errors are both zero. Meanwhile, two saddle circle points P_3 and P_4 have paths approximated by circles (they do not look like circles, but are surely the closest to circles for all coupler curves located in the same valley as for the crank-rocker linkage, shown in Fig. 2.14). The four coupler curves, Γ_1, Γ_2, Γ_3, and Γ_4, are traced by P_1, P_2, P_3, and P_4, respectively.

In the example, 50 initial points are generated on the coupler link of the crank rocker, and 50 solutions or saddle circle points are obtained by the sub-program ArrP. Some of them converge to the same point, shown in Fig. 2.13, in which 16 solutions coincide with the circle point $P_1(B)$, four solutions with the circle point $P_2(C)$, and others with the two saddle circle points P_3 and P_4. On the saddle circle error surface, there exist four obvious valleys. If the given initial points generated randomly just locate on the valley, the optimal results, saddle circle points, lie on the valley bottom. The average iterative step is 26 and the average optimization time is 4.2216 s.

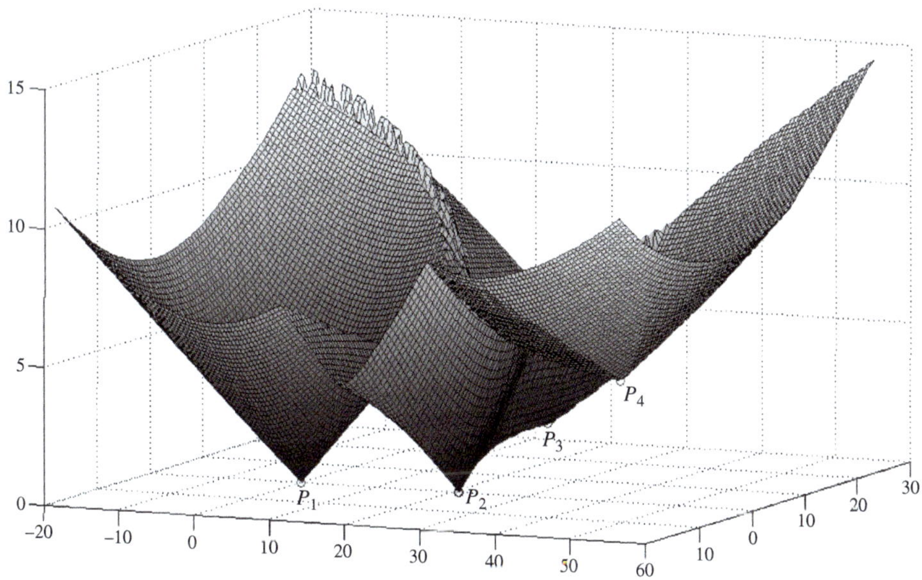

Figure 2.12 The surface of saddle circle errors for a crank-rocker linkage

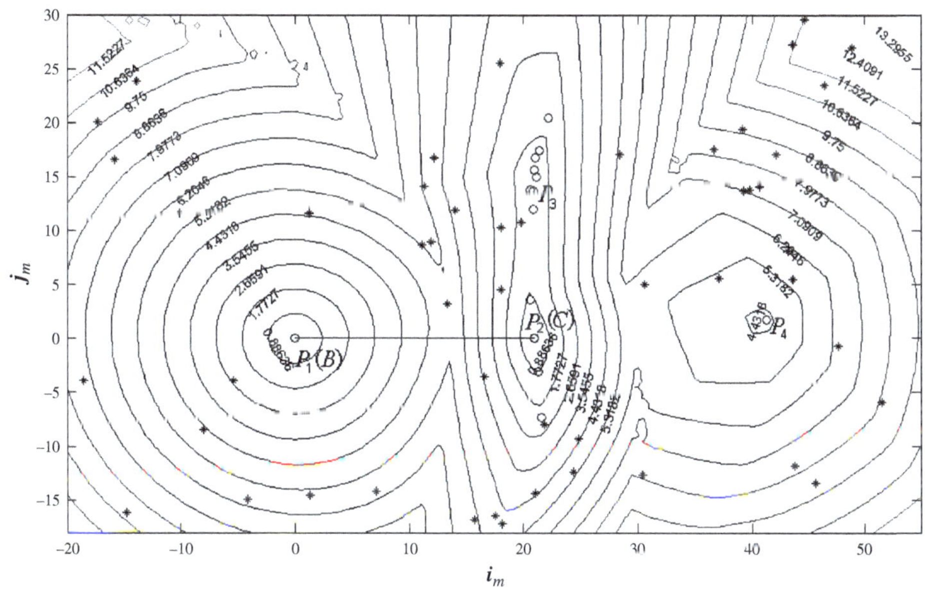

Figure 2.13 The saddle circle errors with contour lines

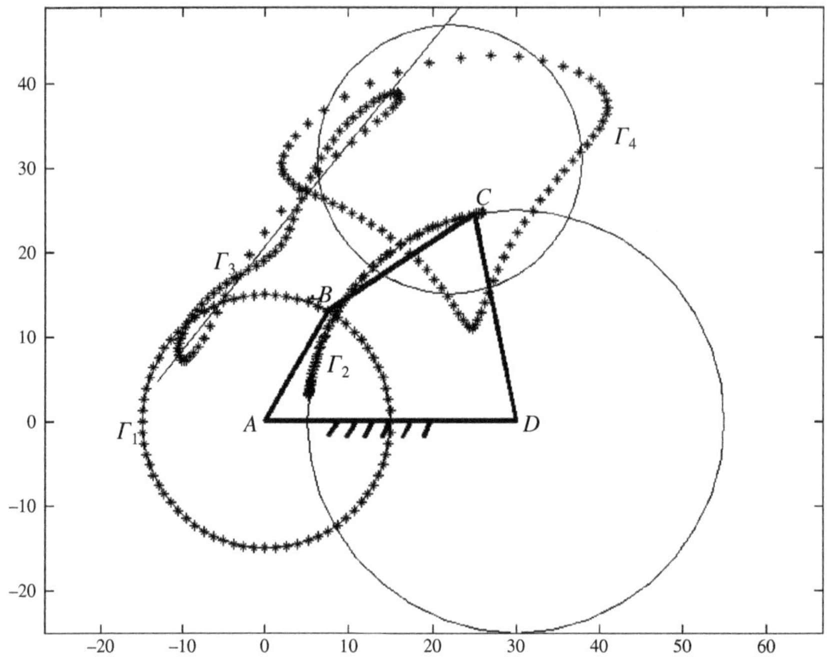

Figure 2.14 The coupler curves of four saddle circle points

Example 2.3 The motion parameters of 10 separated positions of a moving body in planar motion are listed in Table 2.1. Search out the saddle circle points on the moving body.

Solution

50 initial points are randomly generated in the area $x_m \in [-60.66, 60.66], y_m \in [-60.66, 60.66]$ and the corresponding results are obtained by calling the sub-program ArrP, in which 12 saddle circle points with less saddle circle errors are listed in Table 2.2, and numbered with initial letters SCP. The average iteration step is 27 and the average optimization time is 4.8813 s. It should be pointed out that theoretically only one saddle circle point appears at one bottom of the valley of the saddle circle error surface, but in reality several saddle circle points can be obtained for the convergence precision of optimization (see Fig. 2.15).

The saddle circle and the saddle circle error surface for the moving body, whose four or five separated positions are prescribed, have been discussed in Sections 2.3.3 and 2.3.4 and do not need to be repeated here.

2.4 Saddle Sliding Point

For the kinematic synthesis of linkages, a binary link P–R is frequently used to connect the coupler link with an R-pair and the base link with a P-pair for the simplest four-bar linkage, a crank-sliding linkage for instance. The open kinematic chain P–R is a binary link, whose

Table 2.2 The 12 saddle circle points for ten positions

No.	Saddle circle point		Corresponding saddle circle			
			Circle center		Radius	Saddle circle error
	x_{Pm}	y_{Pm}	x_C	y_C	r	δ_{rr}
SCP1	−1.9297	2.2955	16.1933	17.5949	9.0233	0.007458
SCP2	−7.7094	18.4328	30.8980	930.7898	907.3550	0.047845
SCP3	−6.8157	18.2960	30.4512	−712.7540	735.4073	0.093019
SCP4	−8.8986	18.6378	32.4868	310.4826	286.0223	0.101975
SCP5	−5.0106	18.0475	30.7175	−101.6760	122.8157	0.176732
SCP6	−3.8013	17.8970	30.8108	−56.9396	77.0659	0.209690
SCP7	−10.5343	18.7570	30.9165	233.9518	208.0649	0.224472
SCP8	−2.0020	17.8208	31.0452	−24.8049	43.5258	0.232643
SCP9	−0.4301	17.9223	31.2948	−12.0046	29.5739	0.238399
SCP10	−1.4648	17.8349	31.1075	−19.7058	38.0186	0.238029
SCP11	−1.3101	17.8426	31.1296	−18.3961	36.5938	0.238961
SCP12	−2.7296	1.2503	15.0881	18.2065	9.3539	0.368016

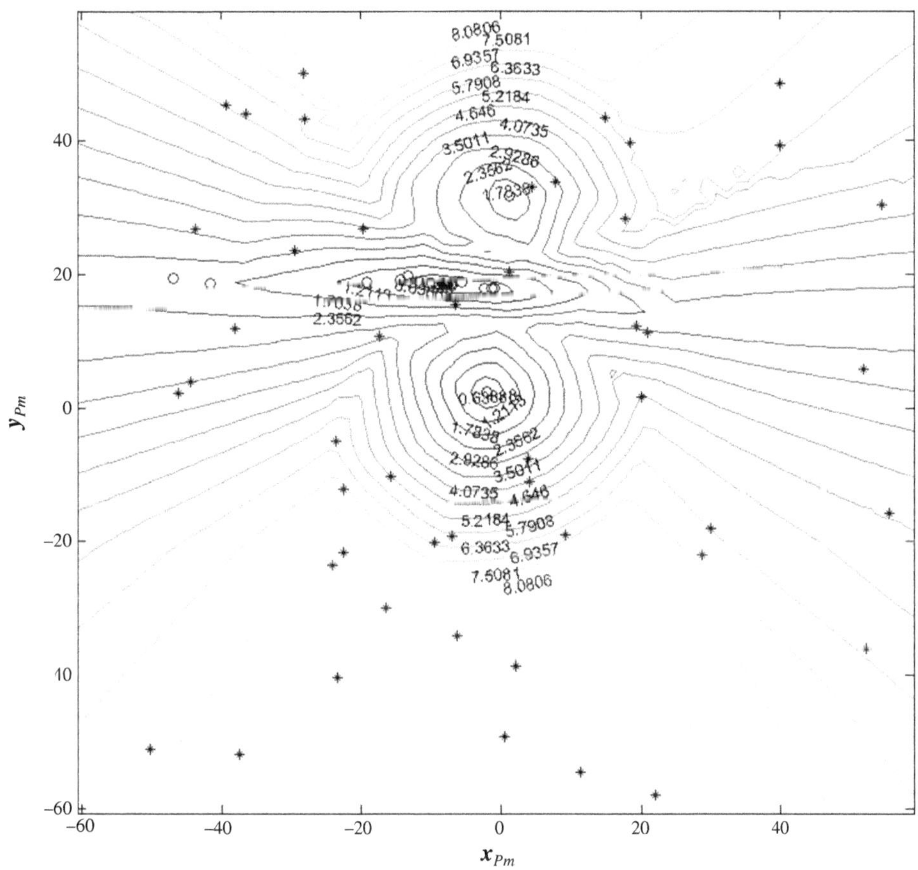

Figure 2.15 The 10-position saddle circle errors

constraint curve is a straight line, which is naturally another basic element in the discrete kinematic geometry of planar linkages. The points of $\sum^*$, whose discrete paths in the base body are globally compared with a straight line, are introduced as follows.

2.4.1 Saddle Line Fitting

For a binary link P–R of a crank-slider linkage, the center point of the moving joint R on the coupler plane coincides with that of the binary link P–R and is defined as a ***sliding point*** since its path in the base link is a line. The path, traced by the point of $\sum^*$, is represented by a discrete point set $\{\boldsymbol{R}_P^{(i)}\}$ in the fixed frame $\{\boldsymbol{O}_f; \boldsymbol{i}_f, \boldsymbol{j}_f\}$. A straight line is adaptively taken to fit the given discrete point set $\{\boldsymbol{R}_P^{(i)}\}$ to estimate the approximation of the line and the discrete path. Similar to a saddle circle, there must be a best fitting line, whose maximum fitting error is minimum.

Definition 2.3 For a given discrete point set $\{\boldsymbol{R}_P^{(i)}\}$, a line, adaptively determined by letting the maximum fitting error be minimum, is called a ***saddle line***, whose corresponding fitting error is defined as the ***saddle line error***.

Given Definition 2.3, we set up the mathematical model of a saddle line via saddle point programming as

$$\begin{cases} \Delta_{\text{pr}} = \min\limits_{\boldsymbol{x}} \max\limits_{1\le i\le n} \left\{\Delta^{(i)}(\boldsymbol{x})\right\} \\ \quad = \min\limits_{\boldsymbol{x}} \max\limits_{1\le i\le n} \left\{ \left| x_P^{(i)} \cos\phi + y_P^{(i)} \sin\phi - h \right| \right\} \\ \text{s.t.} \quad h \in [0, +\infty), \phi \in [0, 2\pi) \\ \boldsymbol{x} = (h, \phi)^T \end{cases} \tag{2.29}$$

where $\{\Delta^{(i)}(\boldsymbol{x})\}$ is the object function, the set of fitting errors for all points in $\{\boldsymbol{R}_P^{(i)}\}$ and the fitting line, Δ_{pr} is the saddle line error. Here, n is the number of discrete points $\{\boldsymbol{R}_P^{(i)}\}$; (h, ϕ) are the optimization variables, or the parameters of the fitting line, in which h is the distance from the original point O_f to the fitting line, or the perpendicular foot E, and ϕ is the inclination angle of the direction $\boldsymbol{l} = [-\sin\phi, \cos\phi]^T$ of the fitting line, as shown in Fig. 2.16.

The saddle line is the line closest to the discrete points set $\{\boldsymbol{R}_P^{(i)}\}$, since there is no other fitting line with less fitting error according to Definition 2.3. In order to solve the optimum

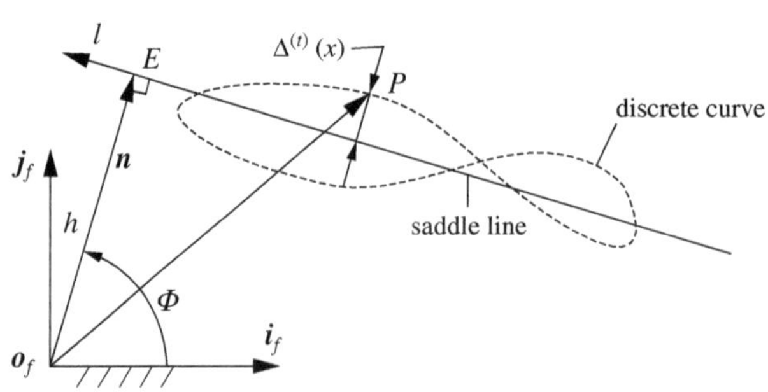

Figure 2.16 The saddle line fitting of a plane discrete curve

mathematical model (2.29), the initial values of the optimization variables (h, ϕ) have to be assigned first. Based on the least-square method, the initial values for a saddle line can be calculated directly and easily. The normal deviation e_i between the discrete point set $\{\boldsymbol{R}_P^{(i)}\}$ and a fitting line, whose vector equation is $\boldsymbol{R}_L = \boldsymbol{R}_E + \lambda \boldsymbol{l} = \{h\cos\phi, h\sin\phi\} + \lambda\{-\sin\phi, \cos\phi\}$, can be represented as

$$e_i = \left| x_P^{(i)} \cos\phi + y_P^{(i)} \sin\phi - h \right| \tag{2.30}$$

The quadratic sum of the normal deviation e_i can be obtained as

$$F = \sum_{i=1}^{n} e_i^2 = \sum_{i=1}^{n} (x_P^{(i)} \cos\phi + y_P^{(i)} \sin\phi - h)^2 \tag{2.31}$$

Differentiating the above equation with respect to ϕ and h, and letting them be zero, we have

$$\begin{cases} \dfrac{\partial F}{\partial \phi} = 2\sum_{i=1}^{n} (x_P^{(i)} \cos\phi + y_P^{(i)} \sin\phi - h) \cdot (-x_P^{(i)} \sin\phi + y_P^{(i)} \cos\phi) = 0 \\ \dfrac{\partial F}{\partial h} = -2\sum_{i=1}^{n} (x_P^{(i)} \cos\phi + y_P^{(i)} \sin\phi - h) = 0 \end{cases} \tag{2.32}$$

The parameters ϕ and h of a fitting line can be calculated by the above equation, or

$$\begin{cases} \cot\phi = -\dfrac{n\sum_{i=1}^{n} x_P^{(i)} y_P^{(i)} - \sum_{i=1}^{n} x_P^{(i)} \sum_{i=1}^{n} y_P^{(i)}}{n\sum_{i=1}^{n} (x_P^{(i)})^2 - \left(\sum_{i=1}^{n} x_P^{(i)}\right)^2} \\ h = \sin\phi \cdot \dfrac{\sum_{i=1}^{n} (x_P^{(i)})^2 \sum_{i=1}^{n} y_P^{(i)} - \sum_{i=1}^{n} x_P^{(i)} \sum_{i=1}^{n} x_P^{(i)} y_P^{(i)}}{n\sum_{i=1}^{n} (x_P^{(i)})^2 - \left(\sum_{i=1}^{n} x_P^{(i)}\right)^2} \end{cases} \tag{2.33}$$

For a given discrete point set $\{\boldsymbol{R}_P^{(i)}\}$ in $\{\boldsymbol{A}; \boldsymbol{i}_f, \boldsymbol{j}_f\}$, equation (2.29), the mathematical model of a saddle line, can be solved by function **fminimax** in Matlab, whose initial values are provided by equation (2.33); the parameters of the function **fminimax** are set in the same way as those of a saddle circle. Like the sub-program ArrF, a ***sub-program AprF*** is constructed for the saddle line fitting of a discrete point set $\{\boldsymbol{R}_P^{(i)}\}$, $i = 1, \ldots, n$. The inputs are the coordinates of the discrete points and the outputs are the parameters (h, ϕ) of the saddle line and the saddle line error Δ_{pr}.

2.4.2 *Saddle Line*

Based on Definition 2.3, a saddle line is located for a given discrete path Γ_P in $\{\boldsymbol{O}_f; \boldsymbol{i}_f, \boldsymbol{j}_f\}$, whose saddle line error Δ_{pr} is the maxima for all the points on Γ_P, and whose maximal error Δ_{pr} is the smallest for all the fitting lines.

For convenience, in discussing the properties of the saddle line error, the given path Γ_P is assumed to be a continuous curve. As shown in Fig. 2.16, the vector equation of the saddle line L is $\boldsymbol{R}_L = \boldsymbol{R}_E + \lambda \boldsymbol{l}$, and $\boldsymbol{R}_E$ is the vector of the reference point of L and $\boldsymbol{l}$ is the unit direction vector. The normal $\boldsymbol{n}$ of the saddle line L is perpendicular to $\boldsymbol{l}$. The path Γ_P has a normal distance Δ to the saddle line L at a point P along the normal vector $\boldsymbol{n}$. The position vector of P can be written as

$$\boldsymbol{R}_P = \boldsymbol{R}_E + \lambda \boldsymbol{l} + \Delta \boldsymbol{n} \tag{2.34}$$

The normal error Δ between Γ_P and the saddle line L is

$$\Delta = (\boldsymbol{R}_P - \boldsymbol{R}_E) \cdot \boldsymbol{n} \tag{2.35}$$

The position vector $\boldsymbol{R}_E$ and the normal vector $\boldsymbol{n}$ of the saddle line L are both constant vectors for the given path Γ_P. Differentiating equation (2.35) with respect to the arc length s of Γ_P, we have

$$\frac{\mathrm{d}\Delta}{\mathrm{d}s} = \frac{\mathrm{d}\boldsymbol{R}_P}{\mathrm{d}s} \cdot \boldsymbol{n} \tag{2.36}$$

The condition for the fitting error Δ to be an extreme value is $\frac{\mathrm{d}\Delta}{\mathrm{d}s} = 0$ at P of Γ_P. By equation (2.36), we have

$$\frac{\mathrm{d}\boldsymbol{R}_P}{\mathrm{d}s} \cdot \boldsymbol{n} = 0 \tag{2.37}$$

This means that when the tangent vector $\frac{\mathrm{d}\boldsymbol{R}_P}{\mathrm{d}s}$ of Γ_P at P is orthogonal to the normal vector $\boldsymbol{n}$ of the saddle line L, or the normal of Γ_P coincides with that of the saddle line, the normal fitting error Δ obtains an extreme value. The extreme values may appear at multiple points of Γ_P but may not be the maximum, such as the points $P^{(k+1)}$, $P^{(k+2)}$, and $P^{(k+3)}$ in Fig. 2.17; the saddle line error Δ_{pr} is only the absolute maxima of all points.

For a given plane curve Γ_P, at least three points on Γ_P should hold equation (2.37) true to determine the two parameters (h, ϕ) of the saddle line and the saddle line error Δ_{pr} of Γ_P. Substituting the position vectors of the three points into equation (2.35), we can solve (h, ϕ) and Δ_{pr}.

Based on saddle point programming [3], the degree of approximation between a given plane curve and the saddle line is estimated by the saddle line fitting, or by letting the maximum

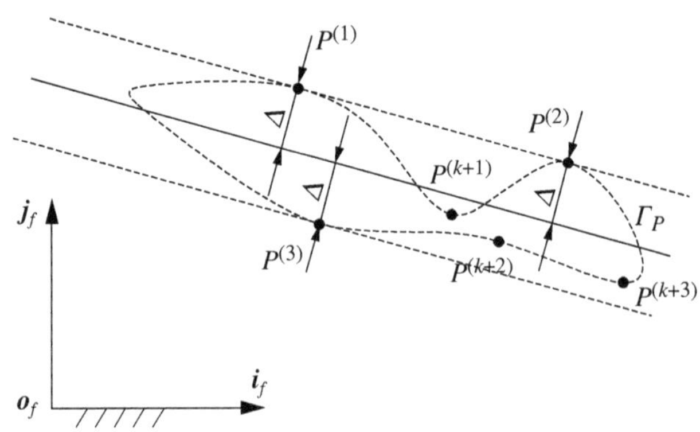

Figure 2.17 The two parallel lines and the normal fitting errors

fitting error be the minima, which is equivalent to the given curve contained by two parallel straight lines. The given plane curve may have at least three tangent points with the two parallel straight lines, leading to the maximum normal fitting errors being the minima with the same value, as shown in Fig. 2.17. The three tangent points may distribute on both sides of the saddle line.

The three tangent points on Γ_P can geometrically determine the saddle line of Γ_P, defined as the ***saddle line fitting points*** on Γ_P. The corresponding three positions of $\sum^*$ are referred to as the ***saddle line fitting positions***. The normal errors at three saddle line fitting points have the same value, just the saddle line error. If the saddle line error is zero, the fitted plane curve must be a straight line.

If the fitted plane curve Γ_P is discretized into a discrete point set $\{\boldsymbol{R}_P^{(i)}\}$, three saddle line fitting points should be selected from $\{\boldsymbol{R}_P^{(i)}\}$ to determine the saddle line. By equation (2.35), the normal fitting error $\Delta^{(i)}$ between a point of $\{\boldsymbol{R}_P^{(i)}\}$ and the saddle line can be written as

$$\Delta^{(i)} = (\boldsymbol{R}_P^{(i)} - \boldsymbol{R}_E) \cdot \boldsymbol{n}, \boldsymbol{R}_P^{(i)} = [\mathbf{M}^{(i)}] \cdot \boldsymbol{R}_{Pm} \tag{2.38}$$

The saddle line error Δ_{pr} is the absolute maximum value of all, or

$$\Delta_{\text{pr}} = \max\{|\Delta^{(i)}|\} = \max\left\{\left|x_P^{(i)}\cos\phi + y_P^{(i)}\sin\phi - h\right|\right\} \tag{2.39}$$

The saddle line error of a discrete point set $\{\boldsymbol{R}_P^{(i)}\}$ can be obtained by numerical calculation and comparison among all the normal errors $\Delta^{(i)}$ of $\{\boldsymbol{R}_P^{(i)}\}$ by equation (2.39).

The relationships among the separated positions of $\sum^*$, the discrete path, and the saddle line can be established by the mathematical model of saddle line fitting.

2.4.3 *Three Positions*

In order to compare a plane discrete point set and a straight line in the global geometrical property, we start with the three separated positions of $\sum^*$, the three fitting positions for the saddle line fitting model. Three separated positions of $\sum^*$ in planar movement are prescribed in $\{\boldsymbol{O}_f; \boldsymbol{i}_f, \boldsymbol{j}_f\}$ and numbered 1, 2, 3; a point $P(x_{Pm}, y_{Pm})$ of $\sum^*$ traces a discrete point set $\{\boldsymbol{R}_P^{(i)}\}$, $i = 1, 2, 3$ in $\{\boldsymbol{O}_f; \boldsymbol{i}_f, \boldsymbol{j}_f\}$, comprised of three points $P^{(1)}$, $P^{(2)}$, and $P^{(3)}$. Based on equation (2.29) of the saddle line fitting model, we call the sub-program AprF and obtain the saddle line of $\{\boldsymbol{R}_P^{(i)}\}$, $i = 1, 2, 3$, referred to as the ***three-point saddle line***. In this case, the three points $P^{(1)}$, $P^{(2)}$, and $P^{(3)}$ are the saddle line fitting points and the three positions of $\sum^*$ are the saddle line fitting positions.

A discrete path traced by a point of $\sum^*$ with three separated positions has three points, and corresponds to a three-point saddle line. According to saddle point programming, the maximum fitting errors at three fitted points obtain the minimum if and only if the fitting errors are the same as an extreme value and the three fitted points $P^{(1)}$, $P^{(2)}$, and $P^{(3)}$ distribute on both sides of the three-point saddle line. There are three distribution cases in total, $P^{(1)}P^{(2)}-P^{(3)}$, $P^{(1)}P^{(3)}-P^{(2)}$, and $P^{(2)}P^{(3)}-P^{(1)}$, and the corresponding three fitting lines, $L_{12,3}$, $L_{13,2}$, and $L_{23,1}$, called the ***distributing lines*** of the three points, are shown in Fig. 2.18. The distributing line with the smallest fitting error, such as $L_{23,1}$ in Fig. 2.18, is the three-point saddle line. The algebraic equation of the distributing line error is discussed in the following.

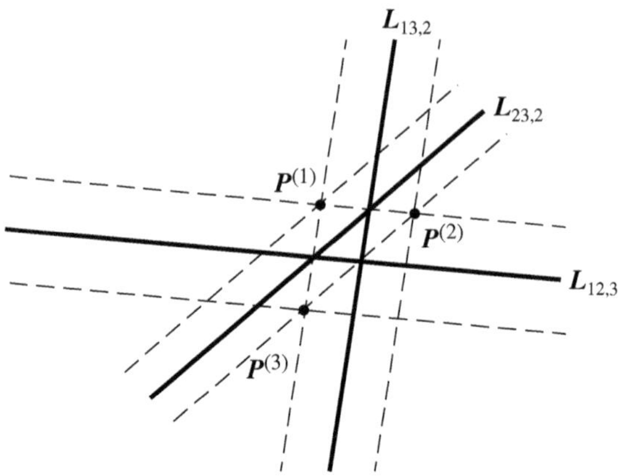

Figure 2.18 Three distributing lines of three points

For the distribution case $P^{(1)}P^{(2)}-P^{(3)}$, two points $P^{(1)}$ and $P^{(2)}$ locate on one side and point $P^{(3)}$ on the other side of $L_{12,3}$. The algebraic equation of line $P^{(1)}P^{(2)}$ is

$$(x_P^{(2)} - x_P^{(1)})y - (y_P^{(2)} - y_P^{(1)})x + y_P^{(2)}x_P^{(1)} - x_P^{(2)}y_P^{(1)} = 0 \tag{2.40}$$

The distance from point $P^{(3)}$ to the line $P^{(1)}P^{(2)}$ is

$$d = \frac{\left|(x_P^{(2)} - x_P^{(1)})y_P^{(3)} - (y_P^{(2)} - y_P^{(1)})x_P^{(3)} + y_P^{(2)}x_P^{(1)} - x_P^{(2)}y_P^{(1)}\right|}{\sqrt{(x_P^{(2)} - x_P^{(1)})^2 + (y_P^{(2)} - y_P^{(1)})^2}} \tag{2.41}$$

Since the three points $P^{(1)}$, $P^{(2)}$, and $P^{(3)}$ have equal distance to the distributing line $L_{12,3}$, the fitting error $\Delta_{12,3}$ of $L_{12,3}$ is

$$\Delta_{12,3} = \frac{|(x_P^{(2)} - x_P^{(1)})y_P^{(3)} - (y_P^{(2)} - y_P^{(1)})x_P^{(3)} + y_P^{(2)}x_P^{(1)} - x_P^{(2)}y_P^{(1)}|}{2\sqrt{(x_P^{(2)} - x_P^{(1)})^2 + (y_P^{(2)} - y_P^{(1)})^2}} \tag{2.42}$$

By equation (2.4), substituting the coordinates of the three points $P^{(1)}$, $P^{(2)}$, and $P^{(3)}$ in $\{\boldsymbol{O}_f; \boldsymbol{i}_f, \boldsymbol{j}_f\}$ into the above equation, we have the quartic algebraic equation of $(x_{Pm}, y_{Pm}, \Delta_{12,3})$ as

$$\begin{aligned} &4(a_1x_{Pm}^2 + a_1y_{Pm}^2 + a_2x_{Pm} + a_3y_{Pm} + a_4)\Delta_{12,3}^2 \\ &-(a_5x_{Pm}^2 + a_5y_{Pm}^2 + a_6x_{Pm} + a_7y_{Pm} + a_8)^2 = 0 \end{aligned} \tag{2.43}$$

where the coefficients a_i, $i = 1, 2, \ldots, 8$ are functions of the position parameters $(x_{Omf}^{(i)}, y_{Omf}^{(i)}, \gamma^{(i)})$, $i = 1, 2, 3$. In a similar way, the distributing line errors $\Delta_{23,1}$ of $L_{23,1}$ and $\Delta_{13,2}$ of $L_{13,2}$ can be obtained. According to Definition 2.3, the three-point saddle line error Δ_{123} of $\{\boldsymbol{R}_P^{(i)}\}$, $i = 1, 2, 3$ is the minima of three distributing line errors, or

$$\Delta_{123} = \min(\Delta_{12,3}, \Delta_{13,2}, \Delta_{23,1}) \tag{2.44a}$$

Another form is

$$\begin{aligned}\Delta_{123} = \frac{1}{2}\left|\left|\frac{1}{2}\left|\Delta_{23,1}+\Delta_{13,2}\right|-\frac{1}{2}\left|\Delta_{23,1}-\Delta_{13,2}\right|\right|+\Delta_{12,3}\right| \\ -\frac{1}{2}\left|\left|\frac{1}{2}\left|\Delta_{23,1}+\Delta_{13,2}\right|-\frac{1}{2}\left|\Delta_{12,1}-\Delta_{13,2}\right|\right|-\Delta_{12,3}\right|\end{aligned} \tag{2.44b}$$

The above equation indicates the saddle line error of the discrete point set $\{\boldsymbol{R}_P^{(i)}\}$, $i = 1, 2, 3$, traced by the point P on $\sum^*$ with three given separated positions. Taking the coordinates (x_{Pm}, y_{Pm}) of the point of $\sum^*$ as the independent variables and the three-point saddle line error Δ_{123} as the dependent variable, or function value, we have the error surface of three-point saddle line, which is mapped onto a plane. All points with the same saddle line errors are connected with the contour lines, as shown in Fig. 2.19, with motion parameters of the three positions from the first three lines of Table 2.1.

In particular, if the saddle line error is zero, we obtain an intersection curve between the error surface and the plane defined by $\Delta_{123} = 0$, the ***sliding point circle***, shown in Fig. 2.19 with the heavy line.

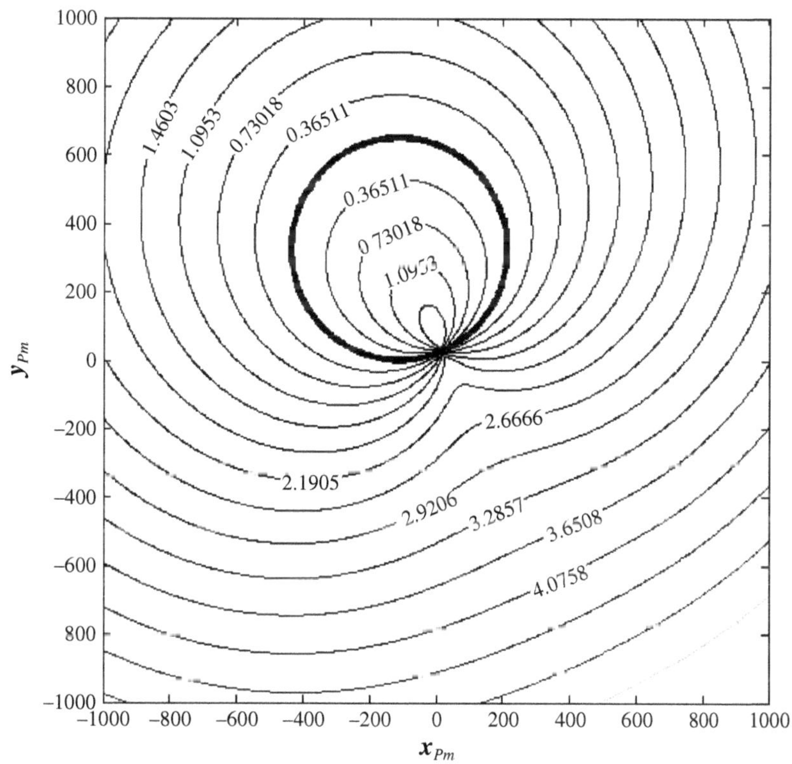

Figure 2.19 The three-position saddle line errors

2.4.4 Four Positions

For the four separated positions of $\sum^*$ numbered 1, 2, 3, 4, a point P of $\sum^*$ traces a discrete point set $\{\boldsymbol{R}_P^{(i)}\}$, $i = 1, 2, 3, 4$, comprised of four points $P^{(1)}$, $P^{(2)}$, $P^{(3)}$, and $P^{(4)}$. Based on equation (2.29) of the saddle line fitting model, we call the sub-program AprF and obtain the saddle line of $\{\boldsymbol{R}_P^{(i)}\}$, $i = 1, 2, 3, 4$, referred to as the ***four-point saddle line***. Similar to the three positions, the error surface of four-point saddle line can be constructed and mapped onto a plane. All points with the same value are connected with contour lines, as shown in Fig. 2.20, with motion parameters of the four positions from the first four lines of Table 2.1.

As previously mentioned, a saddle line of the four points $P^{(1)}$–$P^{(4)}$ is located by the three saddle line fitting points, which correspond to the three saddle line fitting positions. The saddle line of the three points $P^{(1)}$–$P^{(3)}$ has been introduced previously. If another point $P^{(4)}$ is added, we discuss whether it is the saddle line fitting point of the four points in two cases.

(1) The fourth point $P^{(4)}$ is not the saddle line fitting point, and position 4 is not the fitting position. If the fitting error between the point $P^{(4)}$ and the three-point saddle line L_{123} is less than the saddle line error Δ_{123}, the saddle line error Δ_{1234} of the four points $P^{(1)}$–$P^{(4)}$ is determined by the saddle line fitting points $P^{(1)}$–$P^{(3)}$, or $\Delta_{1234} = \Delta_{123}$, as shown in Fig. 2.21. The points of $\sum^*$ with four given separated positions, whose paths have three discrete points ($P^{(1)}$–$P^{(3)}$) as the saddle line fitting points, distribute on the specific area of $\sum^*$ called the ***three-position area*** (123) of $\sum^*$, for example area I in Fig. 2.20. There is also another situation

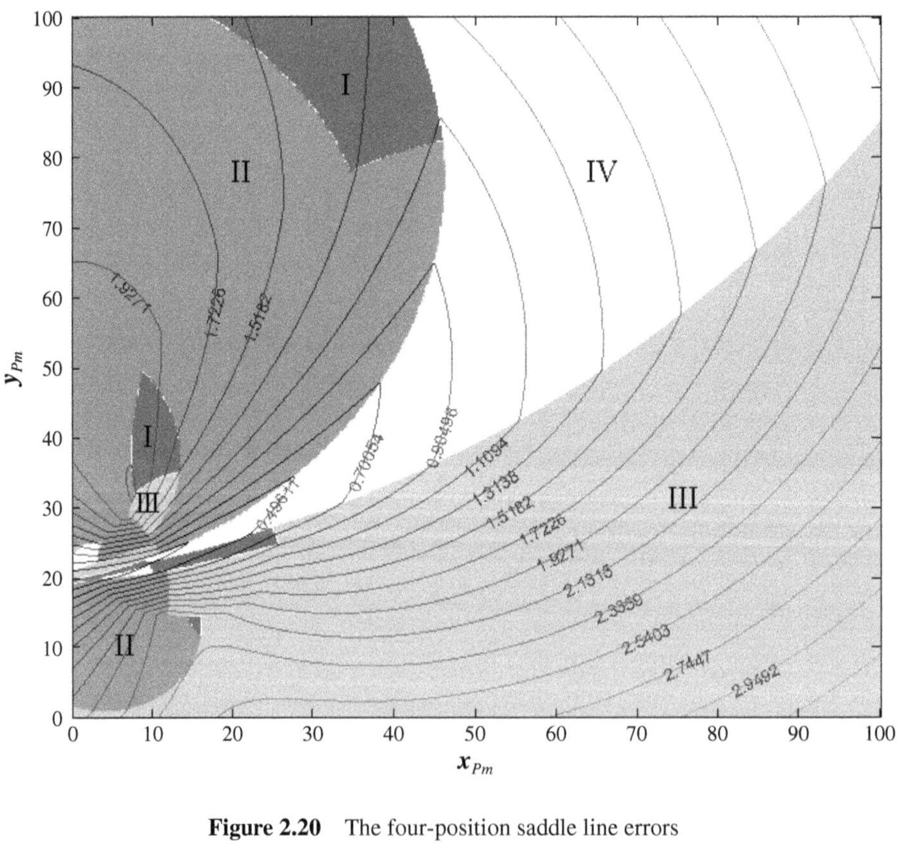

Figure 2.20 The four-position saddle line errors

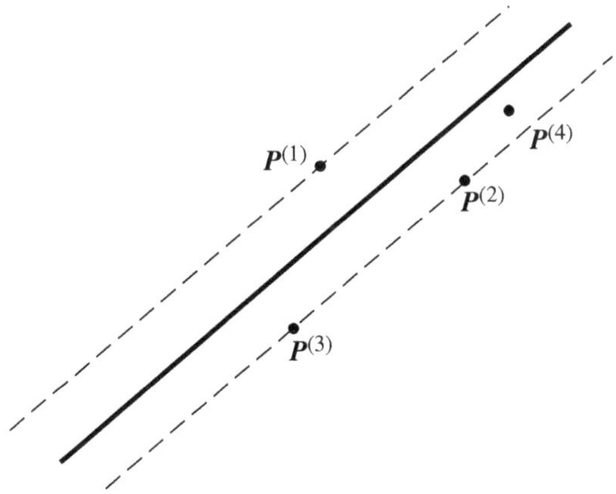

Figure 2.21 The three-point saddle lines and the fourth point

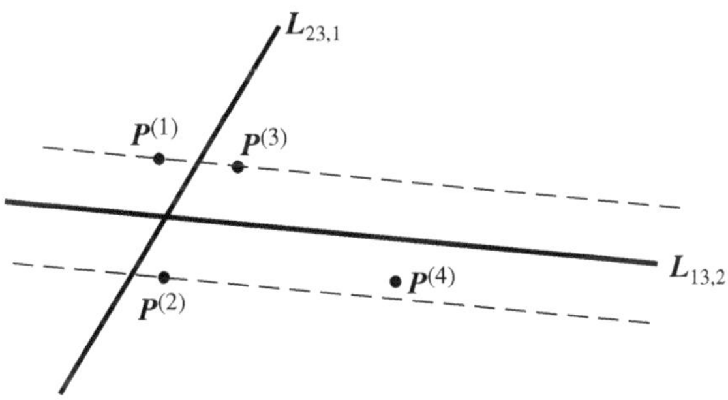

Figure 2.22 The four-point saddle line is the three point distributing line

for the points on the three-position area (123). Although the three points $P^{(1)}$–$P^{(3)}$ are the saddle line fitting points, the saddle line L_{1234} of $P^{(1)}$–$P^{(4)}$ is not the saddle line of $P^{(1)}$–$P^{(3)}$, as shown in Fig. 2.22. The saddle line of $P^{(1)}$–$P^{(3)}$ is the line $L_{23,1}$ in the distribution case $P^{(2)}P^{(3)}$–$P^{(1)}$, or $\Delta_{123} = \Delta_{23,1}$. The saddle line of $P^{(1)}$–$P^{(4)}$ is the line $L_{13,2}$, which is located by the points $P^{(1)}$–$P^{(3)}$ in the distribution case $P^{(1)}P^{(3)}$–$P^{(2)}$, or a distributing line.

(2) The fourth point $P^{(4)}$ is the saddle line fitting point, and position 4 is the fitting position. In this case, both position 4 and two of the three positions 123 may become the saddle line fitting positions, and there exist three possibilities: 124, 134, and 234. Each possible three-position combination has three distributing lines, for example the positions 124 have three distributing lines with error $\Delta_{12,4}$, $\Delta_{14,2}$, and $\Delta_{24,1}$; there exist nine distributing lines for all three possible combinations. The saddle line of the four points $P^{(1)}$–$P^{(4)}$ is one of the nine distributing lines.

For $\sum^*$ with four given separated positions, each point of $\sum^*$ corresponds to a four-point saddle line error, and all points of $\sum^*$ and their four-point saddle line errors correspond to an error surface, as shown in Fig. 2.20. This quartic algebraic surface comprised of several distributing line error surfaces can be divided into several three-position areas, for example area I corresponds to the positions 123, area II to 124, area III to 134, and area IV to 234 in Fig. 2.20. The algebraic equations of the boundaries of the different four-position areas can be derived through equations (2.42) and (2.44). The point locating on the boundaries has the characteristics of the adjacent areas, belonging to both areas, and the four-point saddle line of its path corresponds to two three-point distributing lines with the same errors. If the two three-point distributing lines are coincident with each other, a new characteristic on the boundary of the error surface would appear, such as the Ball point on the rigid body.

As previously mentioned, the error surface of four-position saddle line is theoretically comprised of several error surfaces of three-position distributing lines, and is a quartic-order algebraic surface with many peak valleys. If we compare the error surface of three-position saddle line with that of four-position saddle line, there is a bottom of a trench (sliding point circle for three positions) against a valley bottom (sliding point for four positions).

2.4.5 Multiple Positions

For $\sum^*$ with multiple separated positions, the discrete path, traced by a point of $\sum^*$, is comprised of multiple discrete points. Based on saddle point programming, the saddle line of the discrete path is still located by the three saddle line fitting points, no matter how many points the path has. For all points on $\sum^*$, the saddle lines can be determined in the same way as in the four-position case. Therefore, no matter how many separated positions of $\sum^*$ are given, it can also be viewed as a combination of three positions. There are C_n^3 combinations of three positions for n given positions, and there exist more three-position areas on $\sum^*$. For the 10 separated positions listed in Table 2.1, the contour lines of the saddle line error surface with the three-position areas are shown in Fig. 2.23.

The more separated positions of $\sum^*$ are given, the smaller are the three-position areas of the saddle line. When the multiply separated positions of $\sum^*$ tend to infinity, a discrete path, traced by a point of $\sum^*$, would approximate a continuous curve, whose saddle line is still located by the three fitting points. These fitting points may change continuously as the tracing point varies continuously on the rigid body.

2.4.6 Saddle Sliding Point

The mathematical model of the saddle point program can be applied to line fitting for the path with discrete points $\{\boldsymbol{R}_P^{(i)}\}$, traced by a point of $\sum^*$. All paths have their corresponding saddle lines, while the saddle line errors are a function of the tracing point coordinates on $\sum^*$, described by a quartic algebraic equation. The saddle line error varies as the shapes of the paths change and certainly has a few extreme values at some points.

Definition 2.4 A point of a moving body in planar movement, whose path in the fixed frame corresponds to the saddle line, having minimum saddle line error with respect to other tracing points in its neighborhood, is defined as a ***saddle sliding point***, and the corresponding saddle line error is called the ***saddle sliding point error***.

The saddle sliding point is a sliding point if the saddle sliding point error is zero. The saddle sliding point has similar properties to the saddle circle point, its mathematical model is

$$\begin{cases} \delta_{\text{pr}} = \min \Delta_{\text{pr}}(z) \\ z = (x_{Pm}, y_{Pm})^T \end{cases} \tag{2.45}$$

where $\Delta_{\text{pr}}(z)$ is the objective function, or the saddle line error for a tracing point on $\sum^*$, the optimization variables $z = (x_{Pm}, y_{Pm})^T$ are the coordinates of the tracing point. Similar to the

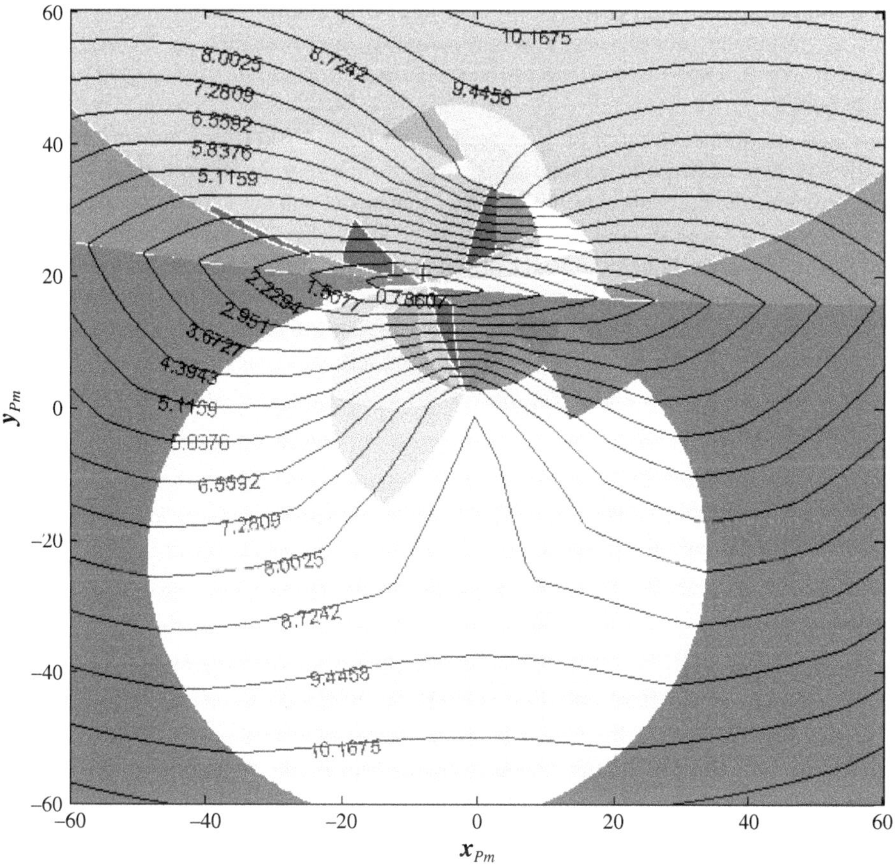

Figure 2.23 The 10-position saddle line errors

optimization model for saddle circle point, we should assign a number of initial values for that of saddle sliding point.

The mathematical model of the saddle sliding point, equation (2.45), can be solved directly by the function **fmincon** in Matlab, whose parameters can be set by referring to the saddle circle points. Similar to the sub-program ArrP, a ***sub-program AprP*** is constructed for the location of saddle sliding points on $\sum^*$ with multiple separated positions. The inputs are the parameters of the prescribed separated positions and the outputs are the parameters and errors of the saddle sliding points.

For all points of $\sum^*$, whose paths correspond to saddle lines and their errors, the saddle line error is a nonlinear function of the tracing point's coordinates, a quartic algebraic equation. Therefore, the saddle line errors must have extreme values at some tracing points of $\sum^*$. Based on Definitions 2.3 and 2.4:

Theorem 2.2 *There exist several saddle sliding points on the moving body in non-degenerate planar motion.*

As mentioned above, the properties of the saddle line and saddle sliding point, such as adaptivity, minimum, uniqueness, and comparability, are described by Definitions 2.3 and 2.4. According to Theorem 2.2, the optimal solutions, the saddle sliding points, exist on a moving body.

For planar motion, each point of $\sum^*$ traces a path in the fixed frame, which corresponds to a saddle line error. The paths change their shape from one to another as their tracing points vary in position, and their corresponding saddle line errors will vary simultaneously to become larger or smaller (of course, taking an extreme value somewhere). Thus, there may be a few saddle sliding points even though they have different values of saddle line errors in the local area. One saddle circle point and one saddle sliding point have to been searched out if a crank-slider linkage is to be synthesized. Needless to say, the values of fitting errors, for both a saddle circle point and a saddle sliding point, depend on the properties of planar motion of $\sum^*$, possibly reproducing the given movement with least error if it cannot be theoretically generated by a crank-slider linkage.

Example 2.4 A crank-slider linkage with dimensions $l_1 = 5$, $l_2 = 15$, $e = 2.5$ is shown in Fig. 2.24. Calculate the saddle line error surface.

Solution

The moving frame $\{\boldsymbol{B}; \boldsymbol{i}_m, \boldsymbol{j}_m\}$ is established on the coupler link BC and the fixed frame $\{\boldsymbol{A}; \boldsymbol{i}_f, \boldsymbol{j}_f\}$ is built on the base link. The rotation angle $\varphi \in [0, 2\pi]$ of the driving link AB is discretized with 5° step. The discrete positions $(x_B^{(i)}, y_B^{(i)}; \gamma^{(i)})$, $i = 1, \ldots, 72$ of the coupler link relative to the base link are obtained by solving the displacement equations of the crank-slider linkage.

In order to show the outline of the saddle line errors with respect to the positions of the coupler, 6161 coupler points are selected in the area $x_m \in [-20, 40]$, $y_m \in [-50, 50]$ of the coupler link with interval step 1; a surface of the saddle line errors is built up in three-dimensional space (Fig. 2.25) and mapped onto the coupler plane. All points with the same values are connected by contour lines, as shown in Fig. 2.26. It is obviously noted that there is only one global minimum point, corresponding to the hinge point $C(15,0)$ in the coupler plane.

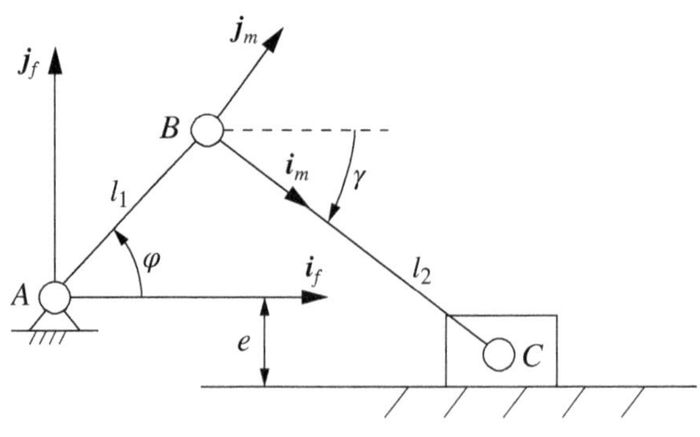

Figure 2.24 A planar crank-slider linkage

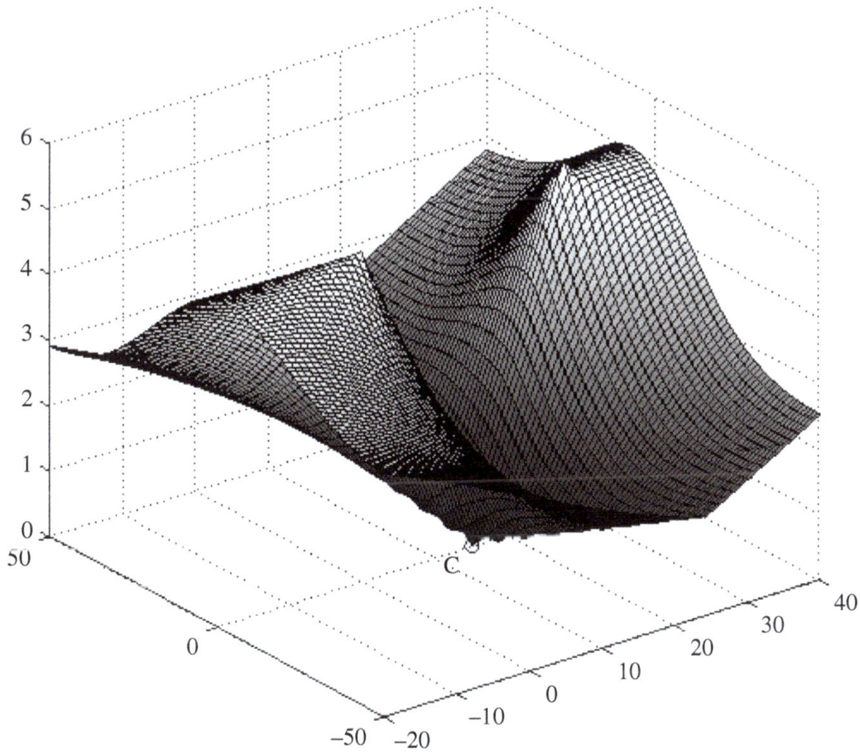

Figure 2.25 The saddle line error surface of a crank-slider linkage

30 initial points are randomly generated by Matlab, and the results, the saddle sliding points, are obtained by calling the sub-program AprP, shown in Fig. 2.26, in which 29 saddle sliding points converge to the sliding point C since there exists only one valley on the surface. The average iterative step is 32 and the average optimization time is 3.725 s.

Example 2.5 The parameters of 10 separated positions of a planar moving body are listed in Table 2.1. Search out the saddle sliding points.

Solution

Based on the optimization model (2.45) of the saddle sliding point, 30 initial points are randomly generated by Matlab in the area $x_m \in [-25, 20]$, $y_m \in [-10, 35]$ of the moving body. The results can be obtained by calling the sub-program AprP. All saddle sliding points converge to the bottom of the valley since there exists only one valley for the saddle line error surface, as shown in Fig. 2.27. The saddle sliding point has rectangular coordinates (−7.5887,18.4669), whose corresponding saddle line has parameters $\phi = 1.5710$, $h = 23.3564$; the saddle line error is 0.056591. The average iterative step is 32 and the average optimization time is 3.8822 s.

For a moving body with three or four separated positions, the saddle lines and the saddle line error surfaces have been shown in Sections 2.4.3 and 2.4.4, and do not need to be repeated here.

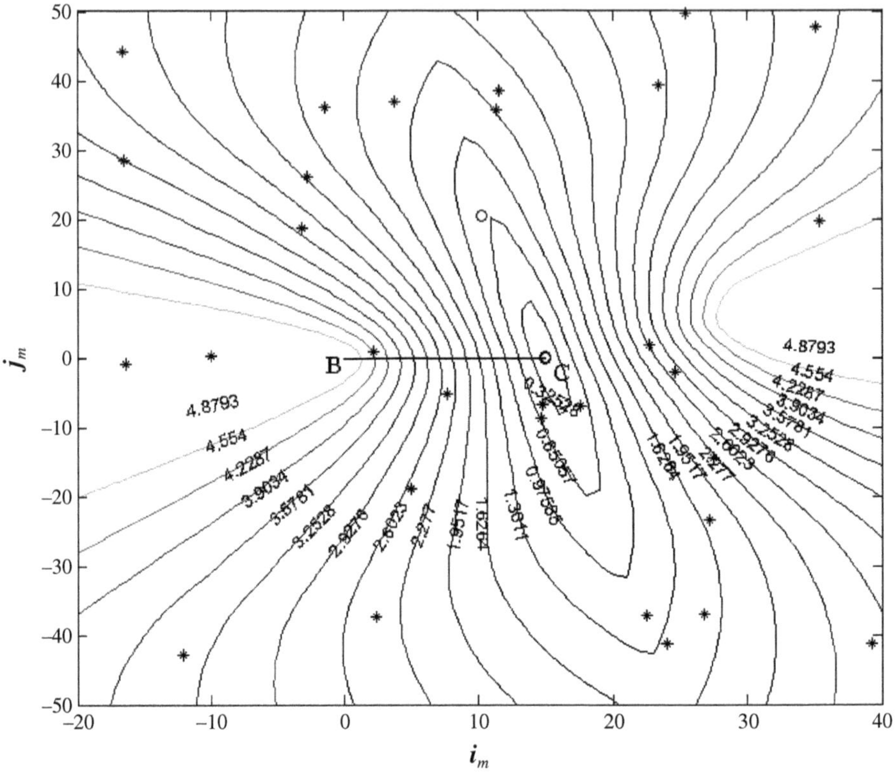

Figure 2.26 The saddle line errors for the coupler of a crank-slider linkage

2.5 The Saddle Kinematic Synthesis of Planar Four-bar Linkages

The kinematic synthesis of planar four-bar linkages, in particular, crank-rocker linkages and crank-slider linkages, is discussed in this section. The synthesis of a crank-rocker linkage has to find two saddle circle points on the moving body with discrete positions, while one saddle circle point and one saddle sliding point of the moving body are needed to synthesize a crank-slider linkage. As mentioned above, for the discrete movement of a moving body, either a few separated positions or multiple positions, we have a unified mathematical model for a saddle circle point and a saddle sliding point; the precise synthesis is just viewed as a precise solution of the discrete kinematic synthesis. In terms of geometrical dimensions and transmission performances, the design requirements of the linkages will be taken into account in the discrete kinematic synthesis of linkages.

Similar to the kinematic geometry of linkages, the kinematic synthesis of linkages can be divided into the geometrical method and the analytical method. The analytical method can be applied to both precise synthesis and approximate synthesis in connection with computing technology. In general, both geometrical and analytical methods have the same theoretical

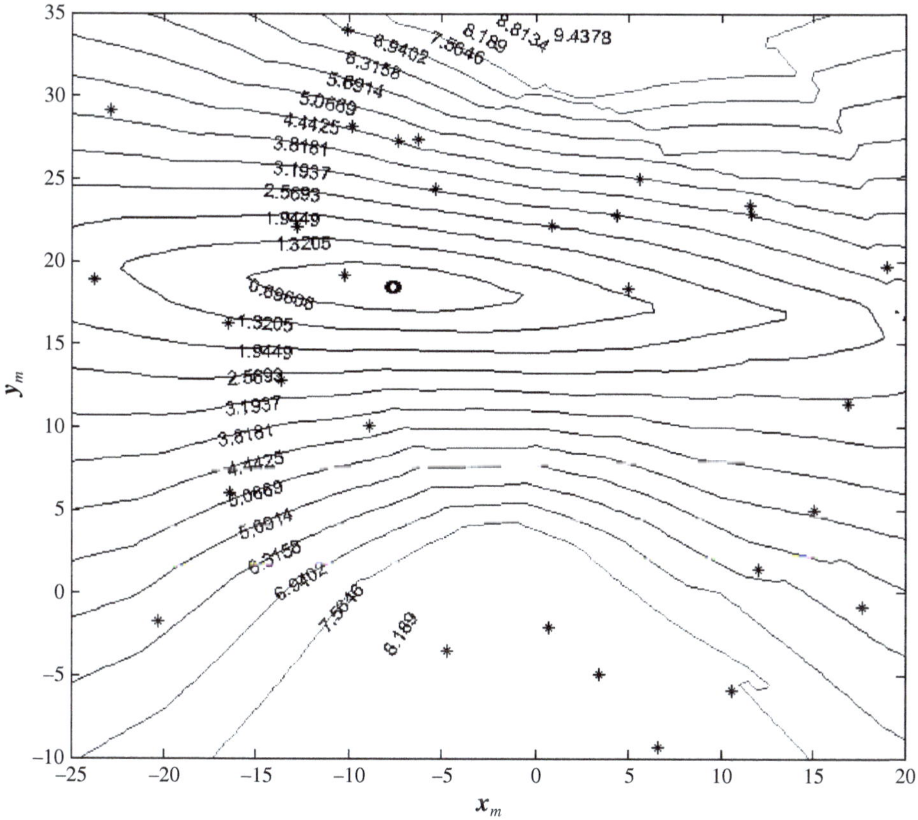

Figure 2.27 The 10-position saddle line errors

foundation, kinematic geometry. They cannot be distinguished from each other in the cases of multi-degree-of-freedom and n-dimensional space. Kinematic geometry of linkages provides the theoretical basis for the kinematic synthesis of linkages; they are the inverse topics of each other. In the kinematic synthesis of linkages, in particular, in the discrete movement, the theoretical foundation is not yet complete; some cases are only treated as special issues of kinematic synthesis. Kinematic synthesis, in fact, either precise or approximate at a finite number of separated positions, can be viewed as saddle curve fitting and locating the saddle circle (sliding) points.

For a good linkage synthesized to meet the requirements in machine design, not only the essential geometrical condition, including position, path, and function generation, but also the geometrical space, kinematic and kinetic properties need to be taken into account, which is beyond the approach of the kinematic synthesis of linkages. We believe it is a key issue to transform the application cases of mechanical engineering into the fundamental topics of kinematic synthesis of linkages. This book only deals with the kinematic synthesis of linkages with discrete multiple positions. The content and approaches of application cases in mechanical engineering will be discussed in the authors' following books.

2.5.1 Kinematic Synthesis

The kinematic synthesis of planar linkages is to determine the type and dimension of the linkages according to the given motion of a rigid body, which may be divided into three types: (1) the positions of a moving body, (2) the path traced by a point of a moving body, (3) the correlation between the input motion and the output motion. These three types of given movement are often referred to as the position synthesis, path synthesis, and function synthesis of linkages. The given positions, the path, and the functional relationship can all be described by continuous analytical functions, or discrete functions. The kinematic synthesis of linkages can be classified into precise synthesis or approximate synthesis depending on whether the prescribed motion is regenerated precisely or approximately by the linkages. With regard to precise synthesis, the classic Burmester theory is useful for finite position synthesis. With the rapid development of numerical procedures and computer technology, the approximate synthesis, also called the optimum synthesis, of linkages has received considerable attention.

Each of the three types of kinematic synthesis of linkages can be converted into another by means of a coordinate system transformation and the additional dyad, which will be introduced briefly. We take position synthesis as the typical topic in the kinematic synthesis of linkages in this book. Meanwhile, the position synthesis of linkages has three representation types: continuous synthesis in higher order, precise synthesis, and approximate synthesis.

2.5.1.1 Continuous Synthesis

The planar movement of a rigid body has been described by adjoint motion in Chapter 1. A moving coordinate system $\{\boldsymbol{O}_m; \boldsymbol{i}_m, \boldsymbol{j}_m\}$ on the moving body $\sum^*$ and a fixed coordinate system $\{\boldsymbol{O}_f; \boldsymbol{i}_f, \boldsymbol{j}_f\}$ on the fixed body $\sum$ are set up respectively. The movement of $\sum^*$, relative to $\sum$, can be described by both the translation along the original curve Γ_{Om} traced by O_m and the relative rotation around the origin point O_m. The original curve Γ_{Om} in $\{\boldsymbol{O}_f; \boldsymbol{i}_f, \boldsymbol{j}_f\}$ and the relative rotation angle are represented as

$$\boldsymbol{R}_{Om}(s) = x_{Omf}(s)\boldsymbol{i}_f + y_{Omf}(s)\boldsymbol{j}_f, \gamma = \gamma(s) \tag{2.46}$$

where (x_{Omf}, y_{Omf}) are the rectangular coordinates of O_m in $\{\boldsymbol{O}_f; \boldsymbol{i}_f, \boldsymbol{j}_f\}$; γ is the angle between the $\boldsymbol{i}_m$ axis and the $\boldsymbol{i}_f$ axis, and s is the arc length of Γ_{Om}. Both (x_{Omf}, y_{Omf}) and γ are continuous functions of s.

In an ideal case, the linkage can be synthesized by the analytical method to regenerate the required movement of a rigid body, described by an analytical curve and a continuous rotation function. A planar 4R linkage is synthesized if there are two circle points on the moving body. A circle point and a sliding point located on the moving body form a crank-slider linkage. The synthesis of these linkages by the analytical method is called the kinematic synthesis of linkages with continuous motion conditions, or the **continuous synthesis of linkages** for short.

There is no acceptable rule to tell us whether a circle point or a sliding point exists on the moving body in continuous movement, or to choose the right linkage to regenerate the given movement of the moving body. Roberts [5] took two algebraic curves, Γ_a with order n_a and Γ_b with order n_b, as two guidance curves in the fixed frame, as shown in Fig. 2.28. Point A and point B of a moving body $\sum^*$ slide along curves Γ_a and Γ_b, respectively. The movement of $\sum^*$ can be completely determined by the two guidance curves. Another point C of $\sum^*$ traces a curve Γ_c in the fixed frame. Roberts proved that the order of curve Γ_c is lower than $2n_a n_b$.

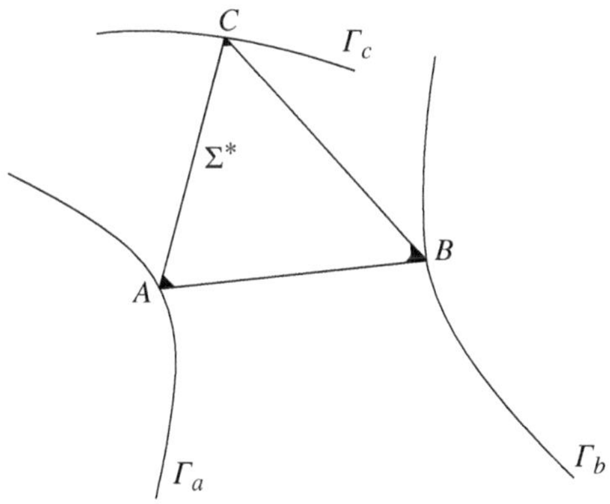

Figure 2.28 The algebraic order of a planar point path

Based on the results, a coupler curve of a planar 4R linkage is a sextic algebraic curve since the two constraint curves, viewed as guidance curves, are circles of order 2. A coupler curve of a slider-crank linkage is a quartic algebraic curve. This tells us the order of a coupler curve for a planar four-bar linkage. The mapping properties of the dimensions and the characteristics, the ranges of the coupler curve, and the kinematic performance of the four-bar linkages have to be studied.

As mentioned in Chapter 1, we take Γ_{Om} as an original curve and Γ_P, traced by a point P of the moving body in the fixed frame, as an adjoint curve. The continuous kinematic synthesis of linkages is to locate the characteristic points by the analytical method. Based on Roberts' work, the guidance curve Γ_a is taken as the original curve. The moving body $\sum^*$ translates on Γ_a and rotates about point A on Γ_a. The movement of $\sum^*$ is completely determined by curve Γ_a and the rotation angle. Another guidance curve Γ_b is viewed as an adjoint curve of Γ_a. In particular, we want the curve Γ_b to be a circle, or the point B on the curve Γ_b will be a circle point. Is there a circle point on the moving body? Where is it if it exists? What happens if it does not exist? In order to answer these questions, we have to give the analytical representation of a point of $\sum^*$ in the fixed frame.

A point P of $\sum^*$, with rectangle coordinates (x_{Pm}, y_{Pm}) in $\{\boldsymbol{O}_m; \boldsymbol{i}_m, \boldsymbol{j}_m\}$, can be represented by a vector equation as

$$\boldsymbol{R}_P = \boldsymbol{R}_{Om} + (x_{Pm}\cos\gamma - y_{Pm}\sin\gamma)\boldsymbol{i}_f + (x_{Pm}\sin\gamma + y_{Pm}\cos\gamma)\boldsymbol{j}_f \tag{2.47}$$

Taking the derivative of the above equation with respect to the arc length s of curve Γ_{Om} and simplifying it by equation (1.74) in Chapter 1, we obtain the following curvature k_P of the path Γ_P traced by P:

$$\begin{cases} k_P = F/G \\ F = (x_{Pm} - a)^2 + (y_{Pm} - b)^2 - D^2/4 \\ G = \{[x_{Pm} + \sin\theta/(k_{Om} - \dot{\theta})]^2 + [y_{Pm} - \cos\theta/(k_{Om} - \dot{\theta})]^2\}^{3/2} \end{cases} \tag{2.48}$$

where k_{Om} is the curvature of the original curve Γ_{Om}, θ is the direction angle of vector $\boldsymbol{\alpha}$ of the Frenet frame $\{\boldsymbol{R}_{Om}; \boldsymbol{\alpha}, \boldsymbol{\beta}\}$ of Γ_{Om} in $\{\boldsymbol{O}_m; \boldsymbol{i}_m, \boldsymbol{j}_m\}$, $D = 1/k^*$ is given in equation (1.75), and a, b

are defined in equation (1.100). Therefore, the continuous synthesis of linkage can be converted into the kinematic geometry analysis of planar motion to confirm whether the characteristic point, which traces a special path Γ_P in the fixed frame, exists on $\sum^*$ or not. If the curvature k_P in equation (2.48) is a constant, $k_P \equiv$ const or $k_P \equiv 0$, this corresponds to the characteristic points being circle points or sliding points. In Chapter 1, the characteristic points – such as the inflection point, the Ball point, the Burmester point, and the Ball-Burmester point – the curvatures of whose paths are instantaneously constant or zero to higher order, have been discussed. According to equation (2.48), the differential equations can be derived for $k_P \equiv$ const and $k_P \equiv 0$ with variable s, which are so complicated that the existence of the solutions and the solving method have never been reported in the literature. Even though the original curve Γ_{Om} is replaced by the constraint curve, a circle or a line for a four-bar linkage, the properties and calculation complexity of the differential equation may not fundamentally change, although equation (2.48) is somewhat simpler.

It is too difficult to synthesize a planar linkage by continuous synthesis, even for a simple four-bar linkage. Just a few successful examples have been reported in the literature. Mechanical engineers may turn their attention to the mechanism with higher pairs for such cases, the cam mechanism and the gear mechanism for instance, even the combined mechanism – although the linkages may be simpler and manufactured conveniently. It is extremely important for mechanical engineers to design linkages easily if the linkages are expected to apply in practice.

2.5.1.2 Discrete Synthesis

As we know, it is difficult to reproduce exactly the given continuous movement of a moving body by the kinematic synthesis of linkages. In mechanical engineering practice, several separated positions of the moving body are usually required to be passed through precisely, or multiple separated positions are required to be passed through approximately, referred to as the ***discrete kinematic synthesis of linkages***, shown in Fig. 2.29. Traditionally, the kinematic synthesis of linkages for the finite precise separated positions, which has precise solutions by solving the algebra equations, is called the ***precise synthesis of linkages***, and that for multiple positions, which can only be solved by mathematical programming, is referred to as the ***approximate synthesis of linkages***, or the optimum kinematic synthesis of linkages. The discrete kinematic synthesis of linkages includes both the precise synthesis and the approximate synthesis, while the results of the precise synthesis are viewed as special cases; just the precise solutions of the discrete kinematic synthesis. The discrete kinematic synthesis of linkages can be described as follows: the type and dimension of a linkage need to be determined for the moving body to occupy the given separated positions precisely or approximately.

A moving body with discrete positions can be represented by the discrete displacement matrix in equation (2.4). This is a traditional topic of the kinematic synthesis of linkages with finite separated positions, such as the circle points curve for four or five separated positions, as special cases of saddle circle points. The circle points may not exist when the number of given positions of a moving body exceeds a certain value. The circle point may be replaced by the saddle circle point, whose path is closest to a circle. A simplest linkage, a four-bar linkage with four revolute joints, is synthesized to reproduce the given positions approximately. The moving body (or coupler link) is guided approximately by two saddle circle points to occupy

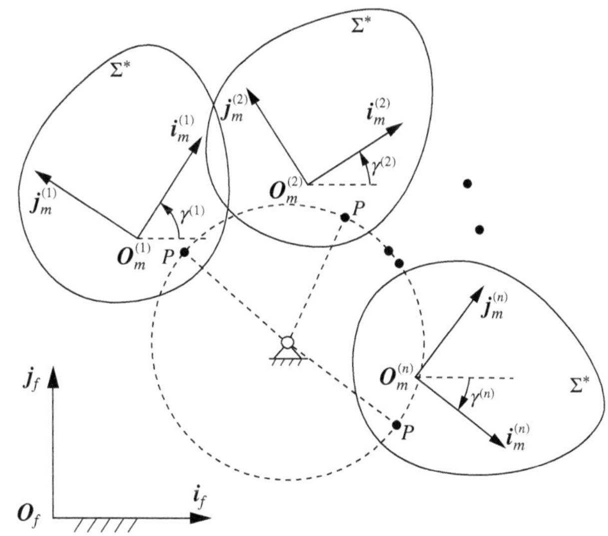

Figure 2.29 The discrete positions of a moving body

given separated positions. Similarly, the saddle sliding point and the saddle circle point have to be searched out if a slide-crank linkage is to be synthesized.

2.5.1.3 Function Synthesis

A linkage to be determined, whose input and output motions should satisfy given transfer functions $\varphi = \varphi(t)$, $\delta = \delta(\varphi)$, or $S = S(\varphi)$, where φ is the rotational angle of the input binary link and δ or S are respectively the rotational angle or linear displacement of the output binary link, is defined as the ***function synthesis of linkages***. The synthesized linkage is a four-bar linkage with four revolute joints if the function is given to connect two angular displacements or rotations, as shown in Fig. 2.30. Three coordinate systems are established, they are respectively $\{\boldsymbol{A};\boldsymbol{i}_f,\boldsymbol{j}_f\}$ on the frame link AD, $\{\boldsymbol{A};\boldsymbol{i}_1,\boldsymbol{j}_1\}$ on the binary link 1, and $\{\boldsymbol{D};\boldsymbol{i}_3,\boldsymbol{j}_3\}$ on the binary link 3. According to the relative motion between the binary links 1 and 3, the function synthesis of linkages can be converted into the position synthesis of linkage if link 1 is assumed to be a relative base link and link 3 be a relative coupler.

Link 1 rotates about hinge point A of the base link with an angle φ, while link 3 rotates about hinge point D with angle δ. The motion of link 3 relative to link 1 can be described as the translation with point D and the rotation about point D with a rotation angle $\gamma = \delta - \varphi$, which is presented by

$$\boldsymbol{R}_D = \cos\varphi \boldsymbol{i}_1 - \sin\varphi \boldsymbol{j}_1,\ \gamma = \delta - \varphi \tag{2.49}$$

The synthesized linkage is a slider crank linkage if the function is given to connect a linear displacement with a rotational displacement, $S = S(\varphi)$, as shown in Fig. 2.31. The motion of link 3 relative to link 1 is presented by

$$\boldsymbol{R}_D = S\cos\varphi \boldsymbol{i}_1 - S\sin\varphi \boldsymbol{j}_1,\ \gamma = -\varphi \tag{2.50}$$

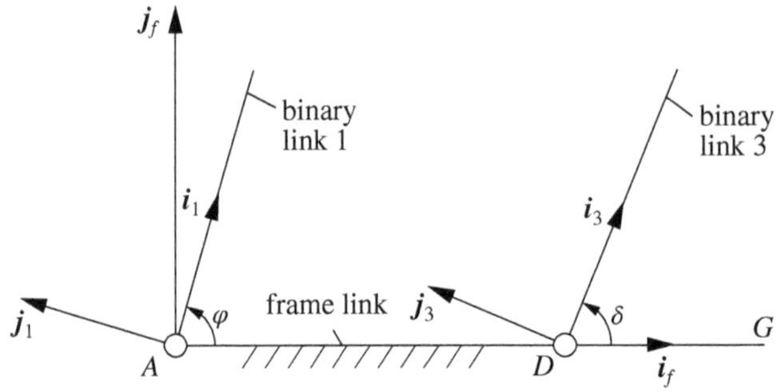

Figure 2.30 The angular displacement function synthesis

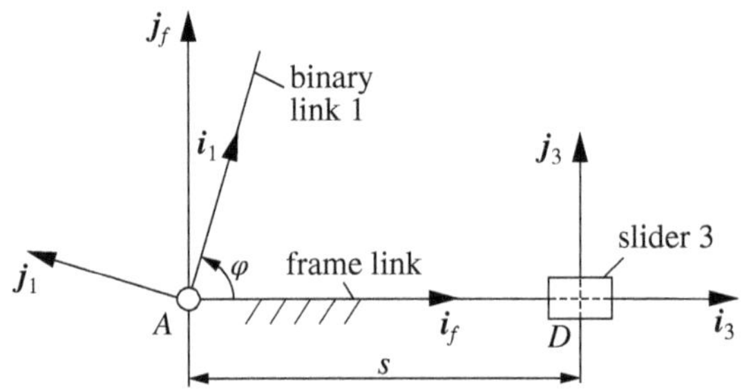

Figure 2.31 The linear displacement function synthesis

As discussed in Chapter 1, both the original curve Γ_{Om} and the angular displacement γ of the moving body $\sum^*$ define completely the kinematic properties of $\sum^*$. The geometrical properties of a path traced by a point of the moving body in the fixed frame $\{\boldsymbol{O}_f; \boldsymbol{i}_f, \boldsymbol{j}_f\}$ depend on Γ_{Om} and γ.

As mentioned above, the given function and the length of the base link can be mapped to the position vector of the original curve and the angular displacement γ of the moving body $\sum^*$ by equation (2.49) or equation (2.50). The function synthesis of linkages can be converted into the position synthesis of linkages.

2.5.1.4 Path Synthesis

The type and dimensions of a linkage need to be determined, whose coupler point traces a curve in the fixed frame, defined by the given function $y = f(x)$ or discrete positions (x_M, y_M), called the ***path synthesis of linkages*** as shown in Fig. 2.32. To synthesize a linkage to regenerate the given curve, mechanical engineers may have some idea of looking up the atlas of coupler curves in a handbook, or designing an experiment to find the linkage. Mathematicians may study the geometrical properties of both the given curve and the coupler curves, and calculate the dimensions of the linkage through complicated algebraic equations. Readers and authors may have an idea from the above position synthesis and function synthesis of linkages, that is, how to map the path synthesis of linkage to position synthesis by converting the given curve into relative motion of a moving body.

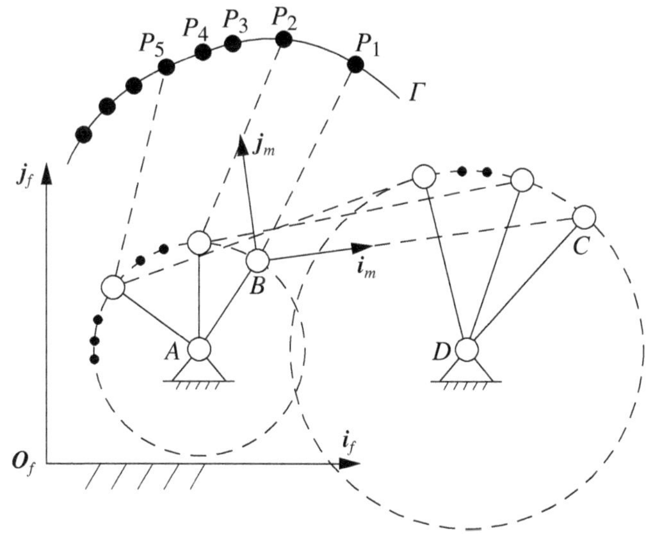

Figure 2.32 The path synthesis of linkage

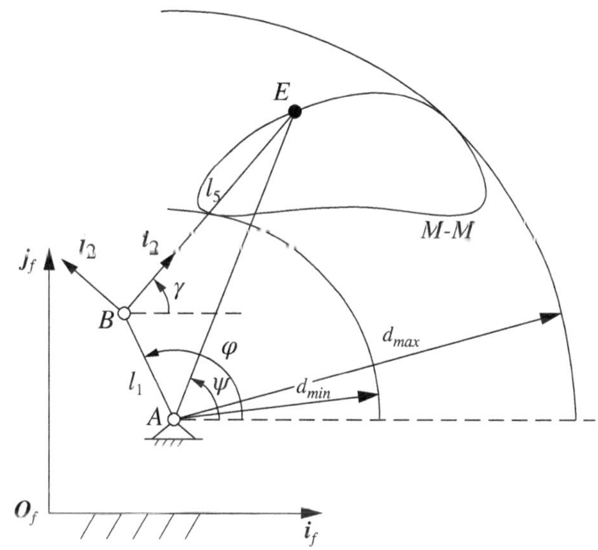

Figure 2.33 A dyad for path generation

As shown in Fig. 2.33, a dyad ABE is given to generate exactly a curve $M-M$. If point E of the dyad moves along the given curve, the floating link BE moves along circle Γ_B traced by point B of the binary link AB, and rotates about point B with a rotational angle γ. In such a case, the kinematic synthesis is to locate the characteristic points (circle point or sliding point) on the floating line BE. If the fixed hinge point A locates outside the given curve $M-M$, and point

E of the dyad has to reach each point of the given curve, the position of point A in the fixed frame $\{\boldsymbol{O}_f, \boldsymbol{i}_f, \boldsymbol{j}_f\}$ and the lengths l_1, l_5 of the links AB and BE would meet some geometrical conditions:

$$l_1 = \frac{d_{\max} - d_{\min}}{2}, \ l_5 = \frac{d_{\max} + d_{\min}}{2} \tag{2.51}$$

where $d_{\max}$ is the maximum distance and $d_{\min}$ is the minimum distance from point A to the given curve. If point A locates inside the given curve, then equation (2.51) can be rewritten as

$$l_1 = \frac{d_{\max} + d_{\min}}{2}, \ l_5 = \frac{d_{\max} - d_{\min}}{2} \tag{2.52}$$

The displacements of the dyad ABE can be determined when point E moves along the given curve M–M. We have

$$\begin{cases} l = \sqrt{(x_M - x_A)^2 + (y_M - y_A)^2} \\ \psi = \begin{cases} \cos^{-1}\dfrac{x_M - x_A}{l}, \ y_M - y_A \geq 0 \\ -\cos^{-1}\dfrac{x_M - x_A}{l}, \ y_M - y_A < 0 \end{cases} \\ \angle BAE = \cos^{-1}\left(\dfrac{l_1^2 + l^2 - l_5^2}{2l_1 l}\right) \\ \varphi = \psi + p_c \angle BAE \end{cases} \tag{2.53}$$

where p_c is the assembly coefficient. Its initial value is 1 if link AB rotates counterclockwise, and -1 if link AB rotates clockwise. Meanwhile, p_c changes its sign when the distance from A to E reaches its extreme value as point E moves along the given curve M–M.

The coordinate system $\{\boldsymbol{B}; \boldsymbol{i}_2, \boldsymbol{j}_2\}$ is established on the floating link BE and the relationship between the frames $\{\boldsymbol{O}_f, \boldsymbol{i}_f, \boldsymbol{j}_f\}$ and $\{\boldsymbol{B}; \boldsymbol{i}_2, \boldsymbol{j}_2\}$ is

$$\begin{cases} \boldsymbol{R}_B = \boldsymbol{R}_A + l_1 \cos\varphi \boldsymbol{i}_f + l_1 \sin\varphi \boldsymbol{j}_f \\ \boldsymbol{i}_2 = \cos\gamma \boldsymbol{i}_f + \sin\gamma \boldsymbol{j}_f \\ \boldsymbol{j}_2 = -\sin\gamma \boldsymbol{i}_f + \cos\gamma \boldsymbol{j}_f \end{cases} \tag{2.54}$$

Based on the above equations, the original curve Γ_B traced by point B in the fixed frame and the rotation angle γ can be written as

$$\begin{cases} x_B = x_A + l_1 \cos\varphi, \ y_B = y_A + l_1 \sin\varphi \\ \gamma = \tan^{-1}\left(y_M - y_B, x_M - x_B\right) \end{cases} \tag{2.55}$$

Needless to say, path synthesis can be converted into position synthesis of linkages, but the position of point A in the fixed frame has to be properly chosen. Chapter 1 may provide the distribution of the coupler curves of four-bar linkages for the path synthesis of linkages.

As mentioned earlier, in the three types of kinematic synthesis – position, function, and path synthesis – or in the property description of the synthesis, like continuous, precise, and approximate synthesis, all given conditions for the kinematic synthesis of planar linkages can be described by the original curve $\Gamma_{Om}(x_{Omf}, y_{Omf})$ and the rotation angle γ. That is, the moving body $\sum^*$ moves along the original curve Γ_{Om} and rotates around the point on Γ_{Om} with a rotational angle γ. The displacement coordinates (x_{Omf}, y_{Omf}) and the rotation angle γ can describe either the continuous functions for the continuous synthesis, or the discrete functions for the precise and approximate synthesis.

The numerical approach for the kinematic synthesis of linkages is systematically introduced in this book. The kinematic synthesis of planar linkages, to some extent, is to search for the saddle circle points and the saddle sliding points, on the moving body by means of the original curve Γ_{Om} and the rotation angle γ. A point P of the moving body $\sum^*$ with rectangular coordinates (x_{Pm}, y_{Pm}) in the moving Cartesian frame traces a path Γ_P in the fixed frame. If the planar motion of the moving body is given by some separated positions, the path Γ_P turns into planar discrete points, whose vector euqation can be written as

$$\boldsymbol{R}_P^{(i)} = f(x_{Omf}^{(i)}, y_{Omf}^{(i)}, \gamma^{(i)}, x_{Pm}, y_{Pm}),\ i = 1, \ldots, n \tag{2.56}$$

The above equation is taken as the typical representation form for discrete movement, or the discrete displacement of $\sum^*$ in planar movement.

2.5.2 *Crank-rocker Linkage*

In this section, based on the discrete kinematic geometry, the mathematical model and process for the kinematic synthesis of planar hinged four-bar linkages are introduced. The numeral examples are discussed.

2.5.2.1 The Mathematical Model

As we know, the kinematic synthesis of a planar hinged four-bar linkage with separated positions is to find two saddle circle points by numerical calculation. In particular, we have to use the optimization program, which includes six terms: the objective function, optimization variables, constraint conditions, initial values of optimization variables, and termination conditions. The essential question is how to establish the optimization variables and objective function. For the kinematic synthesis of planar hinged four-bar linkages, the displacements of the moving body are given as the known conditions in terms of the set of discrete points, and two saddle circle points P_1 and P_2 have to be searched out on the moving body. The mathematical model of the discrete kinematic synthesis of planar hinged four-bar linkages can be set up as

$$\begin{cases} \min F(\mathbf{Z}) = \min(\Delta_{rr}(z_1) + \Delta_{rr}(z_2)) \\ \text{s.t.}\ \ g_j(\mathbf{Z}) \le 0, j = 1, 2, \ldots, k \end{cases} \tag{2.57}$$

In which each term can be explained in detail as follows.

A. **The objective function and optimization variables**

The sum of two saddle circle errors, $\Delta_{rr}(z_1)$ and $\Delta_{rr}(z_2)$, for two different points P_1 and P_2 on the moving body, is taken as the objective function. The definition of saddle circle is presented in Section 2.3.1.

The optimization variables are $\mathbf{Z} = (z_1, z_2)^T$, or $z_1 = (x_{Pm1}, y_{Pm1})^T$ and $z_2 = (x_{Pm2}, y_{Pm2})^T$, which are the rectangular coordinates of two points on the moving body.

The area of optimization variables is defined from minus infinity to plus infinity on the moving body.

B. **The constraint equations**

As mentioned above, the two points $P_1(x_{Pm1}, y_{Pm1})$ and $P_2(x_{Pm2}, y_{Pm2})$ of the moving body trace two curves respectively, and correspond to two saddle circles, of which one has center point $C_1(x_{C1}, y_{C1})$ and radius r_1, while the other has center point $C_2(x_{C2}, y_{C2})$ and radius r_2. We take points P_1 and P_2 as two moving hinge points on the moving body and the center points C_1 and C_2 as two fixed hinge points on the base link. The planar hinged four-bar linkage can be constituted, whose dimensions are

$$\begin{cases} l_1 = r_1 \\ l_2 = \sqrt{(x_{Pm1} - x_{Pm2})^2 + (y_{Pm1} - y_{Pm2})^2} \\ l_3 = r_2 \\ l_4 = \sqrt{(x_{C1} - x_{C2})^2 + (y_{C1} - y_{C2})^2} \end{cases} \tag{2.58}$$

Considering the requirements of geometrical dimensions and transmission performances, some of the constraint conditions can be listed as follows.

1. The geometrical conditions. All link lengths of the linkage should first be larger than a given positive value for the links to be manufactured easily, that is

$$g_1(\mathbf{Z}) = \varepsilon - \min(l_1, l_2, l_3, l_4) \le 0 \tag{2.59}$$

where ε is the permitted minimum positive value for the restriction of the link length.

The linkage should have a suitable ratio of the link length, which can be represented by

$$g_2(\mathbf{Z}) = \max(l_1, l_2, l_3, l_4)/\min(l_1, l_2, l_3, l_4) - [l] \le 0 \tag{2.60}$$

where $[l]$ is the permitted maximum ratio of link lengths.

The condition for the Grashof kinematic chain is

$$g_3(\mathbf{Z}) = 2\{\max(l_1, l_2, l_3, l_4) + \min(l_1, l_2, l_3, l_4)\} - \sum_{i=1}^{4} l_i \le 0 \tag{2.61}$$

If the hinged four-bar linkage has a crank, the link lengths should meet

$$\begin{cases} g_4(\mathbf{Z}) = \min(l_1, l_4) - l_2 \le 0 \\ g_5(\mathbf{Z}) = \min(l_1, l_4) - l_3 \le 0 \end{cases} \tag{2.62}$$

2. The transmission angle

$$g_6(\mathbf{Z}) = [\gamma] - \gamma_{\min} \le 0 \tag{2.63a}$$

where $[\gamma]$ is the minimum allowable value of the transmission angle. $\gamma_{\min}$ is the minimum transmission angle of the four-bar linkage and can be determined by

$$\gamma_{\min} = \min\left(\arccos\left[\frac{l_2^2 + l_3^2 - (l_1 - l_4)^2}{2l_2 l_3}\right], \pi - \arccos\left[\frac{l_2^2 + l_3^2 - (l_1 + l_4)^2}{2l_2 l_3}\right]\right) \tag{2.63b}$$

In practical applications, there exist other constraints for the kinematic synthesis of planar four-bar linkages, such as the position constraint of the joints, the requirement of the movement space, the movement sequence, and the circuit and branch. The additional constraint equations have to be given according to the different application cases.

C. **The optimization algorithm**
Based on the definition of saddle circle point and the optimization model (2.27), whose solving process has been discussed in Section 2.3.6, the optimization model of the kinematic synthesis of a planar hinged four-bar linkage is established as equation (2.57), which is the nonlinear optimum problem under constraint conditions. In this book, the function **fmincon** in Matlab is adopted directly to solve this nonlinear optimum problem.

D. **Initial values of the optimization variables**
The initial values of the optimization variables have a powerful effect on both the computing amount and the convergence of the solving algorithm in an optimization program. It is better to provide good initial values, which are close to the optimal solutions, for the purpose of reducing both the amount of computation and the complexity of the iterative algorithm. Unfortunately, we don't know where the optimal solutions are; we needn't do anything at all if the solutions can be guessed. It is in our interests for the initial points to locate on the same valleys as the positions of the optimal solutions; this is an arduous but necessary stage in the optimization program. In the kinematic synthesis of planar hinged four-bar linkages, initial values should be set three times in: searching for the saddle circles, the saddle circle points, and the synthesized planar four-bar linkage. The former two are introduced above, providing a good background for the third. The saddle circle points of the moving body are naturally chosen to be the initial points for equation (2.57).

E. **Stop conditions**
In general, the values of either the optimization variables or the objective function at two adjacent iterative steps can be taken as the end conditions (or conditions to stop) for an optimization program. The numerical examples in this book are given by means of the optimization tools of the software Matlab, in which the optimum options can be set up in the function **optimset**, such as the maximum number of iterations allowed (**MaxIter**), the maximum number of function evaluations allowed (**MaxFunEvals**), the termination tolerance on the function value (**TolFun**) and the optimization variables (**TolX**).

2.5.2.2 Multi-position Synthesis

To find a planar hinged four-bar linkage, which will guide a lamina to approximately occupy multiple given positions, is called discrete synthesis of planar hinged four-bar linkages, in which the number of separated positions is not limited. There may exist precise solutions if the given separated positions are less than six. This will be discussed later.

The mathematical model (2.57) can be solved directly by Matlab. The function **fmincon** is set up as follows:

$$[\mathrm{x}, \mathrm{fval}, \mathrm{exitflag}] = \mathrm{fmincon}\ \ (\mathrm{fun}, \mathrm{x0}, \mathrm{A}, \mathrm{b}, \mathrm{Aeq}, \mathrm{beq}, \mathrm{lb}, \mathrm{ub}, \mathrm{nonlcon}, \mathrm{options})$$

The optimization conditions of **fmincon** are set up through the function **options**. The medium-scale algorithm is chosen as the optimization scheme. The conditions of convergence are set through MaxFunEvals = 500, MaxIter = 200, TolFun = 0.000001, and TolX = 0.000001.

The outputs of **fmincon** are respectively described as follows.

x: the optimal solution; the Cartesian coordinates of two moving hinge points on the moving body with the given positions.
fval: the value of the objective function, or the sum of two saddle circle errors corresponding to the two moving hinge points.
exitflag: the exit condition, which describes the reason for the termination of the function **fmincon**.

The inputs of the function **fmincon** are respectively

fun: the objective function; the sum of two saddle circle errors of any two points on the moving body, or the discrete point set $\{\boldsymbol{R}_{P1}^{(i)}\}$ and $\{\boldsymbol{R}_{P2}^{(i)}\}$.
x0: the initial values of the optimization variables, which can be provided by a pair of saddle circle points as mentioned in the preparation of initial values.
A,b,Aeq,beq,lb,ub,nonlcon: subject function, in which the linear constraints **A**, **b**, **Aeq**, **beq** are set to null. The lower bound of the optimization variables is $\text{lb} = [x_{Pm1\,\min}, y_{Pm1\,\min}, x_{Pm2\,\min}, y_{Pm2\,\min}]$ and the upper bound is $\text{ub} = [x_{Pm1\,\max}, y_{Pm1\,\max}, x_{Pm2\,\max}, y_{Pm2\,\max}]$. The nonlinear constraints **nonlcon** can be set by calling the constraint function @mycon. For the constraint function @mycon, the relationship among the optimum variables and the dimensions of the synthesized linkage can be established through equation (2.58), and then the nonlinear inequality equations can be constructed by functions g_1, g_2, g_3, g_4, g_5, and g_6 from equations (2.59) to (2.63).

As mentioned above, a **program for kinematic synthesis of planar four-bar linkages,** or **KS-CR** for short, is naturally constructed for multiple separated positions of a moving body, whose input are the separated positions of a moving body and output are hinged four-bar linkages with dimensions and performance.

The process of multi-position synthesis of a planar hinged four-bar linkage by saddle point programming is explained intensively by the following numeral example. As a general rule, the saddle circle point errors, corresponding to the saddle circle points of the moving body, depend on the kinematic properties of the movement of the moving body, which are usually not zero. This means that there exist saddle circle points other than circle points on the moving body for general planar motion, unless it is degenerate.

Example 2.6 Ten positions of the rigid body are described by $(x_{Omf}^{(i)}, y_{Omf}^{(i)}, \gamma^{(i)})$, $i = 1, \ldots, 10$ in Table 2.1. Synthesize a planar hinged four-bar linkage with a crank, whose minimum allowable transmission angle $[\gamma] = 30°$ and maximum permitted ratio of link length $[l] = 5$.

Solution

Based on the above discussion, the mathematical model of the kinematic synthesis of the hinged four-bar linkage with ten positions is established as

$$
\begin{cases}
\min\limits_{Z=(z_1,z_2)^T} (\Delta_{rr}(z_1) + \Delta_{rr}(z_2)) \\
\text{s.t. } g_1(\mathbf{Z}) = 2\{\max(l_1, l_2, l_3, l_4) + \min(l_1, l_2, l_3, l_4)\} - \sum_{i=1}^{4} l_i \le 0 \\
g_2(\mathbf{Z}) = \min(l_1, l_4) - l_2 \le 0 \\
g_3(\mathbf{Z}) = \min(l_1, l_4) - l_3 \le 0 \\
g_4(\mathbf{Z}) = \max(l_1, l_2, l_3, l_4)/\min(l_1, l_2, l_3, l_4) - 5 \le 0 \\
g_5(\mathbf{Z}) = \dfrac{\pi}{6} - \min\left(\arccos\left[\dfrac{l_2^2 + l_3^2 - (l_1 - l_4)^2}{2l_2l_3}\right], \pi - \arccos\left[\dfrac{l_2^2 + l_3^2 - (l_1 + l_4)^2}{2l_2l_3}\right]\right) \le 0 \\
g_6(\mathbf{Z}) = 0.1 - \min(l_1, l_2, l_3, l_4) \le 0
\end{cases}
\tag{E2-6.1}
$$

The letters in the above equations have the same meanings as those in equations (2.57) to (2.63). The kinematic synthesis of a hinged four-bar linkage with 10 separated positions has three stages: the preparation of initial values, the program setting, and the program operation.

1. **Preparation of initial values**
   The saddle circle points in the defined area of the moving body have been searched out in Example 2.3, and 12 saddle circle points (SCP1–SCP12) with lesser errors are listed in Table 2.2. Any two points of the 12 saddle circle points are taken as a group of two moving joints on the moving body, whose corresponding center points on the base link are designated as the fixed joints. A hinged four-bar linkage can then be constituted. In this way, each group of saddle circle points corresponds to a hinged four-bar linkage, whose dimensions can be calculated by equation (2.58) and whose performance can also be calculated by equations (2.59) to (2.63). For these hinged four-bar linkages, some may be good initial linkages for the kinematic synthesis of a hinged four-bar linkage with a crank under 10 prescribed positions if the constraint conditions in equation (E2-6.1) are satisfied. After the judgment of the constraint equations, seven initial hinged four-bar linkages are finally constituted from the 12 saddle circle points (SCP1–SCP12) in Table 2.2 and listed in Table 2.3 (with letter G and roman numeral as sequence number).
2. **Program settings**
   For the kinematic synthesis program **KS-CR**, the parameter settings of the function **fmincon** can refer totally to the previously introduced content, including the optimization algorithm, termination conditions, and constraint conditions setting. The 10 positions of the moving body are described by the parameters $(x_{Omf}^{(i)}, y_{Omf}^{(i)}, \gamma^{(i)})$, $i = 1, \ldots, 10$ in Table 2.1. The nonlinear inequality constraints should be constructed through the inequalities g_1, g_2, g_3, g_4, g_5, and g_6 from equation (E2-6.1). The lower bound is $\text{lb} = [-60.66, -60.66, -60.66, -60.66]$ and the upper bound is $\text{ub} = [60.66, 60.66, 60.66, 60.66]$.
3. **Program operation**
   We run the program **KS-CR** for kinematic synthesis of a crank-rocker linkage. The seven groups of initial values have the corresponding results, or seven synthesized crank-rocker linkages with different dimensions. The average optimization time is 8.2834 s.

Table 2.3 The initial values (combination of saddle circle points)

No.	Combinations	Dimensions of four-bar linkages	Maximum length ratio	Minimum transmission angle	Object function value
GI	SCP1–SCP10	$l_1=9.0233$, $l_2=15.5464$, $l_3=38.0188$, $l_4=40.1720$	4.45	51.67°	0.245487
GII	SCP1–SCP9	$l_1=9.0233$, $l_2=15.6986$, $l_3=29.5733$, $l_4=33.2289$	3.68	44.33°	0.245857
GIII	SCP12–SCP8	$l_1=9.3539$, $l_2=16.5865$, $l_3=43.5257$, $l_4=45.8760$	4.90	52.41°	0.600659
GIV	SCP1–SCP11	$l_1=9.0233$, $l_2=15.5594$, $l_3=36.5932$, $l_4=38.9667$	4.32	50.66°	0.246419
GV	SCP12–SCP10	$l_1=9.3539$, $l_2=16.6328$, $l_3=38.0188$, $l_4=41.1580$	4.40	49.02°	0.606045
GVI	SCP12–SCP11	$l_1=9.3539$, $l_2=16.6529$, $l_3=36.5932$, $l_4=39.9630$	4.27	47.98°	0.606977
GVII	SCP12–SCP9	$l_1=9.3539$, $l_2=16.8298$, $l_3=29.5733$, $l_4=34.2832$	3.67	41.43°	0.606415

In Table 2.4, the synthesized crank-rocker linkages are listed, corresponding to the initial groups (GI–GVII). For groups GI, GII, and GIV, the optimization results are the same as the initial points. However, for group GIII, the optimization results change to be new points P_1 and P_2 within the feasible region. For groups GV, GVI, or GVII, the optimization results include the saddle circle point SCP1 and new points P_3, P_4, and P_5. The rectangular coordinates of these new points P_1 to P_5 and the parameters of their saddle circles are listed in Table 2.5.

The two crank-rocker linkages, corresponding to solutions 1 and 2 respectively, guide their coupler links to occupy the prescribed 10 positions. The deviations between the coupler's positions and the required positions are listed in Tables 2.6 and 2.7 respectively.

2.5.2.3 Five-position Synthesis

To determine a planar hinged four-bar linkage, whose coupler link will guide a lamina to occupy less than six required separated positions, is called the precise synthesis of planar hinged four-bar linkages since there may exist circle points on the lamina, and the number of circle points cannot exceed four. In general, a few circle points with their center points can constitute the planar hinged four-bar linkages in geometry, but they may not satisfy the design requirements, such as the kinematical and transmission conditions. Therefore, the solutions, including the circle points, the saddle circle points obtained by the discrete kinematic synthesis of linkages under constraint conditions may provide much more choice for the designer.

Table 2.4 The seven solutions for 10 given positions

No.	Initial values	Hinged joints	Dimensions of four-bar linkages	Length ratio	Transmission angle	Object function value
Solution 1	GI	SCP1–SCP10	$l_1=9.0233$, $l_2=15.5464$, $l_3=38.0188$, $l_4=40.1720$	4.45	51.67°	0.245487
Solution 2	GII	SCP1–SCP9	$l_1=9.0233$, $l_2=15.6986$, $l_3=29.5733$, $l_4=33.2289$	3.68	44.33°	0.245857
Solution 3	GIII	P_1–P_2	$l_1=9.0247$, $l_2=15.5446$, $l_3=39.0572$, $l_4=41.0577$	4.55	52.38°	0.245407
Solution 4	GIV	SCP1–SCP11	$l_1=9.0233$, $l_2=15.5594$, $l_3=36.5932$, $l_4=38.9667$	4.32	50.66°	0.246419
Solution 5	GV	SCP1–P_3	$l_1=9.0233$, $l_2=15.5390$, $l_3=39.0794$, $l_4=41.0760$	4.55	52.38°	0.244648
Solution 6	GVI	SCP1–P_4	$l_1=9.0233$, $l_2=15.5674$, $l_3=35.8850$, $l_4=38.3723$	4.25	50.13°	0.246791
Solution 7	GVII	SCP1–P_5	$l_1=9.0233$, $l_2=15.6972$, $l_3=29.6143$, $l_4=33.2613$	3.69	44.37°	0.245902

Table 2.5 The new moving joints and their saddle circles

	Coordinates on the moving body		Saddle circle			
			Coordinates of the center point		Radius	Saddle circle error
	x_{Pm}	y_{Pm}	x_C	y_C	r	δ_{rr}
P_1	−1.9288	2.2891	16.1880	17.5939	9.0247	0.007838
P_2	−1.5726	17.8296	31.0909	−20.6637	39.0572	0.237569
P_3	−1.5755	17.8305	31.0928	−20.6835	39.0794	0.237188
P_4	−1.2305	17.8472	31.1417	−17.7460	35.8850	0.239330
P_5	−0.4358	17.9215	31.2935	−12.0411	29.6143	0.238441

Table 2.6 The position deviations for solution 1

Position	$\Delta x_{Omf}^{(i)}$	$\Delta y_{Omf}^{(i)}$	$\Delta\gamma^{(i)}$	Position	$\Delta x_{Omf}^{(i)}$	$\Delta y_{Omf}^{(i)}$	$\Delta\gamma^{(i)}$
1	−0.0465	0.0225	−0.0026	6	0.0000	0.0038	−0.0172
2	−0.0483	0.0399	−0.0186	7	−0.0242	−0.0720	0.0239
3	−0.0199	−0.0284	−0.0050	8	0.0342	0.0180	0.0099
4	0.0364	0.0750	0.0179	9	−0.0283	0.0675	−0.0171
5	0.0532	−0.0333	0.0064	10	−0.0334	0.0295	0.0000

Table 2.7 The position deviations for solution 2

Position	$\Delta x_{Omf}^{(i)}$	$\Delta y_{Omf}^{(i)}$	$\Delta\gamma^{(i)}$	Position	$\Delta x_{Omf}^{(i)}$	$\Delta y_{Omf}^{(i)}$	$\Delta\gamma^{(i)}$
1	−0.0758	−0.0129	0.0115	6	0.0411	−0.0030	−0.0155
2	−0.0825	0.0270	−0.0199	7	0.0035	−0.0542	0.0218
3	−0.0411	−0.0528	−0.0083	8	0.0289	0.0662	0.0000
4	0.0026	0.0596	0.0103	9	0.0015	−0.0686	−0.0100
5	−0.0434	0.0580	0.0018	10	−0.0474	0.0202	0.0128

Table 2.8 Five positions of the rigid body

	$x_{Omf}^{(i)}$	$y_{Omf}^{(i)}$	$\gamma^{(i)}$
1	14.2469	11.7052	0.3671
2	10.8641	13.0941	0.2627
3	6.8561	13.4122	0.2031
4	2.7976	12.4444	0.1677
5	−0.7927	10.2430	0.1506

Example 2.7 Synthesize a planar crank-rocker linkage, which guides the lamina through five positions described by $(x_{Om}^{(i)}, y_{Om}^{(i)}, \gamma^{(i)})$, $i = 1, \ldots, 5$, listed in Table 2.8, whose minimum allowable transmission angle $[\gamma]$ is 30° and maximum permitted ratio of link length $[l]$ is 5.

Solution

For the kinematic synthesis of planar crank-rocker linkages under the condition of five given positions, the process is similar to that of Example 2.6, which includes three steps: the preparation of initial values, the program setting, and the program operation of **KS-CR**.

Firstly, for the preparation of initial values, 50 initial values are generated and the saddle circle points can be located on the moving body by sub-program ArrP for the given five positions, in which eight saddle circle points (SCP1–SCP8) with lesser errors are listed in Table 2.9, and two points (SCP1, SCP2) are just the circle points on the moving body.

Two of the eight saddle circle points can be chosen as a group to form a four-bar linkage. For the eight saddle circle points SCP1–SCP8 in Table 2.9, four combinations GI–GIV of saddle circle points are chosen as the initial points and listed in Table 2.10 after the constraints are judged. Meanwhile, a four-bar linkage, corresponding to the two circle points (SCP1, SCP2), has dimensions $l_1 = 12.1095$, $l_2 = 14.2621$, $l_3 = 15.1426$, $l_4 = 16.2314$, whose maximum ratio of link length is 1.34 and minimum transmission angle is only 15.75°. Unfortunately, it cannot be the final choice; although it is a precise solution in geometry, it does not satisfy the constraint condition $[\gamma] = 30^{\circ}$.

Then, the program settings are similar to those in Example 2.6. The four optimization results corresponding to the four groups of initial points in Table 2.10, four crank-rocker linkages are obtained by calling the program **KS-CR** and listed in Table 2.11. The average optimization time is 17.2732 s.

Table 2.9 The eight saddle circle points for five positions

No.	Saddle circle point		Saddle circle			
			Circle center		Radius	Saddle circle error
	x_{Pm}	y_{Pm}	x_C	y_C	r	δ_{rr}
SCP1	−6.5378	−0.1815	0.5547	−0.1936	12.1095	5.862970×10^{-8}
SCP2	7.7161	0.3034	16.7780	0.3186	15.1426	3.945190×10^{-7}
SCP3	−1.6954	−1.2531	6.2316	−1.3179	13.1853	0.001250
SCP4	−8.9493	0.8557	−2.3463	0.9009	11.5461	0.001304
SCP5	12.4893	3.2531	21.8508	3.4057	16.0771	0.002294
SCP6	18.1774	9.9950	27.4522	10.4837	17.0683	0.007271
SCP7	23.2031	34.7871	28.9100	37.3790	16.0600	0.029510
SCP8	20.9315	−1.3533	37.7935	−11.6698	29.7094	0.123926

Table 2.10 The initial values (combination of saddle circle points)

No.	Combinations	Dimensions of four-bar linkages	Maximum length ratio	Minimum transmission angle	Object function value
GI	SCP4–SCP5	$l_1 = 11.5461, l_2 = 21.5722,$ $l_3 = 16.0772, l_4 = 24.3263$	2.11	35.74°	0.003598
GII	SCP3–SCP7	$l_1 = 13.1853, l_2 = 43.8045,$ $l_3 = 16.0600, l_4 = 44.8527$	3.40	32.12°	0.030760
GIII	SCP1–SCP8	$l_1 = 12.1095, l_2 = 27.4943,$ $l_3 = 29.7094, l_4 = 38.9671$	3.22	53.57°	0.123926
GIV	SCP4–SCP8	$l_1 = 11.5461, l_2 = 29.9623,$ $l_3 = 29.7094, l_4 = 42.0621$	3.64	52.11°	0.125230

Table 2.11 Four solutions for five-position synthesis

No.	Initial values	Hinged joints	Dimensions of four bar linkages	Length ratio	Transmission angle	Object function value
Solution 1	GI	SCP4–SCP5	$l_1 = 11.5461, l_2 = 21.5722,$ $l_3 = 16.0772, l_4 = 24.3263$	2.11	35.74°	0.003598
Solution 2	GII	SCP3–SCP7	$l_1 = 13.1853, l_2 = 43.8045,$ $l_3 = 16.0600, l_4 = 44.8527$	3.40	32.12°	0.030760
Solution 3	GIII	P_1–P_2	$l_1 = 11.2452, l_2 = 16.7368,$ $l_3 = 15.2427, l_4 = 19.6470$	1.75	30°	0.010709
Solution 4	GIV	P_3–P_4	$l_1 = 11.4748, l_2 = 21.3945,$ $l_3 = 17.9436, l_4 = 25.5683$	2.23	39.49°	0.036379

Table 2.12 The new moving joints and their saddle circles

	Coordinates on the moving body		Saddle circle			
			Coordinates of the center point		Radius	Saddle circle error
	x_{Pm}	y_{Pm}	x_C	y_C	r	δ_{rr}
P_1	−10.1950	1.5390	−3.8676	1.6239	11.2452	0.002211
P_2	6.3436	−1.0288	15.5468	−1.3899	15.2427	0.008498
P_3	−8.7829	1.1348	−2.2402	1.2779	11.4748	0.003685
P_4	12.5895	0.1631	23.1887	−1.3887	17.9436	0.032694

Solutions 1 and 2 are the same as their initial group GI and initial group GII. The last two optimization results, solutions 3 and 4, corresponding to the initial groups GIII and GIV have new groups of moving joints, P_1–P_2 and P_3–P_4, whose coordinates and parameters of saddle circles are listed in Table 2.12.

2.5.2.4 Four-position Synthesis

There exists a cubic curve on the moving body with four given positions; all points on the curve are circle points. Theoretically infinite hinged four-bar linkages can be constructed by these circle points to guide the rigid body to occupy the four positions precisely. The optimization synthesis with saddle circle error as the objective function is meaningless. A better kinematic and kinetic performance can be the objective of the synthesis of planar four-bar linkages.

Example 2.8 Synthesize a planar crank-rocker linkage, which guides the lamina through four positions described by $(x^{(i)}_{Om}, y^{(i)}_{Om}, \gamma^{(i)})$, $i = 1, \ldots, 4$ in Table 2.13, whose minimum allowable transmission angle $[\gamma]$ is 30° and maximum permitted ratio of link length $[l]$ is 5.

Solution

200 circle points are located on the moving body by calling the sub-program ArrP, corresponding to 200 initial points generated randomly in the area $x_{Pm} \in [-10, 10]$, $y_{Pm} \in [-10, 10]$ of the moving body.

Table 2.13 Four positions of the rigid body

	$x^{(i)}_{Omf}$	$y^{(i)}_{Omf}$	$\gamma^{(i)}$
1	1.0	1.0	0°
2	2.0	0.5	0°
3	3.0	1.5	45°
4	2.0	2.0	90°

Table 2.14 Three solutions for four-position synthesis

	Circle point		Center point		Link length	Maximum ratio of length	Minimum transmission angle
	x_{Pm}	y_{Pm}	x_C	y_C	l_1, l_2, l_3, l_4		
1	−0.4307 −1.9646	1.7037 1.5283	0.8594 −1.2527	2.0338 0.7021	$l_1 = 0.7299, l_2 = 1.5439,$ $l_3 = 1.8488, l_4 = 2.4968$	3.42	36.15°
2	−0.4307 −1.5123	1.7037 1.7025	0.8594 −0.5156	2.0338 1.4459	$l_1 = 0.7299, l_2 = 1.0815,$ $l_3 = 1.2566, l_4 = 1.4954$	2.05	35.84°
3	0.5249 −0.4041	2.8765 1.0672	1.1121 1.2915	1.8009 2.2084	$l_1 = 2.1163, l_2 = 2.0339,$ $l_3 = 0.7098, l_4 = 0.4453$	4.75	48.29°

Two of the 200 circle points on the moving body are combined as a group to constitute a planar four-bar linkage, whose parameters can be calculated by equation (2.58). After estimating the constraints, a lot of crank-rocker linkages are synthesized, three of which are shown in Table 2.14.

2.5.3 *Crank-slider Linkage*

The discrete kinematic synthesis of planar crank-slider linkages is to search out a saddle circle point and a saddle sliding point on the moving body, while the design requirements of the geometrical dimensions and the transmission performances of the linkage have to be met. In this section, the mathematical model and the process for the kinematic synthesis of planar crank-slider linkages are introduced by saddle point programming. Then, several numeral examples are given, including multiple separated positions, four and three separated positions of the moving body.

2.5.3.1 The Mathematical Model

For the discrete kinematic synthesis of crank-slider linkages, a saddle circle point and a saddle sliding point have to be located on the moving body, and designated as the centers of two R-pairs on the moving body. The saddle circle point P_1 corresponds to a saddle circle, whose center point in the fixed body is assigned as the center of the fixed R-pair. The saddle sliding point P_2 corresponds to a saddle line in the fixed body, which is taken as the guide line of the P-pair in the fixed body. Thus, a four-bar linkage with a slider is constructed. The mathematical model of the discrete kinematic synthesis of planar crank-slider linkages is set up as

$$\begin{cases} \min F(\mathbf{Z}) = \min(\Delta_{\mathrm{rr}}(z_1) + \Delta_{\mathrm{pr}}(z_2)) \\ \text{s.t. } g_j(\mathbf{Z}) \leq 0, j = 1, 2, \ldots, k \end{cases} \tag{2.64}$$

which can be explained in detail as follows.

A. **The objective function and optimization variables**
The sum of the saddle circle error $\Delta_{\mathrm{rr}}(z_1)$ and the saddle line error $\Delta_{\mathrm{pr}}(z_2)$ for two different points P_1 and P_2 on the moving body is taken as the objective function. The definitions of saddle circle and saddle line are presented respectively in Sections 2.3.1 and 2.4.1.

The optimization variables are $\mathbf{Z} = (z_1, z_2)^T$, or $z_1 = (x_{Pm1}, y_{Pm1})^T$ and $z_2 = (x_{Pm2}, y_{Pm2})^T$, which are the rectangular coordinates of two points on the moving body.

B. **The constraint equations**

As mentioned above, the points $P_1(x_{Pm1}, y_{Pm1})$ and $P_2(x_{Pm2}, y_{Pm2})$ are set as the two moving joint R-pairs on the moving body, while the center point $C_1(x_{C1}, y_{C1})$ corresponding to the saddle circle point P_1 is taken as the center of the fixed R-pair and the saddle line to P_2 is taken as the guide line of the fixed P-pair. The dimensions of the so-obtained four-bar linkage with a slider are

$$\begin{cases} l_1 = r_1 \\ l_2 = \sqrt{(x_{Pm1} - x_{Pm2})^2 + (y_{Pm1} - y_{Pm2})^2} \\ e = |h_2 - x_{C1}\cos\phi_2 - y_{C1}\sin\phi_2| \end{cases} \tag{2.65}$$

Considering the requirements of geometrical dimensions and transmission performances, we list the constraint conditions as follows.

1. The geometrical conditions. The length of each link for a linkage has to be larger than a designated positive value first in order to be easily manufactured, that is

$$g_1(\mathbf{Z}) = \varepsilon - \min(l_1, l_2) \le 0 \tag{2.66}$$

where ε is the permitted minimum positive value for the restriction of the link length.

The linkage should have a suitable ratio of the link length, which can be represented by

$$g_2(\mathbf{Z}) = \max(l_1, l_2, e)/\min(l_1, l_2, e) - [l] \le 0 \tag{2.67}$$

where $[l]$ is the permitted maximum ratio of the link lengths.

If the four-bar linkage is a crank-slider linkage, the condition of a Grashof kinematic chain has to be met, or

$$g_3(\mathbf{Z}) = l_1 - l_2 + e \le 0 \tag{2.68}$$

2. The transmission angle should be

$$g_4(\mathbf{Z}) = [\gamma] - \arccos\left(\frac{l_1 + e}{l_2}\right) \tag{2.69}$$

where $[\gamma]$ is the allowable minimum transmission angle.

Similarly, in practical applications, there exist other constraints for the kinematic synthesis of planar four-bar linkages with a slider, such as the position constraint of the joints, the requirement of the movement space, the movement sequence, and the circuit and branch. The additional constraint equations have to be given according to the different application cases.

C. **The optimization algorithm**

For the kinematic synthesis of a planar four-bar linkage with a slider under constraint conditions, the optimization model is established as equation (2.64) and the function **fmincon** in Matlab can be adopted directly to solve the equation.

D. **Initial values of the optimization variables**

In the kinematic synthesis of planar crank-slider linkages, the initial values of the optimization variables, or the coordinates $z_1 = (x_{Pm1}, y_{Pm1})^T$ and $z_2 = (x_{Pm2}, y_{Pm2})^T$ of two points of the moving body have to be inputted at first to run the optimization program. One saddle sliding point and one saddle circle point are chosen as a group of initial points.

E. **Stop conditions**
Similarly, the stop conditions are set by the function **optimset** in Matlab.

2.5.3.2 Multi-position Synthesis

A planar crank-slider linkage is synthesized, which guides the coupler link to occupy approximately multiple given positions. No matter how many separated positions are given, the saddle circle points and the saddle sliding points can respectively be located by calling the sub-programs ArrP and AprP. However, there may exist precise solutions, circle points, and sliding points if the number of given separated positions is less than five.

The mathematical model (2.64) of the discrete kinematic synthesis of crank-slider linkages can be solved directly by Matlab. The function **fmincon** can be set up by referring to the program **KS-CR**. As mentioned above, a **program for kinematic synthesis of a crank-slider linkage under multiple positions**, or **KS-CS** for short, is naturally constructed for multiple separated positions of a moving body, whose input are the separated positions of a moving body and output are crank-slider linkages with dimensions and performance.

One saddle sliding point with its saddle line and one saddle circle point with the center point of its saddle circle are taken as the joint centers of a crank-slider linkage. In general, both the saddle circle point error and the saddle sliding point error, which depends on the motion properties of the moving body, are nonzero. The process of the discrete kinematic synthesis of planar crank-slider linkages is shown by the following numeral example.

Example 2.9 Ten positions of the rigid body are described by $(x_{Omf}^{(i)}, y_{Omf}^{(i)}, \gamma^{(i)})$, $i = 1, \ldots, 10$ and listed in Table 2.1. Synthesize a planar crank-slider linkage with minimum allowable transmission angle $[\gamma] = 40°$ and maximum permitted ratio of link length $[l] = 4$

Solution

Based on equation (2.64), the mathematical model of the discrete kinematic synthesis of crank-slider linkages is established as

$$\begin{cases} \min\limits_{Z=(z_1,z_2)^T} (\Delta_{rr}(z_1) + \Delta_{pr}(z_2)) \\ \text{s.t. } g_1(\mathbf{Z}) = l_1 - l_2 + e \leq 0 \\ \quad g_2(\mathbf{Z}) = \max(l_1, l_2, e)/\min(l_1, l_2, e) - 4 \leq 0 \\ \quad g_3(\mathbf{Z}) = \dfrac{2\pi}{9} - \arccos\left(\dfrac{l_1 + e}{l_2}\right) \leq 0 \\ \quad g_4(\mathbf{Z}) = 0.1 - \min(l_1, l_2, e) \leq 0 \end{cases} \tag{E2-9.1}$$

The letters in the equations above have the same meanings as those in equations (2.64) to (2.69). The process of the discrete kinematic synthesis of crank-slider linkages has three steps: preparation of initial values, program settings, and program operation.

1. **Preparation of initial values**
Firstly, the saddle circle points are located by the sub-program ArrP for the given 10 separated positions of the moving body, such as SCP1–SCP12 with lesser errors in Table 2.2.
There is only one saddle sliding point SSP1 on the moving body in Example 2.5, whose rectangular coordinates are (−7.5887,18.4669) in the moving Cartesian coordinate system; the corresponding saddle line has parameters $(h, \phi) = (23.3564, 1.5710)$ in the base link and its saddle line error Δ_{pr} is 0.056591.
The 12 saddle circle points SCP1–SCP12 in Table 2.2 can be combined with the saddle sliding point SSP1 to be 12 groups, or 12 four-bar linkages with a slider, whose dimensions can be calculated through equation (2.65). After estimating the constraint equations (2.66)–(2.69), the six groups (GI–GVI) are chosen as the initial values and listed in Table 2.15, although they can't satisfy the constraints, but are better than others.
2. **Program settings**
For the kinematic synthesis program **KS-CS**, the parameter settings of the function **fmincon** can refer totally to the previously introduced content, including the optimization algorithm, termination conditions, and constraint conditions setting. The 10 positions of the moving body are described by the parameters $(x_{Omf}^{(i)}, y_{Omf}^{(i)}, \gamma^{(i)})$, $i = 1, \ldots, 10$ in Table 2.1. The nonlinear inequality constraints should be constructed through the inequalities g_1, g_2, g_3, and g_4 from equation (E2-9.1). The lower bound is $\text{lb} = [-60.66, -60.66, -60.66, -60.66]$ and the upper bound is $\text{ub} = [60.66, 60.66, 60.66, 60.66]$.
3. **Program operation**
Based on the six groups of initial values (GI–GVI), we run the program **KS-CS** for the kinematic synthesis of a crank-slider linkage with multi-positions, and have six corresponding solutions listed in Table 2.16. In particular, for the six solutions, new characteristic points are assigned as the moving joints on the moving body, other than the initial points, such as P_1, P_3, P_5, and P_7 instead of SCP1 and SCP8–SCP12, while P_2, P_4, P_6, and P_8 replace SSP1. Despite these new points not being the saddle circle points or the saddle sliding points, they have met the constraint conditions but lost the extreme values of the saddle curve fitting error. The new points and their corresponding parameters are listed in Tables 2.17 and 2.18.

Table 2.15 The initial values(combination of saddle circle points and saddle sliding point)

No.	Groups	Dimensions of the linkages	Maximum length ratio	Minimum transmission angle	Object function value
GI	SCP1–SSP1	$l_1 = 9.0233$, $l_2 = 17.1330$, $e = 5.7655$	2.97	30.33°	0.064049
GII	SCP12–SSP1	$l_1 = 9.3539$, $l_2 = 17.8892$, $e = 5.1535$	3.47	35.81°	0.424607
GIII	SCP8–SSP1	$l_1 = 43.5257$, $l_2 = 5.6239$, $e = 48.1688$	8.56	0°	0.289235
GIV	SCP9–SSP1	$l_1 = 29.5733$, $l_2 = 7.1793$, $e = 35.3681$	4.93	0°	0.294995
GV	SCP10–SSP1	$l_1 = 38.0188$, $l_2 = 6.1564$, $e = 43.0700$	7.00	0°	0.294620
GVI	SCP11–SSP1	$l_1 = 36.5932$, $l_2 = 6.3096$, $e = 41.7595$	6.62	0°	0.295552

Table 2.16 Six synthesized crank-slider linkages

No.	Initial values	Hinged joints	Dimensions of crank-slider linkages	Length ratio	Transmission angle	Object function value
Solution 1	GI	P_1–P_2	$l_1 = 9.0280$, $l_2 = 17.4945$, $e = 4.3736$	4	40°	0.673841
Solution 2	GII	P_3–P_4	$l_1 = 9.3210$, $l_2 = 18.0623$, $e = 4.5156$	4	40°	0.651078
Solution 3	GIII	P_5–P_6	$l_1 = 9.2948$, $l_2 = 18.0115$, $e = 4.5029$	4	40°	2.232560
Solution 4	GIV	P_5–P_6	$l_1 = 9.2948$, $l_2 = 18.0115$, $e = 4.5029$	4	40°	2.232560
Solution 5	GV	P_7–P_8	$l_1 = 12.5413$, $l_2 = 25.4003$, $e = 6.5281$	3.89	41.34°	6.740214
Solution 6	GVI	P_5–P_6	$l_1 = 9.2948$, $l_2 = 18.0115$, $e = 4.5029$	4	40°	2.232560

Table 2.17 The new joints and their saddle circles

	Coordinates		Saddle circle			
			Coordinates of the center		Radius	Saddle circle error
	x_{Pm}	y_{Pm}	x_C	y_C	r	δ_{rr}
P_1	−1.9283	2.2629	16.1628	17.5870	9.0280	0.010796
P_3	−1.8730	1.2129	15.1216	17.4943	9.3210	0.295778
P_5	−1.1714	31.2713	46.4629	19.7389	9.2948	1.883346
P_7	10.5511	40.0710	54.1880	6.7001	12.5413	4.342779

Table 2.18 The new joints and their saddle lines

	Coordinates		Parameters of the saddle line		Saddle line error
	x_{Pm}	y_{Pm}	h	ϕ	δ_{pr}
P_2	−5.3890	19.4117	22.4427	1.5405	0.663045
P_4	−5.8755	18.8262	22.0542	1.5679	0.355300
P_6	−8.4361	17.7899	23.7208	1.5824	0.349233
P_8	11.5603	14.6908	10.0747	1.3849	2.397428

The first two solutions listed in Table 2.16, corresponding to the two crank-slider linkages, are taken as the final results since their objective function values are less. The deviations between the given positions and those guided by the two synthesized crank-slider linkages are listed respectively in Tables 2.19 and 2.20.

Table 2.19 The position deviations for solution 1

Position	$\Delta x_{Omf}^{(i)}$	$\Delta y_{Omf}^{(i)}$	$\Delta\gamma^{(i)}$	Position	$\Delta x_{Omf}^{(i)}$	$\Delta y_{Omf}^{(i)}$	$\Delta\gamma^{(i)}$
1	0.0534	−0.1012	0.0289	6	−0.0601	0.0276	−0.0604
2	0.0621	−0.0975	0.0358	7	−0.0592	0.1072	−0.0598
3	0.0733	−0.0767	0.0367	8	−0.0509	−0.0217	−0.0107
4	0.0463	−0.0197	0.0294	9	0.0277	−0.0748	0.0040
5	−0.0463	0.0465	−0.0068	10	−0.0671	0.0246	0.0152

Table 2.20 The position deviations for solution 2

Position	$\Delta x_{Omf}^{(i)}$	$\Delta y_{Omf}^{(i)}$	$\Delta\gamma^{(i)}$	Position	$\Delta x_{Omf}^{(i)}$	$\Delta y_{Omf}^{(i)}$	$\Delta\gamma^{(i)}$
1	−0.2363	−0.3202	0.0319	6	0.2547	0.2450	−0.0544
2	0.0651	−0.2034	0.0265	7	−0.2607	0.2433	−0.0391
3	−0.1101	0.0445	0.0158	8	0.1198	−0.1448	0.0198
4	−0.2023	−0.2322	0.0325	9	0.2479	0.2781	−0.0019
5	−0.0488	−0.0486	0.0042	10	−0.0788	−0.0036	0.0182

2.5.3.3 Four-position Synthesis

For a moving body with four given separated positions, theoretically, there are infinite circle points on the moving lamina while there exists only one sliding point. This is viewed as a precise kinematic synthesis of four-bar linkages with a slider, geometrically comprised of a circle point and a sliding point, guiding a moving lamina to pass through the four given separated positions.

Sometimes, the four-bar linkage with a slider constituted of a circle point and a sliding point may not meet the design requirements, or the performance may not be good enough. The saddle sliding points would be taken into account for more choice in such a situation. Generally, the saddle kinematic synthesis of crank-slider linkages may be adopted, preferring to lose a little geometrical precision but take the better performance for the constraint conditions. Unfortunately, except for a sliding point, no other saddle sliding points exist on the moving body with four prescribed positions. The points close to the sliding point with less saddle line error would be chosen to constitute the four-bar linkages with a slider to meet the constraint conditions.

Example 2.10 Synthesize a planar crank-slider linkage with minimum allowable transmission angle $[\gamma] = 15°$ and maximum permitted ratio of link length $[l] = 4$, guiding the lamina to pass through four given positions described by $(x_{Om}^{(i)}, y_{Om}^{(i)}, \gamma^{(i)})$, $i = 1, \ldots, 4$ as listed in Table 2.13.

Solution

The kinematic synthesis of a planar crank-slider linkage under four given positions is to search out two characteristic points, a circle point $P_1(x_{Pm1}, y_{Pm1})$ and a sliding point $P_2(x_{Pm2}, y_{Pm2})$

Table 2.21 The obtained crank-slider linkages for four-position synthesis

	Circle point		Center point		Dimensions of the crank-slider linkage	Maximum length ratio	Minimum transmission angle
	x_{Pm}	y_{Pm}	x_C	y_C	l_1, l_2, e		
1	0.8357	3.4099	1.1260	1.7403	$l_1 = 2.7623$, $l_2 = 4.6267$, $e = 1.6672$	2.78	16.79°
2	0.8237	3.3888	1.1255	1.7424	$l_1 = 2.7369$, $l_2 = 4.6033$, $e = 1.6688$	2.76	16.85°
3	0.5249	2.8765	1.1121	1.8009	$l_1 = 2.1163$, $l_2 = 4.0349$, $e = 1.7151$	2.35	18.27°

on the moving body, the summation of whose corresponding saddle circle error and saddle line error reaches a minimum. The optimization model is still the same as equation (E2-9.1).

Based on the optimization model (2.45) of the saddle sliding point and the sub-program AprP, the sliding point can be located on the moving lamina with rectangular coordinates (−2.4728, 0.1757). The parameters (h, ϕ) of its saddle line are (0.3930, 1.1071).

The circle points all distribute on the cubic curve on the moving body with the given four positions, which can be calculated by the sub-program ArrP. A circle point can be chosen and combined with the sliding point to constitute a four-bar linkage with a slider, whose dimensions can be calculated by equation (2.65), and the constraint equations (2.66) to (2.69) are adopted to judge the constraint conditions, such as the existence of the crank, the minimum transmission angle, and the maximum ratio of link lengths. In this way, three circle points are chosen and combined with the sliding point to constitute crank-slider linkages, as listed in Table 2.21.

2.5.3.4 Three-position Synthesis

For the given three positions of a moving lamina, any point on the lamina can be designated as a circle point, or a moving hinged point. Meanwhile, there is a sliding point circle on the lamina, where any point is a sliding point, or another moving joint. A circle point and a sliding point can constitute a four-bar linkage with a slider. In such a situation, there are infinite four-bar linkages with a slider, which can guide the lamina to pass through the three given separated positions precisely. It is unnecessary to discuss the synthesis of a crank-slider linkage with three given separated positions except in terms of higher performance.

2.6 The Saddle Kinematic Synthesis of Planar Six-bar Linkages with Dwell Function

The dwell mechanisms, which convert a continuous motion of the input link into a dwell motion of the output link, have been widely applied in industry, especially in the textile industry. Among the dwell mechanisms, six-bar dwell linkages have many distinct advantages, such as easy manufacture and maintenance, and good performance of dynamics compared with other dwell mechanisms like cams. However, more light needs to be shed on the synthesis approach of six-bar dwell linkages if it is expected to apply in more practical cases instead of cam mechanisms. The saddle point programming of local saddle curve fitting is first applied to the kinematic synthesis of six-bar dwell linkages.

2.6.1 Six-bar Linkages

The function synthesis of linkages is to find a linkage, whose driving (or input) link and driven (or output) link have corresponding displacements described by a given function. For a general function without dwell, a four-bar linkage is chosen preferably, as mentioned in Section 2.5.1.3. We are interested in six-bar dwell linkages for the function with dwell, since this is beyond four-bar linkages. A function with dwell, or a dwell function between independent variable and dependent variable, or input parameter and output parameter, may have different physical meanings for different six-bar dwell linkages, such as 7R, 6RP and 5RPR linkages. This will be briefly discussed in the following.

2.6.1.1 The Dwell Function

A dwell function, the function between independent variable and dependent variable, is divided into three parts, the upper part, dwell part, and down part, as shown in Fig. 2.34. It is

$$F = \begin{cases} F_1(\theta) & \theta_1 \leq \theta < \theta_2 \\ F_0 & \theta_2 \leq \theta \leq \theta_3 \\ F_2(\theta) & \theta_3 < \theta \leq \theta_4 \end{cases} \tag{2.70}$$

where θ is the independent variable and F is the dependent variable, whose period may be 2π or not, while $\theta_1 - \theta_2$, $\theta_2 - \theta_3$, and $\theta_3 - \theta_4$ are the defined area of the independent variable for the upper part, the dwell part, and the down part respectively.

For a six-bar dwell linkage, whose input parameter is generally a rotating angle displacement θ, and output parameter may be a rotating angle displacement ψ or linear sliding displacement S, the letter F in equation (2.70) can be replaced by ψ and S according to the type of six-bar dwell linkage.

2.6.1.2 Six-bar Linkages with Dwell Motion

Planar six-bar linkages with one degree of freedom can be classified into WATT linkages and STEPHENSON linkages. Both WATT and STEPHENSON linkages can be generated by a

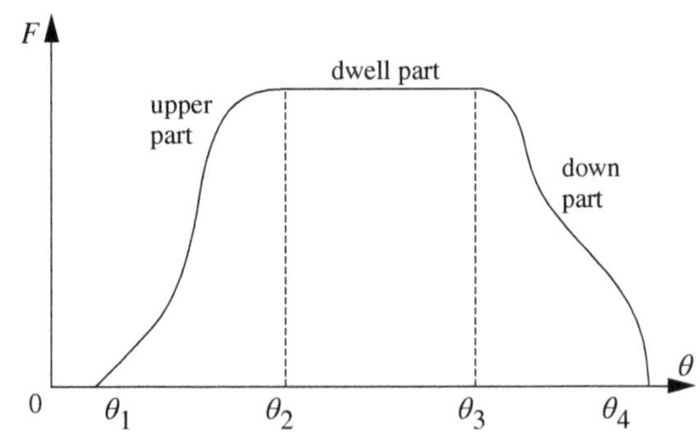

Figure 2.34 A dwell function

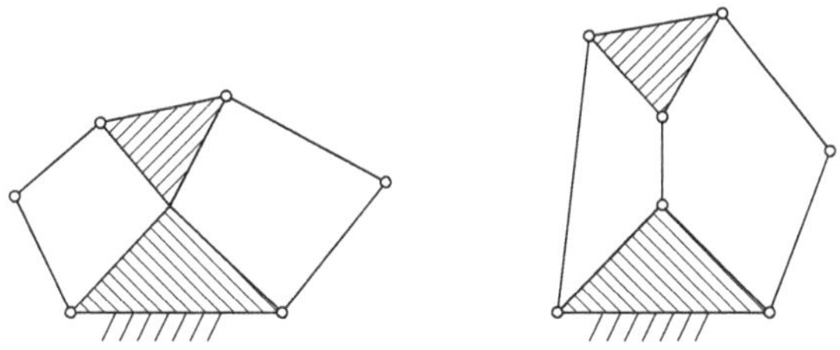

Figure 2.35 Six-bar linkages (a) WATT linkage (b) STEPHENSON linkage

four-bar linkage, called a basic four-bar linkage, and a grade II Assur group, if different positions are chosen to connect them (as shown in Fig. 2.35). A WATT linkage cannot produce a dwell function between its input and output, but a STEPHENSON linkage does since a grade II Assur group connects the coupler point of a basic four-bar linkage to its base link while the coupler curve has special geometric characteristics. Thus, we focus on discussing the kinematic synthesis of dwell mechanisms for STEPHENSON linkages in this section.

Generally, a STEPHENSON linkage is taken to produce a dwell motion between its input and output, whose basic four-bar linkage is designated as a crank-rocker linkage. Either the crank or the rocker is assigned as a driving link or input, while the link of a grade II Assur group connected with the base link is a driven link or output. The performance of the dwell function depends on the geometric characteristics of both coupler curves of a crank-rocker linkage and a grade II Assur group. The coupler curves of crank-rocker linkages are globally discussed in Chapter 1, in particular the distributions of coupler curves with a cusp, a Ball point, a Burmester point, a double point, and the symmetrical coupler curves, etc. A grade II Assur group may have three types – 3R, RRP, and PRR – in which the former two with a rocker or guiding link lead to an output with angular displacement, as shown in Figs 2.36(a) and 2.37, and the latter with a slider leads to a line displacement, as shown in Fig. 2.36(b). The combination of a basic four-bar linkage and a grade II Assur group, by taking advantage of the geometric characteristics of coupler curves, generates different output dwell motions, like linear displacement, angle displacement in either a symmetrical cycle or a non-symmetrical cycle, even in multi-cycles.

As shown in Fig. 2.36, the point P_B on the coupler plane of a four bar linkage, an instantaneous Burmester point, traces a coupler curve Γ_P, which contacts a circle in higher order. Both the coupler point P_B and the curvature center C_B of curve Γ_P are designated as the two hinge points of a link EF, which halts at that instant. If the link EF belongs to the grade II Assur group RRR, we can obtain a planar hinged 7R six-bar linkage with dwell output angular displacement of the link FG when the point G is located at a suitable position fixed in the base link. If the link EF belongs to the grade II Assur group PRR, we can get a planar 6RP six-bar linkage with dwell output linear displacement of the slider when a suitable guiding line is fixed in the base link.

As shown in Fig. 2.37, the coupler point P_b, an instantaneous Ball point, on the coupler plane of a four-bar linkage, traces a coupler curve Γ_P, which has higher-order contact with a line. The coupler point P_b is assigned as a hinge point E of the grade II Assur group RRP and the osculating line of the coupler curve Γ_P is designated as the guiding line of the P-pair. We can

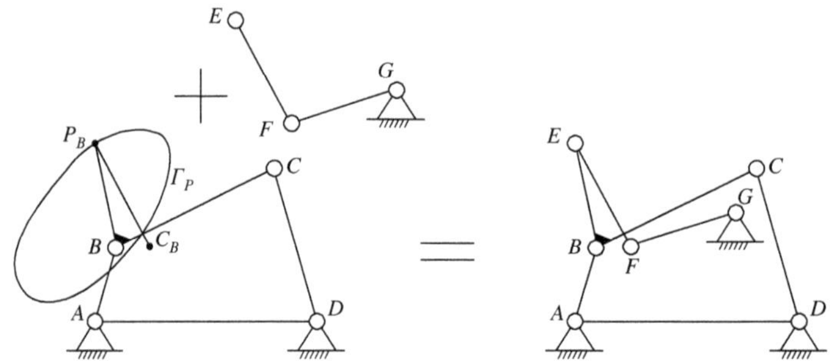

(a)

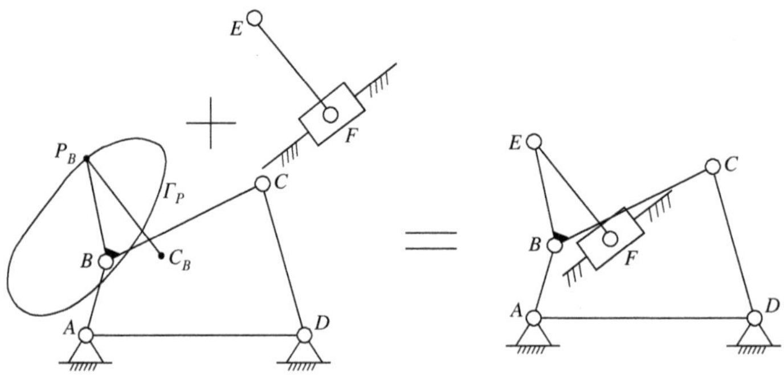

(b)

Figure 2.36 The six-bar linkages with dwell functions (a) A basic four-bar linkage and a grade II Assur group 3R (b) A basic four-bar linkage and a grade II Assur group PRR

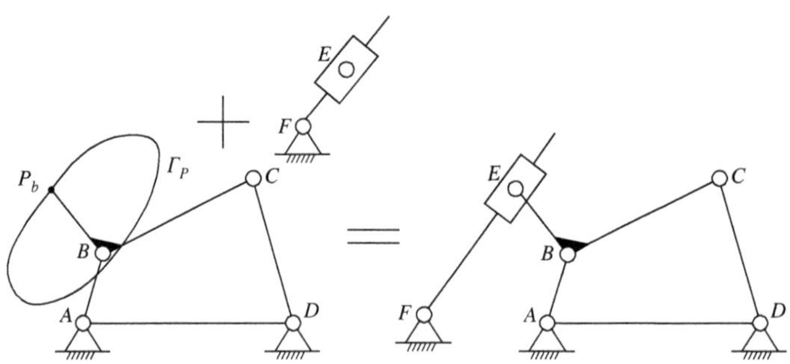

Figure 2.37 A basic four-bar linkage and a grade II Assur group RRP

get the planar 6RP six-bar linkage with dwell output angular displacement when a suitable point F is located in the base link.

Based on a Burmester point and a Ball point on the coupler plane of a four-bar linkage, a six-bar linkage can only produce a dwell output at an instant. However, a large-scale output dwell motion is expected or the coupler curve of a basic four-bar linkage is close to a circle arc or a line segment in large scale. This is a difficult but interesting topic, especially when the dwell output displacement is required to have geometrical characteristics with symmetry and multi-cycling. Hence, the kinematic differential geometry for a four-bar linkage might be expected to have a considerable contribution to the kinematic synthesis of a six-bar linkage with dwell output in large scale.

Based on the distribution of coupler curves for a planar four-bar linkage in Chapter 1, the saddle point programming approach is developed and applied to the local saddle curve fitting. This means that a coupler curve is expected to be close to a circle arc or a line in a definite local segment, whose corresponding coupler points are respectively assigned as a local saddle circle point and a local saddle sliding point. The kinematic synthesis of a planar six-bar linkage with dwell output, or a STEPHENSON linkage with dwell output, is to search out a suitable local saddle circle point or local saddle sliding point on the coupler plane of a basic four-bar linkage and to design both reasonable dimensions and positions for the corresponding grade II Assur group.

2.6.2 *Local Saddle Curve Fitting*

A coupler curve, whose independent variable is the rotational angle φ of the crank, is traced by a coupler point $P(x_{Pm}, y_{Pm})$ of a planar crank-rocker linkage. If the rotational angle φ is discretized into $\varphi^{(i)}$, $i = 1, \dots, n$, the coupler curve is described by the discrete point set $\{\boldsymbol{R}_P^{(i)}\} = \{(x_P^{(i)}, y_P^{(i)})^T\}$, $i = 1, \dots, n$ as

$$\begin{aligned}\boldsymbol{R}_P^{(i)} &= (l_1 \cos\varphi^{(i)} + x_{Pm}\cos\gamma^{(i)} - y_{Pm}\sin\gamma^{(i)})\boldsymbol{i}_f \\ &\quad + (l_1 \sin\varphi^{(i)} + x_{Pm}\sin\gamma^{(i)} + y_{Pm}\cos\gamma^{(i)})\boldsymbol{j}_f \end{aligned} \tag{2.71}$$

where $\gamma^{(i)}$ is an inclination angle of the coupler link with respect to the base link. l_1 is the length of the crank. The discrete points on the segment $\overline{P^{(j)}P^{(j+\Delta j)}}$ of a coupler curve are obtained when the independence variable $\varphi^{(i)}$ varies from $\varphi^{(j)}$ to $\varphi^{(j+\Delta j)}$. We are interested in whether the segment $\overline{P^{(j)}P^{(j+\Delta j)}}$ approximates a circle or not, and how to estimate the degree of approximation for the kinematic synthesis of a planar six-bar linkage with dwell output.

The curve segment $\overline{P^{(j)}P^{(j+\Delta j)}}$ begins at the rotational angle $\varphi^{(j)}$ and ends at $\varphi^{(j+\Delta j)}$, while the length of the curve segment is determined by the interval $\Delta\varphi = \varphi^{(j+\Delta j)} - \varphi^{(j)}$. There is certainly a saddle circle for a given curve segment $\overline{P^{(j)}P^{(j+\Delta j)}}$ with interval $\Delta\varphi$. For a curve with n discrete points, there exist many segments with interval $\Delta\varphi$ if the beginning point j varies from 1 to n, and their corresponding saddle circles and errors $\Delta^{(j)}$ can also be calculated. Among all the errors $\Delta^{(j)}$, there exists a minimum value $\Delta^{(j_0)} = \min\limits_{1<j<n}(\Delta^{(j)})$.

Definition 2.5 A circle, fitting the segments of a discrete curve $\{\boldsymbol{R}_P^{(i)}\}$ within a given interval $\Delta\varphi$ by letting the maximum fitting error be the minimum, is called a ***local saddle circle***.

The saddle curve fitting, which is discussed intensively in Sections 2.3 and 2.4, is developed into ***local saddle curve fitting***, including ***local saddle circle fitting and local***

saddle line fitting. The mathematical model of the local saddle circle fitting is first established as

$$\begin{cases} \Delta^{(j)} = \min\limits_{x} \max\limits_{j \le I \le j+\Delta j} \{\Delta^{(I)}(\boldsymbol{x})\} \\ \quad = \min\limits_{x} \max\limits_{j \le I \le j+\Delta j} \left\{ \left| \sqrt{(x_P^{(I)} - x_C)^2 + (y_P^{(I)} - y_C)^2} - r \right| \right\} \\ \text{s.t.} \quad r \in (0, +\infty) \end{cases} \tag{2.72}$$

where $\{\Delta^{(I)}(\boldsymbol{x})\}$ is the object function, or the fitting errors of the discrete points on the curve segment $\overline{P^{(j)}P^{(j+\Delta j)}}$ to a fitting circle. $\boldsymbol{x} = (x_C, y_C, r)^T$ are the optimization variables, or the rectangular coordinates of the center point and the radius of the fitting circle respectively. $\Delta^{(j)}$ is the maximum normal fitting error for all discrete points on $\overline{P^{(j)}P^{(j+\Delta j)}}$ to the local saddle circle.

On the contrary, we focus on the approximation of a curve segment $\overline{P^{(j)}P^{(j+\Delta j)}}$ to a straight line. The curve segment $\overline{P^{(j)}P^{(j+\Delta j)}}$ begins at the rotational angle $\varphi^{(j)}$ and stops at $\varphi^{(j+\Delta j)}$, while the length of the curve segment is determined by the interval $\Delta\varphi = \varphi^{(j+\Delta j)} - \varphi^{(j)}$. When j varies from 1 to n, every curve segment has a saddle line and corresponding saddle line error.

Definition 2.6 A line, fitting the segments of a discrete curve $\{\boldsymbol{R}_P^{(i)}\}$ within a given interval $\Delta\varphi$ by letting the maximum fitting error be the minimum, is called a ***local saddle line***.

The mathematical model of the local saddle line fitting is established as

$$\begin{cases} \Delta^{(j)} = \min\limits_{x} \max\limits_{j \le I \le j+\Delta j} \{\Delta^{(I)}(\boldsymbol{x})\} \\ \quad = \min\limits_{x} \max\limits_{j \le I \le j+\Delta j} \left\{ \left| x_P^{(I)} \cos\phi + y_P^{(I)} \sin\phi - h \right| \right\} \\ \text{s.t.} \quad h \in [0, +\infty),\ \phi \in [0, 2\pi) \end{cases} \tag{2.73}$$

Here, $\{\Delta^{(I)}(\boldsymbol{x})\}$ is the objective function, or the fitting errors of the discrete points on the curve segment $\overline{P^{(j)}P^{(j+\Delta j)}}$ to a fitting straight line. $\boldsymbol{x} = (h, \phi)^T$ are the optimization variables, or the parameters of the fitting line. $\Delta^{(j)}$ is the maximum normal fitting error for all discrete points on $\overline{P^{(j)}P^{(j+\Delta j)}}$ to the local saddle line.

2.6.3 Dwell Function Synthesis

The dwell function, described as $F = F(\theta)$, $\theta \in (\theta_1, \theta_4)$ in equation (2.70), can be generated by a STEPHENSON six-bar linkage with 7R or 5RPR if both input and output of the dwell function are angular displacements, and by a six-bar linkage with 6RP if the output is a linear displacement. This is key to setting up a mathematical model for synthesizing a STEPHENSON six-bar linkage with a given dwell function output, which includes choosing a suitable basic four-bar linkage, searching out a coupler curve with suitable geometric characteristics, and determining sufficient dimensions of a grade II Assur group, etc. We prefer to focus on the mathematical model for the grade II Assur group, rather than the basic four-bar linkages and coupler curves, which are discussed widely in Chapter 1.

2.6.3.1 7R Six-bar Linkages with Dwell Function

A planar 7R six-bar linkage can be synthesized by the combination of a basic hinged four-bar linkage and a grade II Assur group RRR, shown in Fig. 2.38. The synthesis includes locating the positions of both point E on the coupler plane and point G on the base link, and determining the dimensions of a grade II Assur group RRR if the basic four-bar linkage is given. The locations of the coupler point E and the fixed pivot point G, as well as the optimization model for the Assur group, are introduced respectively as follows.

(1) The location of coupler point E. The coordinates of coupler point E are $(x_{Em}, y_{Em})^T$ in the moving frame $\{\boldsymbol{B}; \boldsymbol{i}_m, \boldsymbol{j}_m\}$ of the basic four-bar linkage $ABCD$, or $l_e = \sqrt{x_{Em}^2 + y_{Em}^2}$, and its coupler curve can be described by discrete points $\{\boldsymbol{R}_E^{(i)}\}$, $i = 1, \ldots, n$. For the rotation angle $\Delta\varphi = \theta_3 - \theta_2$, corresponding to a given dwell interval in equation (2.70), a curve segment $E^{(j)}E^{(j+\Delta j)}$ can be determined by means of the mathematical model of the local saddle circle fitting, or equation (2.72), whose local saddle circle with center point $F(x_{F0}, y_{F0})$ in the base link and radius r can also be calculated. Thus, the radius r is designated as the length l_5 of link EF for the grade II Assur group 3R, and the beginning angle $\varphi^{(j_0)} - \theta_2$ of the dwell motion of a 7R six-bar linkage corresponds to the rotation angle $\varphi^{(j_0)}$ of the basic crank-rocker linkage.

(2) The location of point G. The grade II Assur group 3R has the link FG with two joints F and G, in which the joint F coincides instantaneously with the center point $F(x_{F0}, y_{F0})$ of the local saddle circle in the base link, and the fixed pivot joint G is optimally located by the mathematical model. Hence, the length l_6 of link FG can be determined by the positions of both point F and point G in the base link:

$$l_6 = \sqrt{(x_G - x_{F0})^2 + (y_G - y_{F0})^2} \tag{2.74}$$

(3) The mathematical model. So far, a planar six-bar linkage with basic four-bar linkage $ABCD$ and grade II Assur group EFG has been constructed, but its output should be compared with the given dwell function in equation (2.70). The rectangular coordinates (x_F, y_F) of F in

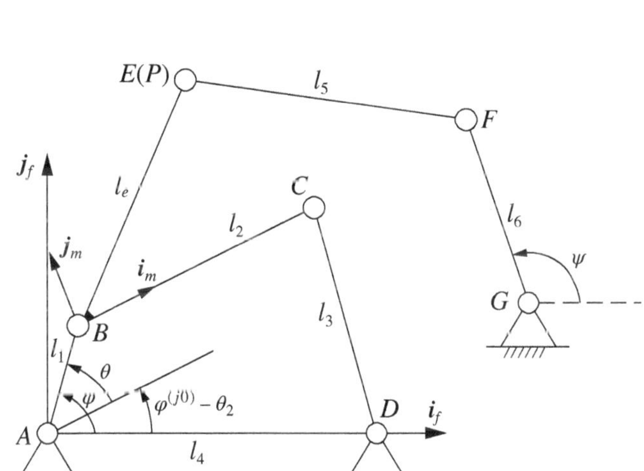

Figure 2.38 A planar 7R six-bar linkage with dwell function

the fixed frame $\{\boldsymbol{A}; \boldsymbol{i}_F, \boldsymbol{j}_F\}$ can be derived by the geometrical equations

$$\begin{cases} (x - x_E)^2 + (y - y_E)^2 = l_5^2 \\ (x - x_G)^2 + (y - y_G)^2 = l_6^2 \end{cases} \tag{2.75}$$

or

$$\begin{cases} x_F = x_E + p(x_G - x_E) - q(y_G - y_E) \\ y_F = y_E + p(y_G - y_E) - q(x_G - x_E) \end{cases} \tag{2.76a}$$

where

$$\begin{cases} p = \dfrac{1}{2} + \dfrac{1}{2} \dfrac{l_5^2 - l_6^2}{(x_G - x_E)^2 + (y_G - y_E)^2} \\ q = M\sqrt{\dfrac{l_5^2}{(x_G - x_E)^2 + (y_G - y_E)^2} - p^2} \end{cases} \tag{2.76b}$$

The parameter M can be chosen as $+1$ or -1, which depends on the initial position determined by (x_{F0}, y_{F0}). If $y_F \geq y_G$, the output angle ψ is

$$\psi = \arccos\left(\frac{x_F - x_G}{\sqrt{(x_F - x_G)^2 + (y_F - y_G)^2}}\right) \tag{2.77}$$

Otherwise, if $y_F < y_G$, we have

$$\psi = 2\pi - \arccos\left(\frac{x_F - x_G}{\sqrt{(x_F - x_G)^2 + (y_F - y_G)^2}}\right) \tag{2.78}$$

When the crank angle $\varphi^{(i)}$ of the basic four-bar linkage belongs to $[\varphi^{(j_0)}, \varphi^{(j_0+\Delta j)}]$, or varies from $\varphi^{(j_0)}$ to $\varphi^{(j_0+\Delta j)}$, the output angle $\psi_{\text{I}}^{(i)}$ of the six-bar linkage derived from equations (2.74) to (2.78) is dwell, otherwise the output angle $\psi_{\Pi}^{(i)}$ is non-dwell. Both dwell output $\psi_{\text{I}}^{(i)}$ and non-dwell output $\psi_{\Pi}^{(i)}$ are compared with the given functions $\overline{\psi}_{I}^{(i)}$ and $\overline{\psi}_{\Pi}^{(i)}$ of equation (2.70), whose corresponding errors are

$$\begin{cases} F_1(\boldsymbol{X}) = \max\left(\left|\psi_{\text{I}}^{(i)} - \overline{\psi}_{\text{I}}^{(i)}\right|\right) \\ F_2(\boldsymbol{X}) = \max\left(\left|\psi_{\Pi}^{(i)} - \overline{\psi}_{\Pi}^{(i)}\right|\right) \end{cases} \tag{2.79}$$

For convenience of expression as a function, a weighting coefficient λ is introduced:

$$F(\boldsymbol{X}) = \lambda F_1(\boldsymbol{X}) + (1 - \lambda)F_2(\boldsymbol{X}) \tag{2.80}$$

Naturally, the maximum error is minimum for the synthesis of a planar 7R six-bar linkage with dwell output. The mathematical model can be set up as

$$\begin{cases} \min\limits_{\boldsymbol{X}} F(\boldsymbol{X}) \\ \text{s.t. } \boldsymbol{X} = (x_{Em}, y_{Em}, x_G, y_G)^T \\ \quad g_k(\boldsymbol{X}) \leq 0,\ k = 1, 2, \ldots, m \end{cases} \tag{2.81}$$

which can be discussed in details as follows.

A. **The objective function and optimization variables**
The objective function $F(\boldsymbol{X})$ is the weighting sum of the maximum errors at both dwell interval and non-dwell interval, which can be obtained by equations (2.74) to (2.80).
The optimization variables in the mathematical model are $\boldsymbol{X} = (x_{Em}, y_{Em}, x_G, y_G)^T$, in which (x_{Em}, y_{Em}) are the rectangular coordinates of coupler point E in $\{\boldsymbol{B}; \boldsymbol{i}_m, \boldsymbol{j}_m\}$ and (x_G, y_G) are the rectangular coordinates of the fixed pivot point G in $\{\boldsymbol{A}; \boldsymbol{i}_f, \boldsymbol{j}_f\}$.

B. **The constraint condition**
Considering the requirements of geometrical dimensions and transmission performances, we list the constraint conditions as follows.

1. The geometrical conditions. The linkage is preferred to have a suitable ratio of the link length, or

$$g_1(\boldsymbol{X}) = \max(l_1, l_2, l_3, l_4, l_e, l_5, l_6) / \min(l_1, l_2, l_3, l_4, l_e, l_5, l_6) - [l] \le 0 \tag{2.82}$$

where $[l]$ is the permitted maximum ratio of link lengths.
All lengths of each link in the six-bar linkage have to be larger than a given positive value in order to be manufactured, that is

$$g_2(\boldsymbol{X}) = \varepsilon - \min(l_1, l_2, l_3, l_4, l_e, l_5, l_6) \le 0 \tag{2.83}$$

where ε is the permitted minimum positive value for the restriction of the link length.
A six-bar linkage with dwell output can work correctly when the coupler point E of the basic four-bar linkage locates on the same area as that of the point E of a grade II Assur group EFG, which means that the dimensions of the grade II Assur group EFG have to meet the constraint as follows:

$$\begin{cases} l_5 + l_6 \ge d_{\max} \\ |l_5 - l_6| \le d_{\min} \end{cases} \tag{2.84}$$

where

$$\begin{cases} d_{PG}^{(i)} = \sqrt{(x_E^{(i)} - x_G)^2 + (y_E^{(i)} - y_G)^2} \\ d_{\max} = \max\limits_i(d_{PG}^{(i)}) \\ d_{\min} = \min\limits_i(d_{PG}^{(i)}) \end{cases} \tag{2.85}$$

2. The transmission angle

$$g_3(\boldsymbol{X}) = [\gamma] - \gamma_{\min} \le 0 \tag{2.86}$$

where $[\gamma]$ is the minimum allowable transmission angle. $\gamma_{\min}$ is the minimum transmission angle of the grade II Assur group and can be calculated by

$$\gamma_{\min} = \min\left(\arccos\left[\frac{l_5^2 + l_6^2 - {d_{\min}}^2}{2l_5 l_6}\right], \pi - \arccos\left[\frac{l_5^2 + l_6^2 - {d_{\max}}^2}{2l_5 l_6}\right]\right) \tag{2.87}$$

C. **Initial values of the optimization variables**
The optimization variables are $\boldsymbol{X} = (x_{Em}, y_{Em}, x_G, y_G)^T$ in the mathematical model of synthesis of planar 7R six-bar linkages with dwell output, which are the coordinates of coupler

point E and fixed pivot point G. For the coupler point E, its coupler curve is locally approximated to a circle and good enough for dwell output, such as a Burmester point. For the fixed pivot point G, it is a possible way to give multiple initial points if you can bear the long computation time, which can be expected to cover the possible area where the global optimal solution is located. The initial point area for locations of coupler point E is recommended to be

$$\begin{cases} a_1 = \min(l_1, l_2, l_3, l_4) \\ a_2 = \max(l_1, l_2, l_3, l_4) \\ r_{in} = a_2/[l],\ r_{out} = a_1[l] \end{cases} \tag{2.88}$$

The above equation defines an annulus area with the moving hinge point B as center, $\frac{a_2}{[l]}$ and $a_1[l]$ as the radii of internal and external circles. The initial values for the fixed pivot point G are recommended to be

$$\begin{cases} a_3 = \min(l_1, l_2, l_3, l_4, l_5, l_e) \\ a_4 = \max(l_1, l_2, l_3, l_4, l_5, l_e) \\ x_G = x_{F0} - \left\{ \frac{a_4}{[l]} + c_1 \left(a_3[l] - \frac{a_4}{[l]} \right) \right\} \cos\psi_0 \\ y_G = y_{F0} - \left\{ \frac{a_4}{[l]} + c_1 \left(a_3[l] - \frac{a_4}{[l]} \right) \right\} \sin\psi_0 \end{cases} \tag{2.89}$$

where c_1 is a random value in the range [0, 1]. ψ_0 is the output rotation angle of the 7R six-bar linkage in the dwell stage.

D. **The optimization algorithm**
In this book, the function **fmincon** in Matlab is adopted directly for the synthesis of planar 7R six-bar linkages with dwell output.

2.6.3.2 6RP Six-bar Linkages with Dwell Function

A planar 6RP six-bar linkage can be synthesized by the combination of a basic hinged four-bar linkage and a grade II Assur group PRR, as shown in Fig. 2.39. The synthesis is to locate the positions of both point E on the coupler plane and the guide line FG on the base link, and then to determine the dimensions of the grade II Assur group PRR. The locations of the coupler point E and the link EF for 6RP linkage are the same to those of the former 7R linkage. We just discuss how to locate the fixed guide line FG and the optimization model for the synthesis in the following.

For the discrete path traced by the coupler point E, the point $F(x_{F0}, y_{F0})$ is instantaneously coincident with the center point of its local saddle circle in the base link, while the length l_5 is the radius of the local saddle circle. It is a key topic to locate the guide line FG for the grade II Assur group PRR. The direction angle of the guide line of the prismatic pair is designated as δ in the fixed frame $\{\boldsymbol{A}; \boldsymbol{i}_f, \boldsymbol{j}_f\}$. The rectangular coordinates (x_F, y_F) of the hinged point F in $\{\boldsymbol{A}; \boldsymbol{i}_f, \boldsymbol{j}_f\}$ can be derived by

$$\begin{cases} (x_F - x_E)^2 + (y_F - y_E)^2 = l_5^2 \\ y_F = (x_F - x_{F0}) \tan\delta + y_{F0} \end{cases} \tag{2.90}$$

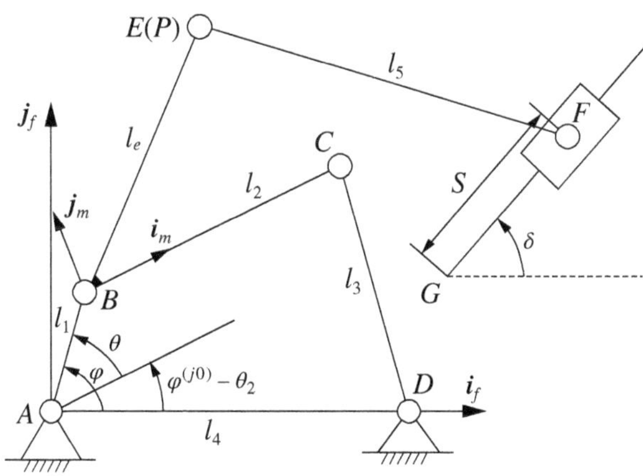

Figure 2.39 A planar 6RP six-bar linkage with dwell function

or

$$\begin{cases} x_F = \frac{-b+M\sqrt{b^2-4ac}}{2a} \\ y_F = \left(x_F - x_{F0}\right)\tan\delta + y_{F0} \end{cases} \tag{2.91a}$$

where

$$\begin{cases} a = 1 + \tan^2\delta \\ b = 2d\tan\delta - 2x_E \\ c = x_E^2 + d^2 - l_5^2 \\ d = y_{F0} - x_{F0}\tan\delta - y_E \end{cases} \tag{2.91b}$$

The parameter M in equation (2.91a) can be chosen as +1 or −1, depending on the position of the center point (x_{F0}, y_{F0}) of the local saddle circle. The output displacement S of the slider is

$$S = S_0 + (x_F - x_{F0})\cos\delta + (x_F - x_{F0})\sin\delta \tag{2.92}$$

As the crank angle $\varphi^{(i)}$ of the four-bar linkage varies from $\varphi^{(j_0)}$ to $\varphi^{(j_0+\Delta j)}$, the joint $F(x_F, y_F)$ has corresponding position coordinates $(x_F^{(i)}, y_F^{(i)})$. Substituting these into equation (2.92), we have the actual output displacement $S_{\mathrm{I}}^{(i)}$ at the dwell interval $[\varphi^{(j_0)}, \varphi^{(j_0+\Delta j)}]$; otherwise, the output displacement $S_{\Pi}^{(i)}$ can be calculated by equations (2.90) to (2.92) at the non-dwell interval. The actual output displacements at both dwell and non-dwell intervals $S_{\mathrm{I}}^{(i)}$ and $S_{\Pi}^{(i)}$ are respectively compared with the given functions $\overline{S}_{\mathrm{I}}^{(i)}$ and $\overline{S}_{\Pi}^{(i)}$ in equation (2.70), so the error functions at the two intervals can be estimated as

$$\begin{cases} F_1(\boldsymbol{X}) = \max(|S_{\mathrm{I}}^{(i)} - \overline{S}_{\mathrm{I}}^{(i)}|) \\ F_2(\boldsymbol{X}) = \max(|S_{\Pi}^{(i)} - \overline{S}_{\Pi}^{(i)}|) \end{cases} \tag{2.93}$$

For convenience in expressing it as a function, a weighting coefficient λ is introduced:

$$F(\boldsymbol{X}) = \lambda F_1(\boldsymbol{X}) + (1 - \lambda)F_2(\boldsymbol{X}) \tag{2.94}$$

The mathematical model for synthesis of planar 6RP six-bar linkage with dwell function can be set up as

$$\begin{cases} \min\limits_{X} F(\boldsymbol{X}) \\ \text{s.t.} \quad \boldsymbol{X} = (x_{Em}, y_{Em}, \delta)^T \\ \qquad g_k(\boldsymbol{X}) \le 0, \ k = 1, 2, \ \dots \ , m \end{cases} \tag{2.95}$$

which can be explained in detail as follows.

A. **The objective function and the optimization variables**
The objective function is the weighted sum of the maximum errors at both dwell and non-dwell intervals, which can be calculated through equations (2.90) to (2.94).
The optimization variables are $\boldsymbol{X} = (x_{Em}, y_{Em}, \delta)^T$, of which (x_{Em}, y_{Em}) are the rectangular coordinates of the coupler point E in $\{\boldsymbol{B}; \boldsymbol{i}_m, \boldsymbol{j}_m\}$, and δ is the directional angle of the guide line of the P-pair of the grade II Assur group PRR in the base link.
B. **The constraint condition**
Considering the requirements of geometrical dimensions and transmission performances, we list the constraint conditions as follows.
1. The geometrical conditions. The linkage is preferred to have a suitable ratio of the link length, or

$$g_1(\boldsymbol{X}) = \max(l_1, l_2, l_3, l_4, l_e, l_5) / \min(l_1, l_2, l_3, l_4, l_e, l_5) - [l] \le 0 \tag{2.96}$$

where $[l]$ is the permitted maximum ratio of link lengths.
All lengths of each link in the six-bar linkage have to be larger than a given positive value in order to be manufactured, that is

$$g_2(\boldsymbol{X}) = \varepsilon - \min(l_1, l_2, l_3, l_4, l_e, l_5) \le 0 \tag{2.97}$$

where ε is the permitted minimum positive value for the restriction of the link length.
The six-bar linkage with dwell output can work correctly when the coupler point E of the basic four-bar linkage passes through the same area as that of the end point E of a grade II Assur group *EFG*, which means that the dimensions of the grade II Assur group *EFG* have to meet the follow constraint:

$$l_5 \ge d_{\max} \tag{2.98}$$

where $d_{\max}$ is the maximum distance from the point E to the guide line:

$$\begin{cases} d^{(i)} = \left| (x_E^{(i)} - x_{F0})(-\sin\delta) + (y_E^{(i)} - y_{F0})\cos\delta \right| \\ d_{\max} = \max(d^{(i)}) \end{cases} \tag{2.99}$$

2. The transmission angle

$$g_3(\boldsymbol{X}) = [\gamma] - \gamma_{\min} \le 0 \tag{2.100}$$

where $\gamma_{\min}$ is the minimum transmission angle of the grade II Assur group and equals $\arccos\left(\frac{d_{\max}}{l_5}\right)$.

C. **Initial values of the optimization variables**
The optimization variables are $\boldsymbol{X} = (x_{Em}, y_{Em}, \delta)^T$ in the mathematical model of synthesis of planar 6RP six-bar linkages with dwell output, which are the coordinates of the coupler point E and the direction angle δ of the guide line FG. The Burmester points on the coupler of the local crank-rocker linkage can also be taken as the initial points for the location of point E. For the direction angle δ, the initial value is recommended to be

$$\begin{cases} c_1 = \text{rand}\,(n) \\ \delta = c_1 \pi \end{cases} \tag{2.101}$$

D. **The optimization algorithm**
Needless to say, the function **fmincon** in Matlab is adopted directly for the synthesis of planar 6RP six-bar linkages with dwell output in this book.

2.6.3.3 5RPR Six-bar Linkages with Dwell Function

A planar 5RPR six-bar linkage can be synthesized by the combination of a local hinged four-bar linkage and a grade II Assur group RPR, shown in Fig. 2.40. The synthesis is to locate the positions of both the coupler point E and the fixed pivot point G, and then to determine the dimensions of a grade II Assur group RPR.

Similar to the synthesis of 7R six-bar linkages with dwell function, the coupler point $E(x_{Em}, y_{Em})$ of the basic four-bar linkage $ABCD$ traces a coupler curve with discrete point set $\{\boldsymbol{R}_E^{(i)}\}$, $i = 1, \ldots, n$ in the fixed frame $\{\boldsymbol{A}; \boldsymbol{i}_f, \boldsymbol{j}_f\}$ by equation (2.71). The curve segment $\overline{E^{(j_0)}E^{(j_0+\Delta j)}}$ and the corresponding local saddle line can be obtained by the mathematical model (2.73). The local saddle line with parameters (h_f, ϕ_f) is designated to coincide with the link EG of the grade II Assur group RPR at the dwell stage. The rectangular coordinates of the fixed hinge point G are supposed to be (x_G, y_G) in $\{A; i_f, j_f\}$, which locates on the local

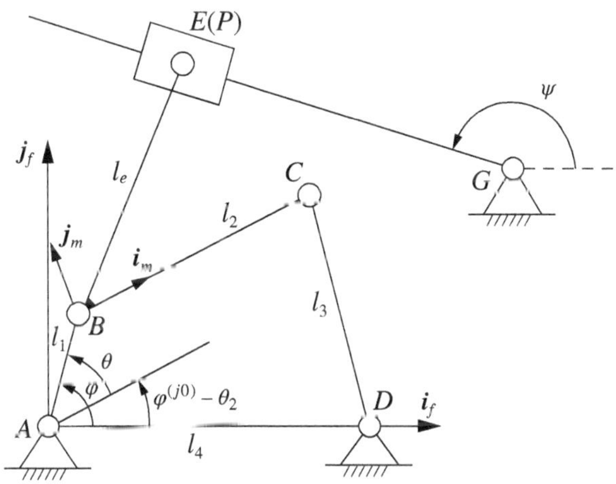

Figure 2.40 A planar 5RPR six-bar linkage with dwell function

saddle line. For the condition $y_E \geq y_G$, the output angle ψ is

$$\psi^{(i)} = \arccos\left(\frac{x_E^{(i)} - x_G}{\sqrt{(x_E^{(i)} - x_G)^2 + (y_E^{(i)} - y_G)^2}}\right) \tag{2.102a}$$

For the condition $y_E < y_G$, the output angle ψ is

$$\psi^{(i)} = 2\pi - \arccos\left(\frac{x_E^{(i)} - x_G}{\sqrt{(x_E^{(i)} - x_G)^2 + (y_E^{(i)} - y_G)^2}}\right) \tag{2.102b}$$

If the crank angle $\varphi^{(i)}$ of the four-bar linkage varies from $\varphi^{(j_0)}$ to $\varphi^{(j_0+\Delta j)}$, we have the dwell output displacement $\psi_{\mathrm{I}}^{(i)}$ at the dwell interval $[\varphi^{(j_0)}, \varphi^{(j_0+\Delta j)}]$ through the above equation, otherwise, the output displacement $\psi_{\mathrm{II}}^{(i)}$ can be calculated at the non-dwell interval. The actual output displacements at both dwell and non-dwell intervals $\psi_{\mathrm{I}}^{(i)}$ and $\psi_{\mathrm{II}}^{(i)}$ are respectively compared with the given functions $\overline{\psi}_{\mathrm{I}}^{(i)}$ and $\overline{\psi}_{\mathrm{II}}^{(i)}$ in equation (2.70), so the error functions at the two intervals can be estimated as follows:

$$\begin{cases} F_1(X) = \max\left(\left|\psi_{\mathrm{I}}^{(i)} - \overline{\psi}_{\mathrm{I}}^{(i)}\right|\right) \\ F_2(X) = \max\left(\left|\psi_{\mathrm{II}}^{(i)} - \overline{\psi}_{\mathrm{II}}^{(i)}\right|\right) \end{cases} \tag{2.103}$$

For convenience in expressing it as a function, a weight coefficient λ is introduced:

$$F(\boldsymbol{X}) = \lambda F_1(\boldsymbol{X}) + (1 - \lambda)F_2(\boldsymbol{X}) \tag{2.104}$$

Then, the mathematical model for the synthesis of the planar 5RPR six-bar linkage with dwell function can be set up as

$$\begin{cases} \min\limits_{\boldsymbol{X}} F(\boldsymbol{X}) \\ \text{s.t.}\ \ \boldsymbol{X} = (x_{Em}, y_{Em}, x_G, y_G)^T \\ \qquad g_k(\boldsymbol{X}) \leq 0,\ k = 1, 2, \ldots, m \end{cases} \tag{2.105}$$

which can be explained in detail as follows.

A. **The objective function and the optimization variables**
The objective function is the weighted sum of the maximum errors at both dwell and non-dwell intervals, which can be calculated by equations (2.102) to (2.104).
The optimization variables are $\boldsymbol{X} = (x_{Em}, y_{Em}, x_G, y_G)^T$, of which (x_{Em}, y_{Em}) are the coordinates of the coupler point E in $\{\boldsymbol{B}; \boldsymbol{i}_m, \boldsymbol{j}_m\}$, and (x_G, y_G) are the coordinates of the fixed pivot point G in $\{\boldsymbol{A}; \boldsymbol{i}_f, \boldsymbol{j}_f\}$. There exists only one independent parameter of (x_G, y_G), since the point G must locate on the local saddle line of the coupler curve traced by point E.

B. **The constraint conditions**

Considering the requirements of geometrical dimensions and transmission performances, we list the constraint conditions as follows.

Similarly, the constraints for the link lengths and the length ratio are

$$g_1(\boldsymbol{X}) = \max(l_1, l_2, l_3, l_4, l_e)/\min(l_1, l_2, l_3, l_4, l_e) - [l] \leq 0 \tag{2.106}$$

$$g_2(\boldsymbol{X}) = \varepsilon - \min(l_1, l_2, l_3, l_4, l_e) \leq 0 \tag{2.107}$$

C. **Initial values of the optimization variables**

The optimization variables $\boldsymbol{X} = (x_{Em}, y_{Em}, x_G, y_G)^T$ are the coordinates of the coupler point E and the fixed pivot point G. Since the coupler curve traced by coupler point E should be locally approximate to a line, the Ball points are natually taken as the inital points. The point G lies on the local saddle line of the coupler curve, and then the inital values are generated by

$$\begin{cases} c_1 = c_0 \cdot [2 \cdot \text{rand}\,(n) - 1] \\ x_G = h_f \cos\phi_f - c_1 \sin\phi_f \\ y_G = h_f \sin\phi_f + c_1 \cos\phi_f \end{cases} \tag{2.108}$$

where c_0 is the searching area coefficient.

D. **The optimization algorithm**

Needless to say, the function **fmincon** in Matlab is adopted directly for the synthesis of planar 5RPR six-bar linkages with dwell output in this book.

2.6.3.4 Kinematic Synthesis of Six-bar Linkages with a Symmetrical Dwell Function

As introduced in Section 1.3 of Chapter 1, there exists a symmetrical circle on the coupler plane for a crank-rocker linkage when the coupler link length is equal to that of the rocker; each point on the circle will trace a symmetrical coupler curve. The two intersection points between a symmetrical coupler curve and its symmetrical axis correspond to the two instants (or positions) when (where) the crank coincides with the base link, or the crank angle φ is 0 or π. If a crank-rocker linkage is taken as the base four-bar linkage for planar six-bar linkages with dwell function, the coupler point on the symmetrical circle is naturally chosen as the hinged joint between the base linkage and the grade II Assur group; the planar six-bar linkage outputs a symmetrical dwell motion.

As shown in Fig. 1.55(a), the symmetrical circle intersects with the Burmester curve on coupler plane of a crank-rocker linkage just at the singularity, such as point P_9. This means the Burmester points E_B certainly locates on the symmetrical axis of the symmetrical coupler curve traced by coupler point E, which contacts with a circle to the fifth order at the Burmester point and the curvature center F lies on the symmetrical axis. The coupler point E and the center F can be regarded as center points of the two R-pairs of the grade II Assur group PRR, and the symmetrical axis is taken as the guide line of the P-pair, shown in Fig. 2.41(a). This Assur group PRR and the basic four-bar linkage can constitute a 6RP six-bar linkage with symmetrical dwell output motion. On the other hand, we might expect there exist two Burmester points on the symmetrical coupler curve, which leads to the tracing point, or the coupler point, must be a Burmester point twice at different time, or the Burmester curve on

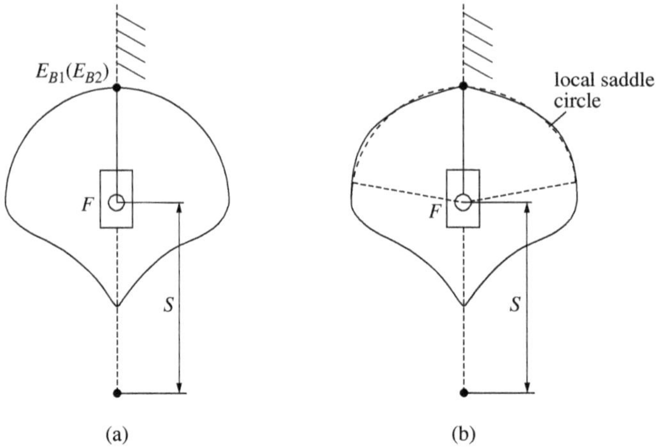

Figure 2.41 Assur group PRR with symmetrical dwell output

the coupler plane intersects itself and the intersection point happens to be on the symmetrical circle. This would be hidden in the kinematic differential geometry of linkages, and the cases will be designed until the symmetrical conditions of coupler curves are derived by differential geometry.

On the contrary, it is a general case for a Ball curve to intersect a symmetrical circle on the coupler plane for a planar four-bar linkage. Logically, a symmetrical coupler curve has two Ball points at two different times for symmetry, which means that a Ball point curve intersects itself and the intersection point happens to be on the symmetrical circle. As shown in Fig. 2.42(a), a coupler point E on the symmetrical circle of the base four-bar linkage traces a symmetrical coupler curve with the instantaneous Ball point E_{b1} and the Ball point E_{b2}, which are symmetric about the symmetrical axis. The coupler point E can be designated as a hinged joint of the grade II Assur group RPR, and the osculating line of the path at the Ball point can be taken as the guide line of the P-pair, which constitute a 5RPR six-bar linkage to produce a symmetrical dwell output.

Based on both a Ball point and a Burmester point, which are instantaneous points on a coupler curve, the instant dwell function can be produced by means of a grade II Assur group. But for a large dwell function with symmetry, we can adopt the local saddle circle or local saddle line fitting to estimate the approximation of the coupler curve and circle or straight line, as is shown in Fig. 2.41(b) and Fig. 2.42(b), and then locate the coupler points through optimization method, whose coupler curves are approximate to circle or line in a large range, while either a Ball point or a Burmester point is usually taken as initial point.

For the kinematic synthesis of six-bar linkages with symmetrical dwell functions, a base crank-rocker linkage has to be designated first, in which the link length of the coupler link is equal to that of the rocker. Then, the characteristic points on the coupler plane are searched

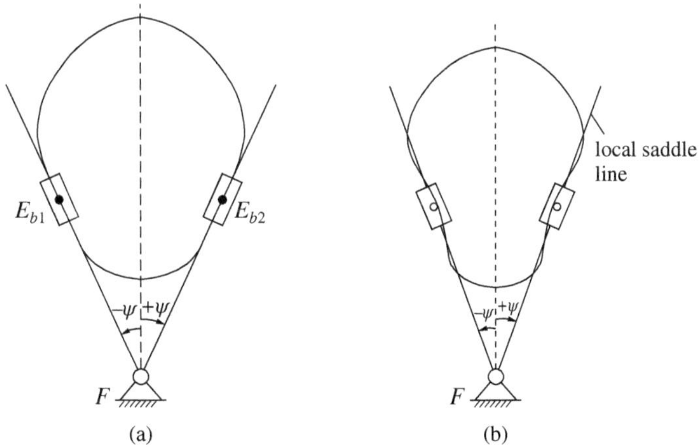

Figure 2.42 Assur group RPR with symmetrical dwell output

out according to the different properties of a symmetrical dwell function, such as points on the symmetrical circle, Ball points, Burmester points, and so on. A grade II Assur group can be synthesized by means of the optimization model and Matlab, as introduced in the previous section. It is believed to be a good idea to take advantage of the distribution of coupler curves, discussed in Chapter 1, which will provide good enough initial points to improve the searching efficiency.

Example 2.11 A given dwell function is

$$\psi = \begin{cases} 25 - 15\cos\left(\dfrac{\theta}{120}\pi\right) & 0 < \theta < 120^{\circ} \\ 40 & 120 \le \theta \le 240^{\circ} \\ 25 + 15\cos\left(\dfrac{\theta}{120}\pi\right) & 240 < \theta < 360^{\circ} \end{cases} \tag{E2-11.1}$$

A planar 7R six-bar dwell linkage is synthesized to regenerate the above dwell function approximately with the minimum allowable transmission angle $[\gamma] = 30^{\circ}$ and the maximum permitted ratio of link length $[l] = 5$.

Solution

The selected basic four-bar linkage $ABCD$ has dimensions l_1=100, l_2=150, l_3=220, l_4=220. The moving frame $\{\boldsymbol{B}, \boldsymbol{i}_m, \boldsymbol{j}_m\}$ is established on the coupler link BC and the fixed frame $\{\boldsymbol{A}; \boldsymbol{i}_f, \boldsymbol{j}_f\}$ is set up on the base link AD.

The mathematical model for the synthesis of planar 7R six-bar linkages with dwell function is set up as

$$\begin{cases} \min_{X} \left(0.8 \max \left(\left| \psi_{\mathrm{I}}^{(i)} - \overline{\psi}_{\mathrm{I}}^{(i)} \right| \right) + 0.2 \max \left(\left| \psi_{\mathrm{II}}^{(i)} - \overline{\psi}_{\mathrm{II}}^{(i)} \right| \right) \right) \\ \text{s.t.} \quad g_1(\boldsymbol{X}) = \dfrac{\max(100, 150, 220, 220, l_e, l_5, l_6)}{\min(100, 150, 220, 220, l_e, l_5, l_6)} - 5 \le 0 \\ g_2(\boldsymbol{X}) = 0.1 - \min(100, 150, 220, 220, l_e, l_5, l_6) \le 0 \\ g_3(\boldsymbol{X}) = \dfrac{\pi}{6} - \min\left(\arccos \dfrac{{l_5}^2 + {l_6}^2 - d_{\min}^2}{2 l_5 l_6}, \pi - \arccos \dfrac{{l_5}^2 + {l_6}^2 - d_{\max}^2}{2 l_5 l_6} \right) \le 0 \\ g_4(\boldsymbol{X}) = d_{\max} - (l_5 + l_6) \le 0 \\ g_5(\boldsymbol{X}) = |l_5 - l_6| - d_{\min} \le 0 \end{cases} \tag{E2-11.2}$$

where $\overline{\psi}_{\mathrm{I}}^{(i)}$ and $\overline{\psi}_{\mathrm{II}}^{(i)}$ respectively correspond to the given functions at both dwell and non-dwell intervals of equation (E2-11.1). $\psi_{\mathrm{I}}^{(i)}$ and $\psi_{\mathrm{II}}^{(i)}$ are the output displacements of the 7R six-bar linkage at the dwell and non-dwell intervals, which can be calculated by equations (2.74) to (2.78). The weight coefficient λ in equation (2.81) is set as 0.8, or the dwell interval is stressed on. The optimization variables $\boldsymbol{X} = (x_{Em}, y_{Em}, x_G, y_G)^T$ are respectively the rectangular coordinates of the coupler point E in $\{\boldsymbol{B}; \boldsymbol{i}_m, \boldsymbol{j}_m\}$ and the fixed pivot point G in $\{\boldsymbol{A}; \boldsymbol{i}_f, \boldsymbol{j}_f\}$. The coupler point E traces a coupler curve with discrete points $\{\boldsymbol{R}_E^{(i)}\} = \{(x_E^{(i)}, y_E^{(i)})^T\}$, $i = 1, \ldots, n$ in $\{\boldsymbol{A}; \boldsymbol{i}_f, \boldsymbol{j}_f\}$ as the movement of the six-bar linkage. Based on equation (2.72), the center point $F(x_{F0}, y_{F0})$ and the radius r of the local saddle circle of $\{\boldsymbol{R}_E^{(i)}\}$ can be obtained. We have the dimensions of the 7R six-bar linkage as

$$\begin{cases} l_e = \sqrt{x_{Em}^2 + y_{Em}^2} \\ l_5 = r \\ l_6 = \sqrt{(x_{F0} - x_G)^2 + (y_{F0} - y_G)^2} \\ d_{\max} = \max\limits_{i}\left(\sqrt{(x_E^{(i)} - x_G)^2 + (y_E^{(i)} - y_G)^2}\right) \\ d_{\min} = \min\limits_{i}\left(\sqrt{(x_E^{(i)} - x_G)^2 + (y_E^{(i)} - y_G)^2}\right) \end{cases} \tag{E2-11.3}$$

The mathematical model (E2-11.2) of the discrete kinematic synthesis of planar 7R six-bar dwell linkages can be solved directly by Matlab. The function **fmincon** can be set up by referring to the program **KS-CR**. As mentioned above, a ***program for synthesis of a 7R six-bar linkage with dwell function***, or ***KS-7R-DWELL*** for short, is naturally constructed, whose input are a given dwell function and a basic crank-rocker linkage and whose output are a six-bar linkage with dimensions and performance.

We run the program **KS-7R-DWELL** with 20 groups of initial values and have the corresponding results, or 7R six-bar dwell linkages with different dimensions. The linkage with the minimum value of objective function is listed in Table 2.22. The difference between the output function of the synthesized linkage and the required function is shown in Fig. 2.43.

Table 2.22 The dimensions of the 7R six-bar dwell linkage

Coordinates of coupler point E		Coordinates of fixed hinge point G		Angle error for dwell	Angle error for non-dwell	Link lengths
x_{Em}	y_{Em}	x_G	y_G	F_1	F_2	l_1=100, l_2=150, l_3=220,
138.2139	65.9793	222.6319	−88.1244	0.339277°	5.532709°	l_4=220, l_e=153.1546, l_5=352.6648, l_6=78.0732

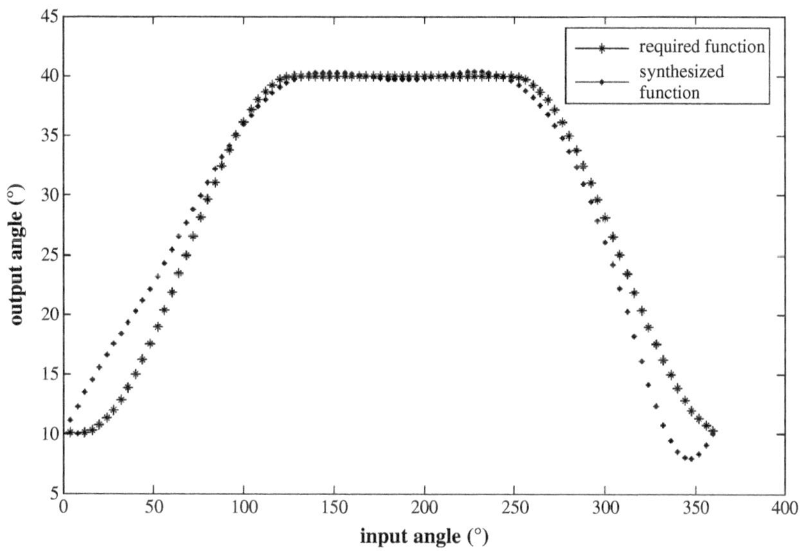

Figure 2.43 The output function of 7R six-bar dwell linkage and the required function

2.7 Discussion

The kinematic synthesis of linkages has been a key content of the kinematics of linkages. The classical kinematic geometry of a rigid body in planar motion, established by Euler, Savary, Ball, Burmester, and Müller [6, 7], provides a theoretical basis for dimensional synthesis of linkages. The geometrical method, presented by Burmester [8], was widely used in the kinematic analysis and synthesis of planar linkages before 1950. Based on the geometrical method, Lichtenheldt [9], Hain [10], and Kiper [11] built a design method suitable for engineering requirements. The geometrical method with intuition is good enough for the analysis and synthesis of planar linkages, but of lower precision and needs more improvement and is commonly believed to be inconvenient and not easily adapted for the analysis and synthesis of spherical and spatial linkages. Kinzel, Schmiedeler, and Pennock [12] presented the kinematic synthesis of finitely separated positions using geometric constraint programming, greatly improving the precision of geometrical methods by means of computer-aided geometrical diagrammatizing.

The application of computer technology has brought rapid development to both the atlas approach and the algebraic method. Alt [13] provided an atlas of coupler curves of four-bar linkages for designers to look for a four-bar linkage with type and dimensions conveniently in order to approximately meet the requirements in practice. Yang [14] and Li [15] set up the performance atlas of four-bar linkages and an atlas of coupler curves for four-bar linkages. Vidosic and Tesar [16–18] selected one Burmester point and two Burmester points to coincide with a Ball point at the instant for a four-bar linkage, called a Ball–Burmester point and a Ball–double Burmester point respectively, and then drew atlases, which provided an efficient tool for the synthesis of four-bar linkages with an approximate straight line since the dimensions of a four-bar linkage and the position of a coupler point can be chosen in the atlas. Based on these atlases, Kota, Erdman, and Riley [19, 20] programmed an expert system for the synthesis of multi-bar linkages with dwell function. By means of the cognate four-bar linkages, Davies [21, 22] presented finite five-dimensional atlases for all planar four-bar linkage coupler curves, which mapped all types of coupler curve for a four-bar linkage into the intersection area between two circles, whose centers locate on both joints of the coupler link with radius of coupler link length, which means that it is necessary to focus on the intersection area without all of the coupler plane for an atlas of coupler curves. Hoeltzel and Chieng [23] presented pattern matching synthesis as an automated approach to the design of mechanisms with the probabilistic method and Hopfield neural networks. Chu and Cao [24] classified the types of coupler curve by fuzzy mathematical methods and synthesized a four-bar linkage by pattern matching of their characteristics. Xie [25] presented an approach for the kinematic synthesis of planar linkages by neural networks. There are more and more scholars [26–32] in the field focusing on the improvement in approaches of the numerical atlas, and there would probably be changes in the approach of this atlas with lower precision and less range chosen with the coming of big data.

In terms of algebraic methods, polynomial theory, first proposed and applied to the function synthesis of linkages by Chebyshev [33], was described and explained further by Denavit and Hartenberg [34], who think that the polynomial equations for closed loop displacements of linkages can be obtained by a coordinate-converting matrix for the precise functional synthesis of linkages. Numerical solutions for the analysis and synthesis of mechanisms were first presented by Freudenstein [35, 36], the pioneer of computing kinematics, and are continuously being developed by more and more researchers, such as Roth [37], Wampler [38], Suh and Radcliffe [39], Erdman and Sandor [40], Waldron and Kinzel [41], Hall [42], McCarthy [43], and so on. Liao [44] and Yang [45] studied the approaches to the precise synthesis of linkages, such as the homotopy approach. Based on both the geometrical and the algebraic method, the software for precise synthesis of planar linkages is declared as programmed in the literature – KINSYN by Kaufman [46], LINKAGES by Erdman [47], and RECSYN by Waldron [48], and so on.

More and more application cases in practical engineering need a larger range, more positions, more transmission performance, etc., which is beyond the boundary of the precise synthesis of planar linkages, although they have been gradually completed in the last decades. Hence, more and more researchers pay attention to the approximate synthesis of planar linkages. Fox and Willmert [49] first introduced the optimum mathematical model for the path synthesis of linkages and searched optimum solutions, in which some requirements for linkages in practical application cases can be described by inequality constraints. Angeles [50, 51] presented the optimum synthesis of linkages for given positions and paths, whose object functions are the minimum construction errors in a least square and whose optimization variables are the coordinates of the joints in both the moving body and the base link. Zhou and Cheung [52] also discussed the optimum path synthesis of linkages, in which the direction construction error is taken as object function by RRR instead of RPR. Further, Refs [53–58] discuss the optimum synthesis of linkages, even multi-bar linkages; they have similar object functions and some of them deal with solving algorithms. Kota and Chiou [59] combined the optimization method with the method of orthogonal experiments, and expected to find good initial values for solving the optimization model, which is a typically difficult problem. However, the results are not satisfied except that they are beneficial for the local convergence of the solving algorithm. On the contrary, the approaches for optimization synthesis of six-bar linkages with dwell function are intensively noted by many more researchers [60–64].

For the optimization synthesis of linkages, there are four topics to be discussed. Firstly, the setup of the object function with optimization variables. In other words, what does a decreasing objective function mean? And how could we reveal the effect caused by the change of each optimization variable? It is the first step, an extremely important one, to set up a mathematical model to describe what linkage is to be synthesized. It is useless for us to synthesize linkages by means of what is thought to be a highly efficient algorithm if it is not described by both a distinct and an accurate mathematical model. Therefore, a suitable object function is good enough to provide the existence of the optimization solution and convergence of the solving algorithm for the optimization synthesis, or approximate synthesis of linkages. The optimization variables are usually taken as the dimensional parameters of linkages to be determined, which are preferred to be less and not redundant. Secondly, the constraint equations are established according to the requirements of the application cases. In general, the constraint equations can be changed as part of the object function by means of penalty functions, which are converted into an optimization model without constraints. Thirdly, the problem of initial values is a puzzling topic but has to be dealt with. Theoretically, it depends on the property of the object function. The object function for the optimization synthesis of linkages is generally the non-convex function with multiple extreme values, which means that the solving algorithm can converge to a local point with extreme value if the initial point is in the same area as the extreme value point, or in the same valley. Thus, it is better to assign the initial points on different areas with extreme value points of the object function for the kinematic synthesis of linkages, which can easily come true when the initial points are assigned several times. Generally, this is normal for the optimization synthesis of linkages whose solutions usually depend on the initial points, because the object function has different extreme value points. Otherwise, unreasonable object functions or redundancy variables lead to cases that must be treated as individual cases; there is only a qualitative change in values between the optimization solution and the initial values. However, the stop or end conditions have to be given for a program, relying on two terms. One is the expected solving accuracy or convergence accuracy. Another is the existence of a solution; it is useless to compute endlessly if there is no solution to meet the conditions. Finally, mathematicians have provided us with a lot of optimization algorithms and software, such as Matlab, which are good enough for the kinematic synthesis of linkages. We suggest that kinematicians or mechanism researchers do not compete with the mathematicians on optimization algorithms, even though they may be competent or interested. In the field, it is not easy to have the necessary skill or experience to apply the optimization algorithms in Matlab, let alone improve them.

Kinematic synthesis can determine both the types and dimensions of linkages, and is an important element of machine design since the linkages' dimensions appear first as the machine is being constructed. Since engineers need both the approaches and the software of kinematic synthesis to design linkages, they have to be convenient, effective, and reliable in practical engineering. So far, the approaches of kinematic synthesis of linkages are still not mature and need theoretical proofs, a theoretical basis of discrete kinematic geometry, and large numerical examples in practical applications to verify the existence of solutions and the convergence of the algorithm.

Based on the curvature of a planar curve, we focus on the curves with special curvature characteristics and take them as basic elements of the kinematic synthesis of linkages, which may avoid redundant variables disturbing the down gradient of the object function and provide a theoretical basis for the existence of the optimization solution and convergence of the solving algorithm [65–71]. On the other hand, based on the saddle point programming, the approach of the kinematic synthesis of linkages is systematically set up from planar, spherical, to spatial linkages [72–75]. The best uniform approximation in multi-dimensional space (the saddle point programming) was presented to test in kinematic synthesis of linkages since 1959 [76–79], which was believed to be more desirable in mechanism synthesis [80]. But due to the numerical difficulty involved, its applications were limited.

The discrete kinematic geometry in planar movement studies the discrete displacements of a rigid body and its points. The geometrical properties of the discrete displacements are compared with those of the constraint curves globally in terms of kinematic invariants; there are multiple degrees of freedom as the computing kinematics develop, whereas the classical topic only deals with a few finite separated positions. In fact, both the discrete kinematic invariants and the global invariants of a discrete curve, like the centrodes of instantaneous planar kinematics of a lamina in Chapter 1 and the global curvature of a curve in Chapter 3, might play a fundamental role in discrete kinematic geometry, from the starting point to searching out the characteristic points of a moving body, and proving the existence of solutions and the convergence of the algorithm. From the viewpoint of the planar, spherical, and spatial linkages, the approach of saddle curve fitting sets up a unified evaluation system of normal errors between two curves. Making a comprehensive survey of all numeral examples, such as the 72 positions of a coupler, the four, five, or 10 positions of a moving body, we may find that the error surfaces of saddle circles or saddle lines look like uninterrupted mountains and weaving rivers. There exist some relationships among the peaks, valleys, and river beds; in particular, the fewer separated positions there are, the more the error surface is like a dry river bed without water. It might forebode that the number of saddle circle points is few and the number of saddle sliding points is one, as discovered in the theory of discrete kinematic geometry in planar movement.

This book presents the approach of saddle point programming for discrete kinematic geometry of linkages, saddle circle fitting, and saddle line fitting for planar linkages, converting the planar discrete kinematic geometry into computing the extreme values of a discrete function, avoiding not only the redundant variables but also the influence of different coordinate systems. The authors have tried to develop the kinematic geometry at less than six separated positions, and the classical issue of the absolute circle points in terms of a pole of two positions of a lamina. The global geometrical properties of discrete paths, saddle circle points, and saddle sliding points have been considered by means of saddle point programming, and the relationships among the points of a lamina, discrete positions, and discrete paths have been discussed, revealing the distributions of the characteristic points on the moving body, linking up the theory of kinematic geometry at infinitesimally successive positions – the differential kinematic geometry of Chapter 1. However, the geometrical properties of both the three-position saddle line and the four-position saddle circle still remain to be revealed completely, and the algebraic properties of their error surfaces have not been demonstrated clearly. In particular, the global geometrical properties and number of points with extreme values on the error surface have not been studied. The error surface of the saddle circle at multiple positions is comprised of the error surfaces for different four-positions; how to combine or overlay them? What are the properties of the boundary curves? All these questions need to be studied further, and solved by algebra. We might expect there to exist discrete kinematic invariants for a four-position saddle circle and a three-position saddle line, or discrete generalized curvature, like the Euler–Savary formula in Chapter 1, corresponding to the centrodes at infinitesimal positions and the poles at two separated positions. Modern differential geometry, such as the differential manifold, the theory of fiber bundles, topology, the moving frame and its differential formula in the classical differential geometry, will lay the theoretical basis for discrete kinematic geometry, particularly in multi-dimensional space and in the kinematics of manipulators with multiple degrees of freedom.

References

[1] Q.X. Zhang, *Analysis and Synthesis of Spatial Mechanisms*, Vol. **1**. China Machine Press, Beijing, 1984 (in Chinese).

[2] K.H. Hunt, *Kinematic Geometry of Mechanism*. Oxford University Press, Oxford, 1978.

[3] J. Liu and X.M. Wang, *Saddle Point Programming and Geometric Error Evaluation*. Dalian University of Technology Press, Dalian, 1996 (in Chinese).

[4] F. Freudenstein, On the variety of motions generated by mechanisms. *ASME Journal of Engineering for Industry*, **84**(1), 156–159, 1962.

[5] S. Roberts, On the motion of a plane under certain conditions. *Proceedings of the London Mathematical Society*, **3**, 286–318, 1871.

[6] H. Nolle, Linkage coupler curve synthesis: A historical review - I. Developments up to 1875. *Mechanism and Machine Theory*, **9**(2), 147–168, 1974.

[7] H. Nolle, Linkage coupler curve synthesis: A historical review - II. Developments after 1875. *Mechanism and Machine Theory*, **9**(3&4), 325–348, 1974.

[8] L. Burmester, *Lehrbuch der Kinematik*. Felix, Leipzig, 1888.

[9] W. Lichtenheldt, *Konstruktionslehre der Getriebe*. Akademie-Verlag, Berlin, 1961.

[10] K. Hain, *Applied Kinematics*, 2nd edn. McGraw-Hill, New York, 1964.

[11] G. Kiper, *Synthese der obenen Gelenkgetriebe*. VDI-Forschungsheft, Berlin, 1952.

[12] E.C. Kinzel, J.P. Schmiedeler, and G.R. Pennock, Kinematic synthesis for finitely separated positions using geometric constraint programming. *ASME Journal of Mechanical Design*, **128**(5), 1070–1079, 2005.

[13] H. Alt, Das Konstruieren von Gelenkvierecken unter Benutzung Einer Kurventafel. *VDI-Z*, **85**, 69–72, 1941.

[14] J.H. Yang and F. Gao, *Space Model and Performance Atlas*. China Machine Press, Beijing, 1989 (in Chinese).

[15] X.R. Li, *Atlas of Coupler Curve*. Chongqing Publishing House, Chongqing, 1993 (in Chinese).

[16] J.P. Vidosic and D. Tesar, Selection of four-bar mechanisms having required approximate straight-line outputs. Part I. The general case of the Ball–Burmester point. *Journal of Mechanisms*, **2**(1), 23–44, 1967.

[17] J.P. Vidosic and D. Tesar, Selection of four-bar mechanisms having required approximate straight-line outputs. Part II. The Ball–Burmester point at the inflection pole. *Journal of Mechanisms*, **2**(1), 45–59, 1967.

[18] J.P. Vidosic and D. Tesar, Selection of four-bar mechanisms having required approximate straight-line outputs. Part III. The Ball–double Burmester point linkage. *Journal of Mechanisms*, **2**(1), 61–76, 1967.

[19] S. Kota, A.G. Erdman, and D.R. Riley, Development of knowledge for designing linkage-type dwell mechanisms. Part 1. Theory. *ASME Journal of Mechanisms, Transmissions, and Automation in Design*, **109**(3), 308–315, 1987.

[20] S. Kota, A.G. Erdman, and D.R. Riley, Development of knowledge for designing linkage-type dwell mechanisms. Part 2. Application. *ASME Journal of Mechanisms, Transmissions, and Automation in Design*, **109**(3), 316–321, 1987.

[21] T.H. Davies and C.J. Yuan, Proposals for finite 5-dimensional atlases for all planar 4-bar linkage coupler curves. *Mechanism and Machine Theory*, **19**(2), 211–221, 1984.

[22] T.H. Davies, Proposals for a finite 5-dimensional atlas of crank-rocker coupler curves. *Mechanism and Machine Theory*, **16**(5), 517–530, 1981.

[23] D.A. Hoeltzel and W.H. Chieng, Pattern matching synthesis as an automated approach to mechanism design. *ASME Journal of Mechanical Design*, **112**(2), 190–199, 1990.

[24] J.K. Chu and W.Q. Cao, Synthesis of coupler curves of planar four-bar linkages through fast Fourier transform. *Chinese Journal of Mechanical Engineering*, **29**(5), 117–122, 1993 (in Chinese).

[25] J. Xie, K.Y. Yan, and Y. Chen, Synthesis of planar motion generation mechanisms using BP neural networks. *Chinese Journal of Mechanical Engineering*, **41**(2), 24–27, 2005 (in Chinese).

[26] Z.X. Wang, Z.B. Chen, and L. Jiang, Study on planar linkages' path synthesis by coupler-angle curve. *Chinese Journal of Mechanical Engineering*, **31**(1), 42–47, 1995 (in Chinese).

[27] J.R. McGarva, Rapid search and selection of path generating mechanisms from a library. *Mechanism and Machine Theory*, **29**(2), 223–235, 1994.

[28] I. Ullah and S. Kota, Optimal synthesis of mechanisms for path generation using Fourier descriptors and global searching methods. *ASME Journal of Mechanical Design*, **119**(4), 504–510, 1997.

[29] Zh.F. Yuan, M.J. Gilmartin, and S.S. Douglas, Optimal mechanism design for path generation and motions with reduced harmonic content. *ASME Journal of Mechanical Design*, **126**(1), 191–196, 2004.

[30] Y. Liu and R.B. Xiao, Optimal synthesis of mechanisms for path generation using refined numerical representation based model and AIS based searching method. *ASME Journal of Mechanical Design*, **127**(4), 688–691, 2004.

[31] Z.H. Lan and H.J. Zou, Parallel optimization of mechanisms based on the local characteristics of the coupler curves. *Chinese Journal of Mechanical Engineering*, **35**(5), 16–19, 1999 (in Chinese).

[32] J.K. Chu and J.W. Sun, *The Method of Harmonic Characteristic Parameters for the Kinematic Synthesis of Linkages*. Science Press, Beijing, 2010 (in Chinese).

[33] P.L. Chebyshev, Sur les parallélogrammes composés de trios éléments quelconques. *Mémoires de l'Académie des Sciences de Saint-Pétersbourg*, **36**(3), 1879.

[34] J. Denavit and R.S. Hartenberg, A kinematic notation for lower-pair mechanisms based on matrices. *ASME Journal of Applied Mechanics*, **22**, 215–221, 1955.

[35] F. Freudenstein, Approximate synthesis of four-bar linkages. *Resonance*, **15**(8), 740–767, 2010. Reprinted with permission from *Transactions of the ASME*, **77**, 853–861, 1955.

[36] F. Freudenstein and G.N. Sandor, Synthesis of path generating mechanisms by means of a programmed digital computer. *ASME Journal of Engineering for Industry*, **81**, 159–168, 1959.

[37] P. Chen and B. Roth, Design equations for the finitely and infinitesimally separated position synthesis of binary links and combined link chains. *ASME Journal of Engineering for Industry*, **91**, 209–219, 1969.

[38] C.W. Wampler, Solving the kinematics of planar mechanisms. *ASME Journal of Mechanical Design*, **121**(3), 387–391, 1999.

[39] C.H. Suh and C.W. Radcliffe, *Kinematics and Mechanisms Design*. Wiley, New York, 1978.

[40] A.G. Erdman and G.N. Sandor, *Mechanism Design: Analysis and Synthesis*, Vol. **1**. Prentice-Hall, Englewood Cliffs, NJ, 1997.

[41] K.J. Waldron and G.L. Kinzel, *Kinematics and Dynamics, and Design of Machinery*. Wiley, New York, 1999.

[42] A.S. Hall Jr.,, *Kinematics and Linkage Design*. Prentice-Hall, Englewood Cliffs, NJ, 1961.

[43] J.M. McCarthy and G.S. Soh, *Geometric Design of Linkages*, 2nd edn. Interdisciplinary Applied Mathematics 11, Springer, New York, 2010.

[44] Q.Z. Liao, Geometry algebra method for solving the kinematics of linkage mechanisms. *Journal of Beijing University of Posts and Telecommunications*, **33**(4), 1–11, 2010 (in Chinese).

[45] T.L. Yang, *Kinematics Design of Mechanical System*. China Petrochemical Press, Beijing, 1999 (in Chinese).

[46] R.E. Kaufman, KINSYN: An interactive kinematic design system. Proceedings of the Third Congress on the Theory of Machines and Mechanisms, Dubrovnik, Yugoslavia, 1971.

[47] A.G. Erdman and J. Gustafson, LINCAGES – A linkage interactive computer analysis and graphically enhanced synthesis package. ASME Paper No. 77-DETC-5, Chicago, IL, 1977.

[48] J.C. Chuang, R.T. Strong, and K.J. Waldron, Implementation of solution rectification techniques in an interactive linkage synthesis program. *ASME Journal of Mechnical Design*, **103**(3), 657–664, 1981.

[49] R.L. Fox and K.D. Willmert, Optimum design of curve-generating linkages with inequality constraints. *ASME Journal of Engineering for Industry*, **89**(1), 144–151, 1967.

[50] J. Angeles, A. Alivizatoss, and R. Akhras, An unconstrained nonlinear least-square method of optimization of RRRR planar path generators. *Mechanism and Machine Theory*, **23**(5), 343–353, 1988.

[51] R. Akhras and J. Angeles, Unconstrained nonlinear least-square optimization of planar linkages for rigid-body guidance. *Mechanism and Machine Theory*, **25**(1), 97–118, 1990.

[52] H. Zhou and E.H.M. Cheung, Optimal synthesis of crank-rocker linkages for path generation using the orientation structural error of the fixed link. *Mechanism and Machine Theory*, **36**(8), 973–982, 2001.

[53] X.S. Hu, R.Y. Wu, *et al.*, Efficient approach for minimax linkage function generation. *Chinese Journal of Mechanical Engineering*, **33**(2), 1–7, 1997 (in Chinese).

[54] H. Diab and A. Smaili, Optimum exact/approximate point synthesis of planar mechanisms. *Mechanism and Machine Theory*, **43**(12), 1610–1624, 2008.

[55] J.E. Holte, T.R. Chase, and A.G. Erdman, Mixed exact–approximate position synthesis of planar mechanisms. *ASME Journal of Mechanical Design*, **122**(3), 278–286, 1998.

[56] G.S. Son and J.M. McCarthy, The synthesis of six-bar linkages as constrained planar 3R chains. *Mechanism and Machine Theory*, **43**(2), 160–170, 2008.

[57] Q. Shen, Y.M. Al-Smadi, P.J. Martin, K. Russell, and R.S. Sodhi, An extension of mechanism design optimization for motion generation. *Mechanism and Machine Theory*, **44**(9), 1759–1767, 2009.

[58] J. Wu, Q.J. Ge, H.J. Su, and F. Gao, Kinematic acquisition of geometric constraints for task-oriented design of planar mechanisms. *ASME Journal of Mechanisms and Robotics*, **5**, 011003-1-7, 2013.

[59] S. Kota and S.-J. Chiou, Use of orthogonal arrays in mechanism synthesis. *Mechanism and Machine Theory*, **28**(6), 777–794, 1993.

[60] C. Bagci, Synthesis of double-crank (drag-link) driven mechanisms with adjustable motion and dwell time ratios. *Mechanism and Machine Theory*, **12**(6), 619–638, 1977.

[61] E. Sandgen, Design of single- and multiple-dwell six-link mechanisms through design optimization. *Mechanism and Machine Theory*, **20**(6), 483–490, 1985.

[62] S. Kota, G.A. Erdman, and R.D. Riley, Development of knowledge base for designing linkage-type dwell mechanisms. Part 1. Theory. *Journal of Mechanisms, Transmissions and Automation in Design*, **109**(3), 308–315, 1987.

[63] S. Kota, G.A. Erdman and R.D. Riley, Development of knowledge base for designing linkage-type dwell mechanisms. Part 2. Application. *Journal of Mechanisms, Transmissions and Automation in Design*, **109**(3), 316–321, 1987.

[64] R.R. Bulatovic, S.R. Dordevic, and V.S. Dordevic, Cuckoo search algorithm: A metaheuristic approach to solving the problem of optimum synthesis of a six-bar double swell linkage. *Mechanism and Machine Theory*, **61**, 1–13, 2013.

[65] D.L. Wang, S.F. Wang, and T. Li, New Approach for Mechanisms Synthesis by Adaptive Saddle-Fitting. *Chinese Journal of Mechanical Engineering*, **37**(12), 21–26, 2001 (in Chinese).

[66] T. Li, Research on the theory and approach of optimal synthesis for planar linkages based on self-adapted approximation of simple curves, Dalian University of Technology, 2000 (in Chinese).

[67] J.C. Zhou, Research on the approximate path generation of planar linkages, Dalian University of Technology, 1999 (in Chinese).

[68] J.J. Zhang, Research on the theory and approach of function synthesis for planar six-bar dwell mechanism, Dalian University of Technology, 2002 (in Chinese).

[69] J. Chai, Research on the approximate function synthesis of planar six-bar dwell mechanism, Dalian University of Technology, 2003 (in Chinese).

[70] C. Ma, Research on the conceptual design and kinematic synthesis of an eight-bar and ten-pair mechanism, Dalian University of Technology, 2009 (in Chinese).

[71] H.M. Dong, D.L. Wang, An Adaptive Approach of Approximate Function Generation for Stephenson III Six-Bar Dwell Linkage. *Journal of Dalian University of Technology*, **52**(**4**), 529–535, 2012 (in Chinese).

[72] B.Y. Zhang, Research on the theory and approach of functional synthesis of spherical four-bar linkage, Dalian University of Technology, 2002 (in Chinese).

[73] P.C. Zheng, Optimal path synthesis of spatial mechanisms, Dalian University of Technology, 1999 (in Chinese).

[74] S.F. Wang, *New Approach for Kinematic Synthesis of Mechanism by Adaptive Curve Fitting*, Dalian University of Technology, 2005 (in Chinese).

[75] D.L. Wang and S.F. Wang, A unified approach to kinematic synthesis of mechanism by adaptive curve fitting. *Science in China Series E, Technological Sciences*, **47**(1), 85–96, 2004.

[76] F. Freudenstein, Structural Error Analysis in Plane Kinematic Synthesis. *ASME Journal of Engineering for Industry*, **81**, 15–22, 1959.
[77] N.L. Levitskii, Y.L. Sarkissyan and G.S. Gekchian, Optimum Synthesis of Four-Bar Function Generating Mechanism. *Mechanism and Machine Theory*, **7**, 387–398, 1972.
[78] R.S. Rose and G.N. Sandor, Direct Analytical Synthesis of Four-bar Function Generators with Optimal Structural Error. *ASME Journal of Engineering for Industry*, May, 563–571, 1973.
[79] V.V. Garbarouk and P.A. Lebedev, Synthesis of Spatial Automatic Operators with the Aid of Electronic Digital Computers. *Mechanism and Machine Theory*, **15**, 9–17, 1980.
[80] A.G. Erdman. *Modern Kinematics: Developments in the Last Forty Years*. New York: John Wiley and Sons, 1993.

3

Differential Geometry of the Constraint Curves and Surfaces

For a spatial four-bar linkage RSCR, the coupler link is connected to the base link by two binary links RS and RC. The center point of the kinematic pair S on the coupler link, called a ***characteristic point***, is constrained by the binary link RS and traces a spatial curve in the fixed frame, defined as a ***constraint curve***. The axis line of the kinematic pair C on the coupler link, designated as a ***characteristic line***, is constrained by the binary link RC and traces a ruled surface in the fixed frame, referred to as a ***constraint surface***. A binary link connecting a coupler link to the base link corresponds to a constraint surface or a constraint curve in the base link, and also corresponds to a characteristic point or a characteristic line on the coupler link, which provides an important bridge between movement of the moving body and a base link for the kinematic synthesis of linkages. The geometrical properties of both constraint curves and constraint surfaces are the topics of geometry and the theoretical basis for the kinematic analysis and synthesis of linkages. The differential geometry, of course, is a powerful approach and naturally chosen to study the geometrical properties of constraint curves and constraint surfaces in terms of invariants and invariance.

For convenience in reading the following sections of this book, the elements of differential geometry are briefly introduced, and their applications in both constraint curves and constraint surfaces of spatial linkages are concisely derived, entitled as the differential geometry of constraint curves and constraint surfaces for this chapter. The proofs of theorems in differential geometry about constraint curves and surfaces, which can be found in Refs [1–3], are not provided.

3.1 Space Curves

3.1.1 Vector Representations

A spatial curve Γ, which is different from a planar curve in Chapter 1 and briefly called a curve in this chapter, can be presented by a point with rectangular coordinates (x, y, z) in the

Kinematic Differential Geometry and Saddle Synthesis of Linkages, First Edition. Delun Wang and Wei Wang.

Cartesian frame $\{\boldsymbol{O};\boldsymbol{i},\boldsymbol{j},\boldsymbol{k}\}$ and has the independent variable t, or the parameter t. The equation of the curve is

$$\begin{cases} x = x(t) \\ y = y(t) \\ z = z(t) \end{cases} \tag{3.1}$$

If the independent variable t is eliminated, the curve Γ can be expressed by

$$\begin{cases} y = y(x) \\ z = z(x) \end{cases} \tag{3.2}$$

The form of implicit function is

$$\begin{cases} F_1(x, y, z) = 0 \\ F_2(x, y, z) = 0 \end{cases} \tag{3.3}$$

In the Cartesian frame $\{\boldsymbol{O};\boldsymbol{i},\boldsymbol{j},\boldsymbol{k}\}$, the vector equation of the curve Γ is

$$\Gamma : \boldsymbol{R} = x(t)\boldsymbol{i} + y(t)\boldsymbol{j} + z(t)\boldsymbol{k} \tag{3.4}$$

It is usually simplified to

$$\boldsymbol{R} = \boldsymbol{R}(t) \tag{3.5}$$

In Chapter 1, the vector function of a unit circle was introduced to describe the planar curve with a concise form. For the vector equation (3.4) of a curve Γ, the two components on the coordinate axes $\boldsymbol{i}$ and $\boldsymbol{j}$ can also be represented by the vector function of a unit circle, as shown in Fig. 3.1, and the vector equation (3.4) can be rewritten as

$$\Gamma : \boldsymbol{R} = r(\varphi)\boldsymbol{e}_{\mathrm{I}(\varphi)} + z(\varphi)\boldsymbol{k} \tag{3.6}$$

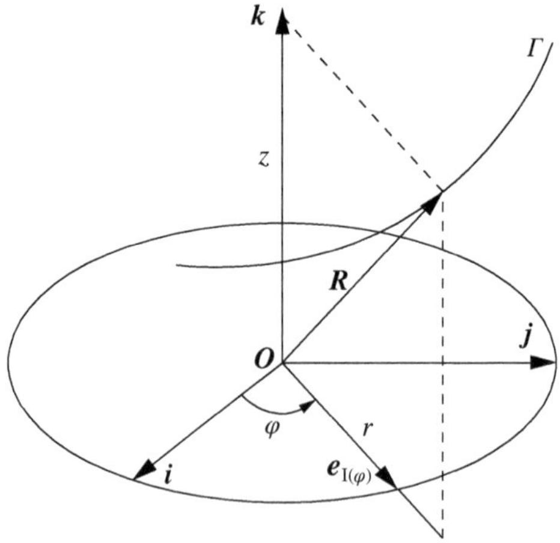

Figure 3.1 The space curve described by vector function of a unit circle

With regard to a curve Γ, the natural parameter is the arc length s, which has the following relationship with the general parameter t:

$$s = \int_{t_a}^{t_b} \left| \frac{\mathrm{d}\boldsymbol{R}}{\mathrm{d}t} \right| \mathrm{d}t,\ \mathrm{d}s = |\mathrm{d}\boldsymbol{R}| = \sqrt{\left(\frac{\mathrm{d}x}{\mathrm{d}t}\right)^2 + \left(\frac{\mathrm{d}y}{\mathrm{d}t}\right)^2 + \left(\frac{\mathrm{d}z}{\mathrm{d}t}\right)^2}\,\mathrm{d}t \tag{3.7}$$

Then, a curve Γ can be written with arc length s as the independent variable, that is

$$\Gamma : \boldsymbol{R} = \boldsymbol{R}(s),\ s_a \le s \le s_b \tag{3.8}$$

Example 3.1 Write the vector equation for a spherical curve, as shown in Fig. 3.2.

Solution

By setting the original point of the Cartesian coordinate system $\{\boldsymbol{O}; \boldsymbol{i}, \boldsymbol{j}, \boldsymbol{k}\}$ at the center of the spherical surface, a spherical curve Γ is represented as

$$\begin{cases} x = x(t),\ y = y(t),\ z = z(t) \\ x^2 + y^2 + z^2 = R^2 \end{cases} \tag{E3-1.1}$$

where R is the radius of the spherical surface and t is the general parameter of curve Γ. The above equations can be rewritten in the form of implicit function when parameter t is eliminated:

$$\begin{cases} z = z(x, y) \\ x^2 + y^2 + z^2 = R^2 \end{cases} \tag{E3-1.2}$$

Since all points of the spherical curve Γ lie on a spherical surface, curve Γ is usually represented by spherical coordinates as

$$\delta = \delta(t),\ \varphi = \varphi(t),\ r = R \tag{E3 1.3}$$

where δ is the included angle between the vector $\boldsymbol{k}$ and the vector $\overrightarrow{OP}$ from the original point $\boldsymbol{O}$ to point P, φ is the included angle between the vector $\boldsymbol{i}$ and the projection vector of vector $\overrightarrow{OP}$

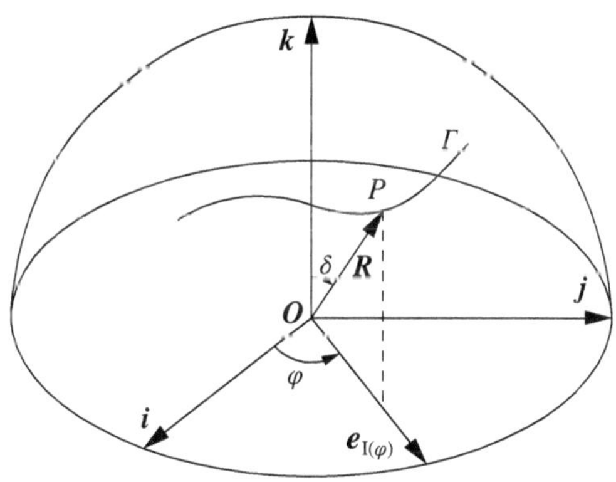

Figure 3.2 Spherical curve

onto the coordinate plane $\boldsymbol{O}-\boldsymbol{ij}$, where $\delta \in [0, \pi]$ and $\varphi \in [0, 2\pi]$. The relationships between the spherical coordinates and the Cartesian coordinates in the frame $\{\boldsymbol{O}; \boldsymbol{i}, \boldsymbol{j}, \boldsymbol{k}\}$ for a point P are

$$\begin{cases} x = R\sin\delta(t)\cos\varphi(t) \\ y = R\sin\delta(t)\sin\varphi(t) \\ z = R\cos\delta(t) \end{cases} \tag{E3-1.4}$$

With t as the general parameter, the vector equation of the spherical curve Γ is

$$\Gamma : \boldsymbol{R} = \boldsymbol{R}(t) = x(t)\boldsymbol{i} + y(t)\boldsymbol{j} + z(t)\boldsymbol{k} \tag{E3-1.5}$$

The vector equation of the spherical curve Γ can be rewritten by the vector function of a unit circle as

$$\boldsymbol{R} = R\sin\delta(\varphi)\boldsymbol{e}_{\mathrm{I}(\varphi)} + R\cos\delta(\varphi)\boldsymbol{k} \tag{E3-1.6}$$

Example 3.2 Write the vector equation for a cylindrical curve, as shown in Fig. 3.3.

Solution

In the Cartesian frame $\{\boldsymbol{O}; \boldsymbol{i}, \boldsymbol{j}, \boldsymbol{k}\}$, the equation of a cylindrical curve is

$$\begin{cases} x = r_0\cos\varphi \\ y = r_0\sin\varphi \\ z = z(\varphi) \end{cases} \tag{E3-2.1}$$

where r_0 is the radius of the cylinder. The vector equation can be given by the vector function of a unit circle as

$$\boldsymbol{R} = r_0\boldsymbol{e}_{\mathrm{I}(\varphi)} + z(\varphi)\boldsymbol{k} \tag{E3-2.2}$$

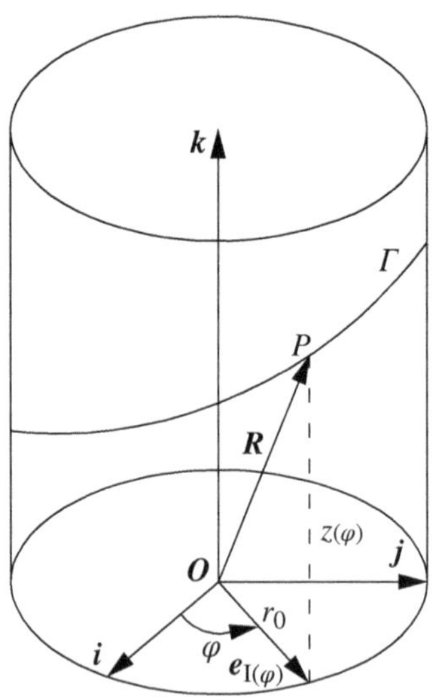

Figure 3.3 Cylindrical curve

3.1.2 Frenet Trihedron

For brevity, a curve Γ is hereinafter presented by a simplified vector equation $\boldsymbol{R} = \boldsymbol{R}(s)$ from the origin point of the fixed frame. A tangent line of curve Γ at point P can be located by point P and its infinitely adjacent point, or the first derivative with respect to arc length s. The unit tangent vector $\boldsymbol{\alpha}(s) = \mathrm{d}\boldsymbol{R}(s)/\mathrm{d}s$ is in the direction of increasing arc length. Differentiating the tangent vector with respect to s, we have

$$\frac{\mathrm{d}\boldsymbol{\alpha}(s)}{\mathrm{d}s} = k(s)\boldsymbol{\beta}(s) \tag{3.9}$$

where the coefficient $k(s)$ is called the ***curvature*** of curve Γ at point P, just the change rate of the inclined angle of the unit tangent vector $\boldsymbol{\alpha}(s)$ relative to the arc length s. The normal vector $\boldsymbol{\beta}(s)$ of the curve Γ at point P directs toward the curvature center. The osculating plane of the curve Γ at point P is determined by both the unit tangent vectors $\boldsymbol{\alpha}(s)$ and $\boldsymbol{\beta}(s)$, or three infinitesimal points including point P. The curvature of a spatial curve reveals the bending of curve Γ at point P on the osculating plane, which is always non-negative, unlike that of planar curve. The reciprocal $\rho(s) = 1/k(s)$ is called the curvature radius if the curvature $k(s)$ is not equal to zero. The position vector of the curvature center C is

$$\boldsymbol{R}_C = \boldsymbol{R}_P + \rho \cdot \boldsymbol{\beta} \tag{3.10}$$

The unit vector $\boldsymbol{\gamma}(s) = \boldsymbol{\alpha}(s) \times \boldsymbol{\beta}(s)$ is a normal of the osculating plane, and called the binormal vector $\boldsymbol{\gamma}(s)$ of the curve Γ at the point P. Thus, the trihedron, or right-handed orthogonal frame $\{\boldsymbol{R}(s); \boldsymbol{\alpha}(s), \boldsymbol{\beta}(s), \boldsymbol{\gamma}(s)\}$, is formed by three orthogonal unit vectors $\boldsymbol{\alpha}(s)$,$\boldsymbol{\beta}(s)$, and $\boldsymbol{\gamma}(s)$, or three base vectors, whose original point $\boldsymbol{R}(s)$ is at the point P of curve Γ. The trihedron is called the ***Frenet trihedron***, or usually the ***Frenet frame***, as shown in Fig. 3.4.

The Frenet trihedron $\{\boldsymbol{R}(s); \boldsymbol{\alpha}(s), \boldsymbol{\beta}(s), \boldsymbol{\gamma}(s)\}$ is also comprised of three planes, the osculating plane $\boldsymbol{\alpha}(s) - \boldsymbol{\beta}(s)$, the normal plane$\boldsymbol{\beta}(s) - \boldsymbol{\gamma}(s)$, and the rectifying plane $\boldsymbol{\alpha}(s) - \boldsymbol{\gamma}(s)$. The Frenet

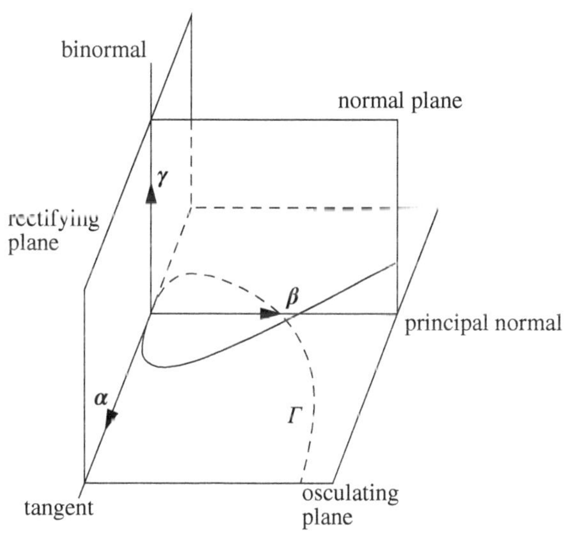

Figure 3.4 Frenet trihedron of a curve

trihedron closely attaches to the curve and moves along it since the three base vectors are determined by the curve, called a ***natural trihedron*** of a curve. Given the differential of the Frenet trihedron $\{\boldsymbol{R}(s); \boldsymbol{\alpha}(s), \boldsymbol{\beta}(s), \boldsymbol{\gamma}(s)\}$ with respect to the arc length s, we have the ***Frenet formulas*** of a curve:

$$\begin{cases} \dfrac{\mathrm{d}\boldsymbol{R}(s)}{\mathrm{d}s} = \boldsymbol{\alpha}(s) \\ \dfrac{\mathrm{d}\boldsymbol{\alpha}(s)}{\mathrm{d}s} = \qquad k(s)\boldsymbol{\beta}(s) \\ \dfrac{\mathrm{d}\boldsymbol{\beta}(s)}{\mathrm{d}s} = -k(s)\boldsymbol{\alpha}(s) \qquad + \tau(s)\boldsymbol{\gamma}(s) \\ \dfrac{\mathrm{d}\boldsymbol{\gamma}(s)}{\mathrm{d}s} = \qquad -\tau(s)\boldsymbol{\beta}(s) \end{cases} \tag{3.11}$$

where $\tau(s)$ is called the ***torsion*** of curve Γ at point P, which measures the change rate of the inclination angle of the binormal vector $\boldsymbol{\gamma}(s)$ with respect to s, or the rate of deviation of the curve relative to the osculating plane.

Based on the Frenet formulas (3.11), the curvature k and torsion τ of a curve Γ can be described by

$$k = \left|\frac{\mathrm{d}^2\boldsymbol{R}}{\mathrm{d}s^2}\right|, \tau = \left(\frac{\mathrm{d}\boldsymbol{R}}{\mathrm{d}s}, \frac{\mathrm{d}^2\boldsymbol{R}}{\mathrm{d}s^2}, \frac{\mathrm{d}^3\boldsymbol{R}}{\mathrm{d}s^3}\right) \Big/ \left|\frac{\mathrm{d}^2\boldsymbol{R}}{\mathrm{d}s^2}\right|^2 \tag{3.12}$$

The curvature k and torsion τ can be expressed as follows if the arc length s is replaced by a general parameter t:

$$k = \left|\frac{\mathrm{d}\boldsymbol{R}}{\mathrm{d}t} \times \frac{\mathrm{d}^2\boldsymbol{R}}{\mathrm{d}t^2}\right| \Big/ \left|\frac{\mathrm{d}\boldsymbol{R}}{\mathrm{d}t}\right|^3, \tau = \left(\frac{\mathrm{d}\boldsymbol{R}}{\mathrm{d}t}, \frac{\mathrm{d}^2\boldsymbol{R}}{\mathrm{d}t^2}, \frac{\mathrm{d}^3\boldsymbol{R}}{\mathrm{d}t^3}\right) \Big/ \left(\frac{\mathrm{d}\boldsymbol{R}}{\mathrm{d}t} \times \frac{\mathrm{d}^2\boldsymbol{R}}{\mathrm{d}t^2}\right)^2 \tag{3.13}$$

The curvature $k(s)$ and torsion $\tau(s)$ are all invariants of the curve. Both $k = k(s)$ and $\tau = \tau(s)$ are called the ***natural equations*** of a curve.

We have the following theorem:

Theorem 3.1 *Given a differentiable function $k(s) > 0$ and $\tau(s)$, $s \in [s_0, s_1]$, with original point $\boldsymbol{R}_0$, and right-handed orthogonal unit vectors $\{\boldsymbol{\alpha}_0, \boldsymbol{\beta}_0, \boldsymbol{\gamma}_0\}$, there must exist only one curve $\boldsymbol{R} = \boldsymbol{R}(s)$, $s \in (s_0, s_1)$ with arc length s, curvature $k(s)$, and torsion $\tau(s)$, whose original point is $\boldsymbol{R}(s_0)$, and the three unit vectors of the Frenet trihedron at point $\boldsymbol{R}(s_0)$ are $\boldsymbol{\alpha}(s_0) = \boldsymbol{\alpha}_0$, $\boldsymbol{\beta}(s_0) = \boldsymbol{\beta}_0$, $\boldsymbol{\gamma}(s_0) = \boldsymbol{\gamma}_0$.*

Let a curve be parameterized by arc length without a singular point, we write the vector equation of the curve in a neighborhood Δs of s, using the Frenet trihedron $\{\boldsymbol{R}(s); \boldsymbol{\alpha}(s), \boldsymbol{\beta}(s), \boldsymbol{\gamma}(s)\}$, and have the Taylor expansion

$$\boldsymbol{R}(s + \Delta s) = \boldsymbol{R}(s) + \frac{\mathrm{d}\boldsymbol{R}(s)}{\mathrm{d}s}\Delta s + \frac{1}{2!}\frac{\mathrm{d}^2\boldsymbol{R}(s)}{\mathrm{d}s^2}(\Delta s)^2 + \cdots \frac{1}{n!}\frac{\mathrm{d}^n\boldsymbol{R}(s)}{\mathrm{d}s^n}(\Delta s)^n + \varepsilon_n(s, \Delta s)(\Delta s)^n \tag{3.14}$$

where $\lim\limits_{\Delta s \to 0} \varepsilon_n(s, \Delta s) = 0$. The all-order deviations of the vector equation $\boldsymbol{R}(s)$ with respect to s can be derived through the Frenet formulas (3.11), such as $\frac{\mathrm{d}\boldsymbol{R}}{\mathrm{d}s} = \boldsymbol{\alpha}$, $\frac{\mathrm{d}^2\boldsymbol{R}}{\mathrm{d}s^2} = k\boldsymbol{\beta}$, $\frac{\mathrm{d}^3\boldsymbol{R}}{\mathrm{d}s^3} = -k^2\boldsymbol{\alpha} + \frac{\mathrm{d}k}{\mathrm{d}s}\boldsymbol{\beta} + k\tau\boldsymbol{\gamma}$.

3.2 Surfaces

3.2.1 Elements of Surfaces

A surface $\sum_P$ with rectangular coordinates (x, y, z) in the Cartesian frame $\{\boldsymbol{O}; \boldsymbol{i}, \boldsymbol{j}, \boldsymbol{k}\}$, which has independent variables, or parameters u, v, can be represented by

$$\begin{cases} x = x(u, v) \\ y = y(u, v) \\ z = z(u, v) \end{cases} \tag{3.15}$$

We have another expression if the parameters u, v are eliminated:

$$z = z(x, y) \tag{3.16}$$

The form of implicit function is

$$F(x, y, z) = 0 \tag{3.17}$$

In the Cartesian frame $\{\boldsymbol{O}; \boldsymbol{i}, \boldsymbol{j}, \boldsymbol{k}\}$, the vector equation of $\sum_P$ can be written as

$$\sum_P : \boldsymbol{R} = x(u, v)\boldsymbol{i} + y(u, v)\boldsymbol{j} + z(u, v)\boldsymbol{k} \tag{3.18}$$

or simplified as

$$\boldsymbol{R} = \boldsymbol{R}(u, v) \tag{3.19}$$

The partial derivatives of $\boldsymbol{R}(u, v)$ are denoted by

$$\boldsymbol{R}_u = \frac{\partial \boldsymbol{R}}{\partial u}, \boldsymbol{R}_v = \frac{\partial \boldsymbol{R}}{\partial v}, \boldsymbol{R}_{uu} = \frac{\partial^2 \boldsymbol{R}}{\partial u^2}, \boldsymbol{R}_{uv} = \frac{\partial^2 \boldsymbol{R}}{\partial u \partial v}, \boldsymbol{R}_{vv} = \frac{\partial^2 \boldsymbol{R}}{\partial v^2} \tag{3.20}$$

On surface $\sum_P$, two curves, named the parameter curve v and the parameter curve u, can respectively be described by equation (3.19) if either u or v is designated as a constant. A curve Γ of $\sum_P$ is expressed as

$$\boldsymbol{R} = \boldsymbol{R}(u(t), v(t)) \tag{3.21}$$

whose differential form is

$$\mathrm{d}\boldsymbol{R} = \boldsymbol{R}_u \mathrm{d}u + \boldsymbol{R}_v \mathrm{d}v \tag{3.22}$$

The arc length of the curve Γ is s, and the square of the line element $\mathrm{d}s$ is

$$\begin{cases} \pi_1 = (\mathrm{d}s)^2 = (\mathrm{d}\boldsymbol{R})^2 = E(\mathrm{d}u)^2 + 2F\mathrm{d}u\mathrm{d}v + G(\mathrm{d}v)^2 \\ E = \boldsymbol{R}_u^2, F = \boldsymbol{R}_u \cdot \boldsymbol{R}_v, G = \boldsymbol{R}_v^2 \end{cases} \tag{3.23}$$

It is called the ***first fundamental form*** of surface $\sum_P$ and is a quadratic differential form, in which the coefficients E, F, and G are the functions of the parameters u, v, and are referred to as the ***coefficients of the first fundamental form***.

The unit normal vector of $\sum_P$ at point $P(u, v)$ can be defined by

$$\boldsymbol{n} = \frac{\boldsymbol{R}_u \times \boldsymbol{R}_v}{|\boldsymbol{R}_u \times \boldsymbol{R}_v|} = \frac{\boldsymbol{R}_u \times \boldsymbol{R}_v}{(EG - F^2)^{1/2}} \tag{3.24}$$

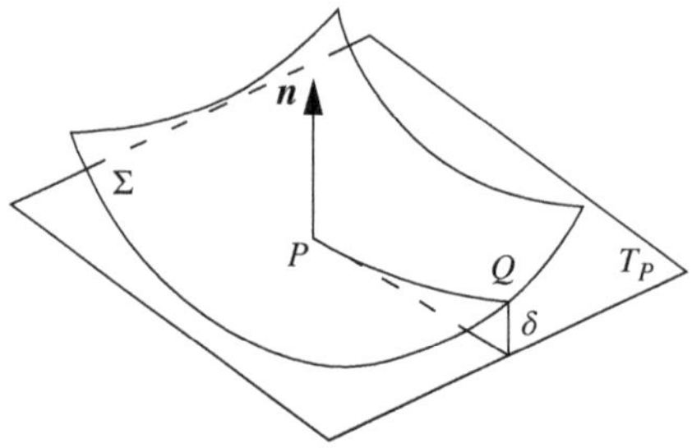

Figure 3.5 Proximity structure of a space surface

The tangent plane of $\sum_P$ at point P is assumed to be T_P with normal $\boldsymbol{n}$, as shown in Fig. 3.5. The distance vector from point $P(u, v)$ to the adjacent point $Q(u + \Delta u, v + \Delta v)$ is described as

$$\begin{aligned}\overrightarrow{PQ} &= \boldsymbol{R}(u + \Delta u, v + \Delta v) - \boldsymbol{R}(u, v)\\ &= \boldsymbol{R}_u \Delta u + \boldsymbol{R}_v \Delta v + \frac{1}{2}[\boldsymbol{R}_{uu}(\Delta u)^2 + 2\boldsymbol{R}_{uv}\Delta u \Delta v + \boldsymbol{R}_{vv}(\Delta v)^2] + \cdots \end{aligned} \tag{3.25}$$

The normal distance from point Q to the tangent plane T_P is

$$\delta_n = \overrightarrow{PQ} \cdot \boldsymbol{n} = \frac{1}{2}\left[\boldsymbol{R}_{uu} \cdot \boldsymbol{n}(\Delta u)^2 + 2\boldsymbol{R}_{uv} \cdot \boldsymbol{n}\Delta u \Delta v + \boldsymbol{R}_{vv} \cdot \boldsymbol{n}(\Delta v)^2\right] + \cdots \tag{3.26}$$

The sign of δ_n reflects the bending direction, while its value reveals the magnitude of $\sum_P$ at P. When $\Delta u, \Delta v \to 0$, the principal part of $2\delta_n$ is

$$\begin{cases} \pi_2 = \mathrm{d}^2\boldsymbol{R} \cdot \boldsymbol{n} = L(\mathrm{d}u)^2 + 2M\mathrm{d}u\mathrm{d}v + N(\mathrm{d}v)^2 \\ L = \boldsymbol{n} \cdot \boldsymbol{R}_{uu}, M = \boldsymbol{n} \cdot \boldsymbol{R}_{uv}, N = \boldsymbol{n} \cdot \boldsymbol{R}_{vv} \end{cases} \tag{3.27}$$

The above equation is defined as the ***second fundamental form*** of surface $\sum_P$. The coefficients L, M, and N are the ***coefficients of the second fundamental form***.

The unit tangent vector $\boldsymbol{\alpha}$ and the normal vector $\boldsymbol{\beta}$ of a curve Γ on the surface $\sum_P$ can be obtained by the Frenet formulas (3.11):

$$\boldsymbol{\alpha} = \frac{\mathrm{d}\boldsymbol{R}}{\mathrm{d}s}, \ k\boldsymbol{\beta} = \frac{\mathrm{d}^2\boldsymbol{R}}{\mathrm{d}s^2} \tag{3.28}$$

Projecting the curvature vector $k\boldsymbol{\beta}$ of Γ to the normal vector $\boldsymbol{n}$ and the binormal vector $\boldsymbol{v} = \boldsymbol{\alpha} \times \boldsymbol{n}$ of $\sum_P$ at point P respectively, we have the ***normal curvature*** k_n and the ***geodesic curvature*** k_g of Γ at P as follows:

$$\begin{cases} k_n = k\boldsymbol{\beta} \cdot \boldsymbol{n} \\ k_g = k\boldsymbol{\beta} \cdot \boldsymbol{v} \end{cases} \tag{3.29}$$

The signs of both k_n and k_g may be positive or negative. The included angle between the vectors $\boldsymbol{\beta}$ and $\boldsymbol{n}$ is designated as θ; we have

$$\begin{cases} k^2 = k_n^2 + k_g^2 \\ k_n = k\cos\theta,\ k_g = k\sin\theta \end{cases} \tag{3.30}$$

The normal plane $\sum_N$ of $\sum_P$ at point P, defined by both the tangent vector $\boldsymbol{\alpha}$ and the normal vector $\boldsymbol{n}$, intersects with $\sum_P$ at a normal section curve $\overline{\Gamma}$, shown in Fig. 3.6(a). The curvature of $\overline{\Gamma}$ is just the normal curvature k_n of Γ on $\sum_P$, which can be derived from equations (3.28) to (3.30). Substituting equation (3.28) into (3.29), and combining them with equations (3.23) and (3.27), we have

$$k_n = \frac{\pi_2}{\pi_1} = \frac{L(\mathrm{d}u)^2 + 2M\mathrm{d}u\mathrm{d}v + N(\mathrm{d}v)^2}{E(\mathrm{d}u)^2 + 2F\mathrm{d}u\mathrm{d}v + G(\mathrm{d}v)^2} \tag{3.31}$$

Similarly, different transversal curves Γ_i on $\sum_P$ can be obtained by intersecting a plane with $\sum_P$; the plane passes through the point P and the tangent vector $\boldsymbol{\alpha}$ and rotates by different angles θ around $\boldsymbol{\alpha}$. Then, each curve Γ_i has its own curvature k_i, but the same normal curvature k_n.

If the curvatures k_i and k_n are replaced by their curvature radii R_i and R_n, another form of $k_n = k\cos\theta$ can be obtained

$$R_i = R_n \cdot \cos\theta \tag{3.32}$$

If the normal curvature k_n of a curve on surface $\sum_P$ at point P is zero, the tangential direction of the curve is called the asymptotic direction of $\sum_P$ at P. Equation (3.32) can be described by ***Meusnier's theorem***.

Meusnier's theorem *All curves on a surface $\sum_P$, passing through a given point P and having the same non-asymptotic tangential direction, have the same normal curvature at P.*

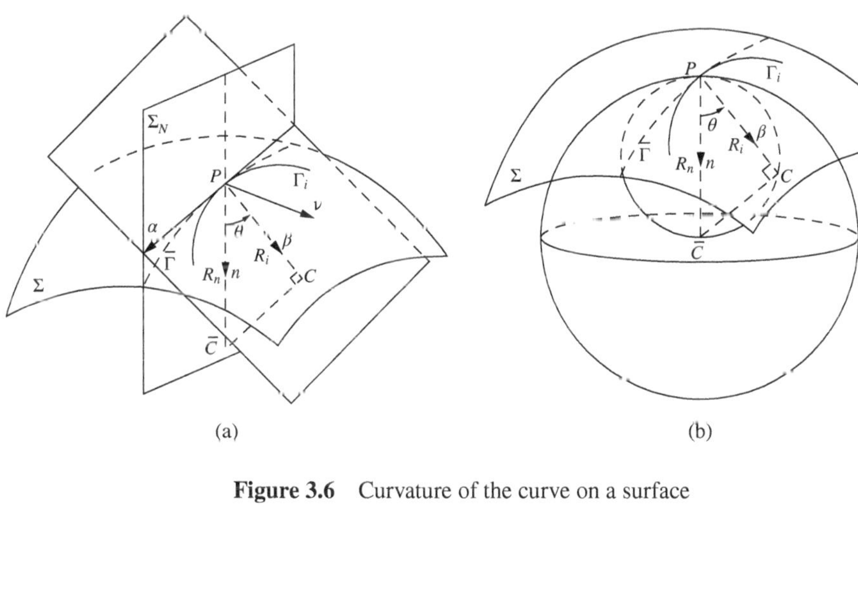

Figure 3.6 Curvature of the curve on a surface

In other words, for all curves on $\sum_P$ at P with the same non-asymptotic tangential direction, their corresponding osculating circles form a sphere, as shown in Fig. 3.6(b).

Based on equation (3.31), the normal curvature k_n of Γ on $\sum_P$ depends not only on the coefficients of first and second fundamental forms, but also on the tangential direction $\mathrm{d}u/\mathrm{d}v$ of Γ. The tangential direction $\mathrm{d}u/\mathrm{d}v$, in which the normal curvature k_n gets an extremum, is defined as the ***principal direction***, and the corresponding extremum of the normal curvature is called the ***principal curvature***.

By differentiating equation (3.31) with respect to $\mathrm{d}u/\mathrm{d}v$, and letting $\mathrm{d}u/\mathrm{d}v$ be zero, we have an expression with variable k_n:

$$\begin{vmatrix} k_nE - L & k_nF - M \\ k_nF - M & k_nG - N \end{vmatrix} = 0 \tag{3.33}$$

Two principal curvatures k_1 and k_2 can be solved from the above equation and two corresponding orthogonal principal directions can be derived by substituting equation (3.31) into the above equation, that is

$$\begin{vmatrix} E\mathrm{d}u + F\mathrm{d}v & F\mathrm{d}u + F\mathrm{d}v \\ L\mathrm{d}u + M\mathrm{d}v & M\mathrm{d}u + N\mathrm{d}v \end{vmatrix} = 0 \tag{3.34}$$

A regular curve is called a ***line of curvature*** on the surface $\sum_P$ if the tangential direction at any point is always the principal direction. Equation (3.34) is also the equation of a line of curvature on a surface. The lines of curvature can be characterized by the following formula of Rodrigues:

$$\mathrm{d}\boldsymbol{n} = -k_n \mathrm{d}\boldsymbol{R} \tag{3.35}$$

where k_n is the normal curvature in the direction $\mathrm{d}\boldsymbol{R}$ of the line of curvature.

The coefficients F and M in equation (3.31) will both be zero if the curvature lines are taken as the parametric lines, and the tangential directions of the two parametric lines $\mathrm{d}v = 0$ and $\mathrm{d}u = 0$ are the principal directions of the surface, which leads to the two principle curvatures being $k_1 = L/E$, $k_2 = N/G$. Any direction $\mathrm{d}u/\mathrm{d}v$ at point P on $\sum_P$ has an inclined angle φ with respect to the principal direction $\mathrm{d}v = 0$, and the normal curvature k_n of $\sum_P$ at P in this direction is

$$k_n = k_1 \cos^2\varphi + k_2 \sin^2\varphi \tag{3.36}$$

The above equation expresses the normal curvature in an arbitrary direction in terms of the two principle curvatures k_1 and k_2, and is known as ***Euler's theorem***.

The points on $\sum_P$, where the coefficients of the first and second fundamental forms are proportional, are called ***umbilical points***. At a non-umbilical point on $\sum_P$, there exist two principal curvatures k_1 and k_2. Then, the ***Gauss curvature*** K and the ***mean curvature*** H can be defined by

$$\begin{cases} K = k_1 k_2 = \dfrac{LN - M^2}{EG - F^2} \\ H = k_1 + k_2 = \dfrac{LG - 2MF + NE}{2\left(EG - F^2\right)} \end{cases} \tag{3.37}$$

The sign of the Gauss curvature K reflects the vicinity structure of $\sum_P$ at the neighborhood of point P.

(i) $K>0$. The two principal curvatures k_1 and k_2 have the same sign at P. The normal sectional curves in any tangential direction have normal curvature k_n between k_1 and k_2, which means that $\sum_P$ bends to the same side in different directions. The point is called an ***elliptic point***.

(ii) $K<0$. Two principal curvatures k_1 and k_2 have opposite sign at P. The normal curvature k_n must change its sign as the tangential direction varies from a principal direction to another one; there exists a tangential direction for $k_n=0$. This point is called a ***hyperbolic point***, or a ***saddle point*** of the surface. Based on equation (3.36), the tangential direction for $k_n = 0$, or the ***asymptotic direction***, can be determined by

$$\tan\varphi = \pm\sqrt{-k_1/k_2} \tag{3.38}$$

For any curve Γ on $\sum_P$, if the tangential direction at any point is always the asymptotic direction, it is called the ***asymptotic line*** on $\sum_P$. Based on equation (3.31), the differential equation of the asymptotic line is

$$L(\mathrm{d}u)^2 + 2M\mathrm{d}u\mathrm{d}v + N(\mathrm{d}v)^2 = 0 \tag{3.39}$$

(iii) $K=0$. The Gauss curvature is zero at point P, which denotes that at least one principal curvature of k_1 and k_2 is zero, and the corresponding principal direction is the asymptotic direction. This point of the surface is called a ***parabolic point***.

The curve of $\sum_P$ is a **geodesic** if its geodesic curvature k_g is always zero. As is known from equations (3.29) and (3.30), the condition for the geodesic curvature $k_g = 0$ at a point is the principal normal vector $\boldsymbol{\beta}$ of Γ coincides with the normal vector $\boldsymbol{n}$ of $\sum_P$ at the point, or the angle between $\boldsymbol{\beta}$ and $\boldsymbol{n}$ is 0 or π. For example, a geodesic on a plane can only be a straight line, and a geodesic on a sphere can only be a great circle, while a helix is a geodesic on a cylindrical surface.

If a curve Γ is a geodesic on $\sum_P$, the ***geodesic torsion*** τ_g can be derived by the Frenet formulas (3.11). That is

$$\tau_g = \frac{\mathrm{d}\boldsymbol{\beta}}{\mathrm{d}s}\cdot\boldsymbol{\gamma} = \frac{\mathrm{d}\boldsymbol{\beta}}{\mathrm{d}s}\cdot(\boldsymbol{\alpha}\times\boldsymbol{\beta}) \tag{3.40}$$

For the geodesic, the condition $\boldsymbol{\beta} = \pm\boldsymbol{n}$ can be substituted into the above equation, and another expression of geodesic torsion τ_g can be written sequentially as

$$\tau_g = \frac{\mathrm{d}\boldsymbol{\beta}}{\mathrm{d}s}\cdot(\boldsymbol{\alpha}\times\boldsymbol{\beta}) = \left(\boldsymbol{\alpha},\boldsymbol{n},\frac{\mathrm{d}\boldsymbol{n}}{\mathrm{d}s}\right) = \left(\frac{\mathrm{d}\boldsymbol{R}}{\mathrm{d}s},\boldsymbol{n},\frac{\mathrm{d}\boldsymbol{n}}{\mathrm{d}s}\right) \tag{3.41}$$

The geodesic torsion τ_g is defined along the direction of vector $\boldsymbol{\alpha}$.

On a curve Γ of $\sum_P$, another moving trihedron is comprised of the three base vectors $\boldsymbol{\alpha}$, $\boldsymbol{n}$, and $\boldsymbol{v} = \boldsymbol{\alpha}\times\boldsymbol{n}$, whose original point locates on the point P of Γ, represented by $\{\boldsymbol{R}_P;\boldsymbol{\alpha},\boldsymbol{n},\boldsymbol{v}\}$ for short. The trihedron is called the ***Darboux frame*** and the differential formulas, or ***Darboux formulas***, are

$$\begin{cases} \dfrac{\mathrm{d}\boldsymbol{\alpha}}{\mathrm{d}s} = \qquad k_n\boldsymbol{n} + k_g\boldsymbol{v} \\ \dfrac{\mathrm{d}\boldsymbol{n}}{\mathrm{d}s} = -k_n\boldsymbol{\alpha} \qquad + \tau_g\boldsymbol{v} \\ \dfrac{\mathrm{d}\boldsymbol{v}}{\mathrm{d}s} = -k_g\boldsymbol{\alpha} - \tau_g\boldsymbol{n} \end{cases} \tag{3.42}$$

As mentioned above, the normal curvature k_n, the geodesic curvature k_g, and the geodesic torsion τ_g are all kinematic invariants of the curve relative to the surface, and the differential coefficients of the Darboux frame. The Darboux frame $\{\boldsymbol{R}_P; \boldsymbol{\alpha}, \boldsymbol{n}, \boldsymbol{v}\}$ moves along a curve of $\sum_P$, revealing the intrinsic geometrical relationship between Γ and $\sum_P$, while the Frenet frame $\{\boldsymbol{R}_P; \boldsymbol{\alpha}, \boldsymbol{\beta}, \boldsymbol{\gamma}\}$ reveals the intrinsic geometrical properties of the curve.

In particular, when $\sum_P$ degenerates to a spherical surface and the original point of the fixed frame $\{\boldsymbol{O}; \boldsymbol{i}, \boldsymbol{j}, \boldsymbol{k}\}$ is placed at the sphere center. The normal vector $\boldsymbol{n}$ of the sphere directs toward the sphere center and the spherical curve Γ can be denoted as

$$\Gamma : \boldsymbol{R}(s) = -R\boldsymbol{n}(s) \tag{3.43}$$

where R is the radius of the sphere and s is the arc length of Γ. Differentiating the above equation with respect to s, we have

$$\frac{\mathrm{d}\boldsymbol{R}}{\mathrm{d}s} = -R\frac{\mathrm{d}\boldsymbol{n}}{\mathrm{d}s} = \boldsymbol{\alpha} \tag{3.44}$$

Combining the above equation and the differential formulas (3.42) of the Darboux frame, the differential formulas of the Darboux frame $\{\boldsymbol{R}; \boldsymbol{\alpha}, \boldsymbol{n}, \boldsymbol{v}\}$ of a spherical curve Γ are

$$\begin{cases} \dfrac{\mathrm{d}\boldsymbol{R}}{\mathrm{d}s} = \boldsymbol{\alpha} \\ \dfrac{\mathrm{d}\boldsymbol{\alpha}}{\mathrm{d}s} = \dfrac{1}{R}\boldsymbol{n} + k_g\boldsymbol{v} \\ \dfrac{\mathrm{d}\boldsymbol{n}}{\mathrm{d}s} = -\dfrac{1}{R}\boldsymbol{\alpha} \\ \dfrac{\mathrm{d}\boldsymbol{v}}{\mathrm{d}s} = -k_g\boldsymbol{\alpha} \end{cases} \tag{3.45}$$

For the unit sphere with $R = 1$, we have

$$\begin{cases} \dfrac{\mathrm{d}\boldsymbol{R}}{\mathrm{d}s} = \boldsymbol{\alpha} \\ \dfrac{\mathrm{d}\boldsymbol{\alpha}}{\mathrm{d}s} = \boldsymbol{n} + k_g\boldsymbol{v} \\ \dfrac{\mathrm{d}\boldsymbol{n}}{\mathrm{d}s} = -\boldsymbol{\alpha} \\ \dfrac{\mathrm{d}\boldsymbol{v}}{\mathrm{d}s} = -k_g\boldsymbol{\alpha} \end{cases} \tag{3.46}$$

Both the frames of a curve Γ on $\sum_P$, the Frenet frame $\{\boldsymbol{R}; \boldsymbol{\alpha}, \boldsymbol{\beta}, \boldsymbol{\gamma}\}$ and the Darboux frame $\{\boldsymbol{R}; \boldsymbol{\alpha}, \boldsymbol{n}, \boldsymbol{v}\}$, have the same vector $\boldsymbol{\alpha}$, so $\{\boldsymbol{R}; \boldsymbol{\alpha}, \boldsymbol{n}, \boldsymbol{v}\}$ can be regarded as the rotation of $\{\boldsymbol{R}; \boldsymbol{\alpha}, \boldsymbol{\beta}, \boldsymbol{\gamma}\}$ around $\boldsymbol{\alpha}$ by angle θ, and the relationships between them are

$$\begin{cases} \boldsymbol{\alpha} = \boldsymbol{\alpha} \\ \boldsymbol{n} = \cos\theta\boldsymbol{\beta} + \sin\theta\boldsymbol{\gamma} \\ \boldsymbol{v} = -\sin\theta\boldsymbol{\beta} + \cos\theta\boldsymbol{\gamma} \end{cases} \tag{3.47}$$

Differentiating the above equations with respect to the arc length s of Γ, and combining the differential formulas (3.11) and (3.42), we have

$$k_n = k\cos\theta,\ k_g = -k\sin\theta,\ \tau_g = \tau + \frac{d\theta}{ds} \tag{3.48}$$

Sequentially differentiating the above equations with respect to s, we get

$$\begin{cases} \dfrac{dk_n}{ds} = \dfrac{dk}{ds}\cos\theta - k\dfrac{d\theta}{ds}\sin\theta,\ \dfrac{dk_g}{ds} = -\dfrac{dk}{ds}\sin\theta - k\dfrac{d\theta}{ds}\cos\theta \\ \dfrac{d^2k_n}{ds^2} = \dfrac{d^2k}{ds^2}\cos\theta - 2\dfrac{dk}{ds}\dfrac{d\theta}{ds}\sin\theta - k\left(\dfrac{d\theta}{ds}\right)^2\cos\theta - k\dfrac{d^2\theta}{ds^2}\sin\theta \\ \dfrac{d^2k_g}{ds^2} = -\dfrac{d^2k}{ds^2}\sin\theta - 2\dfrac{dk}{ds}\dfrac{d\theta}{ds}\cos\theta + k\left(\dfrac{d\theta}{ds}\right)^2\sin\theta - k\dfrac{d^2\theta}{ds^2}\cos\theta \\ \dfrac{d\tau_g}{ds} = \dfrac{d\tau}{ds} + \dfrac{d^2\theta}{ds^2},\ \dfrac{d^2\tau_g}{ds^2} = \dfrac{d^2\tau}{ds^2} + \dfrac{d^3\theta}{ds^3} \end{cases} \tag{3.49}$$

3.2.2 *Ruled Surfaces*

A typical surface, traced by a straight line in spatial motion, is called a ***ruled surface***. The straight line is the generator of the ruled surface, or ruling. Both cylindrical surfaces and conical surfaces can be viewed as special cases of ruled surfaces. The generators are parallel to each other for the former, while the generators intersect at the vertex for the latter.

A curve Γ, which intersects with all the generators on the ruled surface $\sum_L$, is described by

$$\Gamma:\ \boldsymbol{a} = \boldsymbol{a}(u),\quad u_0 \le u \le u_1 \tag{3.50}$$

A unit vector $\boldsymbol{l}(u)$ is taken as the direction vector of generatrices of $\sum_L$, and then vector equation of $\sum_L$ is

$$\sum\nolimits_L :\ \boldsymbol{R}(u,v) = \boldsymbol{a}(u) + v\boldsymbol{l}(u),\ u_0 \le u \le u_1,\ -\infty < v < +\infty \tag{3.51}$$

The curve Γ is called the directrix of $\sum_L$. The above equation of a ruled surface may change as its directrix varies, since any curve can be taken as a directrix if it intersects with all the generatrices. Taking the partial derivatives of the above equation with respect to the parameters u and v respectively, we have

$$\boldsymbol{R}_u = \frac{\partial \boldsymbol{R}}{\partial u} = \frac{d\boldsymbol{a}}{du} + v\frac{d\boldsymbol{l}}{du},\ \boldsymbol{R}_v = \frac{\partial \boldsymbol{R}}{\partial v} = \boldsymbol{l} \tag{3.52}$$

The normal vector $\boldsymbol{N}$ at the point $P(u, v)$ of $\sum_L$ is defined by

$$\boldsymbol{N} = \boldsymbol{R}_u \times \boldsymbol{R}_v = \frac{d\boldsymbol{a}}{du} \times \boldsymbol{l} + v\frac{d\boldsymbol{l}}{du} \times \boldsymbol{l} \tag{3.53}$$

For $\sum_L$, the shortest distance between a generator $\boldsymbol{l}(u)$ and the adjoining generator $\boldsymbol{l}(u + \Delta u)$ is the length of the common normal $\overline{M_1M_2}$, where M_1 is the perpendicular foot on $\boldsymbol{l}(u)$ and M_2 is that on $\boldsymbol{l}(u + \Delta u)$, as shown in Fig. 3.7. The point M_2 will coincide with the point M_1 on

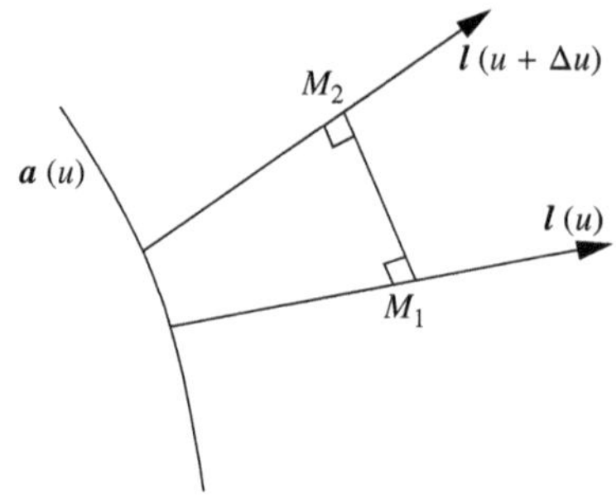

Figure 3.7 Two adjacent generatrices of a ruled surface

$\boldsymbol{l}(u)$ when Δu tends to zero, which is defined as a ***striction point*** on $\boldsymbol{l}(u)$ of $\sum_L$. The distance b_l, from the striction point to a point of the directrix Γ on the generator $\boldsymbol{l}(u)$, or the striction directrix distance for short, is

$$b_l = -\left(\frac{\mathrm{d}\boldsymbol{a}}{\mathrm{d}u} \cdot \frac{\mathrm{d}\boldsymbol{l}}{\mathrm{d}u}\right) / \left(\frac{\mathrm{d}\boldsymbol{l}}{\mathrm{d}u}\right)^2 \tag{3.54}$$

For $\sum_L$, the striction points on all generatrices form a ***striction curve***. Based on equations (3.51) and (3.54), a striction curve can be represented as

$$\boldsymbol{\rho}(u) = \boldsymbol{a}(u) + b_l \boldsymbol{l}(u) = \boldsymbol{a}(u) - \left(\frac{\mathrm{d}\boldsymbol{a}}{\mathrm{d}u} \cdot \frac{\mathrm{d}\boldsymbol{l}}{\mathrm{d}u}\right) \cdot \boldsymbol{l}(u) / \left(\frac{\mathrm{d}\boldsymbol{l}}{\mathrm{d}u}\right)^2 \tag{3.55}$$

A striction curve has the same parameter u in the above equation as that of a directrix in equation (3.50). Obviously, there exists only one striction curve on a ruled surface if the ruled surface is not degenerate, but there are infinite number of directrix on a ruled surface.

Theorem 3.2 *For a ruled surface $\sum_L$ with generator $\boldsymbol{l}(u)$ and directrix $\boldsymbol{a}(u)$, the directrix is the striction curve of $\sum_L$ if and only if $\frac{\mathrm{d}\boldsymbol{a}}{\mathrm{d}u} \cdot \frac{\mathrm{d}\boldsymbol{l}}{\mathrm{d}u} = 0$.*

Hence, the vector equation of $\sum_L$ can be written in standard form by taking its striction curve as a directrix, that is

$$\sum\nolimits_L : \boldsymbol{R}(u, v) = \boldsymbol{\rho}(u) + v\boldsymbol{l}(u), \ u_0 \leq u \leq u_1, \ -\infty < v < +\infty \tag{3.56}$$

To study the geometrical properties of a ruled surface, a trihedron of a ruled surface is set up with three base vectors designated as

$$\boldsymbol{e}_1 = \boldsymbol{l}, \boldsymbol{e}_2 = \frac{\mathrm{d}\boldsymbol{l}}{\mathrm{d}u} / \left|\frac{\mathrm{d}\boldsymbol{l}}{\mathrm{d}u}\right|, \boldsymbol{e}_3 = \boldsymbol{e}_1 \times \boldsymbol{e}_2 = \boldsymbol{l} \times \frac{\mathrm{d}\boldsymbol{l}}{\mathrm{d}u} / \left|\frac{\mathrm{d}\boldsymbol{l}}{\mathrm{d}u}\right| \tag{3.57}$$

The trihedron, whose original point is located on the striction curve $\boldsymbol{\rho}(u)$, is called the ***Frenet trihedron*** of the ruled surface, also the Frenet frame, and can briefly be represented as $\{\boldsymbol{\rho}; \boldsymbol{e}_1, \boldsymbol{e}_2, \boldsymbol{e}_3\}$. It is also named the ***natural trihedron***, since it attaches closely to the ruled surface and moves along it, just similar to the Frenet trihedron of a curve. The differential formulas of the Frenet trihedron or ***Frenet formulas***, are

$$\begin{cases} \dfrac{\mathrm{d}\boldsymbol{\rho}}{\mathrm{d}\sigma} = \alpha \boldsymbol{e}_1 + \gamma \boldsymbol{e}_3 \\ \dfrac{\mathrm{d}\boldsymbol{e}_1}{\mathrm{d}\sigma} = \boldsymbol{e}_2 \\ \dfrac{\mathrm{d}\boldsymbol{e}_2}{\mathrm{d}\sigma} = -\boldsymbol{e}_1 + \beta \boldsymbol{e}_3 \\ \dfrac{\mathrm{d}\boldsymbol{e}_3}{\mathrm{d}\sigma} = -\beta \boldsymbol{e}_2 \end{cases} \tag{3.58}$$

The coefficients $\alpha,\ \beta,\ \gamma$ are defined as the construction parameters, which are all the kinematic invariants of a ruled surface. In some references such as Ref. [4], the natural trihedron $\{\boldsymbol{\rho}; \boldsymbol{e}_1, \boldsymbol{e}_2, \boldsymbol{e}_3\}$ of a ruled surface is written as $\{\boldsymbol{T}, \boldsymbol{G}, \boldsymbol{E}\}$, and the three construction parameters $\alpha,\ \beta,\ \gamma$ are respectively presented as curvature functions $\gamma,\ \Delta,\ \Gamma$.

The geometrical meaning of the kinematic invariants $\sigma,\ \alpha,\ \beta$, and γ of a ruled surface are described as follows.

1. σ is the ***arc length of the spherical image curve***, corresponding to the unit vector of the generator $\boldsymbol{l}(u)$ of $\sum_L$. When the initial point of the generator's unit vector is put at the center of a unit sphere, the end point of the unit vector will draw a curve on the unit sphere, which is called the ***spherical image curve of the generator***. The arc length σ has a relation with the parameter u as

$$\mathrm{d}\sigma = \left|\frac{\mathrm{d}\boldsymbol{l}}{\mathrm{d}u}\right| \mathrm{d}u \tag{3.59}$$

2. β is the geodesic curvature of the spherical image curve of the generator $\boldsymbol{l}(u)$ of a ruled surface $\sum_L$. Based on the differential formulas (3.58), we have

$$\beta = \left(\boldsymbol{l}, \frac{\mathrm{d}\boldsymbol{l}}{\mathrm{d}\sigma}, \frac{\mathrm{d}^2\boldsymbol{l}}{\mathrm{d}\sigma^2}\right) = \left(\boldsymbol{l}, \frac{\mathrm{d}\boldsymbol{l}}{\mathrm{d}u}, \frac{\mathrm{d}^2\boldsymbol{l}}{\mathrm{d}u^2}\right) \Big/ \left|\frac{\mathrm{d}\boldsymbol{l}}{\mathrm{d}u}\right|^3 \tag{3.60}$$

3. γ is the ***distribution parameter***, which indicates the variation of the normal along the generator of $\sum_L$, and can be derived from the Frenet formulas (3.58). That is

$$\gamma = \frac{\mathrm{d}\boldsymbol{\rho}}{\mathrm{d}\sigma} \cdot \boldsymbol{e}_3 = \left(\frac{\mathrm{d}\boldsymbol{\rho}}{\mathrm{d}\sigma}, \boldsymbol{l}, \frac{\mathrm{d}\boldsymbol{l}}{\mathrm{d}\sigma}\right) = \left(\frac{\mathrm{d}\boldsymbol{\rho}}{\mathrm{d}u}, \boldsymbol{l}, \frac{\mathrm{d}\boldsymbol{l}}{\mathrm{d}u}\right) \Big/ \left|\frac{\mathrm{d}\boldsymbol{l}}{\mathrm{d}u}\right|^2 \tag{3.61}$$

   When γ is equal to zero, the ruled surface $\sum_L$ is called a ***developable surface***.

4. α is an angle between tangent line of the striction curve and the generator. Based on the Frenet formulas (3.58), we have

$$\alpha = \frac{\mathrm{d}\boldsymbol{\rho}}{\mathrm{d}\sigma} \cdot \boldsymbol{e}_1 = \frac{\mathrm{d}\boldsymbol{\rho}}{\mathrm{d}u} \cdot \boldsymbol{l} \Big/ \left|\frac{\mathrm{d}\boldsymbol{l}}{\mathrm{d}u}\right| \tag{3.62}$$

Example 3.3 Write the vector equation of a cylindrical surface (see Fig. 3.8(a)).

Solution

Given a spatial curve $\Gamma: \boldsymbol{a} = \boldsymbol{a}(u),\ \ u_0 \le u \le u_1$ and a straight line with unit direction vector $\boldsymbol{l}$, the straight line moves along the curve Γ and keeps parallel, whose trajectory is a cylindrical

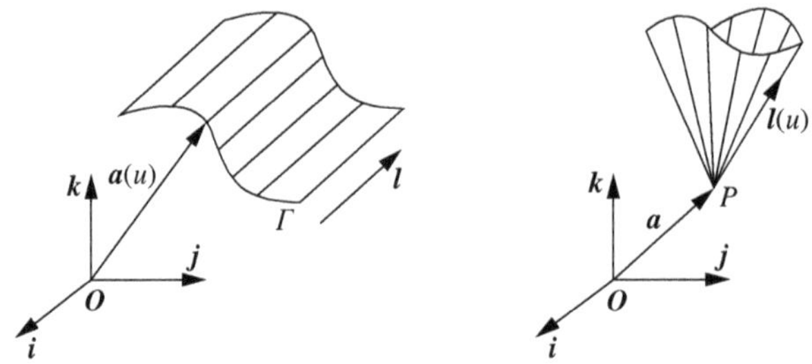

Figure 3.8 A cylindrical surface and a conical surface (a) A cylindrical surface (b) A conical surface

surface and can be expressed as

$$\boldsymbol{R}(u, v) = \boldsymbol{a}(u) + v\boldsymbol{l},\ u_0 \leq u \leq u_1,\ -\infty < v < \infty \tag{E3-3.1}$$

By means of equation (3.24), the unit normal vector of the cylindrical surface is

$$\boldsymbol{n} = \frac{\boldsymbol{R}_u \times \boldsymbol{R}_v}{|\boldsymbol{R}_u \times \boldsymbol{R}_v|} = \frac{\mathrm{d}\boldsymbol{a}}{\mathrm{d}u} \times \boldsymbol{l} \Big/ \left|\frac{\mathrm{d}\boldsymbol{a}}{\mathrm{d}u} \times \boldsymbol{l}\right| \tag{E3-3.2}$$

Any directrix of the cylindrical surface can be viewed as its striction curve since all generatrices are parallel to each other.

Example 3.4 Write the vector equation of a conical surface (see Fig. 3.8(b)).

Solution

Given a fixed point P with position vector $\boldsymbol{a}$ and a straight line with a unit direction vector $\boldsymbol{l}$, the straight line rotates around the fixed point P, whose trajectory is a conical surface, with equation

$$\boldsymbol{R}(u, v) = \boldsymbol{a} + v\boldsymbol{l}(u),\ u_0 \leq u \leq u_1,\ -\infty < v < \infty \tag{E3-4.1}$$

The unit normal vector of the conical surface can be expressed by

$$\boldsymbol{n} = \frac{\boldsymbol{R}_u \times \boldsymbol{R}_v}{|\boldsymbol{R}_u \times \boldsymbol{R}_v|} = \frac{\mathrm{d}\boldsymbol{l}}{\mathrm{d}u} \times \boldsymbol{l} \Big/ \left|\frac{\mathrm{d}\boldsymbol{l}}{\mathrm{d}u} \times \boldsymbol{l}\right| \tag{E3-4.2}$$

It is viewed as both the directrix curve and the striction curve of the conical surface degenerate to the fixed point P.

3.2.3 Adjoint Approach

To facilitate the study of the geometrical properties of constraint curves and surfaces of linkages by differential geometry, the displacements of points or lines can be examined intensively by means of the adjoint motion to the moving frame, or an ***adjoint approach,*** including a curve adjoint to another curve with Frenet trihedron, a curve adjoint to a surface with Darboux frame, and a ruled surface adjoint to another ruled surface with Frenet trihedron. We expand the contents usually introduced in a textbook of differential geometry to the general cases, and

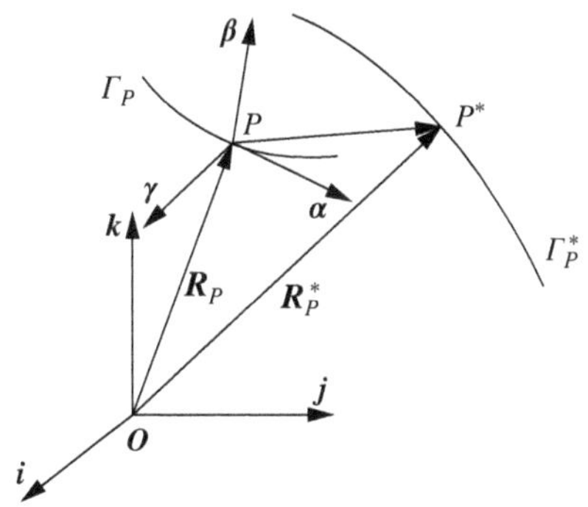

Figure 3.9 A spatial curve adjoining another spatial curve

discuss some important results and formulas for the kinematic geometry of linkages, which can readily be referred to in later chapters.

3.2.3.1 A Spatial Curve Adjoining Another Spatial Curve

As shown in Fig. 3.9, for a spatial curve Γ_P in the fixed coordinate system $\{\boldsymbol{O};\boldsymbol{i},\boldsymbol{j},\boldsymbol{k}\}$, a point P moves along Γ_P and another point P^*, which does not belong to Γ_P, also moves and traces another curve Γ_P^* in the same fixed coordinate system $\{\boldsymbol{O};\boldsymbol{i},\boldsymbol{j},\boldsymbol{k}\}$. Meanwhile, each position of P^* at Γ_P^* always corresponds to a position of P at Γ_P. The curve Γ_P is defined as an ***original curve*** and Γ_P^* as an ***adjoint curve***.

The Frenet trihedron $\{\boldsymbol{R}_P;\boldsymbol{\alpha},\boldsymbol{\beta},\boldsymbol{\gamma}\}$ of the original curve Γ_P is set up in the way mentioned above, and the adjoint curve Γ_P^* can be expressed as

$$\Gamma_P^* : \boldsymbol{R}_P^* = \boldsymbol{R}_P + u_1\boldsymbol{\alpha} + u_2\boldsymbol{\beta} + u_3\boldsymbol{\gamma} \tag{3.63}$$

where (u_1, u_2, u_3) are the coordinates of P^* in $\{\boldsymbol{R}_P;\boldsymbol{\alpha},\boldsymbol{\beta},\boldsymbol{\gamma}\}$. Based on the Frenet formulas (3.11), the first derivative of the above equation with respect to arc length s of Γ_P is

$$\begin{cases} \dfrac{\mathrm{d}\boldsymbol{R}_P^*}{\mathrm{d}s} = A_1\boldsymbol{\alpha} + A_2\boldsymbol{\beta} + A_3\boldsymbol{\gamma} \\ A_1 = 1 + \dfrac{\mathrm{d}u_1}{\mathrm{d}s} - ku_2 \\ A_2 = ku_1 + \dfrac{\mathrm{d}u_2}{\mathrm{d}s} - \tau u_3 \\ A_3 = \dfrac{\mathrm{d}u_3}{\mathrm{d}s} + \tau u_2 \end{cases} \tag{3.64}$$

Similar to the planar adjoint approach, $\mathrm{d}\boldsymbol{R}_P^*/\mathrm{d}s$ is the tangent vector of Γ_P^*. The absolute motion of point P^* in $\{\boldsymbol{O};\boldsymbol{i},\boldsymbol{j},\boldsymbol{k}\}$ is expressed by the Frenet trihedron $\{\boldsymbol{R}_P;\boldsymbol{\alpha},\boldsymbol{\beta},\boldsymbol{\gamma}\}$ of Γ_P. In particular, if point P^* is a fixed point in $\{\boldsymbol{O};\boldsymbol{i},\boldsymbol{j},\boldsymbol{k}\}$, the absolute coordinates of P^* do not change with s of Γ_P, or $\mathrm{d}\boldsymbol{R}_P^*/\mathrm{d}s = 0$. By equation (3.64), we have detailed expressions of $\mathrm{d}\boldsymbol{R}_P^*/\mathrm{d}s = 0$:

$$\begin{cases} A_1 = 1 + \dfrac{\mathrm{d}u_1}{\mathrm{d}s} - ku_2 = 0 \\ A_2 = ku_1 + \dfrac{\mathrm{d}u_2}{\mathrm{d}s} - \tau u_3 = 0 \\ A_3 = \dfrac{\mathrm{d}u_3}{\mathrm{d}s} + \tau u_2 = 0 \end{cases} \tag{3.65}$$

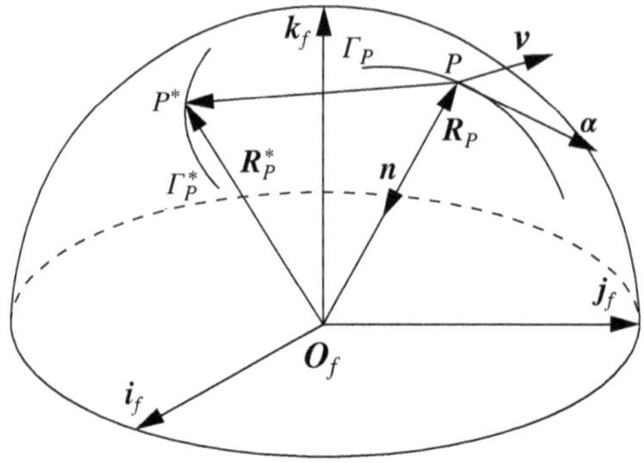

Figure 3.10 Two adjoint spherical curves

The above equations are called the ***fixed point conditions of an adjoint spatial curve***. The phrase "spatial curve" is used here to differentiate it from a planar curve with a Frenet frame. For point P^*, described in the Frenet trihedron $\{\boldsymbol{R}_P; \boldsymbol{\alpha}, \boldsymbol{\beta}, \boldsymbol{\gamma}\}$, to be an absolute fixed point in the fixed coordinate system $\{\boldsymbol{O}; \boldsymbol{i}, \boldsymbol{j}, \boldsymbol{k}\}$ at an instant, the parameters have to meet the equations (3.65).

As a special case, the original curve Γ_P lies on a sphere with radius R, or Γ_P is a spherical curve. It is convenient to take the Darboux frame $\{\boldsymbol{R}; \boldsymbol{\alpha}, \boldsymbol{n}, \boldsymbol{v}\}$ as the moving frame of Γ_P, other than the Frenet trihedron $\{\boldsymbol{R}_P; \boldsymbol{\alpha}, \boldsymbol{\beta}, \boldsymbol{\gamma}\}$, shown in Fig. 3.10. Then, another spherical curve Γ_P^* is an adjoint curve of Γ_P, whose vector equation is

$$\Gamma_P^* : \boldsymbol{R}_P^* = \boldsymbol{R}_P + u_1\boldsymbol{\alpha} + u_2\boldsymbol{n} + u_3\boldsymbol{v} \tag{3.66}$$

where (u_1, u_2, u_3) are the coordinates of P^* in the Darboux frame $\{\boldsymbol{R}; \boldsymbol{\alpha}, \boldsymbol{n}, \boldsymbol{v}\}$ of Γ_P. Differentiating the above equation with respect to the arc length s of Γ_P, and substituting the Darboux formulas (3.45) into it, we have

$$\begin{cases} \dfrac{\mathrm{d}\boldsymbol{R}_P^*}{\mathrm{d}s} = A_1\boldsymbol{\alpha} + A_2\boldsymbol{n} + A_3\boldsymbol{v} \\ A_1 = 1 + \dfrac{\mathrm{d}u_1}{\mathrm{d}s} - \dfrac{u_2}{R} - k_g u_3 \\ A_2 = \dfrac{u_1}{R} + \dfrac{\mathrm{d}u_2}{\mathrm{d}s} \\ A_3 = k_g u_1 + \dfrac{\mathrm{d}u_3}{\mathrm{d}s} \end{cases} \tag{3.67}$$

When the point P^* is a fixed point in the fixed coordinate system $\{\boldsymbol{O}; \boldsymbol{i}, \boldsymbol{j}, \boldsymbol{k}\}$ at an instant, we have

$$\begin{cases} A_1 = 1 + \dfrac{\mathrm{d}u_1}{\mathrm{d}s} - \dfrac{u_2}{R} - k_g u_3 = 0 \\ A_2 = \dfrac{u_1}{R} + \dfrac{\mathrm{d}u_2}{\mathrm{d}s} = 0 \\ A_3 = k_g u_1 + \dfrac{\mathrm{d}u_3}{\mathrm{d}s} = 0 \end{cases} \tag{3.68}$$

The above equations are the ***fixed point conditions of an adjoint spherical curve***.

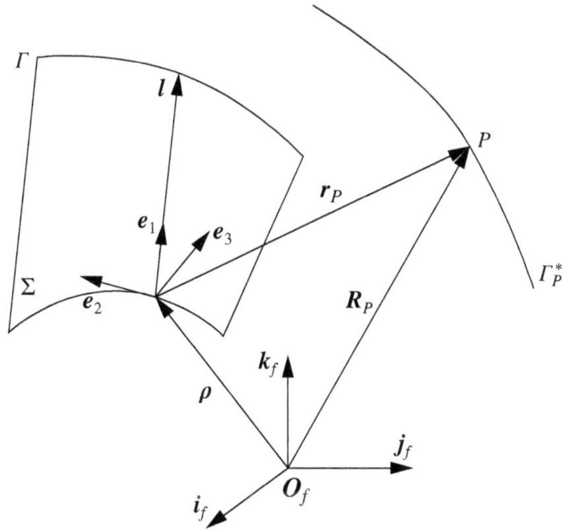

Figure 3.11 A spatial curve adjoint to a ruled surface

3.2.3.2 A Spatial Curve Adjoining a Ruled Surface

For a ruled surface $\sum_L$ in the fixed frame $\{\boldsymbol{O}_f;\boldsymbol{i}_f,\boldsymbol{j}_f,\boldsymbol{k}_f\}$, the vector expressions are written as equation (3.56) and the Frenet trihedron $\{\boldsymbol{\rho};\boldsymbol{e}_1,\boldsymbol{e}_2,\boldsymbol{e}_3\}$ is set up as equation (3.57), as shown in Fig. 3.11. A point P, which doesn't belong to $\sum_L$, moves and traces a spatial curve Γ_P^* in the same fixed coordinate system $\{\boldsymbol{O}_f;\boldsymbol{i}_f,\boldsymbol{j}_f,\boldsymbol{k}_f\}$. Each position of P at Γ_P^* is adjoint to the generator L of $\sum_L$, hence, the curve Γ_P^* is adjoint to the ruled surface $\sum_L$. Γ_P^* is defined as an adjoint curve of $\sum_L$, while $\sum_L$ is referred to as an original surface.

The adjoint curve Γ_P^* can be described as

$$\Gamma_P^*:\ \boldsymbol{R}_P=\boldsymbol{\rho}+\boldsymbol{r}_P=\boldsymbol{\rho}+u_1\boldsymbol{e}_1+u_2\boldsymbol{e}_2+u_3\boldsymbol{e}_3 \tag{3.69}$$

where (u_1,u_2,u_3) are the coordinates of P in $\{\boldsymbol{\rho};\boldsymbol{e}_1,\boldsymbol{e}_2,\boldsymbol{e}_3\}$ and $\boldsymbol{\rho}$ is the vector of the striction curve of $\sum_L$. Based on the Frenet formulas (3.58), the first derivative of the above equation with respect to arc length σ of the generatrices' spherical image curve of $\sum_L$ is

$$\begin{cases}\dfrac{\mathrm{d}\boldsymbol{R}_P}{\mathrm{d}\sigma}=A_1\boldsymbol{e}_1+A_2\boldsymbol{e}_2+A_3\boldsymbol{e}_3\\[2mm] A_1=\dfrac{\mathrm{d}u_1}{\mathrm{d}\sigma}-u_2+\alpha\\[2mm] A_2=u_1+\dfrac{\mathrm{d}u_2}{\mathrm{d}\sigma}-\beta u_3\\[2mm] A_3=\beta u_3+\dfrac{\mathrm{d}u_3}{\mathrm{d}\sigma}+\gamma\end{cases} \tag{3.70}$$

where A_i, $i = 1, 2, 3$, represented in $\{\boldsymbol{\rho}; \boldsymbol{e}_1, \boldsymbol{e}_2, \boldsymbol{e}_3\}$, are the rates of change of the absolute coordinates of P in $\{\boldsymbol{O}_f; \boldsymbol{i}_f, \boldsymbol{j}_f, \boldsymbol{k}_f\}$, while $\mathrm{d}u_i/\mathrm{d}\sigma$, $i = 1, 2, 3$ are the rates of change of the relative coordinates of P in $\{\boldsymbol{\rho}; \boldsymbol{e}_1, \boldsymbol{e}_2, \boldsymbol{e}_3\}$. If P is a fixed point in $\{\boldsymbol{O}_f; \boldsymbol{i}_f, \boldsymbol{j}_f, \boldsymbol{k}_f\}$, which means $\mathrm{d}\boldsymbol{R}_P/\mathrm{d}\sigma = 0$, the last three expressions of equation (3.70) are zero:

$$\begin{cases} A_1 = \dfrac{\mathrm{d}u_1}{\mathrm{d}\sigma} - u_2 + \alpha = 0 \\ A_2 = u_1 + \dfrac{\mathrm{d}u_2}{\mathrm{d}\sigma} - \beta u_3 = 0 \\ A_3 = \beta u_3 + \dfrac{\mathrm{d}u_3}{\mathrm{d}\sigma} + \gamma = 0 \end{cases} \tag{3.71}$$

The above equations are called the ***fixed point conditions for a curve adjoint to a ruled surface***.

3.2.3.3 A Ruled Surface Adjoining Another Ruled Surface

A ruled surface $\sum_L$ is presented as equation (3.56) in the fixed frame $\{\boldsymbol{O}_f; \boldsymbol{i}_f, \boldsymbol{j}_f, \boldsymbol{k}_f\}$, and its Frenet trihedron $\{\boldsymbol{\rho}; \boldsymbol{e}_1, \boldsymbol{e}_2, \boldsymbol{e}_3\}$ is the same as equation (3.57), shown in Fig. 3.12. A straight line L moves along a striction curve Γ of $\sum_L$. Another straight line L^* with unit direction vector $\boldsymbol{l}^*$, not belonging to $\sum_L$, moves along the directrix Γ_P^* and traces another ruled surface $\sum_L^*$ in frame $\{\boldsymbol{O}_f; \boldsymbol{i}_f, \boldsymbol{j}_f, \boldsymbol{k}_f\}$. Each position of line L^* of $\sum_L^*$ is adjoint to the line L hence the ruled surface $\sum_L^*$ is adjoint to the ruled surface $\sum_L$. We call $\sum_L^*$ an ***adjoint ruled surface*** of $\sum_L$, while $\sum_L$ is referred to as an ***original ruled surface***.

The adjoint ruled surface $\sum_L^*$ can be represented as

$$\begin{cases} \sum_L^*: \boldsymbol{R}_l = \boldsymbol{\rho} + \boldsymbol{r}_P + \lambda \boldsymbol{l}^* \\ \boldsymbol{r}_P = u_1\boldsymbol{e}_1 + u_2\boldsymbol{e}_2 + u_3\boldsymbol{e}_3, \ \boldsymbol{l}^* = l_1^*\boldsymbol{e}_1 + l_2^*\boldsymbol{e}_2 + l_3^*\boldsymbol{e}_3 \end{cases} \tag{3.72}$$

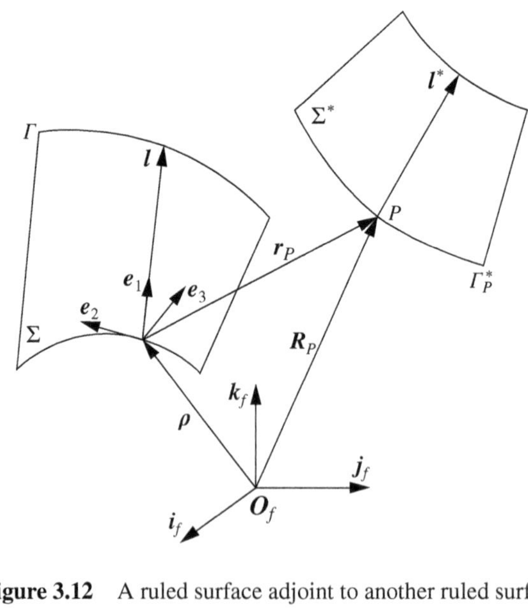

Figure 3.12 A ruled surface adjoint to another ruled surface

where $\boldsymbol{\rho}$ is the vector of the striction curve of $\sum_L$, $\boldsymbol{r}_P$ is the vector from the striction point of $\sum_L$ to the point P on L^*, and u_i, $i = 1, 2, 3$ are the components of $\boldsymbol{r}_P$ in the Frenet frame $\{\boldsymbol{\rho}; \boldsymbol{e}_1, \boldsymbol{e}_2, \boldsymbol{e}_3\}$. l_i^*, $i = 1, 2, 3$ are the components of $\boldsymbol{l}^*$ in $\{\boldsymbol{\rho}; \boldsymbol{e}_1, \boldsymbol{e}_2, \boldsymbol{e}_3\}$. λ is a parameter along the generator of $\sum_L^*$. The vector $\boldsymbol{\rho} + \boldsymbol{r}_P$ represents the directrix Γ_P^* of $\sum_L^*$. Taking the differential of equation (3.72) with respect to the arc length σ of the generatrices' spherical image curve of $\sum_L$, we have

$$\begin{cases} \mathrm{d}\boldsymbol{R}_l = \left(\dfrac{\mathrm{d}\boldsymbol{\rho}}{\mathrm{d}\sigma} + \dfrac{\mathrm{d}\boldsymbol{r}_P}{\mathrm{d}\sigma} + \lambda\dfrac{\mathrm{d}\boldsymbol{l}^*}{\mathrm{d}\sigma}\right)\mathrm{d}\sigma + \boldsymbol{l}^*\mathrm{d}\lambda \\ \dfrac{\mathrm{d}\boldsymbol{\rho}}{\mathrm{d}\sigma} + \dfrac{\mathrm{d}\boldsymbol{r}_P}{\mathrm{d}\sigma} = \left(\alpha + \dfrac{\mathrm{d}u_1}{\mathrm{d}\sigma} - u_2\right)\boldsymbol{e}_1 + \left(u_1 + \dfrac{\mathrm{d}u_2}{\mathrm{d}\sigma} - \beta u_3\right)\boldsymbol{e}_2 + \left(\beta u_2 + \dfrac{\mathrm{d}u_3}{\mathrm{d}\sigma} + \gamma\right)\boldsymbol{e}_3 \\ \dfrac{\mathrm{d}\boldsymbol{l}^*}{\mathrm{d}\sigma} = \left(\dfrac{\mathrm{d}l_1^*}{\mathrm{d}\sigma} - l_2^*\right)\boldsymbol{e}_1 + \left(l_1^* + \dfrac{\mathrm{d}l_2^*}{\mathrm{d}\sigma} - \beta l_3^*\right)\boldsymbol{e}_2 + \left(\beta l_2^* + \dfrac{\mathrm{d}l_3^*}{\mathrm{d}\sigma}\right)\boldsymbol{e}_3 \\ \boldsymbol{l}^* = l_1^*\boldsymbol{e}_1 + l_2^*\boldsymbol{e}_2 + l_3^*\boldsymbol{e}_3 \end{cases} \tag{3.73}$$

where α, β, and γ are the construction parameters of $\sum_L$. In particular, when the straight line L^* is a fixed line in the fixed frame $\{\boldsymbol{O}_f; \boldsymbol{i}_f, \boldsymbol{j}_f, \boldsymbol{k}_f\}$, the directrix Γ_P^* will be a fixed point on L^*, or $\frac{\mathrm{d}\boldsymbol{\rho}}{\mathrm{d}\sigma} + \frac{\mathrm{d}\boldsymbol{r}_P}{\mathrm{d}\sigma} = 0$ and $\frac{\mathrm{d}\boldsymbol{l}^*}{\mathrm{d}\sigma} = 0$. Based on equation (3.73), we have

$$\begin{cases} \dfrac{\mathrm{d}u_1}{\mathrm{d}\sigma} = u_2 - \alpha, \ \dfrac{\mathrm{d}u_2}{\mathrm{d}\sigma} = \beta u_3 - u_1, \ \dfrac{\mathrm{d}u_3}{\mathrm{d}\sigma} = -\beta u_2 - \gamma \\ \dfrac{\mathrm{d}l_1^*}{\mathrm{d}\sigma} = l_2^*, \ \dfrac{\mathrm{d}l_2^*}{\mathrm{d}\sigma} = \beta l_3^* - l_1^*, \ \dfrac{\mathrm{d}l_3^*}{\mathrm{d}\sigma} = -\beta l_2^* \end{cases} \tag{3.74}$$

Equations (3.74), in which the first to third expressions are the fixed point conditions while the other expressions are the fixed vector conditions, are called the absolute fixed line conditions for a line L^* adjoint to the ruled surface $\sum_L$, or the ***absolute fixed line conditions*** for short, and L^* is referred to as the ***absolute fixed line***.

On the other hand, if L^* is always collinear with a fixed line in $\{\boldsymbol{O}_f; \boldsymbol{i}_f, \boldsymbol{j}_f, \boldsymbol{k}_f\}$, but slides along the fixed line, L^* would satisfy the following equations:

$$\left(\frac{\mathrm{d}\boldsymbol{\rho}}{\mathrm{d}\sigma} + \frac{\mathrm{d}\boldsymbol{r}_P}{\mathrm{d}\sigma}\right) \times \boldsymbol{l}^* = 0, \ \frac{\mathrm{d}\boldsymbol{l}^*}{\mathrm{d}\sigma} = 0 \tag{3.75}$$

The line L^* is defined as the ***quasi-fixed line***. Substituting equation (3.73) into the above equation, we have

$$\begin{cases} \dfrac{\mathrm{d}l_1^*}{\mathrm{d}\sigma} = l_2^*, \ \dfrac{\mathrm{d}l_2^*}{\mathrm{d}\sigma} = \beta l_3^* - l_1^*, \ \dfrac{\mathrm{d}l_3^*}{\mathrm{d}\sigma} = -\beta l_2^* \\ l_3\left(\alpha + \dfrac{\mathrm{d}u_1}{\mathrm{d}\sigma} - u_2\right) - l_1\left(\beta u_2 + \dfrac{\mathrm{d}u_3}{\mathrm{d}\sigma} + \gamma\right) = 0 \\ l_3\left(u_1 + \dfrac{\mathrm{d}u_2}{\mathrm{d}\sigma} - \beta u_3\right) - l_2\left(\beta u_2 + \dfrac{\mathrm{d}u_3}{\mathrm{d}\sigma} + \gamma\right) = 0 \end{cases} \tag{3.76}$$

The above equations are the ***quasi-fixed line conditions***, in which the first to third expressions are the fixed vector conditions while the others are the conditions for the rectilinear sliding.

3.3 Constraint Curves and Surfaces

For planar linkages, the planar four-bar linkages RRRP and RRRR, as mentioned in Chapters 1 and 2, the characteristic points of a moving body are taken as the moving hinge points, corresponding to binary links with fixed joints and constraint curves. There are three kinds of binary link, R–R, R–P, and P–R, which are listed in Table 3.1. The binary link, the characteristic point or line, and a constraint curve, have an equivalent meaning since they play the same role in constraining the movement of the coupler link, or they correspond to each other.

For spherical linkages, the kinematic pairs are currently all revolute pairs R, either a moving joint or a fixed joint. There is only one kind of binary link R–R, in which the axes of two R joints intersect with each other at the center point of the sphere, the corresponding constraint ruled surface $\sum_{RR}$ is a conical surface. The axis of the moving joint R is a characteristic line L_{RR} in three-dimensional space. Needless to say, it is easily treated as a special case in the spatial kinematic geometry of a rigid body. However, for convenience of reading and systematic understanding, we will study it in Chapters 4 and 5 by mapping it to a characteristic point on the spherical surface (like the case in planar movement). It provides a bridge from planar to spatial for the kinematic geometry of a rigid body. The characteristic line L_{RR} and the ***constraint ruled surface*** $\sum_{RR}$ for a spherical binary link R–R are examined in two-dimensional space. The former is a characteristic point P_{RR} on the sphere of the coupler link, while the latter is a circle, or constraint curve C_{RR} on the sphere of the base link, as listed in Table 3.1.

For spatial linkages, a binary link kinematically connects a moving body to a base link, or fixed frame, by two kinematic pairs. The point or line of the moving body, which is the center point or axis of a moving joint, is defined as the characteristic point or the characteristic line. The trajectory of a characteristic point or line in the fixed frame is called a ***constraint curve*** or ***constraint ruled surface***. As shown in Fig. 3.13(a), the moving joint is a cylindrical pair C and the fixed joint is a revolute pair R. The constraint ruled surface $\sum_{RC}$ is a ***hyperboloid of one sheet***, traced by the characteristic line L_{RC}, the axis of the cylindrical pair C. In Fig. 3.13(b), the moving joint is a spherical pair S and the fixed joint is a revolute pair R. The constraint curve C_{RS} is a circle, traced by the characteristic point P_{RS}, the center of the spherical pair S. Hereinafter, all constraint curves are represented by a capital letter C with subscripts, while all

Table 3.1 Binary links and their constraint curves

	Binary links	Constraint curves	Characteristic points	Constraint curves
Planar linkage	R–R	Circle	P_{RR}	C_{RR}
	P–R	Line	P_{PR}	C_{PR}
	R–P	Envelope circle	L_{RP}	C_{RP}
Spherical linkage	R–R	Spherical circle (Conical surface)	P_{RR} (L_{RR})	C_{RR} ($\sum_{RR}$)
Spatial linkage	C–S	Cylindrical curve	P_{CS}	C_{CS}
	H–S	Cylindrical spiral	P_{HS}	C_{HS}
	R–S	Circle	P_{RS}	C_{RS}
	P–S	Straight line	P_{PS}	C_{PS}
	S–S	Spherical curve	P_{SS}	C_{SS}

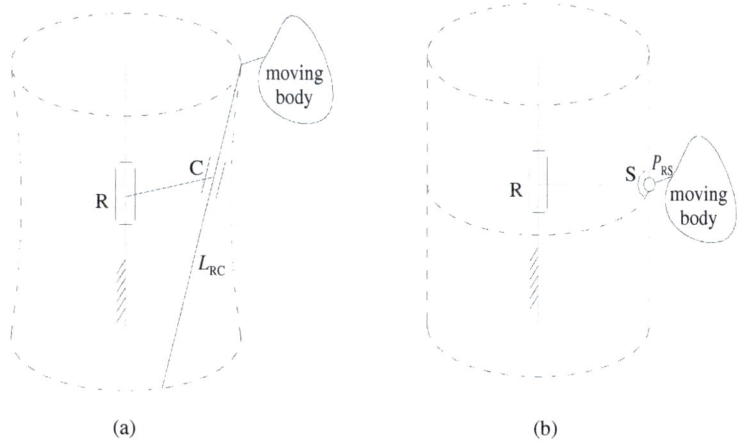

Figure 3.13 (a) Binary link R–C (b) Binary link R–S

constraint ruled surfaces are represented by a symbol $\sum$ with subscripts. For example, $\sum_{RC}$ stands for the constraint ruled surface of a binary link R–C.

Obviously, the geometrical properties, such as the shapes, of both the constraint curves and the constraint ruled surfaces depend on the joints at both ends of a binary link. Thus, a binary link, a characteristic point or a characteristic line, and a constraint curve or a constraint ruled surface correspond to each other. That is, one can be derived by any of the others, or vice versa, since they play the same roles in kinematical constraint of a moving body but in different views. For example, the constraint ruled surface $\sum_{RC}$, the characteristic line L_{RC}, and a binary link R–C can be constructed from any of the others. In kinematic analysis and synthesis of spatial linkages, we are interested in the trajectories of both a point and a line and their geometrical properties, and estimate them to be (or approximately to be) a constraint curve or a constraint ruled surface of spatial linkages. We want to find the geometrical properties of the characteristic points or characteristic lines of the moving body, which are the theoretical basis for the synthesis of spatial linkages.

We restrict our analysis to five common kinematic joints, such as the cylindrical pair C, the revolute pair R, the helical pair H, the prismatic pair P, and the spherical pair S, which do not cover all the kinematic joints in the spatial linkages but are most frequently used. Taking two of the five kinematic joints as the moving and fixed joints of the binary link respectively, or connecting the moving body and the fixed frame, we have 25 different kinds of binary link with two kinematic pairs, each corresponding to a constraint curve or a constraint ruled surface. Of course, there are many more kinds of binary link, but these 25 ones are of sufficient interest to us so that we study the global geometrical properties of them in this chapter. More can be done with the help of mathematicians. If the generalized binary links with combinations of more than two kinematic joints are also discussed, there are more cases to be taken into account. These also form the theoretical basis of the robot mechanisms with multiple degrees of freedom, including the parallel mechanisms.

A characteristic line of a moving body traces a constraint ruled surface in the fixed frame, and can be designated as the axis of a moving joint. In some binary links, the axes of the moving joints are the same characteristic lines on both the moving body and the binary link, such as the characteristic lines of R–C, R–R, R–H, and R–P trace the same constraint ruled surface, a hyperboloid of one sheet. In fact, the four moving joints, C, R, H, and P, may be identified by kinematic constraint in Chapter 6. Therefore, a cylindrical pair C can be taken as a typical moving joint for a binary link with moving joints C, R, H, and P, corresponding to the same constraint ruled surface. Then, the constraint ruled surface can be extended to the combinations of three kinematic pairs, such as C–P–C, called a ***generalized binary link***, or an ***open kinematic chain***, in which the sliding displacement in the P pair should be designated as a given function of the rotational displacement of the fixed pair C. In Table 3.1, nine kinds of binary link and their corresponding constraint curves are listed, including three ones for planar linkages and one for spherical linkages. In Table 3.2, five kinds of binary link and three generalized binary links are listed. The geometrical properties of both constraint curves and constraint ruled surfaces will be discussed in the following sections.

We focus on discussing both the constraint curves and the constraint ruled surfaces, which are respectively traced by characteristic points and characteristic lines when they move with one degree of freedom (DOF), or the corresponding binary link moves with one independent parameter. For binary links such as C–C or C–S, the sliding displacement should be correlated with the rotation displacement of the fixed joint C, just like it is applied in a spatial linkage with one DOF. Three open chains C–P–C, H–P–C, and R–P–C have the same problem and need to be degenerated to chains with one DOF. All letters of the kinematic pair whose multi-DOF is to be degenerated to one DOF are represented by a prime, such as the binary link C′–C for C–C, C′–S for C–S, and the open chain C′–P′–C for C–P–C, etc. Otherwise, a characteristic line moves with two DOFs and will trace a congruence of lines, other than a ruled surface, while a characteristic point moves with two DOFs and will trace a surface, not a curve. Moreover, two kinematic pairs at both ends of a binary link have a general relative position. Taking a binary link C–C for instance, the axes of the two cylindrical pairs C should be neither parallel nor intersecting.

Table 3.2 Binary (generalized) links and their constraint surfaces

	Binary links (generalized)	Constraint surfaces	Characteristic lines	Constraint ruled surfaces
Spatial linkage	C–C	Constant axis ruled surface	L_{CC}	Σ_{CC}
	H–C	helicoid	L_{HC}	Σ_{HC}
	R–C	hyperboloid	L_{RC}	Σ_{RC}
	P–C	plane	L_{PC}	Σ_{PC}
	S–C	Constant distance ruled surface	L_{SC}	Σ_{SC}
	C–P–C	Constant inclination ruled surface	L_{CPC}	Σ_{CPC}
	H–P–C	Constant inclination helical ruled surface	L_{HPC}	Σ_{HPC}
	R–P–C	Constant inclination revolving ruled surface	L_{RPC}	Σ_{RPC}

3.4 Spherical and Cylindrical Curves

The constraint curves, listed in Table 3.1, can be classified into three types, planar curves, spherical curves, and cylindrical curves. The planar constraint curves have been introduced in Chapters 1 and 2. We will discuss the geometrical characteristics of both the spherical constraint curves and the cylindrical constraint curves in this section.

3.4.1 Spherical Curves (S–S)

Spherical constraint curves correspond to both the spherical binary link R–R and the spatial binary link S–S. The former is just a circle on the spherical surface while the latter is an arbitrary curve on the spherical surface. Hence, we focus on the latter while the former is taken as a special case.

3.4.1.1 Properties of Spherical Curves

A spherical curve is expressed by equations (3.43) and (3.44), whose Darboux frame $\{\boldsymbol{R};\boldsymbol{\alpha},\boldsymbol{n},\boldsymbol{v}\}$ as described in Section 3.2.1 has the differential formulas (3.45). Based on the differential formulas of the Darboux frame, we have the following properties for a spherical curve:

1. The geodesic torsion τ_g of a spherical curve is zero, and the normal curvature k_n is constant and equal to the reciprocal of the radius of the sphere.
2. The curvature of the spherical curve is $k = [1/R^2 + k_g^2]^{1/2}$.
3. The geodesic curvature k_g totally reflects the geometrical properties of the spherical curve, revealing the curvature degree of the spherical curve on the tangent plane. It has the equivalent meaning to the curvature k of a planar curve.

From the third property, similar to the condition for a planar curve to be a circle, we have the following conclusion:

A spherical curve Γ is a circle if and only if its geodesic curve k_g is constant.

For the sufficient condition, or when we know that the geodesic curve k_g of a spherical curve Γ is a constant, we need to prove that it is a circle. Based on equations (3.43) and (3.44), the first to third derivatives of Γ can be given as

$$\frac{\mathrm{d}\boldsymbol{R}}{\mathrm{d}s} = \boldsymbol{\alpha},\ \frac{\mathrm{d}^2\boldsymbol{R}}{\mathrm{d}s^2} = \frac{1}{R}\boldsymbol{n} + k_g\boldsymbol{v},\ \frac{\mathrm{d}^3\boldsymbol{R}}{\mathrm{d}s^3} = -\left(\frac{1}{R^2} + k_g^2\right)\boldsymbol{\alpha} + \frac{\mathrm{d}k_g}{\mathrm{d}s}\boldsymbol{v} \tag{3.77}$$

From the above equations, it is easy to verify that the equation $\left(\frac{\mathrm{d}\boldsymbol{R}}{\mathrm{d}s}, \frac{\mathrm{d}^2\boldsymbol{R}}{\mathrm{d}s^2}, \frac{\mathrm{d}^3\boldsymbol{R}}{\mathrm{d}s^3}\right) = 0$ holds true. Then, substituting them into equation (3.12), we have $\tau = 0$, which means the spherical curve Γ is a planar curve, and further it must be a circle if a spherical curve is also a planar curve.

On the contrary, for the necessary condition or when we know that the spherical curve Γ is a circle, we need to prove that the geodesic curve k_g is a constant. Substituting equation (3.77) into equation (3.12), we have the torsion of Γ as $\tau = \frac{\mathrm{d}k_g}{\mathrm{d}s}/R$, which should be zero since curve Γ is a planar curve, a circle. That is, $\frac{\mathrm{d}k_g}{\mathrm{d}s} = 0$, which leads to the inference that the geodesic curve k_g of Γ is a constant.

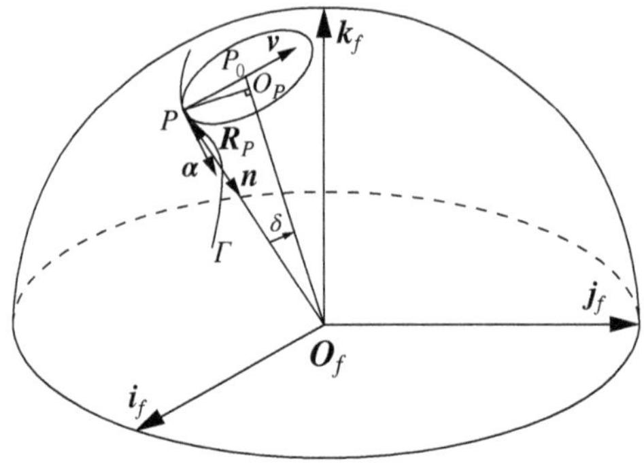

Figure 3.14 The geodesic curvature center of a spherical curve

The position of the geodesic curvature center P_0 can be located by vector $\boldsymbol{R}_{P_0} = \boldsymbol{R}_P + \rho_g \boldsymbol{v}$ if the radius $\rho_g \neq 0$, in which the radius ρ_g of the geodesic curvature is the reciprocal of the geodesic curvature k_g of Γ at P, and $|\rho_g| = R \tan \delta$, in which δ is the included angle between the line $\overline{O_f P_0}$ and $\overline{O_f P}$, as shown in Fig. 3.14. Then, a circle on the tangent plane $\boldsymbol{R} - \boldsymbol{\alpha v}$ of Γ at P, whose center is at point P_0 with radius $|\rho_g|$, is called a ***geodesic curvature circle***. The curvature center O_P just locates at line $\overline{O_f P_0}$, and the curvature radius ρ is $|PO_P|$. It can be seen from equation (3.45) that the curvature k of Γ satisfies the following equations:

$$\begin{cases} k^2 = k_g^2 + \dfrac{1}{R^2} \\ k\dfrac{\mathrm{d}k}{\mathrm{d}s} = k_g \dfrac{\mathrm{d}k_g}{\mathrm{d}s} \end{cases} \tag{3.78}$$

The curvature k of Γ cannot be zero, or the inflection point cannot be found on the spherical curve. If the geodesic curvature k_g of Γ at point P is zero, the curvature is equal to $1/R$. So, the three consecutive points in the neighborhood of point P locate on a great circle (geodesic) on the sphere. If the geodesic curvature k_g of Γ at P meets the equation $\mathrm{d}k_g/\mathrm{d}s = 0$, we have $\mathrm{d}k/\mathrm{d}s = 0$, and the four consecutive points including point P all lie on the curvature circle. Further, if $\mathrm{d}k_g/\mathrm{d}s = 0$ and $\mathrm{d}^2 k_g/\mathrm{d}s^2 = 0$, the five consecutive points including point P locate on the curvature circle.

3.4.1.2 Necessary and Sufficient Conditions of Spherical Curves

For the kinematic analysis and synthesis of spatial linkages, the global properties of a spherical curve have to be studied, in particular, the necessary and sufficient conditions for a spatial curve to be a spherical curve form the theoretical basis for estimating the spherical constraint curves.

Firstly, a spatial curve Γ is defined as Γ: $\boldsymbol{R} = \boldsymbol{R}(s)$, whose Darboux frame is $\{\boldsymbol{R}; \boldsymbol{\alpha}, \boldsymbol{n}, \boldsymbol{v}\}$. The equidistant curve Γ^* of Γ is described by

$$\Gamma^* : \boldsymbol{R}^* = \boldsymbol{R} + h\boldsymbol{n} \tag{3.79}$$

where h is the distance constant and $\boldsymbol{n}$ is the unit vector of normal of Γ. In fact, curve Γ^* can be viewed as an adjoint curve of Γ. Differentiating the above equation with respect to the arc length s of Γ, and combining the differential formulas (3.42) of the Darboux frame, we have

$$\begin{cases} \dfrac{\mathrm{d}\boldsymbol{R}^*}{\mathrm{d}s} = A\boldsymbol{\alpha} + B\boldsymbol{v},\ A = 1 - hk_n,\ B = h\tau_g \\ \dfrac{\mathrm{d}^2\boldsymbol{R}^*}{\mathrm{d}s^2} = \left(\dfrac{\mathrm{d}A}{\mathrm{d}s} - Bk_g\right)\boldsymbol{\alpha} + \left(\dfrac{\mathrm{d}B}{\mathrm{d}s} + Ak_g\right)\boldsymbol{v} + (Ak_n - B\tau_g)\boldsymbol{n} \end{cases} \tag{3.80}$$

The tangent vector $\frac{\mathrm{d}\boldsymbol{R}^*}{\mathrm{d}s}$ of Γ^* holds $\frac{\mathrm{d}\boldsymbol{R}^*}{\mathrm{d}s} \cdot \boldsymbol{n} = 0$ true, which indicates that the vector $\boldsymbol{n}$ is the common normal and h is the shortest distance between Γ and Γ^*.

Particularly, curve Γ^* degenerates to a fixed point, or the distance h from each point of Γ to the fixed point is a constant, which means that Γ should be on a spherical surface with radius h, or a spherical curve. Hence, the conditions for Γ^* to be a fixed point are just the fixed point conditions of an adjoint spherical curve, or the variables A and B in equations (3.80) are zero. We have

$$\frac{\mathrm{d}k_n}{\mathrm{d}s} = 0,\ \tau_g = 0 \tag{3.81}$$

The above equations are called the ***invariants of spherical curves***. Based on the above deriving process and the first property of the spherical curve, we have the following theorem:

Theorem 3.3 *For a spatial curve Γ to be a spherical curve, the necessary and sufficient conditions are that there exists a moving frame $\{\boldsymbol{R}; \boldsymbol{a}, \boldsymbol{n}, \boldsymbol{v}\}$ on Γ with arc length s, which leads to $\tau_g = 0$ and $\frac{\mathrm{d}k_n}{\mathrm{d}s} = 0$.*

The proof for the necessary condition in Theorem 3.3 has been provided by the deriving process for the equidistant curve of a spatial curve to be a fixed point, and the sufficient condition can be deduced from the first property of spherical curve.

By equation (3.79), if the spatial curve Γ degenerates to a spherical curve, the position vector $\boldsymbol{R}_0$ of the sphere center and the radius r_0 are

$$\begin{cases} \boldsymbol{R}_0 = \boldsymbol{R} + r_0\boldsymbol{n} \\ r_0 = 1/k_n \end{cases} \tag{3.82}$$

3.4.2 *Cylindrical Curves (C–S)*

The constraint curve, corresponding to the binary link C–S is a cylindrical curve. The constraint curves corresponding to binary links H–S and R–S can be viewed as special cases when the fixed joint C degenerates to R and H, which are respectively circles and cylindrical helices on the cylindrical surface. The properties of cylindrical curves, especially the global properties, can provide the theoretical basis for the kinematic synthesis of spatial linkages with binary links C–S, R–S, and H–S.

3.4.2.1 Properties of Cylindrical Curves

A cylindrical surface, shown in Fig. 3.15, can be expressed as

$$\Gamma : \boldsymbol{R} = r_0\boldsymbol{e}_{\mathrm{I}(\varphi)} + \lambda\boldsymbol{k} \tag{3.83}$$

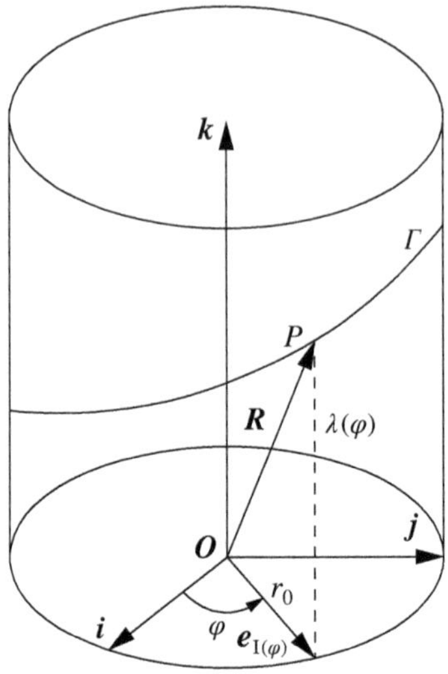

Figure 3.15 Cylindrical curve

Both parameters λ and φ are independent variables of a cylindrical surface. Based on equation (3.83), the unit normal vector $\boldsymbol{n}$ of the cylindrical surface at point $P(\varphi, \lambda)$ is

$$\boldsymbol{n} = \frac{\mathrm{d}\boldsymbol{R}}{\mathrm{d}\lambda} \times \frac{\mathrm{d}\boldsymbol{R}}{\mathrm{d}\varphi} \Big/ \left| \frac{\mathrm{d}\boldsymbol{R}}{\mathrm{d}\lambda} \times \frac{\mathrm{d}\boldsymbol{R}}{\mathrm{d}\varphi} \right| = \boldsymbol{k} \times \boldsymbol{e}_{\mathrm{II}(\varphi)} = -\boldsymbol{e}_{\mathrm{I}(\varphi)} \tag{3.84}$$

A curve Γ on the cylindrical surface can be represented by letting parameter λ be a function of φ, whose equation is given in Example 3.2. Differentiating equation (3.83) to φ, we have

$$\frac{\mathrm{d}\boldsymbol{R}}{\mathrm{d}\varphi} = r_0 \boldsymbol{e}_{\mathrm{II}(\varphi)} + \frac{\mathrm{d}\lambda}{\mathrm{d}\varphi}\boldsymbol{k} \tag{3.85}$$

The line element $\mathrm{d}s$ of the cylindrical curve Γ can be derived as

$$\mathrm{d}s = \left| \frac{\mathrm{d}\boldsymbol{R}}{\mathrm{d}\varphi} \right| \mathrm{d}\varphi = \sqrt{r_0^2 + \left(\frac{\mathrm{d}\lambda}{\mathrm{d}\varphi} \right)^2} \mathrm{d}\varphi \tag{3.86}$$

The Darboux frame $\{\boldsymbol{R}; \boldsymbol{\alpha}, \boldsymbol{n}, \boldsymbol{v}\}$ can be established on Γ as

$$\begin{cases} \boldsymbol{\alpha} = \frac{\mathrm{d}\boldsymbol{R}}{\mathrm{d}\varphi} \Big/ \left| \frac{\mathrm{d}\boldsymbol{R}}{\mathrm{d}\varphi} \right| = \left[r_0 \boldsymbol{e}_{\mathrm{II}(\varphi)} + \frac{\mathrm{d}\lambda}{\mathrm{d}\varphi}\boldsymbol{k} \right] \Big/ \sqrt{r_0^2 + \left(\frac{\mathrm{d}\lambda}{\mathrm{d}\varphi} \right)^2} \\ \boldsymbol{n} = -\boldsymbol{e}_{\mathrm{I}(\varphi)} \\ \boldsymbol{v} = \boldsymbol{\alpha} \times \boldsymbol{n} = \left[r_0 \boldsymbol{k} - \frac{\mathrm{d}\lambda}{\mathrm{d}\varphi} \boldsymbol{e}_{\mathrm{II}(\varphi)} \right] \Big/ \sqrt{r_0^2 + \left(\frac{\mathrm{d}\lambda}{\mathrm{d}\varphi} \right)^2} \end{cases} \tag{3.87}$$

The curvature invariants k_n, k_g, and τ_g of Γ can be derived through the differential formulas (3.42) of the Darboux frame as

$$\begin{cases} k_n = -\dfrac{\mathrm{d}\boldsymbol{n}}{\mathrm{d}s}\cdot\boldsymbol{\alpha} = \dfrac{r_0}{r_0^2 + (\mathrm{d}\lambda/\mathrm{d}\varphi)^2} \\ k_g = \dfrac{\mathrm{d}\boldsymbol{\alpha}}{\mathrm{d}s}\cdot\boldsymbol{v} = \dfrac{r_0\cdot\mathrm{d}^2\lambda/\mathrm{d}\varphi^2}{[r_0^2 + (\mathrm{d}\lambda/\mathrm{d}\varphi)^2]^{3/2}} \\ \tau_g = \dfrac{\mathrm{d}\boldsymbol{n}}{\mathrm{d}s}\cdot\boldsymbol{v} = \dfrac{\mathrm{d}\lambda/\mathrm{d}\varphi}{r_0^2 + (\mathrm{d}\lambda/\mathrm{d}\varphi)^2} \end{cases} \tag{3.88}$$

The parameter φ in the above expressions of k_n, k_g, and τ_g can be replaced by the natural parameter s by

$$\frac{\mathrm{d}\lambda}{\mathrm{d}\varphi} = \frac{\mathrm{d}\lambda}{\mathrm{d}s}\frac{\mathrm{d}s}{\mathrm{d}\varphi}, \left(\frac{\mathrm{d}s}{\mathrm{d}\varphi}\right)^2 = r_0^2 + \left(\frac{\mathrm{d}\lambda}{\mathrm{d}\varphi}\right)^2 \tag{3.89}$$

The first and second derivatives of λ with respect to φ are solved from the above equations, and they are

$$\frac{\mathrm{d}\lambda}{\mathrm{d}\varphi} = \frac{r_0\mathrm{d}\lambda/\mathrm{d}s}{\sqrt{1-(\mathrm{d}\lambda/\mathrm{d}s)^2}}, \frac{\mathrm{d}^2\lambda}{\mathrm{d}\varphi^2} = \frac{r_0^2\mathrm{d}^2\lambda/\mathrm{d}s^2}{[1-(\mathrm{d}\lambda/\mathrm{d}s)^2]^2} \tag{3.90}$$

Substituting the above equations into equation (3.88), and simplifying, we have

$$\begin{cases} k_n = \dfrac{1-(\mathrm{d}\lambda/\mathrm{d}s)^2}{r_0} \\ k_g = \dfrac{\mathrm{d}^2\lambda/\mathrm{d}s^2}{\sqrt{1-(\mathrm{d}\lambda/\mathrm{d}s)^2}} \\ \tau_g = \dfrac{(\mathrm{d}\lambda/\mathrm{d}s)\sqrt{1-(\mathrm{d}\lambda/\mathrm{d}s)^2}}{r_0} \end{cases} \tag{3.91}$$

Eliminating λ and its derivatives from equation (3.91), we get the important properties of the cylindrical curve as

$$\begin{cases} \dfrac{\mathrm{d}k_n}{\mathrm{d}s} = -2k_g\tau_g \\ \dfrac{\mathrm{d}\tau_g}{\mathrm{d}s} = -\dfrac{k_g\left(\tau_g^2 - k_n^2\right)}{k_n} \\ r_0 = \dfrac{k_n}{k_n^2 + \tau_g^2} = \text{const.} \end{cases} \tag{3.92}$$

The above equations are called the ***invariants of cylindrical curves***. Two expressions of the above equation are independent, since the third expression can be derived from the first two expressions. In particular, for a cylindrical helix, the geodesic curvature k_g is zero since it is the geodesic of a cylindrical surface. Then the invariants of cylindrical helix are $k_n = \text{const.}$, $\tau_g = \text{const.}$, and $k_g = 0$. For a circle on the cylindrical surface, the geodesic curvature k_g and the geodesic torsion τ_g are both zero, since the circle is both the geodesic and the line of curvature. The invariants are $k_g = 0$, $\tau_g = 0$, and $k_n = \text{const.}$

3.4.2.2 Necessary and Sufficient Conditions of Cylindrical Curves

For a spatial curve to be a cylindrical curve, the necessary and sufficient conditions need to be discussed for the kinematic synthesis of spatial linkages.

A spatial curve Γ is given by Γ: $\boldsymbol{R} = \boldsymbol{R}(s)$, whose Darboux frame is $\{\boldsymbol{R}; \boldsymbol{\alpha}, \boldsymbol{n}, \boldsymbol{v}\}$, while the equidistant curve Γ^* of Γ is expressed as (3.79), whose derivatives can also be obtained as equation (3.80) by the differential formulas (3.42) of the Darboux frame.

Particularly, when the equidistant curve Γ^* of Γ degenerates to a straight line, other than a fixed point for a spherical curve, or the distance from curve Γ to the straight line is a constant h, Γ is a cylindrical curve whose axis is the straight line Γ^* and radius is the constant distance h.

The equidistant curve Γ^* is a straight line if and only if its curvature is identically equal to zero, or $\frac{d\boldsymbol{R}^*}{ds} \times \frac{d^2\boldsymbol{R}^*}{ds^2} = 0$, which can be rewritten as follows by means of equations (3.13) and (3.80):

$$-B\tau_g + Ak_n = 0, A\frac{dB}{ds} - B\frac{dA}{ds} + (A^2 + B^2)k_g = 0 \tag{3.93}$$

To simplify the above equations, we have

$$\begin{cases} h = k_n / \left(k_n^2 + \tau_g^2\right) = \text{const.} \\ \dfrac{dk_n}{ds} = -2k_g\tau_g \\ \dfrac{d\tau_g}{ds} = -k_g(\tau_g^2 - k_n^2)/k_n \end{cases} \tag{3.94}$$

These equations are identical to equation (3.92), so we get the following theorem:

Theorem 3.4 *For a spatial curve Γ to be a cylindrical curve, the necessary and sufficient conditions are that there exists a moving frame $\{\boldsymbol{R}; \boldsymbol{a}, \boldsymbol{n}, \boldsymbol{v}\}$ on Γ with arc length s, and the curvature invariants k_n, k_g, and τ_g hold equation (3.94) true.*

The proof for the necessary condition in Theorem 3.4 is provided by the analytical process of properties of a cylindrical curve, such as equation (3.92), and that for the sufficient condition is provided by the deriving process of the equidistant straight line of a cylindrical curve.

Considering a cylindrical helix and a circle as special cylindrical curves, we have the following two corollaries:

Corollary 3.1 *For a spatial curve Γ to be a cylindrical helix, k_g is zero and both k_n and τ_g are constant, which are the necessary and sufficient conditions.*

Corollary 3.2 *For a spatial curve Γ to be a circle, both k_g and τ_g are zero and k_n is a constant, which are the necessary and sufficient conditions.*

When a cylindrical curve is degenerated from a spatial curve Γ, its axis and radius r_0 can be determined by

$$\begin{cases} \boldsymbol{R}_a = \boldsymbol{R} + r_0\boldsymbol{n} + \mu\boldsymbol{a} \\ r_0 = k_n / \left(k_n^2 + \tau_g^2\right) \\ \boldsymbol{a} = \dfrac{\tau_g \boldsymbol{\alpha} + k_n \boldsymbol{v}}{\sqrt{k_n^2 + \tau_g^2}} \end{cases} \tag{3.95}$$

where $\boldsymbol{R}_a$ is the vector equation of the axis of the cylindrical surface, whose unit direction vector is $\boldsymbol{a}$.

If a circle is degenerated by a spatial curve Γ, the center and radius can also be determined by equation (3.82).

If a cylindrical helix is degenerated by a spatial curve Γ, its axis and radius can also be calculated by equation (3.95), while the helical lead is

$$p = \frac{\mathrm{d}\lambda}{\mathrm{d}\varphi} = \frac{r_0\tau_g}{k_n} = \frac{\tau_g}{k_n^2 + \tau_g^2} \tag{3.96}$$

3.5 Constraint Ruled Surfaces

In Section 3.3 some constraint surfaces, the line trajectories traced by the characteristic lines of the binary links with one DOF, were introduced. The differential geometry of the constraint ruled surfaces is discussed in this section.

3.5.1 *Constant Inclination Ruled Surfaces ($C'-P'-C$)*

A ruled surface $\sum_L$ is defined as a ***constant inclination ruled surface***, whose generatrices have a constant inclination angle with respect to a fixed line, such as a constraint ruled surface $\sum_{\mathrm{C'P'C}}$, corresponding to the combined links with three kinematic pairs C′–P′–C. The characteristic line $L_{\mathrm{C'P'C}}$ is the generator of $\sum_{\mathrm{C'P'C}}$ and the axis of the fixed joint C′ is designated as the fixed line L_A. The constant inclination angle between $L_{\mathrm{C'P'C}}$ and L_A is δ, as shown in Fig. 3.16.

3.5.1.1 Construction Parameters of a Constant Inclination Ruled Surface

The constant inclination ruled surface $\sum_{\mathrm{C'P'C}}$ can be described in a fixed coordinate system $\{\boldsymbol{O};\boldsymbol{i},\boldsymbol{j},\boldsymbol{k}\}$ (see Fig. 3.16). The fixed line L_A with a unit direction vector $\boldsymbol{d}$ is supposed to pass through a fixed point A_0, whose position vector is $\boldsymbol{R}_0$ in $\{\boldsymbol{O};\boldsymbol{i},\boldsymbol{j},\boldsymbol{k}\}$. The normal distance between L_A and $L_{\mathrm{C'P'C}}$ is denoted by η. The points A and B are the feet of their perpendicular line $\overline{AB}$ on L_A and $L_{\mathrm{C'P'C}}$ respectively.

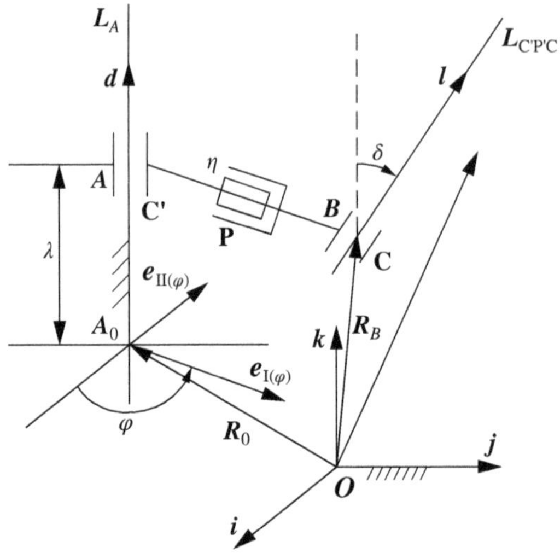

Figure 3.16 The constant inclination ruled surface and an open chain C′–P′–C

The generator $L_{C'P'C}$ rotates around the fixed line L_A by an angle φ. Both unit direction vectors $\boldsymbol{l}$ of the generator and $\boldsymbol{d}$ of the fixed line L_A can be described by the vector function of a unit circle:

$$\boldsymbol{d} = \boldsymbol{e}_{\mathrm{I}(\varphi)} \times \boldsymbol{e}_{\mathrm{II}(\varphi)},\ \boldsymbol{l} = \cos\delta \boldsymbol{d} + \sin\delta \boldsymbol{e}_{\mathrm{II}(\varphi)} \tag{3.97}$$

Let σ be the arc length of the spherical image curve of $\boldsymbol{l}$. The differential of σ can be obtained by the above equation in terms of

$$\mathrm{d}\sigma = \left|\frac{\mathrm{d}\boldsymbol{l}}{\mathrm{d}\varphi}\right| \mathrm{d}\varphi = \sin\delta \cdot \mathrm{d}\varphi \tag{3.98}$$

In $\{\boldsymbol{O};\boldsymbol{i},\boldsymbol{j},\boldsymbol{k}\}$, the point B will trace a curve Γ_B, expressed by

$$\boldsymbol{R}_B = \boldsymbol{R}_0 + \lambda_{(\varphi)}\boldsymbol{d} + \eta_{(\varphi)}\boldsymbol{e}_{\mathrm{I}(\varphi)} \tag{3.99}$$

The curve Γ_B is taken as a directrix of $\sum_{C'P'C}$, and the vector equation of $\sum_{C'P'C}$ is

$$\sum\nolimits_{C'P'C} : \boldsymbol{R} = \boldsymbol{R}_B + \mu\boldsymbol{l} \tag{3.100}$$

Actually, by equation (3.55), the striction curve of $\sum_{C'P'C}$ can be written as

$$\boldsymbol{\rho} = \boldsymbol{R}_0 + \lambda_{(\varphi)}\boldsymbol{d} + \eta_{(\varphi)}\boldsymbol{e}_{\mathrm{I}(\varphi)} + \frac{\mathrm{d}\eta}{\mathrm{d}\sigma}\boldsymbol{l} \tag{3.101}$$

where $\frac{\mathrm{d}\eta}{\mathrm{d}\sigma}$ describes the distance from the point B of the directrix Γ_B to the striction point. This means striction point is not a fixed point of the generator $L_{C'P'C}$.

The Frenet frame $\{\boldsymbol{\rho};\boldsymbol{e}_1,\boldsymbol{e}_2,\boldsymbol{e}_3\}$ of $\sum_{C'P'C}$ is established as

$$\boldsymbol{e}_1 = \boldsymbol{l},\ \boldsymbol{e}_2 = \frac{\mathrm{d}\boldsymbol{l}}{\mathrm{d}\sigma},\ \boldsymbol{e}_3 = \boldsymbol{l} \times \frac{\mathrm{d}\boldsymbol{l}}{\mathrm{d}\sigma} \tag{3.102}$$

Based on equation (3.58), the construction parameters of $\sum_{C'PC}$ can be derived as

$$\begin{cases} \alpha = \dfrac{d\lambda}{d\sigma}\cos\delta + \eta + \dfrac{d^2\eta}{d\sigma^2} \\ \beta = \cot\delta \\ \gamma = \dfrac{d\lambda}{d\sigma}\sin\delta - \eta\cot\delta \end{cases} \tag{3.103}$$

Based on equation (3.102), the unit vector $\boldsymbol{d}$ can be expressed in $\{\boldsymbol{\rho}; \boldsymbol{e}_1, \boldsymbol{e}_2, \boldsymbol{e}_3\}$ as

$$\boldsymbol{d} = \cos\delta \boldsymbol{e}_1 + \sin\delta \boldsymbol{e}_3, \ \cot\delta = \beta \tag{3.104}$$

In particular, the general kinematic pair C′ is actually a revolute pair R; $\sum_{C'P'C}$ degenerates to a ***constant inclination revolving ruled surface*** $\sum_{RP'C}$, or $\frac{d\lambda}{d\sigma} = \frac{d\lambda}{d\varphi} = 0$, whose construction parameters can be derived from equation (3.103):

$$\alpha = \eta + \frac{d^2\eta}{d\sigma^2}, \ \beta = \cot\delta, \ \gamma = -\eta\cot\delta \tag{3.105}$$

The fixed axis L_A can be represented in $\{\boldsymbol{\rho}; \boldsymbol{e}_1, \boldsymbol{e}_2, \boldsymbol{e}_3\}$ of $\sum_{RP'C}$ as

$$\begin{cases} \boldsymbol{R}_A = \boldsymbol{\rho} + \dfrac{d\gamma/d\sigma}{\beta}\boldsymbol{e}_1 - \dfrac{\gamma}{\beta}\boldsymbol{e}_2 \\ \boldsymbol{d} = \cos\delta \boldsymbol{e}_1 + \sin\delta \boldsymbol{e}_3, \ \cot\delta = \beta \end{cases} \tag{3.106}$$

On the contrary, the general kinematic pair C′ is a helical pair H, $\sum_{C'P'C}$ will be a ***constant inclination helical ruled surface*** $\sum_{HP'C}$, or $\frac{d\lambda}{d\sigma} = \frac{d\lambda}{d\varphi}\frac{d\varphi}{d\sigma} = \frac{p}{\sin\delta}$, whose construction parameters can be obtained from equation (3.103):

$$\alpha = p\cot\delta + \eta + \frac{d^2\eta}{d\sigma^2}, \ \beta = \cot\delta, \ \gamma = p - \eta\cot\delta \tag{3.107}$$

The fixed axis L_A can be represented in $\{\boldsymbol{\rho}; \boldsymbol{e}_1, \boldsymbol{e}_2, \boldsymbol{e}_3\}$ of $\sum_{HP'C}$ as

$$\begin{cases} \boldsymbol{R}_A = \boldsymbol{\rho} + \dfrac{d\gamma/d\sigma}{\beta}\boldsymbol{e}_1 + \dfrac{d^2\gamma/d\sigma^2 + \alpha\beta - \beta^2\gamma}{\beta(1+\beta^2)}\boldsymbol{e}_2 \\ \boldsymbol{d} = \cos\delta \boldsymbol{e}_1 + \sin\delta \boldsymbol{e}_3, \ \cot\delta = \beta \end{cases} \tag{3.108}$$

The helix parameter p can be obtained from equation (3.107) as

$$p = \frac{\alpha\beta + \gamma + d^2\gamma/d\sigma^2}{1+\beta^2} \tag{3.109}$$

3.5.1.2 Necessary and Sufficient Conditions of a Constant Inclination Ruled Surface

According to the definition of the constant inclination ruled surface $\sum_{C'P'C}$, the included angle δ between the ruling of $\sum_{C'P'C}$ and the fixed line L_A is constant. Thus, the geodesic curvature β of the spherical image curve of the unit vector $\boldsymbol{l}$ of the generator $L_{C'P'C}$ is also constant according to equation (3.103). Obviously, this is a necessary condition for a ruled surface to be a constant inclination ruled surface $\sum_{C'P'C}$.

On the contrary, the sufficient condition is that a general ruled surface with constant β must be a constant inclination ruled surface $\sum_{C'P'C}$. The spherical image curve is a circle on a unit sphere if the construction parameter β is identically constant by virtue of its geometrical definition, as shown in Fig. 3.17. Connecting the center O' of the circle and the center O of the unit sphere with a line $\overline{OO\prime}$, we have a cone with half vertex angle δ and fixed axis $\overline{OO\prime}$. Since the rulings of the cone have the same unit vectors as those of the ruled surface, there must exist a fixed line, which corresponds to the fixed axis $\overline{OO\prime}$, and forms a constant inclination angle δ with all the rulings of the ruled surface. Of course, the ruled surface must be the constant inclination ruled surface $\sum_{C'P'C}$ by its definition.

As a result, we have the following theorem:

Theorem 3.5 *A general ruled surface is a constant inclination ruled surface if and only if its construction parameter β is constant.*

3.5.2 Constant Axis Ruled Surfaces ($C'-C$)

A ruled surface $\sum_L$ is defined as a ***constant axis ruled surface*** if not only the inclination angle δ between the ruling of $\sum_L$ and a fixed line is always constant, but also the normal distance η between the two lines is constant. According to such a definition, the constraint ruled surface $\sum_{C'C}$ of a binary link C′–C is a constant axis ruled surface, as shown in Fig. 3.18.

3.5.2.1 Construction Parameters of a Constant Axis Ruled Surface

According to the definition of the constant axis ruled surface, it has an additional condition to the constant inclination ruled surface, namely the normal distance between the ruling of the ruled surface and the fixed line is constant, or the variable $\eta_{(\varphi)}$ in equation (3.100) has

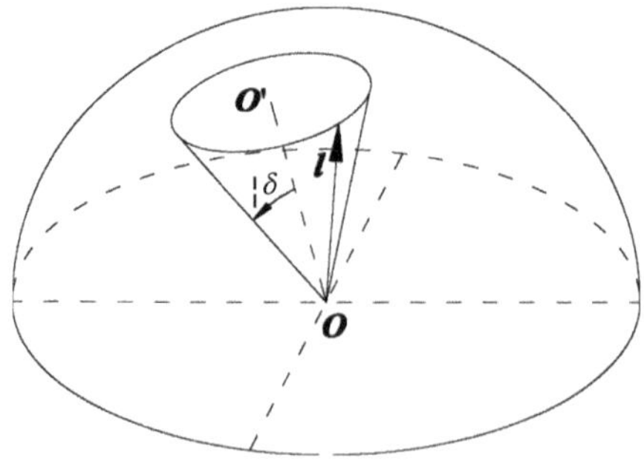

Figure 3.17 The spherical image curve of a constant inclination ruled surface

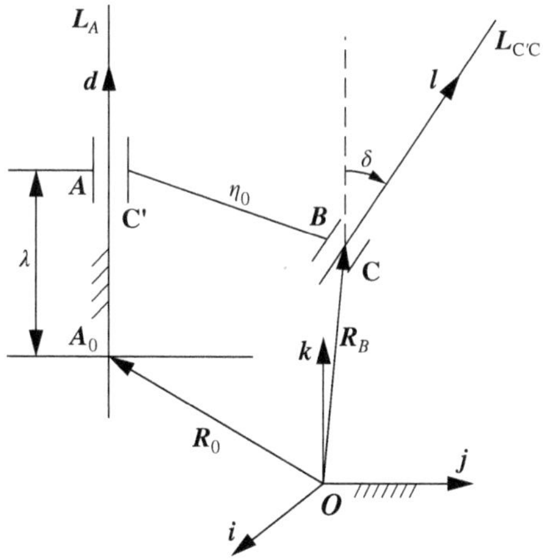

Figure 3.18 The constant axis ruled surface and a binary link C′–C

constant value η_0. The vector equation of the constant axis ruled surface $\sum_{C'C}$ is

$$\sum\nolimits_{C'C}: \boldsymbol{R} = \boldsymbol{R}_0 + \lambda_{(\varphi)}\boldsymbol{d} + \eta_0 \frac{\boldsymbol{l} \times \boldsymbol{d}}{|\boldsymbol{l} \times \boldsymbol{d}|} + \mu \boldsymbol{l},\ \boldsymbol{d} \cdot \boldsymbol{l} = \cos\delta \tag{3.110}$$

in which $\boldsymbol{R}_B = \boldsymbol{R}_0 + \lambda_{(\varphi)}\boldsymbol{d} + \eta_0 \frac{\boldsymbol{l}\times\boldsymbol{d}}{|\boldsymbol{l}\times\boldsymbol{d}|}$ is the vector of the directrix, $\boldsymbol{R}_0$, $\boldsymbol{d}$ are the constant vectors and η_0 is a constant. Based on equations (3.54) and (3.101), the distance between the striction point and the point of the directrix is zero, or the perpendicular foot point B is always the striction point; certainly it is a fixed point of the characteristic line $L_{C'C}$ of a binary link, but not a fixed point of a moving body. We rewrite the vector equation of the striction curve $\Gamma_{C'C}$ of $\sum_{C'C}$ as

$$\boldsymbol{\rho} = \boldsymbol{R}_B = \boldsymbol{R}_0 + \lambda_{(\varphi)}\boldsymbol{d} + \eta_0 \frac{\boldsymbol{l} \times \boldsymbol{d}}{|\boldsymbol{l} \times \boldsymbol{d}|} \tag{3.111}$$

The Frenet frame of the constant axis ruled surface $\sum_{C'C}$ has an identical form to equation (3.102), whose origin vector is replaced by $\boldsymbol{\rho}$ of the above equation. By letting $\eta_{(\varphi)}$ in equation (3.103) be constant and replaced by η_0, the construction parameters of $\sum_{C'C}$ can be obtained:

$$\begin{cases} \alpha = \dfrac{\mathrm{d}\boldsymbol{\rho}}{\mathrm{d}\sigma} \cdot \boldsymbol{l} - \dfrac{\mathrm{d}\lambda}{\mathrm{d}\sigma}\cos\delta + \eta_0 \\ \beta = \left(\boldsymbol{l}, \dfrac{\mathrm{d}\boldsymbol{l}}{\mathrm{d}\sigma}, \dfrac{\mathrm{d}^2\boldsymbol{l}}{\mathrm{d}\sigma^2}\right) = \cot\delta \\ \gamma = \left(\dfrac{\mathrm{d}\boldsymbol{\rho}}{\mathrm{d}\sigma}, \boldsymbol{l}, \dfrac{\mathrm{d}\boldsymbol{l}}{\mathrm{d}\sigma}\right) - \dfrac{\mathrm{d}\lambda}{\mathrm{d}\sigma}\sin\delta - \eta_0\cot\delta \end{cases} \tag{3.112}$$

Based on the above equations, we have the invariants of $\sum_{C'C}$ as

$$\begin{cases} \dfrac{\mathrm{d}\beta}{\mathrm{d}\sigma} = 0 \text{ or } \beta = \text{const.} \\ \dfrac{\mathrm{d}(\alpha - \beta\gamma)}{\mathrm{d}\sigma} = 0 \text{ or } \alpha - \beta\gamma = \text{const.} \end{cases} \tag{3.113}$$

These imply that a constant axis ruled surface $\sum_{C'C}$ has two important properties:

1. The spherical image curve of the rulings is a circle on a unit sphere.
2. The shortest distance between the generator $L_{C'P'C}$ and the fixed line L_A is the radius $\eta_0 = \frac{\alpha-\beta\gamma}{1+\beta^2}$ of a cylindrical surface, while the perpendicular foot on the generator is the striction point, or the striction curve is a cylindrical curve.

By means of equations (3.104), (3.111), and (3.112), the direction and position of the fixed axis L_A of the constant axis ruled surface $\sum_{C'C}$ can be represented by the Frenet frame as

$$\begin{cases} \boldsymbol{R}_A = \boldsymbol{\rho} + \eta_0 \boldsymbol{e}_2 = \boldsymbol{\rho} + \dfrac{\alpha - \beta\gamma}{1+\beta^2}\boldsymbol{e}_2 \\ \boldsymbol{d} = \cos\delta \boldsymbol{e}_1 + \sin\delta \boldsymbol{e}_3, \ \cot\delta = \beta \end{cases} \tag{3.114}$$

3.5.2.2 Necessary and Sufficient Conditions of a Constant Axis Ruled Surface

A constant axis ruled surface $\sum_{C'C}$ has two global properties just mentioned above, which reveal the relationship between a constant axis ruled surface and the corresponding two curves. The spherical image curve of the rulings and the striction curve, in particular, provide a solid theoretical basis for the kinematic geometry of lines both at infinitesimally separated positions in Chapter 6 and finitely separated positions in Chapter 7. Currently, we may estimate a ruled surface, or a line trajectory in spatial movement, to be a constant axis ruled surface $\sum_{C'C}$ or not, by means of the necessary and sufficient conditions. The first property of a constant axis ruled surface has been proved in Theorem 3.5, as that for a ruled surface to be a constant inclination ruled surface $\sum_{C'P'C}$. Hence, we focus on proving the necessary and sufficient conditions for a constant inclination ruled surface $\sum_{C'P'C}$ to be a constant axis ruled surface $\sum_{C'C}$. To identify the striction curve as a cylindrical curve, the following lemma is adopted:

Lemma 3.1 *A constant inclination ruled surface is a constant axis ruled surface $\sum_{C'C}$ if and only if its invariant $\alpha - \beta\gamma$ is constant.*

Firstly, we prove the necessary condition, or the invariant $\alpha - \beta\gamma$ of a constant axis ruled surface $\sum_{C'C}$ must be constant (or its striction curve is a cylindrical curve). In general, it can be verified directly by equations (3.112), or property (2), but it is not clear how to reveal the striction curve $\Gamma_{C'C}$ of $\sum_{C'C}$ as a cylindrical curve. Based on the Frenet formulas (3.58), other than equation (3.111) – since the latter is more complex than the former – the differential of arc length s for the striction curve of $\sum_{C'C}$ can be derived as

$$\mathrm{d}s = \sqrt{\alpha^2 + \gamma^2}\mathrm{d}\sigma \tag{3.115}$$

The Darboux frame $\{\boldsymbol{\rho}; \boldsymbol{\alpha}, \boldsymbol{n}, \boldsymbol{v}\}$ is established on the striction curve $\Gamma_{C'C}$ and has the following relation with the Frenet frame $\{\boldsymbol{\rho}; \boldsymbol{e}_1, \boldsymbol{e}_2, \boldsymbol{e}_3\}$:

$$\begin{cases} \boldsymbol{\alpha} = \dfrac{\alpha\boldsymbol{e}_1 + \gamma\boldsymbol{e}_3}{\sqrt{\alpha^2+\gamma^2}} \\ \boldsymbol{n} = \boldsymbol{e}_2 \\ \boldsymbol{v} = \boldsymbol{\alpha} \times \boldsymbol{n} = \dfrac{-\gamma\boldsymbol{e}_1 + \alpha\boldsymbol{e}_3}{\sqrt{\alpha^2+\gamma^2}} \end{cases} \tag{3.116}$$

Based on the differential formulas (3.42) of $\{\boldsymbol{\rho};\boldsymbol{\alpha},\boldsymbol{n},\boldsymbol{v}\}$, the curvatures k_n, k_g, τ_g of the striction curve can be represented in terms of the construction parameters α, β, γ of $\sum_{C'C}$, which are

$$\begin{cases} k_n = \dfrac{\alpha - \beta\gamma}{\alpha^2 + \gamma^2} \\ k_g = \dfrac{\alpha \mathrm{d}\gamma/\mathrm{d}\sigma - \gamma \mathrm{d}\alpha/\mathrm{d}\sigma}{\left(\alpha^2 + \gamma^2\right)^{3/2}} \\ \tau_g = \dfrac{\alpha\beta + \gamma}{\alpha^2 + \gamma^2} \end{cases} \tag{3.117}$$

Differentiating the above equations with respect to s, and combining with the fact that the parameter β is constant, we have

$$\begin{cases} \dfrac{\mathrm{d}k_n}{\mathrm{d}s} = \left[\left(\alpha^2 + \gamma^2\right)\left(\dfrac{\mathrm{d}\alpha}{\mathrm{d}\sigma} - \beta\dfrac{\mathrm{d}\gamma}{\mathrm{d}\sigma}\right) - 2(\alpha - \beta\gamma)\left(\alpha\dfrac{\mathrm{d}\alpha}{\mathrm{d}\sigma} + \gamma\dfrac{\mathrm{d}\gamma}{\mathrm{d}\sigma}\right)\right] / (\alpha^2 + \gamma^2)^{5/2} \\ \dfrac{\mathrm{d}\tau_g}{\mathrm{d}s} = \left[\left(\alpha^2 + \gamma^2\right)\left(\beta\dfrac{\mathrm{d}\alpha}{\mathrm{d}\sigma} + \dfrac{\mathrm{d}\gamma}{\mathrm{d}\sigma}\right) - 2(\gamma + \alpha\beta)\left(\alpha\dfrac{\mathrm{d}\alpha}{\mathrm{d}\sigma} + \gamma\dfrac{\mathrm{d}\gamma}{\mathrm{d}\sigma}\right)\right] / (\alpha^2 + \gamma^2)^{5/2} \end{cases} \tag{3.118}$$

It is known from equation (3.94) that the necessary and sufficient conditions for the striction curve to be a cylindrical curve are

$$\begin{cases} \dfrac{\mathrm{d}k_n}{\mathrm{d}s} + 2k_g\tau_g = -\dfrac{\left(\dfrac{\mathrm{d}\alpha}{\mathrm{d}\sigma} - \beta\dfrac{\mathrm{d}\gamma}{\mathrm{d}\sigma}\right)}{(\alpha^2 + \gamma^2)^{3/2}} = 0 \\ \dfrac{\mathrm{d}\tau_g}{\mathrm{d}s} + \dfrac{k_g(\tau_g^2 - k_n^2)}{k_n} = -\dfrac{(\gamma + \alpha\beta)\left(\dfrac{\mathrm{d}\alpha}{\mathrm{d}\sigma} - \beta\dfrac{\mathrm{d}\gamma}{\mathrm{d}\sigma}\right)}{(\alpha^2 + \gamma^2)^{3/2}(\alpha - \beta\gamma)} = 0 \end{cases} \tag{3.119}$$

Based on the above equations, we get the necessary and sufficient condition for the striction curve of the constant inclination ruled surface to be a cylindrical curve:

$$\frac{\mathrm{d}\alpha}{\mathrm{d}\sigma} - \beta\frac{\mathrm{d}\gamma}{\mathrm{d}\sigma} = 0 \tag{3.120}$$

The above equation implies that the invariant $\alpha - \beta\gamma$ is constant. According to Theorem 3.4, the striction curve of a constant axis ruled surface $\sum_{C'C}$ is certainly a cylindrical curve.

On the contrary, we prove the sufficient condition – namely, a constant inclination ruled surface $\sum_{C'P'C}$, whose invariant $\alpha - \beta\gamma$ is constant, must be a constant axis ruled surface $\sum_{C'C}$. For a constant inclination ruled surface $\sum_{C'P'C}$, whose invariant $\alpha - \beta\gamma$ is constant, we can rewrite the differentiating equation since the construction parameter β is also constant. That is

$$\frac{\mathrm{d}(\alpha - \beta\gamma)}{\mathrm{d}\sigma} = 0 \tag{3.121}$$

Substituting equation (3.103), the construction parameters of the constant inclination ruled surface $\sum_{C'P'C}$, into the above equation and simplifying, we have another expression:

$$\frac{d(\alpha - \beta\gamma)}{d\sigma} = -(1+\beta^2)\frac{d\eta}{d\sigma} - \frac{d^3\eta}{d\sigma^3} = 0 \tag{3.122}$$

In order to solve the above differentiating equation, the arc length σ has to be replaced by the rotation angle φ of the generator of $\sum_{C'P'C}$. Substituting equations (3.98) and (3.103) into the above equation, we have

$$\frac{d\eta}{d\varphi} + \frac{d^3\eta}{d\varphi^3} = 0 \tag{3.123}$$

The general solution of this differentiating equation is

$$\eta = c_1 \cos(\varphi + \varphi_0) + c_2 \tag{3.124}$$

where c_1, c_2, and φ_0 are constants of integration and can be determined by the initial conditions. By substituting the general solution and equation (3.99) into equation (3.101), the vector equation of the striction curve of a constant inclination ruled surface with $\alpha - \beta\gamma =$ const. can be written as

$$\begin{aligned}\boldsymbol{\rho} &= \boldsymbol{R}_0 + [c_1\cos(\varphi+\varphi_0) + c_2]\boldsymbol{e}_{\mathrm{I}}(\varphi) - c_1\sin(\varphi+\varphi_0)\boldsymbol{e}_{\mathrm{II}}(\varphi) + [\lambda + \beta c_1 \sin(\varphi+\varphi_0)]\boldsymbol{d}\\ &= \boldsymbol{R}_0 + c_1\boldsymbol{e}_{\mathrm{I}}(\varphi_0) + c_2\boldsymbol{e}_{\mathrm{I}}(\varphi) + [\lambda + \beta c_1\sin(\varphi+\varphi_0)]\boldsymbol{d}\end{aligned} \tag{3.125}$$

where $\boldsymbol{R}_0$, $c_1\boldsymbol{e}_{\mathrm{I}}(\varphi_0)$, and $\boldsymbol{d}$ are all constant vectors. They locate the initial position of the fixed line L_A or point A_0, regardless of the choice of the fixed frame $\{\boldsymbol{O};\boldsymbol{i},\boldsymbol{j},\boldsymbol{k}\}$, as shown in Fig. 3.16. Hence, the normal distance between the fixed line L_A and the rulings of $\sum_{C'P'C}$ with $\alpha - \beta\gamma =$ const. can be expressed as

$$[\boldsymbol{\rho} - \boldsymbol{R}_0 - c_1\boldsymbol{e}_{\mathrm{I}}(\varphi_0)] \cdot \frac{\boldsymbol{d}\times\boldsymbol{l}}{|\boldsymbol{d}\times\boldsymbol{l}|} = \{c_2\boldsymbol{e}_{\mathrm{I}}(\varphi) + [\lambda + \beta c_1\sin(\varphi+\varphi_0)]\boldsymbol{d}\}\cdot\boldsymbol{e}_{\mathrm{I}}(\varphi) = c_2 \tag{3.126}$$

The result of the above equation is a constant, of course – the radius of the cylindrical surface, on which the striction curve locates. Therefore, the constant inclination ruled surface $\sum_{C'P'C}$ with $\alpha - \beta\gamma =$ const. is a constant axis ruled surface $\sum_{C'C}$.

Combining Theorem 3.5 and the above lemma, we have

Theorem 3.6 *A ruled surface is a constant axis ruled surface if and only if (a)* $\beta =$ *const., (b)* $\alpha - \beta\gamma =$ *const.*

3.5.3 Constant Parameter Ruled Surfaces (H–C, R–C)

A ruled surface $\sum_L$ is defined as a ***constant parameter ruled surface*** if its three construction parameters are all constant. The constraint ruled surfaces of the binary links H–C and R–C are both constant parameter ruled surfaces, as shown in Figs 3.19 and 3.20.

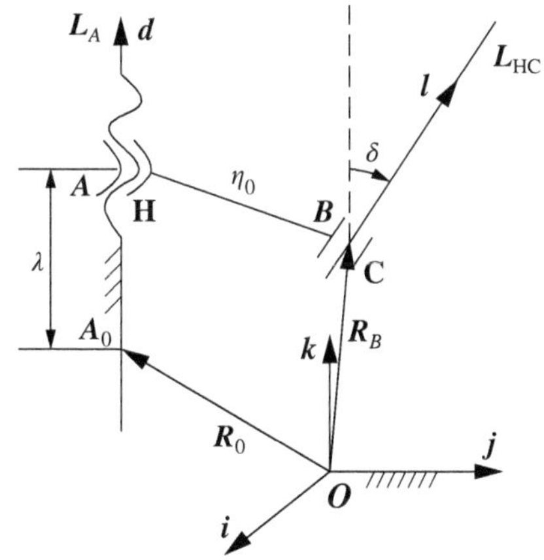

Figure 3.19 The constant helicoids and a binary link H–C

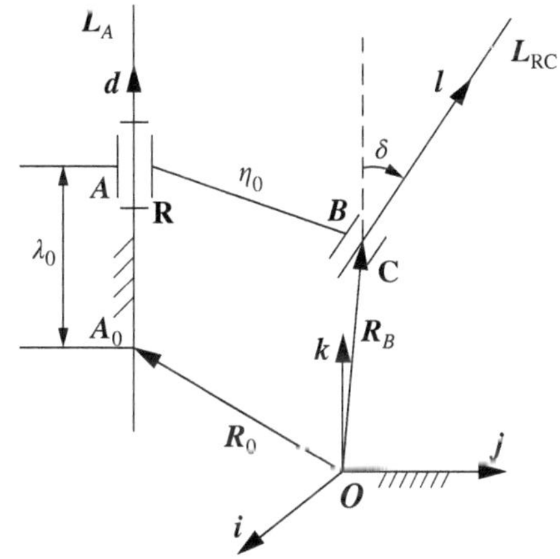

Figure 3.20 The hyperboloid of one sheet and a binary link R–C

3.5.3.1 Geometrical Characteristics of Constant Parameter Ruled Surfaces

Since three construction parameters are all constant, a constant parameter ruled surface has more special geometrical characteristics, especially in the striction curve.

Based on the definition of a constant parameter ruled surface, and the constant axis ruled surface $\sum_{C'C}$, the equation of the constant parameter ruled surface can be obtained by letting $d\lambda/d\sigma$ be constant in equation (3.110). That is

$$\sum : \boldsymbol{R} = \boldsymbol{R}_0 + \lambda \boldsymbol{d} + \eta_0 \frac{\boldsymbol{l} \times \boldsymbol{d}}{|\boldsymbol{l} \times \boldsymbol{d}|} + \mu \boldsymbol{l},\ \boldsymbol{d} \cdot \boldsymbol{l} = \cos\delta,\ \frac{d\lambda}{d\sigma} = \text{const.} \tag{3.127}$$

Here, the vector of the striction curve is

$$\boldsymbol{\rho} = \boldsymbol{R}_0 + \lambda \boldsymbol{d} + \eta_0 \frac{\boldsymbol{l} \times \boldsymbol{d}}{|\boldsymbol{l} \times \boldsymbol{d}|} \tag{3.128}$$

The Frenet frame of a constant parameter ruled surface can be set up in the same way as that of the constant axis ruled surface $\sum_{C'C}$, whose construction parameters can be obtained by letting $d\lambda/d\sigma$ be constant in equation (3.112) of $\sum_{C'C}$:

$$\begin{cases} \alpha = \dfrac{d\boldsymbol{\rho}}{d\sigma} \cdot \boldsymbol{l} = \dfrac{d\lambda}{d\sigma} \cos\delta + \eta_0 \\ \beta = \left(\boldsymbol{l}, \dfrac{d\boldsymbol{l}}{d\sigma}, \dfrac{d^2\boldsymbol{l}}{d\sigma^2} \right) = \cot\delta \\ \gamma = \left(\dfrac{d\boldsymbol{\rho}}{d\sigma}, \boldsymbol{l}, \dfrac{d\boldsymbol{l}}{d\sigma} \right) = \dfrac{d\lambda}{d\sigma} \sin\delta - \eta_0 \cot\delta \end{cases} \tag{3.129}$$

Based on the differential formulas (3.58) of the Frenet frame of a ruled surface and the definition of the constant parameter ruled surface, we can write the first to third derivatives of the vector $\boldsymbol{\rho}$ of the striction curve with respect to σ as

$$\begin{cases} \dfrac{d\boldsymbol{\rho}}{d\sigma} = \alpha \boldsymbol{e}_1 + \gamma \boldsymbol{e}_3, \ \dfrac{d^2\boldsymbol{\rho}}{d\sigma^2} = (\alpha - \beta\gamma)\boldsymbol{e}_2 \\ \dfrac{d^3\boldsymbol{\rho}}{d\sigma^3} = (\alpha - \beta\gamma)(-\boldsymbol{e}_1 + \beta \boldsymbol{e}_3) \end{cases} \tag{3.130}$$

From equation (3.13), the curvature k_y and torsion τ_y of the striction curve of a constant parameter ruled surface can be expressed by

$$\begin{cases} k_y = (\alpha - \beta\gamma)/(\alpha^2 + \gamma^2) \\ \tau_y = (\gamma + \alpha\beta)/(\alpha^2 + \gamma^2) \end{cases} \tag{3.131}$$

According to Theorem 3.6 and the above equations, some geometrical characteristics of a constant parameter ruled surface can be described as follows.

1. A constant parameter ruled surface is a constant axis ruled surface.
2. The tangent line of the striction curve forms a constant angle with respect to the ruling.
3. The curvature k_y and torsion τ_y of the striction curve are constant.

3.5.3.2 Classification of the Constant Parameter Ruled Surfaces

By substituting equation (3.129) into (3.131), the constant parameter ruled surface can be classified into three types according to the shapes of their striction curves.

1. A cone
   A constant parameter ruled surface becomes a cone if the striction curve is a point, or the vector $\boldsymbol{\rho}$ of the striction point is a constant vector. In this case, the three construction parameters are

$$\alpha = 0, \ \beta = \cot\delta, \ \gamma = 0 \tag{3.132}$$

The position of the cone can be located by

$$\begin{cases} \boldsymbol{R}_A = \boldsymbol{\rho} \\ \boldsymbol{d} = \cos\delta \boldsymbol{e}_1 + \sin\delta \boldsymbol{e}_3, \cot\delta = \beta \end{cases} \tag{3.133}$$

2. The hyperboloid $\sum_{RC}$
   The constant parameter ruled surface is a ***hyperboloid*** $\sum_{RC}$ if the striction curve is a circle, which corresponds to a binary link R–C. From equation (3.131), the three construction parameters satisfy

$$\gamma + \alpha\beta = 0,\ \alpha - \beta\gamma \neq 0 \tag{3.134}$$

Meanwhile, $\frac{d\lambda}{d\sigma} = 0$ or $\lambda = \text{const.}$ is derived by substituting equation (3.129) into equation (3.134), and the three construction parameters can be rewritten as

$$\alpha = \eta_0,\ \beta = \cot\delta,\ \gamma = -\beta\eta_0 \tag{3.135}$$

The position of the hyperboloid $\sum_{RC}$ can be determined by

$$\begin{cases} \boldsymbol{R}_A = \boldsymbol{\rho} + \alpha\boldsymbol{e}_2 \\ \boldsymbol{d} = \cos\delta \boldsymbol{e}_1 + \sin\delta \boldsymbol{e}_3,\ \cot\delta = \beta \end{cases} \tag{3.136}$$

3. The constant helicoid $\sum_{HC}$
   A constant parameter ruled surface is a ***constant helicoid*** $\sum_{HC}$ when its striction curve is a circular helix, which corresponds to a binary link H–C. Both the curvature k_y and the torsion τ_y of the striction curve must be constant. According to equation (3.131), the three construction parameters have to satisfy

$$\gamma + \alpha\beta \neq 0, \alpha - \beta\gamma \neq 0 \tag{3.137}$$

The helix parameter of the constant helicoid $\sum_{HC}$ is

$$p = \frac{\alpha\beta + \gamma}{\sqrt{1 + \beta^2}} \tag{3.138}$$

The position of the hyperboloid $\sum_{HC}$ can be presented as

$$\begin{cases} \boldsymbol{R}_A = \boldsymbol{\rho} + \dfrac{\alpha - \beta\gamma}{1 + \beta^2}\boldsymbol{e}_2 \\ \boldsymbol{d} = \cos\delta \boldsymbol{e}_1 + \sin\delta \boldsymbol{e}_3,\ \cot\delta = \beta \end{cases} \tag{3.139}$$

Especially, the constant helicoid $\sum_{HC}$ is an ***Archimedes helicoid*** when the striction curve is a straight line if $\alpha - \beta\gamma = 0$, and $\sum_{HC}$ is a ***right helicoid*** if $\alpha = 0,\ \beta = 0$. Finally, a constant helicoid $\sum_{HC}$ with $\gamma = 0$ is an ***involute helicoid***; it is the only developable surface of all constant helicoids.

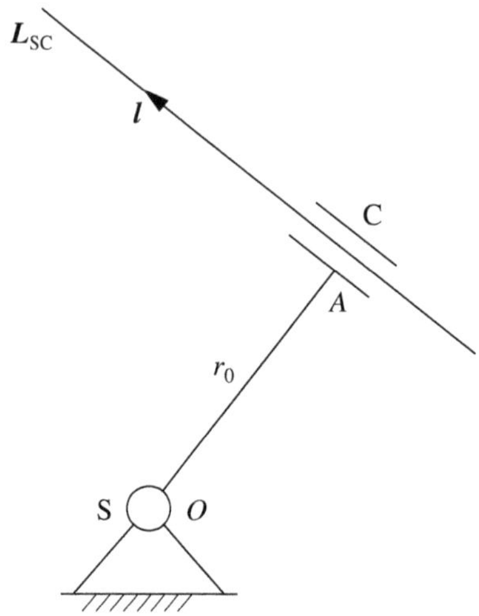

Figure 3.21 The constant distance ruled surface and a binary link S−C

3.5.4 *Constant Distance Ruled Surfaces (S′−C)*

A ruled surface $\sum_L$ is called a ***constant distance ruled surface*** if the distance between its rulings and a fixed point keeps constant. The constant distance ruled surface is certainly traced by a straight line, which rotates around a fixed point. The constraint ruled surface $\sum_{S'C}$ of the binary link S′−C is a constant distance ruled surface, as shown in Fig. 3.21.

3.5.4.1 Construction Parameters of a Constant Distance Ruled Surface

Based on the definition of the constant distance ruled surface $\sum_{S'C}$, the distance, naturally the normal distance, from a fixed point O to the rulings of $\sum_{S'C}$ is identically a constant value r_0. From the viewpoint of geometry, the ruling of $\sum_{S'C}$ should contact tangentially at point A of a spherical surface with fixed center point O and radius r_0, which is the normal distance. All the contacting points A with position vector $\boldsymbol{R}_A$ form a spherical curve $\Gamma_{S'C}$ on the spherical surface with arc length s. The Darboux frame $\{\boldsymbol{R}_A; \boldsymbol{\alpha}, \boldsymbol{n}, \boldsymbol{v}\}$ can be established on $\Gamma_{S'C}$, as shown in Fig. 3.22. The vector equation of $\Gamma_{S'C}$ is written as

$$\Gamma_{S'C}: \boldsymbol{R}_A(\mathrm{s}) = -r_0\boldsymbol{n}(\mathrm{s}) \tag{3.140}$$

A generator, or a ruling, must lie on the tangent plane $\boldsymbol{R}_A - \boldsymbol{\alpha v}$. Since the rulings of $\sum_{S'C}$ are always tangent to the spherical surface along the curve $\Gamma_{S'C}$, its unit direction vector $\boldsymbol{l}$ can be expressed by

$$\boldsymbol{l} = \cos\delta\boldsymbol{\alpha} + \sin\delta\boldsymbol{v} \tag{3.141}$$

where δ is the included angle between the vectors $\boldsymbol{l}$ and $\boldsymbol{\alpha}$ and the function of the arc length s. The curve $\Gamma_{S'C}$ is taken as a directrix of $\sum_{S'C}$, and the vector equation of $\sum_{S'C}$ can be written as

$$\sum\nolimits_{S'C}: \boldsymbol{R} = -r_0\boldsymbol{n} + \mu(\cos\delta\boldsymbol{\alpha} + \sin\delta\boldsymbol{v}) \tag{3.142}$$

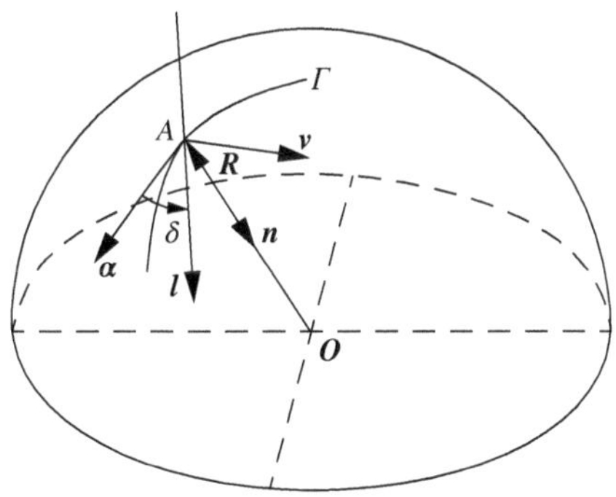

Figure 3.22 The Darboux frame of $\sum_{SC}$

Differentiating equations (3.141) and (3.142) with respect to s, we have

$$\begin{cases} \dfrac{\mathrm{d}\boldsymbol{R}_A}{\mathrm{d}s} = \boldsymbol{\alpha} \\ \dfrac{\mathrm{d}\boldsymbol{l}}{\mathrm{d}s} = \left(k_g + \dfrac{\mathrm{d}\delta}{\mathrm{d}s}\right)(\cos\delta\boldsymbol{v} - \sin\delta\boldsymbol{\alpha}) + \dfrac{\cos\delta}{r_0}\boldsymbol{n} \end{cases} \tag{3.143}$$

By means of equation (3.55), the equation of the striction curve of $\sum_{S'C}$ is represented as

$$\begin{aligned} \boldsymbol{\rho} &= \boldsymbol{R}_A - \frac{(\mathrm{d}\boldsymbol{R}_A/\mathrm{d}s)\cdot(\mathrm{d}\boldsymbol{l}/\mathrm{d}s)}{(\mathrm{d}\boldsymbol{l}/\mathrm{d}s)^2}\boldsymbol{l} \\ &= -r_0\boldsymbol{n} + \frac{r_0^2\left(k_g + \dfrac{\mathrm{d}\delta}{\mathrm{d}s}\right)\sin\delta}{\cos^2\delta + r_0^2\left(k_g + \dfrac{\mathrm{d}\delta}{\mathrm{d}s}\right)^2}(\cos\delta\boldsymbol{\alpha} + \sin\delta\boldsymbol{v}) \end{aligned} \tag{3.144}$$

The above equation indicates that the directrix $\Gamma_{S'C}$ is not the striction curve of $\sum_{S'C}$ unless $\delta = 0$ or $k_g + \frac{\mathrm{d}\delta}{\mathrm{d}s} = 0$.

The parameters μ and s in equation (3.142) are respectively designated as a constant, or the parametric curves μ =const. and s = const. on $\sum_{S'C}$ can be obtained. The distance from the fixed point O to a point on $\sum_{S'C}$ is

$$h = |\boldsymbol{R}| = \sqrt{r_0^2 + \mu^2} \tag{3.145}$$

The distance h is a constant if μ is constant, which implies that any parametric curve μ =const. is a spherical curve with radius h and center point O. But only on the directrix $\Gamma_{S'C}(\mu = 0)$, $\sum_{S'C}$ tangentially contacts the spherical surface.

3.5.4.2 Properties of a Constant Distance Ruled Surface

The constant distance ruled surface $\sum_{S'C}$, whose directrix $\Gamma_{S'C}$ is a spherical curve, naturally the line of curvature of the spherical surface, tangentially contacts the spherical surface with radius r_0 along $\Gamma_{S'C}$. The normal lines of both the spherical surface and $\sum_{S'C}$ coincide with each other at points along $\Gamma_{S'C}$. Hence, the curvature vector $k\boldsymbol{\beta}$ of $\Gamma_{S'C}$ is projected onto the normal vector $\boldsymbol{n}$ of both the spherical surface and $\sum_{S'C}$, which have the same value, or the normal curvature of $\sum_{S'C}$ is equal to that of the spherical surface, $k_n = 1/r_0$.

By differentiating equation (3.140) with respect to the arc length s, and simplifying it by the differential formulas (3.45) of the Darboux frame of the spherical curve, we have

$$\frac{\mathrm{d}\boldsymbol{n}}{\mathrm{d}s} = -\frac{1}{r_0}\boldsymbol{\alpha} = -\frac{1}{r_0}\frac{\mathrm{d}\boldsymbol{R}_A}{\mathrm{d}s} \tag{3.146}$$

According to the formula of Rodrigues (3.35), the directrix $\Gamma_{S'C}$ is the line of curvature of $\sum_{S'C}$, whose curvature k_n is the principal curvature of both $\sum_{S'C}$ and the spherical surface, and identically the constant $1/r_0$. Thus, we have the following theorem:

Theorem 3.7 *A ruled surface $\sum_L$ is a constant distance ruled surface $\sum_{S'C}$ if and only if there is a line of curvature on $\sum_L$, whose principal curvature is constant.*

Proof. The necessary conditions can be proved by the above derivation process. Now we prove the sufficient conditions.

Since the curve Γ is the line of curvature of the ruled surface, it must satisfy the formulas of Rodrigues (3.35). That is

$$\mathrm{d}\boldsymbol{n} = -k_n \mathrm{d}\boldsymbol{R} \tag{3.147}$$

where $\boldsymbol{R}$ is the vector of the curve Γ, $\boldsymbol{n}$ is the normal vector of the ruled surface $\sum_L$, and the principal curvature k_n of Γ on $\sum_L$ is constant. Integrating the above equation, we have

$$\boldsymbol{n} = -k_n(\boldsymbol{R} - \boldsymbol{R}_0) \tag{3.148}$$

where $\boldsymbol{R}_0$ is a fixed vector. Taking the dot product of both ends of the above equation by itself, we have

$$(\boldsymbol{R} - \boldsymbol{R}_0)^2 = \frac{1}{k_n^2} \tag{3.149}$$

This indicates that the curve Γ is located on the spherical surface, whose radius r_0 is $1/k_n$ and whose vector of center point is $\boldsymbol{R}_0$. The normal vectors of both the ruled surface $\sum_L$ and the spherical surface coincide with each other along the curve Γ according to equation (3.148). This means that the rulings of the ruled surface $\sum_L$ are tangent to the spherical surface along the curve Γ. Thus, by the definition of a constant distance ruled surface, the ruled surface $\sum_L$ with a line of curvature, whose principal curvature is constant, is a constant distance ruled surface.

3.6 Generalized Curvature of Curves

In differential geometry, as mentioned in Section 3.1, at a point P of a space curve, the tangent line is defined as a straight line comprised of two consecutive points, and the osculating plane is constituted by three consecutive points. Then, the local structure of a curve at the neighborhood of a point can be described by its higher-order curvature, such as equation (3.14).

In Section 3.4, the necessary and sufficient conditions for a space curve to be both a spherical curve and a cylindrical curve were respectively discussed, providing global properties of a space curve, rather than local properties for the kinematic synthesis of spatial linkages. In planar kinematics of the moving body, the local properties of a curve, traced by a point of a moving body, are studied intensively. The Ball points and Burmester points, corresponding to binary link P–R and R–R, can be located on the moving body, whose trajectories are approximate to straight lines or circles in higher-order. It is a natural idea to expand the planar kinematics to spatial kinematics, but we need to discuss how to estimate the degree of approximation between a space curve and a constraint curve, or a space curve and a constraint surface, as listed in Table 3.1, or the contacting orders at infinitesimal positions. In this section, we expand the contact order between two curves into that between curves and constraint surfaces, in terms of invariants of curves, such as curvatures and torsions, and present the new concept of ***generalized curvature*** to discuss the degree of approximation between a curve and a spherical surface, or a curve and a cylindrical surface.

3.6.1 Generalized Curvature of Space Curves

The contact orders between two curves have been discussed in differential geometry [1–3]. The deviation ε of two curves at the neighborhood of the contact point is adopted to evaluate the degree of approximation. Two curves intersect or have a common point if the deviation ε is a first-order infinitely small quantity, and they are tangent or have two common points when ε is a second-order quantity, while the contact order of the two curves is naturally one. The two curves have three common points, or contact in second order if ε is third-order, and so on. The contact order of two curves is referred to as a higher order if it exceeds two.

Now we discuss the contact order for a space curve to contact a surface at the neighborhood of the contact point. The deviation ε, an infinitely small quantity, between a curve Γ_1 and a surface $\sum_2$ can be taken to define the contact order at the contact point P.

As is shown in Fig. 3.23, a curve Γ_1 tangentially contacts a surface $\sum_2$ at the point P, or two infinitesimal consecutive points of Γ_1, including point P, lie on $\sum_2$. Supposing the arc length of Γ_1 is s, we can establish the moving frame $\{\boldsymbol{R}_1; \boldsymbol{\alpha}, \boldsymbol{n}, \boldsymbol{\nu}\}$ of Γ_1, in which $\boldsymbol{R}_1$ is the vector

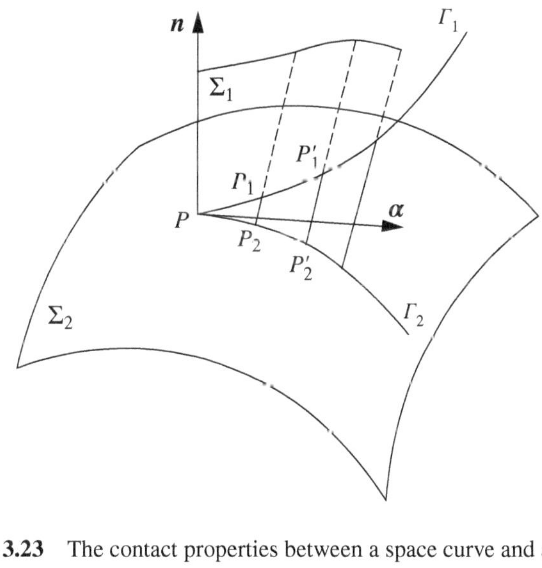

Figure 3.23 The contact properties between a space curve and a surface

of Γ_1, $\boldsymbol{\alpha}$ is the common tangent vector of both Γ_1 and $\sum_2$, and $\boldsymbol{n}$ is the unit normal vector of $\sum_2$ at point P. Taking the Taylor expansion of Γ_1 at point P, we have

$$\begin{aligned}\boldsymbol{R}_1(s+\Delta s)=\boldsymbol{R}_1(s)+\frac{\mathrm{d}\boldsymbol{R}_1(s)}{\mathrm{d}s}\Delta s+\frac{1}{2!}\frac{\mathrm{d}^2\boldsymbol{R}_1(s)}{\mathrm{d}s^2}(\Delta s)^2\\ +\cdots\frac{1}{n!}\frac{\mathrm{d}^n\boldsymbol{R}_1(s)}{\mathrm{d}s^n}(\Delta s)^n+\varepsilon_n(s,\Delta s)(\Delta s)^n\end{aligned}\tag{3.150}$$

where

$$\begin{cases}\dfrac{\mathrm{d}\boldsymbol{R}_1(s)}{\mathrm{d}s}=\boldsymbol{\alpha},\ \dfrac{\mathrm{d}^2\boldsymbol{R}_1(s)}{\mathrm{d}s^2}=k_{n1}\boldsymbol{n}+k_{g1}\boldsymbol{v}\\ \dfrac{\mathrm{d}^3\boldsymbol{R}_1(s)}{\mathrm{d}s^3}=-{k_1}^2\boldsymbol{\alpha}+\left(\dfrac{\mathrm{d}k_{n1}}{\mathrm{d}s}-k_{g1}\tau_{g1}\right)\boldsymbol{n}+\left(\dfrac{\mathrm{d}k_{g1}}{\mathrm{d}s}+k_{n1}\tau_{g1}\right)\boldsymbol{v}\\ \dfrac{\mathrm{d}^4\boldsymbol{R}_1(s)}{\mathrm{d}s^4}=-3k_1\dfrac{\mathrm{d}k_1}{\mathrm{d}s}\boldsymbol{\alpha}+\left[\dfrac{\mathrm{d}^2k_{n1}}{\mathrm{d}s^2}-2\dfrac{\mathrm{d}k_{g1}}{\mathrm{d}s}\tau_{g1}-k_{g1}\dfrac{\mathrm{d}\tau_{g1}}{\mathrm{d}s}-\left(k_1^2+{\tau_{g1}}^2\right)k_{n1}\right]\boldsymbol{n}\\ +\left[\dfrac{\mathrm{d}^2k_{g1}}{\mathrm{d}s^2}+\dfrac{\mathrm{d}k_{n1}}{\mathrm{d}s}\tau_{g1}+k_{n1}\dfrac{\mathrm{d}\tau_{g1}}{\mathrm{d}s}-\left(k_1^2+{\tau_{g1}}^2\right)k_{g1}\right]\boldsymbol{v}\end{cases}$$

In the above equation, k_1 is the curvature of Γ_1 and its components are the normal curvature k_{n1} and the geodesic curvature k_{g1}, or $k_1^2={k_{n1}}^2+{k_{g1}}^2$.

Taking each consecutive point P_1, P_1', P_1'', … at the neighborhood of the point P of Γ_1, and connecting them to the corresponding points P_2, P_2', P_2'', … on $\sum_2$, the normal lines of $\sum_2$, $\overline{P_1P_2}$, $\overline{P_1'P_2'}$, …, are formed and a ruled surface $\sum_1$ is constituted. The surface $\sum_1$ orthogonally intersects $\sum_2$ along Γ_2. It is obvious that curves Γ_1 and Γ_2 are both on $\sum_1$, but Γ_2 is just on $\sum_2$. The distance vector $\boldsymbol{\varepsilon}=\boldsymbol{R}_1-\boldsymbol{R}_2$ between the point P_1 of Γ_1 and the point P_2 of Γ_2, at the neighborhood of point P, is projected to the normal of $\sum_2$, or $h=\boldsymbol{\varepsilon}\cdot\boldsymbol{n}=(\boldsymbol{R}_1-\boldsymbol{R}_2)\cdot\boldsymbol{n}$, which describes the normal deviation between Γ_1 and $\sum_2$. The curve Γ_2 naturally has the same moving frame $\{\boldsymbol{R}_1;\boldsymbol{\alpha},\boldsymbol{n},\boldsymbol{v}\}$ as that of Γ_1 at point P, and Γ_2 have the same Taylor expressions as equation (3.150), only with the subscript 1 replaced by 2 of the variables, such as the curvature invariants k_2, k_{n2}, k_{g2}, and τ_{g2}. The geodesic torsion of Γ_2 on $\sum_1$ is equal to that of Γ_2 on $\sum_2$ [3]. Since $\sum_1$ and $\sum_2$ orthogonally intersect along the curve Γ_2, and curve Γ_1 tangentially contacts curve Γ_2 at point P, we have

$$\tau_{g1}=\tau_{g2}\tag{3.151}$$

Then, the normal deviation between Γ_1 and $\sum_2$ can be expressed as

$$\begin{aligned}h&=[\boldsymbol{R}_1(s)-\boldsymbol{R}_2(s)]\cdot\boldsymbol{n}\\ &=\left[(\boldsymbol{R}_1-\boldsymbol{R}_2)+\left(\frac{\mathrm{d}\boldsymbol{R}_1}{\mathrm{d}s}-\frac{\mathrm{d}\boldsymbol{R}_2}{\mathrm{d}s}\right)\Delta s+\left(\frac{\mathrm{d}^2\boldsymbol{R}_1}{\mathrm{d}s^2}-\frac{\mathrm{d}^2\boldsymbol{R}_2}{\mathrm{d}s^2}\right)(\Delta s)^2\right.\\ &\left.+\left(\frac{\mathrm{d}^3\boldsymbol{R}_1}{\mathrm{d}s^3}-\frac{\mathrm{d}^3\boldsymbol{R}_2}{\mathrm{d}s^3}\right)(\Delta s)^3+\cdots\right]\cdot\boldsymbol{n}\end{aligned}$$

$$= (k_{n1} - k_{n2})(\Delta s)^2 + \left[\left(\frac{\mathrm{d}k_{n1}}{\mathrm{d}s} - k_{g1}\tau_{g1}\right) - \left(\frac{\mathrm{d}k_{n2}}{\mathrm{d}s} - k_{g2}\tau_{g2}\right)\right](\Delta s)^3$$
$$+ \left\{\left[\frac{\mathrm{d}^2k_{n1}}{\mathrm{d}s^2} - 2\frac{\mathrm{d}k_{g1}}{\mathrm{d}s}\tau_{g1} - k_{g1}\frac{\mathrm{d}\tau_{g1}}{\mathrm{d}s} - \left(k_{n1}^2 + \tau_{g1}^2\right)k_{n1}\right]\right.$$
$$\left. - \left[\frac{\mathrm{d}^2k_{n2}}{\mathrm{d}s^2} - 2\frac{\mathrm{d}k_{g2}}{\mathrm{d}s}\tau_{g2} - k_{g2}\frac{\mathrm{d}\tau_{g2}}{\mathrm{d}s} - \left(k_{n2}^2 + \tau_{g2}^2\right)k_{n2}\right]\right\}(\Delta s)^4 + \cdots \tag{3.152}$$

Here, the normal deviation h is a second-order infinitely small quantity, which indicates that the curve Γ_1 contacts surface $\sum_2$ in the first order.

When the normal deviation h is a third-order infinitely small quantity, or Γ_1 contacts $\sum_2$ in the second order, based on equation (3.152), the conditions are

$$k_{n1} = k_{n2} \tag{3.153}$$

Similarly, the conditions for third-order contact between Γ_1 and $\sum_2$ can be derived as

$$\left(\frac{\mathrm{d}k_{n1}}{\mathrm{d}s} - k_{g1}\tau_{g1}\right) - \left(\frac{\mathrm{d}k_{n2}}{\mathrm{d}s} - k_{g2}\tau_{g2}\right) = 0 \tag{3.154}$$

The conditions for fourth-order contact between Γ_1 and $\sum_2$ are

$$\frac{\mathrm{d}^2k_{n1}}{\mathrm{d}s^2} - 2\frac{\mathrm{d}k_{g1}}{\mathrm{d}s}\tau_{g1} - k_{g1}\frac{\mathrm{d}\tau_{g1}}{\mathrm{d}s} - \left(k_{n1}^2 + \tau_{g1}^2\right)k_{n1} - \frac{\mathrm{d}^2k_{n2}}{\mathrm{d}s^2}$$
$$+ 2\frac{\mathrm{d}k_{g2}}{\mathrm{d}s}\tau_{g2} + k_{g2}\frac{\mathrm{d}\tau_{g2}}{\mathrm{d}s} + \left(k_{n2}^2 + \tau_{g2}^2\right)k_{n2} = 0 \tag{3.155}$$

The contact order between curve Γ_1 and surface $\sum_2$ is n, which geometrically means that Γ_1 and $\sum_2$ have n+1 consecutive common points at the neighborhood of point P. The additional conditions will be added as the contact order increases. Referring to the curvature of plane curves, we give a definition for generalized curvature of curve attached to surfaces:

Definition 3.1 For a given smooth curve Γ_1, if there exists a regular surface $\sum_2$ contacting Γ_1 at the neighborhood of point P in higher order, whose contact conditions are described in terms of invariants, and $\sum_2$ is completely determined, the invariants are defined as ***generalized curvature*** of Γ_1 attached to $\sum_2$, and $\sum_2$ is referred to as a ***generalized curvature surface***.

For a given curve, there are infinite number of surfaces which contain the given curve, just like there are certainly infinite number of curves on a given surface that pass through a given point. It is just for several special constraint surfaces that we can deduce the global geometrical properties of a generalized curvature surface from the local geometrical properties of a curve, such as a plane (F–S), a spherical surface (S–S), and a cylindrical surface (C–S), whose adjacent structures in the neighborhood of any point on the surface are totally the same. Each of these three surfaces has only one kind of point, a flat point, a ball point, and a parabolic point respectively. For some other constraint surfaces, a helicoid (H–P–S) surface, a helix torus (H–R–S) surface, and a toroidal surface (R–R–S), whose global properties need to be studied, we expect the theoretical support provided by mathematicians.

3.6.2 *Spherical Curvature and Cylindrical Curvature*

There are two constraint surfaces, a spherical surface (S–S) and a cylindrical surface (C–S). The points on the constraint surfaces are ball point for the former, and parabolic point for the latter. Based on their local geometrical properties, the generalized curvature of a curve at a point, the ***spherical curvature*** or the ***cylindrical curvature***, can be derived respectively by the contact conditions between a curve and a spherical surface, or a curve and a cylindrical surface. Particularly, either a circle (R–S) or a helix (H–S) is viewed as a special case of a cylindrical surface.

3.6.2.1 Spherical Curvature of a Curve and its Higher Order

If a curve Γ_1 contacts a spherical surface $\sum_2$, the contact orders can be analyzed by the contact conditions in terms of invariants of both the curve and the surface. The invariants of a spherical curve (mentioned in Section 3.4.1) are k_{n2} = const. and $\tau_{g2} = 0$ and the contact conditions are described by equations (3.153) to (3.155).

First order: A curve Γ_1 contacts a spherical surface $\sum_2$ at the contact point, with two consecutive common points at its neighborhood, which means that they are tangent and the zero-order contact is naturally satisfied.

Second order: A Frenet frame $\{\boldsymbol{R}_1; \boldsymbol{\alpha}, \boldsymbol{\beta}, \boldsymbol{\gamma}\}$ of curve Γ_1 at point P can naturally be determined by three consecutive points in the neighborhood of point P, and the curvature k_1, or the corresponding curvature circle, can be obtained. For the curve Γ_1 to contact a spherical surface $\sum_2$ in second order, the contact conditions are described by equation (3.153), or they have the same normal curvature, $k_{n1} = k_{n2}$. It is easy to obtain the normal curvature k_{n1} by equation (3.48) after the Frenet frame $\{\boldsymbol{R}_1; \boldsymbol{\alpha}, \boldsymbol{\beta}, \boldsymbol{\gamma}\}$ is rotated around the vector $\boldsymbol{\alpha}$ with an arbitrary angle θ, to be a Darboux frame $\{\boldsymbol{R}_1; \boldsymbol{\alpha}, \boldsymbol{n}, \boldsymbol{v}\}$. Then, a spherical surface $\sum_2$ is constructed by the normal curvature k_{n1}, whose center and radius changes with the variation of the rotation angle θ.

Third order: For a curve Γ_1 with four consecutive points including point P, the Frenet frame $\{\boldsymbol{R}_1; \boldsymbol{\alpha}, \boldsymbol{\beta}, \boldsymbol{\gamma}\}$, the curvature k_1 and its derivative $\mathrm{d}k_1/\mathrm{d}s$, and the torsion τ_1, can naturally be determined. If the curve Γ_1 contacts a spherical surface $\sum_2$ at the point P in third order along a curve Γ_2 of $\sum_2$, Γ_2 has the properties of a spherical curve, or k_{n2} = const. and $\tau_{g2} = 0$. Taking into account equations (3.151) and (3.154), we simplify the contact conditions as

$$\frac{\mathrm{d}k_{n1}}{\mathrm{d}s} = 0,\ \tau_{g1} = 0 \tag{3.156}$$

Substituting equations (3.48) and (3.49) into the above equations, we have a rotation angle θ for the Frenet frame $\{\boldsymbol{R}_1; \boldsymbol{\alpha}, \boldsymbol{\beta}, \boldsymbol{\gamma}\}$ to rotate around the vector $\boldsymbol{\alpha}$ and become a Darboux frame $\{\boldsymbol{R}_1; \boldsymbol{\alpha}, \boldsymbol{n}, \boldsymbol{v}\}$, or

$$\tan\theta = -\frac{\mathrm{d}k_1/\mathrm{d}s}{k_1\tau_1},\ \frac{\mathrm{d}\theta}{\mathrm{d}s} = -\tau_1 \tag{3.157}$$

Thus, the spherical surface $\sum_2$, which contacts the curve Γ_1 in third order, can be uniquely determined. It is called ***curvature sphere***, just the ***osculating sphere*** in differential geometry [3]. The invariants in equations (3.156), $\frac{\mathrm{d}k_{n1}}{\mathrm{d}s}$ and τ_{g1}, are defined as the ***spherical curvature*** of a curve Γ_1 at a contact point.

Based on equation (3.82), the curvature sphere $\sum_2$ of Γ_1, or its center and radius, can be represented as

$$\begin{cases} \boldsymbol{R}_0 = \boldsymbol{R}_1 + r_0 \boldsymbol{n} \\ r_0 = 1/k_{n1} \end{cases} \tag{3.158}$$

Substituting equations (3.47), (3.48), and (3.157) into the above equation, we can get another expression in the Frenet frame $\{\boldsymbol{R}_1; \boldsymbol{\alpha}, \boldsymbol{\beta}, \boldsymbol{\gamma}\}$ as

$$\boldsymbol{R}_0 = \boldsymbol{R}_1 + \frac{1}{k_{n1}}\boldsymbol{n} = \boldsymbol{R}_1 + \frac{1}{k_1}\boldsymbol{\beta} - \frac{\mathrm{d}k_1/\mathrm{d}s}{{k_1}^2\tau_1}\boldsymbol{\gamma} \tag{3.159}$$

The radius of the curvature sphere $\sum_2$ is $r_0 = \sqrt{\frac{1}{k_1^2} + \frac{(\mathrm{d}k_1/\mathrm{d}s)^2}{k_1^4\tau_1^2}}$.

Higher order: A space curve Γ_1 contacting a spherical surface $\sum_2$ in fourth order or higher is referred to as higher-order contact. The contact conditions increase as the contact order goes up. For a curve Γ_1 with five consecutive points including point P, the Frenet frame $\{\boldsymbol{R}_1; \boldsymbol{\alpha}, \boldsymbol{\beta}, \boldsymbol{\gamma}\}$, the curvature k_1, $\mathrm{d}k_1/\mathrm{d}s$, $\mathrm{d}^2k_1/\mathrm{d}s^2$, and the torsion τ_1, $\mathrm{d}\tau_1/\mathrm{d}s$ can naturally be obtained. Γ_1 contacts a spherical surface $\sum_2$ along a curve Γ_2 in fourth order. Γ_2 has the properties $\frac{\mathrm{d}k_{n2}}{\mathrm{d}s} = 0$, $\frac{\mathrm{d}^2k_{n2}}{\mathrm{d}s^2} = 0$, $\tau_{g2} = 0$, and $\frac{\mathrm{d}\tau_{g2}}{\mathrm{d}s} = 0$. Based on equations (3.151) to (3.155), all contact conditions for fourth order can be simplified as

$$\frac{\mathrm{d}k_{n1}}{\mathrm{d}s} = 0,\ \frac{\mathrm{d}^2k_{n1}}{\mathrm{d}s^2} = 0,\ \tau_{g1} = 0,\ \frac{\mathrm{d}\tau_{g1}}{\mathrm{d}s} = 0 \tag{3.160}$$

The Frenet frame $\{\boldsymbol{R}_1; \boldsymbol{\alpha}, \boldsymbol{\beta}, \boldsymbol{\gamma}\}$ can be rotated around the vector $\boldsymbol{\alpha}$ to become the Darboux frame $\{\boldsymbol{R}_1; \boldsymbol{\alpha}, \boldsymbol{n}, \boldsymbol{v}\}$ by a rotation angle θ. The normal curvature k_{n1}, geodesic curvature k_{g1}, and the geodesic torsion τ_{g1} of Γ_1 at point P can be obtained, and have to satisfy equations (3.160). Substituting equation (3.48) into equation (3.160), we have

$$\begin{cases} \dfrac{\mathrm{d}k_{n1}}{\mathrm{d}s} = \dfrac{\mathrm{d}k_1}{\mathrm{d}s}\cos\theta - k_1\dfrac{\mathrm{d}\theta}{\mathrm{d}s}\sin\theta = 0,\ \tau_{g1} = \tau_1 + \dfrac{\mathrm{d}\theta}{\mathrm{d}s} = 0 \\ \dfrac{\mathrm{d}^2k_{n1}}{\mathrm{d}s^2} = \dfrac{\mathrm{d}^2k_1}{\mathrm{d}s^2}\cos\theta - 2\dfrac{\mathrm{d}k_1}{\mathrm{d}s}\dfrac{\mathrm{d}\theta}{\mathrm{d}s}\sin\theta - k_1\left(\dfrac{\mathrm{d}\theta}{\mathrm{d}s}\right)^2\cos\theta - k_1\dfrac{\mathrm{d}^2\theta}{\mathrm{d}s^2}\sin\theta = 0 \\ \dfrac{\mathrm{d}\tau_{g1}}{\mathrm{d}s} = \dfrac{\mathrm{d}\tau_1}{\mathrm{d}s} + \dfrac{\mathrm{d}^2\theta}{\mathrm{d}s^2} = 0 \end{cases} \tag{3.161}$$

in which three variables, θ, $\frac{\mathrm{d}\theta}{\mathrm{d}s}$, and $\frac{\mathrm{d}^2\theta}{\mathrm{d}s^2}$, have to hold the above four equations true simultaneously, or they can be eliminated, that is

$$k_1\tau_1\frac{\mathrm{d}^2k_1}{\mathrm{d}s^2} - 2\left(\frac{\mathrm{d}k_1}{\mathrm{d}s}\right)^2\tau_1 - k_1\frac{\mathrm{d}k_1}{\mathrm{d}s}\frac{\mathrm{d}\tau_1}{\mathrm{d}s} - k_1^2\tau_1^3 = 0 \tag{3.162}$$

This indicates the condition for a curve Γ_1 to contact a spherical surface $\sum_2$ in fourth order, which means that a curve can't contact a spherical surface at an arbitrary point in fourth order unless the curve holds equation (3.162) true at the contact point. This is just the condition $\frac{\tau_1}{k_1} + \frac{\mathrm{d}}{\mathrm{d}s}\left(\frac{1}{\tau_1 k_1^2}\frac{\mathrm{d}k_1}{\mathrm{d}s}\right) = 0$ for a curve to be a spherical curve in terms of invariants, curvature, and torsion, by the Frenet frame $\{\boldsymbol{R}_1; \boldsymbol{\alpha}, \boldsymbol{\beta}, \boldsymbol{\gamma}\}$ [3], while Theorem 3.3 gives another expression in terms of invariants, the geodesic curvature, and the geodesic torsion by the Darboux frame $\{\boldsymbol{R}_1; \boldsymbol{\alpha}, \boldsymbol{n}, \boldsymbol{v}\}$. They are equivalent, but the former is beneficial in analyzing global properties and the latter is suitable for analyzing the local properties of the curve at a contacting point.

3.6.2.2 Cylindrical Curvature of a Curve and its Higher-order Curvatures

A curve Γ_1 contacts a cylindrical surface $\sum_2$ at a point with several consecutive common points, or they contact along a curve Γ_2 on $\sum_2$, which implies that there are three preconditions, the local properties of both curves Γ_1 and Γ_2, and the contacting conditions. The local properties of curve Γ_1 are expressed by invariants in its Frenet frame $\{\boldsymbol{R}_1; \boldsymbol{\alpha}, \boldsymbol{\beta}, \boldsymbol{\gamma}\}$, such as the curvature k_1, the torsion τ_1, and their derivatives. The local properties of curve Γ_2 are described by invariants in its Darboux frame $\{\boldsymbol{R}_1; \boldsymbol{\alpha}, \boldsymbol{n}, \boldsymbol{v}\}$, k_{n2}, k_{g2}, τ_{g2}, and their derivatives, which satisfy the condition equation (3.92) of a cylindrical curve, mentioned in Section 3.4.2. The contacting conditions are revealed as equations (3.153) to (3.155) in terms of invariants of curves, which are currently not sufficient for the curve Γ_2 or a cylindrical surface $\sum_2$ to be determined, although they set up a correlation of two curves Γ_1 and Γ_2. In particular, they are not as concise and convenient as those for a curve contacting a spherical surface, which leads to more complex and broad solutions.

For convenience of solving the contacting conditions, only the special case, the normal vector $\boldsymbol{\beta}_1$ of Γ_1 coincides with the normal vector $\boldsymbol{\beta}_2$ of Γ_2 at the contact point, is taken into account and discussed as follows. In such a situation, both vectors $\boldsymbol{\beta}_1$ and $\boldsymbol{\beta}_2$ have the same inclination angle θ with the normal vector $\boldsymbol{n}$ of the cylindrical surface $\sum_2$, or the Frenet frames $\{\boldsymbol{R}_1; \boldsymbol{\alpha}, \boldsymbol{\beta}, \boldsymbol{\gamma}\}$ of both curve Γ_1 and curve Γ_2 coincide with each other. The geodesic curvature of curve Γ_2 and its derivative can be derived from equation (3.48) as

$$k_{g2} = k_{n2} \cdot \tan\theta, \quad \frac{\mathrm{d}k_{g2}}{\mathrm{d}s} = \frac{\mathrm{d}k_{n2}}{\mathrm{d}s}\tan\theta + \frac{k_{n2}}{\cos^2\theta}\frac{\mathrm{d}\theta}{\mathrm{d}s} \tag{3.163}$$

Based on the condition for second order contact, or equation (3.153), we have

$$k_{g2} = k_{n2}\tan\theta = k_{n1}\tan\theta = k_{g1} \tag{3.164}$$

Combining equation (3.151), we simplify the condition equation (3.154) for third order contact, that is

$$\frac{\mathrm{d}k_{n2}}{\mathrm{d}s} = \frac{\mathrm{d}k_{n1}}{\mathrm{d}s} \tag{3.165}$$

Substituting the above equation and equation (3.164) into equation (3.49), and combining it with equation (3.151), we get

$$\frac{\mathrm{d}k_{g2}}{\mathrm{d}s} = \frac{\mathrm{d}k_{n2}}{\mathrm{d}s}\tan\theta + \frac{k_{n2}}{\cos^2\theta}\frac{\mathrm{d}\theta}{\mathrm{d}s} = \frac{\mathrm{d}k_{n1}}{\mathrm{d}s}\tan\theta + \frac{k_{n1}}{\cos^2\theta}\frac{\mathrm{d}\theta}{\mathrm{d}s} = \frac{\mathrm{d}k_{g1}}{\mathrm{d}s} \tag{3.166}$$

The condition equation (3.155) for fourth order contact can be simplified as

$$\frac{\mathrm{d}^2 k_{n2}}{\mathrm{d}s^2} = \frac{\mathrm{d}^2 k_{n1}}{\mathrm{d}s^2} \tag{3.167}$$

Hence, the conditions for the third and fourth order contact have been converted into simple equations; or the normal curvature and its derivatives of curve Γ_1 are equal to those of curve Γ_2 at the corresponding contact order, and their application cases are discussed as follows.

It is not necessary to discuss the zero order, first order and second order contact of a curve Γ_1 and a cylindrical surface $\sum_2$ for the corresponding cylindrical surface $\sum_2$ at the contact point can be arbitrarily determined.

Third order: A curve Γ_1 at point P, with four consecutive points, and the Frenet frame $\{\boldsymbol{R}_1; \boldsymbol{\alpha}, \boldsymbol{\beta}, \boldsymbol{\gamma}\}$, k_1, τ_1, and $\mathrm{d}k_1/\mathrm{d}s$, contacts a cylindrical surface $\sum_2$ in the third order. Combining the contact conditions, equations (3.151), (3.153), (3.165), and (3.92), we have four condition equations for the third order contact:

$$k_{n1} = k_{n2}, \quad \frac{\mathrm{d}k_{n1}}{\mathrm{d}s} = \frac{\mathrm{d}k_{n2}}{\mathrm{d}s}, \quad \tau_{g2} = \tau_{g1}, \quad \frac{\mathrm{d}k_{n2}}{\mathrm{d}s} = -2k_{g2}\tau_{g2} \tag{3.168}$$

Five parameters, θ, $\frac{\mathrm{d}\theta}{\mathrm{d}s}$, k_{n2}, $\frac{\mathrm{d}k_{n2}}{\mathrm{d}s}$, and τ_{g2}, have to be solved for a cylindrical surface $\sum_2$. A rotation angle θ is freely given, or the Frenet frame $\{\boldsymbol{R}_1; \boldsymbol{\alpha}, \boldsymbol{\beta}, \boldsymbol{\gamma}\}$ is rotated around the vector $\boldsymbol{\alpha}$ by angle θ to be the moving frame $\{\boldsymbol{R}_1; \boldsymbol{\alpha}, \boldsymbol{n}, \boldsymbol{v}\}$, and $\frac{\mathrm{d}\theta}{\mathrm{d}s}$ can be derived from the last expression of equation (3.168) as

$$\frac{\mathrm{d}\theta}{\mathrm{d}s} = \frac{\cot\theta \cdot \mathrm{d}k_1/\mathrm{d}s - 2k_1\tau_1}{3k_1} \tag{3.169}$$

The other parameters, k_{n2}, $\frac{\mathrm{d}k_{n2}}{\mathrm{d}s}$, and τ_{g2}, are obtained by equations (3.48) and (3.49) and the first three expression of equation (3.168). As a result, a lot of cylindrical surfaces can be determined by equation (3.95), which contacts the curve Γ_1 in third order.

Fourth order: A curve Γ_1 at point P, with five consecutive points, and the Frenet frame $\{\boldsymbol{R}_1; \boldsymbol{\alpha}, \boldsymbol{\beta}, \boldsymbol{\gamma}\}$, k_1, τ_1, and $\mathrm{d}k_1/\mathrm{d}s$, $\mathrm{d}^2k_1/\mathrm{d}s^2$, contacts a cylindrical surface $\sum_2$ in the fourth order. All the eight condition equations for the fourth-order contact, through equations (3.92), (3.151), (3.167), and (3.168), are

$$\begin{cases} k_{n1} = k_{n2}, \quad \dfrac{\mathrm{d}k_{n1}}{\mathrm{d}s} = \dfrac{\mathrm{d}k_{n2}}{\mathrm{d}s}, \quad \dfrac{\mathrm{d}^2k_{n1}}{\mathrm{d}s^2} = \dfrac{\mathrm{d}^2k_{n2}}{\mathrm{d}s^2}, \quad \tau_{g1} = \tau_{g2}, \quad \dfrac{\mathrm{d}\tau_{g1}}{\mathrm{d}s} = \dfrac{\mathrm{d}\tau_{g2}}{\mathrm{d}s} \\ \dfrac{\mathrm{d}k_{n2}}{\mathrm{d}s} = -2k_{g2}\tau_{g2}, \quad \dfrac{\mathrm{d}^2k_{n2}}{\mathrm{d}s^2} = -2\left(\dfrac{\mathrm{d}k_{g2}}{\mathrm{d}s}\tau_{g2} + \dfrac{\mathrm{d}\tau_{g2}}{\mathrm{d}s}k_{g2}\right), \quad \dfrac{\mathrm{d}\tau_{g2}}{\mathrm{d}s} = -\dfrac{k_{g2}(\tau_{g2}^2 - k_{n2}^2)}{k_{n2}} \end{cases} \tag{3.170}$$

in which there are eight parameters, k_{n2}, $\frac{\mathrm{d}k_{n2}}{\mathrm{d}s}$, $\frac{\mathrm{d}^2k_{n2}}{\mathrm{d}s^2}$, τ_{g2}, $\frac{\mathrm{d}\tau_{g2}}{\mathrm{d}s}$, θ, $\frac{\mathrm{d}\theta}{\mathrm{d}s}$, and $\frac{\mathrm{d}^2\theta}{\mathrm{d}s^2}$ to be determined. Based on equation (3.169) and the last expression of the above equation, we have

$$\frac{\mathrm{d}^2\theta}{\mathrm{d}s^2} = \frac{\left(\dfrac{\mathrm{d}k_1}{\mathrm{d}s}\right)^2\cos^2\theta + k_1^2\tau_1^2\sin^2\theta + 2k_1\tau_1\dfrac{\mathrm{d}k_1}{\mathrm{d}s}\sin\theta\cos\theta - 9k_1^4\sin^2\theta\cos^2\theta}{9k_1^2\sin\theta\cos\theta} - \frac{\mathrm{d}\tau_1}{\mathrm{d}s} \tag{3.171}$$

Substituting equations (3.169) and (3.171) into the seventh expression of equation (3.170), we get a sextic algebraic equation of the rotation angle θ as

$$
\begin{aligned}
-9k_1^4\sin^4\theta\cos^2\theta - 3k_1^2\frac{d\tau_1}{ds}\sin^3\theta\cos\theta + \left[4\left(\frac{dk_1}{ds}\right)^2 - 3k_1\frac{d^2k_1}{ds^2}\right]\sin^2\theta\cos^2\theta \\
+ k_1^2\tau_1^2\sin^4\theta - 2\frac{dk_1}{ds}k_1\tau_1\sin\theta\cos^3\theta + \left(\frac{dk_1}{ds}\right)^2\cos^2\theta = 0
\end{aligned}
\tag{3.172}
$$

Theoretically, at most six rotation angles θ can be solved, corresponding to six cylindrical surfaces, which contact the given curve at the contact point with five common consecutive points. The invariants in equation (3.170) are referred to as the ***cylindrical curvature*** of the curve Γ_1 at the point P, and the corresponding cylinder surfaces are called ***curvature cylindrical surfaces***. In Chapter 6 for the infinitesimally separated five positions and Chapter 7 for the finitely separated five positions, the numerical examples demonstrate that there are no more than six curvature cylinder surfaces. This is identified from the usually unique curvature of a curve at a point, such as the curvature circle and the curvature sphere.

For any rotation angle θ solved from equation (3.172), the normal curvature k_{n1}, geodesic curvature k_{g1}, and the geodesic torsion τ_{g1} of the curve Γ_1 at the point P can be obtained. By equation (3.95), the cylindrical surface $\sum_2$ with the axis and the radius, which contacts Γ_1 to fourth order, can be represented by the moving frame $\{\boldsymbol{R}_1; \boldsymbol{\alpha}, \boldsymbol{n}, \boldsymbol{v}\}$ as

$$
\begin{cases}
\boldsymbol{R}_a = \boldsymbol{R}_1 + r_0\boldsymbol{n} + \mu\boldsymbol{a} \\
r_0 = k_{n1}/\left(k_{n1}^2 + \tau_{g1}^2\right) \\
\boldsymbol{a} = \dfrac{\tau_{g1}\boldsymbol{\alpha} + k_{n1}\boldsymbol{v}}{\sqrt{k_{n1}^2 + \tau_{g1}^2}}
\end{cases}
\tag{3.173}
$$

Combining equations (3.47), (3.48), and (3.173), we can describe the direction and position of the axis and the radius of the cylindrical surface $\sum_2$ by the Frenet frame $\{\boldsymbol{R}_1; \boldsymbol{\alpha}, \boldsymbol{\beta}, \boldsymbol{\gamma}\}$ as

$$
\begin{cases}
\boldsymbol{R}_a = \boldsymbol{R}_1 + r_0(\cos\theta\cdot\boldsymbol{\beta} + \sin\theta\cdot\boldsymbol{\gamma}) + \mu\boldsymbol{a} \\
r_0 = \dfrac{9k_1^3\sin^2\theta\cos\theta}{(\cos\theta\cdot dk_1/ds + \sin\theta\cdot k_1\tau_1)^2 + 9k_1^4\sin^2\theta\cos^2\theta} \\
\boldsymbol{a} = \dfrac{(\cos\theta\cdot dk_1/ds + \sin\theta\cdot k_1\tau_1)\boldsymbol{\alpha} + 3k_1^2\sin\theta\cos\theta(-\sin\theta\cdot\boldsymbol{\beta} + \cos\theta\cdot\boldsymbol{\gamma})}{\sqrt{(\cos\theta\cdot dk_1/ds + \sin\theta\cdot k_1\tau_1)^2 + 9k_1^4\sin^2\theta\cos^2\theta}}
\end{cases}
\tag{3.174}
$$

Higher order: A space curve Γ_1 contacting a cylindrical surface $\sum_2$ in fifth or higher order is referred to as higher-order contact. According to the contact order, all of the equations, including the simplified equations of the contact conditions, two constraint equations of a cylindrical curve, and equation (3.151), can be solved for the invariants. For instance, when the curve Γ_1 contacts a cylindrical surface $\sum_2$ in fifth order, based on condition equation (3.170) for fourth order contact, the additional equations are

$$\begin{cases} \dfrac{d^3k_{n1}}{ds^3} = \dfrac{d^3k_{n2}}{ds^3}, \dfrac{d^2\tau_{g1}}{ds^2} = \dfrac{d^2\tau_{g2}}{ds^2}, \dfrac{d^3k_{n1}}{ds^3} = -2\left(\dfrac{d^2k_{g1}}{ds^2}\tau_{g1} + \dfrac{d^2\tau_{g1}}{ds^2}k_{g1} + 2\dfrac{dk_{g1}}{ds}\dfrac{d\tau_{g1}}{ds}\right) \\ \dfrac{d^2\tau_{g1}}{ds^2} = -\dfrac{1}{k_{n1}}\left[\dfrac{dk_{g1}}{ds}\left(\tau_{g1}^2 - k_{n1}^2\right) + k_{g1}\left(2\tau_{g1}\dfrac{d\tau_{g1}}{ds} - 2k_{n1}\dfrac{dk_{n1}}{ds}\right)\right] + \dfrac{k_{g1}}{k_{n1}^2}(\tau_{g1}^2 - k_{n1}^2)\dfrac{dk_{n1}}{ds} \end{cases} \tag{3.175}$$

in which three parameters are newly introduced, $\frac{d^3k_{n2}}{ds^3}$, $\frac{d\tau^2{}_{g2}}{ds^2}$, and $\frac{d^3\theta}{ds^3}$, that can be eliminated by the above equations, leading to a compatible equation. It is only when the curve satisfies the compatible equation that it contacts a cylindrical surface in fifth order. When the contact order increases by one, there will be three more parameters and four more equations, while the derivative of angle θ goes up by one order. This is not discussed any further, since its complexity increases rapidly.

In particular, a curve Γ_1 contacts the cylindrical surface $\sum_2$ along the curve Γ_2, while the curve Γ_2 is just a cylindrical helix on $\sum_2$. Based on Corollary 1 to Theorem 3.4, the cylindrical helix Γ_2 has special geometrical properties, such as k_{g2} is zero, k_{n2} and τ_{g2} are both constant. Based on equations (3.153) to (3.155), the contact conditions can be listed as follows.

Second order: $k_{n1} = k_{n2},\ k_{g1} = 0$

Third order: $k_{n2} = k_{n1},\ k_{g1} = 0,\ \dfrac{dk_{g1}}{ds} = 0,\ \dfrac{dk_{n1}}{ds} = 0,\ \tau_{g2} = \tau_{g1}$

Fourth order: $k_{n1} = k_{n2},\ k_{g1} = 0,\ \dfrac{dk_{g1}}{ds} = 0,\ \dfrac{dk_{n1}}{ds} = 0,\ \dfrac{d^2k_{g1}}{ds^2} = 0,\ \dfrac{d^2k_{n1}}{ds^2} = 0,$
$\frac{d\tau_{g1}}{ds} = 0$

The position and parameters of Γ_2 can be determined by equations (3.95) and (3.96).

As another special case, the curve Γ_2, is just a circle on the cylindrical surface. Based on Corollary 2 to Theorem 3.4, the geodesic curvature k_{g2} and the geodesic torsion τ_{g2} are both zero, the normal curvature k_{n2} is constant, and the contact conditions are given as follows.

Second order: $k_{n1} = k_{n2},\ k_{g1} = 0$

Third order: $k_{n1} = k_{n2},\ k_{g1} = 0,\ \dfrac{dk_{g1}}{ds} = 0,\ \dfrac{dk_{n1}}{ds} = 0,\ \tau_{g1} = 0$

Fourth order: $k_{n1} = k_{n2},\ k_{g1} = 0,\ \dfrac{dk_{g1}}{ds} = 0,\ \dfrac{dk_{n1}}{ds} = 0,\ \tau_{g1} = 0,\ \dfrac{d^2k_{g1}}{ds^2} = 0,$
$\frac{d^2k_{n1}}{ds^2} = 0,\ \frac{d\tau_{g1}}{ds} = 0$

As is well known, a curve Γ_1 contacts a circle Γ_2 in the third or higher order only when it has a stationary curvature at the contact point.

3.7 Generalized Curvature of Ruled Surfaces

In Section 3.5, the global properties of the ruled surfaces (or the necessary and sufficient conditions for a ruled surface to be a constraint ruled surface) are discussed. For a ruled surface, which is not a constraint ruled surface globally but locally, how do we evaluate the approximation? Based on the conception of the generalized curvature of a curve, the local approximation between a ruled surface and a constraint ruled surface can be analyzed by their consecutive common rulings. The generalized curvature of the ruled surface is presented for the approximation of two ruled surfaces.

As mentioned in the previous section, two curves are tangent to each other if they have two consecutive common points. We call this case "first-order contact," or contact in the first order. Then, nth-order contact is defined as two curves having $n+1$ common points at infinitesimally separated positions. The concepts will be extended to the contact order between two ruled surfaces. A ruled surface is traced by the straight line in spatial movement; the ruling is an element of a ruled surface in kinematics, just like a point of a curve. Consequently, two ruled surfaces contact along a ruling in nth-order when they have $n+1$ common rulings at infinitesimally separated positions. This is defined as nth-order contact. If two ruled surfaces have only one common ruling, they intersect each other or they contact to zero order, there is no need to discuss this case in the book.

3.7.1 Tangent Conditions

If two ruled surfaces have two common rulings at infinitesimally separated positions, or they are tangent along the common ruling, this means that the two ruled surfaces have the same normal at each point along the common ruling. It is a very strict condition for two arbitrary ruled surfaces.

Let two ruled surfaces be $\sum_L$ and $\sum_L^*$ respectively, we write their equations as

$$\sum\nolimits_L : \boldsymbol{R} = \boldsymbol{\rho} + \mu \boldsymbol{l},\ \sum\nolimits_L^* : \boldsymbol{R}^* = \boldsymbol{\rho}^* + \mu^* \boldsymbol{l}^* \tag{3.176}$$

where $\boldsymbol{\rho}$ and $\boldsymbol{\rho}^*$ are the position vectors of the striction points of $\sum_L$ and $\sum_L^*$ respectively. Then, two Frenet frames $\{\boldsymbol{\rho}; \boldsymbol{e}_1, \boldsymbol{e}_2, \boldsymbol{e}_3\}$ and $\{\boldsymbol{\rho}^*; \boldsymbol{e}_1{}^*, \boldsymbol{e}_2{}^*, \boldsymbol{e}_3{}^*\}$ are set up on $\sum_L$ and $\sum_L^*$. As a result, the construction parameters α, β, γ and α^*, β^*, γ^* can be derived through equation (3.58), and they are functions of σ or σ^*. By equation (3.53), the unit normal vectors of $\sum_L$ and $\sum_L^*$ at any point (σ, μ) or (σ^*, μ^*) are respectively expressed as

$$\boldsymbol{n} = \frac{\mu \boldsymbol{e}_3 - \gamma \boldsymbol{e}_2}{\sqrt{\mu^2 + \gamma^2}},\ \boldsymbol{n}^* = \frac{\mu^* \boldsymbol{e}_3{}^* - \gamma^* \boldsymbol{e}_2{}^*}{\sqrt{\mu^{*2} + \gamma^{*2}}} \tag{3.177}$$

The tangent conditions between $\sum_L$ and $\sum_L^*$ can be described by the following theorem.

Theorem 3.8 *Two non-developable ruled surfaces $\sum_L$ and $\sum_L^*$ contact along a common ruling to the first order if and only if (1) $\gamma = \gamma^*$, (2) $\boldsymbol{\rho} = \boldsymbol{\rho}^*$, and (3) each axis of the Frenet frame of $\sum_L$ is collinear with the corresponding axis of $\sum_L^*$, that is $\boldsymbol{e}_i \times \boldsymbol{e}_i^* = 0$ $(i = 1, 2, 3)$.*

Proof. Necessity: Suppose two non-developable ruled surfaces $\sum_L$ and $\sum_L^*$ contact along a common ruling corresponding to σ and σ^*, we have $\boldsymbol{e}_1 \times \boldsymbol{e}_1{}^* = 0$. $\boldsymbol{\rho}$ and $\boldsymbol{\rho}^*$ are respectively the position vectors of the striction points of $\sum_L$ and $\sum_L^*$. Let the distance between the two striction points of $\sum_L$ and $\sum_L^*$ be h. That is

$$\mu^* = \mu - h \tag{3.178}$$

$\boldsymbol{e}_2$ and $\boldsymbol{e}_3$ ($\boldsymbol{e}_2{}^*$ and $\boldsymbol{e}_3{}^*$) are perpendicular to $\boldsymbol{e}_1(\boldsymbol{e}_1{}^*)$, thus

$$\boldsymbol{e}_2{}^* = \cos\theta \boldsymbol{e}_2 + \sin\theta \boldsymbol{e}_3,\ \boldsymbol{e}_3{}^* = -\sin\theta \boldsymbol{e}_2 + \cos\theta \boldsymbol{e}_3 \tag{3.179}$$

Since $\sum_L$ contacts $\sum_L^*$ along the common ruling in the first order, they are tangent along the common ruling, or $\boldsymbol{n} \times \boldsymbol{n}^* = 0$. Substituting equation (3.177) into this vector equation, we have

$$\mu\mu^* \boldsymbol{e}_3 \times \boldsymbol{e}_3{}^* - \gamma\gamma^* \boldsymbol{e}_2 \times \boldsymbol{e}_2{}^* - \mu\gamma^* \boldsymbol{e}_3 \times \boldsymbol{e}_2{}^* - \mu^*\gamma \boldsymbol{e}_2 \times \boldsymbol{e}_3{}^* = 0 \tag{3.180}$$

By substituting equations (3.178) and (3.179) into the above equation, it can be simplified as

$$\tan\theta[\mu(\mu - h) + \gamma\gamma^*] + \mu(\gamma - \gamma^*) - h\gamma = 0 \tag{3.181}$$

The above equation must hold true for any value μ, which leads to

$$\gamma = \gamma^*,\ h = \mu - \mu^* = 0,\ \theta = 0 \text{ or } \pi \tag{3.182}$$

These three expressions are equivalent to the three conditions of Theorem 3.8.

Sufficiency: Suppose the three conditions of Theorem 3.8 hold true, that is $\gamma = \gamma^*,\ \mu = \mu^*$, $\boldsymbol{e}_i \times \boldsymbol{e}_i{}^* = 0,\ i = 1,\ 2,\ 3$. Then substituting them into $\boldsymbol{n} \times \boldsymbol{n}^*$, we have

$$\boldsymbol{n} \times \boldsymbol{n}^* = \frac{\mu \boldsymbol{e}_3 - \gamma \boldsymbol{e}_2}{\sqrt{\mu^2 + \gamma^2}} \times \frac{\mu^* \boldsymbol{e}_3{}^* - \gamma^* \boldsymbol{e}_2{}^*}{\sqrt{\mu^{*2} + \gamma^{*2}}} \tag{3.183}$$

The result of the above equation is zero for any value μ, which implies that $\sum_L$ and $\sum_L^*$ contact to the first order along the ruling.

Deriving again in the same way, but now for two developable surfaces ($\gamma = \gamma^* = 0$), we have:

Corollary 3.3 *Two developable surfaces contact to first order along a common ruling if and only if* $\boldsymbol{e}_i \times \boldsymbol{e}_i^* = 0$ ($i = 1,\ 2,\ 3$).

Corollary 3.4 *A developable surface and a non-developable surface cannot contact to first order along a common ruling.*

3.7.2 Generalized Curvature

Two ruled surfaces have the same principal normal at each point along the common ruling if they contact to first order. They also satisfy this condition for higher-order contact, but the distances between the two ruled surfaces in the normal direction change with the order of contact. Therefore, we take the maximum relative differences in the principal normal direction to describe higher-order contact conditions of two ruled surfaces, which is similar to that for two planar curves. Suppose the standard vector equation of a ruled surface $\sum_L$ is

$$\sum\nolimits_L : \boldsymbol{R} = \boldsymbol{\rho} + \mu \boldsymbol{l} \tag{3.184}$$

In order to get a difference vector between two points, which belongs to two ruled surfaces $\sum_L$ and $\sum_L^*$ in the neighborhood of a contact point along the common ruling, we take the increment vector $\Delta\boldsymbol{R}$ of a point $\boldsymbol{R}_{(\sigma,\mu)}$ in the neighborhood of one ruled surface $\sum_L$. That is

$$\Delta\boldsymbol{R} = \boldsymbol{R}_{(\sigma+\Delta\sigma,\mu+\Delta\mu)} - \boldsymbol{R}_{(\sigma,\mu)} \tag{3.185}$$

By use of the Taylor series, we expand the two-element vector function

$$\Delta\boldsymbol{R} = \boldsymbol{\varepsilon}_1 + \frac{1}{2}\boldsymbol{\varepsilon}_2 + \frac{1}{6}\boldsymbol{\varepsilon}_3 + \cdots \frac{1}{n!}\boldsymbol{\varepsilon}_n + \cdots \tag{3.186}$$

where, up to third order, these vectors can be written as follows:

$$\boldsymbol{\varepsilon}_1 = (\alpha \mathrm{d}\sigma + \mathrm{d}\mu)\boldsymbol{e}_1 + \mu \mathrm{d}\sigma \boldsymbol{e}_2 + \gamma \mathrm{d}\sigma \boldsymbol{e}_3$$

$$\boldsymbol{\varepsilon}_2 = \left(\frac{\mathrm{d}\alpha}{\mathrm{d}\sigma} - \mu\right)\mathrm{d}\sigma^2 \boldsymbol{e}_1 + [(\alpha - \beta\gamma)\mathrm{d}\sigma^2 + 2\mathrm{d}\sigma \mathrm{d}\mu]\boldsymbol{e}_2 + \left(\frac{\mathrm{d}\gamma}{\mathrm{d}\sigma} + \beta\mu\right)\mathrm{d}\sigma^2 \boldsymbol{e}_3$$

$$\boldsymbol{\varepsilon}_3 = \left[\left(\frac{\mathrm{d}^2\alpha}{\mathrm{d}\sigma^2} - \alpha + \beta\gamma\right)\mathrm{d}\sigma^3 - 3\mathrm{d}\sigma^2\mathrm{d}\mu\right]\boldsymbol{e}_1 + \left[2\left(\frac{\mathrm{d}\alpha}{\mathrm{d}\sigma} - \beta\frac{\mathrm{d}\gamma}{\mathrm{d}\sigma}\right) - \frac{\mathrm{d}\beta}{\mathrm{d}\sigma}\gamma - (1+\beta^2)\mu\right]\mathrm{d}\sigma^3 \boldsymbol{e}_2$$
$$+ \left[\left(\frac{\mathrm{d}^2\gamma}{\mathrm{d}\sigma^2} + \frac{\mathrm{d}\beta}{\mathrm{d}\sigma}\mu + \alpha\beta - \beta^2\gamma\right)\mathrm{d}\sigma^3 + 3\beta \mathrm{d}\sigma^2 \mathrm{d}\mu\right]\boldsymbol{e}_3$$

From the above equations, $\boldsymbol{\varepsilon}_1, \boldsymbol{\varepsilon}_2, \boldsymbol{\varepsilon}_3, \ldots, \boldsymbol{\varepsilon}_n$ are the first-order, second-order, third-order, …, nth-order infinitesimal vectors respectively. They depend on the construction parameters α, β, γ. In the same way, the increment vector $\Delta\boldsymbol{R}^*$ of a point $\boldsymbol{R}^*_{(\sigma^*,\mu^*)}$ on another ruled surface $\sum_L^*$ can also be represented by a Taylor expansion and infinitesimal vectors $\boldsymbol{\varepsilon}_1{}^*, \boldsymbol{\varepsilon}_2{}^*, \boldsymbol{\varepsilon}_3{}^*, \ldots, \boldsymbol{\varepsilon}_n{}^*$ at each order can be derived. They have the same forms as equation (3.186) and depend on α^*, β^*, γ^*.

$\sum_L$ and $\sum_L^*$ have the same Frenet frame at the common ruling, by Theorem 3.8, if they contact to first order or more. Therefore, the infinitesimal relative vectors are

$$\Delta\boldsymbol{\varepsilon}_1 = \boldsymbol{\varepsilon}_1 - \boldsymbol{\varepsilon}_1{}^*,\ \Delta\boldsymbol{\varepsilon}_2 = \boldsymbol{\varepsilon}_2 - \boldsymbol{\varepsilon}_2{}^*,\ \Delta\boldsymbol{\varepsilon}_3 = \boldsymbol{\varepsilon}_3 - \boldsymbol{\varepsilon}_3{}^*,\ \ldots \tag{3.187}$$

$\sum_L$ and $\sum_L^*$ possess the same unit normal vector $\boldsymbol{n}$ at the points of the common ruling, that is

$$\boldsymbol{n} = \frac{\mu\boldsymbol{e}_3 - \gamma\boldsymbol{e}_2}{\sqrt{\mu^2 + \gamma^2}} \tag{3.188}$$

Hence, the relative normal differences can be obtained by projecting each order infinitesimal relative vector onto the normal vector $\boldsymbol{n}$:

$$\varepsilon_{1n} = \Delta\boldsymbol{\varepsilon}_1 \cdot \boldsymbol{n},\ \varepsilon_{2n} = \Delta\boldsymbol{\varepsilon}_2 \cdot \boldsymbol{n},\ \varepsilon_{3n} = \Delta\boldsymbol{\varepsilon}_3 \cdot \boldsymbol{n}, \cdots \tag{3.189}$$

Thus, $\sum_L$ and $\sum_L^*$ contact to first order if $\boldsymbol{\varepsilon}_{1n} = 0$, $\sum_L$ and $\sum_L^*$ contact to second order if $\boldsymbol{\varepsilon}_{1n} = 0$ and $\boldsymbol{\varepsilon}_{2n} = 0$ simultaneously, and so forth.

Substituting equations (3.186) to (3.188) into equation (3.189) and simplifying, we can write the contact conditions for two ruled surfaces to any order along a common ruling.

$$\text{Second order}: \alpha = \alpha^*,\ \beta = \beta^*,\ \gamma = \gamma^*,\ \frac{d\gamma}{d\sigma} = \frac{d\gamma^*}{d\sigma} \tag{3.190}$$

$$\text{Third order}: \ \alpha = \alpha^*,\ \beta = \beta^*,\ \gamma = \gamma^*,\ \frac{d\alpha}{d\sigma} = \frac{d\alpha^*}{d\sigma},\ \frac{d\beta}{d\sigma} = \frac{d\beta^*}{d\sigma},\ \frac{d\gamma}{d\sigma} = \frac{d\gamma^*}{d\sigma},\ \frac{d^2\gamma}{d\sigma^2} = \frac{d^2\gamma^*}{d\sigma^2} \tag{3.191}$$

$$\begin{aligned} &\alpha = \alpha^*,\ \beta = \beta^*,\ \gamma = \gamma^*,\ \frac{d\alpha}{d\sigma} = \frac{d\alpha^*}{d\sigma},\ \frac{d\beta}{d\sigma} = \frac{d\beta^*}{d\sigma},\ \frac{d\gamma}{d\sigma} = \frac{d\gamma^*}{d\sigma} \\ \text{Fourth order}: \ &\frac{d^2\alpha}{d\sigma^2} = \frac{d^2\alpha^*}{d\sigma^2},\ \frac{d^2\beta}{d\sigma^2} = \frac{d^2\beta^*}{d\sigma^2},\ \frac{d^2\gamma}{d\sigma^2} = \frac{d^2\gamma^*}{d\sigma^2},\ \frac{d^3\gamma}{d\sigma^3} = \frac{d^3\gamma^*}{d\sigma^3} \end{aligned} \tag{3.192}$$

The contact condition for any higher order can be derived similarly.

It is noted that the above contact conditions are described for two ruled surfaces, rather than constraint ruled surfaces in spatial linkages. For convenience in applying the contact conditions in the kinematic geometry of spatial linkages, we combine them with the geometrical properties of constraint ruled surfaces and define the generalized curvature of a ruled surface.

Definition 3.2 For a ruled surface $\sum_L$ with ruling L, a regular ruled surface $\sum_L^*$ contacts $\sum_L$ along the ruling L in higher order, and is finitely determined by the ***invariants***, which are defined as the ***generalized curvature***, and $\sum_L^*$ is called the ***ruled surface of generalized curvature***.

For a given ruled surface $\sum_L$, a lot of ruled surfaces can be taken to approximate it at the neighborhood of a ruling L. The local properties of a ruled surface at the neighborhood of a ruling would deduce the global properties with the theoretical support of the global differential geometry. In this book, we just discuss a few cases for the contact between a ruled surface and constraint ruled surfaces of spatial linkages, such as the constant inclination ruled surface $\sum_{C'P'C}$, the constant axis ruled surface $\sum_{C'C}$, and the constant parameter ruled surface. For other more constraint ruled surfaces, such as the constant distance ruled surface $\sum_{S'C}$, their global properties need to be studied. We expect that the mathematicians can also provide theoretical foundations.

3.7.3 Constant Inclination Curvature

For a given ruled surface $\sum_L$ with ruling L, a constant inclination ruled surface $\sum_{C'PC}$ is taken to contact it along the ruling. As mentioned in Section 3.5.1, the construction parameter $\beta_{C'PC}$ is constant, or the spherical image curve of the ruling is a spherical circle, which means that the derivatives of $\beta_{C'PC}$ are zero. The contact orders can be analyzed along the common rulings by means of the contact conditions, from equations (3.190) to (3.192).

Second order: For a given ruled surface $\sum_L$ with construction parameters at a ruling L and its neighborhood, another ruled surface, a constant inclination ruled surface $\sum_{C'P'C}$, is constructed to contact $\sum_L$ along the ruling L in second order. Based on condition equation (3.190) for second order contact, the construction parameters of $\sum_{C'P'C}$ are identically equal to those of the given ruled surface $\sum_L$, or the constant inclination ruled surface $\sum_{C'P'C}$ is uniquely determined by the invariant, the equation (3.190), that is $\beta = \beta_{C'P'C}$, called a ***constant inclination curvature*** of the ruled surface $\sum_L$; $\sum_{C'P'C}$ is referred to as a ***constant inclination curvature surface*** of $\sum_L$.

Third and higher order: From the condition equation (3.191) for third order contact, the construction parameters and their derivatives of the given ruled surface $\sum_L$ are equal to the corresponding ones of the constant inclination ruled surface $\sum_{C'P'C}$. Particularly, the construction parameter $\beta_{C'P'C}$ of $\sum_{C'P'C}$ is constant, which leads to $\frac{d\beta}{d\sigma} = 0$ for a given ruled surface $\sum_L$. The higher derivatives of β should be zero for higher-order contact of $\sum_L$ and $\sum_{C'P'C}$ higher-order contact.

3.7.4 Constant Axis Curvature

In Section 3.5.2, a constant axis ruled surface $\sum_{C'C}$ has two properties, the construction parameter $\beta_{C'C}$ is constant and the invariant $\alpha_{C'C} - \beta_{C'C}\gamma_{C'C}$ is also constant. Based on the contact conditions, equations (3.190) to (3.192), we can discuss the cases for a ruled surface $\sum_L$ contacting the constant axis ruled surface $\sum_{C'C}$.

Second order: According to the local structure of a given ruled surface $\sum_L$ at the neighborhood of a ruling L, a constant axis ruled surface $\sum_{C'C}$ is built up to contact the given ruled surface $\sum_L$ along the ruling L in second order. Based on the contact conditions, equation (3.190), the constant axis ruled surface $\sum_{C'C}$ is uniquely determined by the invariants, $\beta = \beta_{C'C}$ and $\alpha - \beta\gamma = \alpha_{C'C} - \beta_{C'C}\gamma_{C'C}$, which is called the ***constant axis curvature*** of the ruled surface $\sum_L$; the constant axis ruled surface $\sum_{C'C}$ is referred to as the ***constant axis curvature surface***.

Third and higher order: For contact in third and higher order, by the contact conditions, the construction parameters and their derivatives of the given ruled surface $\sum_L$ are equal to the corresponding ones of the constant axis ruled surface $\sum_{C'C}$, so the construction parameters have to meet equations: $\frac{d\beta}{d\sigma} = 0$ and $\frac{d\alpha}{d\sigma} - \beta\frac{d\gamma}{d\sigma} = 0$ for third order; additional two equations $\frac{d^2\beta}{d\sigma^2} = 0$ and $\frac{d^2\alpha}{d\sigma^2} - \beta\frac{d^2\gamma}{d\sigma^2} = 0$ for fourth order. When the contact order increases by one, the number of additional equations increases by two. More cases in the kinematic geometry of linkages are discussed intensively in Chapter 6.

3.8 Discussion

The properties of curves and surfaces, both algebraic and geometrical properties, are the theoretical basis of the kinematic geometry of mechanisms. The algebraic properties of curves and surfaces, the classic and traditional topics, have abundant results, the well-known facts in kinematic geometry of mechanisms for instance – the degree of a curve or a surface, the intersection points of two curves, a curve and a surface, the transversal of two surfaces. The geometrical properties of curves and surfaces, classic and traditional topics too, study shapes and variations of figures, curves, and surfaces. The geometrical figures may be described in terms of algebraic equations, whose coordinate system is naturally introduced. The complexity of algebraic equations for a figure may be different since the coordinate system is chosen differently; even a simple figure will have a complicated algebraic equation if the coordinate system is not properly taken. Modern mathematics cannot distinguish clearly between classical algebra and traditional geometry. The differential geometry, based on vector algebra and vector analysis, takes the moving frame as a basic approach, and studies the shapes of a geometrical figure. In particular, the geometrical figures are viewed as the trajectories traced by points or lines, and are described in the moving frames, such as the Frenet frame $\{\boldsymbol{R}_1; \boldsymbol{\alpha}, \boldsymbol{\beta}, \boldsymbol{\gamma}\}$ and the Darboux frame $\{\boldsymbol{R}; \boldsymbol{\alpha}, \boldsymbol{n}, \boldsymbol{v}\}$, which build up the relationship among the geometrical figures, the movements, and the points or lines. The invariants and invariance of the geometrical figure, or kinematic invariants, which do not vary as the figure (rigid body) totally has a displacement, can be taken to describe the geometrical properties of a figure without the external factors. Both the curvature and the torsion of a curve, three construction parameters of a ruled surface, are all the invariants. The equations of figures comprised of invariants, such as a spherical curve and a cylindrical surface, are the invariance, introduced in differential geometry [1–3].

The geometrical properties of a curve can be described by its curvature and torsion. The geometrical properties of surfaces can also be expressed by the curvatures, such as the geodesic curvature, the normal curvature, and the geodesic torsion. These curvatures are used to reveal the local properties of curves and surfaces, referred to as the curvature theories and a long-term topic in the kinematic geometry of mechanisms, although they have been well studied in textbooks on classical differential geometry. The local geometrical properties for a curve contacting a surface are studied by Lagrange [5] and Cauchy [6]; Struik [2] introduced them by a curve contacting a spherical surface. Lee, Ravani, and Yang [7] presented contact theory for geometric continuity of parametric curves by the pole and evolute curve. As for a special kind of surface, the ruled surface, Schaaf and Ravani [8] studied the geometric continuity of a ruled surface by dual numbers. Hunt [9], Bottema and Roth [10] researched the geometrical properties of lines' trajectories by line geometry and screw theory. Veldkamp [11] studied the geometrical properties of the ruled surface and its evolute ruled surface based on the dual vector method. Yang, Kirson, and Roth [12] defined characteristic scalars, which are independent of stretch/rotation transformations, to determine the local geometrical properties of ruled surface. Köse [13] presented a method of determining a developable ruled surface through dual vector calculus. Gürsoy [14] and Köse [15] gave the pitch and angle of pitch of a closed ruled surface in terms of the integral invariants. McCarthy [16] re-derived the scalar and dual formulations of the curvature theory of a ruled surface and exposed the fundamental curvature functions that characterize the shape of a ruled surface. Park [17] provided the developable surface design problem in an optimal control setting. Ge and Ravani [18] presented a method for constructing rational Bezier line congruence and ruled surfaces suitable for CAGD based on line geometry.

The moving frame, as a significant tool in differential geometry, can be used to derive the kinematic invariants of a ruled surface [19]. McCarthy and Roth [4] derived three curvature parameters to describe the local shape of a ruled surface, which is the first reference to apply differential geometry in kinematics. Wang [20] studied extensively the invariants of some typical constraint ruled surface by means of the moving frame method of differential geometry. The sufficient and necessary conditions were given for a ruled surface to be one of the constraint ruled surfaces.

The curves and surfaces are usually formed in two ways, topology and movement. The former studies a curve or a surface from one to another, which is the field that mathematicians are interested in. The latter is the field that kinematics researchers have more interest in. A point or a line in spatial movement traces a trajectory, a curve, or a ruled surface. In particular, the movement and geometrical properties of the trajectory of a point (line) have a certain relationship, which has been a key topic researched in the kinematic geometry of mechanisms.

The constraint curves or surfaces may be expected to come from the trajectories of points or lines in spatial movement, which is a general idea to synthesize a spatial linkage. In the classical kinematic geometry of planar mechanisms, the paths or curves traced by points are compared with a line or a circle, and estimated by their corresponding curvatures. Chapter 1 introduced Burmester points and Ball points, which correspond to the constraint curve of the binary link R–R and that of the binary link P–R. It is easy and convenient to do so for the curvature and its variation of simple curves, such as a line and a circle, and is locally and instantaneously valid. Can we compare a line's trajectories with a constraint ruled surface? For example, spherical curves, cylindrical curves, and ruled surfaces? It is so complex that both the curves and the surfaces cannot be expressed by a simple curve. Therefore, for some simple constraint surfaces of spatial linkages, authors develop the local curvature of a curve at a contact point to the generalized curvature, the global properties of a surface, such as the spherical curvature, the cylindrical curvature, the constant inclination curvature, the constant axis curvature, etc., which are described by the invariance of the curves and surfaces. The generalized curvatures will be applied to analyze the kinematic geometry of both points and lines in Chapter 6. However, for the kinematic synthesis of spatial linkages, we may pay attention to the global properties of both the points' and lines' trajectories compared with the constraint curves or constraint surfaces. Here, the global properties in differential geometry are involved, so the contents are interesting but difficult. There is little literature to describe them. The authors try to study the global properties of two special constraint surfaces [21], a spherical surface and a cylindrical surface, whose adjacent structures of a point with infinitesimal consecutive points are totally the same on the surface. Each of these surfaces has only one kind of point, a ball point or a parabolic point respectively. For a ruled surface to be a constraint ruled surface, such as a constant inclination ruled surface, a constant axis ruled surface, or a constant parameters ruled surface, the sufficient and necessary conditions in terms of invariance are derived and proved, which will be applied to the kinematic synthesis of spatial linkages in Chapter 7.

Obviously, in the local properties for generalized curvature or the global properties for necessary and sufficient conditions, the current study is only an attempt and is applicable to only a few simple constraint curves and simple constraint surfaces in the linkages. More light needs to be shed on the constraint curves and constraint surfaces for kinematic synthesis of spatial linkages. Further studies are expected with more effort put in by mathematicians, and this is believed to be attractive enough if the topics are vividly described.

References

[1] M.P. Do Carmo, *Differential Geometry of Curves and Surfaces*. Prentice-Hall, Englewood Cliffs, NJ, 1976.

[2] D.J. Struik, *Lectures on Classical Differential Geometry*. Dover Publications, London, 1961.

[3] D.R. Wu, *Differential Geometry*. Higher Education Press, Beijing, 1981 (in Chinese).

[4] J.M. McCarthy and B. Roth, The curvature theory of line trajectories in spatial kinematics. *ASME Journal of Mechanical Design*, **103**(4), 718–724, 1981.

[5] J.L. Lagrange, *Theorie Des Fonctions Analytiques*. L'Imprimerie de la Republique, Paris, 1797.

[6] A.L. Cauchy, *Lecons Sur Les Applications Du Calcul Infinitesimal a la Geometrie*, V1-2. Kessinger Publishing, Whitefish, MT, 1826.

[7] C. Lee, B. Ravani, and A.T. Yang, Theory of contact for geometric continuity of parametric curves. *The Visual Computer*, **8**, 338–350, 1992.

[8] J.A. Schaaf and B. Ravani, Geometric continuity of ruled surface. *Computer Aided Geometric Design*, **15**, 289–310, 1998.

[9] K.H. Hunt, *Kinematic Geometry of Mechanisms*. Clarendon Press, Oxford, 1978.

[10] O. Bottema and B. Roth, *Theoretical Kinematics*. North-Holland, New York, 1979.
[11] G.R. Veldkamp, On the use of dual numbers, vectors and matrices in instantaneous, spatial kinematics. *Mechanism and Machine Theory*, **11**(2), 141–156, 1976.
[12] A.T. Yang, Y. Kirson, and B. Roth, On a kinematic curvature theory for ruled surfaces. Proceedings of the Fourth World Congress on the Theory of Machines and Mechanisms, Newcastle-upon-Tyne, pp. 737–742, 1975.
[13] Ö. Köse, A method of the determination of a developable ruled surface. *Mechanism and Machine Theory*, **34**(8), 1187–1193, 1999.
[14] O. Gürsoy, The dual angle of pitch of a closed ruled surface. *Mechanism and Machine Theory*, **25**(2), 131–140, 1990.
[15] Ö. Köse, Contributions to the theory of integral invariants of a closed ruled surface. *Mechanism and Machine Theory*, **32**(2), 261–277, 1997.
[16] J.M. McCarthy, On the scalar and dual formulations of the curvature theory of line trajectories. *Journal of Mechanisms, Transmissions, and Automation*, **109**(1), 101–106, 1987.
[17] F.C. Park, J. Yu, C. Chun, *et al.* Design of developable surfaces using optimal control. *ASME Journal of Mechanical Design*, **124**, 602–608, 2002.
[18] Q.J. Ge and B. Ravani, Geometric design of rational Bezier line congruences and ruled surfaces using line geometry. *Geometric Modelling*, **13**, 101–120, 1998.
[19] S. Sasaki, *Differential Geometry*. Kyolitsu Press, Tokyo, 1956 (in Japanese).
[20] D.L. Wang, J. Liu, and D.Z. Xiao, Geometrical characteristics of some typical spatial constraints. *Mechanism and Machine Theory*, **35**(10), 1413–1430, 2000.
[21] D.L. Wang, *Kinematic Differential Geometry of Mechanisms*. Dalian University of Technology, Dalian 1995 (in Chinese).

4

Spherical Kinematic Differential Geometry

The spherical movement, a special case of space movement, is usually studied by the approach of space kinematic geometry, and has three independent parameters, or three degrees of freedom in the kinematic properties, like that of plane movement, while the space movement has six independent parameters. Spherical movement can be viewed as the extension of plane movement to the spherical surface, and the spherical kinematic geometry in both form and content surely corresponds to the plane kinematic geometry. Hence, the kinematic geometry in spherical movement is a bridge from plane to space movement.

In this chapter, the spherical kinematic differential geometry is briefly introduced in the same way as the plane one, so as to not only reveal the systematicness of the approach of differential geometry and the contents of kinematic geometry of linkages, but also ensure convenience of reading by arranging the contents gradually from planar to spatial movement.

4.1 Spherical Displacement

4.1.1 General Expression

For a spherical displacement of a moving body $\sum^*$ relative to a fixed body $\sum$ from one position to another, a moving Cartesian coordinate system $\{\boldsymbol{O}_m;\boldsymbol{i}_m,\boldsymbol{j}_m,\boldsymbol{k}_m\}$ is set up in $\sum^*$ and a fixed Cartesian coordinate system $\{\boldsymbol{O}_f;\boldsymbol{i}_f,\boldsymbol{j}_f,\boldsymbol{k}_f\}$ is established in $\sum$. For convenience in describing the spherical movement, the original points of both the moving and the fixed frames are put at the center O of the spherical surface, a fixed point which $\sum^*$ just rotates around. For a rigid body in spherical motion, it rotates around three axes by three successive angular displacements, which can usually be described in two ways – RPY angles and Euler angles, as shown in Fig. 4.1. For the former, the moving body rotates around axes $\boldsymbol{i}_f$, $\boldsymbol{j}_f$, and $\boldsymbol{k}_f$ respectively by three angles α_3, α_2, and α_1, while for the later, it rotates around axes $\boldsymbol{k}_m$, $\boldsymbol{j}_m$, and $\boldsymbol{i}_m$ by three angles β_1, β_2, and β_3 in sequence. We prefer to use the dihedral angles to describe the relationship between the moving coordinate system and the fixed coordinate system by means of another coordinate system, a Cartesian coordinate system $\{\boldsymbol{O};\boldsymbol{i}_t,\boldsymbol{j}_t,\boldsymbol{k}_m\}$ whose coordinate vector $\boldsymbol{i}_t$ is

---

Kinematic Differential Geometry and Saddle Synthesis of Linkages, First Edition. Delun Wang and Wei Wang.

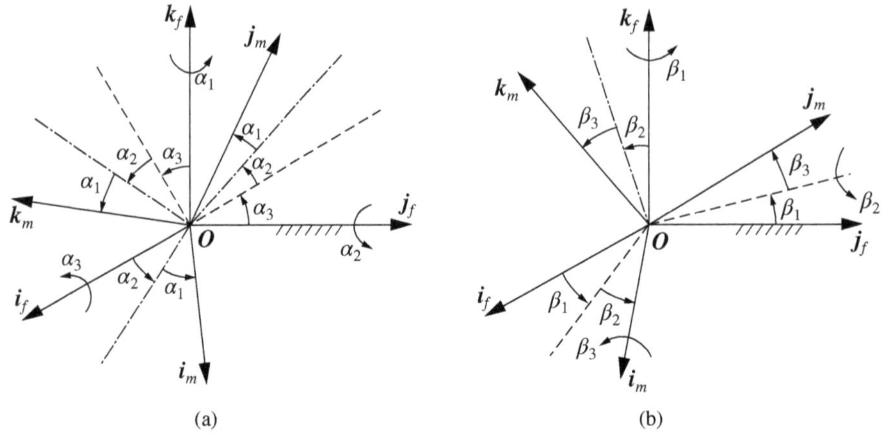

Figure 4.1 The RPY angles and Euler angles for spherical movement (a) RPY angles (b) Euler angles

determined by $\boldsymbol{i}_t = \boldsymbol{k}_m \times \boldsymbol{k}_f$ and $\boldsymbol{j}_t = \boldsymbol{k}_m \times \boldsymbol{i}_t$, or a right-hand rectangular coordinate system $\boldsymbol{i}_t$, $\boldsymbol{j}_t$, $\boldsymbol{k}_m$. The dihedral angles are defined as: two planes $\boldsymbol{i}_t - \boldsymbol{k}_m$ and $\boldsymbol{i}_t - \boldsymbol{k}_f$, or axes $\boldsymbol{k}_m$ and $\boldsymbol{k}_f$, having angle θ_1; two planes $\boldsymbol{k}_m - \boldsymbol{k}_f$ and $\boldsymbol{k}_f - \boldsymbol{i}_f$ having angle θ_2; two planes $\boldsymbol{k}_m - \boldsymbol{i}_t$ and $\boldsymbol{k}_m - \boldsymbol{i}_m$ having angle θ_3, as shown in Fig. 4.2. Hence, the relationship between $\{\boldsymbol{O};\boldsymbol{i}_m,\boldsymbol{j}_m,\boldsymbol{k}_m\}$ and $\{\boldsymbol{O};\boldsymbol{i}_f,\boldsymbol{j}_f,\boldsymbol{k}_f\}$ is

$$\begin{cases} \boldsymbol{i}_m = \left(\mathrm{c}\theta_3\mathrm{s}\theta_2 + \mathrm{c}\theta_1\mathrm{s}\theta_3\mathrm{c}\theta_2\right)\boldsymbol{i}_f + \left(-\mathrm{c}\theta_3\mathrm{c}\theta_2 + \mathrm{c}\theta_1\mathrm{s}\theta_3\mathrm{s}\theta_2\right)\boldsymbol{j}_f - \mathrm{s}\theta_1\mathrm{s}\theta_3\boldsymbol{k}_f \\ \boldsymbol{j}_m = \left(-\mathrm{s}\theta_3\mathrm{s}\theta_2 + \mathrm{c}\theta_1\mathrm{c}\theta_3\mathrm{c}\theta_2\right)\boldsymbol{i}_f + \left(\mathrm{s}\theta_3\mathrm{c}\theta_2 + \mathrm{c}\theta_1\mathrm{c}\theta_3\mathrm{s}\theta_2\right)\boldsymbol{j}_f - \mathrm{s}\theta_1\mathrm{c}\theta_3\boldsymbol{k}_f \\ \boldsymbol{k}_m = \mathrm{s}\theta_1\mathrm{c}\theta_2\boldsymbol{i}_f + \mathrm{s}\theta_1\mathrm{s}\theta_2\boldsymbol{j}_f + \mathrm{c}\theta_1\boldsymbol{k}_f \end{cases} \tag{4.1}$$

where s and c are abbreviations of the sine and cosine functions respectively.

Based on the above equations, a point P (x_{Pm}, y_{Pm}, z_{Pm}) of $\sum^*$ in $\{\boldsymbol{O};\boldsymbol{i}_m,\boldsymbol{j}_m,\boldsymbol{k}_m\}$ has rectangular coordinates (x_{Pf}, y_{Pf}, z_{Pf}) in $\{\boldsymbol{O};\boldsymbol{i}_f,\boldsymbol{j}_f,\boldsymbol{k}_f\}$, which can usually be expressed by the displacement matrix as

$$\begin{bmatrix} x_{Pf} \\ y_{Pf} \\ z_{Pf} \end{bmatrix} = \begin{bmatrix} \mathrm{c}\theta_3\mathrm{s}\theta_2 + \mathrm{c}\theta_1\mathrm{s}\theta_3\mathrm{c}\theta_2 & -\mathrm{s}\theta_3\mathrm{s}\theta_2 + \mathrm{c}\theta_1\mathrm{c}\theta_3\mathrm{c}\theta_2 & \mathrm{s}\theta_1\mathrm{c}\theta_2 \\ -\mathrm{c}\theta_3\mathrm{c}\theta_2 + \mathrm{c}\theta_1\mathrm{s}\theta_3\mathrm{s}\theta_2 & \mathrm{s}\theta_3\mathrm{c}\theta_2 + \mathrm{c}\theta_1\mathrm{c}\theta_3\mathrm{s}\theta_2 & \mathrm{s}\theta_1\mathrm{s}\theta_2 \\ -\mathrm{s}\theta_1\mathrm{s}\theta_3 & -\mathrm{s}\theta_1\mathrm{c}\theta_3 & \mathrm{c}\theta_1 \end{bmatrix} \begin{bmatrix} x_{Pm} \\ y_{Pm} \\ z_{Pm} \end{bmatrix} \tag{4.2}$$

Similar to the plane movement, a point B of $\sum^*$ is chosen as the reference point on a spherical surface of $\sum^*$, and the movement of $\sum^*$ can be described as both the movement along with the point B and the movement relative to B on the spherical surface. In particular, the reference point B is located on the coordinate axis $\boldsymbol{k}_m$, with distance R to the original point O, or its coordinates are (0, 0, R) in $\{\boldsymbol{O};\boldsymbol{i}_m,\boldsymbol{j}_m,\boldsymbol{k}_m\}$, whose coordinates in $\{\boldsymbol{O};\boldsymbol{i}_f,\boldsymbol{j}_f,\boldsymbol{k}_f\}$ are

$$x_{Bf} = R\sin\theta_1\cos\theta_2,\ y_{Bf} = R\sin\theta_1\sin\theta_2,\ z_{Bf} = R\cos\theta_1 \tag{4.3}$$

The spherical movements of $\sum^*$ relative to $\sum$ can be given in other forms, such as the spherical displacements of two points, etc. We can convert them into the form of the displacement matrix, equation (4.2). The spherical movement of $\sum^*$ may be given in continuous functions,

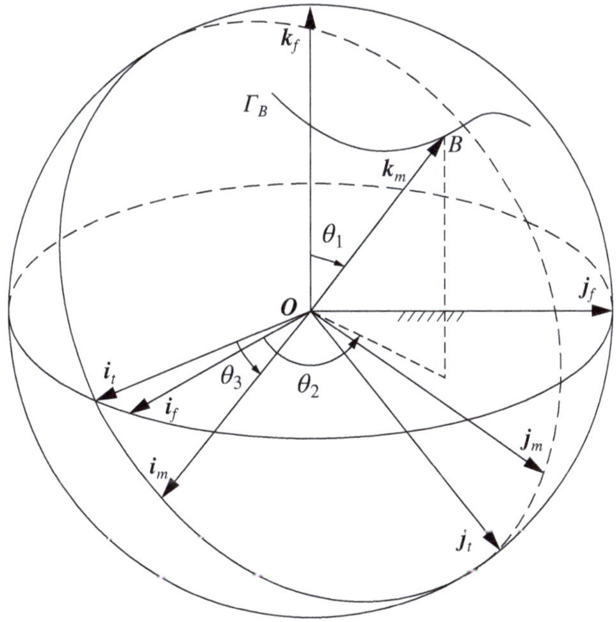

Figure 4.2 Cartesian coordinate systems in spherical movement

or discrete ones. The former corresponds to the spherical differential kinematic geometry at infinitesimally separated positions, while the latter to the spherical discrete kinematic geometry at finite separated positions.

4.1.2 Adjoint Expression

As mentioned above, the point B with rectangular coordinates $(0, 0, R)$ in $\{\boldsymbol{O}; \boldsymbol{i}_m, \boldsymbol{j}_m, \boldsymbol{k}_m\}$ traces a spherical curve Γ_B in $\{\boldsymbol{O}; \boldsymbol{i}_f, \boldsymbol{j}_f, \boldsymbol{k}_f\}$ as $\sum^*$ moves, and the curve Γ_B can be represented as

$$\Gamma_B : \boldsymbol{R}_B = x_{Bf}\boldsymbol{i}_f + y_{Bf}\boldsymbol{j}_f + z_{Bf}\boldsymbol{k}_f = -R\boldsymbol{n}(s) \tag{4.4}$$

where s is the arc length of Γ_B. $\boldsymbol{n}$ is the unit normal vector of the sphere at B and directs toward the sphere center. Based on the properties of spherical curves, the Darboux frame $\{\boldsymbol{R}_B; \boldsymbol{\alpha}, \boldsymbol{n}, \boldsymbol{v}\}$ on Γ_B, shown in Fig. 4.3, is set up as

$$\boldsymbol{\alpha} = \frac{\mathrm{d}\boldsymbol{R}_B}{\mathrm{d}s}, \ \boldsymbol{n} = -\frac{1}{R}\boldsymbol{R}_B, \ \boldsymbol{v} = -R\frac{\mathrm{d}\boldsymbol{R}_B}{\mathrm{d}s} \times \boldsymbol{R}_B \tag{4.5}$$

According to equation (3.45), the differential formulas of $\{\boldsymbol{R}_B; \boldsymbol{\alpha}, \boldsymbol{n}, \boldsymbol{v}\}$ are

$$\frac{\mathrm{d}\boldsymbol{\alpha}}{\mathrm{d}s} = \frac{1}{R}\boldsymbol{n} + k_{gB}\boldsymbol{v}, \ \frac{\mathrm{d}\boldsymbol{n}}{\mathrm{d}s} = -\frac{1}{R}\boldsymbol{\alpha}, \ \frac{\mathrm{d}\boldsymbol{v}}{\mathrm{d}s} = -k_{gB}\boldsymbol{\alpha} \tag{4.6}$$

where the invariant k_{gB} is the geodesic curvature of Γ_B.

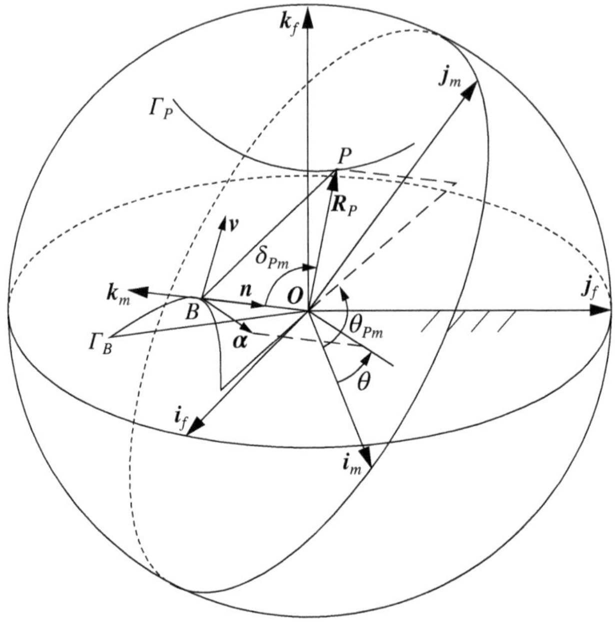

Figure 4.3 The adjoint expression of spherical movement

A point P on the same sphere as B of $\sum^*$, with rectangular coordinates (x_{Pm}, y_{Pm}, z_{Pm}) or spherical coordinates $(R, \delta_{Pm}, \theta_{Pm})$ in $\{\boldsymbol{O}; \boldsymbol{i}_m, \boldsymbol{j}_m, \boldsymbol{k}_m\}$, can be represented as

$$\begin{aligned}\boldsymbol{R}_{Pm} &= x_{Pm}\boldsymbol{i}_m + y_{Pm}\boldsymbol{j}_m + z_{Pm}\boldsymbol{k}_m \\ &= R\sin\delta_{Pm}\cos\theta_{Pm}\boldsymbol{i}_m + R\sin\delta_{Pm}\sin\theta_{Pm}\boldsymbol{j}_m + R\cos\delta_{Pm}\boldsymbol{k}_m\end{aligned} \tag{4.7}$$

A spherical curve Γ_P traced by point P in $\{\boldsymbol{O}; \boldsymbol{i}_f, \boldsymbol{j}_f, \boldsymbol{k}_f\}$ is viewed as an adjoint curve of Γ_B, while the curve Γ_B is the original curve. According to equation (3.66), Γ_P can be expressed as

$$\boldsymbol{R}_P = \boldsymbol{R}_B + u_1\boldsymbol{\alpha} + u_2\boldsymbol{n} + u_3\boldsymbol{v} \tag{4.8}$$

The relationship between the Darboux frame $\{\boldsymbol{R}_B; \boldsymbol{\alpha}, \boldsymbol{n}, \boldsymbol{v}\}$ and $\{\boldsymbol{O}; \boldsymbol{i}_m, \boldsymbol{j}_m, \boldsymbol{k}_m\}$ is

$$\begin{cases}\boldsymbol{R}_B = R\boldsymbol{k}_m \\ \boldsymbol{\alpha} = \cos\theta\boldsymbol{i}_m + \sin\theta\boldsymbol{j}_m \\ \boldsymbol{n} = -\boldsymbol{R}_B \\ \boldsymbol{v} = -\sin\theta\boldsymbol{i}_m + \cos\theta\boldsymbol{j}_m\end{cases} \tag{4.9}$$

where θ is the angle between the projective vector of $\boldsymbol{\alpha}$ on the plane $\boldsymbol{O} - \boldsymbol{i}_m\boldsymbol{j}_m$ and the unit vector $\boldsymbol{i}_m$, and is a function of arc length s, or $\theta = \theta(s)$. Combining equations (4.7), (4.8), and (4.9), we have the coordinates (u_1, u_2, u_3) of P in $\{\boldsymbol{R}_B; \boldsymbol{\alpha}, \boldsymbol{n}, \boldsymbol{v}\}$ as

$$\begin{cases} u_1 = R\sin\delta_{Pm}\cos\left(\theta_{Pm}-\theta\right) \\ u_2 = R\left(1-\cos\delta_{Pm}\right) \\ u_3 = R\sin\delta_{Pm}\sin\left(\theta_{Pm}-\theta\right) \end{cases} \tag{4.10}$$

Differentiating equation (4.8) of $\varGamma_P$ with respect to s (not the arc length s_P of $\varGamma_P$), and substituting equation (4.6) into it, we have

$$\begin{cases} \dot{\boldsymbol{R}}_P = A_1\boldsymbol{\alpha} + A_2\boldsymbol{n} + A_3\boldsymbol{v} \\ A_1 = 1 + \dot{u}_1 - \dfrac{u_2}{R} - k_{gB}u_3 = 1 - \dfrac{u_2}{R} - \left(k_{gB}-\dot{\theta}\right)u_3 \\ A_2 = \dfrac{u_1}{R} + \dot{u}_2 = \dfrac{u_1}{R} \\ A_3 = k_{gB}u_1 + \dot{u}_3 = \left(k_{gB}-\dot{\theta}\right)u_1 \end{cases} \tag{4.11}$$

where the variable with an overdot represents the derivative with respect to s. The geodesic curvature k_{gB} of the original curve $\varGamma_B$ is a function of s. Based on the conditions of a spherical fixed point, or equation (3.67), the coordinates (u_1, u_2, u_3) of the spherical fixed point P_0 of $\sum^*$ can be solved by letting equation (4.11) be zero, or

$$\begin{cases} u_2 + R\left(k_{gB}-\dot{\theta}\right)u_3 = R \\ u_1 = 0 \\ \left(k_{gB}-\dot{\theta}\right)u_1 = 0 \end{cases} \tag{4.12}$$

Substituting equation (4.10) into equation (4.12), we have

$$\begin{cases} R\sin\delta_{Pm}\cos\left(\theta_{Pm}-\theta\right) = 0 \\ \cos\delta_{Pm} - R\left(k_{gB}-\dot{\theta}\right)\sin\delta_{Pm}\sin\left(\theta_{Pm}-\theta\right) = 0 \end{cases} \tag{4.13}$$

Then, the spherical coordinates $(\delta_{P0}, \theta_{P0})$ of P_0 in $\{\boldsymbol{O};\boldsymbol{i}_m,\boldsymbol{j}_m,\boldsymbol{k}_m\}$ are solved as

$$\begin{cases} \theta_{P0} = (n+1/2)\pi + \theta, \ n = 0, 1 \\ \tan\delta_{P0} = \pm\dfrac{1}{R(k_{gB}-\dot{\theta})} \end{cases} \tag{4.14}$$

Converting $\dot{\boldsymbol{R}}_P = 0$ into $\frac{\mathrm{d}\boldsymbol{R}_P}{\mathrm{d}t}\cdot\frac{\mathrm{d}t}{\mathrm{d}s} = 0$, we find $\boldsymbol{V}_P = \frac{\mathrm{d}\boldsymbol{R}_P}{\mathrm{d}t} = 0$, or the spherical fixed point is an instantaneous velocity center on the spherical surface of $\sum^*$. Based on equation (4.12), the instant center P_0 locates on the normal plane ($u_1 = 0$) of $\varGamma_B$, or the great circle of the spherical surface, and its position can be determined exactly by equation (4.14), or

$$\boldsymbol{R}_{P0} = \boldsymbol{R}_B + u_2\boldsymbol{n} + u_3\boldsymbol{v} = -R\cos\delta_{P0}\boldsymbol{n} + R\sin\delta_{P0}\boldsymbol{v} \tag{4.15}$$

For a different instant, or a different arc length s of $\varGamma_B$, the position of the instant center P_0 on the spherical surface will change with the spherical movement of $\sum^*$. Hence, for the instant velocity centers described in the Darboux frame $\{\boldsymbol{R}_B;\boldsymbol{\alpha},\boldsymbol{n},\boldsymbol{v}\}$ of $\varGamma_B$, if they are examined in

$\{\boldsymbol{O};\boldsymbol{i}_m,\boldsymbol{j}_m,\boldsymbol{k}_m\}$, a moving centrode π_m on the spherical surface of $\sum^*$ is formed; when examined in $\{\boldsymbol{O};\boldsymbol{i}_f,\boldsymbol{j}_f,\boldsymbol{k}_f\}$, a fixed centrode π_f on the spherical surface of a fixed body is formed. If the original spherical curve Γ_B is on the unit spherical surface, $\boldsymbol{R}_B$ in equation (4.15) is a unit vector.

Example 4.1 A spherical four-bar linkage *ABCD* is shown in Fig. 4.4. Write the vector equation of the coupler curve and the instant velocity center.

Solution

For the spherical four-bar linkage *ABCD*, the driving link *AB*, coupler link *BC*, driven link *CD*, and base link *AD* can be viewed as segments of great circle on a spherical surface with radius *R*, and their central angles of great circular arcs are a_1, a_2, a_3, and a_4 respectively. The fixed Cartesian coordinate system $\{\boldsymbol{O};\boldsymbol{i}_f,\boldsymbol{j}_f,\boldsymbol{k}_f\}$ is established on the base link *AD*, whose coordinate plane $\boldsymbol{O}-\boldsymbol{i}_f\boldsymbol{k}_f$ is coplanar with plane *OAD*. The moving Cartesian coordinate system $\{\boldsymbol{O};\boldsymbol{i}_m,\boldsymbol{j}_m,\boldsymbol{k}_m\}$ is set up on the coupler link *BC* with coordinate axis vector $\boldsymbol{k}_m$ coaxial with the axis *OB* of revolute joint *B* and coordinate plane $\boldsymbol{O}-\boldsymbol{j}_m\boldsymbol{k}_m$ coplanar with plane *OBC*. The original points of both $\{\boldsymbol{O};\boldsymbol{i}_m,\boldsymbol{j}_m,\boldsymbol{k}_m\}$ and $\{\boldsymbol{O};\boldsymbol{i}_f,\boldsymbol{j}_f,\boldsymbol{k}_f\}$ are placed at the center of the sphere. The input rotational angle φ of link *AB* around axis *OA* is defined by the dihedral angle between planes *OAB* and *OAD*. The rotational angle ψ of link *BC* around axis *OB* is defined by the dihedral angle between planes *OBC* and *OAB*.

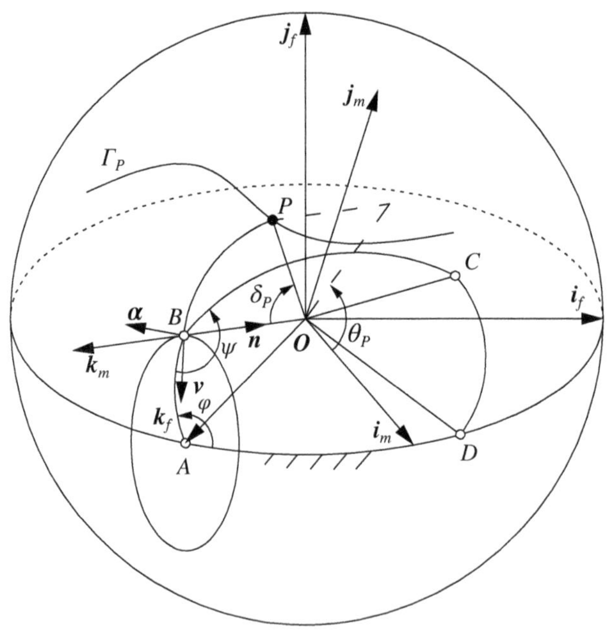

Figure 4.4 A spherical four-bar linkage

For the spherical four-bar linkage, joint B traces a spherical curve Γ_B in the frame $\{\boldsymbol{O};\boldsymbol{i}_f,\boldsymbol{j}_f,\boldsymbol{k}_f\}$, or

$$\Gamma_B : \boldsymbol{R}_B = R\sin a_1 \cos\varphi \boldsymbol{i}_f + R\sin a_1 \sin\varphi \boldsymbol{j}_f + R\cos a_1 \boldsymbol{k}_f \tag{E4-1.1}$$

The Darboux frame $\{\boldsymbol{R}_B;\boldsymbol{\alpha}_B,\boldsymbol{n}_B,\boldsymbol{v}_B\}$ of Γ_B is set up as

$$\begin{cases} \boldsymbol{\alpha}_B = -\sin\varphi \boldsymbol{i}_f + \cos\varphi \boldsymbol{j}_f \\ \boldsymbol{n}_B = -\sin a_1 \cos\varphi \boldsymbol{i}_f - \sin a_1 \sin\varphi \boldsymbol{j}_f - \cos a_1 \boldsymbol{k}_f \\ \boldsymbol{v}_B = -\cos a_1 \cos\varphi \boldsymbol{i}_f - \cos a_1 \sin\varphi \boldsymbol{j}_f + \sin a_1 \boldsymbol{k}_f \end{cases} \tag{E4-1.2}$$

The arc length s of Γ_B can be derived by the above equation, or $\mathrm{d}s = R\sin a_1 \mathrm{d}\varphi$, and the geodesic curvature k_{gB} of Γ_B is

$$k_{gB} = \frac{\mathrm{d}\boldsymbol{\alpha}_B}{\mathrm{d}s} \cdot \boldsymbol{v}_B = \frac{\mathrm{d}\varphi}{\mathrm{d}s}\frac{\mathrm{d}\boldsymbol{\alpha}_B}{\mathrm{d}\varphi} \cdot \boldsymbol{v}_B = \frac{1}{R}\cot a_1 \tag{E4-1.3}$$

Similar to that of a plane four-bar linkage, a coupler point of a spherical four-bar linkage is always adjoint to joint B as the spherical linkage moves. The curve Γ_B, traced by joint B in $\{\boldsymbol{O};\boldsymbol{i}_f,\boldsymbol{j}_f,\boldsymbol{k}_f\}$, is logically taken as an original spherical curve to study the geometrical properties of coupler curves since it is just a circle on the spherical surface. Hence, the Darboux frame $\{\boldsymbol{R}_B;\boldsymbol{\alpha}_B,\boldsymbol{n}_B,\boldsymbol{v}_B\}$ is built up on Γ_B, and the relationship between $\{\boldsymbol{O};\boldsymbol{i}_m,\boldsymbol{j}_m,\boldsymbol{k}_m\}$ and $\{\boldsymbol{R}_B;\boldsymbol{\alpha}_B,\boldsymbol{n}_B,\boldsymbol{v}_B\}$ is:

$$\begin{cases} \boldsymbol{R}_B = R\boldsymbol{k}_m \\ \boldsymbol{\alpha}_B = \cos\psi \boldsymbol{i}_m - \sin\psi \boldsymbol{j}_m \\ \boldsymbol{n}_B = -\boldsymbol{k}_m \\ \boldsymbol{v}_B = \sin\psi \boldsymbol{i}_m + \cos\psi \boldsymbol{j}_m \end{cases} \tag{E4-1.4}$$

Comparing the above equation with equation (4.9), we replace the parameter θ by the direction angle ψ of vector $\boldsymbol{\alpha}_B$ on the coordinate plane $\boldsymbol{O}-\boldsymbol{i}_m\boldsymbol{j}_m$, or $\theta = 2\pi - \psi$. As in Fig. 4.4, a point P of coupler link BC, whose spherical coordinates are (R, δ_P, θ_P) in $\{\boldsymbol{O};\boldsymbol{i}_m,\boldsymbol{j}_m,\boldsymbol{k}_m\}$, traces a spherical coupler curve Γ_P, and its vector equation is

$$\boldsymbol{R}_P = \boldsymbol{R}_B + u_1\boldsymbol{\alpha}_B + u_2\boldsymbol{n}_B + u_3\boldsymbol{v}_B \tag{E4-1.5}$$

where (u_1, u_2, u_3) are the coordinates of P in the Darboux frame $\{\boldsymbol{R}_B;\boldsymbol{\alpha}_B,\boldsymbol{n}_B,\boldsymbol{v}_B\}$, and can be obtained by equations (E4-1.4) and (E4-1.5):

$$\begin{cases} u_1 = R\sin\delta_P \cos\left(\theta_P + \psi\right) \\ u_2 = R\left(1 - \cos\delta_P\right) \\ u_3 = R\sin\delta_P \sin\left(\theta_P + \psi\right). \end{cases} \tag{E4-1.6}$$

The geodesic curvature of Γ_B is a constant, or $k_{gB} = \frac{1}{R}\cot a_1$, and all-order derivatives of k_{gB} are equal to zero by equation (E4-1.3). For a spherical four-bar linkage, the fixed conditions of spherical movement, or equation (4.14), can be solved as

$$\begin{cases}\theta_{P0} = (n + 1/2)\pi - \psi, n = 0, 1 \\ \tan\delta_{P0} = \pm\dfrac{1}{\cot a_1 + R\psi}\end{cases} \tag{E4-1.7}$$

which describes the position of the instant velocity center P_0 on the coupler link at an instant $s(s = R\sin a_1 \cdot \varphi)$.

4.2 Spherical Differential Kinematics

4.2.1 Spherical Centrodes (Axodes)

Giulio Mozzi [1] first introduced the concept of ***instantaneous screw axis (ISA)***, and Chasles [2] definitely pointed out the spatial motion of a rigid body is the composite of the rotation about the ISA and the translation along the ISA, or the screw motion. Spherical motion of a rigid body is rotation motion about a fixed point and can be viewed as a special case of spatial motion, whose ISA always passes through a fixed point, or whose ISA generates a ***moving axode*** in the moving frame $\{\boldsymbol{O}; \boldsymbol{i}_m, \boldsymbol{j}_m, \boldsymbol{k}_m\}$ and a ***fixed axode*** in the fixed frame $\{\boldsymbol{O}; \boldsymbol{i}_f, \boldsymbol{j}_f, \boldsymbol{k}_f\}$. Both axodes are conical surfaces with their vertexes on the sphere center. The moving axode intersects the spherical surface of the moving body with a spherical curve, the ***spherical moving centrode*** π_m, and the fixed axode intersects the spherical surface of the base body with a spherical curve, the ***spherical fixed centrode*** π_f, derived as equation (4.15). As mentioned in Chapter 3, the ruled surfaces of both the moving and the fixed axodes can be mapped onto a unit sphere to study their geometrical properties. The spherical moving axode and the spherical fixed axode are respectively mapped onto two unit spheres, whose center points are designated to coincide with each other. We then have two spherical image curves, the spherical moving centrode π_m and the spherical fixed centrode π_f on the unit sphere, as shown in Fig. 4.5. For demonstrating the systematic approach of differential geometry in this book, particularly in studying the geometric properties of a line trajectory in spatial movement, or ruled surfaces, we prefer to use two spherical image curves, or two spherical centrodes on two unit sphere surfaces, to represent two axodes of a rigid body in spherical movement. Hereinafter, we no longer declare that the spherical movement is a unit spherical movement and let the radius of the spherical surface be a unit.

1. Spherical moving centrode π_m

   The moving spherical centrode π_m of $\sum^*$ in spherical movement is examined in the moving frame $\{\boldsymbol{O}; \boldsymbol{i}_m, \boldsymbol{j}_m, \boldsymbol{k}_m\}$ by means of equations (4.15) as follows:

$$\pi_m : \boldsymbol{R}_m = \boldsymbol{R}_B + u_2\boldsymbol{n} + u_3\boldsymbol{v} \tag{4.16}$$

   The original spherical curve Γ_B in $\{\boldsymbol{O}; \boldsymbol{i}_f, \boldsymbol{j}_f, \boldsymbol{k}_f\}$, traced by point B of $\sum^*$, has its Darboux frame $\{\boldsymbol{R}_B; \boldsymbol{\alpha}, \boldsymbol{n}, \boldsymbol{v}\}$, where the vector $\boldsymbol{n}$ is a fixed vector and vectors $\boldsymbol{\alpha}$, $\boldsymbol{v}$ are vector functions of unit circle on the plane $\boldsymbol{O} - \boldsymbol{i}_m\boldsymbol{j}_m$, or $\boldsymbol{\alpha} = \boldsymbol{e}_{\mathrm{I}(\theta)}$, $\boldsymbol{v} = \boldsymbol{e}_{\mathrm{II}(\theta)}$ with inclination angle θ as the parameter. Based on the properties of vector function of a unit circle in Section 1.1, the

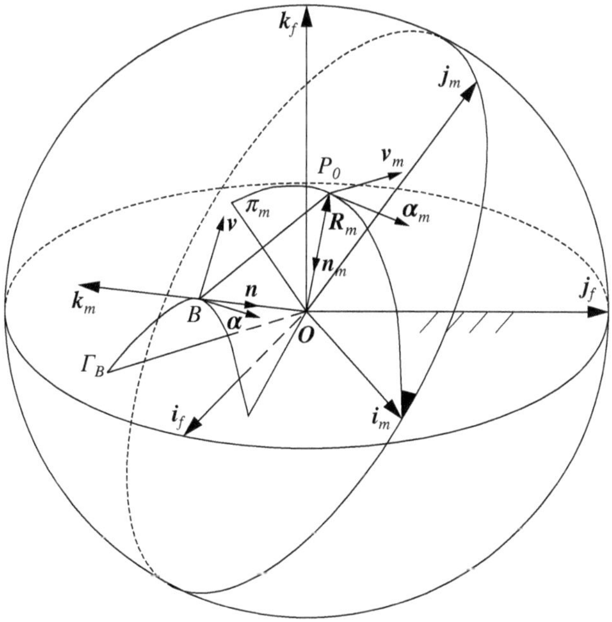

Figure 4.5 The spherical centrodes in the spherical movement

derivatives of $\boldsymbol{\alpha}$, $\boldsymbol{v}$ with respect to θ are $\frac{\mathrm{d}\boldsymbol{\alpha}}{\mathrm{d}\theta} = \boldsymbol{v}$, $\frac{\mathrm{d}\boldsymbol{v}}{\mathrm{d}\theta} = -\boldsymbol{\alpha}$. The tangent vector of the moving centrode π_m can be written as

$$\frac{\mathrm{d}\boldsymbol{R}_m}{\mathrm{d}s} = -\dot{\theta}\sin\delta_{P0}\boldsymbol{\alpha} + \dot{\delta}_{P0}\sin\delta_{P0}\boldsymbol{n} + \dot{\delta}_{P0}\cos\delta_{P0}\boldsymbol{v} \tag{4.17}$$

where s is the arc length of Γ_B, not the natural parameter of π_m. The arc length σ_m of π_m can be derived from equation (4.17), and the differential relationship between σ_m and s is

$$\mathrm{d}\sigma_m = \left|\mathrm{d}\boldsymbol{R}_m\right| = \sqrt{\dot{\theta}^2\sin^2\delta_{P0} + \dot{\delta}_{P0}^{\,2}}\,\mathrm{d}s \tag{4.18}$$

As in Fig. 4.5, the Darboux frame $\{\boldsymbol{R}_m; \boldsymbol{\alpha}_m, \boldsymbol{n}_m, \boldsymbol{v}_m\}$ of π_m is set up as

$$\begin{cases} \boldsymbol{\alpha}_m = \left(-\dot{\theta}\sin\delta_{P0}\boldsymbol{\alpha} + \dot{\delta}_{P0}\sin\delta_{P0}\boldsymbol{n} + \dot{\delta}_{P0}\cos\delta_{P0}\boldsymbol{v}\right)\cdot\dfrac{\mathrm{d}s}{\mathrm{d}\sigma_m} \\ \boldsymbol{n}_m = -\boldsymbol{R}_m = \cos\delta_{P0}\boldsymbol{n} - \sin\delta_{P0}\boldsymbol{v} \\ \boldsymbol{v}_m = \boldsymbol{\alpha}_m \times \boldsymbol{n}_m = \left(-\dot{\delta}_{P0}\boldsymbol{\alpha} - \dot{\theta}\sin^2\delta_{P0}\boldsymbol{n} - \dot{\theta}\sin\delta_{P0}\cos\delta_{P0}\boldsymbol{v}\right)\cdot\dfrac{\mathrm{d}s}{\mathrm{d}\sigma_m} \end{cases} \tag{4.19}$$

The geodesic curvature k_{gm} of π_m can be derived from the differential formulas (4.6):

$$k_{gm} = \frac{\mathrm{d}\boldsymbol{\alpha}_m}{\mathrm{d}\sigma_m}\cdot\boldsymbol{v}_m = \frac{\ddot{\theta}\dot{\delta}_{P0}\sin\delta_{P0} + 2\dot{\theta}\dot{\delta}_{P0}^{\,2}\cos\delta_{P0} - \dot{\theta}\ddot{\delta}_{P0}\sin\delta_{P0} + \dot{\theta}^3\sin^2\delta_{P0}\cos\delta_{P0}}{\left(\dot{\theta}^2\sin^2\delta_{P0} + \dot{\delta}_{P0}^{\,2}\right)^{3/2}} \tag{4.20}$$

2. Spherical fixed centrode π_f
   The instant velocity center of $\sum^*$ in spherical movement forms a curve on the unit sphere in the fixed frame $\{\boldsymbol{O}; \boldsymbol{i}_f, \boldsymbol{j}_f, \boldsymbol{k}_f\}$ for all instants, the spherical fixed centrode π_f. By means of equation (4.15), it can be viewed as the adjoint curve of Γ_B and has the following vector equation

$$\pi_f : \boldsymbol{R}_f = -\cos\delta_{P0}\boldsymbol{n} + \sin\delta_{P0}\boldsymbol{v} \tag{4.21}$$

Deriving equation (4.21) with respect to the arc length s of Γ_B and combining it with the differential formulas (4.6), we have the tangent vector of π_f as

$$\begin{aligned}\frac{\mathrm{d}\boldsymbol{R}_f}{\mathrm{d}s} &= \left(\cos\delta_{P0} - k_{gB}\sin\delta_{P0}\right)\boldsymbol{\alpha} + \dot{\delta_{P0}}\sin\delta_{P0}\boldsymbol{n} + \dot{\delta_{P0}}\cos\delta_{P0}\boldsymbol{v} \\ &= -\dot{\theta}\sin\delta_{P0}\boldsymbol{\alpha} + \dot{\delta_{P0}}\sin\delta_{P0}\boldsymbol{n} + \dot{\delta_{P0}}\cos\delta_{P0}\boldsymbol{v}\end{aligned} \tag{4.22}$$

Based on the above equation, the differential relationship between s and the arc length σ_f of π_f is

$$\mathrm{d}\sigma_f = \left|\mathrm{d}\boldsymbol{R}_f\right| = \sqrt{\dot{\theta}^2\sin^2\delta_{P0} + \dot{\delta_{P0}}^2}\,\mathrm{d}s \tag{4.23}$$

Comparing equation (4.18) with equation (4.23), we find that the differential arc length of π_f is equal to that of π_m, $\mathrm{d}\sigma_f = \mathrm{d}\sigma_m$, or $\mathrm{d}\sigma$ for short, which means that the spherical moving centrode π_m rolls on the spherical fixed centrode π_f on the unit sphere without sliding in spherical movement. The Darboux frame $\{\boldsymbol{R}_f; \boldsymbol{\alpha}_f, \boldsymbol{n}_f, \boldsymbol{v}_f\}$ of π_f is established as

$$\begin{cases}\boldsymbol{\alpha}_f = \left(-\dot{\theta}\sin\delta_{P0}\boldsymbol{\alpha} + \dot{\delta_{P0}}\sin\delta_{P0}\boldsymbol{n} + \dot{\delta_{P0}}\cos\delta_{P0}\boldsymbol{v}\right)\cdot\dfrac{\mathrm{d}s}{\mathrm{d}\sigma} \\ \boldsymbol{n}_f = -\boldsymbol{R}_f = \cos\delta_{P0}\boldsymbol{n} - \sin\delta_{P0}\boldsymbol{v} \\ \boldsymbol{v}_f = \boldsymbol{\alpha}_f \times \boldsymbol{n}_f = \left(-\dot{\delta_{P0}}\boldsymbol{\alpha} - \dot{\theta}\sin^2\delta_{P0}\boldsymbol{n} - \dot{\theta}\sin\delta_{P0}\cos\delta_{P0}\boldsymbol{v}\right)\cdot\dfrac{\mathrm{d}s}{\mathrm{d}\sigma}\end{cases} \tag{4.24}$$

Comparing the above equation with equation (4.19), we can see that the Darboux frame $\{\boldsymbol{R}_f; \boldsymbol{\alpha}_f, \boldsymbol{n}_f, \boldsymbol{v}_f\}$ of π_f coincides with that of π_m at the instant. By taking derivative of the first expression of above equation with respect to σ, we have the geodesic curvature k_{gf} of π_f as

$$\begin{aligned}k_{gf} &= \frac{\mathrm{d}\boldsymbol{\alpha}_f}{\mathrm{d}\sigma_f}\cdot\boldsymbol{v}_f \\ &= \frac{\ddot{\theta}\dot{\delta_{P0}}\sin\delta_{P0} + \dot{\theta}\dot{\delta_{P0}}^2\cos\delta_{P0} - \dot{\theta}\ddot{\delta_{P0}}\sin\delta_{P0} + \left(\dot{\theta}^2\sin^2\delta_{P0} + \dot{\delta_{P0}}^2\right)\left(\sin\delta_{P0} + k_{gB}\cos\delta_{P0}\right)}{\left(\dot{\theta}^2\sin^2\delta_{P0} + \dot{\delta_{P0}}^2\right)^{3/2}}\end{aligned} \tag{4.25}$$

A new and significant invariant, the ***induced geodesic curvature*** k_g^* of $\sum^*$ in the spherical movement, is derived by the geodesic curvature of π_f minus that of π_m at the instant center, or subtracting equation (4.20) from equation (4.25) as follows:

$$k_g^* = k_{gf} - k_{gm} = \frac{\sin\delta_{P0} + \left(k_{gB} - \dot{\theta}\right)\cos\delta_{P0}}{\left(\dot{\theta}^2\sin^2\delta_{P0} + \dot{\delta_{P0}}^2\right)^{1/2}}$$

$$= \frac{1}{\sin\delta_{P0}}\frac{1}{\left(\dot{\theta}^2\sin^2\delta_{P0} + \dot{\delta_{P0}}^2\right)^{1/2}} = \frac{1}{\sin\delta_{P0}}\frac{\mathrm{d}s}{\mathrm{d}\sigma} \tag{4.26}$$

From equations (4.18) and (4.23), (4.19) and (4.24), we have the conclusions for a moving body in spherical movement:

For a moving body $\sum^*$ in spherical movement, there exist a spherical moving centrode π_m of $\sum^*$ and a spherical fixed centrode π_f of $\sum$, respectively, with π_m rolling on π_f without sliding and their Darboux frames being coincident with each other.

The above conclusions can be expressed as

$$\mathrm{d}\sigma_f = \mathrm{d}\sigma_m,\ \boldsymbol{R}_f = \boldsymbol{R}_m,\ \boldsymbol{\alpha}_f//\boldsymbol{\alpha}_m,\ \boldsymbol{n}_f//\boldsymbol{n}_m,\ \boldsymbol{v}_f//\boldsymbol{v}_m \tag{4.27}$$

Note that the above results are true only at an instant.

Example 4.2 A spherical four-bar linkage is given in Example 4.1. Write its spherical moving and fixed centrodes.

Solution

For a spherical crank-rocker linkage, shown in Fig. 4.4, its displacements can be solved for the dihedral angle ψ_1 of plane *OAB* and *OBD* by applying the spherical triangle cosine theorem to the spherical triangle *ABD*. Similarly, the dihedral angle ψ_2 between the plane *OBD* and the plane *OBC* can be derived by

$$\begin{cases} \cos\psi_1 = \left(\cos a_4 - \cos\angle BOD \cdot \cos a_1\right) / \left(\sin\angle BOD \cdot \sin a_1\right) \\ \cos\psi_2 = \left(\cos a_3 - \cos\angle BOD \cdot \cos a_2\right) / \left(\sin\angle BOD \cdot \sin a_2\right) \end{cases} \tag{E4-2.1}$$

where $\cos\angle BOD = \cos a_1 \cdot \cos a_4 + \sin a_1 \cdot \sin a_4 \cdot \cos\varphi$. The dihedral angle ψ between the plane *OAB* and the plane *OBC* is

$$\begin{cases} \psi = \psi_1 + \psi_2, 0 \le \varphi < \pi \\ \psi = -\psi_1 + \psi_2, \pi \le \varphi < 2\pi \end{cases} \tag{E4-2.2}$$

The driving link *AB* of the spherical four-bar linkage rotates around joint *A* for $\varphi \in (0,2\pi)$, the velocity center P_0 traces a spherical moving centrode π_m in the coupler link *BC*. Based on equations (E4-1.4) and (4.16), we get the equation of π_m in the coupler link as

$$\pi_m : \boldsymbol{R}_m = -R\cos\delta_{P0}\boldsymbol{n}_B + R\sin\delta_{P0}\boldsymbol{v}_B$$

$$= R\sin\delta_{P0}\sin\psi\boldsymbol{i}_m + R\sin\delta_{P0}\cos\psi\boldsymbol{j}_m + R\cos\delta_{P0}\boldsymbol{k}_m \tag{E4-2.3}$$

Substituting equation (E4-1.7) and $\theta = 2\pi - \psi$ into equation (4.20), we can derive the geodesic curvature k_{gm} of π_m.

Based on equations (E4-1.2) and (4.21), the velocity center P_0 traces a spherical fixed centrode π_f in the base link:

$$\begin{aligned}\pi_f : \boldsymbol{R}_f &= -R\cos\delta_{P0}\boldsymbol{n}_B + R\sin\delta_{P0}\boldsymbol{v}_B \\ &= \left(\cos\delta_{P0}\sin a_1 - \sin\delta_{P0}\cos a_1\right)\left(R\cos\varphi\boldsymbol{i}_f + R\sin\varphi\boldsymbol{i}_f\right) \\ &\quad + \left(R\cos\delta_{P0}\cos a_1 + R\sin\delta_{P0}\sin a_1\right)\boldsymbol{k}_f \end{aligned} \tag{E4-2.4}$$

Substituting equation (E4-1.7) and $\theta = 2\pi - \psi$, $k_{gB} = \frac{1}{R}\cot a_1$ into equation (4.25), we have the geodesic curvature k_{gf} of π_f. The above equation (E4-2.2) denotes $\psi - \varphi$. $\dot{\psi} = \mathrm{d}\psi/\mathrm{d}s$ can also be obtained by differentiating this equation with respect to the arc length s of Γ_B. A spherical crank-rocker linkage with $a_1 = 20°$, $a_2 = 37.5°$, $a_3 = 37.5°$, $a_4 = 50°$ is given as an example. We can calculate the rectangular coordinates of the velocity center in both $\{\boldsymbol{O};\boldsymbol{i}_m,\boldsymbol{j}_m,\boldsymbol{k}_m\}$ and $\{\boldsymbol{O};\boldsymbol{i}_f,\boldsymbol{j}_f,\boldsymbol{k}_f\}$ by means of equations (E4-2.3) and (E4-2.4). For all instants φ, the spherical moving centrode π_m and the spherical fixed centrode π_f are shown in Fig. 4.6.

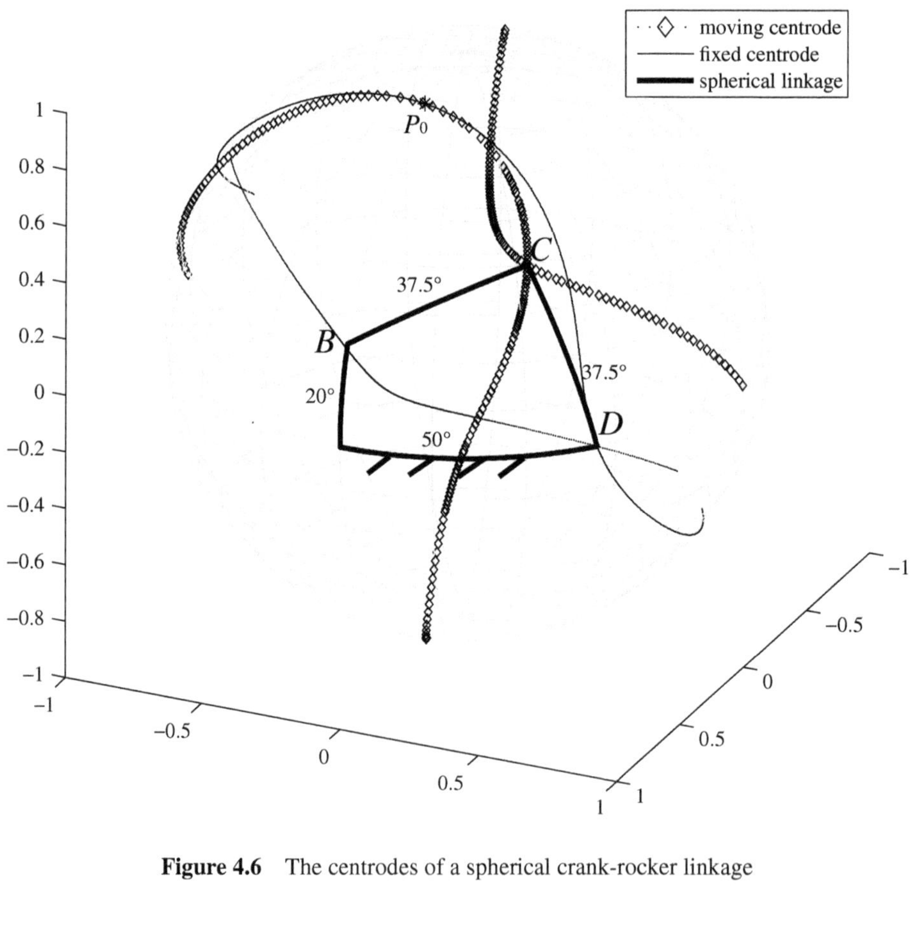

Figure 4.6 The centrodes of a spherical crank-rocker linkage

At the instant $\varphi = \pi/2$, π_m and π_f contact each other tangentially at the spherical instant center P_0, whose spherical coordinates $(\delta_{P0}, \theta_{P0})$ are (50.7948°,146.3707°) in $\{\boldsymbol{O}; \boldsymbol{i}_m, \boldsymbol{j}_m, \boldsymbol{k}_m\}$ and rectangular coordinates are (0, 0.9443, 0.3289) in $\{\boldsymbol{O}; \boldsymbol{i}_f, \boldsymbol{j}_f, \boldsymbol{k}_f\}$. By equations (4.20) and (4.25), the geodesic curvatures k_{gf} of π_f and k_{gm} of π_m are 0.6960 and 1.1565 respectively, and their induced geodesic curvature k_g^* is −0.4605.

4.2.2 *Curvature and Euler–Savary Formula*

As to the well-known Euler–Savary formula in planar movement, both the moving centrode and the fixed centrode reveal the intrinsic information and geometrical properties of a moving body. For spherical movement of a rigid body, the spherical centrodes based curvature theory would be similar to that in planar movement, both in content and form, and presented in an intuitive and concise way.

A point P of $\sum^*$ in spherical movement traces a spherical curve Γ_P in $\{\boldsymbol{O}; \boldsymbol{i}_f, \boldsymbol{j}_f, \boldsymbol{k}_f\}$, which can be designated as a spherical curve adjoint to the spherical fixed centrode π_f. Referring to Fig. 4.7, we have

$$\Gamma_P : \boldsymbol{R}_P = \boldsymbol{R}_f + v_1\boldsymbol{\alpha}_f + v_2\boldsymbol{n}_f + v_3\boldsymbol{v}_f \tag{4.28}$$

where (v_1, v_2, v_3) are the coordinates of P in the Darboux frame $\{\boldsymbol{R}_f; \boldsymbol{\alpha}_f, \boldsymbol{n}_f, \boldsymbol{v}_f\}$ of π_f. Differentiating the above equation with respect to the arc length σ and simplifying it by means of the differential formula (4.6), we have

$$\frac{\mathrm{d}\boldsymbol{R}_P}{\mathrm{d}\sigma} = \left(1 + \frac{\mathrm{d}v_1}{\mathrm{d}\sigma} - v_2 - k_{gf}v_3\right)\boldsymbol{\alpha}_f + \left(v_1 + \frac{\mathrm{d}v_2}{\mathrm{d}\sigma}\right)\boldsymbol{n}_f + \left(k_{gf}v_1 + \frac{\mathrm{d}v_3}{\mathrm{d}\sigma}\right)\boldsymbol{v}_f \tag{4.29}$$

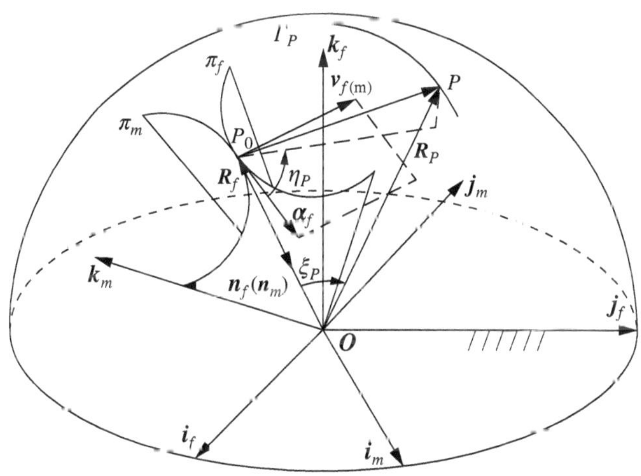

Figure 4.7 The point trajectory adjoint to centrodes in spherical movement

In $\{\boldsymbol{O};\boldsymbol{i}_m,\boldsymbol{j}_m,\boldsymbol{k}_m\}$ of $\sum^*$, the moving centrode π_m has Darboux frame $\{\boldsymbol{R}_m;\boldsymbol{\alpha}_m,\boldsymbol{n}_m,\boldsymbol{v}_m\}$. The point P of $\sum^*$ may be examined in $\{\boldsymbol{R}_m;\boldsymbol{\alpha}_m,\boldsymbol{n}_m,\boldsymbol{v}_m\}$, and viewed as a curve adjoint to π_m:

$$\boldsymbol{R}_{Pm} = \boldsymbol{R}_m + u_1\boldsymbol{\alpha}_m + u_2\boldsymbol{n}_m + u_3\boldsymbol{v}_m \tag{4.30}$$

where (u_1, u_2, u_3) are the coordinates of P in $\{\boldsymbol{R}_m;\boldsymbol{\alpha}_m,\boldsymbol{n}_m,\boldsymbol{v}_m\}$. Differentiating the above equation with respect to the arc length σ and simplifying it with the differential formula (4.6), we have

$$\frac{\mathrm{d}\boldsymbol{R}_{Pm}}{\mathrm{d}\sigma} = \left(1 + \frac{\mathrm{d}u_1}{\mathrm{d}\sigma} - u_2 - k_{gm}u_3\right)\boldsymbol{\alpha}_m + \left(u_1 + \frac{\mathrm{d}u_2}{\mathrm{d}\sigma}\right)\boldsymbol{n}_m + \left(k_{gm}u_1 + \frac{\mathrm{d}u_3}{\mathrm{d}\sigma}\right)\boldsymbol{v}_m \tag{4.31}$$

Obviously, the point P does not change its position at all in $\{\boldsymbol{O};\boldsymbol{i}_m,\boldsymbol{j}_m,\boldsymbol{k}_m\}$ and remains a fixed point of $\sum^*$. Hence, it meets the spherical fixed point condition in equation (3.68), or

$$\begin{cases} 1 + \dfrac{\mathrm{d}u_1}{\mathrm{d}\sigma} - u_2 - k_{gm}u_3 = 0 \\ u_1 + \dfrac{\mathrm{d}u_2}{\mathrm{d}\sigma} = 0 \\ k_{gm}u_1 + \dfrac{\mathrm{d}u_3}{\mathrm{d}\sigma} = 0 \end{cases} \tag{4.32}$$

That is

$$\frac{\mathrm{d}u_1}{\mathrm{d}\sigma} = u_2 + k_{gm}u_3 - 1,\ \frac{\mathrm{d}u_2}{\mathrm{d}\sigma} = -u_1,\ \frac{\mathrm{d}u_3}{\mathrm{d}\sigma} = -k_{gm}u_1 \tag{4.33}$$

The above equation indicates the change rates of the relative coordinates of point P in the Darboux frame $\{\boldsymbol{R}_m;\boldsymbol{\alpha}_m,\boldsymbol{n}_m,\boldsymbol{v}_m\}$. As mentioned above, the Darboux frames of both π_m and π_f are coincident with each other at the instant and π_m rolls on π_f without sliding, or equation (4.27) holds true. Therefore, point P has the same coordinates and their change rates in the Darboux frame $\{\boldsymbol{R}_m;\boldsymbol{\alpha}_m,\boldsymbol{n}_m,\boldsymbol{v}_m\}$ of both π_f and π_m:

$$u_1 = v_1,\ u_2 = v_2,\ u_3 = v_3,\ \frac{\mathrm{d}u_1}{\mathrm{d}\sigma} = \frac{\mathrm{d}v_1}{\mathrm{d}\sigma},\ \frac{\mathrm{d}u_2}{\mathrm{d}\sigma} = \frac{\mathrm{d}v_2}{\mathrm{d}\sigma},\ \frac{\mathrm{d}u_3}{\mathrm{d}\sigma} = \frac{\mathrm{d}v_3}{\mathrm{d}\sigma} \tag{4.34}$$

Substituting equation (4.33) into equation (4.29), we have the first derivative of the curve Γ_P as

$$\frac{\mathrm{d}\boldsymbol{R}_P}{\mathrm{d}\sigma} = k_g^*\left(-v_3\boldsymbol{\alpha}_f + v_1\boldsymbol{v}_f\right) \tag{4.35}$$

The velocity vector of P in $\{\boldsymbol{O};\boldsymbol{i}_f,\boldsymbol{j}_f,\boldsymbol{k}_f\}$ is

$$\boldsymbol{V}_P = \frac{\mathrm{d}\boldsymbol{R}_P}{\mathrm{d}t} = \frac{\mathrm{d}\boldsymbol{R}_P}{\mathrm{d}\sigma}\frac{\mathrm{d}\sigma}{\mathrm{d}t} \tag{4.36}$$

The spherical motion of $\sum^*$ can be regarded as the relative rotation about the ISA at an instant, so the velocity of P can also be deduced in another way:

$$\begin{aligned} \boldsymbol{V}_P &= \boldsymbol{\omega} \times \left(\boldsymbol{R}_P - \boldsymbol{R}_f\right) = -\omega\boldsymbol{n}_f \times \left(v_1\boldsymbol{\alpha}_f + v_2\boldsymbol{n}_f + v_3\boldsymbol{v}_f\right) \\ &= \omega\left(-v_3\boldsymbol{\alpha}_f + v_1\boldsymbol{v}_f\right) \end{aligned} \tag{4.37}$$

where ω is the angular velocity of rotation of $\sum^*$ about the ISA and can be calculated through equations (4.35) to (4.37) as

$$\omega = k_g^* \frac{d\sigma}{dt} \tag{4.38}$$

The above equation demonstrates the kinematic meaning of the induced geodesic curvature k_g^*.

The point P always moves on a unit spherical surface, and its position in the Darboux frame $\{\boldsymbol{R}_f; \boldsymbol{\alpha}_f, \boldsymbol{n}_f, \boldsymbol{v}_f\}$ of π_f can be represented by parameters (ξ_P, η_P), where ξ_P is the angle between vectors $\overrightarrow{OP}$ and $\overrightarrow{OP_0}$, and η_P is the angle between the projection of vector $\overrightarrow{P_0P}$ on the plane $\boldsymbol{R}_f - \boldsymbol{\alpha}_f\boldsymbol{v}_f$ and the vector $\boldsymbol{\alpha}_f$, as shown in Fig. 4.7. The Cartesian coordinates of P in $\{\boldsymbol{R}_f; \boldsymbol{\alpha}_f, \boldsymbol{n}_f, \boldsymbol{v}_f\}$ are

$$\begin{cases} v_1 = u_1 = \sin\xi_P \cos\eta_P \\ v_2 = u_2 = 1 - \cos\xi_P \\ v_3 = u_3 = \sin\xi_P \sin\eta_P \end{cases} \tag{4.39}$$

Substituting the above equation into equation (4.33), we obtain

$$\begin{cases} \dfrac{d\eta_P}{d\sigma} = \cot\xi_P \sin\eta_P - k_{gm} \\ \dfrac{d\xi_P}{d\sigma} = -\cos\eta_P \end{cases} \tag{4.40}$$

The Darboux frame $\{\boldsymbol{R}_P; \boldsymbol{\alpha}_P, \boldsymbol{n}_P, \boldsymbol{v}_P\}$ of Γ_P is set up by

$$\begin{cases} \boldsymbol{\alpha}_P = -\sin\eta_P\boldsymbol{\alpha}_f + \cos\eta_P\boldsymbol{v}_f \\ \boldsymbol{n}_P = -\sin\xi_P\cos\eta_P\boldsymbol{\alpha}_f + \cos\xi_P\boldsymbol{n}_f - \sin\xi_P\sin\eta_P\boldsymbol{v}_f \\ \boldsymbol{v}_P = -\cos\xi_P\cos\eta_P\boldsymbol{\alpha}_f - \sin\xi_P\boldsymbol{n}_f - \cos\xi_P\sin\eta_P\boldsymbol{v}_f \end{cases} \tag{4.41}$$

The relationship between the arc length σ_P of Γ_P and σ of π_f can be deduced from the above equation as

$$d\sigma_P = k_g^* \sin\xi_P d\sigma \tag{4.42}$$

Differentiating $\boldsymbol{\alpha}_P$ of equation (4.41) with respect to σ_P, and dotting the second equation, we can write the geodesic curvature k_{gP} of Γ_P as

$$k_{gP} = \frac{d\boldsymbol{\alpha}_P}{d\sigma_P} \cdot \boldsymbol{v}_P = \frac{\sin\eta_P + k_g^* \sin\xi_P\cos\xi_P}{k_g^* \sin^2\xi_P} = \frac{\sin\xi_P\cos\xi_P + \sin\eta_P/k_g^*}{\sin^2\xi_P} \tag{4.43a}$$

If k_{gP} is replaced by the reciprocal of the geodesic curvature radius ρ_g, the above equation can be rewritten as

$$\rho_g\left(\cos\xi_P\sin\xi_P + \frac{1}{k_g^*}\sin\eta_P\right) = \sin^2\xi_P \tag{4.43b}$$

As shown in Fig. 4.8, the tangent plane $\sum_P$ of Γ_P at point P is determined by the tangent vector $\boldsymbol{\alpha}_P$ and the binormal vector $\boldsymbol{v}_P$ of Γ_P. The curve Γ_P is projected onto $\sum_P$ and we have a corresponding plane curve C_P. The geodesic curvature of Γ_P at P is just the curvature of C_P, and the geodesic curvature center is the point O_{Pg}.

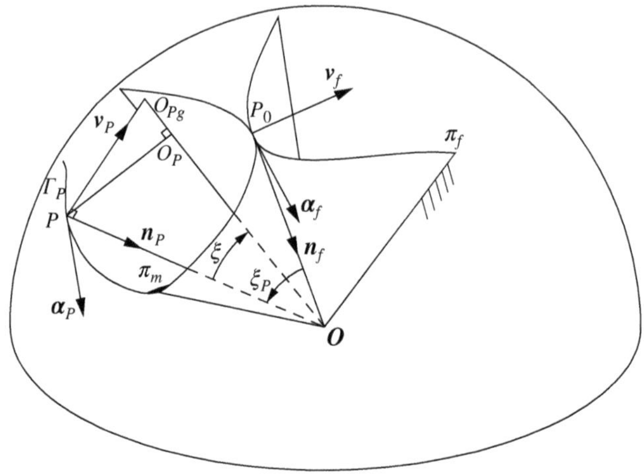

Figure 4.8 The curvature center and geodesic curvature center of the spherical curve

Hence, equations (4.43a) and (4.43b) indicate the relationships among the position of the point, the geodesic curvature center, and the induced geodesic curvature k_g^* of $\sum^*$ in the spherical movement, which can be called the ***geodesic Euler–Savary formulas*** for a point of a moving body in spherical movement, similar to equation (1.74) in planar movement.

Based on equations (3.29) and (3.30) in Chapter 3, the curvature vector of a curve is the vector sum of the normal curvature vector and the geodesic curvature vector. The spherical curve Γ_P traced by the point P has geodesic curvature center point O_{Pg} and normal curvature center point O, the center point of the sphere. The three points P, O_{Pg}, and O form a right triangle. The curvature center O_P of Γ_P is just the foot of the perpendicular line $\overline{PO_P}$ of P to the hypotenuse $\overline{OO_{Pg}}$ based on the expression $k_P^2 = k_{nP}^2 + k_{gP}^2$. It denotes that the normal curvature center O, the geodesic curvature center O_{Pg}, and the curvature center O_P of Γ_P all locate on the line $\overline{OO_{Pg}}$. By letting $\tan\xi = \rho_g$, we can rewrite equation (4.43b) as

$$\cot\left(\xi - \xi_P\right) + \cot\xi_P = -k_g^*/\sin\eta_P \tag{4.43c}$$

From the viewpoint of the central angle of the unit sphere, the above equation can be called the ***spherical Euler–Savary formula*** of both curvature and geodesic curvature for a point of a moving body in spherical movement.

By letting the geodesic curvature k_{gP} of Γ_P in equation (4.43a) be zero, we have

$$\sin 2\xi_P + \frac{2\sin\eta_P}{k_g^*} = 0 \tag{4.44}$$

The above equation indicates that there exists a spherical curve of $\sum^*$ at an instant, which is referred to as a ***spherical geodesic inflection curve***. Each point of the curve will trace a curve in the fixed frame, which contacts with a great circle at three infinitesimally separated positions. Based on the symmetry of the spherical movement about the spherical center, the domains of definition for the parameters η_P and ξ_P of P in $\{\boldsymbol{R}_m; \boldsymbol{\alpha}_m, \boldsymbol{n}_m, \boldsymbol{v}_m\}$ are respectively given as

$\xi_P \in [0, \pi/2)$ and $\eta_P \in [0, 2\pi)$. According to equation (4.44), two corresponding values of ξ_P for the given η_P can be obtained as

$$\begin{cases} \xi_P^{(1)} = \dfrac{1}{2}\arcsin\left(-\dfrac{2\sin\eta_P}{k_g^*}\right) \\ \xi_P^{(2)} = \dfrac{\pi}{2} - \dfrac{1}{2}\arcsin\left(-\dfrac{2\sin\eta_P}{k_g^*}\right) \end{cases} \tag{4.45}$$

If $k_g^{*2} \geq 4\sin^2\eta_P$ or $|k_g^*| \geq 2|\sin\eta_P|$, the above equation has solutions, as shown in Fig. 4.9. ξ_P can't reach all the values in the interval $[0, \pi/2)$ when $|k_g^*| > 2$, which means the spherical geodesic inflection curve Γ_1 has two branches. One branch is a symmetrical closed curve, which starts from point P_1 and closes at point P_2. The parameter ξ_{P1} of P_1 is equal to zero and the parameter ξ_{P2} of P_2 is $\frac{1}{2}\arcsin(2/|k_g^*|)$. Another branch is a symmetrical open curve, which starts from point P_4, whose parameter ξ_{P4} is $\frac{\pi}{2} - \frac{1}{2}\arcsin(2/|k_g^*|)$.

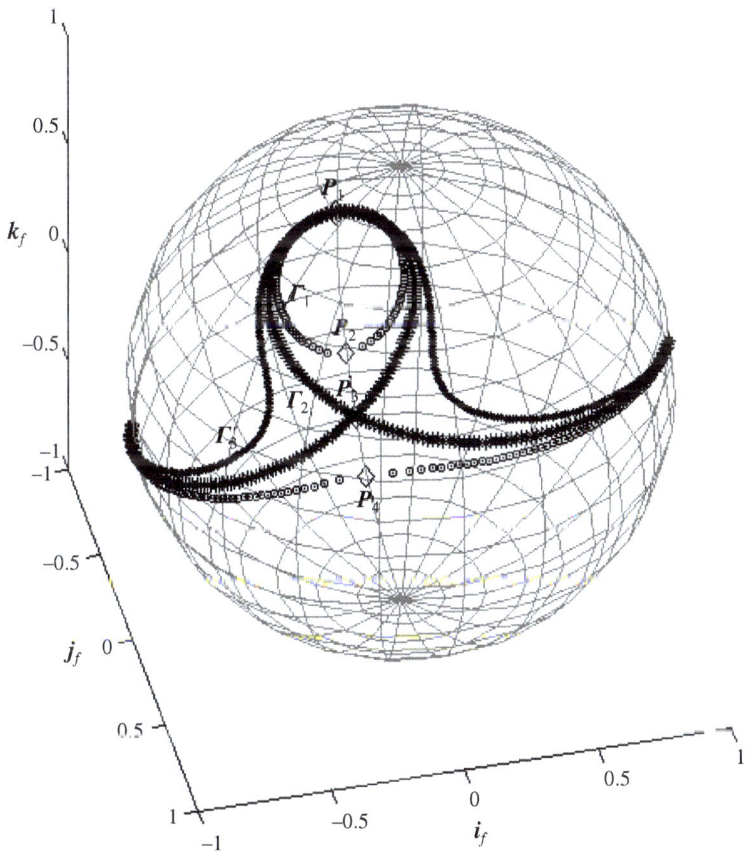

Figure 4.9 The spherical geodesic inflection curve of spherical movement

When the induced geodesic curvature k_g^* equals 2 or -2, we have $\xi_P^{(1)} = \xi_P^{(2)} = \frac{\pi}{4}$ for the given value $\frac{3\pi}{2}$ or $\frac{\pi}{2}$ of the parameter η_P. The spherical geodesic inflection curve Γ_2 is a symmetrical self-intersecting curve at point P_3, whose parameter ξ_{P3} equals $\pi/4$.

When $|k_g^*| < 2$, we have $|\sin\eta_P| \leq |k_g^*|/2$. Thus, the parameter η_P is defined in the range $\left[-\arcsin\frac{|k_g^*|}{2}, \arcsin\frac{|k_g^*|}{2}\right] \cup \left[\pi - \arcsin\frac{|k_g^*|}{2}, \pi + \arcsin\frac{|k_g^*|}{2}\right]$. The spherical geodesic inflection curve Γ_3 starts from the point P_1 and forms a symmetrical open curve.

From equation (4.43a), the first derivative of the geodesic curvature k_{gP} of Γ_P is derived with respect to σ, which is

$$\frac{\mathrm{d}k_{gP}}{\mathrm{d}\sigma} = \frac{\left(k_g^* - k_{gm}\right)\sin\xi_P\cos\eta_P - \dfrac{1}{k_g^*}\dfrac{\mathrm{d}k_g^*}{\mathrm{d}\sigma}\sin\xi_P\sin\eta_P + 3\cos\xi_P\sin\eta_P\cos\eta_P}{k_g^*\sin^3\xi_P} \tag{4.46}$$

When the normal curvature k_{nP} of Γ_P is constant, the variation of the geodesic curvature k_{gP} can be reflected by that of the curvature k_P through $k_P^2 = k_{nP}^2 + k_{gP}^2$. If the curvature k_P does not change instantaneously, we let $\mathrm{d}k_{gP}/\mathrm{d}\sigma = 0$ in the above equation and simplify it as

$$\cot\xi_P = \frac{1}{M\sin\eta_P} + \frac{1}{N\cos\eta_P} \tag{4.47a}$$

where

$$\frac{1}{M} = \frac{\left(k_{gm} - k_g^*\right)}{3}, \quad \frac{1}{N} = \frac{\mathrm{d}k_g^*/\mathrm{d}\sigma}{3k_g^*} \tag{4.47b}$$

Obviously, equation (4.47a) is similar to equation (1.79) in Chapter 1 for the cubic curve of stationary curvature of a moving body in planar movement. The spherical curve determined by equation (4.47a) is called the ***spherical cubic curve of stationary curvature*** of a moving body in spherical movement. A point of a moving body, belonging to the cubic curve, will trace a trajectory Γ_P in a fixed frame with $\mathrm{d}k_{gP}/\mathrm{d}\sigma = 0$ or $\mathrm{d}k_P/\mathrm{d}\sigma = 0$ at the instant σ, which contacts a circle in third order, or has four common points with a circle at infinitesimal separated positions at this instant.

Letting $k_{gP} = 0$ and $\mathrm{d}k_{gP}/\mathrm{d}\sigma = 0$ simultaneously, we have the intersection points between the spherical geodesic inflection curves and the cubic curve of stationary curvature, called ***spherical Ball points***. A spherical Ball point of $\sum^*$ traces a trajectory in the fixed frame, which has four common points with a spherical great circle at four infinitesimal separated positions at the instant. According to equations (4.44) and (4.47a), the positions of spherical Ball points can be located by

$$\begin{cases} \sin 2\xi_P + \dfrac{2\sin\eta_P}{k_g^*} = 0 \\ \cot\xi_P = \dfrac{1}{M\sin\eta_P} + \dfrac{1}{N\cos\eta_P} \end{cases} \tag{4.48a}$$

The above equation is equivalent to

$$\begin{cases} \sin 2\xi_P + \dfrac{2\sin\eta_P}{k_g^*} = 0 \\ \mathrm{d}\left(\sin 2\xi_P + \dfrac{2\sin\eta_P}{k_g^*}\right)/\mathrm{d}\sigma = 0 \end{cases} \tag{4.48b}$$

It indicates that a spherical Ball curve is just a curve enveloped by the spherical geodesic inflection curves.

Eliminating parameter η_P in equations (4.48a) and simplifying, we have

$$f_1\cos^8\xi_P + f_2\cos^6\xi_P + f_3\cos^4\xi_P + f_4\cos^2\xi_P + f_5 = 0 \tag{4.49a}$$

where

$$\begin{cases} f_1 = -k_g^{*4} \\ f_2 = k_g^{*4} - 2k_g^{*3}/M \\ f_3 = 2k_g^{*3}/M - k_g^{*2}/N^2 - k_g^{*2}/M^2 - k_g^{*2} \\ f_4 = k_g^{*2}/N^2 + k_g^{*2}/M^2 - 2k_g^*/M \\ f_5 = -1/M^2 \end{cases} \tag{4.49b}$$

The above equation is an eight-order algebraic equation about $\cos\xi_P$ or perhaps more precisely a quartic algebraic equation about $\cos^2\xi_P$. Equation (4.49a) may have eight solutions for the variable ξ_P, which theoretically correspond to the eight spherical Ball points of $\sum^*$. But for the symmetry of spherical movement, half of them locate on the opposite semi-sphere, or actually four spherical Ball points at most may be found at an instant.

Differentiating equation (4.46) with respect to σ once more, and letting both the first and second derivatives of k_{gP} be zero, or $\mathrm{d}k_{gP}/\mathrm{d}\sigma = 0$ and $\mathrm{d}^2k_{gP}/\mathrm{d}\sigma^2 = 0$, we simplify them by substituting equation (4.47a) and have

$$\begin{aligned} &\left(\tan^2\eta_P + 1\right)\left[\frac{1}{M^2}\left(2 - k_{gm}M\right) + \frac{1}{M^2}\left(\frac{3M}{N} + \frac{\mathrm{d}M}{\mathrm{d}\sigma}\right)\tan\eta_P + \frac{1}{N^2}\left(1 + \frac{\mathrm{d}N}{\mathrm{d}\sigma}\right)\tan^2\eta_P \right. \\ &\left. + \frac{1}{N}\left(k_{gm} - \frac{1}{M}\right)\tan^3\eta_P - \frac{1}{N^2}\tan^4\eta_P\right] + \tan^2\eta_P = 0 \end{aligned} \tag{4.50}$$

The above equation is a sextic algebraic equation about $\tan\eta_P$. For an instant, the parameters M, N and their derivatives can be determined by the kinematic parameters of $\sum^*$, and the roots of equation (4.50) are definitely not more than six. The corresponding points of $\sum^*$ are called the Burmester points of $\sum^*$ in spherical movement, or ***spherical Burmester points*** for short. The trajectories in a fixed frame, traced by spherical Burmester points, have five common points with a circle at infinitesimal separated positions at the instant.

The geometrical properties of spherical curves are studied by means of the approach of the moving frame in differential geometry, and some results can be applied in the bevel

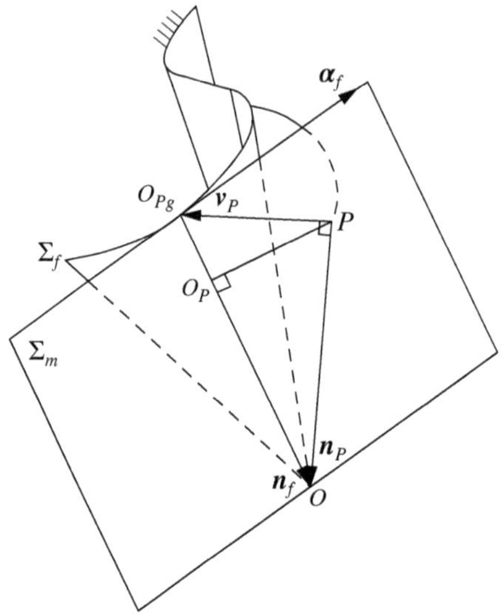

Figure 4.10 Spherical involute

gearing. Particularly, a point P of $\sum^*$ in spherical movement traces a spherical involute Γ_P in $\{\boldsymbol{O}_f; \boldsymbol{i}_f, \boldsymbol{j}_f, \boldsymbol{k}_f\}$ if it always locates on the tangent plane $\boldsymbol{R}_f - \boldsymbol{\alpha}_f \boldsymbol{v}_f$ of the moving centrode π_m (or the moving axode), which means that the coordinate parameter v_3 is zero in equation (4.28), or $\eta_P = 0$ in equation (4.39) for all instants. Hence, equation (4.43a) of the geodesic curvature can be simplified into $k_{gP} = \cot \xi_P$, which reveals that both the geodesic curvature center O_{Pg} and the curvature center O_P locate on the ISA, as shown in Fig. 4.10.

For an involute bevel gear, the base cone is regarded as the fixed axode and the generating plane is taken as the moving axode. The generating plane rolls on the base cone without sliding, and a point P of the generating plane traces a spherical curve Γ_P, which is just the spherical involute and has analogue properties to the plane involute. Based on the spherical adjoint approach, the geometrical properties of the spherical bevel gears with tooth profile of spherical involute can be discussed by the axodes, pitch cones, as shown in Fig. 4.11, which are similar to those of involute cylindrical gears, and can be developed into two conjugating involute surfaces by kinematic geometry in spherical movement.

Example 4.3 Given a spherical four-bar linkage with dimensions $a_1 = 30^\circ$, $a_2 = 45^\circ$, $a_3 = 60^\circ$, $a_4 = 15^\circ$, reveal the instantaneous kinematics of its coupler, including the spherical moving and fixed centrodes, Euler–Savary formula of a coupler curve, geodesic inflection curves, and spherical Ball curve.

Solution

The displacement solutions of a spherical four-bar linkage are illustrated in Example 4.2, and the vector equations of the spherical moving and fixed centrodes for the coupler link relative to the base link are given.

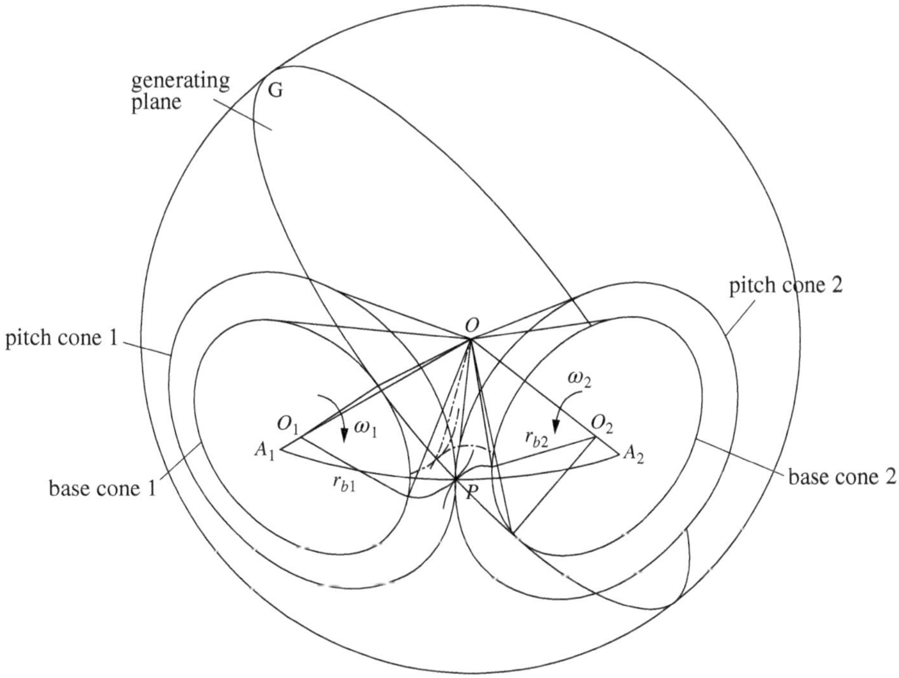

Figure 4.11 The meshing of two spherical involute bevel gears

1. **The spherical moving centrode and the spherical fixed centrode.**
   Based on equations (E4-2.1) and (E4-2.2) in Example 4.2, the spherical moving centrode π_m and the spherical fixed centrode π_f for the spherical double-crank linkage are shown in Fig. 4.12. At the instant $\varphi = \pi/2$, π_m and π_f contact tangentially at the spherical instant center P_0 ($\delta_{P0} = 55.6183^{\circ}, \theta_{P0} = 318.1524^{\circ}$), whose rectangular coordinates are $(0, -0.4324, 0.9017)$ in $\{\boldsymbol{O}; \boldsymbol{i}_f, \boldsymbol{j}_f, \boldsymbol{k}_f\}$. The geodesic curvatures of π_f and π_m are respectively $k_{gf} = 2.4959$ and $k_{gm} = 1.1023$ by equations (4.20) and (4.25), and then the induced geodesic curvature k_g^* is 1.3936.
2. **The spherical Euler–Savary equation for spherical coupler curves.**
   From equations (E4-1.4) and (4.19), the relationship between the Darboux frame $\{\boldsymbol{R}_m; \boldsymbol{\alpha}_m, \boldsymbol{n}_m, \boldsymbol{v}_m\}$ of π_m and $\{\boldsymbol{O}; \boldsymbol{i}_m, \boldsymbol{j}_m, \boldsymbol{k}_m\}$ is

$$\begin{cases} x_m = \dfrac{\mathrm{d}s}{\mathrm{d}\sigma}\left(\dot{\psi}\sin\delta_{P0}v_1 - \dot{\delta}_{P0}v_3\right)\cos\psi \\ \qquad + \left[\dot{\delta}_{P0}\cos\delta_{P0}\dfrac{\mathrm{d}s}{\mathrm{d}\sigma}v_1 - \left(v_2 - 1\right)\sin\delta_{P0} + \dot{\psi}\sin\delta_{P0}\cos\delta_{P0}\dfrac{\mathrm{d}s}{\mathrm{d}\sigma}v_3\right]\sin\psi \\ y_m = -\dfrac{\mathrm{d}s}{\mathrm{d}\sigma}\left(\dot{\psi}\sin\delta_{P0}v_1 - \dot{\delta}_{P0}v_3\right)\sin\psi \\ \qquad + \left[\dot{\delta}_{P0}\cos\delta_{P0}\dfrac{\mathrm{d}s}{\mathrm{d}\sigma}v_1 - \left(v_2 - 1\right)\sin\delta_{P0} + \dot{\psi}\sin\delta_{P0}\cos\delta_{P0}\dfrac{\mathrm{d}s}{\mathrm{d}\sigma}v_3\right]\cos\psi \\ z_m = -\dot{\delta}_{P0}\sin\delta_{P0}\dfrac{\mathrm{d}s}{\mathrm{d}\sigma}v_1 - \left(v_2 - 1\right)\cos\delta_{P0} - \dot{\psi}\sin^2\delta_{P0}\dfrac{\mathrm{d}s}{\mathrm{d}\sigma}v_3 \end{cases} \tag{E4-3.1}$$

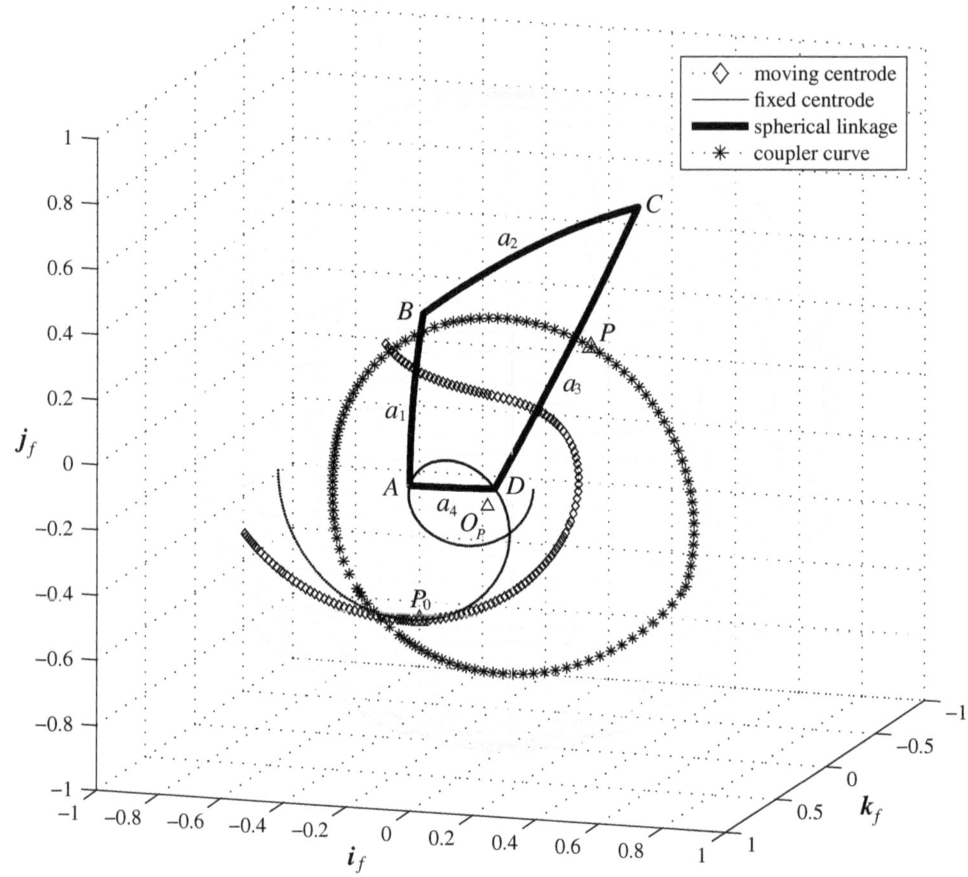

Figure 4.12 The moving and fixed centrodes for a double-crank linkage

where $\mathrm{d}s/\mathrm{d}\sigma = 1/\sqrt{\dot{\psi}^2\sin^2\delta_{P0} + \dot{\delta}_{P0}^{2}}$. The coupler point $P(x_m, y_m, z_m)$ has coordinates (v_1, v_2, v_3) in the Darboux frame $\{\boldsymbol{R}_m; \boldsymbol{\alpha}_m, \boldsymbol{n}_m, \boldsymbol{v}_m\}$, defined as $v_1 = \sin\xi_P\cos\eta_P$, $v_2 = 1 - \cos\xi_P$, and $v_3 = \sin\xi_P\sin\eta_P$.

A coupler point $P(0.3536, 0.3536, 0.8660)$ in $\{\boldsymbol{O}; \boldsymbol{i}_m, \boldsymbol{j}_m, \boldsymbol{k}_m\}$, or parameters $\xi_P = 1.0336$ and $\eta_P = 0.8466$ in $\{\boldsymbol{R}_m; \boldsymbol{\alpha}_m, \boldsymbol{n}_m, \boldsymbol{v}_m\}$, traces a spherical trajectory Γ_P, shown in Fig. 4.12, whose geodesic curvature k_{gP} is 1.3238 by equation (4.43a) at the instant $\varphi = \pi/2$. The included angle ξ of the lines $\overline{OP}$ and $\overline{OO_{Pg}}$ or $\overline{OO_P}$ is 0.6469, where O_{Pg} and O_P are respectively the geodesic curvature center and the curvature center of Γ_P at the instant. The rectangular coordinates of the curvature center O_P are $(0.1748, -0.0987, 0.7722)$ in $\{\boldsymbol{O}; \boldsymbol{i}_f, \boldsymbol{j}_f, \boldsymbol{k}_f\}$, and the radius of the curvature circle is 0.6027.

3. **The instantaneous geodesic inflection curves on the coupler link.**
At the instant $\varphi = 143.2394°$, the induced geodesic curvature k_g^* is 1.6869. By equation (4.44), the equation of the geodesic inflection curve on the coupler link is

$$\sin 2\xi_P + \frac{2\sin\eta_P}{1.6869} = 0 \tag{E4-3.2}$$

The geodesic inflection curve Γ_1 can be described in the moving frame $\{\boldsymbol{O};\boldsymbol{i}_m,\boldsymbol{j}_m,\boldsymbol{k}_m\}$, as shown in Fig.4.13.
Similarly, at the instant φ =167.7620°, the induced geodesic curvature of π_m and π_f is 2, the equation of the spherical geodesic inflection curve on the coupler link is

$$\sin 2\xi_P + \sin \eta_P = 0 \tag{E4-3.3}$$

The spherical geodesic inflection curve Γ_2 crosses itself at point $P_3(\xi_{P3}=45°, \eta_{P3}=-90°)$, as shown in Fig. 4.13. The point P_3 has rectangular coordinates (0.9607, 0.2758, 0.0317) in the frame $\{\boldsymbol{O};\boldsymbol{i}_m,\boldsymbol{j}_m,\boldsymbol{k}_m\}$.
At the instant $\varphi = 177.6169°$, k_g^* is 2.1810, and the equation of the spherical geodesic inflection curve on the coupler link is

$$\sin 2\xi_P + \frac{2\sin\eta_P}{2.1810} = 0 \tag{E4-3.4}$$

The spherical geodesic inflection curve Γ_3 has two branches. One branch is a closed curve, which starts from initial point ($\xi_P = 0$) and stops at point $P_1(\xi_{P1} = 33.2487°, \eta_{P1} = -90°)$. Another is an open curve, which starts from $P_2(\xi_{P2} = 56.7515°, \eta_{P2} = -90°)$. The Cartesian coordinates of P_1 and P_2 in $\{\boldsymbol{O};\boldsymbol{i}_m,\boldsymbol{j}_m,\boldsymbol{k}_m\}$ are respectively (0.9186, 0.2914, 0.2638) and (0.8831, 0.4566, –0.1080).

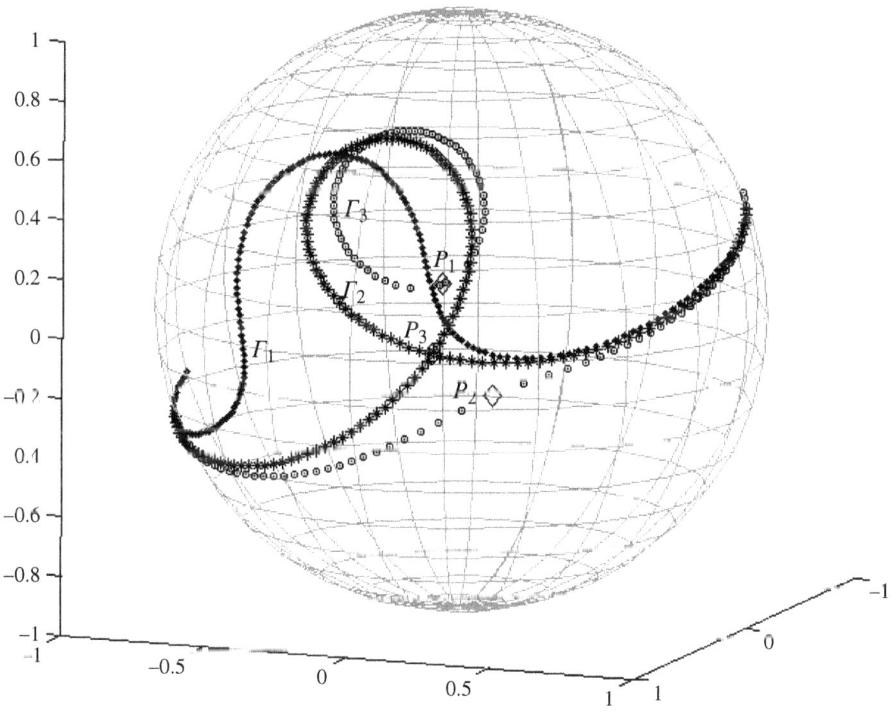

Figure 4.13 The geodesic inflection curve of a spherical double-crank linkage

4. **Spherical Ball curve on the coupler link.**
At the instant $\varphi = 90^\circ$, the induced geodesic curvature k_g^* is 1.3936. Based on equation (4.46), the equations of the spherical Ball points on the coupler link are

$$\begin{cases} \sin 2\xi_P + \dfrac{2\sin\eta_P}{1.3936} = 0 \\ \cot\xi_P = -\dfrac{1}{10.2980\sin\eta_P} + \dfrac{1}{12.2980\cos\eta_P} \end{cases} \quad \text{(E4-3.5)}$$

We obtain two spherical Ball points P_{B1} and P_{B2} on the coupler link, whose coordinate parameters are $P_{B1}(103.0407^\circ, 17.8362^\circ)$ and $P_{B2}(108.0427^\circ, 155.7700^\circ)$ in $\{\boldsymbol{R}_m; \boldsymbol{\alpha}_m, \boldsymbol{n}_m, \boldsymbol{v}_m\}$, or $P_{B1}(0.2905, 0.9363, 0.1971)$ and $P_{B2}(-0.9182, -0.3898, 0.0710)$ in $\{\boldsymbol{O}; \boldsymbol{i}_m, \boldsymbol{j}_m, \boldsymbol{k}_m\}$.
For $\varphi \in [0, 2\pi)$, the spherical moving centrode π_m, the geodesic inflection curves, and the ***spherical Ball curve*** can be drawn, as shown in Fig. 4.14. Similar to plane linkage, spherical geodesic inflection curves envelope the spherical Ball curves and the moving centrode π_m.

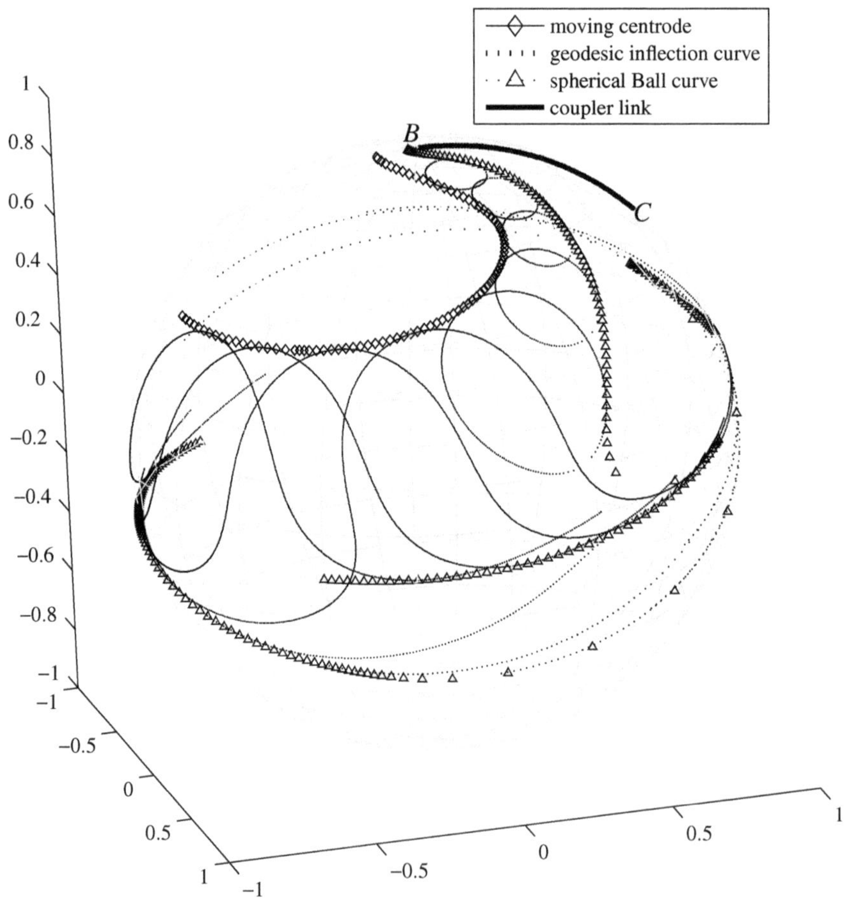

Figure 4.14 The spherical Ball curve of the spherical double-crank linkage

4.3 Spherical Coupler Curves

Spherical four-bar linkages have been discussed intensively in Ref. [3]. We intend to apply the discussions in the above section to spherical four-bar linkages to reveal the systematical approach of differential geometry, and the contents, except for the global property and double points of coupler curves, bear a likeness to those of planar four-bar linkages.

4.3.1 Basic Equation

For a unit spherical four-bar linkage *ABCD*, the moving coordinate system $\{\boldsymbol{O}; \boldsymbol{i}_m, \boldsymbol{j}_m, \boldsymbol{k}_m\}$ and the fixed coordinate system $\{\boldsymbol{O}; \boldsymbol{i}_f, \boldsymbol{j}_f, \boldsymbol{k}_f\}$ are set up respectively on the coupler link *BC* and base link *AB*, as shown in Fig. 4.4. A coupler point *P* with spherical coordinates $(1, \delta_P, \theta_P)$ in $\{\boldsymbol{O}; \boldsymbol{i}_m, \boldsymbol{j}_m, \boldsymbol{k}_m\}$ traces a spherical curve $\varGamma_P$ in the fixed frame, and is always adjoint to joint *B*. $\varGamma_P$ is called an adjoint curve while curve $\varGamma_B$ of joint *B*, a circle on the sphere, is referred to as an original curve. The Darboux frame $\{\boldsymbol{R}_B; \boldsymbol{\alpha}_B, \boldsymbol{n}_B, \boldsymbol{v}_B\}$ of the original curve $\varGamma_B$ is established. The equation of the coupler curve $\varGamma_P$ can be expressed in $\{\boldsymbol{R}_B; \boldsymbol{\alpha}_B, \boldsymbol{n}_B, \boldsymbol{v}_B\}$ as

$$\boldsymbol{R}_P = \boldsymbol{R}_B + u_1\boldsymbol{\alpha}_B + u_2\boldsymbol{n}_B + u_3\boldsymbol{v}_B \tag{4.51}$$

where (u_1, u_2, u_3) are the coordinates of *P* in $\{\boldsymbol{R}_B; \boldsymbol{\alpha}_B, \boldsymbol{n}_B, \boldsymbol{v}_B\}$, or

$$\begin{cases} u_1 = \sin\delta_P \cos\left(\theta_P + \psi\right) \\ u_2 = 1 - \cos\delta_P \\ u_3 = \sin\delta_P \sin\left(\theta_P + \psi\right) \end{cases} \tag{4.52}$$

where ψ is the rotation angle (dihedral angle) of the coupler link *BC* relative to the driving link *AB*.

4.3.2 Double Point

For a spherical linkage, a ***spherical double point*** (SDP) of the spherical coupler curve in the fixed frame relates to the coupler point, or the tracing point of the coupler link, that passes through it twice at two different instants, like that for planar linkages, as shown in Fig. 4.15. There are also two types of SDP: ***spherical crunode*** and ***spherical tacnode***. For a spherical double point on a spherical coupler curve, the tracing point must lie in an identical position at two different instants (or positions). That is

$$\boldsymbol{R}_P^{(1)} = \boldsymbol{R}_P^{(2)} \tag{4.53}$$

where $\boldsymbol{R}_P^{(1)}$ and $\boldsymbol{R}_P^{(2)}$ represent the position vectors of the tracing point at the spherical double point at two different instants. Equation (4.53) may be replaced by identical moduli and arguments of two vectors.

$$\begin{cases} \left|\boldsymbol{R}_P^{(1)}\right| = \left|\boldsymbol{R}_P^{(2)}\right| \\ \boldsymbol{R}_P^{(1)} \times \boldsymbol{R}_P^{(2)} = 0 \end{cases} \tag{4.54}$$

Since the point *P* moves along the unit sphere, the first equation holds all the time if the center point of the sphere is taken as the original reference point. Hence, we can adopt joint *A* as

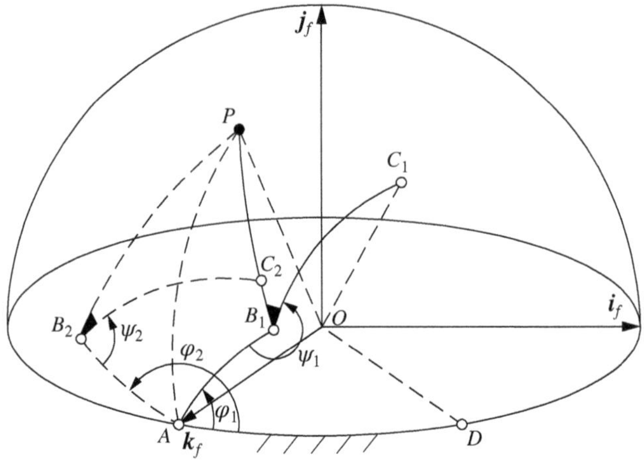

Figure 4.15 The double point of the spherical coupler curve

the reference point, and then the equality of the moduli of the vector $\overrightarrow{AP}$ is taken as the first condition, that is $|\overrightarrow{AP^{(1)}}| = |\overrightarrow{AP^{(2)}}|$. The vector $\overrightarrow{AP}$ can be represented by the Darboux frame $\{\boldsymbol{R}_B; \boldsymbol{\alpha}_B, \boldsymbol{n}_B, \boldsymbol{v}_B\}$ of the curve $\varGamma_B$ of joint B:

$$\overrightarrow{AP} = \overrightarrow{AB} + \overrightarrow{BP} = u_1\boldsymbol{\alpha}_B + \left(u_2 + \cos a_1 - 1\right)\boldsymbol{n}_B + \left(u_3 - \sin a_1\right)\boldsymbol{v}_B \tag{4.55}$$

Substituting equation (E4-1.6) and the above equation into $|\overrightarrow{AP^{(1)}}| = |\overrightarrow{AP^{(2)}}|$, we have

$$\sin\left(\theta_P + \psi_1\right) = \sin\left(\theta_P + \psi_2\right) \tag{4.56}$$

which leads to

$$\theta_P + \psi_1 = n\pi - \left(\theta_P + \psi_2\right) \tag{4.57}$$

where n is an odd number.

In the fixed frame $\{\boldsymbol{O}; \boldsymbol{i}_f, \boldsymbol{j}_f, \boldsymbol{k}_f\}$, the position vector of the point P of the coupler link is

$$\boldsymbol{R}_P = \sin\angle AOP\cos\left(\angle PAB + \varphi\right)\boldsymbol{i}_f + \sin\angle AOP\sin\left(\angle PAB + \varphi\right)\boldsymbol{j}_f + \cos\angle AOP\boldsymbol{k}_f \tag{4.58}$$

where $\angle AOP$ is the center angle of the great circular arc $\widehat{PA}$ and $\angle PAB$ is the spherical included angle (dihedral angle) of the great circular arc $\widehat{PA}$ and $\widehat{AB}$. Substituting equation (4.58) into the second expression of equation (4.54), we have

$$\angle PAB_1 + \varphi_1 = \angle PAB_2 + \varphi_2 \tag{4.59}$$

As shown in Fig. 4.15, $\angle PAB$ keeps constant, $\angle PAB_1 = -\angle PAB_2$, so equation $\angle PAB_1 = (\varphi_2 - \varphi_1)/2 = \Delta\varphi/2$ can be obtained from equation (4.59). Applying the spherical sine/cosine law to the spherical triangle PAB_1, we obtain

$$\sin\delta_P = \frac{\sin a_1 \sin\dfrac{\Delta\varphi}{2}}{\sin\angle APB} \tag{4.60a}$$

where

$$\cos\angle APB = -\cos\frac{\Delta\varphi}{2}\cos\left(\theta_P + \psi_1 - \frac{\pi}{2}\right) - \sin\frac{\Delta\varphi}{2}\sin\left(\theta_P + \psi_1 - \frac{\pi}{2}\right)\cos a_1 \quad (4.60b)$$

Equations (4.57) and (4.60a) demonstrate the corresponding positions of the spherical linkage and the tracing point when an SDP occurs on the coupler curve. Differentiating equation (4.57) with respect to the arc length s of Γ_B, we have

$$\dot{\psi}_1 + \dot{\psi}_2\left(\sin a_1 \frac{\mathrm{d}\Delta\varphi}{\mathrm{d}s} + 1\right) = 0 \quad (4.61)$$

in which $\mathrm{d}(\Delta\varphi)/\mathrm{d}s = -2/\sin a_1$ when $\Delta\varphi \to 0$, $\varphi_2 \to \varphi_1$, $\psi_2 \to \psi_1$, $\dot{\psi}_2 \to \dot{\psi}_1$, and we get

$$\theta_P + \psi_1 = \frac{n}{2}\pi \quad (4.62)$$

Differentiating equation (4.60b) with respect to s, we obtain

$$\frac{\mathrm{d}\angle APB}{\mathrm{d}s} = \mp\frac{\sqrt{1 + \sin^2 a_1 \dot{\psi}^2 + 2\sin a_1 \cos a_1 \dot{\psi}}}{\sin a_1} \quad (4.63)$$

Taking the limit of equation (4.60a) as $\Delta\varphi$ approaches zero:

$$\sin\delta_P = \lim_{\Delta\varphi\to 0} \sin a_1 \frac{\cos\frac{\Delta\varphi}{2}\,\frac{1}{2}\frac{\mathrm{d}\Delta\varphi}{\mathrm{d}s}}{\cos\angle APB\,\frac{\mathrm{d}\angle APB}{\mathrm{d}s}} = -\frac{1}{\frac{\mathrm{d}\angle APB}{\mathrm{d}s}} = \pm\frac{\sin a_1}{\sqrt{1 + \sin^2 a_1 \dot{\psi}^2 + 2\sin a_1 \cos a_1 \dot{\psi}}} \quad (4.64)$$

which leads to

$$\tan\delta_P = \pm\frac{\sin a_1}{\cos a_1 + \dot{\psi}\sin a_1} = \pm\frac{1}{\cot a_1 + \dot{\psi}} \quad (4.65)$$

Comparing equation (E4-1.7) with equations (4.62) and (4.65), we find that the spherical double point of the coupler curve occurs at the instant center.

For $\varphi \in [0, 2\pi]$, different coupler points on a great circle with the same θ_P of the coupler link can be located by equations (4.57) and (4.60a), which respectively locate at SDPs of spherical coupler curves, while the position parameter δ_P of these coupler points varies within a certain limited area. If the variation of δ_P is not monotonic, an extreme value on the limited area must exist. Differentiating equation (4.60a) with respect to s and letting $\frac{\mathrm{d}\delta_P}{\mathrm{d}s} = 0$, we have

$$\cos\frac{\Delta\varphi}{2}\frac{\mathrm{d}\Delta\varphi}{\mathrm{d}s}\sin\angle APB - 2\sin\frac{\Delta\varphi}{2}\cos\angle APB\frac{\mathrm{d}\angle APB}{\mathrm{d}s} = 0 \quad (4.66)$$

By combining equations (4.57) and (4.60a) with equation (4.66), the extreme value of δ_P can be calculated for a given value of θ_P. Similar to plane linkage, the conditions of the self-tangent point or the tacnode for a spherical linkage are

$$\boldsymbol{R}_P^{(1)} = \boldsymbol{R}_P^{(2)}, \quad \frac{\mathrm{d}\boldsymbol{R}_P^{(1)}}{\mathrm{d}s} = \frac{\mathrm{d}\boldsymbol{R}_P^{(2)}}{\mathrm{d}s} \quad (4.67)$$

which is actually a parallel condition for the tangent vectors (the second expression) in addition to the condition (4.53) for an SDP. From equations (4.67), (4.57), (4.60a), and (4.66), a coupler point with parameter δ_P in the limited area will trace a coupler curve with an SDP, and further, the SDP is a spherical tacnode if δ_P has an extreme value in the limited area.

As for the plane linkages, it is difficult to determine whether a coupler point locates at an SDP of its coupler curve or not, due to the different cases of planar linkage. Theoretically, the spherical linkages are viewed as an extension of the plane linkages to the spherical surface. The distributions of the DPs of the coupler curves for planar linkages naturally correspond to those for spherical linkages, although they have different condition equations of double points and different variation areas of their coordinate parameters. Therefore, the distributions of coupler points, tracing coupler curves with double points, for planar four-bar linkages can also be expanded to those for the spherical four-bar linkages, and the calculation procedure is similar.

Example 4.4 The spherical crank-rocker linkage $ABCD$ with dimensions $a_1 = 20°$, $a_2 = 37.5°$, $a_3 = 37.5°$, $a_4 = 50°$ is shown in Example 4.2. Reveal the spherical double points of the coupler curves.

Solution

Three great circles are located on the coupler link, the great circle BC of the coupler link, the great circle BG_1 with spherical included angle (dihedral angle) $\tau_1 = 0.85$ to BC, and the great circle BG_2 with spherical included angle (dihedral angle) $\tau_2 = -0.5$ to BC. There are two intersection points between the great circle BG_1 and two branches of the moving centrode, $P_0^{(1)}(0.8666, 2.4208)$ and $P_0^{(2)}(1.2139, 2.4208)$, which correspond to the instants $\varphi_0^{(1)} = 1.4615$ and $\varphi_0^{(2)} = 5.3763$ respectively.

At instants $\varphi_0^{(1)}$ and $\varphi_0^{(2)}$, the tracing points are just the instant centers $P_0^{(1)}$ and $P_0^{(2)}$ respectively, and also coincident with the spherical double points on their coupler curves. As mentioned in Section 1.3.3, we focus on the great circle BG_1 with $\tau_1 = 0.85$ and the interval of rotation angle $[\varphi_0^{(1)}, \varphi_0^{(2)}]$, to examine where the coupler points locate on BG_1 of the coupler link when they trace coupler curves with spherical double points. As shown in Fig. 4.16(a), the coupler points, between the two instant center points $P_0^{(1)}(\delta_P^{(1)})$ and $P_0^{(2)}(\delta_P^{(2)})$, trace coupler curves with SDPs since the instant center is a special SDP. Hence, the coupler points, whose parameter δ_P varies from $\delta_P^{(1)}$ to $\delta_P^{(2)}$, will be the tracing point of the SDP. If the variation of the parameter δ_P is monotonic, the points on the great circle, whose parameters δ_P are between $[\delta_P^{(1)}, \delta_P^{(2)}]$, trace coupler curves with just an SDP. If the variation is not monotonic, the parameter δ_P gets an extreme value δ_{P_2} at point P_2 but δ_{P_2} is beyond the area $[\delta_P^{(1)}, \delta_P^{(2)}]$, or $\delta_{P_2} < \delta_P^{(1)} < \delta_P^{(2)}$. Thus, the tracing point of SDPs with parameter δ_P on this great circle will change position from $P_0^{(1)}$ to $P_0^{(2)}$ via P_2, or the parameter δ_P will cover the interval $[\delta_{P_2}, \delta_P^{(1)}]$ twice and the interval $[\delta_P^{(1)}, \delta_P^{(2)}]$ only once, which indicates that the coupler points between $[\delta_{P_2}, \delta_P^{(1)}]$ will trace the coupler curves with two SDPs, while the coupler points between $[\delta_P^{(1)}, \delta_P^{(2)}]$ will trace the coupler curves with only one SDP. The extreme value of parameter δ_P can be calculated by equations (4.60) and (4.66). For different values of $\theta_P \in (0, 2\pi)$, all the coupler points with extreme values of the parameter δ_P will constitute a curve of the coupler link, called a ***spherical self-tangent tracer curve***, or a ***spherical transition curve*** for a spherical linkage, as shown in Fig. 4.16(a), of which both ends coalesce with two branches of the spherical moving centrode. Each point on the spherical transition curve will trace a spherical coupler curve

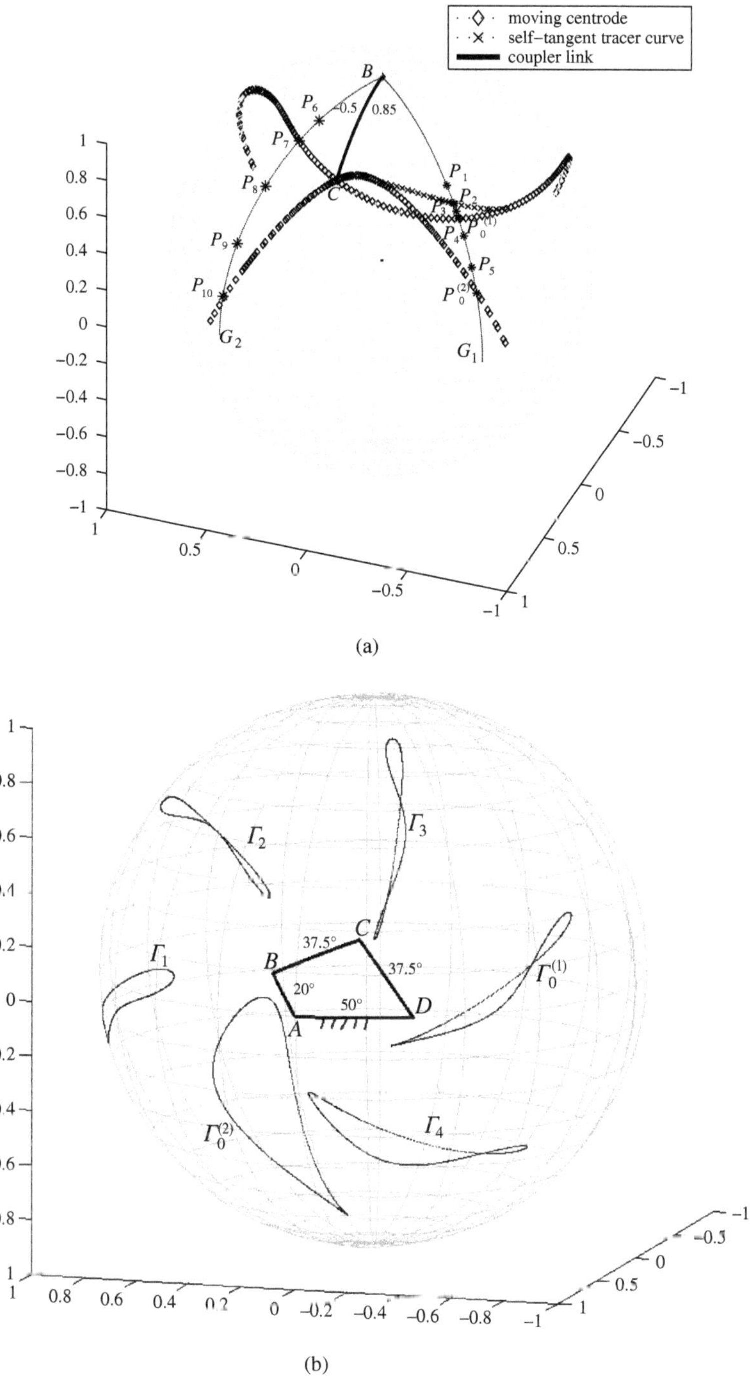

(a)

(b)

Figure 4.16 Double points for spherical crank-rocker linkages

with a spherical tacnode. Therefore, a coupler point located between the spherical transition curve and the spherical moving centrode will trace a curve with two spherical crunodes, while a point located between the two branches of the spherical moving centrode will trace a curve with only one spherical crunode. These can be seen from the coupler curves in Fig. 4.16(b).

For the coupler points on the great circle BG_2 with $\tau_2 = -0.5$ from P_6 to P_{10}, or from one branch to another branch of the spherical moving centrode, the parameter δ_P varies monotonically from P_6 to P_{10}, or from one branch to another branch of the spherical moving centrode. Thus, it is only the points in between the two branches of the spherical moving centrode that trace spherical coupler curves with one SDP. The spherical coupler curves selected are shown in Fig. 4.16(b).

4.3.3 Distribution

As mentioned above, the kinematic geometry of spherical linkages can be viewed as an extension of that of planar linkages. The properties of planar curves can be described by their curvatures, while spherical curves are expressed by their geodesic curvatures. In particular, a plane curve is a straight line if its curvature is zero, and a spherical curve is a great circle if its geodesic curvature is zero. Hence, both the local characteristics and distribution of spherical coupler curves for spherical four-bar linkages correspond to those of coupler curves for planar four-bar linkages, with only some differences in the expression form. In this section, the distribution of spherical coupler curves with both local and global characteristics is briefly introduced, corresponding to that of plane linkages.

1. *The cusp of a spherical coupler curve.* The spherical coupler curve of a spherical four-bar linkage has a cusp point if the geodesic curvature approaches infinity at a point, or the denominator of equation (4.43a) is zero. At this instant, the coupler point coincides with the instant center. All the coupler points locate on the spherical moving centrode of the coupler link if these coupler points trace spherical coupler curves with a cusp.
2. *Geodesic inflection points.* A geodesic inflection point of a spherical coupler curve appears if the geodesic curvature is zero at the point. There is a geodesic inflection curve on the coupler link at an instant, which can be described by equation (4.44). Unlike that of planar linkages, the spherical geodesic inflection curve appears in different forms and changes with the variation of the induced geodesic curvatures, which may envelop two kinds of curve on the coupler link, the spherical Ball curve and the spherical moving centrode.
3. *Spherical Ball points and Ball curves.* At any instant, a spherical Ball point on the coupler link traces a spherical coupler curve with $k_{gP} = 0$ and $\mathrm{d}k_{gP}/\mathrm{d}\sigma = 0$, which contacts a great circle in third order at this point. There may exist four spherical Ball points on the moving body in spherical movement at an instant, but only one in planar movement. All the spherical Ball points form spherical Ball curves on the coupler link.
4. *Spherical Burmester points and Burmester curves.* A spherical coupler curve with $\mathrm{d}k_{gP}/\mathrm{d}\sigma = 0$ and $\mathrm{d}^2k_{gP}/\mathrm{d}\sigma^2 = 0$ has five common points with a great circle at infinitesimally separated positions in the neighborhood of the point. According to equation (4.50), there exist less than six spherical Burmester points on the moving body, of which two are the hinged points on the coupler link for a spherical four-bar linkage. With the continuous movement of the coupler link, the spherical Burmester points form curves on the coupler link, referred to as ***spherical Burmester curves***.

For a spherical four-bar linkage – a spherical crank-rocker linkage – the coupler curves may correspond to those of a plane crank-rocker linkage, and appear in different shapes, such as eggs, pears, raindrops, bananas, figures-of-eight, double figures-of-eight, etc. on the sphere. The shapes of spherical coupler curves depend on their local and global characteristics, or the positions of the tracing point in the coupler link. For a spherical crank-rocker linkage with given dimensions, the spherical moving centrode, the spherical self-tangent tracer curve, the spherical Ball curves can be depicted based on the equations in Section 4.2, as shown in Fig. 4.17. The point P_1 on the spherical Ball curve of the coupler link traces a spherical coupler curve, which is widely approximated to a segment of a great circle. The point P_2 on the self-tangent tracer curve traces a coupler curve with a spherical tacnode. The point P_5 on the spherical moving centrode traces a spherical coupler curve with a cusp.

The point P_3, in between the spherical self-tangent tracer curve and the spherical moving centrode, will trace a coupler curve with two spherical crunodes shaped like a double figure-of-eight, while the point P_4, in between the two branches of the two spherical centrodes, will trace a coupler curve with a single spherical crunode shaped like a single figure-of-eight.

4.4 Discussion

The spherical movement of a rigid body, like the planar movement, has three degrees of freedom, or three independent parameters. It is currently viewed as an extension of the planar movement, and the kinematic geometry in spherical movement is usually studied by comparing both the contents and approaches in plane movement. The kinematic properties were previously studied by Dobrovolskii [4], Meyer zur Capellen, Dittrich, and Janssen [5], and Müller [6]. Applying the right angle theorem [i.e., when the vertex of a right angle moves along a fixed curve, if one side of a right angle remains tangential to the curve at the vertex (in fact, the right angle is generally the Frenet frame of the curve), the pole of a right angle in the motion coincides with the center of curvature of the curve] to the spherical kinematics, Veldkamp [7] derived the spherical Euler–Savary equation and Bobillier theorem, etc. Kamphuis [8] studied the instantaneous spherical kinematics based on the instantaneous angular velocity vector and its derivatives. A spherical slider-crank motion was analyzed to obtain the properties of the Ball point and Burmester point on the coupler link. Based on the description of the spherical motion by Kamphuis, Yang and Roth [9] defined the characteristic numbers and described the spherical curve to a given order, and then studied the loci of all points of a rigid body in spherical motion, which have the same characteristic numbers. The research results provide the theoretical basis for the analysis and synthesis of spherical linkages. Sodhi and Shoup [10] derived the algebraic equations for the axodes of the spherical 4R mechanism, whose order is 16. Ting and Soni [11] studied the plane envelope in spherical motion, and described its geometrical characteristics by the curvature-like characteristic numbers, so that the plane with the same characteristic numbers can be identified in the families of planes. McCarthy [12,13] defined a kinematic mapping from the spherical displacement to the curve on a unit hypersphere by the Euler parameters of the rotation of a rigid body. The instantaneous kinematics of a rigid body in spherical motion can be shown by the curves on the hypersphere and studied by differential geometry. Ting [14] studied the spherical curvature theory of point, plane, and circle paths. The research results revealed the duality between point and plane paths. The works of Hunt [15], Bottema and Roth [16] covered the curvature theories of spherical motion by means of kinematic invariants. The book about the kinematic analysis and synthesis

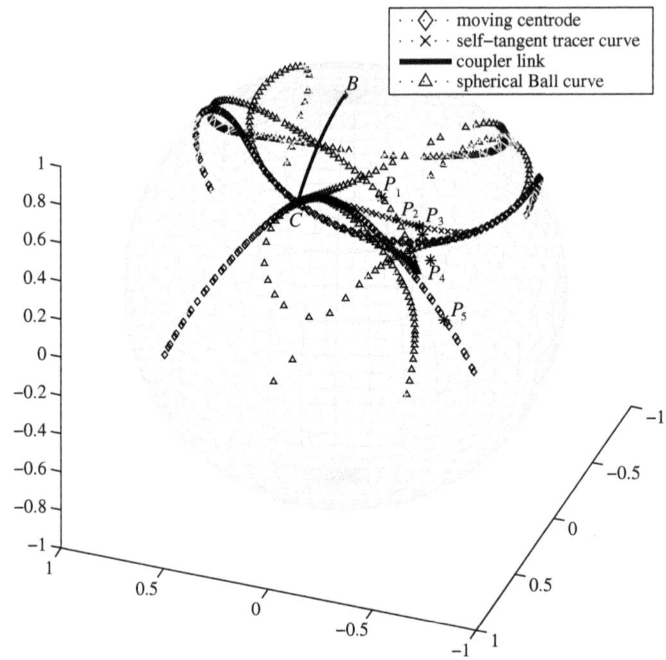

(a)

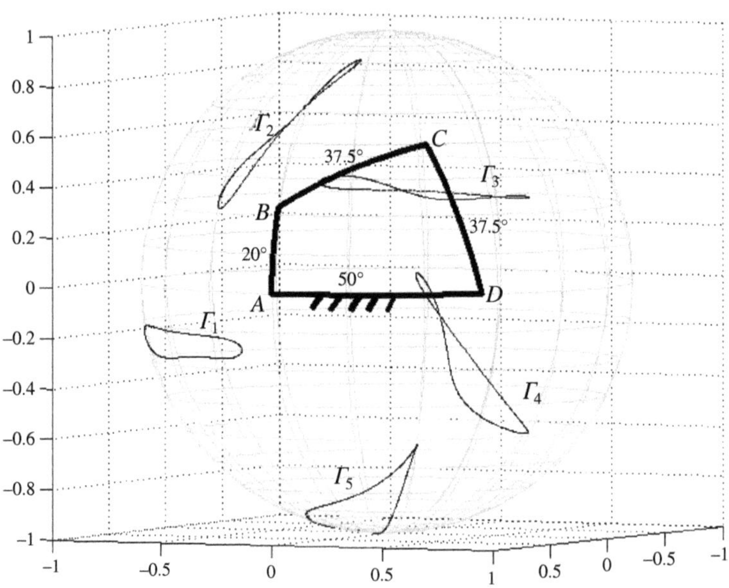

(b)

Figure 4.17 Coupler curve distribution for a spherical crank-rocker linkage

of spherical mechanisms by Chiang [3] developed this from planar motion to spherical motion by the vector and matrix method, and illustrated their close connections.

The local geometrical characteristics of the coupler curves for spherical linkages can be described totally by the parameters of instantaneous kinematics, but their global properties have not yet been revealed sufficiently compared with those of planar movement. Dobrovolskii [17] gave the primary discussion on the properties of the spherical coupler curve and demonstrated that the coupler curve of a spherical four-bar linkage has eighth degree. Lu and Hwang [18] analyzed the symmetrical coupler curves of spherical four-bar linkages, which need an additional condition relative to that of planar four-bar linkages, and pointed out that the lengths of both the coupler link and the driven link are 90°. Lu [19] and Hwang [20,21] gave the triangular nomograms for spherical crank-rocker linkages, double-crank linkages, and Grashof/non-Grashof double-rocker linkages with symmetrical coupler curves. Shirazi [22] studied the properties of the coupler curve for planar and spherical inverted crank-slider linkages, and derived the relationship between numbers of double points on the coupler curves and the lengths of links.

The global properties of coupler curves for spherical linkages, of course, can be studied more directly and concisely by the approach of differential geometry, but they are more complex than those for planar linkages. As mentioned above for the double points of the spherical coupler curves, we believe that the conditions of symmetrical spherical coupler curves should be described by simple analyzed equations, which can be derived by the approach of differential geometry. Similarly, the oval spherical coupler curves could be derived based on the geodesic curvatures and the simple spherical coupler curve, and the distributions of their tracing points are certainly located on the coupler link.

The kinematic geometry in spherical movement, in both contents and expressional forms, corresponds to that in planar movement, and is currently viewed as a bridge from planar to spatial movement. Unfortunately, we believe that the bridge is not set up yet until you have read Chapters 6 and 7 of this book – including the curvature theories for point and line trajectories in spatial movement, or the spatial curves and ruled surfaces, which are expected to be similar to those in planar movement. Yang and Kirson discussed the properties of the ruled surface by the involute ruled surface and the characteristic scalars. McCarthy studied the ruled surface by the natural trihedron and the curvature functions. So far, fewer constraint ruled surfaces in spatial linkages have been systematically studied in terms of three construction parameters in Chapter 3 of this book, or a ruled surface is decomposed into both the spherical image curve of the rulings and the striction curve for discussion, which provides the connective condition for moving from the spherical kinematics to the spatial kinematics. However, the axodes in spatial movement are the key ruled surfaces and don't correspond to those in spherical movement yet. Needless to say, this isn't easy in the kinematic analysis and synthesis of spatial linkages, yet these are discussed respectively in Chapters 6 and 7. Comparing the kinematics in spherical movement with that in planar movement, especially for the local and global properties of the spherical image curves of the rulings for ruled surfaces, we believe that it is just the beginning of the research into spatial movement, which may signify the development of a new research field in spherical kinematics. On the contrary, a curve of a moving body in spherical movement will produce a complex surface in a fixed frame, although Ting has carried out innovative work. The study of geometrical properties by differential geometry will certainly reveal the fantastic contents of spherical kinematics, and highlight the charm of differential geometry [23, 24].

References

[1] G. Mozzi, Discorso matematico sopra il rotamento momentaneo dei corpi. Stamperia di Donato Campo, Napoli, 1763.

[2] M. Chasles, Note sur les Proprietes Generales du Systeme de Deux Corps Semblables entr'eux. Bullettin de Sciences Mathematiques, Astronomiques Physiques et Chimiques, Baron de Ferussac, Paris, 1830, pp. 321–326.

[3] C.H. Chiang, *Kinematics of Spherical Mechanisms*, 2nd edn. Krieger Publishing, Malabar, FL, 2000.

[4] V.V. Dobrovolskii, *Theory of Spherical Mechanisms*. Akadamiia Nauk, Moscow, 1947 (in Russian).

[5] W. Meyer zur Capellen, G. Dittrich, and B. Janssen, *Systematik und Kinematik ebener und sphärischer Viergelenkgetriebe*. Westdeutscher Verlag, Köln und Opladen, 1966.

[6] H.R. Müller, *Sphärische Kinematik*. VEB-Verlag der Wissenschaften, Berlin, 1962.

[7] G.R. Veldkamp, An approach to spherical kinematics using tools suggested by plane kinematics. *Journal of Mechanisms*, **2**(4), 437–450, 1967.

[8] H.J. Kamphuis, Application of spherical instantaneous kinematics to the spherical slider-crank mechanism. *Journal of Mechanisms*, **4**(1), 43–56, 1969.

[9] A.T. Yang and B. Roth, Higher-order path curvature in spherical kinematics. *ASME Journal of Engineering for Industry*, **95**(2), 612–616, 1973.

[10] R. Sodhi and T.E. Shoup, Axodes for the four-revolute spherical mechanism. *Mechanism and Machine Theory*, **17**(3), 173–178, 1982.

[11] K.L. Ting and A.H. Soni, Instantaneous kinematics of a plane in spherical motion. *ASME Journal of Mechanisms, Transmissions, and Automation in Design*, **105**(3), 560–567, 1983.

[12] J.M. McCarthy and B. Ravani, Differential kinematics of spherical and spatial motions using kinematic mapping. *ASME Journal of Applied Mechanics*, **53**(1), 15–22, 1986.

[13] J.M. McCarthy, The differential geometry of curves in an image space of spherical kinematics. *Mechanism and Machine Theory*, **22**(3), 205–211, 1987.

[14] K.L. Ting and R. Bunduwongse, Unified spherical curvature theory of point-, plane-, and circle-paths. *ASME Journal of Mechanical Design*, **113**(2), 142–149, 1991.

[15] K.H. Hunt, *Kinematic Geometry of Mechanism*. Oxford University Press, Oxford, 1978.

[16] O. Bottema and B. Roth, *Theoretical Kinematics*. North-Holland, New York, 1979.

[17] V.V. Dobrovolskii, On spherical coupler curves. *Prikl. Math. Mekh.*, **VIII**, 475–477, 1944 (in Russian).

[18] D.M. Lu and W.M. Hwang, Spherical four-bar linkages with symmetrical coupler-curves. *Mechanism and Machine Theory*, **31**(1), 1–10, 1996.

[19] D.M. Lu, A triangular nomogram for spherical symmetric coupler curves and its application to mechanism design. *ASME Journal of Mechanical Design*, **121**(2), 323–326, 1999.

[20] W.M. Hwang and K.H. Chen, Triangular nonograms for symmetrical spherical four-bar linkages generating symmetrical coupler curves. *Transactions of the Canadian Society for Mechanical Engineering*, **29**(1), 109–128, 2005.

[21] W.M. Hwang and K.H. Chen, Triangular nomograms for symmetrical spherical non-Grashof double-rockers generating symmetrical coupler curves. *Mechanism and Machine Theory*, **42**, 871–888, 2007.

[22] K.H. Shirazi, Symmetrical coupler curve and singular point classification in planar and spherical swinging-block linkages. *ASME Journal of Mechanical Design*, **128**, 436–443, 2006.

[23] T.J. Li, *Kinematics Geometry of Spherical Mechanisms*. Dalian University of Technology, Dalian, 2000 (in Chinese).

[24] W. Wang, D.L. Wang, Curvature Theory for Point-Path and Plane-Envelope in Spherical Kinematics by New Adjoint Approach, *Chinese Journal of Mechanical Engineering*, **27**(6), 1157–1168, 2014.

5

Discrete Kinematic Geometry and Saddle Synthesis of Spherical Linkages

Spherical linkages are simplex in both the kinematic pairs and the types relative to planar and spatial linkages, but the expressional form belongs to three-dimensional space, which means it is inconvenient to synthesize a spherical linkage in practical engineering by the geometrical graphic method. The kinematic synthesis of spherical linkages is presented by means of analytical and optimal methods as computing technology develops. Based on the kinematical differential geometry in spherical movement in Chapter 4, the saddle point program for kinematic synthesis of planar linkages in Chapter 2 is developed into that of the spherical linkages in this chapter, which may set up a bridge from planar to spatial linkages, and provides a theoretical basis for the kinematic synthesis of spatial linkages in Chapter 7.

5.1 Matrix Representation

The spherical movement of the rigid body $\sum^*$ can be described by the three rotational angles $\theta_1, \theta_2, \theta_3$ of the moving Cartesian coordinate system $\{\boldsymbol{O}; \boldsymbol{i}_m, \boldsymbol{j}_m, \boldsymbol{k}_m\}$ of $\sum^*$ relative to the fixed Cartesian coordinate system $\{\boldsymbol{O}; \boldsymbol{i}_f, \boldsymbol{j}_f, \boldsymbol{k}_f\}$ established in the fixed body $\sum$ (as shown in Fig. 4.2). θ_1 is the included angle of the coordinate axes $\boldsymbol{k}_m$ and $\boldsymbol{k}_f$, θ_2 is the dihedral angle of the coordinate planes $\boldsymbol{O}-\boldsymbol{k}_m\boldsymbol{k}_f$ and $\boldsymbol{O}-\boldsymbol{k}_f\boldsymbol{i}_f$, while θ_3 is the dihedral angle of the coordinate planes $\boldsymbol{O}-\boldsymbol{k}_m\boldsymbol{i}_f$ and $\boldsymbol{O}-\boldsymbol{k}_m\boldsymbol{i}_m$. The displacement transformation matrix is taken to represent spherical movement, and is discretized and applied in the position synthesis.

For a point P on the unit sphere of $\sum^*$ with Cartesian coordinates (x_{Pm}, y_{Pm}, z_{Pm}), or spherical coordinates $(1, \delta_{Pm}, \theta_{Pm})$ in $\{\boldsymbol{O}; \boldsymbol{i}_m, \boldsymbol{j}_m, \boldsymbol{k}_m\}$, the vector equation is

$$\begin{aligned}\boldsymbol{R}_{Pm} &= x_{Pm}\boldsymbol{i}_m + y_{Pm}\boldsymbol{j}_m + z_{Pm}\boldsymbol{k}_m \\ &= \sin\delta_{Pm}\cos\theta_{Pm}\boldsymbol{i}_m + \sin\delta_{Pm}\sin\theta_{Pm}\boldsymbol{j}_m + \cos\delta_{Pm}\boldsymbol{k}_m\end{aligned} \tag{5.1}$$

Kinematic Differential Geometry and Saddle Synthesis of Linkages, First Edition. Delun Wang and Wei Wang.

The three rotation angles $\theta_1, \theta_2, \theta_3$ of $\{\boldsymbol{O}; \boldsymbol{i}_m, \boldsymbol{j}_m, \boldsymbol{k}_m\}$ to $\{\boldsymbol{O}; \boldsymbol{i}_f, \boldsymbol{j}_f, \boldsymbol{k}_f\}$ are the same as those in Chapter 4. The position vector $\boldsymbol{R}_P = (x_{Pf}, y_{Pf}, z_{Pf})^T$ of P in $\{\boldsymbol{O}; \boldsymbol{i}_f, \boldsymbol{j}_f, \boldsymbol{k}_f\}$ can be obtained by the following coordinate transformation:

$$\begin{bmatrix} x_{Pf} \\ y_{Pf} \\ z_{Pf} \end{bmatrix} = [\mathbf{R}] \cdot \begin{bmatrix} x_{Pm} \\ y_{Pm} \\ z_{Pm} \end{bmatrix} = [\mathbf{R}] \cdot \begin{bmatrix} \sin\delta_{Pm}\cos\theta_{Pm} \\ \sin\delta_{Pm}\sin\theta_{Pm} \\ \cos\delta_{Pm} \end{bmatrix} \tag{5.2}$$

where

$$[\mathbf{R}] = \begin{bmatrix} \mathrm{c}\theta_3\mathrm{s}\theta_2 + \mathrm{c}\theta_1\mathrm{s}\theta_3\mathrm{c}\theta_2 & -\mathrm{s}\theta_3\mathrm{s}\theta_2 + \mathrm{c}\theta_1\mathrm{c}\theta_3\mathrm{c}\theta_2 & \mathrm{s}\theta_1\mathrm{c}\theta_2 \\ -\mathrm{c}\theta_3\mathrm{c}\theta_2 + \mathrm{c}\theta_1\mathrm{s}\theta_3\mathrm{s}\theta_2 & \mathrm{s}\theta_3\mathrm{c}\theta_2 + \mathrm{c}\theta_1\mathrm{c}\theta_3\mathrm{s}\theta_2 & \mathrm{s}\theta_1\mathrm{s}\theta_2 \\ -\mathrm{s}\theta_1\mathrm{s}\theta_3 & -\mathrm{s}\theta_1\mathrm{c}\theta_3 & \mathrm{c}\theta_1 \end{bmatrix}$$

Here, s and c are respectively abbreviations of the sine and cosine functions. The ***rotation transformation matrix*** $[\mathbf{R}]$ sets up the connection between $\sum^*$ and $\sum$, which can be determined through the above equation if the three kinematic parameters $\theta_1, \theta_2, \theta_3$ are known.

If the spherical movement of $\sum^*$ is represented by the displacements of two reference points A and B of $\sum^*$, or (x_A, y_A, z_A) and (x_B, y_B, z_B) in $\{\boldsymbol{O}; \boldsymbol{i}_f, \boldsymbol{j}_f, \boldsymbol{k}_f\}$, we can also obtain the rotation transformation matrix. On the normal of the plane determined by points A, B and the original point O, the third point C has a unit distance to O. The displacements (x_C, y_C, z_C) of C in $\{\boldsymbol{O}; \boldsymbol{i}_f, \boldsymbol{j}_f, \boldsymbol{k}_f\}$ are

$$\begin{cases} x_C = y_A z_B - y_B z_A \\ y_C = z_A x_B - z_B x_A \\ z_C = x_A y_B - x_B y_A \end{cases} \tag{5.3a}$$

The displacements of the three points A, B, and C can be represented by the arrays (A_{I}), (B_{I}), (C_{I}) for position I and (A_{II}), (B_{II}), (C_{II}) for position II, and the rotation transformation matrix $[\mathbf{R}]$ can be obtained as

$$\left[\mathbf{R}_{\mathrm{I\,II}}\right] = \begin{bmatrix} x_{A\mathrm{II}} & x_{B\mathrm{II}} & x_{C\mathrm{II}} \\ y_{A\mathrm{II}} & y_{B\mathrm{II}} & y_{C\mathrm{II}} \\ z_{A\mathrm{II}} & z_{B\mathrm{II}} & z_{C\mathrm{II}} \end{bmatrix} \cdot \begin{bmatrix} x_{A\mathrm{I}} & x_{B\mathrm{I}} & x_{C\mathrm{I}} \\ y_{A\mathrm{I}} & y_{B\mathrm{I}} & y_{C\mathrm{I}} \\ z_{A\mathrm{I}} & z_{B\mathrm{I}} & z_{C\mathrm{I}} \end{bmatrix}^{-1} \tag{5.3b}$$

At the original position, $\{\boldsymbol{O}; \boldsymbol{i}_m, \boldsymbol{j}_m, \boldsymbol{k}_m\}$ coincides with $\{\boldsymbol{O}; \boldsymbol{i}_f, \boldsymbol{j}_f, \boldsymbol{k}_f\}$, or $\left[\mathbf{R}_{\mathrm{I\,II}}\right]$ is a unit matrix. The rotation angles $\theta_1, \theta_2, \theta_3$ can be obtained by letting the corresponding elements be equal of $[\mathbf{R}]$ in equation (5.2) and $[\mathbf{R}_{\mathrm{I\,II}}]$ in equation (5.3b). Equation (5.2), including the rotation transformation matrix $[\mathbf{R}]$, is taken as the general representation of the spherical movement of $\sum^*$.

If the spherical movement of $\sum^*$ is described by either three rotation parameters $(\theta_1^{(i)}, \theta_2^{(i)}, \theta_3^{(i)})$ or discrete displacements of two points of $\sum^*$, the point P of $\sum^*$ traces a discrete point set in $\{\boldsymbol{O}; \boldsymbol{i}_f, \boldsymbol{j}_f, \boldsymbol{k}_f\}$, which can be represented by

$$\boldsymbol{R}_P^{(i)} = \begin{bmatrix} x_{Pf}^{(i)} \\ y_{Pf}^{(i)} \\ z_{Pf}^{(i)} \end{bmatrix} = \left[\mathbf{R}^{(i)}\right] \cdot \begin{bmatrix} x_{Pm} \\ y_{Pm} \\ z_{Pm} \end{bmatrix} \tag{5.4}$$

where

$$[\mathbf{R}^{(i)}] = \begin{bmatrix} c\theta_3^{(i)}s\theta_2^{(i)} + c\theta_1^{(i)}s\theta_3^{(i)}c\theta_2^{(i)} & -s\theta_3^{(i)}s\theta_2^{(i)} + c\theta_1^{(i)}c\theta_3^{(i)}c\theta_2^{(i)} & s\theta_1^{(i)}c\theta_2^{(i)} \\ -c\theta_3^{(i)}c\theta_2^{(i)} + c\theta_1^{(i)}s\theta_3^{(i)}s\theta_2^{(i)} & s\theta_3^{(i)}c\theta_2^{(i)} + c\theta_1^{(i)}c\theta_3^{(i)}s\theta_2^{(i)} & s\theta_1^{(i)}s\theta_2^{(i)} \\ -s\theta_1^{(i)}s\theta_3^{(i)} & -s\theta_1^{(i)}c\theta_3^{(i)} & c\theta_1^{(i)} \end{bmatrix}$$

So far, the discrete spherical movement of $\sum^*$ can be defined by either the discrete rotation parameters $(\theta_1^{(i)}, \theta_2^{(i)}, \theta_3^{(i)})$ or the ***discrete rotation transformation matrix*** $[\mathbf{R}^{(i)}]$. The discrete trajectory $\{\boldsymbol{R}_P^{(i)}\}$ of the point P of $\sum^*$ in $\{\boldsymbol{O}; \boldsymbol{i}_f, \boldsymbol{j}_f, \boldsymbol{k}_f\}$ can be described by equation (5.4). Therefore, for convenience of description hereinafter, the spherical movement of $\sum^*$ is prescribed by the discrete rotation parameters $(\theta_1^{(i)}, \theta_2^{(i)}, \theta_3^{(i)})$ or even the discrete trajectory $\{\boldsymbol{R}_P^{(i)}\}$, which will not be explained further.

5.2 Saddle Spherical Circle Point

The spherical linkages have only one kind of kinematic pair, the R-pair, and all axes of R-pairs intersect at the center point of the sphere. There merely exists one kind of binary link R–R for spherical linkages, whose constraint curve is a circle on the sphere. Hence, the kinematic synthesis of spherical linkages is just to locate the characteristic points, or the spherical circle points on the moving body. The approximation between a discrete trajectory, a point trajectory in spherical movement, and a spherical circle is discussed for the global geometrical property.

5.2.1 Saddle Spherical Circle Fitting

For the spherical positions of $\sum^*$ described by equation (5.4), the kinematic synthesis of a spherical four-bar linkage is to locate two spherical circle points of $\sum^*$ to constitute two spherical binary links R–R, by comparing the point trajectory with the constraint curve of spherical linkages. The point trajectory – a set of spherical discrete points $\{\boldsymbol{R}_P^{(i)}\}$ in $\{\boldsymbol{O}; \boldsymbol{i}_f, \boldsymbol{j}_f, \boldsymbol{k}_f\}$ – is traced by a point P of $\sum^*$ with coordinates (x_{Pm}, y_{Pm}, z_{Pm}) or $(1, \delta_{Pm}, \theta_{Pm})$ in $\{\boldsymbol{O}; \boldsymbol{i}_m, \boldsymbol{j}_m, \boldsymbol{k}_m\}$. The constraint curve of a spherical four-bar linkage is certainly a spherical curve, a circle on the spherical surface. As mentioned in Chapter 3, the geodesic curvature of a circle on a spherical surface is constant, and the normal curvature is just the radius of the sphere (constant). We estimate the approximation of the discrete point set $\{\boldsymbol{R}_P^{(i)}\}$ and a spherical circle by saddle point programming.

A spherical circle and the sphere center constitute a cone, whose position and dimension depend on the direction angles (ξ, η) of the axis and the half-cone angle δ of the cone. The angles (ξ, η) are adaptively determined by letting the maximum normal fitting error be minimum for a spherical circle to fit the discrete point set $\{\boldsymbol{R}_P^{(i)}\}$ if the half-cone angle δ is given a definite value. The spherical circle with a given half-cone angle δ, shown in Fig. 5.1, is called a ***floating fitting spherical circle***. Different half-cone angles may correspond to different floating fitting spherical circles. Among them, there must exist a spherical circle for all half-cone angles, whose maximum fitting error is minimum.

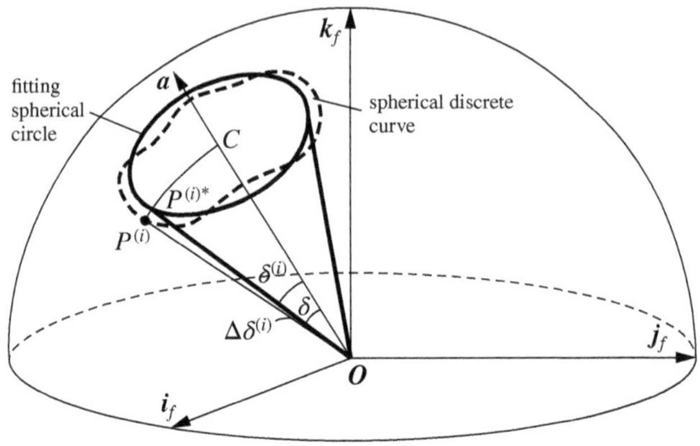

Figure 5.1 The saddle spherical circle fitting of a spherical curve

Definition 5.1 For a spherical discrete point set $\{\boldsymbol{R}_P^{(i)}\}$, a spherical circle, adaptively located by letting the maximum fitting error be minimum, is defined as a ***saddle spherical circle***, while the corresponding maximum fitting error is referred to as the ***saddle spherical circle error***.

For the discrete point set $\{\boldsymbol{R}_P^{(i)}\}$, the saddle spherical circle and the center of the sphere constitute a cone, whose axis $\boldsymbol{a}$ in $\{\boldsymbol{O}; \boldsymbol{i}_f, \boldsymbol{j}_f, \boldsymbol{k}_f\}$ can be represented as

$$\boldsymbol{a} = (\sin\xi\cos\eta, \sin\xi\sin\eta, \cos\xi)^T \tag{5.5}$$

Each point $\boldsymbol{R}_P^{(i)} = (x_P^{(i)}, y_P^{(i)}, z_P^{(i)})^T$ of $\{\boldsymbol{R}_P^{(i)}\}$ is connected to the sphere center O and forms the corresponding angle $\delta^{(i)}$ with axis $\boldsymbol{a}$, that is

$$\delta^{(i)} = \arccos\left(\sin\xi\cos\eta\cdot x_P^{(i)} + \sin\xi\sin\eta\cdot y_P^{(i)} + \cos\xi\cdot z_P^{(i)}\right) \tag{5.6}$$

Hence, the mathematical model of the saddle spherical circle fitting is set up as

$$\begin{cases} \Delta_{\text{rR}} = \min\limits_{\boldsymbol{x}} \max\limits_{1\le i\le n} \left\{\Delta^{(i)}(\boldsymbol{x})\right\} \\ = \min\limits_{\boldsymbol{x}} \max\limits_{1\le i\le n} \left\{\left|\arccos\left(\sin\xi\cos\eta\cdot x_P^{(i)} + \sin\xi\sin\eta\cdot y_P^{(i)} + \cos\xi\cdot z_P^{(i)}\right) - \delta\right|\right\} \\ \text{s.t. } \xi\in[0,\pi),\ \eta\in[0,2\pi),\ \delta\in\left[0,\pi/2\right] \\ \boldsymbol{x} = (\xi,\eta,\delta)^T \end{cases} \tag{5.7}$$

in which the objective function $\{\Delta^{(i)}(\boldsymbol{x})\}$ is the set of normal fitting errors between the fitting spherical circle and all points in $\{\boldsymbol{R}_P^{(i)}\}$. The optimization variables $\boldsymbol{x} = (\xi, \eta, \delta)^T$ have the same meanings as mentioned above. The number of discrete points is n, and the saddle spherical circle error Δ_{rR} is the maximum fitting error of the discrete points to the saddle spherical circle.

The optimal mathematical model (5.7) of the saddle spherical circle can be solved directly by the function **fminimax** in the software Matlab®, whose solving process and parameter setting will be introduced in detail in the following numerical examples. The initial values of the optimization variables can also be determined by the least-square method and are discussed as follows.

A circle fitting a spherical curve is equivalent to a plane fitting a spherical curve since the plane can intersect the sphere at a circle. Thus, based on the discrete spherical point set $\{\boldsymbol{R}_P^{(i)}\} = \{(x_P^{(i)}, y_P^{(i)}, z_P^{(i)})^T\}$, the fitting plane can be determined by the least-square method, or by letting the quadratic sum of fitting errors be zero. If a plane with equation $A \cdot x + B \cdot y + C \cdot z + 1 = 0$ is adopted to fit $\{\boldsymbol{R}_P^{(i)}\}$, the deviations e_i between the point set $\{\boldsymbol{R}_P^{(i)}\}$ and the fitting plane are

$$e_i = A \cdot x_P^{(i)} + B \cdot y_P^{(i)} + C \cdot z_P^{(i)} + 1,\ i = 1,\ \ldots\ , n \tag{5.8}$$

The quadratic sum of the normal deviations e_i can be written as

$$F = \sum_{i=1}^{n} e_i^2 = \sum_{i=1}^{n} \left(A \cdot x_P^{(i)} + B \cdot y_P^{(i)} + C \cdot z_P^{(i)} + 1\right)^2 \tag{5.9}$$

Differentiating the above equation with respect to the coefficients A, B, and C respectively and letting them be zero, we have

$$\begin{cases} A \cdot \sum_{i=1}^{n} x_i^2 + B \cdot \sum_{i=1}^{n} x_i y_i + C \cdot \sum_{i=1}^{n} x_i z_i + \sum_{i=1}^{n} x_i = 0 \\ A \cdot \sum_{i=1}^{n} x_i y_i + B \cdot \sum_{i=1}^{n} y_i^2 + C \cdot \sum_{i=1}^{n} y_i z_i + \sum_{i=1}^{n} y_i = 0 \\ A \cdot \sum_{i=1}^{n} x_i z_i + B \cdot \sum_{i=1}^{n} y_i z_i + C \cdot \sum_{i=1}^{n} z_i^2 + \sum_{i=1}^{n} z_i = 0 \end{cases} \tag{5.10}$$

The coefficients A, B, and C for the fitting plane can be solved uniquely by the above equation, or

$$\begin{bmatrix} A \\ B \\ C \end{bmatrix} = \begin{bmatrix} \sum_{i=1}^{n} x_i^2 & \sum_{i=1}^{n} x_i y_i & \sum_{i=1}^{n} x_i z_i \\ \sum_{i=1}^{n} x_i y_i & \sum_{i=1}^{n} y_i^2 & \sum_{i=1}^{n} y_i z_i \\ \sum_{i=1}^{n} x_i z_i & \sum_{i=1}^{n} y_i z_i & \sum_{i=1}^{n} z_i^2 \end{bmatrix}^{-1} \cdot \begin{bmatrix} -\sum_{i=1}^{n} x_i \\ -\sum_{i=1}^{n} y_i \\ -\sum_{i=1}^{n} z_i \end{bmatrix} \tag{5.11}$$

According to the intersection circle between the fitting plane and the sphere, the position of the intersection circle, or the fitting circle, can be located by the direction angles (ξ_0, η_0) and the half-cone angle δ_0:

$$\begin{cases} \sin \xi_0 \cos \eta_0 = -A/\sqrt{A^2 + B^2 + C^2} \\ \sin \xi_0 \sin \eta_0 = -B/\sqrt{A^2 + B^2 + C^2} \\ \cos \xi_0 = -C/\sqrt{A^2 + B^2 + C^2} \\ \tan \delta_0 = \sqrt{A^2 + B^2 + C^2 - 1} \end{cases} \tag{5.12}$$

The values of $(\xi_0, \eta_0, \delta_0)$ can be designated as the initial values of the optimization variables $\boldsymbol{x} = (\xi, \eta, \delta)^T$.

Based on the above definition and the optimization model (5.7), we determine the saddle spherical circle by the function **fminimax** in Matlab. All parameters of **fminimax** can be set up in the same way as those for a saddle circle to fit a planar curve in Chapter 2. The ***sub-program ArRF of saddle spherical circle*** is naturally programmed for the location of the saddle spherical circle. The inputs of the sub-program ArRF are the spherical coordinates of the discrete point set $\{\boldsymbol{R}_P^{(i)}\}$ and the outputs are the direction angles (ξ, η) of the axis, the half-cone angle δ of the saddle spherical circle (cone), and the saddle spherical circle error Δ_{rR}.

The saddle spherical circle of $\{\boldsymbol{R}_P^{(i)}\}$ is optimally determined by solving the mathematic model or the saddle point program. This means that there is no other spherical circle which can fit $\{\boldsymbol{R}_P^{(i)}\}$ to make the maximum fitting error less in the neighborhood of the parameters of the saddle spherical circle. Like the saddle circle in Chapter 2, the saddle spherical circle also has the same properties of adaptivity, uniqueness, and comparability.

5.2.2 Saddle Spherical Circle

For a spherical trajectory Γ_P traced by a point P of $\sum^*$ in $\{\boldsymbol{O}; \boldsymbol{i}_f, \boldsymbol{j}_f, \boldsymbol{k}_f\}$, the saddle spherical circle error Δ_{rR} is the maxima for all the points on Γ_P. However, for all the fitting spherical circles, the maximal error Δ_{rR} is the smallest.

For convenience in discussing the properties of the fitting error of the saddle spherical circle, the given trajectory Γ_P is assumed to be continuous and to lie on a unit sphere. The unit direction vector of the axis of the saddle spherical circle (cone) of Γ_P is $\boldsymbol{a}$, and the half-cone angle is δ. For a point P of Γ_P, whose position vector is $\boldsymbol{R}_P$, the normal fitting error to the saddle spherical circle is Δ. Thus we have

$$\boldsymbol{R}_P \cdot \boldsymbol{a} = \cos(\delta + \Delta) \tag{5.13}$$

The sign of Δ is positive if point P locates outside the saddle spherical circle (cone) and negative if P is inside, and the normal fitting error Δ is

$$\Delta = \arccos\left(\boldsymbol{R}_P \cdot \boldsymbol{a}\right) - \delta \tag{5.14}$$

For a saddle spherical circle of Γ_P, the vector $\boldsymbol{a}$ is a constant vector and the half-cone angle δ is a constant. By differentiating equation (5.13) with respect to the arc length s of Γ_P, we have

$$\frac{\mathrm{d}\boldsymbol{R}_P}{\mathrm{d}s} \cdot \boldsymbol{a} = -\sin(\delta + \Delta) \cdot \frac{\mathrm{d}\Delta}{\mathrm{d}s} \tag{5.15}$$

The condition for Δ to be an extreme value is $\frac{\mathrm{d}\Delta}{\mathrm{d}s} = 0$ at P of Γ_P, or

$$\frac{\mathrm{d}\boldsymbol{R}_P}{\mathrm{d}s} \cdot \boldsymbol{a} = 0 \tag{5.16}$$

When the normal plane of Γ_P at P passes through the axis of the saddle spherical circle (cone), the normal fitting error Δ gets an extreme value. Similar to the saddle circle of the planar curve,

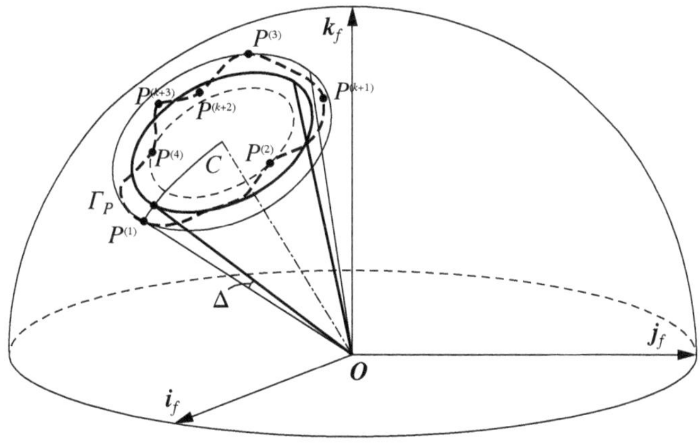

Figure 5.2 The saddle spherical circle and two coaxial cones

the extreme values appears at multiple points of Γ_P but not all of them may be the maxima, such as the points $P^{(k+1)}$, $P^{(k+2)}$, and $P^{(k+3)}$ in Fig. 5.2. For a given spherical trajectory Γ_P, the position vector $\boldsymbol{R}_P$ and the tangent vector $\frac{\mathrm{d}\boldsymbol{R}_P}{\mathrm{d}s}$ are both known. Four points have to be located on Γ_P at least, which hold equation (5.16) true for determining the direction angles (ξ, η) of the axis, the half-cone angle δ, and the error Δ_{rR} of the saddle spherical circle by substituting the position vectors of the points into equation (5.13) or equation (5.14).

Applying saddle point programming to the saddle spherical circle, we can estimate the approximation of a given spherical curve and a spherical circle. The spherical circle fitting by letting the maximum fitting error be the minima is equivalent to the given spherical curve being contained by two spherical circles (or two coaxial cones). The given spherical curve may have four tangent points at least with the two spherical circles, which leads to the maximum normal fitting errors being the minima with the same value, as shown in Fig. 5.2. The four tangent points distribute on both sides of the saddle spherical circle (cone), two are inside and two are outside the saddle spherical circle.

Similar to the saddle circle fitting, the four tangent points of the given curve Γ_P that can totally determine the corresponding saddle spherical circle of Γ_P are defined as the ***saddle spherical circle fitting points***. The corresponding positions of $\sum^*$ are referred to as the ***saddle spherical circle fitting positions***. If the saddle spherical circle error is zero, the given spherical curve must be a spherical circle.

We now focus on the discrete point set $\{\boldsymbol{R}_P^{(i)}\}$ traced by P of $\sum^*$ with separated positions. The four saddle spherical circle fitting points have to be selected from $\{\boldsymbol{R}_P^{(i)}\}$. By equation (5.14), the normal fitting error $\Delta^{(i)}$ of a point of $\{\boldsymbol{R}_P^{(i)}\}$ and the saddle spherical circle can be written as

$$\begin{cases} \Delta^{(i)} = \arccos\left(\boldsymbol{R}_P^{(i)} \cdot \boldsymbol{a}\right) - \delta = \arccos\left(\sin\xi\cos\eta \cdot x_P^{(i)} + \sin\xi\sin\eta \cdot y_P^{(i)} + \cos\xi \cdot z_P^{(i)}\right) - \delta \\ \boldsymbol{R}_P^{(i)} - \left[\mathbf{R}^{(i)}\right] \cdot \boldsymbol{R}_{Pm} \end{cases} \tag{5.17}$$

The saddle spherical circle error Δ_{rR} is the maximum absolute value of all $\Delta^{(i)}$:

$$\begin{aligned}\Delta_{rR} &= \max\{|\Delta^{(i)}|\}\\ &= \max\{|\arccos(\sin\xi\cos\eta\cdot x_P^{(i)}+\sin\xi\sin\eta\cdot y_P^{(i)}+\cos\xi\cdot z_P^{(i)})-\delta|\},\ i=1,\ \ldots,n\end{aligned} \tag{5.18}$$

It indicates that the saddle spherical circle error of $\{\boldsymbol{R}_P^{(i)}\}$ can be obtained by numerical calculation and comparison among all the errors $\Delta^{(i)}$ of $\{\boldsymbol{R}_P^{(i)}\}$ by equation (5.18).

5.2.3 *Four Positions*

Four separated positions of $\sum^*$ in spherical movement are given in $\{\boldsymbol{O};\boldsymbol{i}_f,\boldsymbol{j}_f,\boldsymbol{k}_f\}$ and numbered 1, 2, 3, 4. A point P of $\sum^*$ with spherical coordinates $(1,\delta_{Pm},\theta_{Pm})$ traces a spherical discrete point set $\{\boldsymbol{R}_P^{(i)}\}$, $i=1,\ 2,\ 3,\ 4$, comprised of four points $P^{(1)}$, $P^{(2)}$, $P^{(3)}$, and $P^{(4)}$ in $\{\boldsymbol{O};\boldsymbol{i}_f,\boldsymbol{j}_f,\boldsymbol{k}_f\}$. Based on the mathematical model (5.7), we call the sub-program ArRF and obtain the saddle spherical circle of $\{\boldsymbol{R}_P^{(i)}\}$, or a ***four-point saddle spherical circle***. In this case, the four points $P^{(1)}$, $P^{(2)}$, $P^{(3)}$, and $P^{(4)}$ are just the saddle spherical circle fitting points and the four positions of $\sum^*$ are the saddle spherical circle fitting positions.

A spherical discrete trajectory, comprised of four points, corresponds to a four-point saddle spherical circle. According to the saddle point program, the maximum of the fitting errors is minimum if and only if the four fitting errors are identical to an extreme value at each point. The four fitted points $P^{(1)}$, $P^{(2)}$, $P^{(3)}$, and $P^{(4)}$ distribute on both sides of the four-point saddle spherical circle. There are three distribution cases $P^{(1)}P^{(2)}–P^{(3)}P^{(4)}$, $P^{(1)}P^{(3)}–P^{(2)}P^{(4)}$, $P^{(1)}P^{(4)}–P^{(2)}P^{(3)}$ in total, corresponding to three cases of spherical circle fitting. The obtained three spherical circle are referred to as the ***distributing spherical circles*** of the four points. For the distribution case $P^{(1)}P^{(2)}–P^{(3)}P^{(4)}$, two points $P^{(1)}$ and $P^{(2)}$ are inside the fitting spherical circle and two points $P^{(3)}$ and $P^{(4)}$ are outside. By equation (5.17), the fitting error Δ meets

$$\begin{cases}\boldsymbol{R}_P^{(i)}\cdot\boldsymbol{a}=\cos(\delta-\Delta),\ i=1,\ 2\\ \boldsymbol{R}_P^{(i)}\cdot\boldsymbol{a}=\cos(\delta+\Delta),\ i=3,\ 4\end{cases},\qquad \boldsymbol{R}_P^{(i)}=\left[\mathbf{R}^{(i)}\right]\cdot\boldsymbol{R}_{Pm} \tag{5.19}$$

The unit direction vector $\boldsymbol{a}$ of the axis is determined by

$$\boldsymbol{a}=\frac{\left(\boldsymbol{R}_P^{(1)}-\boldsymbol{R}_P^{(2)}\right)\times\left(\boldsymbol{R}_P^{(3)}-\boldsymbol{R}_P^{(4)}\right)}{\left|\left(\boldsymbol{R}_P^{(1)}-\boldsymbol{R}_P^{(2)}\right)\times\left(\boldsymbol{R}_P^{(3)}-\boldsymbol{R}_P^{(4)}\right)\right|} \tag{5.20}$$

The distributing spherical circle error for the distribution case $P^{(1)}P^{(2)}–P^{(3)}P^{(4)}$ is

$$\Delta_{12,34}=\frac{1}{2}\left|\arccos\left(\boldsymbol{R}_P^{(1)}\cdot\boldsymbol{a}\right)-\arccos\left(\boldsymbol{R}_P^{(3)}\cdot\boldsymbol{a}\right)\right| \tag{5.21}$$

Substituting equation (5.20) into the above equation, we can obtain the higher order algebraic equation of δ_{Pm}, θ_{Pm}, $\Delta_{12,34}$, or

$$f\left(\Delta_{12,34},\delta_{Pm},\theta_{Pm}\right)=0 \tag{5.22}$$

In a similar way, the distributing spherical circle errors $\Delta_{13,24}$ for the case $P^{(1)}P^{(3)}–P^{(2)}P^{(4)}$ and $\Delta_{14,23}$ for the case $P^{(1)}P^{(4)}–P^{(2)}P^{(3)}$ can be obtained. According to Definition 5.1, the four-point saddle spherical circle error Δ_{1234} of $\{\boldsymbol{R}_P^{(i)}\}$, $i = 1, 2, 3, 4$ is the minimum of the three distributing spherical circle errors, or

$$\Delta_{1234} = \min\left(\Delta_{12,34}, \Delta_{13,24}, \Delta_{14,23}\right) \tag{5.23a}$$

Another form is

$$\begin{aligned}\Delta_{1234} = \frac{1}{2}\left|\left|\frac{1}{2}\left|\Delta_{12,34} + \Delta_{13,24}\right| - \frac{1}{2}\left|\Delta_{12,34} - \Delta_{13,24}\right|\right| + \Delta_{14,23}\right| \\ - \frac{1}{2}\left|\left|\frac{1}{2}\left|\Delta_{12,34} + \Delta_{13,24}\right| - \frac{1}{2}\left|\Delta_{12,34} - \Delta_{13,24}\right|\right| - \Delta_{14,23}\right|\end{aligned} \tag{5.23b}$$

For the continuity of the point coordinates in the moving body, the surface of the four-point saddle spherical circle error Δ_{1234} is a higher-order piecewise algebraic surface.

For the four separated positions of $\sum^*$ in spherical movement, which are defined by the motion parameters of the first four lines listed in Table 5.1, the spherical coordinates $(\delta_{Pm}, \theta_{Pm})$ are taken as the independent variables and the four-point saddle spherical circle error Δ_{1234} as the function value, the error surface $\Sigma_{1234}^{\text{rR}}$ of Δ_{1234} is constructed, as shown in Fig. 5.3(a). The bottom of the valleys of $\Sigma_{1234}^{\text{rR}}$ is comprised of the points of $\sum^*$, whose corresponding saddle spherical circle errors are all zero. All these points constitute the ***spherical circle point curve***. To reveal it clearly, the error surface $\Sigma_{1234}^{\text{rR}}$ is mapped onto the coordinate plane $\boldsymbol{O}$-$\boldsymbol{i}_m\,\boldsymbol{j}_m$ and all points with the same values of the saddle spherical circle error are connected with contour lines, as shown in Fig. 5.3(b).

5.2.4 Five Positions

Five separated spherical positions of $\sum^*$ are given in $\{\boldsymbol{O}; \boldsymbol{i}_f, \boldsymbol{j}_f, \boldsymbol{k}_f\}$ and numbered 1, 2, 3, 4, 5. A point P of $\sum^*$ traces a discrete point set $\{\boldsymbol{R}_P^{(i)}\}$, $i = 1, 2, 3, 4, 5$, comprised of five points $P^{(1)}$, $P^{(2)}$, $P^{(3)}$, $P^{(4)}$, and $P^{(5)}$ in $\{\boldsymbol{O}; \boldsymbol{i}_f, \boldsymbol{j}_f, \boldsymbol{k}_f\}$. Based on the mathematical model (5.7), we call the sub-program ArRF and obtain the saddle spherical circle of $\{\boldsymbol{R}_P^{(i)}\}$, referred to as the ***five-point saddle spherical circle***, whose fitting error is Δ_{12345}. Similar to the case of four positions, by taking the spherical coordinates $(\delta_{Pm}, \theta_{Pm})$ as the independent variables and the five-point saddle spherical circle errors Δ_{12345} as the dependent variables, the error surface

Table 5.1 Nine spherical positions (unit: radian)

	$\theta_1^{(i)}$	$\theta_2^{(i)}$	$\theta_3^{(i)}$
1	0.16	0.75	1.16
2	0.30	0.90	0.39
3	0.31	1.08	−0.18
4	0.28	1.35	−0.70
5	0.15	1.72	−1.13
6	0.12	2.21	−1.59
7	0.04	5.20	2.83
8	0.08	6.25	2.24
9	0.16	0.60	1.47

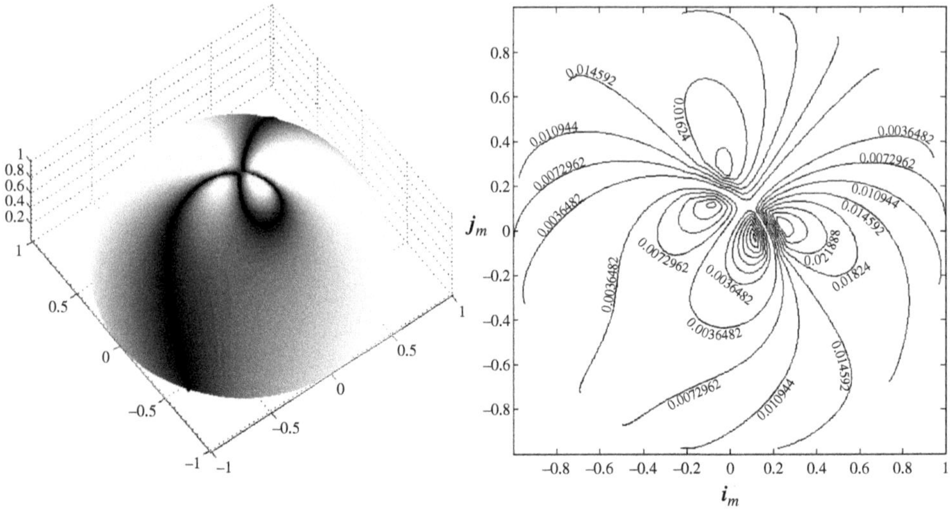

(a) (b)

Figure 5.3 The saddle spherical circle errors for four positions

Σ^{rR}_{12345} can be constructed and mapped onto the plane $\boldsymbol{O}$–$\boldsymbol{i}_m\,\boldsymbol{j}_m$, in which all points with the same values of the saddle spherical circle error are connected with contour lines, as shown in Fig. 5.4.

The five-point saddle spherical circle of $P^{(1)}$–$P^{(5)}$ is also determined by the four saddle spherical circle fitting points, which correspond to the four saddle spherical circle fitting positions of $\sum^*$. The four-point saddle spherical circle of $P^{(1)}$–$P^{(4)}$ has been discussed in the above section, and the point $P^{(5)}$ is an additional point. Whether it is a fitting point of the four-point saddle spherical circle or not is discussed in the following two cases.

1. The fifth point $P^{(5)}$ is not the saddle spherical circle fitting point, and position 5 is not the fitting position. If the fitting error between point $P^{(5)}$ and the four-point saddle spherical circle is less than the four-point saddle spherical circle error Δ_{1234}, the five-point saddle spherical circle error Δ_{12345} of points $P^{(1)}$–$P^{(5)}$ is determined by the fitting points $P^{(1)}$–$P^{(4)}$, or $\Delta_{12345} = \Delta_{1234}$. Hence, all points of $\sum^*$ with five prescribed separated positions, whose four points ($P^{(1)}$–$P^{(4)}$) are the saddle spherical circle fitting points locate on the specific area of $\sum^*$, called the ***spherical four-position area*** (1234) of $\sum^*$. Area I is the projection of the spherical four-position area (1234) on the coordinate plane $\boldsymbol{O}$–$\boldsymbol{i}_m\,\boldsymbol{j}_m$ in Fig. 5.4. As mentioned in Section 2.3.4, there is also another situation for the points on the spherical four-position area (1234). Although the four points $P^{(1)}$–$P^{(4)}$ are fitting points, the saddle spherical circle of $P^{(1)}$–$P^{(5)}$ is merely a distributing spherical circle of $P^{(1)}$–$P^{(4)}$, rather than the saddle spherical circle of $P^{(1)}$–$P^{(4)}$.
2. The fifth point $P^{(5)}$ is the saddle spherical circle fitting point, and position 5 is the fitting position. In this case, position 5 and any three of the four positions (1234) become the saddle spherical circle positions, which leads to four combinations: 1235, 1245, 1345, and 2345. Each combination has three distributing spherical circles, so there exist 12

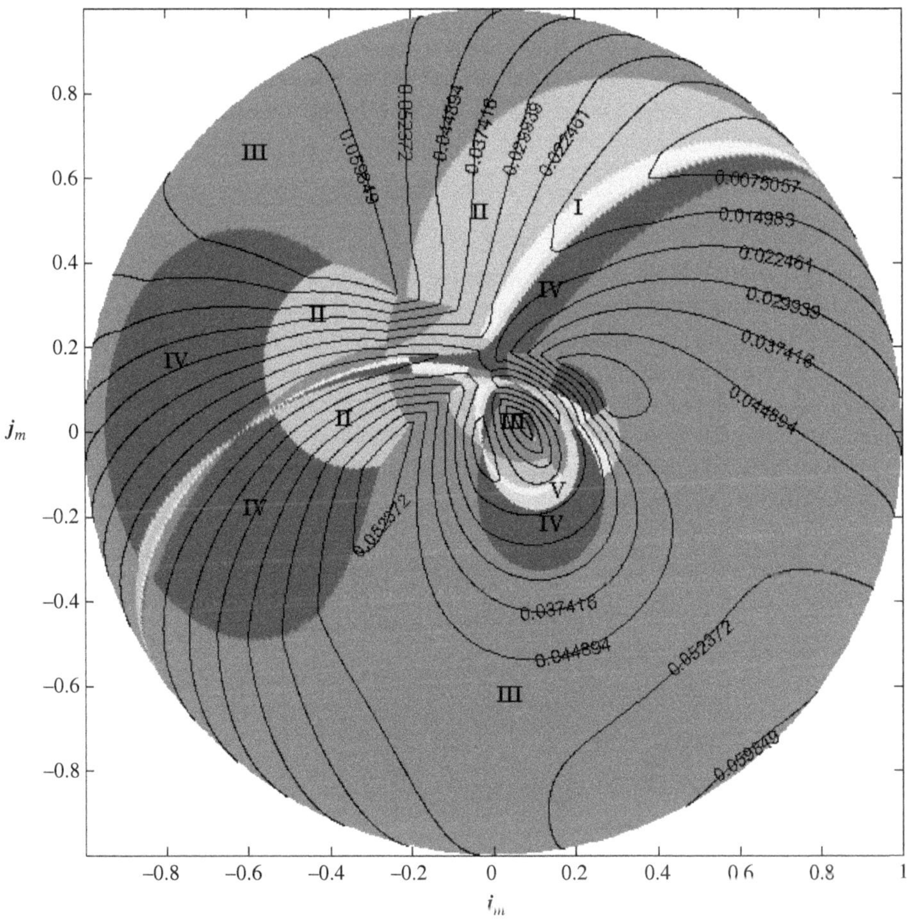

Figure 5.4 The saddle spherical circle errors for five positions

distributing spherical circles for all four combinations. The five-point saddle spherical circle of $P^{(1)}–P^{(5)}$ is one of the 12 distributing spherical circles.

The unit sphere of $\sum^*$ can be divided into several spherical four-position areas, with area I corresponding to positions 1234, area II to 1235, area III to 1245, area IV to 1345, and area V to 2345, as shown in Fig. 5.4. The points located on the boundaries of the areas of $\sum^*$, whose coordinate parameters can be solved through equations (5.22) and (5.23a), have the characteristics of the adjacent areas, and the five-point saddle spherical circle of any point path may correspond to two four-point distributing spherical circles with the same distributing spherical circle error. If the two four-point distributing spherical circles coincide with each other, there would be some special points on the boundary such as the spherical Burmester points on the rigid body $\sum^*$ with five separated positions.

As mentioned above, the error surface of the saddle spherical circle for five positions theoretically can be divided into several distributing spherical circle error surfaces. It is a higher-order algebraic surface and has many peaks/valleys. If the error surface of four-position saddle spherical circle is compared with that of five-position one, we can find a valley bottom, a ***spherical circle point curve*** for four positions, but several valley bottoms, ***spherical circle points*** for five positions, and the saddle spherical circle error of five positions is usually larger than that of four positions.

5.2.5 Multiple Positions

A point of $\sum^*$ with multiple spherical separated positions traces a discrete trajectory. Based on the saddle point program, the saddle spherical circle of the discrete trajectory is still determined by the four saddle spherical circle fitting points, no matter how many points the discrete trajectory has. Therefore, no matter how many separated positions of $\sum^*$ are given, it can also be viewed as combinations of four positions, such as C_n^4 combinations of four positions if n positions are given. There exist more four-position areas on $\sum^*$. For the nine separated positions listed in Table 5.1, the error surface of the saddle spherical circle is mapped as the contour lines, with different four-position areas, as shown in Fig. 5.5.

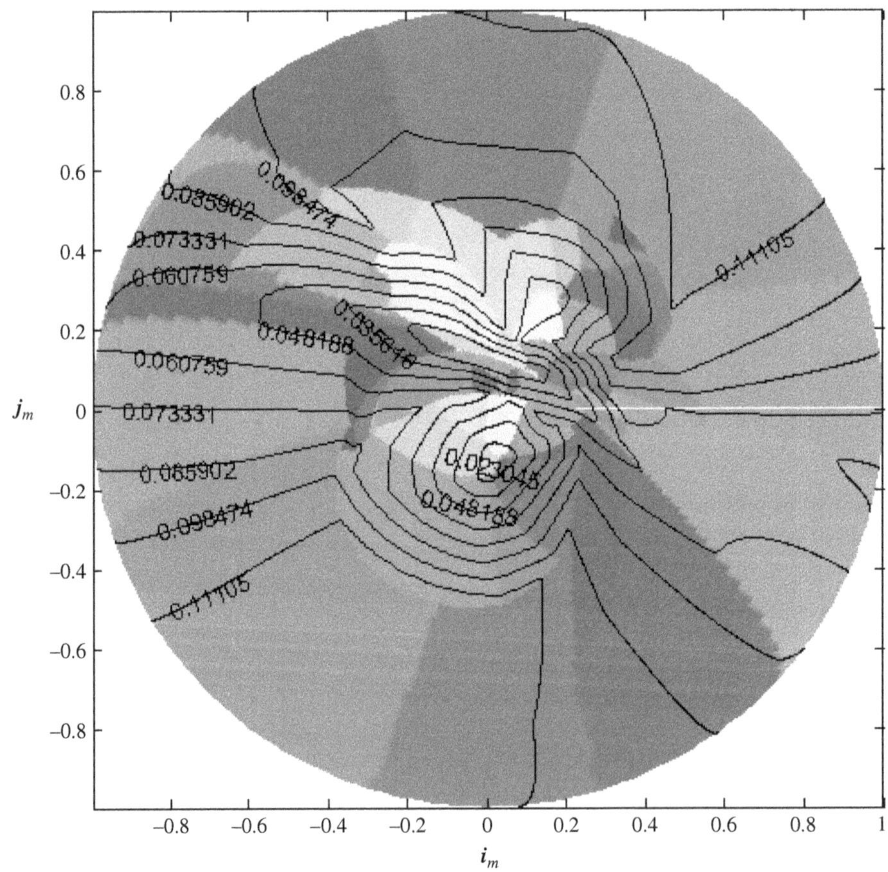

Figure 5.5 The saddle spherical circle errors for nine positions

The more separated positions of $\sum^*$ are given, the smaller are the four-position areas of the saddle spherical circle. When the multiply separated positions of $\sum^*$ tend to be infinite, a discrete trajectory would be a continuous curve, whose saddle spherical circle is still determined by the four fitting points. Those fitting points may vary continuously as the point of $\sum^*$ moves continuously.

5.2.6 Saddle Spherical Circle Point

A point of $\sum^*$ in spherical movement traces a spherical trajectory, which corresponds to a saddle spherical circle and its error. Different points of $\sum^*$ trace different spherical trajectories with different saddle spherical circle errors. In other words, the saddle spherical circle error is a nonlinear function of the point coordinates and may have maximum or minimum value at some points.

Definition 5.2 A point of a moving body in spherical movement is defined as a ***saddle spherical circle point*** if its corresponding saddle spherical circle error achieves a minimum with respect to other points in its neighborhood, which is referred to as the ***saddle spherical circle point error***.

If the saddle spherical circle error is zero, the saddle spherical circle point is just the spherical circle point. The above definition indicates that the saddle spherical circle point of $\sum^*$ is a special point, whose trajectory is closest to a spherical circle relative to other points of $\sum^*$ in its neighborhood. Based on the definition, the mathematical model of the saddle spherical circle point can be expressed by

$$\begin{cases} \delta_{\text{rR}} = \min \Delta_{\text{rR}}(z) \\ \text{s.t.}\ \ \delta_{Pm} \in \left[0, \pi/2\right],\ \theta_{Pm} \in [0, 2\pi] \\ z = \left(\delta_{Pm}, \theta_{Pm}\right)^T \end{cases} \tag{5.24}$$

in which the objective function $\Delta_{\text{rR}}(z)$ is the saddle spherical circle error, determined by equation (5.7). The optimization variables $z = (\delta_{Pm}, \theta_{Pm})^T$ are the spherical coordinates of a point of $\sum^*$. The saddle spherical circle point of $\sum^*$ is located using the saddle point programming twice, or estimating which fitting spherical circle is the saddle spherical circle first, then identifying which point of $\sum^*$ is the saddle spherical circle point second.

The optimal model (5.24) for the saddle spherical circle point is an unconstrained optimization problem, which can be solved directly by the function **fmincon** of Matlab. In the optimizing search for the saddle spherical circle points, the search area of the optimization variables, the spherical coordinates $(\delta_{Pm}, \theta_{Pm})$ are the hemisphere defined by $\delta_{Pm} \in [0, \pi/2]$ and $\theta_{Pm} \in [0, 2\pi]$, since the two symmetrical points about the sphere center have the same spherical trajectory. Because the mathematical model of the saddle spherical circle point is also an optimization problem with multiple extreme values, it is necessary to randomly assign enough initial points on the hemisphere or a group of initial values $(\delta_{Pm}, \theta_{Pm})$ on different convex regions.

Based on the optimization model (5.24), the function **fmincon** is used to search the saddle spherical circle points on the hemisphere area $\delta_{Pm} \in [0, \pi/2], \theta_{Pm} \in [0, 2\pi)$. All the parameters of **fmincon** are set in the same way as those in the sub-program ArrP for saddle circle point in Chapter 2. By the parameter setting, a sub-program ArRP is naturally programmed for the location of the saddle spherical circle points in $\sum^*$. The inputs of the sub-program ArRP are the position parameters of $\sum^*$ with separated spherical positions and the outputs are the spherical coordinates of the saddle spherical circle points and their errors δ_{rR}.

For a general spherical movement of $\sum^*$, the saddle spherical circle error is a nonlinear function of the point coordinates in $\sum^*$. The extreme values of the saddle spherical circle error may appear at some points if the spherical movement is not degenerate. Based on Definitions 5.1 and 5.2, we have the existence theorem of the saddle spherical circle points:

Theorem 5.1 *There exist several saddle spherical circle points in a moving body if the spherical movement is not degenerate.*

The kinematic synthesis of spherical linkages is to locate characteristic points, or saddle spherical circle points in the moving body, whose trajectories are closest to the spherical circles. The definitions of both the saddle spherical circle and the saddle spherical circle point are given in Definitions 5.1 and 5.2, while Theorem 5.1 points out the existence of saddle spherical circle point. Hence, for $\sum^*$ in spherical movement, the saddle spherical circle points can certainly be located in $\sum^*$, whose saddle spherical circle error may depend on the properties of the spherical movement. Based on the properties of the spherical curve, the existence and distributions of the saddle spherical circle points are expected to be theoretically proved by readers and scholars in the field.

Similar to the kinematic synthesis of planar linkages, a saddle spherical circle point of $\sum^*$ is generally not a spherical circle point, unless the saddle spherical circle point errors happen to be zero. We view a spherical circle point as a special saddle spherical circle point in the saddle point programming approach. For five separated spherical positions of $\sum^*$, there are no more than six spherical circle points. It is a general case that there are infinite spherical circle points, located on the spherical circle point curve, for the given four separated positions. Some numeral examples are given to illustrate the discrete kinematic geometry of a spherical moving body with multiple positions, five positions, and four positions, and the error surfaces of the saddle spherical circle are mapped onto the planes as contour lines with the same values.

Example 5.1 The spherical crank-rocker *ABCD* with dimensions $l_1 = 20°$, $l_2 = 37.5°$, $l_3 = 37.5°$, $l_4 = 50°$ is shown in Fig. 5.6. Determine the error surface of the saddle spherical circle and the saddle spherical circle points.

Solution

The moving Cartesian coordinate system $\{\boldsymbol{O}; \boldsymbol{i}_m, \boldsymbol{j}_m, \boldsymbol{k}_m\}$ is established in the coupler link *BC* and the fixed Cartesian coordinate system $\{\boldsymbol{O}; \boldsymbol{i}_f, \boldsymbol{j}_f, \boldsymbol{k}_f\}$ is set up in the base link *AD*, as shown in Fig. 5.6. The rotation angle $\varphi \in [0, 2\pi]$ of the driving link *AB* is discretized with interval 5°, or $\varphi^{(i)}$, $i = 1, \dots, 72$, and 72 discrete positions $(l_1, \varphi^{(i)}, \psi^{(i)})$, $i = 1, \dots, 72$ of the coupler link *BC* relative to the base link *AD* are generated by the displacement solution of the spherical crank-rocker linkage.

4032 coupler points on the hemisphere defined by $\delta_{Pm} \in [0, \pi/2]$, $\theta_{Pm} \in [0, 2\pi)$ are selected with step 0.05 in the two coordinate parameters. The spherical coupler curves traced by these coupler points correspond to saddle spherical circle errors Δ_{rR}, which can be obtained by calling the sub-program ArRF. The error surface can be constructed and mapped onto the plane $\boldsymbol{O}\text{–}\boldsymbol{i}_m\,\boldsymbol{j}_m$ as contour lines with the same values, shown in Fig. 5.7.

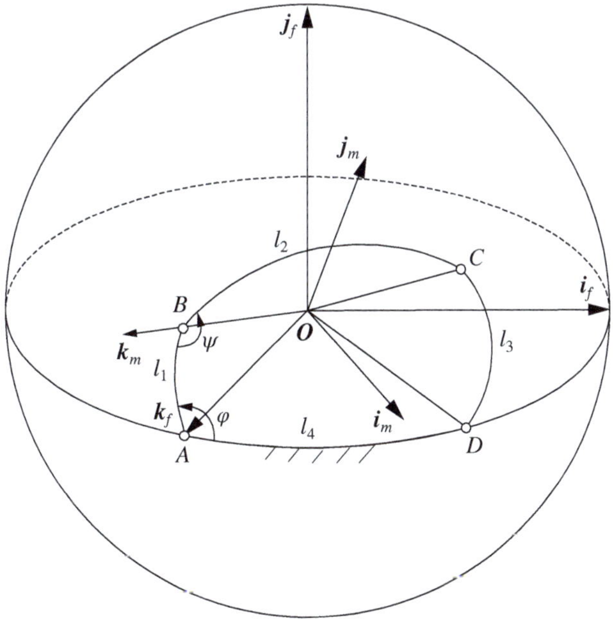

Figure 5.6 A spherical crank-rocker linkage

Obviously, there are a few extreme points $(\delta_{Pm}, \theta_{Pm})$, or saddle spherical circle points, shown in Fig. 5.7, such as four points, $P_1(0,0)$, $P_2(0.6545,1.5708)$, $P_3(1.2953,1.6579)$, and $P_4(0.7693, 2.2672)$ in $\{\boldsymbol{O}; \boldsymbol{i}_m, \boldsymbol{j}_m, \boldsymbol{k}_m\}$, among which points P_1 and P_2 are the spherical circle points or two hinge points B and C on the coupler link. The points P_1, P_2, P_3, and P_4 trace the spherical coupler curves Γ_1, Γ_2, Γ_3, and Γ_4 respectively, and their corresponding saddle spherical circles are shown in Fig. 5.8.

In this example, 50 initial points, denoted "*," are randomly generated in the coupler link, and 50 solutions or saddle spherical circle points, denoted "o," are obtained by the sub-program ArRP. 35 solutions are the spherical circle point P_1, or the joint B, two solutions are the spherical circle point P_2, or the joint C, as shown in Fig. 5.7. The average iterative step is 10 and the average optimization time is 3.0715 s.

Example 5.2 Nine separated spherical positions of a moving body are listed in Table 5.1. Determine the error surface of the saddle spherical circle and the saddle spherical circle points.

Solution

Based on the sub-program ArRF, the error surface of the saddle spherical circle can be constructed and mapped onto the plane $\boldsymbol{O}-\boldsymbol{i}_m\boldsymbol{j}_m$ as contour lines with the same values, shown in Fig. 5.9. 50 random initial points are generated and the corresponding results, the saddle spherical circle points of the moving body, can be obtained by calling the sub-program ArRP, shown in Fig. 5.9, and nine saddle spherical circle points with less errors are listed in Table 5.2. The average iterative steps are 14 and the average optimization time is 2.4937 s.

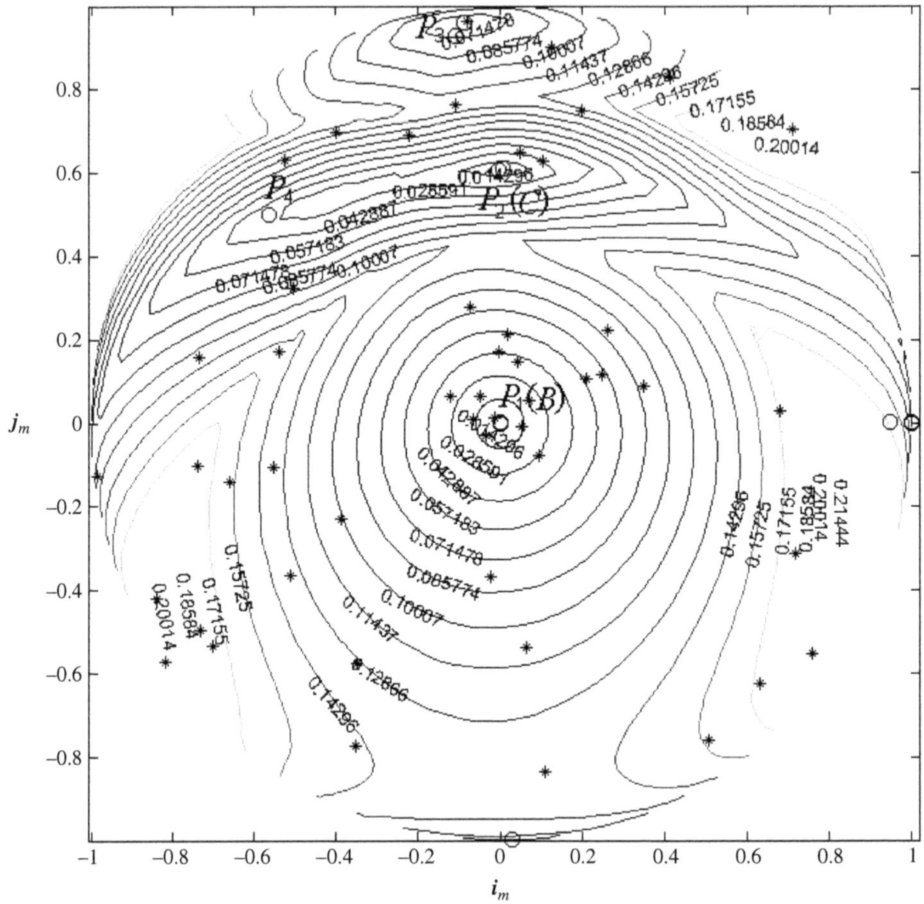

Figure 5.7 The saddle spherical circle error surface for a spherical crank-rocker linkage

The saddle spherical circles and the error surfaces of the saddle spherical circle for the points of the moving body with four and five separated positions have been discussed intensively in Sections 5.2.3 and 5.2.4, and we will not go any further here.

5.3 The Saddle Kinematic Synthesis of Spherical Four-bar Linkages

We now discuss the discrete kinematic synthesis of spherical four-bar linkages, which is to locate two saddle spherical circle points (spherical circle points) in the moving body, and similar to the discrete kinematic synthesis of planar linkages, including rigid-body guidance synthesis or position synthesis, function synthesis, and path synthesis according to the variety of motion requirements, which can be unified into position synthesis. The mathematical model for the kinematic synthesis of spherical four-bar linkages is discussed, while numerical examples are introduced in the last two sections, including multi-position approximate synthesis and precision synthesis for both four and five positions.

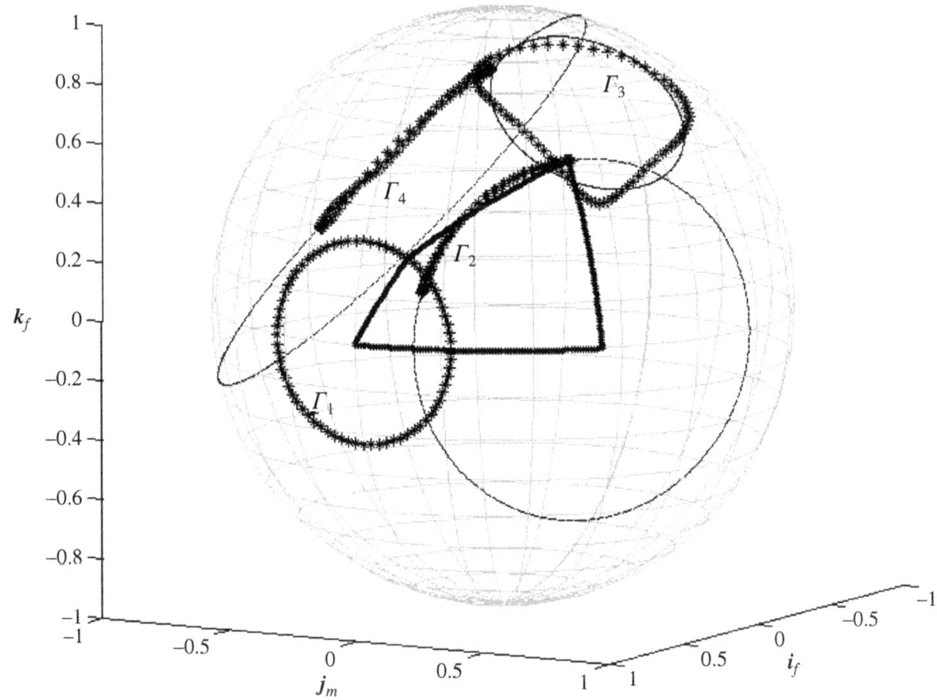

Figure 5.8 The saddle spherical circle points and their spherical coupler curves

5.3.1 *Kinematic Synthesis*

The kinematic synthesis of spherical linkages, which is the same as that of planar linkages, has three types: position synthesis, path synthesis, and function synthesis. The given positions, path, and functional relationship can all be described by continuous analytical functions, or discrete functions.

5.3.1.1 Spherical Continuous Synthesis

The continuous spherical motion of $\sum^*$, described by three rotation angles θ_1, θ_2, and θ_3 of $\{\boldsymbol{O};\boldsymbol{i}_m,\boldsymbol{j}_m,\boldsymbol{k}_m\}$ relative to $\{\boldsymbol{O};\boldsymbol{i}_f,\boldsymbol{j}_f,\boldsymbol{k}_f\}$, are designated as the known conditions. A point O_k on the axis $\boldsymbol{k}_m$, whose Cartesian coordinates are (0,0,1) in $\{\boldsymbol{O};\boldsymbol{i}_m,\boldsymbol{j}_m,\boldsymbol{k}_m\}$, is chosen as the reference point. By equation (5.2), the vector equation of the spherical trajectory Γ_{Ok} traced by point O_k in $\{\boldsymbol{O};\boldsymbol{i}_f,\boldsymbol{j}_f,\boldsymbol{k}_f\}$ is

$$\boldsymbol{R}_{Ok} = s\theta_1 c\theta_2 \boldsymbol{i}_f + s\theta_1 s\theta_2 \boldsymbol{j}_f + c\theta_1 \boldsymbol{k}_f \tag{5.25}$$

By the arc length s of Γ_{Ok}, or $\mathrm{d}s = \left|\frac{\mathrm{d}\boldsymbol{R}_{Ok}}{\mathrm{d}t}\right|\mathrm{d}t$, the Darboux frame $\{\boldsymbol{R}_{Ok};\boldsymbol{\alpha},\boldsymbol{n},\boldsymbol{\nu}\}$ is set up on Γ_{Ok}, or

$$\begin{cases} \boldsymbol{\alpha} = \dfrac{\mathrm{d}\boldsymbol{R}_{Ok}}{\mathrm{d}s} = \cos\theta \boldsymbol{i}_m + \sin\theta \boldsymbol{j}_m \\ \boldsymbol{n} = -\boldsymbol{R}_{Ok} = -\boldsymbol{k}_m \\ \boldsymbol{\nu} = \boldsymbol{\alpha}\times\boldsymbol{n} = -\sin\theta \boldsymbol{i}_m + \cos\theta \boldsymbol{j}_m \end{cases} \tag{5.26}$$

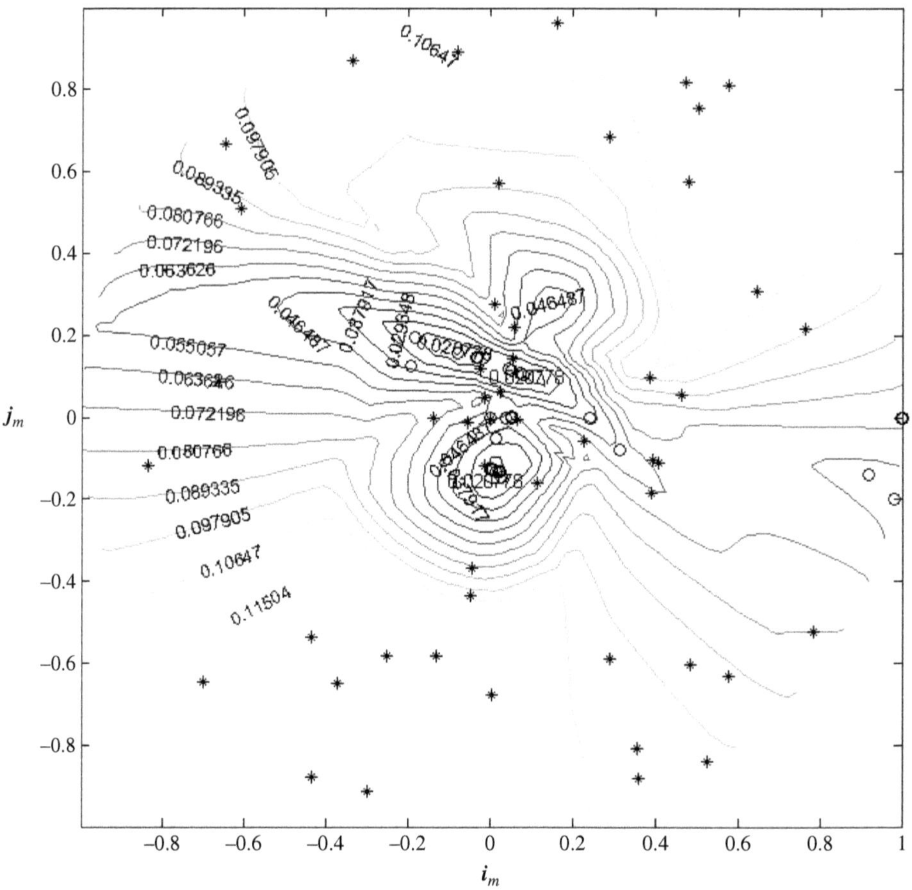

Figure 5.9 The saddle spherical circle errors for nine positions

Table 5.2 Nine saddle spherical circle points (unit: radian)

No.	Spherical coordinates		Saddle spherical circle (cone)			
			Direction angles of the axis		Half-cone angle	Saddle spherical circle point error
	δ_{Pm}	θ_{Pm}	ξ	η	δ	δ_{rR}
SrRP1	0.1316	0.9251	0.5160	6.0488	0.4787	0.010456
SrRP2	0.1262	1.1561	0.4982	6.1087	0.4728	0.010958
SrRP3	0.1749	2.0349	1.0494	5.8044	1.1635	0.011061
SrRP4	0.1256	1.1913	0.4925	6.1210	0.4688	0.011065
SrRP5	0.1497	1.7957	0.8579	5.8621	0.9301	0.012333
SrRP6	0.1319	4.8240	0.0193	0.0302	0.1721	0.016876
SrRP7	0.1289	4.7414	0.0171	0.7600	0.1698	0.017382
SrRP8	0.1494	4.8933	0.0300	5.7125	0.1831	0.017461
SrRP9	0.2718	2.3252	0.7968	5.8722	1.0091	0.021601

where θ is the direction angle of the vector $\boldsymbol{\alpha}$ in $\{\boldsymbol{O};\boldsymbol{i}_m,\boldsymbol{j}_m\}$. A point $P(\delta_{Pm},\theta_{Pm})$ on the unit sphere of $\sum^*$ traces another spherical trajectory Γ_P, which is regarded as the spherical adjoint curve of Γ_{Ok}, while Γ_{Ok} is taken as an original curve. The vector equation of Γ_P is

$$\boldsymbol{R}_P = \boldsymbol{R}_{Ok} + u_1\boldsymbol{\alpha} + u_2\boldsymbol{n} + u_3\boldsymbol{v} \tag{5.27a}$$

where

$$\begin{cases} u_1 = \sin\delta_{Pm}\cos\left(\theta_{Pm} - \theta\right) \\ u_2 = 1 - \cos\delta_{Pm} \\ u_3 = \sin\delta_{Pm}\sin\left(\theta_{Pm} - \theta\right) \end{cases} \tag{5.27b}$$

By the above equation, the geodesic curvature k_{gP} of Γ_P can be derived.

As mentioned in Chapter 4, a point P of $\sum^*$, tracing a trajectory Γ_P in $\{\boldsymbol{O};\boldsymbol{i}_f,\boldsymbol{j}_f,\boldsymbol{k}_f\}$, is referred to as a point of the spherical cubic of stationary curvature, a spherical Ball point, or even a Burmester point at an instant if the higher-order derivatives of geodesic curvature k_{gP} of Γ_P equal zero instantaneously. But if the geodesic curvature k_{gP} is consistently equal to a constant for all instants, we should deal with the existence of solutions and the solving method of differential equations, which is a considerably complicated problem and cannot generally be solved so far, except for $\sum^*$ in degenerate spherical movement.

5.3.1.2 Spherical Discrete Synthesis

The discrete synthesis of spherical linkages is to locate the saddle spherical circle points of $\sum^*$, including finite and multiple separate position synthesis according to the given separated spherical positions of $\sum^*$ with three discrete independent parameters.

The finite separated position synthesis is usually called the precise synthesis, whose theoretical basis is the kinematic geometry at separated positions (logically corresponding to the kinematic geometry at infinitesimally successive positions in Chapter 4). For instance, there exist at most six spherical Burmester points in the moving body in spherical movement at an instant, whose trajectories contact the spherical circle to fourth order. Consequently, there are only at most six spherical circle points in the moving body at five separated positions.

The discrete synthesis of spherical linkages for multiple positions, traditionally called approximate synthesis, is to determine the saddle spherical circle points in the moving body with multiple positions. For general spherical motion, a spherical circle point does not exist in the moving body if the number of given positions exceeds five.

In fact, we systematically build up the mathematical model for the kinematic synthesis of spherical linkages with both finite and multiple separated positions, uniformly called the ***discrete spherical kinematic synthesis of linkages***. The precise spherical synthesis is just the special case of the discrete spherical synthesis, which will be introduced in the next section.

5.3.1.3 Spherical Function Synthesis

For the function synthesis of spherical linkages, the function $\psi = \psi(\varphi)$ and the dimension (central angle) d of the frame link AD are first prescribed, as shown in Fig. 5.10. A spherical linkage is synthesized to regenerate the function, where ψ and φ are the rotational angles of the

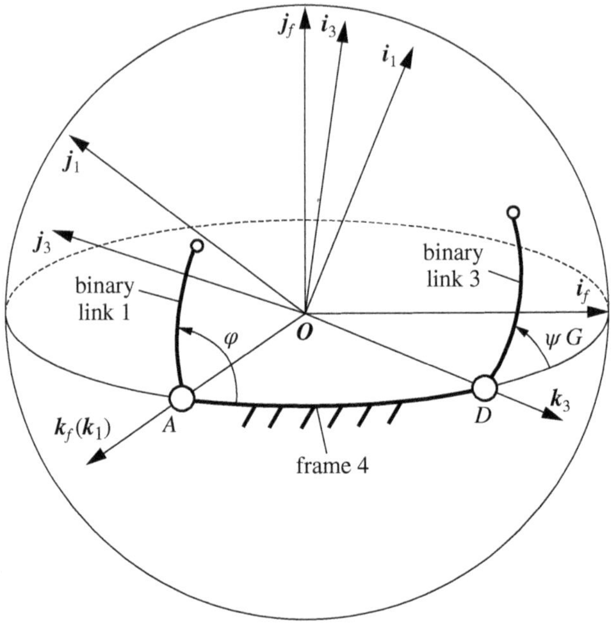

Figure 5.10 The function synthesis of spherical linkages

driven link and the driving link respectively. The function synthesis is also to locate a saddle spherical circle point on either the driving link 1 or the driven link 3, since one binary link makes relative spherical movement to another. The Cartesian coordinate systems $\{\boldsymbol{O}; \boldsymbol{i}_f, \boldsymbol{j}_f, \boldsymbol{k}_f\}$ on the base link *AD*, $\{\boldsymbol{O}; \boldsymbol{i}_1, \boldsymbol{j}_1, \boldsymbol{k}_1\}$ on link 1, and $\{\boldsymbol{O}; \boldsymbol{i}_3, \boldsymbol{j}_3, \boldsymbol{k}_3\}$ on link 3 are respectively set up. The movement of binary link 3 relative to link 1 can be described by the displacement of $\{\boldsymbol{O}; \boldsymbol{i}_3, \boldsymbol{j}_3, \boldsymbol{k}_3\}$ relative to $\{\boldsymbol{O}; \boldsymbol{i}_1, \boldsymbol{j}_1, \boldsymbol{k}_1\}$:

$$\begin{cases} \boldsymbol{i}_3 = (\mathrm{cdc}\psi\mathrm{c}\varphi + \mathrm{s}\psi\mathrm{s}\varphi)\boldsymbol{i}_1 + (-\mathrm{cdc}\psi\mathrm{s}\varphi + \mathrm{s}\psi\mathrm{c}\varphi)\boldsymbol{j}_1 - \mathrm{sdc}\psi\boldsymbol{k}_1 \\ \boldsymbol{j}_3 = (-\mathrm{cds}\psi\mathrm{c}\varphi + \mathrm{c}\psi\mathrm{s}\varphi)\boldsymbol{i}_1 + (\mathrm{cds}\psi\mathrm{s}\varphi + \mathrm{c}\psi\mathrm{c}\varphi)\boldsymbol{j}_1 + \mathrm{sds}\psi\boldsymbol{k}_1 \\ \boldsymbol{k}_3 = \mathrm{sdc}\varphi\boldsymbol{i}_1 - \mathrm{sds}\varphi\boldsymbol{j}_1 + \mathrm{cd}\boldsymbol{k}_1 \end{cases} \tag{5.28}$$

By the above equation, the three rotation angles $(\theta_1, \theta_2, \theta_3)$ of binary link 3, relative to binary link 1, can be expressed by the three angles (φ, ψ, d) of the given function:

$$\theta_1 = d,\ \theta_2 = -\varphi,\ \theta_3 = \psi + \pi/2 \tag{5.29}$$

Hence, the function generation of spherical linkage can be converted into position synthesis, that is to locate the saddle spherical circle points in the binary link 3.

5.3.1.4 Spherical Path Synthesis

Based on a required spherical curve $\Gamma : F(x_E, y_E, z_E) = 0$ in $\{\boldsymbol{O}; \boldsymbol{i}_f, \boldsymbol{j}_f, \boldsymbol{k}_f\}$, a spherical linkage is synthesized to reproduce it, referred to as the ***spherical path synthesis of spherical linkages***.

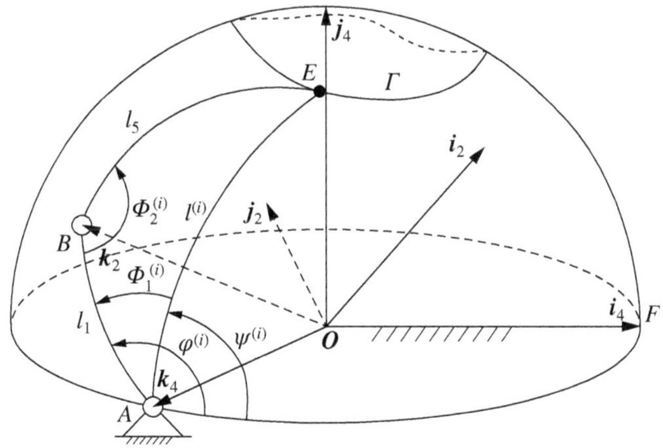

Figure 5.11 A dyad for path generation of spherical linkages

We may develop the approach from that of planar linkages to that of spherical linkages, which converts the path synthesis into position synthesis by means of a ***floating spherical dyad*** *ABE*, as shown in Fig. 5.11. As the end point *E* of the dyad *ABE* moves along the given spherical curve Γ, the floating link *BE* makes a definite spherical movement. Thus, we only need to locate a saddle spherical circle point in the floating link *BE*, to form a spherical four-bar linkage. The path generation is then converted into the location of saddle spherical circle points in the floating link *BE*, or position synthesis of spherical linkages.

The given spherical curve $\Gamma : F(x_E, y_E, z_E) = 0$ is described in $\{\boldsymbol{O}; \boldsymbol{i}_f, \boldsymbol{j}_f, \boldsymbol{k}_f\}$ with original point at the sphere center. A fixed point *A* with spherical coordinates $(1, \alpha_A, \beta_A)$ in $\{\boldsymbol{O}; \boldsymbol{i}_f, \boldsymbol{j}_f, \boldsymbol{k}_f\}$ is designated as a fixed pivot point. For convenience of description, another base frame $\{\boldsymbol{O}; \boldsymbol{i}_4, \boldsymbol{j}_4, \boldsymbol{k}_4\}$ is set up on the fixed pivot point *A*, shown in Fig. 5.11. The relationship between $\{\boldsymbol{O}; \boldsymbol{i}_f, \boldsymbol{j}_f, \boldsymbol{k}_f\}$ and $\{\boldsymbol{O}; \boldsymbol{i}_4, \boldsymbol{j}_4, \boldsymbol{k}_4\}$ is

$$\begin{cases} \boldsymbol{i}_4 = \sin\beta_A \boldsymbol{i}_f - \cos\beta_A \boldsymbol{j}_f \\ \boldsymbol{j}_4 = \cos\alpha_A \cos\beta_A \boldsymbol{i}_f + \cos\alpha_A \sin\beta_A \boldsymbol{j}_f - \sin\alpha_A \boldsymbol{k}_f \\ \boldsymbol{k}_4 = \sin\alpha_A \cos\beta_A \boldsymbol{i}_f + \sin\alpha_A \sin\beta_A \boldsymbol{j}_f + \cos\alpha_A \boldsymbol{k}_f \end{cases} \tag{5.30}$$

Based on the above equation, the given spherical curve Γ can also be represented in $\{\boldsymbol{O}; \boldsymbol{i}_4, \boldsymbol{j}_4, \boldsymbol{k}_4\}$. In order that point *E* of the dyad *ABE* moves along the curve completely, the specific conditions have to be satisfied by the dimensions l_1 of link *AB* and l_5 of link *BE*. The maximum and minimum spherical distances from point *A* to Γ are $d_{\max}$ and $d_{\min}$ respectively. If point *A* lies outside the closed curve Γ, the dimensions of links *AB* and *BE* are

$$l_1 = \frac{d_{\max} - d_{\min}}{2},\ l_5 = \frac{d_{\max} + d_{\min}}{2} \tag{5.31}$$

Under the condition that point *A* lies inside the closed curve Γ, the dimensions of the links of *AB* and *BE* are

$$l_1 = \frac{d_{\max} + d_{\min}}{2},\ l_5 = \frac{d_{\max} - d_{\min}}{2} \tag{5.32}$$

In $\{\boldsymbol{O};\boldsymbol{i}_4,\boldsymbol{j}_4,\boldsymbol{k}_4\}$, the spherical curve Γ can be discretized into the point set $\{(x_E^{(i)}, y_E^{(i)}, z_E^{(i)})^T\}$. The Cartesian coordinates of points A and F are respectively (0,0,1) and (1,0,0), and the dimensions of the spherical great circular segments $\widehat{AE}$ and $\widehat{EF}$ are

$$\begin{cases} l^{(i)} = 2\text{arc}\sin\dfrac{\sqrt{\left(x_E^{(i)}\right)^2 + \left(y_E^{(i)}\right)^2 + \left(z_E^{(i)} - 1\right)^2}}{2} \\ \widehat{EF} = 2\text{arc}\sin\dfrac{\sqrt{\left(x_E^{(i)} - 1\right)^2 + \left(y_E^{(i)}\right)^2 + \left(z_E^{(i)}\right)^2}}{2} \end{cases} \tag{5.33}$$

Based on the spherical triangle AEF and the spherical cosine theorem, we have

$$\psi^{(i)} = \arccos\frac{\cos(\widehat{EF}) - \cos\left(l^{(i)}\right)\cos(\pi/2)}{\sin\left(l^{(i)}\right)\sin(\pi/2)} = \arccos\frac{\cos(\widehat{EF})}{\sin\left(l^{(i)}\right)} \tag{5.34}$$

By applying the spherical cosine theorem to the spherical triangle ABE, we can solve the angles $\phi_1^{(i)}$ and $\phi_2^{(i)}$ as

$$\begin{cases} \phi_1^{(i)} = \arccos\dfrac{\cos\left(l_5\right) - \cos\left(l_1\right)\cos\left(l^{(i)}\right)}{\sin\left(l_1\right)\sin\left(l^{(i)}\right)} \\ \phi_2^{(i)} = \arccos\dfrac{\cos\left(l^{(i)}\right) - \cos\left(l_1\right)\cos\left(l_5\right)}{\sin\left(l_1\right)\sin\left(l_5\right)} \end{cases} \tag{5.35}$$

Hence, the rotational angle of the link AB is

$$\varphi^{(i)} = \psi^{(i)} + p_c\phi_1^{(i)} \tag{5.36}$$

where p_c as an assembly coefficient is 1 for the link AB to rotate anticlockwise, and -1 for the link AB to rotate clockwise.

So far, the movement of the dyad ABE can be completely determined. In order to describe the spherical movement of the link BE, we establish the moving Cartesian coordinate system $\{\boldsymbol{O};\boldsymbol{i}_2,\boldsymbol{j}_2,\boldsymbol{k}_2\}$ on the link BE, which has the following relationship with $\{\boldsymbol{O};\boldsymbol{i}_4,\boldsymbol{j}_4,\boldsymbol{k}_4\}$:

$$\begin{cases} \boldsymbol{i}_2 = \left(-\text{c}\phi_2^{(i)}\text{c}l_1\text{c}\varphi^{(i)} + \text{s}\phi_2^{(i)}\text{s}\varphi^{(i)}\right)\boldsymbol{i}_4 + \left(-\text{c}\phi_2^{(i)}\text{c}l_1\text{s}\varphi^{(i)} - \text{s}\phi_2^{(i)}\text{c}\varphi^{(i)}\right)\boldsymbol{j}_4 + \text{c}\phi_2^{(i)}\text{s}l_1\boldsymbol{k}_4 \\ \boldsymbol{j}_2 = \left(\text{s}\phi_2^{(i)}\text{c}l_1\text{c}\varphi^{(i)} + \text{c}\phi_2^{(i)}\text{s}\varphi^{(i)}\right)\boldsymbol{i}_4 + \left(\text{s}\phi_2^{(i)}\text{c}l_1\text{s}\varphi^{(i)} - \text{c}\phi_2^{(i)}\text{c}\varphi^{(i)}\right)\boldsymbol{j}_4 - \text{s}\phi_2^{(i)}\text{s}l_1\boldsymbol{k}_4 \\ \boldsymbol{k}_2 = \text{s}l_1\text{c}\varphi^{(i)}\boldsymbol{i}_4 + \text{s}l_1\text{s}\varphi^{(i)}\boldsymbol{j}_4 + \text{c}l_1\boldsymbol{k}_4 \end{cases} \tag{5.37}$$

The three rotation angles of the link BE in $\{\boldsymbol{O};\boldsymbol{i}_4,\boldsymbol{j}_4,\boldsymbol{k}_4\}$ are

$$\theta_1 = l_1,\ \theta_2 = \varphi,\ \theta_3 = \phi_2 - \pi/2 \tag{5.38}$$

Hence, the path generation of spherical linkage can be converted into spherical position synthesis, that is to locate the saddle spherical circle points of the coupler link BE in the base frame $\{\boldsymbol{O};\boldsymbol{i}_4,\boldsymbol{j}_4,\boldsymbol{k}_4\}$.

For the kinematic synthesis of spherical linkages, regardless of type (position synthesis, function generation or path generation, continuous position synthesis, discrete synthesis, including both finite separated position precise synthesis and multi-position approximate synthesis), the given conditions can be unified into spherical movement of $\sum^*$ with $\{\boldsymbol{O};\boldsymbol{i}_m,\boldsymbol{j}_m,\boldsymbol{k}_m\}$ relative to $\sum$ with $\{\boldsymbol{O};\boldsymbol{i}_f,\boldsymbol{j}_f,\boldsymbol{k}_f\}$, or three rotation angles $(\theta_1,\theta_2,\theta_3)$. In continuous position synthesis, $(\theta_1,\theta_2,\theta_3)$ are continuous functions, while in separate position synthesis, $(\theta_1,\theta_2,\theta_3)$ are discrete functions. Based on equation (5.4), the discrete point set $\{\boldsymbol{R}_P^{(i)}\}$ traced by any point P of $\sum^*$ is uniformly represented by

$$\boldsymbol{R}_P^{(i)} = f\left(\theta_1^{(i)},\theta_2^{(i)},\theta_3^{(i)},x_{Pm},y_{Pm},z_{Pm}\right),\ i=1,\ \ldots\ ,n \tag{5.39}$$

It can be written briefly as $\{\boldsymbol{R}_P^{(i)}\}$ in this chapter.

5.3.2 Saddle Kinematic Synthesis of Spherical Four-bar Linkages

Each point of $\sum^*$ has a discrete trajectory in $\{\boldsymbol{O};\boldsymbol{i}_f,\boldsymbol{j}_f,\boldsymbol{k}_f\}$, corresponding to a saddle spherical circle and its error. Two points P_1 and P_2 in $\sum^*$ are optimally located by the mathematical model for the kinematic synthesis of spherical hinged four-bar linkages:

$$\begin{cases}\min F(\boldsymbol{Z}) = \min\left(\Delta_{\mathrm{rR}}(z_1)+\Delta_{\mathrm{rR}}(z_2)\right)\\ \text{s.t. } \delta_{Pmk}\in\left[0,\pi/2\right],\ \theta_{Pmk}\in[0,2\pi],\ k=1,\ 2\\ \quad g_j(\boldsymbol{Z})\le 0,\ j=1,2,\ \ldots\ ,n\end{cases} \tag{5.40}$$

A. **Objective function and Optimization variables**
The objective function is the summation of two saddle spherical circle errors $\Delta_{\mathrm{rR}}(z_1)$ and $\Delta_{\mathrm{rR}}(z_2)$ corresponding to two different points P_1 and P_2. The definition of saddle spherical circle is given in Section 5.2.1.
The optimization variables $Z=(z_1,z_2)^T$ include $z_1=(\delta_{Pm1},\theta_{Pm1})^T$ and $z_2=(\delta_{Pm2},\theta_{Pm2})^T$, which are the spherical coordinates of two points in $\sum^*$, and the domain of definition for the optimization variables is $\delta_{Pmk}\in[0,\pi/2]$, $\theta_{Pmk}\in[0,2\pi]$, $k=1,\ 2$.
B. **Constraint equations**
The two points $P_1(\delta_{Pm1},\theta_{Pm1})$ and $P_2(\delta_{Pm2},\theta_{Pm2})$ of $\sum^*$ are designated as the hinge points of two revolute joints of the coupler link, and the axes of the two saddle spherical circles are taken as those of the two fixed revolute joints, while the half-cone angles are taken as the lengths of the binary links R-R. The dimensions of the synthesized spherical four-bar linkages are

$$\begin{cases}l_1=\delta_1\\ l_2=\arccos(\sin\delta_{Pm1}\sin\delta_{Pm2}\cos\theta_{Pm1}\cos\theta_{Pm2}\\ \qquad +\sin\delta_{Pm1}\sin\delta_{Pm2}\sin\theta_{Pm1}\sin\theta_{Pm2}+\cos\delta_{Pm1}\cos\delta_{Pm2})\\ l_3=\delta_2\\ l_4=\arccos(\sin\xi_1\sin\xi_2\cos\eta_1\cos\eta_2\\ \qquad +\sin\xi_1\sin\xi_2\sin\eta_1\sin\eta_2+\cos\xi_1\cos\xi_2)\end{cases} \tag{5.41}$$

Considering the kinematic requirements, we may have some constraint conditions for a spherical four-bar linkage.

1. Grashof kinematic chain condition

$$g_1(z) = 2\left\{\max\left(l_1, l_2, l_3, l_4\right) + \min\left(l_1, l_2, l_3, l_4\right)\right\} - \sum_{i=1}^{4} l_i \leq 0 \tag{5.42}$$

2. Existence conditions of a crank

$$\begin{cases} g_2(z) = \min\left(l_1, l_4\right) - l_2 \leq 0 \\ g_3(z) = \min\left(l_1, l_4\right) - l_3 \leq 0 \end{cases} \tag{5.43}$$

3. Dimension condition of links

$$g_4(z) = \varepsilon - \min\left(l_1, l_2, l_3, l_4\right) \leq 0 \tag{5.44}$$

4. Length ratio of links

$$g_5(z) = \max\left(l_1, l_2, l_3, l_4\right) / \min\left(l_1, l_2, l_3, l_4\right) - [l] \leq 0 \tag{5.45}$$

   where $[l]$ is the permissible maximum length ratio.
5. Conditions for spherical linkages, or the sum of the lengths of two links

$$\begin{cases} l_1 + l_2 \leq \pi, \; l_1 + l_3 \leq \pi, \; l_1 + l_4 \leq \pi \\ l_2 + l_3 \leq \pi, \; l_2 + l_4 \leq \pi, \; l_3 + l_4 \leq \pi \end{cases} \tag{5.46}$$

6. Allowable transmission angle

$$g_6(z) = [\gamma] - \gamma_{\min} \leq 0 \tag{5.47a}$$

   where $[\gamma]$ is the permissible minimum transmission angle. $\gamma_{\min}$ is the actual minimum transmission of the synthesized linkage, whose value is

$$\gamma_{\min} = \min\left(\arccos\frac{\cos\left(l_4 - l_1\right) - \cos\left(l_2\right)\cos\left(l_3\right)}{\sin\left(l_2\right)\sin\left(l_3\right)}, \; \pi - \arccos\frac{\cos\left(l_4 + l_1\right) - \cos\left(l_2\right)\cos\left(l_3\right)}{\sin\left(l_2\right)\sin\left(l_3\right)}\right) \tag{5.47b}$$

In practical applications, there exist other constraints for the kinematic synthesis of spherical four-bar linkages, such as the position requirement of the fixed joints, the requirement of the movement space, the movement sequence, and the circuit and branch. The additional constraint equations have to be given according to the different application cases, which is beyond the scope of this book.

C. **Optimization algorithm**

The mathematical model (5.40) for the kinematic synthesis of spherical four-bar linkages can be solved directly by the function **fmincon** of Matlab, although it is a nonlinear optimum problem under constraint conditions.

D. **Initial values of the optimization variables**

The selection of the initial values for the mathematical model (5.40) is also a complicated and difficult problem due to the constraint conditions. Even a simple constraint, such as the proper length ratio will greatly increase the difficulty of the optimization solution. In this book, the saddle spherical circle points of $\sum^*$ are geometrically chosen as the initial points for the optimal model (5.40) if they locate on the feasible region.

E. **Stop condition**
In this book, we set the termination criterion through the **optimset** function and judge the stop conditions of the optimization based on the exit condition (**exitflag**) of the optimization function.

Based on the above content, the mathematical model (5.40) is solved directly by the function **fmincon** of Matlab. Like the program KS-CR for kinematic synthesis of a planar four-bar linkage, a **program for kinematic synthesis of spherical four-bar linkages**, or **KS-CR-Sphere** for short, is naturally constructed for discrete position synthesis. The inputs are the parameters of the prescribed separated positions of a moving body and the outputs are a spherical four-bar linkage with dimensions and performance.

5.3.2.1 Multiple Positions

In general, the saddle spherical circle point error is not zero, or spherical circle points may not exist in a moving body with multiple positions unless the spherical movement is degenerate. The process for the kinematic synthesis of spherical hinged four-bar linkages is demonstrated by the following numeral examples. By means of the mathematical model of equation (5.24), some saddle spherical circle points can be optimized in the moving body, and two of them constitute a spherical four-bar linkage, which can be taken as a group of initial values for the kinematic synthesis of spherical four-bar linkages, the optimization model (5.40).

Example 5.3 Given the moving body with nine spherical separated positions as listed in Table 5.1, synthesize a spherical four-bar linkage with a crank, minimum allowable transmission angle $[\gamma] = 30^{\circ}$, and maximum allowable length ratio $[l] = 5$.

Solution

According to equation (5.40) and the application requirements, the mathematical model for the kinematic synthesis of a spherical hinged four-bar linkage is

$$\left\{\begin{array}{l}
\min\limits_{Z=(z_1,z_2)^T} \left(\Delta_{\mathrm{rR}}(z_1)+\Delta_{\mathrm{rR}}(z_2)\right) \\
\text{s.t. } \delta_{Pmk} \in [0,\pi/2],\ \theta_{Pmk} \in [0,2\pi],\ k=1,\ 2 \\
\quad g_1(\mathbf{Z}) = 2\left\{\max\left(l_1,l_2,l_3,l_4\right)+\min\left(l_1,l_2,l_3,l_4\right)\right\} - \sum\limits_{i=1}^{4} l_i \le 0 \\
\quad g_2(\mathbf{Z}) = \min\left(l_1,l_4\right) - l_2 \le 0, g_3(\mathbf{Z}) = \min\left(l_1,l_4\right) - l_3 \le 0 \\
\quad g_4(\mathbf{Z}) = \max\left(l_1,l_2,l_3,l_4\right) / \min\left(l_1,l_2,l_3,l_4\right) - 5 \le 0 \\
\quad g_5(\mathbf{Z}) = 0.1 - \min\left(l_1,l_2,l_3,l_4\right) \le 0 \\
\quad g_6(\mathbf{Z}) = l_1+l_2-\pi \le 0,\ g_7(\mathbf{Z}) = l_1+l_3-\pi \le 0 \\
\quad g_8(\mathbf{Z}) = l_1+l_4-\pi \le 0,\ g_9(\mathbf{Z}) = l_2+l_3-\pi \le 0 \\
\quad g_{10}(\mathbf{Z}) = l_2+l_4-\pi \le 0,\ g_{11}(\mathbf{Z}) = l_3+l_4-\pi \le 0 \\
\quad g_{12}(\mathbf{Z}) = \dfrac{\pi}{6} - \min\left(\arccos\dfrac{\cos\left(l_4-l_1\right)-\cos\left(l_2\right)\cos\left(l_3\right)}{\sin\left(l_2\right)\sin\left(l_3\right)},\right. \\
\qquad\qquad\qquad \left.\pi - \arccos\dfrac{\cos\left(l_4+l_1\right)-\cos\left(l_2\right)\cos\left(l_3\right)}{\sin\left(l_2\right)\sin\left(l_3\right)}\right) \le 0
\end{array}\right. \tag{E5-3.1}$$

All letters of the above equation have the same meanings as those in equations (5.40) to (5.47). The process for kinematic synthesis of a spherical hinged four-bar linkage with nine separated positions can be treated in three parts: the preparation of initial values, the parameter setting of the function **fmincon**, and the program operation.

1. **Preparation of initial values**
   For the rigid body with nine given separated positions, the saddle spherical circle points are searched out, in which nine points (SrRP1-SrRP9) with less errors are listed in Table 5.2. Any two of the nine spherical circle points can be taken as a group of initial points, or a initial spherical four-bar linkage, whose dimensions and performance can be calculated through equations (5.41) to (5.47). In this way, the nine saddle spherical circle points correspond to 36 groups of initial values, but just a few (or even none) of them meet the constraint conditions. The six groups of initial values, coming from the nine saddle spherical circle points (SrRP1-SrRP9), are listed in Table 5.3, where a letter G with roman numerals is taken as the sequence number.
2. **Parameter setting of the function fmincon**
   The parameter setting of the function **fmincon** can refer totally to the previously introduced content, including the optimization algorithm, termination conditions, and constraint condition setting. The nonlinear inequality constraints should be constructed through the inequalities $g_1 - g_{12}$ of equation (E5-3.1).
3. **Program operation**
   We run the program **KS-CR-Sphere** for discrete kinematic synthesis of spherical four-bar linkages with a crank by means of the six groups of initial values, and have the corresponding results, or six spherical four-bar linkages with different dimensions, listed in Table 5.4. The average optimization time is 18.3174 s. Obviously not all the optimization results from the program **KS-CR-Sphere** convergence to their initial points (saddle spherical circle points), and the new characteristic points are listed in Table 5.5, which make the value of the objective function reach a minimum.

Table 5.3 Initial values (combination of saddle spherical circle points)

No.	Combination	Dimensions of the linkage	Maximum length ratio	Minimum transmission angle	Value of the objective function
GI	SrRP6–SrRP4	$l_1=0.1721, l_2=0.2497$ $l_3=0.4690, l_4=0.4738$	2.75	35.91°	0.027943
GII	SrRP6–SrRP1	$l_1=0.1721, l_2=0.2447$ $l_3=0.4787, l_4=0.4974$	2.89	39.27°	0.027332
GIII	SrRP7–SrRP1	$l_1=0.1698, l_2=0.2457$ $l_3=0.4787, l_4=0.5069$	2.98	42.35°	0.027836
GIV	SrRP7–SrRP4	$l_1=0.1698, l_2=0.2492$ $l_3=0.4690, l_4=0.4826$	2.84	38.67°	0.028447
GV	SrRP8–SrRP1	$l_1=0.1830, l_2=0.2573$ $l_3=0.4787, l_4=0.4878$	2.67	35.40°	0.027913
GVI	SrRP8–SrRP4	$l_1=0.1830, l_2=0.2643$ $l_3=0.4690, l_4=0.4654$	2.56	32.70°	0.028524

Table 5.4 Six synthesized spherical four-bar linkages for nine positions

No.	Initial values	Moving points	Dimensions of the linkage	Length ratio	Transmission angle	Objective function
Solution 1	GI	P_1–P_2	$l_1 = 0.1746$, $l_2 = 0.2514$ $l_3 = 0.4967$, $l_4 = 0.4911$	2.84	33.54°	0.027768
Solution 2	GII	P_3–P_4	$l_1 = 0.1748$, $l_2 = 0.2450$ $l_3 = 0.4790$, $l_4 = 0.4901$	2.80	36.58°	0.026700
Solution 3	GIII	P_5–P_6	$l_1 = 0.1699$, $l_2 = 0.2460$ $l_3 = 0.4796$, $l_4 = 0.5078$	2.99	42.39°	0.027816
Solution 4	GIV	P_7–P_8	$l_1 = 0.1699$, $l_2 = 0.2492$ $l_3 = 0.4689$, $l_4 = 0.4824$	2.84	38.66°	0.028429
Solution 5	GV	P_9–P_{10}	$l_1 = 0.1748$, $l_2 = 0.2450$ $l_3 = 0.4794$, $l_4 = 0.4906$	2.81	36.63°	0.026992
Solution 6	GVI	P_{11}–P_{12}	$l_1 = 0.1748$, $l_2 = 0.2443$ $l_3 = 0.4732$, $l_4 = 0.4835$	2.77	36.10°	0.027082

Table 5.5 Positions of the moving joints

Moving hinge points	Spherical coordinates		Saddle spherical circle (cone)			
			Direction angles of the axis		Half-cone angle	Saddle spherical circle error
	δ_{Pm}	θ_{Pm}	ξ	η	δ	δ_{rR}
P_1	0.1351	4.8874	0.0267	5.9260	0.1746	0.016575
P_2	0.1269	1.1754	0.5175	6.0841	0.4967	0.011193
P_3	0.1355	4.8855	0.0266	5.9123	0.1748	0.016549
P_4	0.1317	0.9242	0.5164	6.0483	0.4790	0.010450
P_5	0.1290	4.7415	0.0170	0.7544	0.1699	0.017369
P_6	0.1317	0.9251	0.5169	6.0480	0.4796	0.010447
P_7	0.1289	4.7418	0.0171	0.7562	0.1699	0.017364
P_8	0.1256	1.1912	0.4926	6.1209	0.4689	0.011065
P_9	0.1355	4.8854	0.0266	5.9121	0.1748	0.016550
P_{10}	0.1318	0.9219	0.5169	6.0474	0.4794	0.010442
P_{11}	0.1355	4.8852	0.0265	5.9132	0.1748	0.016550
P_{12}	0.1305	0.9326	0.5097	6.0555	0.4732	0.010532

For the six spherical crank-rocker linkages listed in Table 5.4, the values of the objective function are close to each other, and any one of them can be taken as the final result. But if taking into account the transmission performance of the linkage, we may preferentially choose the third solution, whose minimum transmission angle is maximal. For the first solution in Table 5.4, the deviations between the spherical separated positions guided by the spherical four-bar linkage and the given positions are listed in Table 5.6.

Table 5.6 The position deviations for solution 1

Position	$\Delta\theta_1^{(i)}$	$\Delta\theta_2^{(i)}$	$\Delta\theta_3^{(i)}$	Position	$\Delta\theta_1^{(i)}$	$\Delta\theta_2^{(i)}$	$\Delta\theta_3^{(i)}$
1	0.03	−0.02	−0.02	6	−0.02	0.04	0.04
2	−0.01	0.03	0.02	7	0.02	0.04	0.07
3	0.00	0.02	0.01	8	−0.00	−0.06	−0.04
4	−0.02	−0.01	0.01	9	−0.02	−0.03	−0.02
5	0.02	−0.01	−0.02				

5.3.2.2 Five Positions

Five-position Synthesis

If the five separated spherical positions of a moving body are given, the kinematic synthesis of spherical four-bar linkages is traditionally called five-position synthesis, or precise synthesis, that is to locate two spherical circle points or saddle spherical circle points in the moving body. Based on the mathematical model (5.24) and the optimization procedure, at most six saddle spherical circle points, can be located in the moving body. Two optional spherical circle points and their spherical center points constitute a spherical four-bar linkage, which geometrically guides the moving body to occupy the five given positions precisely. Owing to the kinematic constraints, the results may not be good enough to satisfy the design requirements, so the saddle spherical circle points would be preferred as the moving joints instead of the spherical circle points. The five-position synthesis can also be treated as the approximate kinematic synthesis, although it is theoretically a classic precise synthesis problem, which can be indicated by a the following numerical example.

Example 5.4 The five separated spherical positions of the moving body are listed in Table 5.7. Synthesize a spherical four-bar linkage with a crank, minimum allowable transmission angle $[\gamma]=35°$, and maximum allowable length ratio $[l]=5$.

Solution

The process of the kinematic synthesis of spherical four-bar linkages for the five given positions is similar to that of Example 5.3, which includes three steps: the preparation of initial values, the parameter setting of the function **fmincon**, and program operation.

Firstly, for the preparation of initial values, saddle spherical circle points need to be located in the moving body by sub-program ArRP for the given five positions. 50 initial points of the

Table 5.7 Five spherical positions (unit: radian)

	$\theta_1^{(i)}$	$\theta_2^{(i)}$	$\theta_3^{(i)}$
1	0.1802	0.7589	1.1566
2	0.2884	0.9021	0.3919
3	0.3005	1.0656	−0.2087
4	0.2545	1.3323	−0.6920
5	0.1844	1.6951	−1.1317

moving body are generated randomly on the hemisphere defined by $\delta_{Pm} \in [0, \pi/2]$, $\theta_{Pm} \in [0, 2\pi)$. The corresponding results, the saddle spherical circle points, are located by calling the sub-program ArRP, which is shown in Fig. 5.12, and six saddle spherical circle points (SrRP1-SrRP6) with less errors are listed in Table 5.8, in which two points P_1 and P_2 are spherical circle points.

Any two of the six saddle spherical circle points can be taken as a group of initial points, and there are a total of 15 groups, or 15 initial spherical four-bar linkages, whose dimensions and performances can be calculated through equations (5.41) to (5.47). Five groups of initial values, from the six saddle spherical circle points SrRP1-SrRP6 in Table 5.8, are listed in Table 5.9. Group GI, comprised of the two spherical circle points (SrRP1, SrRP2), is a precise solution in geometry, whose minimum transmission angle is 31.60°. Unfortunately, it can't be the final choice since it does not satisfy the constraint condition $[\gamma] = 35^{\circ}$.

Then, the parameters of the function **fmincon** are set up in the same way as in Example 5.3. Based on the five groups of initial values in Table 5.9, we run the optimization program **KS-CR-Sphere** and have five spherical four-bar linkages listed in Table 5.10, whose moving joints and fixed joints are given in Table 5.11.

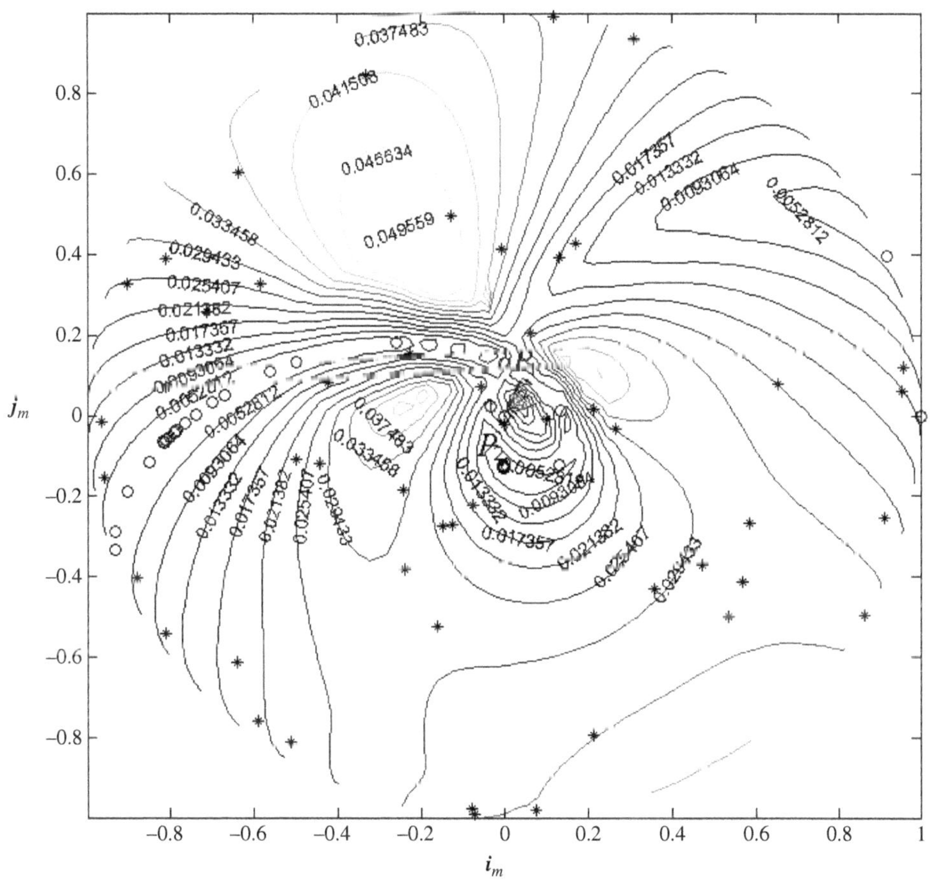

Figure 5.12 The saddle spherical circle errors for five positions

Table 5.8 Six saddle spherical circle points

No.	Spherical coordinates		Saddle spherical circle (cone)			
			Direction angles of the axis		Half-cone angle	Saddle spherical circle point error
	δ_{Pm}	θ_{Pm}	ξ	η	δ	δ_{rR}
SrRP1	0.1308	4.7115	0.0002	1.7493	0.1744	5.436217×10^{-8}
SrRP2	0.1312	1.5769	0.3840	0.0012	0.3844	1.905142×10^{-8}
SrRP3	0.6717	3.0103	0.7312	0.0071	1.2566	0.001065
SrRP4	0.5894	2.9308	0.7277	6.2440	1.1828	0.001074
SrRP5	0.1472	1.8021	0.3911	0.0309	0.4102	0.001183
SrRP6	0.1813	2.0539	0.5216	6.0950	0.6147	0.001700

Table 5.9 Five groups of initial values for five-position synthesis

No.	Combination	Dimensions of the linkage	Maximum length ratio	Minimum transmission angle	Value of the objective function
GI	SrRP1–SrRP2	$l_1=0.1745, l_2=0.2620$ $l_3=0.3843, l_4=0.3840$	2.20	31.60°	3.050083×10^{-5}
GII	SrRP1–SrRP6	$l_1=0.1744, l_2=0.3032$ $l_3=0.6147, l_4=0.5216$	3.52	21.08°	0.001704
GIII	SrRP2–SrRP4	$l_1=0.3843, l_2=0.5739$ $l_3=1.1827, l_4=0.3443$	3.44	0°	0.001078
GIV	SrRP1–SrRP5	$l_1=0.1744, l_2=0.2761$ $l_3=0.4101, l_4=0.3911$	2.35	29.84°	0.001186
GV	SrRP2–SrRP3	$l_1=0.3843, l_2=0.6646$ $l_3=1.2566, l_4=0.3472$	3.62	0°	0.001066

Table 5.10 Five synthesized spherical four-bar linkages

No.	Initial values	Moving points	Dimensions of the linkage	Length ratio	Transmission angle	Objective function
Solution 1	GI	P_1–P_2	$l_1=0.1585, l_2=0.2470$ $l_3=0.3810, l_4=0.3844$	2.43	35.00°	0.000990
Solution 2	GII	P_3–P_4	$l_1=0.1650, l_2=0.2706$ $l_3=0.3292, l_4=0.3515$	2.13	35.00°	0.009712
Solution 3	GIII	P_5–P_6	$l_1=0.1664, l_2=0.2395$ $l_3=0.4035, l_4=0.4130$	2.48	35.00°	0.002619
Solution 4	GIV	P_1–P_2	$l_1=0.1585, l_2=0.2470$ $l_3=0.3810, l_4=0.3844$	2.43	35.00°	0.000990
Solution 5	GV	P_7–P_8	$l_1=0.1227, l_2=0.1957$ $l_3=0.3285, l_4=0.3238$	2.68	35.00°	0.015279

Table 5.11 Parameters of the moving joints

Moving hinge points	Spherical coordinates		Saddle spherical circle (cone)			
			Direction angles of the axis		Half-cone angle	Saddle spherical circle error
	δ_{Pm}	θ_{Pm}	ξ	η	δ	δ_{rR}
P_1	0.1166	4.5915	0.0368	1.5951	0.1585	0.000990
P_2	0.1309	1.5715	0.3821	0.0065	0.3810	0.000005
P_3	0.1251	4.4686	0.0413	2.1304	0.1650	0.003712
P_4	0.1647	2.0639	0.3402	0.3378	0.3292	0.006000
P_5	0.1239	4.6541	0.0184	1.6181	0.1664	0.000461
P_6	0.1190	1.1782	0.4088	6.1203	0.4035	0.002158
P_7	0.0816	3.8062	0.1521	1.5861	0.1227	0.009242
P_8	0.1648	2.0656	0.3400	0.3407	0.3285	0.006037

It can be seen from Table. 5.10 that the moving hinged points of the synthesized spherical four-bar linkage are no longer those of the initial spherical linkage. The average optimization time is 11.6760 s.

Four-position Synthesis

If the four separated spherical positions of a moving body are given, the kinematic synthesis of spherical four-bar linkages is to choose two spherical circle points on the spherical circle point curve and constitute a spherical four-bar linkage. We can obtain a lot of spherical linkages to guide the rigid body to occupy the given four positions precisely. After judgment by the constraint conditions (5.44) to (5.47), the final spherical linkages can be chosen to meet the design requirements.

Example 5.5 The four spherical positions of the moving body are listed in the first four lines of Table 5.7. Synthesize a spherical four-bar linkage with a crank, minimum allowable transmission angle $[\gamma] = 30°$, and maximum allowable length ratio $[l] = 5$.

Solution

Based on the optimal model (5.24), 200 initial points are randomly generated on the unit hemisphere $\delta_{Pm} \in [0, \pi/2]$ and $\theta_{Pm} \in [0, 2\pi]$ of the moving body, and the spherical circle points can be obtained by calling the sub-program ArRP. Two spherical circle points can be combined as a group of moving joints for a spherical four-bar linkage, whose dimensions can be calculated through equation (5.41). After judgment by the constraint conditions (5.44) to (5.47), such as the existence of the crank, the minimum allowable transmission angle $[\gamma] = 30^o$, and the maximum permissible link length ratio $[l] = 5$, the spherical four-bar linkages can easily be obtained, of which two solutions are listed in Table 5.12.

Table 5.12 Spherical crank-rocker linkages for four positions (unit: radian)

	Spherical coordinates		Direction angles of the axis		Half-cone angle	Dimensions of the linkage	Maximum length ratio	Minimum transmission angle
1	δ_{Pm1}	θ_{Pm1}	ξ_1	η_1	δ_1	$l_1 = 0.0611$	2.80	30.31°
	0.0420	1.0764	0.2728	0.9604	0.0611	$l_2 = 0.1203$		
	δ_{Pm2}	θ_{Pm2}	ξ_2	η_2	δ_2	$l_3 = 0.1202$		
	0.0802	4.5837	0.1061	1.1866	0.1202	$l_4 = 0.1710$		
2	δ_{Pm1}	θ_{Pm1}	ξ_1	η_1	δ_1	$l_1 = 0.0622$	1.81	30.34°
	0.0358	1.0690	0.2666	0.9720	0.0622	$l_2 = 0.0839$		
	δ_{Pm2}	θ_{Pm2}	ξ_2	η_2	δ_2	$l_3 = 0.0974$		
	0.0492	4.5427	0.1580	1.1237	0.0974	$l_4 = 0.1128$		

5.4 Discussion

Both graphics and analytics are conventional methods for the kinematic synthesis of linkages, but the former is rarely used for spherical linkages due to its complexity and inconvenience in three-dimensional representation. Bagci [1] presented an approach of conformal projection and construction for the synthesis of spherical linkages and cam-follower mechanisms for position, function, and path generation by the pole, as well as inversion and overlay techniques. With the rapid development of computing technology, algebraic and optimal methods are widely adopted in the kinematic synthesis of spherical linkages. Shirazi [2] presented the computer modeling and geometric construction of a Burmester curve for four-position synthesis of a 4R spherical linkage by the symbolic mathematical software Maple. The atlas of coupler curves by both traditional and modern numerical methods is applied to the kinematic synthesis of spherical linkages, which is less extensive than that of plane linkages. Chu and Sun [3] built up a numerical atlas database, comprised of more than 6 million harmonic characteristic parameters of coupler curves, produced by Fourier series for the basic dimensional types of spherical four-bar linkages. Mullineux [4] set up a database of the atlas of spherical four-bar linkages by Fourier techniques too, which provided a useful aid in the path generation of spherical four-bar linkages. Dobrovolskii [5] was considered the first to study the kinematic synthesis theory of spherical linkages. Roth [6] described the finitely separated positions of a rigid body by linear coordinate transformation and screw motion, which leads to the algebraic equation for points to lie on a sphere with several separated positions. Based on the vector and matrix method, Suh and Radcliffe [7,8] derived the constraint equations for the binary links and solved them for the dimensions of the spherical linkages. Dowler, Duffy, and Tesar [9,10] applied a similar method to the kinematic synthesis of a spherical linkage with three to five positions. Based on the relative pole theory, Chiang [11–13] systematically studied the position, function, and path generation of spherical linkages at both infinitesimal and finitely separated positions. McCarthy [14] improved the structure of the constraint equation of the spherical RR binary link by means of the quarternion, which facilitated the solution of the constraint equation by the algebraic method for kinematic synthesis of four-bar and six-bar spherical linkages. Ruth and McCarthy [15] introduced the functions of the software SphinxPC for the position synthesis of spherical linkages, and pointed out that a lot of spherical four-bar linkages with the characteristic atlas can be generated if the four position of the moving body is given. Alizade [16,17] presented the algebraic equations between the input and output angular displacements for the function synthesis of spherical four-bar linkages, analyzed and compared three approaches (such as the interpolation approximation, least-square approximations and Chebyshev approximation), which revealed that the Chebyshev approximation has minimum values in both the maximum error and the fluctuation range.

Similar to those of planar linkages, the solutions for the kinematic synthesis of spherical linkages depend on the position numbers of the moving body by means of the algebraic method, and the calculation complexity increases with the degree of the algebraic equations, which indicates that the structure of the constraint equations needs to be improved in many ways. With regard to the approximately kinematic synthesis of the spherical linkage, Angeles and Liu [18] presented the optimization method for the path generation of the spherical 4R linkages, whose objective function is the quadratic sum of the error between the path-generating point of the coupler curve and the prescribed points. In a following reference [19], they presented the optimization method for the function generation of the spherical 4R linkage, whose objective function is the quadratic sum of the error between the actual function and the ideal output function of the spherical linkages synthesized. Bodduluri and McCarthy [20] mapped a spherical four-bar linkage to a curve in the four-dimensional image space, through which the spherical four-bar position synthesis was converted into the image curve fitting of the given points by least squares. Ge and Larochelle [21] described approximately the spherical motion by NURBS motion, and suggested that the image curve in the four-dimensional image space can be replaced by the rational B-spline curve and the position synthesis of the spherical linkage can be resolved into the design of the rational B-spline curve. Tse and Larochelle [22] mapped the prescribed positions of the rigid body to the four-dimensional Euclidian space and defined the metric by the bi-quaternion, and proved that a spherical linkage can be synthesized by sphere fitting to the given multiple positions, leading to the conversion of the position synthesis of spherical linkages into curve and surface fitting in the image space. Farhang and Zargar [23] analyzed a spherical 4R linkage, whose crank has much shorter length than the other three links, and whose output rotational displacement can be viewed approximately as a linear combination of the crank angle, with a simple function in both the first and the second harmonics, including lengths of the 4R spherical linkage. Alizade, Can, and Kilit [24] presented the displacement equation of the spherical RR binary link, which can be solved by the linear superposition method for the precise position synthesis and the least-square method for the approximate position synthesis.

The kinematic synthesis of spherical four-bar linkages, in both precision and approximate synthesis, whose theoretical base is the discrete kinematic geometry, has fewer results than those of the plane four-bar linkages, in terms of both contents and depth. On the contrary, the precision synthesis at finitely separated positions has the corresponding properties of the kinematic geometry at infinitesimal positions, which may provide a referring clue to the solutions for the approximate synthesis with multiple positions. Of course, the objective function is the key fact for the optimization synthesis of spherical linkages. Most of the objective functions in the references so far are limited to the structure error directly or indirectly, not properly reflecting the intrinsic connections between the objective function and the optimization variables. Hence, the discovery of the optimal solution and the convergence of the optimal algorithm are still difficult problems to solve for the optimal synthesis of spherical linkages.

The kinematic synthesis of spherical linkages is not only a bridge from plane to spatial, but also a special form for spatial linkages. Unfortunately, researchers in the field might have paid much more attention to the latter than the former, or the approach for kinematic synthesis from plane, spherical to spatial linkage should be systematically taken into account. Especially, the kinematic synthesis of spherical linkages would provide a theoretical basis for the kinematic synthesis of a spatial linkage with ruled surfaces. Based on the curvature theories of the spherical curve, the authors presented the approach to the kinematic synthesis of spherical linkages with multiple positions by saddle spherical circle fitting [25–28], whose form and content correspond to those of both plane linkages and spatial linkages.

References

[1] C. Bagci, Geometric methods for the synthesis of spherical mechanisms for the generation of functions, paths and rigid-body positions using conformal projections. *Mechanism and Machine Theory*, **19**(1), 113–127, 1984.

[2] K.H. Shirazi, Computer modelling and geometric construction for four-point synthesis of 4R spherical linkages. *Applied Mathematical Modelling*, **31**(9), 1874–1888, 2007.

[3] J.K. Chu and J.W. Sun, Numerical atlas method for path generation of spherical four-bar mechanism. *Mechanism and Machine Theory*, **45**(6), 867–879, 2010.

[4] G. Mullineux, Atlas of spherical four-bar mechanism. *Mechanism and Machine Theory*, **46**(11), 1811–1823, 2011.

[5] V.V. Dobrovolskii, Synthesis of spherical mechanisms. *Akadamiia Nauk, SSSR Trudi Seminara po Teorii Ashin, Mechanizmov*, 5–20, 1943 (in Russian).

[6] B. Roth, The kinematics of motion through finitely separated positions. *ASME Journal of Applied Mechanics*, **34**(3), 591–598, 1967.

[7] C.H. Suh and C.W. Radcliffe, Synthesis of spherical mechanisms with the use of the displacement matrix. *ASME Journal of Engineering for Industry*, **89**(2), 215–222, 1967.

[8] C.H. Suh and C.W. Radcliffe, *Kinematics and Mechanism Design*. Wiley, New York, 1978.

[9] H.J. Dowler, J. Duffy, and D. Tesar, A generalized study of three multiply separated positions in spherical kinematics. *Mechanism and Machine Theory*, **11**(6), 395–410, 1976.

[10] H.J. Dowler, J. Duffy, and D. Tesar, A generalized study of four and five multiply separated positions in spherical kinematics-II. *Mechanism and Machine Theory*, **13**(4), 409–435, 1978.

[11] C.H. Chiang, *Kinematics of Spherical Mechanisms*. Cambridge University Press, New York, 1988.

[12] C.H. Chiang, Synthesis of spherical four-bar path generators. *Mechanism and Machine Theory*, **21**(2), 135–143, 1986.

[13] C.H. Chiang, Design of spherical and planar crank-rockers and double-rockers as function generators. *Mechanism and Machine Theory*, **21**(4), 287–305, 1986.

[14] J.M. McCarthy and G.S. Soh, *Geometric Design of Linkages*, 2nd edn. Interdisciplinary Applied Mathematics 11. Springer, New York, 2010.

[15] D.A. Ruth and J.M. McCarthy, The design of spherical 4R linkages for four specified orientations. *Mechanism and Machine Theory*, **34**(5), 677–692, 1999.

[16] R. Alizade and Ö. Kilit, Analytical synthesis of function generating spherical four-bar mechanism for the five precision points. *Mechanism and Machine Theory*, **40**(7), 863–878, 2005.

[17] R. Alizade, Synthesis of function generating spherical four bar mechanism for the six independent parameters. *Mechanism and Machine Theory*, **46**(9), 1316–1326, 2011.

[18] J. Angeles and Z. Liu, The constrained least-square optimization of spherical four-bar path generators. *ASME Journal of Mechanical Design*, **114**(3), 394–405, 1992.

[19] Z. Liu and J. Angeles, Least-square optimization of planar and spherical four-bar function generator under mobility constraint. *ASME Journal of Mechanical Design*, **114**(4), 569–573, 1992.

[20] R.M.C. Bodduluri and J.M. McCarthy, Finite position synthesis using the image curve of a spherical four-bar motion. *ASME Journal of Mechanical Design*, **114**(1), 55–60, 1992.

[21] Q.J. Ge and P.M. Larochelle, Algebraic motion approximation with NURBS motions and its application to spherical mechanism synthesis. *ASME Journal of Mechanical Design*, **121**, 529–532, 1999.

[22] D.M. Tse and P.M. Larochelle, Approximating spatial locations with spherical orientations for spherical mechanism design. *ASME Journal of Mechanical Design*, **122**(4), 457–463, 1998.

[23] K. Farhang and Y.S. Zargar, Design of spherical 4R mechanisms: Function generation for the entire motion cycle. *ASME Journal of Mechanical Design*, **121**(4), 521–528, 1999.
[24] R. Alizade, F.C. Can, and Ö. Kilit, Least square approximate motion generation synthesis of spherical linkages by using Chebyshev and equal spacing. *Mechanism and Machine Theory*, **61**, 123–135, 2013.
[25] D.L. Wang and S.F. Wang, A unified approach to kinematic synthesis of mechanism by adaptive curve fitting. *Science in China Series E: Technological Sciences*, **47**(1), 85–96, 2004.
[26] S.F. Wang, *New Approach for Kinematic Synthesis of Mechanism by Adaptive Curve Fitting*. Dalian University of Technology, Dalian, 2005 (in Chinese).
[27] D.L. Wang, S.F. Wang, and B.Y. Zhang, New approach to function synthesis of spherical mechanisms. *Chinese Journal of Mechanical Engineering*, **40**(2), 45–49, 2004 (in Chinese).
[28] B.Y. Zhang, *Study on Theory and Method of the Function Synthesis of Spherical Four-Bar Linkages*. Dalian University of Technology, Dalian, 2000 (in Chinese).

6

Spatial Kinematic Differential Geometry

The spatial kinematic geometry, particularly at infinitesimally consecutive positions, is to study the instantaneous spatial kinematics of a rigid body and the geometry of graphics, which are theoretical bases of the kinematic synthesis of spatial linkages. The graphics, including space curves and surfaces, produced by the points and lines in spatial movement are so diverse and complex that general approaches such as algebra or geometry are invalid. It is an essential topic for differential geometry to study the geometric properties of graphics by the moving frame and kinematic invariants, in which the differential of the moving frame is viewed as the spatial movement of a rigid body, and the invariants are used to describe the geometry of the graphics, in particular compare them with the regular graphics of linkages. Hence, the spatial kinematics and space graphic geometry are naturally combined for spatial linkages in this chapter, termed "spatial kinematic differential geometry." In this chapter, a line trajectory is first taken as the original ruled surface to describe the space movement of a rigid body, which leads to the axodes–trajectories of the ***instantaneous screw axis (ISA)*** in the moving and fixed body–being derived. Then, both the point trajectories and the line trajectories are described concisely in the moving frame of the axodes and their local properties are discussed and compared with those of the constraint surfaces in terms of invariants and invariance.

6.1 Displacement Equation

There are many ways to describe the spatial movement of a rigid body. The Cartesian coordinate system is currently used for the convenience of calculation, but may lead to additional complexity if not properly chosen. Hence, the moving frame of a reference line trajectory, being adjoint to the movement of the moving body, is introduced to connect the Cartesian coordinate system and the movement of the rigid body in analytical form, revealing the geometrical meaning and convenience of calculation that are part of the adjoint approach.

Kinematic Differential Geometry and Saddle Synthesis of Linkages, First Edition. Delun Wang and Wei Wang.

6.1.1 General Description

In order to describe the spatial movement of a rigid body $\sum^*$ relative to the fixed body $\sum$, the moving Cartesian coordinate system $\{\boldsymbol{O}_m; \boldsymbol{i}_m, \boldsymbol{j}_m, \boldsymbol{k}_m\}$ is established on $\sum^*$ and the fixed system $\{\boldsymbol{O}_f; \boldsymbol{i}_f, \boldsymbol{j}_f, \boldsymbol{k}_f\}$ on $\sum$. Six parameters are required to define the displacements of $\sum^*$ relative to $\sum$, including both the three linear displacements of a reference point to locate the position of a moving body and three angular displacements to define the rotation angles of the moving body with respect to the reference point. If the movement belongs to a plane, there are three parameters, two linear displacements of a reference point of $\sum^*$ and one angular displacement of $\sum^*$ about the reference point. The spherical movement of the rigid body can be viewed as an extension of the plane motion on the sphere; likewise, three parameters define the spherical displacements. If two points of $\sum^*$ in spatial movement are taken as reference points, a reference line L_k through the two points can be determined. The spatial displacements of $\sum^*$ can be described by the movement along with the line L_k and the rotation of $\sum^*$ about L_k. In particular, the reference line L_k is chosen to coincide with the coordinate axis $\boldsymbol{k}_m$ of $\{\boldsymbol{O}_m; \boldsymbol{i}_m, \boldsymbol{j}_m, \boldsymbol{k}_m\}$, as shown in Fig. 6.1. The position vector of the original point O_m in $\{\boldsymbol{O}_f; \boldsymbol{i}_f, \boldsymbol{j}_f, \boldsymbol{k}_f\}$ is $\boldsymbol{R}_{Om}$, and the relative inclination angles of the corresponding axes between $\{\boldsymbol{O}_m; \boldsymbol{i}_m, \boldsymbol{j}_m, \boldsymbol{k}_m\}$ and $\{\boldsymbol{O}_f; \boldsymbol{i}_f, \boldsymbol{j}_f, \boldsymbol{k}_f\}$ are θ_1, θ_2, and θ_3, defined in Chapter 4. The relationship between $\{\boldsymbol{O}_m; \boldsymbol{i}_m, \boldsymbol{j}_m, \boldsymbol{k}_m\}$ and $\{\boldsymbol{O}_f; \boldsymbol{i}_f, \boldsymbol{j}_f, \boldsymbol{k}_f\}$ is

$$\begin{cases} \boldsymbol{R}_{Om} = x_{Omf}\boldsymbol{i}_f + y_{Omf}\boldsymbol{j}_f + z_{Omf}\boldsymbol{k}_f \\ \boldsymbol{i}_m = (\mathrm{c}\theta_3\mathrm{s}\theta_2 + \mathrm{c}\theta_1\mathrm{s}\theta_3\mathrm{c}\theta_2)\boldsymbol{i}_f + (-\mathrm{c}\theta_3\mathrm{c}\theta_2 + \mathrm{c}\theta_1\mathrm{s}\theta_3\mathrm{s}\theta_2)\boldsymbol{j}_f - \mathrm{s}\theta_1\mathrm{s}\theta_3\boldsymbol{k}_f \\ \boldsymbol{j}_m = (-\mathrm{s}\theta_3\mathrm{s}\theta_2 + \mathrm{c}\theta_1\mathrm{c}\theta_3\mathrm{c}\theta_2)\boldsymbol{i}_f + (\mathrm{s}\theta_3\mathrm{c}\theta_2 + \mathrm{c}\theta_1\mathrm{c}\theta_3\mathrm{s}\theta_2)\boldsymbol{j}_f - \mathrm{s}\theta_1\mathrm{c}\theta_3\boldsymbol{k}_f \\ \boldsymbol{k}_m = \mathrm{s}\theta_1\mathrm{c}\theta_2\boldsymbol{i}_f + \mathrm{s}\theta_1\mathrm{s}\theta_2\boldsymbol{j}_f + \mathrm{c}\theta_1\boldsymbol{k}_f \end{cases} \tag{6.1}$$

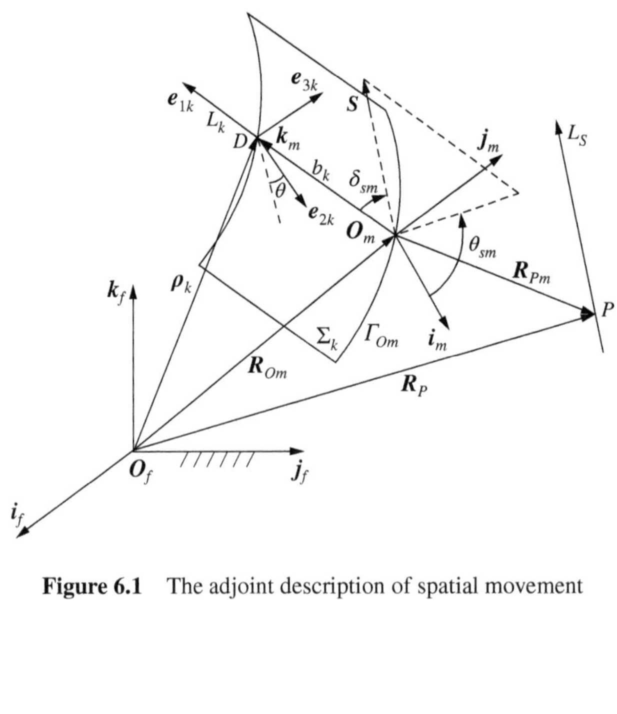

Figure 6.1 The adjoint description of spatial movement

where $c\theta_i = \cos\theta_i$, $s\theta_i = \sin\theta_i$, $i = 1, 2, 3$ (and the same hereinafter in this chapter). $(\theta_1, \theta_2, \theta_3)$ and $(x_{Omf}, y_{Omf}, z_{Omf})$ are the six independent kinematic parameters (or six DOFs) and will be functions of one independent parameter if $\sum^*$ moves with a single DOF. Hereinafter, we only discuss the spatial movement of a moving body with one DOF.

A point P of $\sum^*$ has Cartesian coordinates (x_{Pm}, y_{Pm}, z_{Pm}) in $\{\boldsymbol{O}_m; \boldsymbol{i}_m, \boldsymbol{j}_m, \boldsymbol{k}_m\}$. The position vector $\boldsymbol{R}_P(x_{Pf}, y_{Pf}, z_{Pf})$ of P in $\{\boldsymbol{O}_f; \boldsymbol{i}_f, \boldsymbol{j}_f, \boldsymbol{k}_f\}$ can be obtained by the following coordinate transformation:

$$\begin{bmatrix} x_{Pf} \\ y_{Pf} \\ z_{Pf} \\ 1 \end{bmatrix} = [\mathbf{T}_{mf}] \cdot [\mathbf{R}_{mf}] \cdot \begin{bmatrix} x_{Pm} \\ y_{Pm} \\ z_{Pm} \\ 1 \end{bmatrix} \tag{6.2a}$$

where

$$[\mathbf{T}_{mf}] = \begin{bmatrix} 1 & 0 & 0 & x_{Omf} \\ 0 & 1 & 0 & y_{Omf} \\ 0 & 0 & 1 & z_{Omf} \\ 0 & 0 & 0 & 1 \end{bmatrix} \tag{6.2b}$$

$$[\mathbf{R}_{mf}] = \begin{bmatrix} c\theta_3 s\theta_2 + c\theta_1 s\theta_3 c\theta_2 & -s\theta_3 s\theta_2 + c\theta_1 c\theta_3 c\theta_2 & s\theta_1 c\theta_2 & 0 \\ -c\theta_3 c\theta_2 + c\theta_1 s\theta_3 s\theta_2 & s\theta_3 c\theta_2 + c\theta_1 c\theta_3 s\theta_2 & s\theta_1 s\theta_2 & 0 \\ -s\theta_1 s\theta_3 & -s\theta_1 c\theta_3 & c\theta_1 & 0 \\ 0 & 0 & 0 & 1 \end{bmatrix} \tag{6.2c}$$

The above chosen parameters are not unique to express the spatial movement of $\sum^*$. We can also turn to other description forms such as the displacements of three points in $\sum^*$ which are not collinear. These forms can also be converted into the above description and represented by the coordinate transformation formula (6.2a).

The vector equation of P in $\{\boldsymbol{O}_m; \boldsymbol{i}_m, \boldsymbol{j}_m, \boldsymbol{k}_m\}$ is written in Cartesian coordinates (x_{Pm}, y_{Pm}, z_{Pm}) or cylindrical coordinates $(r_{Pm}, \theta_{Pm}, z_{Pm})$ as

$$\boldsymbol{R}_{Pm} = x_{Pm}\boldsymbol{i}_m + y_{Pm}\boldsymbol{j}_m + z_{Pm}\boldsymbol{k}_m = r_{Pm}\boldsymbol{e}_{\mathrm{I}(\theta_{Pm})} + z_{Pm}\boldsymbol{k}_m \tag{6.3}$$

The point P traces a spatial curve Γ_P in $\{\boldsymbol{O}_f; \boldsymbol{i}_f, \boldsymbol{j}_f, \boldsymbol{k}_f\}$, whose vector equation is

$$\Gamma_P : \boldsymbol{R}_P = \boldsymbol{R}_{Om} + [\mathbf{R}_{mf}] \cdot \boldsymbol{R}_{Pm} \tag{6.4}$$

Taking the derivative of the above vector equation with respect to time t, we have the absolute velocity vector of P as

$$\boldsymbol{V}_P = \frac{\mathrm{d}\boldsymbol{R}_P}{\mathrm{d}t} = \frac{\mathrm{d}\boldsymbol{R}_{Om}}{\mathrm{d}t} + \frac{\mathrm{d}[\mathbf{R}_{mf}]}{\mathrm{d}t}\boldsymbol{R}_{Pm} + [\mathbf{R}_{mf}]\frac{\mathrm{d}\boldsymbol{R}_{Pm}}{\mathrm{d}t} \tag{6.5}$$

The spatial movement of $\sum^*$ can be regarded as a combination of the following velocity with the relative velocity: the following velocity of P as the original point $\boldsymbol{O}_m$ of $\{\boldsymbol{O}_m; \boldsymbol{i}_m, \boldsymbol{j}_m, \boldsymbol{k}_m\}$ and the relative velocity of P with respect to $\boldsymbol{O}_m$. The selection of the moving frame $\{\boldsymbol{O}_m; \boldsymbol{i}_m, \boldsymbol{j}_m, \boldsymbol{k}_m\}$ affects the property and complexity of the following motion and the relative motion. In general, $\{\boldsymbol{O}_m; \boldsymbol{i}_m, \boldsymbol{j}_m, \boldsymbol{k}_m\}$ and P are both fixed on $\sum^*$. The relative displacement of P to O_m is zero, or $\frac{\mathrm{d}\boldsymbol{R}_{Pm}}{\mathrm{d}t} = 0$ in equation (6.5). The absolute velocity of P can be determined by equations (6.2) and (6.5).

6.1.2 Adjoint Description

The properties of the trajectories, traced by the points or lines of $\sum^*$ in spatial movement, can be analytically studied, but the complexity of the analytical equations may change as different $\{\boldsymbol{O}_m;\boldsymbol{i}_m,\boldsymbol{j}_m,\boldsymbol{k}_m\}$ of $\sum^*$ are chosen, which do not adequately take advantage of the analytical properties of both the reference point trajectory (or line trajectory) and the spatial movement of the rigid body. In this chapter, the adjoint approach for spatial movement, a space curve adjoint to a ruled surface and a ruled surface adjoint to another ruled surface, is introduced, or the trajectory traced by the reference line is first designated as the original surface and the points or lines of $\sum^*$ are adjoint to the reference line and trace the adjoint space curves or the adjoint ruled surfaces.

The spatial movement of $\sum^*$ with moving frame $\{\boldsymbol{O}_m;\boldsymbol{i}_m,\boldsymbol{j}_m,\boldsymbol{k}_m\}$ is given: a line L_k of $\sum^*$ traces a line trajectory, a ruled surface $\sum_k$ in $\{\boldsymbol{O}_f;\boldsymbol{i}_f,\boldsymbol{j}_f,\boldsymbol{k}_f\}$, whose vector equation can be derived by equations in the above section. For a general expression, the line L_k is just the coordinate axis $\boldsymbol{k}_m$ of $\{\boldsymbol{O}_m;\boldsymbol{i}_m,\boldsymbol{j}_m,\boldsymbol{k}_m\}$, which can naturally be chosen as the reference line. The original point O_m on L_k traces a spatial curve $\varGamma_{Om}$ in $\{\boldsymbol{O}_f;\boldsymbol{i}_f,\boldsymbol{j}_f,\boldsymbol{k}_f\}$, as shown in Fig. 6.1, whose vector equation is

$$\varGamma_{Om}\ :\ \boldsymbol{R}_{Om}(s) = x_{Omf}(s)\boldsymbol{i}_f + y_{Omf}(s)\boldsymbol{j}_f + z_{Omf}(s)\boldsymbol{k}_f \tag{6.6}$$

where s is the arc length of $\varGamma_{Om}$ and the three displacements $(x_{Omf}, y_{Omf}, z_{Omf})$ can be viewed as the functions of s. Taking $\varGamma_{Om}$ as a directrix of $\sum_k$, we have the vector equation of $\sum_k$ in $\{\boldsymbol{O}_f;\boldsymbol{i}_f,\boldsymbol{j}_f,\boldsymbol{k}_f\}$:

$$\sum\nolimits_k\ :\ \boldsymbol{R}(s,\lambda) = \boldsymbol{R}_{Om}(s) + \lambda\boldsymbol{l}_k(s),\ \boldsymbol{l}_k(s) = l_{k1}\boldsymbol{i}_f + l_{k2}\boldsymbol{j}_f + l_{k3}\boldsymbol{k}_f \tag{6.7}$$

where the unit direction vector of the line L_k is $\boldsymbol{l}_k(s) = (l_{k1}, l_{k2}, l_{k3})^T = [\mathbf{R}_{mf}]\cdot(0,0,1)^T$. The three rotation angles $(\theta_1, \theta_2, \theta_2)$ in $[\mathbf{R}_{mf}]$ are also the functions of s. Based on equation (3.55), the vector equation of the striction curve of $\sum_k$ is derived as

$$\begin{aligned}\rho_k &= \boldsymbol{R}_{Om} + b_k\boldsymbol{l}_k\\ &= \boldsymbol{R}_{Om} + \left(\frac{\mathrm{d}\boldsymbol{R}_{Om}}{\mathrm{d}s}\cdot\frac{\mathrm{d}\boldsymbol{l}_k}{\mathrm{d}s}\right)\boldsymbol{l}_k/\left(\frac{\mathrm{d}\boldsymbol{l}_k}{\mathrm{d}s}\right)^2\end{aligned} \tag{6.8}$$

where the striction directrix distance b_k is the distance between the striction point and the original point O_m on L_k. The striction point has different positions on L_k at different instants. The vector equation of $\sum_k$ is rewritten as

$$\sum\nolimits_k\ :\ \boldsymbol{R}(s,\lambda) = \rho_k(s) + \lambda\boldsymbol{l}_k(s) \tag{6.9}$$

According to equation (3.57), the Frenet frame $\{\rho_k;\boldsymbol{e}_{1k},\boldsymbol{e}_{2k},\boldsymbol{e}_{3k}\}$ of $\sum_k$ is set up as

$$\begin{cases}\rho_k = \boldsymbol{R}_{Om} + b_k\boldsymbol{l}_k\\ \boldsymbol{e}_{1k} = \boldsymbol{l}_k\\ \boldsymbol{e}_{2k} = \dfrac{\mathrm{d}\boldsymbol{e}_{1k}}{\mathrm{d}\sigma_k} = \dfrac{\mathrm{d}\boldsymbol{l}_k}{\mathrm{d}\sigma_k}\\ \boldsymbol{e}_{3k} = \boldsymbol{e}_{1k}\times\boldsymbol{e}_{2k} = \boldsymbol{l}_k\times\dfrac{\mathrm{d}\boldsymbol{l}_k}{\mathrm{d}\sigma_k}\end{cases} \tag{6.10}$$

where σ_k is the arc length of the spherical image curve of the generatrices of $\sum_k$, which can be expressed by s as $\mathrm{d}\sigma_k = |\mathrm{d}\boldsymbol{l}_k| = \left|\frac{\mathrm{d}\boldsymbol{l}_k}{\mathrm{d}s}\right| \mathrm{d}s$. The relationship between $\{\boldsymbol{O}_m; \boldsymbol{i}_m, \boldsymbol{j}_m, \boldsymbol{k}_m\}$ and $\{\boldsymbol{\rho}_k; \boldsymbol{e}_{1k}, \boldsymbol{e}_{2k}, \boldsymbol{e}_{3k}\}$ is

$$\begin{cases} \boldsymbol{\rho}_k = \boldsymbol{R}_{Om} + b_k \boldsymbol{l}_k \\ \boldsymbol{e}_{1k} = \boldsymbol{k}_m \\ \boldsymbol{e}_{2k} = \cos\theta \boldsymbol{i}_m + \sin\theta \boldsymbol{j}_m \\ \boldsymbol{e}_{3k} = -\sin\theta \boldsymbol{i}_m + \cos\theta \boldsymbol{j}_m \end{cases} \tag{6.11}$$

where θ is the direction angle of the axis $\boldsymbol{e}_{2k}$ on the plane $\boldsymbol{O}_m - \boldsymbol{i}_m \boldsymbol{j}_m$. Based on the differential formulas (3.58) of the Frenet frame of the ruled surface, the construction parameters α_k, β_k, and γ_k of $\sum_k$ are

$$\begin{cases} \alpha_k = (\dot{\boldsymbol{\rho}}_k, \boldsymbol{e}_{1k}) = \left(\dfrac{\mathrm{d}\boldsymbol{\rho}_k}{\mathrm{d}s}, \boldsymbol{l}_k\right) \Big/ \left|\dfrac{\mathrm{d}\boldsymbol{l}_k}{\mathrm{d}s}\right| \\ \beta_k = (\boldsymbol{l}_k, \dot{\boldsymbol{l}}_k, \ddot{\boldsymbol{l}}_k) = \left(\boldsymbol{l}_k, \dfrac{\mathrm{d}\boldsymbol{l}_k}{\mathrm{d}s}, \dfrac{\mathrm{d}^2\boldsymbol{l}_k}{\mathrm{d}s^2}\right) \Big/ \left|\dfrac{\mathrm{d}\boldsymbol{l}_k}{\mathrm{d}s}\right|^3 \\ \gamma_k = (\dot{\boldsymbol{\rho}}_k, \boldsymbol{l}_k, \dot{\boldsymbol{l}}_k) = \left(\dfrac{\mathrm{d}\boldsymbol{\rho}_k}{\mathrm{d}s}, \boldsymbol{l}_k, \dfrac{\mathrm{d}\boldsymbol{l}_k}{\mathrm{d}s}\right) \Big/ \left|\dfrac{\mathrm{d}\boldsymbol{l}_k}{\mathrm{d}s}\right|^2 \end{cases} \tag{6.12}$$

where the letters with one and two overdots denote their first and second derivatives with respect to σ_k (and the same hereinafter in this chapter).

A point P of $\sum^*$, not belonging to L_k, traces a point trajectory, a space curve Γ_P in $\{\boldsymbol{O}_f; \boldsymbol{i}_f, \boldsymbol{j}_f, \boldsymbol{k}_f\}$. Point P is examined in the Frenet frame $\{\boldsymbol{\rho}_k; \boldsymbol{e}_{1k}, \boldsymbol{e}_{2k}, \boldsymbol{e}_{3k}\}$ of $\sum_k$, or P is adjoint to L_k. Hence, $\sum_k$ is viewed as an original ruled surface and Γ_P is naturally treated as the adjoint curve of $\sum_k$. The vector equation of Γ_P is

$$\Gamma_P : \boldsymbol{R}_P = \boldsymbol{\rho}_k + u_1 \boldsymbol{e}_{1k} + u_2 \boldsymbol{e}_{2k} + u_3 \boldsymbol{e}_{3k} \tag{6.13}$$

By equations (6.4) and (6.11), the coordinates (u_1, u_2, u_3) of P in $\{\boldsymbol{\rho}_k; \boldsymbol{e}_{1k}, \boldsymbol{e}_{2k}, \boldsymbol{e}_{3k}\}$ are

$$\begin{cases} u_1 = z_{Pm} - b_k \\ u_2 = r_{Pm} \cos(\theta_{Pm} - \theta) \\ u_3 = r_{Pm} \sin(\theta_{Pm} - \theta) \end{cases} \tag{6.14}$$

Taking the derivative of equation (6.13) with respect to σ_k, we have

$$\begin{cases} \dot{\boldsymbol{R}}_P = A_1 \boldsymbol{e}_{1k} + A_2 \boldsymbol{e}_{2k} + A_3 \boldsymbol{e}_{3k} \\ A_1 = \alpha_k + \dot{u}_1 \quad u_2 = \alpha_k - \dot{b}_k - u_2 \\ A_2 = u_1 + \dot{u}_2 - \beta_k u_3 = u_1 - \left(\beta_k - \dot{\theta}\right) u_3 \\ A_3 = \gamma_k + \beta_k u_2 + \dot{u}_3 = \gamma_k + (\beta_k - \dot{\theta}) u_2 \end{cases} \tag{6.15}$$

The absolute velocity vector $\boldsymbol{V}_P$ of P can be obtained by $\boldsymbol{V}_P = \frac{\mathrm{d}\boldsymbol{R}_P}{\mathrm{d}t} = \frac{\mathrm{d}\boldsymbol{R}_P}{\mathrm{d}\sigma_k}\frac{\mathrm{d}\sigma_k}{\mathrm{d}t} = \frac{\mathrm{d}\sigma_k}{\mathrm{d}t}\dot{\boldsymbol{R}}_P$.

For a straight line L_S of $\sum^*$ passing through P, its unit direction vector is $\boldsymbol{s}_m(s_{m1}, s_{m2}, s_{m3})$ in $\{\boldsymbol{O}_m; \boldsymbol{i}_m, \boldsymbol{j}_m, \boldsymbol{k}_m\}$. We have

$$\boldsymbol{s}_m = s_{m1}\boldsymbol{i}_m + s_{m2}\boldsymbol{j}_m + s_{m3}\boldsymbol{k}_m = \sin\delta_{sm}\cos\theta_{sm}\boldsymbol{i}_m + \sin\delta_{sm}\sin\theta_{sm}\boldsymbol{j}_m + \cos\delta_{sm}\boldsymbol{k}_m \tag{6.16a}$$

where δ_{sm} is the angle between $\boldsymbol{s}_m$ and $\boldsymbol{k}_m$, and θ_{sm} is the direction angle of the projection vector of $\boldsymbol{s}_m$ onto the plane $\boldsymbol{O}_m - \boldsymbol{i}_m\boldsymbol{j}_m$, as shown in Fig. 6.1. By equation (6.2), the unit direction vector of L_S can be described in $\{\boldsymbol{O}_f; \boldsymbol{i}_f, \boldsymbol{j}_f, \boldsymbol{k}_f\}$ as

$$\boldsymbol{s} = (s_{f1}, s_{f2}, s_{f3})^T = [\mathbf{R}_{mf}] \cdot (s_{m1}, s_{m2}, s_{m3})^T \tag{6.16b}$$

If L_S is expressed in $\{\boldsymbol{\rho}_k; \boldsymbol{e}_{1k}, \boldsymbol{e}_{2k}, \boldsymbol{e}_{3k}\}$ of $\sum_k$, the unit vector $\boldsymbol{s}$ is

$$\begin{cases} \boldsymbol{s}(\sigma_k) = s_1\boldsymbol{e}_{1k} + s_2\boldsymbol{e}_{2k} + s_3\boldsymbol{e}_{3k} \\ s_1 = \cos\delta_{sm} \\ s_2 = \sin\delta_{sm}\cos(\theta_{sm} - \theta) \\ s_3 = \sin\delta_{sm}\sin(\theta_{sm} - \theta) \end{cases} \tag{6.17}$$

where (s_1, s_2, s_3) are the components of $\boldsymbol{s}$ in $\{\boldsymbol{\rho}_k; \boldsymbol{e}_{1k}, \boldsymbol{e}_{2k}, \boldsymbol{e}_{3k}\}$, which can be obtained by equations (6.11) and (6.16).

Taking the first derivative of $\boldsymbol{s}$ in equation (6.17) with respect to σ_k, we have

$$\begin{aligned} \dot{\boldsymbol{s}} &= (\dot{s_1} - s_2)\boldsymbol{e}_{1k} + (s_1 + \dot{s_2} - \beta_k s_3)\boldsymbol{e}_{2k} + (\beta_k s_2 + \dot{s_3})\boldsymbol{e}_{3k} \\ &= -s_2\boldsymbol{e}_{1k} + [s_1 - (\beta_k - \dot{\theta})s_3]\boldsymbol{e}_{2k} + [(\beta_k - \dot{\theta})s_2]\boldsymbol{e}_{3k} \end{aligned} \tag{6.18}$$

As the spatial movement of $\sum^*$, the line L_S is adjoint to the reference line L_k and traces another ruled surface $\sum_S$ in $\{\boldsymbol{O}_f; \boldsymbol{i}_f, \boldsymbol{j}_f, \boldsymbol{k}_f\}$. $\sum_k$ is taken as the original ruled surface and $\sum_S$ is the adjoint ruled surface of $\sum_k$. The vector equation of $\sum_S$ is

$$\sum\nolimits_S : \boldsymbol{R}_s(\sigma_k, \mu) = \boldsymbol{R}_P(\sigma_k) + \mu\boldsymbol{s}(\sigma_k) \tag{6.19}$$

Any point $P_i(\mu = \mu_i)$ on L_S with distance μ_i, other than the point $P(\mu = 0)$, has its vector equation

$$\boldsymbol{R}_{Pi} = \boldsymbol{R}_P + \mu_i\boldsymbol{s} \tag{6.20}$$

The absolute velocity of P_i can be obtained by differentiating the above equation with respect to t, or

$$\boldsymbol{V}_{Pi} = \frac{\mathrm{d}\boldsymbol{R}_{Pi}}{\mathrm{d}t} = (\dot{\boldsymbol{R}}_P + \mu_i\dot{\boldsymbol{s}})\frac{\mathrm{d}\sigma_k}{\mathrm{d}t} \tag{6.21}$$

where $\dot{\boldsymbol{R}}_P$ is presented in equation (6.15).

In particular, the velocity vectors $\boldsymbol{V}_{Pi}$ of all points P_i, $\mu_i \in (-\infty, +\infty)$, on the line L_S, have the same magnitude and direction vector $\boldsymbol{s}$, which means that $\boldsymbol{V}_{Pi}$ in equation (6.21) is independent of μ_i, or

$$\dot{\boldsymbol{s}} = 0,\ \dot{\boldsymbol{R}}_{Pi} \times \boldsymbol{s} = 0 \tag{6.22}$$

This leads to the above equations being the conditions for the line L_S to be the ISA in $\sum^*$. Substituting equation (6.18) into $\dot{s} = 0$, we get the unit direction vector $\boldsymbol{s}$ of the ISA:

$$\begin{cases} \boldsymbol{s} = s_1\boldsymbol{e}_{1k} + s_3\boldsymbol{e}_{3k} \\ s_1 = \dfrac{\beta_k - \dot{\theta}}{\omega_0}, \ s_2 = 0, \ s_3 = \dfrac{1}{\omega_0} \\ \omega_0 = \left[1 + \left(\beta_k - \dot{\theta}\right)^2\right]^{1/2} \end{cases} \tag{6.23a}$$

From the above equation, the vector $\boldsymbol{s}$ of the ISA just lies on the normal plane $\boldsymbol{\rho}_k - \boldsymbol{e}_{1k}\boldsymbol{e}_{3k}$ of the spherical image curve of $\boldsymbol{l}_k$ and has an included angle δ_{sm} with $\boldsymbol{l}_k$, which satisfies the equation $\tan\delta_{sm} = \frac{s_3}{s_1} = \frac{1}{\beta_k - \dot{\theta}}$, identical to equation (4.14). Substituting equations (6.15) and (6.18) into $\dot{\boldsymbol{R}}_{Pi} \times \boldsymbol{s} = 0$, we get the position vector of P of the ISA:

$$\begin{cases} \boldsymbol{R}_{PS} = \boldsymbol{\rho}_k + u_1\boldsymbol{e}_{1k} + u_2\boldsymbol{e}_{2k} + u_3\boldsymbol{e}_{3k} \\ u_1 = (\beta_k - \dot{\theta})u_3 \\ u_2 = \dfrac{\alpha_k - \dot{b}_k - \gamma_k(\beta_k - \dot{\theta})}{\omega_0^2} \end{cases} \tag{6.23b}$$

By equation (6.23b), we designate another point Q in $\{\boldsymbol{\rho}_k; \boldsymbol{e}_{1k}, \boldsymbol{e}_{2k}, \boldsymbol{e}_{3k}\}$ with coordinates $u_1 = 0$, $u_2 = [\alpha_k - \dot{b}_k - \gamma_k(\beta_k - \dot{\theta})]/\omega_0^2$, $u_3 = 0$. At any instant, the ISA orthogonally intersects with the axis $\boldsymbol{e}_{2k}$ at the point Q and the position vector of Q in $\{\boldsymbol{O}_f; \boldsymbol{i}_f, \boldsymbol{j}_f, \boldsymbol{k}_f\}$ is

$$\boldsymbol{R}_Q = \boldsymbol{\rho}_k + [\alpha_k - \dot{b}_k - \gamma_k(\beta_k - \dot{\theta})]\boldsymbol{e}_{2k}/\omega_0^2 \tag{6.23c}$$

The vector equation of the ISA in $\{\boldsymbol{O}_f; \boldsymbol{i}_f, \boldsymbol{j}_f, \boldsymbol{k}_f\}$ is

$$\boldsymbol{R}_S = \boldsymbol{R}_Q + \mu\boldsymbol{s} \tag{6.24}$$

where the position vector $\boldsymbol{R}_Q$ and the direction vector $\boldsymbol{s}$ are described in the Frenet frame $\{\boldsymbol{\rho}_k; \boldsymbol{e}_{1k}, \boldsymbol{e}_{2k}, \boldsymbol{e}_{3k}\}$ of $\sum_k$. As the movement of $\sum^*$, the ISA produces a ruled surface in $\{\boldsymbol{O}_f; \boldsymbol{i}_f, \boldsymbol{j}_f, \boldsymbol{k}_f\}$, the ***fixed axode*** $\sum_f$. The ISA traces another ruled surface in $\{\boldsymbol{O}_m; \boldsymbol{i}_m, \boldsymbol{j}_m, \boldsymbol{k}_m\}$, the ***moving axode*** $\sum_m$.

A point D of $\sum^*$ instantaneously coincides with the striction point of the original ruled surface $\sum_k$ (not the original point O_m), whose coordinates in $\{\boldsymbol{O}_m; \boldsymbol{i}_m, \boldsymbol{j}_m, \boldsymbol{k}_m\}$ are all zero, as shown in Fig. 6.2. The velocity vector of point D is

$$\boldsymbol{V}_D = \frac{\mathrm{d}\boldsymbol{R}_D}{\mathrm{d}t} = \frac{\mathrm{d}\boldsymbol{R}_D}{\mathrm{d}\sigma_k}\frac{\mathrm{d}\sigma_k}{\mathrm{d}t} = [(\alpha_k - \dot{b}_k)\boldsymbol{e}_{1k} + \gamma_k\boldsymbol{e}_{3k}]\frac{\mathrm{d}\sigma_k}{\mathrm{d}t} \tag{6.25}$$

By the definition of ISA, the velocity vector of D can also be obtained by

$$\boldsymbol{V}_D = \omega_s\boldsymbol{s} \times (\boldsymbol{\rho}_k - \boldsymbol{R}_Q) + V_s\boldsymbol{s} \tag{6.26}$$

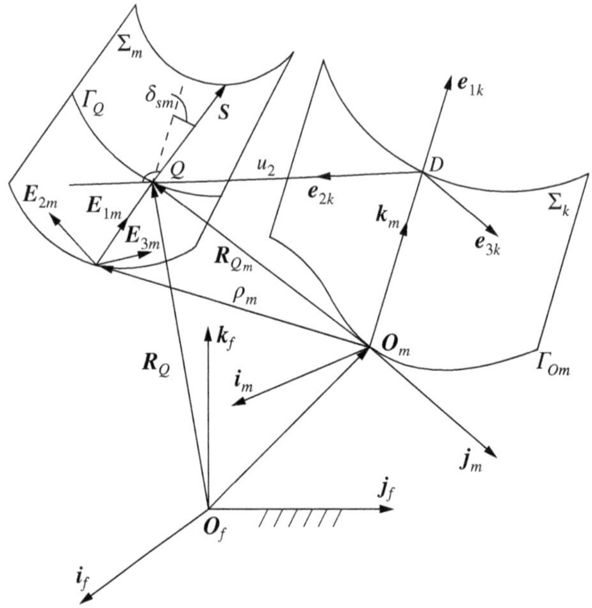

Figure 6.2 The adjoint approach and ISA of the spatial motion

where ω_s and V_s are respectively the angular velocity and translational velocity of $\sum^*$, when $\sum^*$ rolls about and slides along the ISA.

Substituting equation (6.23) into equation (6.26) and combining it with equation (6.25), we have

$$\begin{cases} V_s = V_{s0}\dfrac{\mathrm{d}\sigma_k}{\mathrm{d}t}, \ V_{s0} = \left(\alpha_k - \dot{b}_k\right) s_1 + \gamma_k s_3 \\ \omega_s = \omega_{s0}\dfrac{\mathrm{d}\sigma_k}{\mathrm{d}t}, \ \omega_{s0} = [1 + (\beta_k - \dot{\theta})^2]^{1/2} \end{cases} \tag{6.27}$$

Equation (6.27) indicates the kinematic meanings of the geometrical parameters of the original ruled surface $\sum_k$.

6.2 Axodes

For the spatial movement of $\sum^*$, the ISA traces the moving axode $\sum_m$ in $\{\boldsymbol{O}_m; \boldsymbol{i}_m, \boldsymbol{j}_m, \boldsymbol{k}_m\}$ and the fixed axode $\sum_f$ in $\{\boldsymbol{O}_f; \boldsymbol{i}_f, \boldsymbol{j}_f, \boldsymbol{k}_f\}$, whose geometrical properties essentially reveal the spatial kinematics. Like the centrodes in planar kinematics, the axodes will play an important role in the spatial kinematic geometry.

6.2.1 Fixed Axode

The fixed axode $\sum_f$ can be viewed as an adjoint ruled surface of $\sum_k$ in $\{\boldsymbol{O}_f; \boldsymbol{i}_f, \boldsymbol{j}_f, \boldsymbol{k}_f\}$, whose vector equation is

$$\sum\nolimits_f : \boldsymbol{R}_f = \boldsymbol{R}_Q + \mu \boldsymbol{s} \tag{6.28}$$

where $\boldsymbol{R}_Q$ is the position vector of the directrix $\varGamma_Q$ and $\boldsymbol{s}$ is the unit vector of the generator of $\sum_f$:

$$\begin{cases} \boldsymbol{R}_Q = \boldsymbol{\rho}_k + u_2\boldsymbol{e}_{2k} \\ \boldsymbol{s} = s_1\boldsymbol{e}_{1k} + s_3\boldsymbol{e}_{3k} \end{cases} \tag{6.29}$$

We have the derivatives of equation (6.29) with respect to σ_k

$$\begin{cases} \dot{\boldsymbol{R}}_Q = (\alpha_k - u_2)\boldsymbol{e}_{1k} + \dot{u}_2\boldsymbol{e}_{2k} + (\gamma_k + \beta_k u_2)\boldsymbol{e}_{3k} \\ \dot{\boldsymbol{s}} = \dot{s}_1\boldsymbol{e}_{1k} + (s_1 - \beta_k s_3)\boldsymbol{e}_{2k} + \dot{s}_3\boldsymbol{e}_{3k} \end{cases} \tag{6.30}$$

The spherical image curve of the generatrices of $\sum_f$ is denoted by C_f in $\{\boldsymbol{O}_f; \boldsymbol{i}_f, \boldsymbol{j}_f, \boldsymbol{k}_f\}$ and the arc length of C_f is σ_f. From the above equation, we obtain the differential relationship between σ_f and σ_k as

$$\mathrm{d}\sigma_f = |\dot{\boldsymbol{s}}|\mathrm{d}\sigma_k = \sqrt{\dot{s}_1{}^2 + \dot{s}_3{}^2 + (s_1 - \beta_k s_3)^2}\mathrm{d}\sigma_k = \frac{\sqrt{\dot{\theta}^2\omega_{s0}^2 + (\dot{\beta}_k - \ddot{\theta})^2}}{\omega_{s0}^2}\mathrm{d}\sigma_k \tag{6.31}$$

The striction directrix distance b_f of $\sum_f$ can be calculated by equations (3.54) and (6.30):

$$b_f = -\frac{\dot{\boldsymbol{R}}_Q \cdot \dot{\boldsymbol{s}}}{\dot{\boldsymbol{s}}^2} = -\frac{\dot{s}_1(\alpha_k - u_2) + \dot{u}_2(s_1 - \beta_k s_3) + \dot{s}_3(\gamma_k + \beta_k u_2)}{\dot{\boldsymbol{s}}^2} \tag{6.32}$$

The vector equation of the striction curve of $\sum_f$ and its first derivative with respect to σ_f are

$$\begin{cases} \boldsymbol{\rho}_f = \boldsymbol{R}_Q + b_f\boldsymbol{s} \\ \dfrac{\mathrm{d}\boldsymbol{\rho}_f}{\mathrm{d}\sigma_f} = (\dot{\boldsymbol{R}}_Q + \dot{b}_f\boldsymbol{s} + b_f\dot{\boldsymbol{s}})/|\dot{\boldsymbol{s}}| \end{cases} \tag{6.33}$$

The Frenet frame $\{\boldsymbol{\rho}_f; \boldsymbol{E}_{1f}, \boldsymbol{E}_{2f}, \boldsymbol{E}_{3f}\}$ of $\sum_f$ can be established as

$$\begin{cases} \boldsymbol{E}_{1f} = \boldsymbol{s} = s_1\boldsymbol{e}_{1k} + s_3\boldsymbol{e}_{3k} \\ \boldsymbol{E}_{2f} = \dot{\boldsymbol{s}}/|\dot{\boldsymbol{s}}| = (\dot{s}_1\boldsymbol{e}_{1k} - \dot{\theta}s_3\boldsymbol{e}_{2k} + \dot{s}_3\boldsymbol{e}_{3k})/|\dot{\boldsymbol{s}}| \\ \boldsymbol{E}_{3f} = \boldsymbol{s} \times \dot{\boldsymbol{s}}/|\dot{\boldsymbol{s}}| = [s_3^2\dot{\theta}\boldsymbol{e}_{1k} + (\dot{s}_1 s_3 - s_1\dot{s}_3)\boldsymbol{e}_{2k} - s_1 s_3\dot{\theta}\boldsymbol{e}_{3k}]/|\dot{\boldsymbol{s}}| \end{cases} \tag{6.34}$$

The construction parameters of $\sum_f$ can be derived by the Frenet formulas (3.58):

$$\begin{cases} \alpha_f = \dfrac{\mathrm{d}\boldsymbol{\rho}_f}{\mathrm{d}\sigma_f} \cdot \boldsymbol{E}_{1f} = \dfrac{s_1(\alpha_k - u_2) + s_3(\gamma_k + \beta_k u_2) + \dot{b}_f}{|\dot{\boldsymbol{s}}|} \\ \beta_f = \dfrac{\mathrm{d}\boldsymbol{E}_{2f}}{\mathrm{d}\sigma_f} \cdot \boldsymbol{E}_{3f} = \dfrac{s_3^2\dot{\theta}(s_3\dot{\theta} + \ddot{s}_1) + (\dot{s}_1 s_3 - s_1\dot{s}_3)(\dot{s}_1 - s_3\ddot{\theta} - \dot{s}_3\dot{\theta} - \beta_k\dot{s}_3) - s_1 s_3\dot{\theta}(\ddot{s}_3 - \beta_k s_3\dot{\theta})}{|\dot{\boldsymbol{s}}|^3} \\ \gamma_f = \dfrac{\mathrm{d}\boldsymbol{\rho}_f}{\mathrm{d}\sigma_f} \cdot \boldsymbol{E}_{3f} = \dfrac{s_3^2\dot{\theta}(\alpha_k - u_2) + \dot{u}_2(\dot{s}_1 s_3 - s_1\dot{s}_3) - s_1 s_3\dot{\theta}(\gamma_k + \beta_k u_2)}{|\dot{\boldsymbol{s}}|^2} \end{cases} \tag{6.35}$$

The construction parameter β_f is the geodesic curvature of C_f and can be determined by $\beta_f = \left(s, \frac{ds}{d\sigma_f}, \frac{d^2 s}{d\sigma_f^2}\right)$ from equations (6.34) and (6.35). It has similar geometrical meanings to the geodesic curvature k_{gf} of the spherical fixed centrode π_f for spherical motion.

6.2.2 *Moving Axode*

Based on equations (6.11) and (6.23), we have the vector expression of the moving axode $\sum_m$ with directrix Γ_Q in $\{\boldsymbol{O}_m; \boldsymbol{i}_m, \boldsymbol{j}_m, \boldsymbol{k}_m\}$:

$$\sum\nolimits_m : \boldsymbol{R}_m = \boldsymbol{R}_{Qm} + \mu \boldsymbol{s}_m \tag{6.36a}$$

where

$$\begin{cases} \boldsymbol{R}_{Qm} = u_2 \cos\theta \boldsymbol{i}_m + u_2 \sin\theta \boldsymbol{j}_m + b_k \boldsymbol{k}_m \\ \boldsymbol{s}_m = -s_3 \sin\theta \boldsymbol{i}_m + s_3 \cos\theta \boldsymbol{j}_m + s_1 \boldsymbol{k}_m \end{cases} \tag{6.36b}$$

In order to study the geometrical properties of $\sum_m$, we take the derivative of $\boldsymbol{s}_m$ in equation (6.36b) with respect to σ_k:

$$\dot{\boldsymbol{s}_m} = (-\dot{s_3} \sin\theta - s_3\dot{\theta} \cos\theta)\boldsymbol{i}_m + (\dot{s_3} \cos\theta - s_3\dot{\theta} \sin\theta)\boldsymbol{j}_m + \dot{s_1}\boldsymbol{k}_m \tag{6.37}$$

where $\dot{s_1} = (\dot{\beta}_k - \ddot{\theta})/\omega_{s0}^3$ and $\dot{s_3} = -(\beta_k - \dot{\theta})(\dot{\beta}_k - \ddot{\theta})/\omega_{s0}^3$. The spherical image curve of the generatrices of $\sum_m$ is denoted by C_m in $\{\boldsymbol{O}_m; \boldsymbol{i}_m, \boldsymbol{j}_m, \boldsymbol{k}_m\}$ and the arc length of C_m is σ_m. From the above equation, we obtain the differential relationship between σ_m and σ_k:

$$d\sigma_m = |\dot{\boldsymbol{s}_m}| d\sigma_k = \sqrt{\dot{s_1}^2 + \dot{s_3}^2 + s_3^2\dot{\theta}^2} d\sigma_k = \frac{\sqrt{\dot{\theta}^2\omega_{s0}^2 + (\dot{\beta}_k - \ddot{\theta})^2}}{\omega_{s0}^2} d\sigma_k \tag{6.38}$$

Comparing equations (6.31) with (6.38), we find $d\sigma_f = d\sigma_m$, or $d\sigma$ for short. For a rigid body in spatial motion, the differential arc length of C_m of $\sum_m$ equals that of C_f of $\sum_f$, or the spherical image curve C_m rolls on C_f without sliding.

Differentiating $\boldsymbol{R}_{Qm}$ in equation (6.36b) with respect to σ_k, we have

$$\dot{\boldsymbol{R}_{Qm}} = (\dot{u_2} \cos\theta - u_2\dot{\theta} \sin\theta)\boldsymbol{i}_m + (\dot{u_2} \sin\theta + u_2\dot{\theta} \cos\theta)\boldsymbol{j}_m + \dot{b_k}\boldsymbol{k}_m \tag{6.39a}$$

where

$$\dot{u_2} = \frac{\dot{\alpha_k} - \dot{b_k} - \gamma_k(\dot{\beta}_k - \ddot{\theta}) - \dot{\gamma_k}(\beta_k - \dot{\theta}) - 2(\beta_k - \dot{\theta})(\dot{\beta}_k - \ddot{\theta})u_2}{\omega_{s0}^2} \tag{6.39b}$$

By equation (3.54), we have the striction directrix distance b_m of $\sum_m$:

$$b_m = -\frac{\dot{\boldsymbol{R}_{Qm}} \cdot \dot{\boldsymbol{s}_m}}{\dot{\boldsymbol{s}_m}^2} = -\frac{(\dot{s_3}u_2 - s_3\dot{u_2})\dot{\theta} + \dot{s_1}\dot{b_k}}{\dot{\boldsymbol{s}_m}^2} \tag{6.40}$$

Simplifying equations (6.32) and (6.40), we have

$$\begin{cases} \dot{s_1}\alpha_k + \dot{s_3}\gamma_k = (\omega_{s0}^2 u_2 + \dot{b_k})\dot{s_1} \\ -\dot{s_1}u_2 + \dot{u_2}(s_1 - \beta_k s_3) + \dot{s_3}\beta_k u_2 = (\dot{s_3}u_2 - s_3\dot{u_2})\dot{\theta} - \omega_{s0}^2 u_2 \dot{s_1} \end{cases} \tag{6.41}$$

The striction directrix distances b_f and b_m have the same expression, or $b_f = b_m$. This implies that the striction points of both $\sum_m$ and $\sum_f$ are coincident at any instant.

The vector equation of the striction curve of $\sum_m$ and its first derivative with respect to σ_m are

$$\begin{cases} \boldsymbol{\rho}_m = \boldsymbol{R}_{Qm} + b_m \boldsymbol{s}_m \\ \dfrac{\mathrm{d}\boldsymbol{\rho}_m}{\mathrm{d}\sigma_m} = (\dot{\boldsymbol{R}}_{Qm} + \dot{b}_m \boldsymbol{s}_m + b_m \dot{\boldsymbol{s}}_m)/|\dot{\boldsymbol{s}}_m| \end{cases} \tag{6.42}$$

The Frenet frame $\{\boldsymbol{\rho}_m; \boldsymbol{E}_{1m}, \boldsymbol{E}_{2m}, \boldsymbol{E}_{3m}\}$ of $\sum_m$ can be set up as

$$\boldsymbol{E}_{1m} = \boldsymbol{s}_m,\ \boldsymbol{E}_{2m} = \dot{\boldsymbol{s}}_m/|\dot{\boldsymbol{s}}_m|,\ \boldsymbol{E}_{3m} = \boldsymbol{s}_m \times \dot{\boldsymbol{s}}_m/|\dot{\boldsymbol{s}}_m| \tag{6.43}$$

The Frenet frame $\{\boldsymbol{\rho}_m; \boldsymbol{E}_{1m}, \boldsymbol{E}_{2m}, \boldsymbol{E}_{3m}\}$ can be represented by $\{\boldsymbol{\rho}_k; \boldsymbol{e}_{1k}, \boldsymbol{e}_{2k}, \boldsymbol{e}_{3k}\}$ of $\sum_k$, that is

$$\begin{cases} \boldsymbol{E}_{1m} = s_1 \boldsymbol{e}_{1k} + s_3 \boldsymbol{e}_{3k} \\ \boldsymbol{E}_{2m} = (\dot{s}_1 \boldsymbol{e}_{1k} - s_3 \dot{\theta} \boldsymbol{e}_{2k} + \dot{s}_3 \boldsymbol{e}_{3k})/|\dot{\boldsymbol{s}}_m| \\ \boldsymbol{E}_{3m} = [s_3^2 \dot{\theta} \boldsymbol{e}_{1k} + (\dot{s}_1 s_3 - s_1 \dot{s}_3) \boldsymbol{e}_{2k} - s_1 s_3 \dot{\theta} \boldsymbol{e}_{3k}]/|\dot{\boldsymbol{s}}_m| \end{cases} \tag{6.44}$$

Comparing equations (6.34) with (6.44), we find that the Frenet frame of $\sum_m$ coincides with that of $\sum_f$.

By combining equations (6.11), (6.36) with the above equation, the relationships between the Frenet frame $\{\boldsymbol{\rho}_m; \boldsymbol{E}_{1m}, \boldsymbol{E}_{2m}, \boldsymbol{E}_{3m}\}$ and $\{\boldsymbol{O}_m; \boldsymbol{i}_m, \boldsymbol{j}_m, \boldsymbol{k}_m\}$ are

$$\begin{cases} \boldsymbol{\rho}_m = (u_2 \cos\theta - s_3 b_m \sin\theta)\boldsymbol{i}_m + (u_2 \sin\theta + s_3 b_m \cos\theta)\boldsymbol{j}_m + (b_k + s_1 b_m)\boldsymbol{k}_m \\ \boldsymbol{E}_{1m} = -s_3 \sin\theta \boldsymbol{i}_m + s_3 \cos\theta \boldsymbol{j}_m + s_1 \boldsymbol{k}_m \\ \boldsymbol{E}_{2m} = \dfrac{-(s_3\dot{\theta}\cos\theta + \dot{s}_3 \sin\theta)\boldsymbol{i}_m + (\dot{s}_3 \cos\theta - s_3\dot{\theta}\sin\theta)\boldsymbol{j}_m + \dot{s}_1 \boldsymbol{k}_m}{|\dot{\boldsymbol{s}}_m|} \\ \boldsymbol{E}_{3m} = \dfrac{[(\dot{s}_1 s_3 - s_1 \dot{s}_3)\cos\theta + s_1 s_3 \dot{\theta}\sin\theta]\boldsymbol{i}_m + [(\dot{s}_1 s_3 - s_1 \dot{s}_3)\sin\theta - s_1 s_3\dot{\theta}\cos\theta]\boldsymbol{j}_m + s_3^2\dot{\theta}\boldsymbol{k}_m}{|\dot{\boldsymbol{s}}_m|} \end{cases} \tag{6.45}$$

In $\{\boldsymbol{O}_m; \boldsymbol{i}_m, \boldsymbol{j}_m, \boldsymbol{k}_m\}$, $\boldsymbol{e}_{1k}(\boldsymbol{k}_m)$ is a fixed vector. $\boldsymbol{e}_{2k}$ and $\boldsymbol{e}_{3k}$ can be viewed as vector functions of a unit circle on the plane $\boldsymbol{O}_m - \boldsymbol{i}_m \boldsymbol{j}_m$, or $\boldsymbol{e}_{2k} = \boldsymbol{e}_{\mathrm{I}(\theta)}$ and $\boldsymbol{e}_{3k} = \boldsymbol{e}_{\mathrm{II}(\theta)}$ with included angle θ as the independent variable. Based on the properties of the vector function of a unit circle, the derivatives of $\boldsymbol{e}_{2k}$ and $\boldsymbol{e}_{3k}$ with respect to θ are obtained as $\frac{\mathrm{d}\boldsymbol{e}_{2k}}{\mathrm{d}\theta} = \boldsymbol{e}_{3k}$ and $\frac{\mathrm{d}\boldsymbol{e}_{3k}}{\mathrm{d}\theta} = -\boldsymbol{e}_{2k}$. The construction parameters of $\sum_m$ are

$$\begin{cases} \alpha_m = \dfrac{\mathrm{d}\boldsymbol{\rho}_m}{\mathrm{d}\sigma_m} \cdot \boldsymbol{E}_{1m} = \dfrac{s_3 u_2 \dot{\theta} + s_1 \dot{b}_k + \dot{b}_m}{|\dot{\boldsymbol{s}}_m|} \\ \beta_m = \dfrac{\mathrm{d}\boldsymbol{E}_{2m}}{\mathrm{d}\sigma_m} \cdot \boldsymbol{E}_{3m} = \dfrac{s_1 s_3 \dot{\theta}(s_3 \dot{\theta}^2 - \ddot{s}_3) + \ddot{s}_1 s_3^2 \dot{\theta} - (2\dot{s}_3 \dot{\theta} + s_3 \ddot{\theta})(\dot{s}_1 s_3 - s_1 \dot{s}_3)}{|\dot{\boldsymbol{s}}_m|^3} \\ \gamma_m = \dfrac{\mathrm{d}\boldsymbol{\rho}_m}{\mathrm{d}\sigma_m} \cdot \boldsymbol{E}_{3m} = \dfrac{s_3^2 \dot{b}_k \dot{\theta} + \dot{u}_2(\dot{s}_1 s_3 - s_1 \dot{s}_3) - s_1 s_3 u_2 \dot{\theta}^2}{|\dot{\boldsymbol{s}}_m|^2} \end{cases} \tag{6.46a}$$

where

$$\ddot{s}_1 = \frac{(\ddot{\beta}_k - \dddot{\theta})\omega_{s0}^2 - 3(\dot{\beta}_k - \ddot{\theta})^2(\beta_k - \dot{\theta})}{\omega_{s0}^5},\ \ddot{s}_3 = -(\dot{\beta}_k - \ddot{\theta})\dot{s}_1 - (\beta_k - \dot{\theta})\ddot{s}_1 \tag{6.46b}$$

Based on equations (6.43) and (6.46), the construction parameter β_m, the geodesic curvature of C_m, can be determined by $\beta_m = \left(\boldsymbol{s}_m, \frac{\mathrm{d}\boldsymbol{s}_m}{\mathrm{d}\sigma_m}, \frac{\mathrm{d}^2\boldsymbol{s}_m}{\mathrm{d}\sigma_m{}^2}\right)$. It has similar geometrical meanings to the geodesic curvature k_{gm} of the spherical moving centrode π_m for spherical motion.

The ***induced construction parameters*** α^*, β^*, and γ^*, as very important kinematic invariants for spatial movement, can be obtained by subtracting equation (6.46) from equation (6.35):

$$\begin{cases} \alpha^* = \alpha_f - \alpha_m = \dfrac{s_1(\alpha_k - \dot{b}_k) + s_3\gamma_k}{\mathrm{d}\sigma/\mathrm{d}\sigma_k} \\ \beta^* = \beta_f - \beta_m = \dfrac{\dot{\theta}^2\omega_{s0}^2 + (\dot{\beta}_k - \ddot{\theta})^2}{\omega_{s0}^3|\dot{\boldsymbol{s}}|^3} \\ \gamma^* = \gamma_f - \gamma_m = \dfrac{s_3^2\dot{\theta}(\alpha_k - u_2 - \dot{b}_k) - s_1s_3\dot{\theta}(\gamma_k + \beta_k u_2 - u_2\dot{\theta})}{|\dot{\boldsymbol{s}}|^2} = 0 \end{cases} \tag{6.47}$$

Comparing the above equation with equation (6.27), we have the kinematic meanings of the induced construction parameters α^* and β^*: $\alpha^* = \frac{V_{s0}}{\mathrm{d}\sigma/\mathrm{d}\sigma_k} = \frac{V_s}{\mathrm{d}\sigma/\mathrm{d}t}$ reveals the instantaneous translation movement and $\beta^* = \frac{\omega_{s0}}{\mathrm{d}\sigma/\mathrm{d}\sigma_k} = \frac{\omega_s}{\mathrm{d}\sigma/\mathrm{d}t}$ represents the instantaneous rotation of $\sum^*$. From equations (6.31), (6.38), (6.34), and (6.45), we have the following conclusions for spatial movement of a rigid body:

For spatial motion of a moving body relative to a fixed body, there exists a moving axode $\sum_m$ in the moving body and a fixed axode $\sum_f$ in the fixed body respectively. The moving and fixed axodes have the same distribution parameter and their striction points are coincident, as well as the Frenet frame of $\sum_m$ coinciding with that of $\sum_f$.

The moving axode $\sum_m$ contacts the fixed axode $\sum_f$ along the ISA tangentially at any instant, and the spherical image curve C_m of $\sum_m$ rolls on the spherical image curve C_f of $\sum_f$ without sliding.

The above conclusions can be expressed as

$$\mathrm{d}\sigma_f = \mathrm{d}\sigma_m,\ \boldsymbol{\rho}_f = \boldsymbol{R}_{Om} + \boldsymbol{\rho}_m,\ \boldsymbol{E}_{1m}//\boldsymbol{E}_{1f},\ \boldsymbol{E}_{2m}//\boldsymbol{E}_{2f},\ \boldsymbol{E}_{3m}//\boldsymbol{E}_{3f},\ \gamma_f = \gamma_m = \gamma \tag{6.48}$$

The conclusions lay the groundwork for the following spatial kinematics.

6.3 Differential Kinematics of Points

For a spatial trajectory traced by a point of a moving body in spatial movement, the kinematic invariants – curvature and torsion, are the current content of the differential geometry, introduced in Chapter 3. For the kinematic geometry of linkages, more attention is paid to comparing the point trajectories with constraint curves or surfaces in order to locate the characteristic points on the moving body.

The fixed and moving centrodes form a logical starting point of the instantaneous planar kinematic geometry of a moving body [1]. We may also expect the axodes to occupy a similar important fundamental place in the spatial kinematic geometry of linkages and develop a unified axode-based curvature theory, which is expressible in terms of the curvature invariants. The intrinsic connections among a point trajectory or a line trajectory, the axodes and the constraint curves (or surfaces), can be fully discussed by means of the adjoint approach in this section.

6.3.1 Point Trajectory

As mentioned in Section 6.1, the Cartesian coordinate system $\{\boldsymbol{O}_m; \boldsymbol{i}_m, \boldsymbol{j}_m, \boldsymbol{k}_m\}$ is established on a moving body $\sum^*$ in spatial movement. A point $P(x_{Pm}, y_{Pm}, z_{Pm})$ of $\sum^*$ traces a spatial curve Γ_P in the fixed Cartesian coordinate system $\{\boldsymbol{O}_f; \boldsymbol{i}_f, \boldsymbol{j}_f, \boldsymbol{k}_f\}$, as shown in Fig. 6.3. The ISA of $\sum^*$ at all instants forms the moving axode $\sum_m$ in $\{\boldsymbol{O}_m; \boldsymbol{i}_m, \boldsymbol{j}_m, \boldsymbol{k}_m\}$, whose vector equation (6.36) and Frenet frame (6.44) are obtained in Section 6.2. Now we take $\sum_m$ as an original ruled surface to examine the point P, or the point P is adjoint to the ISA, the generator of $\sum_m$ at an instant. In $\{\boldsymbol{O}_m; \boldsymbol{i}_m, \boldsymbol{j}_m, \boldsymbol{k}_m\}$, the vector equation of P can be written as

$$\boldsymbol{R}_{Pm} = \boldsymbol{\rho}_m + v_1\boldsymbol{E}_{1m} + v_2\boldsymbol{E}_{2m} + v_3\boldsymbol{E}_{3m} \tag{6.49}$$

where (v_1, v_2, v_3) are the coordinates of P in the Frenet frame $\{\boldsymbol{\rho}_m; \boldsymbol{E}_{1m}, \boldsymbol{E}_{2m}, \boldsymbol{E}_{3m}\}$ of $\sum_m$ and can be obtained from equation (6.45). By the Frenet formulas (3.58), the first derivative of equation (6.49) with respect to the arc length σ is

$$\frac{\mathrm{d}\boldsymbol{R}_{Pm}}{\mathrm{d}\sigma} = \left(\frac{\mathrm{d}v_1}{\mathrm{d}\sigma} - v_2 + \alpha_m\right)\boldsymbol{E}_{1m} + \left(v_1 + \frac{\mathrm{d}v_2}{\mathrm{d}\sigma} - \beta_m v_3\right)\boldsymbol{E}_{2m} + \left(\beta_m v_2 + \frac{\mathrm{d}v_3}{\mathrm{d}\sigma} + \gamma_m\right)\boldsymbol{E}_{3m} \tag{6.50}$$

Since point P is a fixed point in $\sum^*$, the absolute velocity of P in $\{\boldsymbol{O}_m; \boldsymbol{i}_m, \boldsymbol{j}_m, \boldsymbol{k}_m\}$ is zero, or $\frac{\mathrm{d}\boldsymbol{R}_{Pm}}{\mathrm{d}\sigma} = 0$. Based on the fixed point conditions (3.71) for a curve being adjoint to a ruled surface, we have

$$\begin{cases} \dfrac{\mathrm{d}v_1}{\mathrm{d}\sigma} - v_2 + \alpha_m = 0 \\ v_1 + \dfrac{\mathrm{d}v_2}{\mathrm{d}\sigma} - \beta_m v_3 = 0 \\ \beta_m v_2 + \dfrac{\mathrm{d}v_3}{\mathrm{d}\sigma} + \gamma_m = 0 \end{cases} \tag{6.51}$$

or

$$\begin{cases} \dfrac{\mathrm{d}v_1}{\mathrm{d}\sigma} = v_2 - \alpha_m \\ \dfrac{\mathrm{d}v_2}{\mathrm{d}\sigma} = \beta_m v_3 - v_1 \\ \dfrac{\mathrm{d}v_3}{\mathrm{d}\sigma} = -\beta_m v_2 - \gamma_m \end{cases} \tag{6.52}$$

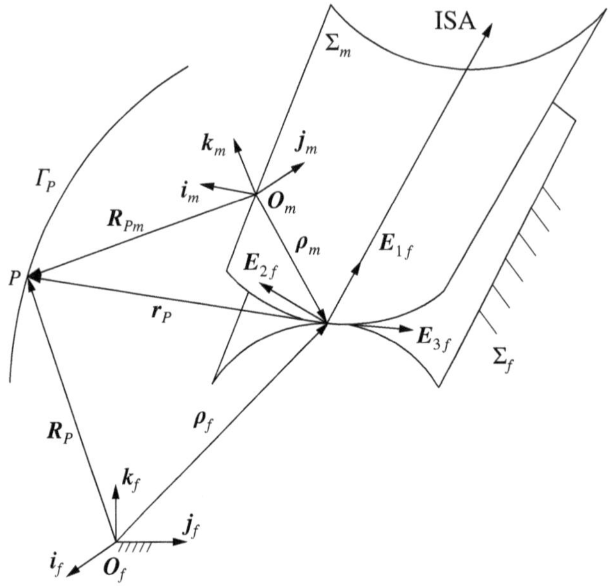

Figure 6.3 The point adjoint to the ISA

On the contrary, the ISA at all instants forms the fixed axode $\sum_f$ in $\{\boldsymbol{O}_f;\boldsymbol{i}_f,\boldsymbol{j}_f,\boldsymbol{k}_f\}$, whose vector equation (6.28) and Frenet frame (6.34) are obtained in the previous section. $\sum_f$ is taken as an original ruled surface and the trajectory Γ_P in $\{\boldsymbol{O}_f;\boldsymbol{i}_f,\boldsymbol{j}_f,\boldsymbol{k}_f\}$ is visualized as an adjoint curve of $\sum_f$. The vector equation of trajectory Γ_P is

$$\Gamma_P : \boldsymbol{R}_P = \boldsymbol{\rho}_f + u_1\boldsymbol{E}_{1f} + u_2\boldsymbol{E}_{2f} + u_3\boldsymbol{E}_{3f} \tag{6.53}$$

where (u_1, u_2, u_3) are the coordinates of P in the Frenet frame $\{\boldsymbol{\rho}_f;\boldsymbol{E}_{1f},\boldsymbol{E}_{2f},\boldsymbol{E}_{3f}\}$ of $\sum_f$. Based on the Frenet formulas (3.58), we take the first derivative of equation (6.53) with respect to σ as

$$\frac{\mathrm{d}\boldsymbol{R}_P}{\mathrm{d}\sigma} = \left(\frac{\mathrm{d}u_1}{\mathrm{d}\sigma} - u_2 + \alpha_f\right)\boldsymbol{E}_{1f} + \left(u_1 + \frac{\mathrm{d}u_2}{\mathrm{d}\sigma} - \beta_f u_3\right)\boldsymbol{E}_{2f} + \left(\beta_f u_2 + \frac{\mathrm{d}u_3}{\mathrm{d}\sigma} + \gamma_f\right)\boldsymbol{E}_{3f} \tag{6.54}$$

Since the Frenet frame of $\sum_m$ is coincident with that of $\sum_f$ and the spherical image curve of the generatrices of $\sum_m$ rolls on that of $\sum_f$ without sliding, as mentioned in equation (6.48), the point P has the same coordinates and their change rates in both $\{\boldsymbol{\rho}_m;\boldsymbol{E}_{1m},\boldsymbol{E}_{2m},\boldsymbol{E}_{3m}\}$ and $\{\boldsymbol{\rho}_f;\boldsymbol{E}_{1f},\boldsymbol{E}_{2f},\boldsymbol{E}_{3f}\}$, or

$$u_1 = v_1,\ u_2 = v_2,\ u_3 = v_3,\ \frac{\mathrm{d}u_1}{\mathrm{d}\sigma} = \frac{\mathrm{d}v_1}{\mathrm{d}\sigma},\ \frac{\mathrm{d}u_2}{\mathrm{d}\sigma} = \frac{\mathrm{d}v_2}{\mathrm{d}\sigma},\ \frac{\mathrm{d}u_3}{\mathrm{d}\sigma} = \frac{\mathrm{d}v_3}{\mathrm{d}\sigma} \tag{6.55}$$

Substituting equations (6.52) and (6.55) into equation (6.54), we have

$$\frac{\mathrm{d}\boldsymbol{R}_P}{\mathrm{d}\sigma} = \alpha^*\boldsymbol{E}_{1f} + \beta^*(-u_3\boldsymbol{E}_{2f} + u_2\boldsymbol{E}_{3f}) \tag{6.56}$$

Differentiating the above equation with respect to σ continuously, and simplifying them by the Frenet formulas (3.58) and the fixed point condition (6.52), we have

$$\begin{cases} \dfrac{\mathrm{d}^2\boldsymbol{R}_P}{\mathrm{d}\sigma^2} = \left(\dfrac{\mathrm{d}\alpha^*}{\mathrm{d}\sigma} + \beta^* u_3\right)\boldsymbol{E}_{1f} + \left(\alpha^* - \dfrac{\mathrm{d}\beta^*}{\mathrm{d}\sigma}u_3 - \beta^{*2}u_2 + \beta^*\gamma\right)\boldsymbol{E}_{2f} + \left(\dfrac{\mathrm{d}\beta^*}{\mathrm{d}\sigma}u_2 - \beta^* u_1 - \beta^{*2}u_3\right)\boldsymbol{E}_{3f} \\ \dfrac{\mathrm{d}^3\boldsymbol{R}_P}{\mathrm{d}\sigma^3} = \left[\dfrac{\mathrm{d}^2\alpha^*}{\mathrm{d}\sigma^2} - \alpha^* + \dfrac{\mathrm{d}\beta^*}{\mathrm{d}\sigma}(2u_3 - \gamma) + \beta^*(\beta^* - \beta_m)u_2 - \beta^*\gamma\right]\boldsymbol{E}_{1f} + \\ \left[2\dfrac{\mathrm{d}\alpha^*}{\mathrm{d}\sigma} - \dfrac{\mathrm{d}\beta^*}{\mathrm{d}\sigma}u_3 - \dfrac{\mathrm{d}\beta^*}{\mathrm{d}\sigma}(\beta^* u_2 + 2u_2 - 2\gamma) + \beta^*(u_3 - \beta^*)^2 u_3 + \beta^* u_1 + \beta_f u_1 + \dfrac{\mathrm{d}\gamma}{\mathrm{d}\sigma}\right]\boldsymbol{E}_{2f} + \\ \left[\beta_f\alpha^* + \dfrac{\mathrm{d}\beta^*}{\mathrm{d}\sigma}(\beta_f\gamma - 2u_1 - 3\beta^* u_3) + \dfrac{\mathrm{d}^2\beta^*}{\mathrm{d}\sigma^2}u_2 + \beta^*(\beta^*\gamma + \beta_m u_3 - \beta^{*2}u_2 - u_2)\right]\boldsymbol{E}_{3f} \end{cases} \tag{6.57}$$

Equations (6.56) and (6.57) indicate that the all-order derivatives of the vector equation of Γ_P deal with the induced construction parameters α^*, β^*, and γ. By taking the first and second derivatives of the vector equation (6.53) with respect to time t, the velocity and acceleration vector equations of P are

$$\begin{cases} \boldsymbol{V}_P = \dfrac{\mathrm{d}\boldsymbol{R}_P}{\mathrm{d}t} = \dfrac{\mathrm{d}\boldsymbol{R}_P}{\mathrm{d}\sigma}\dfrac{\mathrm{d}\sigma}{\mathrm{d}t} \\ \boldsymbol{a}_P = \dfrac{\mathrm{d}^2\boldsymbol{R}_P}{\mathrm{d}t^2} = \dfrac{\mathrm{d}^2\boldsymbol{R}_P}{\mathrm{d}\sigma^2}\left(\dfrac{\mathrm{d}\sigma}{\mathrm{d}t}\right)^2 + \dfrac{\mathrm{d}\boldsymbol{R}_P}{\mathrm{d}\sigma}\dfrac{\mathrm{d}^2\sigma}{\mathrm{d}t^2} \end{cases} \tag{6.58}$$

The spatial movement of $\sum^*$ can be viewed as a screw movement about the ISA $(\boldsymbol{s}, \boldsymbol{E}_{1f})$, whose rotation angular velocity vector $\boldsymbol{\omega}_s$ and sliding velocity vector $\boldsymbol{V}_s$ are

$$\boldsymbol{\omega}_s = \omega_s\boldsymbol{E}_{1f},\ \boldsymbol{V}_s = V_s\boldsymbol{E}_{1f} \tag{6.59}$$

The velocity vector of P is

$$\boldsymbol{V}_P = \boldsymbol{V}_s + \boldsymbol{\omega}_s \times \boldsymbol{r}_P = V_s\boldsymbol{E}_{1f} + \omega_s(u_2\boldsymbol{E}_{3f} - u_3\boldsymbol{E}_{2f}) \tag{6.60}$$

where $\boldsymbol{r}_P$ is the radius vector from the striction point of $\sum_f$ to point P, or $\boldsymbol{r}_P = u_1\boldsymbol{E}_{1f} + u_2\boldsymbol{E}_{2f} + u_3\boldsymbol{E}_{3f}$. The acceleration vector equation of P is

$$\begin{aligned} \boldsymbol{a}_P = \dfrac{\mathrm{d}\boldsymbol{V}_P}{\mathrm{d}t} &= \left(\dfrac{\mathrm{d}V_s}{\mathrm{d}t} + \omega_s u_3\dfrac{\mathrm{d}\sigma}{\mathrm{d}t}\right)\boldsymbol{E}_{1f} + \left(V_s\dfrac{\mathrm{d}\sigma}{\mathrm{d}t} - \dfrac{\mathrm{d}\omega_s}{\mathrm{d}t}u_3 + \omega_s\gamma\dfrac{\mathrm{d}\sigma}{\mathrm{d}t} - \omega_s\beta^* u_2\dfrac{\mathrm{d}\sigma}{\mathrm{d}t}\right)\boldsymbol{E}_{2f} \\ &+ \left(\dfrac{\mathrm{d}\omega_s}{\mathrm{d}t}u_2 - \omega_s\dfrac{\mathrm{d}\sigma}{\mathrm{d}t}u_1 - \omega_s\beta^* u_3\dfrac{\mathrm{d}\sigma}{\mathrm{d}t}\right)\boldsymbol{E}_{3f} \end{aligned} \tag{6.61}$$

where $\frac{\mathrm{d}V_s}{\mathrm{d}t}$ and $\frac{\mathrm{d}\omega_s}{\mathrm{d}t}$ are the translational acceleration and angular acceleration of $\sum^*$ about the ISA. Substituting equations (6.56) and (6.57) into equation (6.58), and comparing them with equations (6.60) and (6.61) respectively, we have the following relationships among α^*, β^*, V_s, and ω_s:

$$\begin{cases} V_s = \alpha^*\dfrac{\mathrm{d}\sigma}{\mathrm{d}t},\ \omega_s = \beta^*\dfrac{\mathrm{d}\sigma}{\mathrm{d}t} \\ \dfrac{\mathrm{d}V_s}{\mathrm{d}t} = \dfrac{\mathrm{d}\alpha^*}{\mathrm{d}\sigma}\left(\dfrac{\mathrm{d}\sigma}{\mathrm{d}t}\right)^2 + \alpha^*\dfrac{\mathrm{d}^2\sigma}{\mathrm{d}t^2} \\ \dfrac{\mathrm{d}\omega_s}{\mathrm{d}t} = \dfrac{\mathrm{d}\beta^*}{\mathrm{d}\sigma}\left(\dfrac{\mathrm{d}\sigma}{\mathrm{d}t}\right)^2 + \beta^*\dfrac{\mathrm{d}^2\sigma}{\mathrm{d}t^2} \end{cases} \tag{6.62}$$

The above equations also reveal the kinematic meanings of the induced construction parameters α^* and β^*, like equations (6.27) and (6.47).

The points of $\sum^*$ will have some special characteristics in kinematics, such as the special values of velocity and acceleration, which can be readily located by equations (6.60) and (6.61).

1. *The acceleration center*. The point of $\sum^*$ with zero acceleration at an instant is called the ***acceleration center***, whose coordinates (u_1, u_2, u_3) in $\{\boldsymbol{\rho}_m; \boldsymbol{E}_{1m}, \boldsymbol{E}_{2m}, \boldsymbol{E}_{3m}\}$ can be solved by letting $\boldsymbol{a}_P$ in equation (6.61) be zero, or

$$\begin{cases} u_1 = \left\{\dfrac{\mathrm{d}\omega_s}{\mathrm{d}t}\left[V_s\dfrac{\mathrm{d}\sigma}{\mathrm{d}t} + \omega_s\dfrac{\mathrm{d}\sigma}{\mathrm{d}t}\gamma + \dfrac{\mathrm{d}V_s}{\mathrm{d}t}\dfrac{\mathrm{d}\omega_s}{\mathrm{d}t} \bigg/ \left(\omega_s\dfrac{\mathrm{d}\sigma}{\mathrm{d}t}\right)\right] \bigg/ \left(\beta^*\omega_s\dfrac{\mathrm{d}\sigma}{\mathrm{d}t}\right) + \beta^*\dfrac{\mathrm{d}V_s}{\mathrm{d}t}\right\} \bigg/ \left(\omega_s\dfrac{\mathrm{d}\sigma}{\mathrm{d}t}\right) \\ u_2 = \left[V_s\dfrac{\mathrm{d}\sigma}{\mathrm{d}t} + \omega_s\gamma\dfrac{\mathrm{d}\sigma}{\mathrm{d}t} + \dfrac{\mathrm{d}V_s}{\mathrm{d}t}\dfrac{\mathrm{d}\omega_s}{\mathrm{d}t} \bigg/ \left(\omega_s\dfrac{\mathrm{d}\sigma}{\mathrm{d}t}\right)\right] \bigg/ \left(\beta^*\omega_s\dfrac{\mathrm{d}\sigma}{\mathrm{d}t}\right) \\ u_3 = -\dfrac{\mathrm{d}V_s}{\mathrm{d}t} \bigg/ \left(\omega_s\dfrac{\mathrm{d}\sigma}{\mathrm{d}t}\right) \end{cases} \tag{6.63}$$

This indicates that there exists only one acceleration center in $\sum^*$ at an instant. Of course, the acceleration center can also be located through equations (6.56), (6.57), and (6.58).

2. *The Bresse hyperboloid*. The ***Bresse hyperboloid*** in $\sum^*$ can be located by the vector equation $\boldsymbol{a}_P \cdot \boldsymbol{V}_P = 0$. By substituting equations (6.60) and (6.61) into $\boldsymbol{a}_P \cdot \boldsymbol{V}_P = 0$, the equation of the Bresse hyperboloid in $\{\boldsymbol{\rho}_m; \boldsymbol{E}_{1m}, \boldsymbol{E}_{2m}, \boldsymbol{E}_{3m}\}$ is represented by

$$V_s\frac{\mathrm{d}V_s}{\mathrm{d}t} - \omega_s{}^2\gamma\frac{\mathrm{d}\sigma}{\mathrm{d}t}u_3 - \omega_s{}^2\frac{\mathrm{d}\sigma}{\mathrm{d}t}u_1u_3 + \omega_s\frac{\mathrm{d}\omega_s}{\mathrm{d}t}(u_2^2 + u_3^2) = 0 \tag{6.64}$$

3. *The inflection point*. The ***inflection points*** in $\sum^*$ can be determined by the vector equation $\boldsymbol{a}_P \times \boldsymbol{V}_P = 0$. By equations (6.56), (6.57), and (6.58), we have three equations of the inflection points in $\{\boldsymbol{\rho}_m; \boldsymbol{E}_{1m}, \boldsymbol{E}_{2m}, \boldsymbol{E}_{3m}\}$:

$$\begin{cases} \alpha^*(\alpha^* + \beta^*\gamma) + \left(\dfrac{\mathrm{d}\alpha^*}{\mathrm{d}\sigma}\beta^* - \dfrac{\mathrm{d}\beta^*}{\mathrm{d}\sigma}\alpha^*\right)u_3 - \alpha^*\beta^{*2}u_2 + \beta^{*2}u_3^2 = 0 \\ \alpha^*\beta^*u_1 + \left(\dfrac{\mathrm{d}\alpha^*}{\mathrm{d}\sigma}\beta^* - \dfrac{\mathrm{d}\beta^*}{\mathrm{d}\sigma}\alpha^*\right)u_2 + \alpha^*\beta^{*2}u_3 + \beta^{*2}u_2u_3 = 0 \\ -(\alpha^* + \beta^*\gamma)u_2 + \beta^*u_1u_3 + \beta^{*2}(u_2^2 + u_3^2) = 0 \end{cases} \tag{6.65}$$

Among the above three equations, there are only two independent equations, which implies that the inflection points all lie on a spatial curve.

In particular, $\sum^*$ will be in spherical motion if α_m, γ_m of $\sum_m$ and α_f, γ_f of $\sum_f$ are all identically zero, and equations (6.63) to (6.65) can be simplified as

1. The acceleration center: $u_1 = u_2 = u_3 = 0$.
2. The Bresse hyperboloid: $-\omega_s\frac{\mathrm{d}\sigma}{\mathrm{d}t}u_1u_3 + \frac{\mathrm{d}\omega_s}{\mathrm{d}t}(u_2^2 + u_3^2) = 0$.
3. The inflection point: $u_2 = u_3 = 0$.

6.3.2 Darboux Frame

According to equation (6.56), the tangent vector of $\varGamma_P$ is independent of the coordinate u_1. As shown in Fig. 6.4, the normal plane $\sum_P$ of $\varGamma_P$ at P is defined by

$$\frac{\mathrm{d}\boldsymbol{R}_P}{\mathrm{d}\sigma}\cdot\left(\boldsymbol{R}_\Sigma-\boldsymbol{R}_P\right)=0 \tag{6.66}$$

where $\boldsymbol{R}_\Sigma=\boldsymbol{\rho}_f+x\boldsymbol{E}_{1f}+y\boldsymbol{E}_{2f}+z\boldsymbol{E}_{3f}$ is the position vector of any point on $\sum_P$. Substituting equation (6.56) into the above equation, we get the algebraic equation of $\sum_P$ in $\{\boldsymbol{\rho}_f;\boldsymbol{E}_{1f},\boldsymbol{E}_{2f},\boldsymbol{E}_{3f}\}$:

$$\alpha^*(x-u_1)-\beta^*u_3(y-u_2)+\beta^*u_2(z-u_3)=0 \tag{6.67}$$

The normal plane $\sum_P$ intersects the ISA ($\boldsymbol{E}_{1f}$) at point O' with coordinates $x=u_1$, $y=0$, $z=0$. A normal of $\varGamma_P$ can be obtained by connecting points P and O', whose direction vector $\boldsymbol{N}$ is

$$\boldsymbol{N}=\overrightarrow{PO'}=-u_2\boldsymbol{E}_{2f}-u_3\boldsymbol{E}_{3f} \tag{6.68}$$

The vector $\boldsymbol{N}$ is perpendicular to $\boldsymbol{E}_{1f}$, or $\boldsymbol{N}\cdot\boldsymbol{E}_{1f}=0$, which means a normal of the curve $\varGamma_P$ is orthogonal to the ISA at an instant. As a special case, the normal of the plane point path always passes through the instant center.

In order to examine the geometrical properties of $\varGamma_P$, we set up a Darboux frame $\{\boldsymbol{R}_P;\boldsymbol{E}_1,\boldsymbol{E}_2,\boldsymbol{E}_3\}$ on $\varGamma_P$ at first, whose original point is at point P on $\varGamma_P$, and axis vector $\boldsymbol{E}_3$ is collinear with the normal $\boldsymbol{N}$. The Darboux frame is expressed as

$$\boldsymbol{E}_1=\frac{\mathrm{d}\boldsymbol{R}_P}{\mathrm{d}\sigma}\Big/\left|\frac{\mathrm{d}\boldsymbol{R}_P}{\mathrm{d}\sigma}\right|,\ \boldsymbol{E}_2=\boldsymbol{E}_3\times\boldsymbol{E}_1,\ \boldsymbol{E}_3=\boldsymbol{N}/|\boldsymbol{N}| \tag{6.69}$$

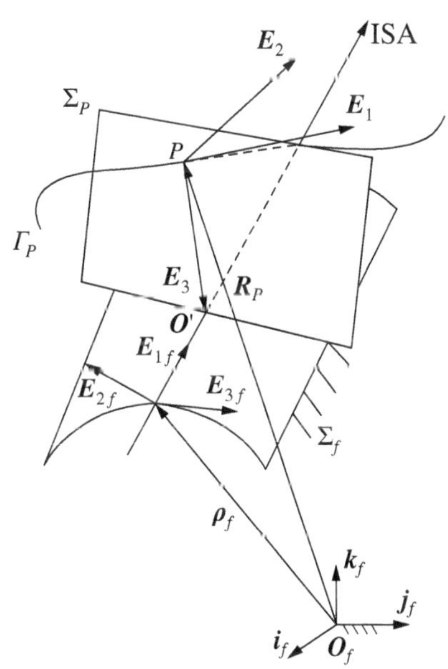

Figure 6.4 Darboux frame of the point trajectory

Substituting equations (6.56) and (6.68) into the above equation, we have

$$\begin{cases} \boldsymbol{E}_1 = \dfrac{\alpha^* \boldsymbol{E}_{1f} + \beta^*(-u_3\boldsymbol{E}_{2f} + u_2\boldsymbol{E}_{3f})}{R} \\ \boldsymbol{E}_2 = -\dfrac{\beta^* r^2 \boldsymbol{E}_{1f} + \alpha^*(u_3\boldsymbol{E}_{2f} - u_2\boldsymbol{E}_{3f})}{Rr} \\ \boldsymbol{E}_3 = -\dfrac{u_2\boldsymbol{E}_{2f} + u_3\boldsymbol{E}_{3f}}{r} \\ R = (\alpha^{*2} + \beta^{*2} r^2)^{1/2}, \ r^2 = u_2^2 + u_3^2 \end{cases} \tag{6.70}$$

The differential of the arc length σ_P of Γ_P can be derived from the first expression of the above equation as $\mathrm{d}\sigma_P = R\mathrm{d}\sigma$. Taking the derivative of the vectors $\boldsymbol{E}_1$ and $\boldsymbol{E}_3$ with respect to σ_P, we have

$$\begin{cases} \dfrac{\mathrm{d}\boldsymbol{E}_1}{\mathrm{d}\sigma_P} = \left[\left(\dfrac{\mathrm{d}\alpha^*}{\mathrm{d}\sigma} + \beta^* u_3\right)\boldsymbol{E}_{1f} + \left(\alpha^* - \dfrac{\mathrm{d}\beta^*}{\mathrm{d}\sigma}u_3 - \beta^{*2}u_2 + \beta^*\gamma\right)\boldsymbol{E}_{2f} + \right. \\ \qquad \left. \left(\dfrac{\mathrm{d}\beta^*}{\mathrm{d}\sigma}u_2 - \beta^* u_1 - \beta^{*2}u_3\right)\boldsymbol{E}_{3f} - \dfrac{\mathrm{d}R}{\mathrm{d}\sigma}\boldsymbol{E}_1\right] / R^2 \\ \dfrac{\mathrm{d}\boldsymbol{E}_3}{\mathrm{d}\sigma_P} = -\left[-u_2\boldsymbol{E}_{1f} - (u_1 + \beta^* u_3)\boldsymbol{E}_{2f} + (\beta^* u_2 - \gamma)\boldsymbol{E}_{3f} + \dfrac{\mathrm{d}r}{\mathrm{d}\sigma}\boldsymbol{E}_3\right] / (Rr) \end{cases} \tag{6.71}$$

Based on the differential formulas (3.42) of the Darboux frame, the normal curvature k_n, geodesic curvature k_g, and geodesic torsion τ_g can be derived as

$$\begin{cases} k_n = \dfrac{\mathrm{d}\boldsymbol{E}_1}{\mathrm{d}\sigma_P}\cdot\boldsymbol{E}_2 = -\dfrac{r^2\left(\dfrac{\mathrm{d}\alpha^*}{\mathrm{d}\sigma}\beta^* - \alpha^*\dfrac{\mathrm{d}\beta^*}{\mathrm{d}\sigma} + \beta^{*2}u_3\right) + \alpha^*(\alpha^* u_3 + \beta^*\gamma u_3 + \beta^* u_1 u_2)}{R^3 r} \\ k_g = -\dfrac{\mathrm{d}\boldsymbol{E}_3}{\mathrm{d}\sigma_P}\cdot\boldsymbol{E}_1 = \dfrac{-\alpha^* u_2 + \beta^{*2}r^2 + \beta^*(u_1u_3 - u_2\gamma)}{R^2 r} \\ \tau_g = -\dfrac{\mathrm{d}\boldsymbol{E}_3}{\mathrm{d}\sigma_P}\cdot\boldsymbol{E}_2 = \dfrac{\beta^* r^2(u_2 + \alpha^*) + \alpha^*(u_1u_3 - u_2\gamma)}{R^2 r^2} \end{cases} \tag{6.72}$$

6.3.3 *Euler–Savary Analogue*

The normal curvature k_n and the geodesic curvature k_g are respectively the projections of the curvature vector $k\boldsymbol{\beta}$ of Γ_P on the normal vectors $\boldsymbol{E}_2$ and $\boldsymbol{E}_3$, and we have $k^2 = k_n^2 + k_g^2$. Since the vector $\boldsymbol{E}_3$ is perpendicular to the vector $\boldsymbol{E}_{1f}$ (ISA), the principal normal $\boldsymbol{\beta}$ of Γ_P is orthogonal to the ISA in the case $k_n = 0$. At this instant, the curvature k equals the geodesic curvature k_g:

$$\begin{cases} r^2\left(\dfrac{\mathrm{d}\alpha^*}{\mathrm{d}\sigma}\beta^* - \alpha^*\dfrac{\mathrm{d}\beta^*}{\mathrm{d}\sigma} + \beta^{*2}u_3\right) + \alpha^*(\alpha^* u_3 + \beta^*\gamma u_3 + \beta^* u_1u_2) = 0 \\ k = \dfrac{-\alpha^* u_2 + \beta^{*2}r^2 + \beta^*(u_1u_3 - u_2\gamma)}{R^2 r} \end{cases} \tag{6.73}$$

Some replacements are taken in the above equation:

$$\begin{cases} u_2 = r\cos\theta_g,\ u_3 = r\sin\theta_g \\ \sin\theta_{g0} = \dfrac{u_1}{D_{g0}\beta^*},\ \cos\theta_{g0} = \dfrac{\alpha^* + \beta^*\gamma}{D_{g0}\beta^{*2}},\ D_{g0} = \dfrac{\left[\beta^{*2}u_1^2 + (\alpha^* + \beta^*\gamma)^2\right]^{1/2}}{\beta^{*2}} \end{cases} \tag{6.74}$$

Equation (6.73) can be rewritten as the following two equations:

$$r\left[r\frac{\mathrm{d}}{\mathrm{d}\sigma}\left(\frac{\alpha^*}{\beta^*}\right) + r^2\sin^2\theta_g + \alpha^* D_{g0}\sin(\theta_g + \theta_{g0})\right] = 0 \tag{6.75}$$

$$k = \frac{r - D_{g0}\cos(\theta_g + \theta_{g0})}{r^2 + (\alpha^*/\beta^*)^2} \tag{6.76}$$

Equation (6.75) or the first expression of equation (6.73) clearly defines a surface in $\sum^*$ at an instant, called the ***geodesic curvature distribution surface***. Any point of the surface traces a spatial trajectory, whose principal normal is orthogonal to the ISA at this instant. Two expressions from equation (6.76) are obtained by replacing the curvature k with the curvature radius ρ, or

$$\rho[r - D_{g0}\cos(\theta_g + \theta_{g0})] = r^2 + \left(\frac{\alpha^*}{\beta^*}\right)^2 \tag{6.77a}$$

$$\frac{1}{r} + \frac{1}{\rho - r} = \frac{\rho}{(\alpha^*/\beta^*)^2 - \rho D_{g0}\cos(\theta_g + \theta_{g0})} \tag{6.77b}$$

The above equations, called the ***Euler–Savary analogue of a spatial point trajectory*** in spatial motion, indicate the relationships among the point's position in the moving body $\sum^*$, the curvature radius of the point trajectory, and the induced construction parameters of the axodes. They are identical to the Euler–Savary equation (1.74) of a point path in planar movement and the geodesic Euler–Savary analogue (4.43) of a point trajectory in spherical movement, except for the additional formula $(\alpha^*/\beta^*)^2$ generated due to the sliding of $\sum_m$ relative to $\sum_f$. Like the Euler–Savary equations in both planar and spherical movement, equation (6.77a) can be geometrically represented as

$$\overrightarrow{PO_P}\cdot\overrightarrow{PJ_P} = \overrightarrow{PO'}^2 + (\alpha^*/\beta^*)^2 \tag{6.77c}$$

As shown in Fig. 6.5, O_P is the curvature center of $\varGamma_P$ at point P, and O' is the intersection point between the principal normal of $\varGamma_P$ and the ISA ($\boldsymbol{E}_{1f}$), while J_P is the intersection point between the principal normal and the geodesic inflection circle, which lies on the plane perpendicular to the ISA. The coordinates (x_0, y_0, z_0) of the geodesic inflection circle's center in $\{\boldsymbol{\rho}_m; \boldsymbol{E}_{1m}, \boldsymbol{E}_{2m}, \boldsymbol{E}_{3m}\}$ are

$$x_0 = u_1,\ y_0 = \frac{\alpha^* + \beta^*\gamma}{2\beta^{*2}},\ z_0 = -\frac{u_1}{2\beta^*} \tag{6.78}$$

The radius of the geodesic inflection circle is $D_{g0}/2$, as shown in Fig. 6.6.

It should be pointed out that all the special points of $\sum^*$, whose trajectories meet the Euler–Savary analogue (6.77), locate on the distribution surface defined by equation (6.75). For other points out of the surface, the normal curvature k_n is not equal to zero and the

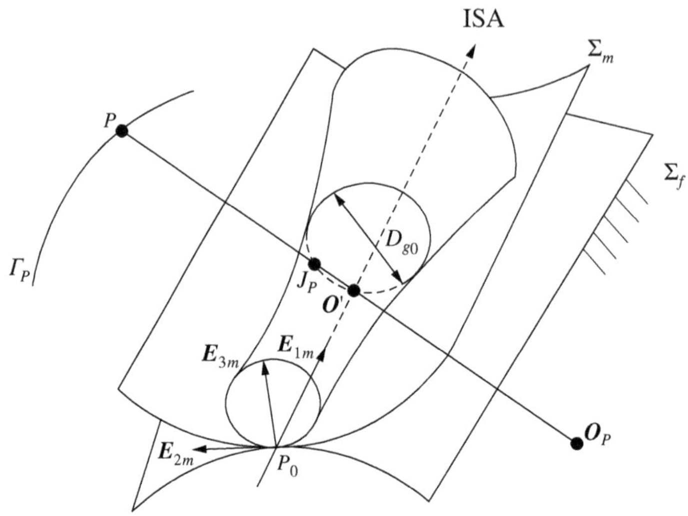

Figure 6.5 Euler–Savary analogue of point's trajectory

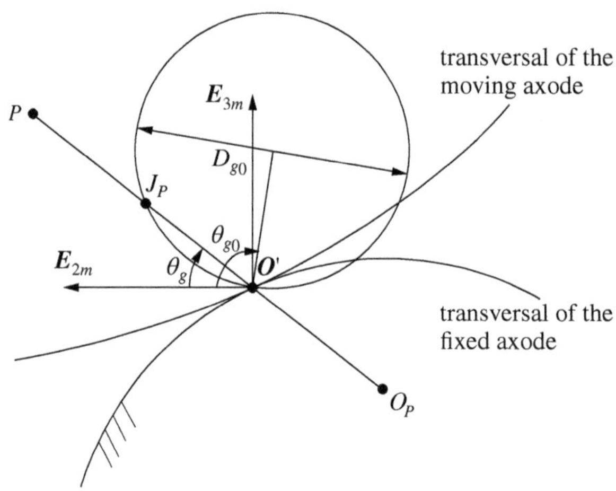

Figure 6.6 Geodesic inflection circle

Euler–Savary analogue (6.77) is inapplicable. Based on equation (6.74), the geodesic curvature k_g of trajectories of all points of $\sum^*$ can be represented by

$$k_g = \frac{r - D_{g0}\cos(\theta_g + \theta_{g0})}{r^2 + (\alpha^*/\beta^*)^2} \tag{6.79}$$

If the geodesic curvature k_g is replaced by the radius ρ_g of the geodesic curvature, the above equation has the following two forms:

$$\rho_g[r - D_{g0}\cos(\theta_g + \theta_{g0})] = r^2 + \left(\frac{\alpha^*}{\beta^*}\right)^2 \tag{6.80a}$$

$$\frac{1}{r} + \frac{1}{\rho_g - r} = \frac{\rho_g}{(\alpha^*/\beta^*)^2 - \rho_g D_{g0}\cos(\theta_g + \theta_{g0})} \tag{6.80b}$$

Equation (6.80a) is referred to as the ***geodesic Euler–Savary analogue of a point trajectory*** in spatial motion. By equation (6.79), the instantaneous geodesic inflection points all lie on a ***geodesic inflection surface***, defined by

$$r - D_{g0}\cos(\theta_g + \theta_{g0}) = 0 \tag{6.81}$$

According to equation (6.74), D_{g0} and θ_{g0} in the above equation are instantaneously constant for a given value of u_1, and equation (6.81) is the equation of the geodesic inflection circle. In view of a series of values of $u_1 \in (-\infty, +\infty)$, equation (6.81) may be regarded as the equation of a family of geodesic inflection circles, or a geodesic inflection surface. The geodesic inflection surface has four properties:

1. On the plane perpendicular to the ISA, the cross-section curve of the geodesic inflection surface is just the geodesic inflection circle, which passes through the ISA.
2. On the plane defined by $u_1 = 0$, the center of the geodesic inflection circle lies on the normal of the moving axode at the striction point.
3. The geodesic inflection surface is a conical surface whose vertex is located at the striction point of the moving axode when the equation $\alpha^* + \beta^*\gamma = 0$ is satisfied.
4. If u_1 tends to infinity, the diameter D_{g0} tends to zero and the geodesic inflection circle degenerates to a straight line.

In particular, the inflection point of the moving body in spatial motion at an instant is defined by $k_g = 0$ and $k_n = 0$, or both the geodesic curvature and the normal curvature are zero simultaneously. We know that this point locates on an ***inflection curve***, determined by equations (6.75) and (6.81), since $k^2 = k_n^2 + k_g^2$.

6.3.4 Generalized Curvature

The (geodesic) Euler–Savary analogue of a point trajectory in spatial movement describes the point's position in the moving body, the (geodesic) curvature of its trajectory, and the induced construction parameters of the axodes, which are the traditional topics in kinematic geometry. For the application in kinematic synthesis of spatial linkages, the curvatures of the point trajectories have to be developed into ***generalized curvatures***, including the ***spherical curvature*** and the ***cylindrical curvature*** discussed in Chapter 3, which typically correspond to the spherical constraint curve and the cylindrical constraint surface.

A spatial trajectory Γ_P in $\{\boldsymbol{O}_f; \boldsymbol{i}_f, \boldsymbol{j}_f, \boldsymbol{k}_f\}$ with arc length σ_P, traced by a point P of $\sum^*$, has the Darboux frame $\{\boldsymbol{R}_P; \boldsymbol{E}_1, \boldsymbol{E}_2, \boldsymbol{E}_3\}$ defined by equation (6.70). Its normal curvature k_n, geodesic curvature k_g, and geodesic torsion τ_g are described by equation (6.72). For convenience in analyzing the generalized curvature later, the Frenet frame $\{\boldsymbol{R}_P; \boldsymbol{\alpha}, \boldsymbol{\beta}, \boldsymbol{\gamma}\}$ of Γ_P is

$$\begin{cases} \boldsymbol{\beta} = \dfrac{1}{k}\left(k_n \boldsymbol{E}_2 + k_g \boldsymbol{E}_3\right) = \cos\varphi \cdot \boldsymbol{E}_2 + \sin\varphi \cdot \boldsymbol{E}_3 \\ \boldsymbol{\gamma} = \dfrac{1}{k}(-k_g \boldsymbol{E}_2 + k_n \boldsymbol{E}_3) = -\sin\varphi \cdot \boldsymbol{E}_2 + \cos\varphi \cdot \boldsymbol{E}_3 \end{cases} \tag{6.82}$$

where φ is the angle between the vectors $\boldsymbol{\beta}$ and $\boldsymbol{E}_2$, and $\tan\varphi = k_g/k_n$. Based on equations (3.47) to (3.49), the curvature k and torsion τ of Γ_P can be expressed by k_n, k_g, τ_g as

$$\begin{cases} k^2 = k_n^2 + k_g^2 \\ \tau = \dfrac{k_n \mathrm{d}k_g/\mathrm{d}\sigma_P - k_g \mathrm{d}k_n/\mathrm{d}\sigma_P}{k_n^2 + k_g^2} + \tau_g \end{cases} \tag{6.83}$$

The curve Γ_P has an osculating plane at the point P, whose normal vector $\boldsymbol{n}$ is just the binormal vector γ of Γ_P at P, or

$$\boldsymbol{n} = \boldsymbol{\gamma} = \frac{1}{k}(-k_g \boldsymbol{E}_2 + k_n \boldsymbol{E}_3) = \frac{1}{\sqrt{k_g{}^2 + k_n{}^2}}(-k_g \boldsymbol{E}_2 + k_n \boldsymbol{E}_3) \tag{6.84}$$

Here, the vectors $\boldsymbol{E}_2$, $\boldsymbol{E}_3$ are determined by equation (6.70) and the curvatures k_n, k_g are obtained by equation (6.72).

The curve Γ_P contacts its osculating plane at P to second order. The torsion τ of Γ_P at P is zero if the contact order is three, or

$$\frac{\mathrm{d}k_g}{\mathrm{d}\sigma_P} k_n - \frac{\mathrm{d}k_n}{\mathrm{d}\sigma_P} k_g + \tau_g (k_n^2 + k_g^2) = 0 \tag{6.85}$$

Substituting k_n, k_g, and τ_g of Γ_P in equation (6.72) into the above equation, we have the constraint equation about the point's coordinates (u_1, u_2, u_3) in $\{\boldsymbol{\rho}_m; \boldsymbol{E}_{1m}, \boldsymbol{E}_{2m}, \boldsymbol{E}_{3m}\}$. If the curve Γ_P and its osculating plane contact in higher order, the higher derivatives of the torsion τ should be zero besides $\tau = 0$.

If the curvature k and torsion τ of Γ_P at the point P are not zero, we can discuss the contact conditions of Γ_P and the spherical or cylindrical surface.

6.3.4.1 Spherical Curvature and Sphere Point

If the spatial curve Γ_P has third-order contact with a spherical surface at the point P, the spherical surface is referred to as the ***osculating sphere***, or ***curvature sphere***. By means of equations (3.156) and (3.157), a new Darboux frame $\{\boldsymbol{R}_P; \boldsymbol{\alpha}, \boldsymbol{n}, \boldsymbol{v}\}$ can be obtained by rotating the Frenet frame $\{\boldsymbol{R}_P; \boldsymbol{\alpha}, \boldsymbol{\beta}, \boldsymbol{\gamma}\}$ about the vector $\boldsymbol{\alpha}$. Substituting equations (6.82) and (6.83) into equation (3.159), we can locate the position of the curvature sphere. The position vector of the center is

$$\boldsymbol{R}_S = \boldsymbol{R}_P + \frac{\left(\dfrac{\mathrm{d}k_g}{\mathrm{d}\sigma_P} + \tau_g k_n\right)\boldsymbol{E}_2 + \left(-\dfrac{\mathrm{d}k_n}{\mathrm{d}\sigma_P} + \tau_g k_g\right)\boldsymbol{E}_3}{\dfrac{\mathrm{d}k_g}{\mathrm{d}\sigma_P} k_n - \dfrac{\mathrm{d}k_n}{\mathrm{d}\sigma_P} k_g + \tau_g (k_n^2 + k_g^2)} \tag{6.86}$$

Furthermore, if Γ_P has fourth-order contact with a spherical surface at the point P, the curvature k and torsion τ have to meet the contact condition equation (3.162), and of course, the spherical surface is the curvature sphere. Substituting equation (6.82) into equation (3.162), we have

$$\left(\frac{\mathrm{d}k_g}{\mathrm{d}\sigma_P}+\tau_g k_n\right)\left(\frac{\mathrm{d}^2k_n}{\mathrm{d}\sigma_P^2}-2\tau_g\frac{\mathrm{d}k_g}{\mathrm{d}\sigma_P}-\tau_g^2k_n-k_g\frac{\mathrm{d}\tau_g}{\mathrm{d}\sigma_P}\right)$$
$$-\left(\frac{\mathrm{d}k_n}{\mathrm{d}\sigma_P}-\tau_g k_g\right)\left(\frac{\mathrm{d}^2k_g}{\mathrm{d}\sigma_P^2}+2\tau_g\frac{\mathrm{d}k_n}{\mathrm{d}\sigma_P}-\tau_g^2k_g-k_n\frac{\mathrm{d}\tau_g}{\mathrm{d}\sigma_P}\right)=0 \tag{6.87}$$

Substituting k_n, k_g, and τ_g of Γ_P in equation (6.72) into the above equation, we have the constraint equation about the point's coordinates (u_1, u_2, u_3) in $\{\boldsymbol{\rho}_m; \boldsymbol{E}_{1m}, \boldsymbol{E}_{2m}, \boldsymbol{E}_{3m}\}$. The equation defines a surface in $\sum^*$ at an instant, which is referred to as the ***surface of the stationary spherical curvature***. All points (∞^2) on the surface, called ***stationary points of spherical curvature***, trace spatial curves in $\{\boldsymbol{O}_f; \boldsymbol{i}_f, \boldsymbol{j}_f, \boldsymbol{k}_f\}$, which contact with the curvature sphere at five infinitesimally consecutive positions.

If the trajectory Γ_P traced by point P of $\sum^*$ contacts the curvature sphere at six infinitesimally consecutive positions, or has fifth-order contact with the curvature sphere at an instant, the point is called the ***point of second-order stationary spherical curvature***. Besides equation (6.87), the curvatures of Γ_P should meet another equation. There exist infinite points (∞^1) of second-order stationary sphere curvature on a curve in $\sum^*$, which is defined by the two contact condition equations and referred to as the ***curve of second-order stationary spherical curvature***. By that analogy, a point of $\sum^*$, whose trajectory contacts the curvature sphere at seven infinitesimally consecutive positions, is called the ***sphere point***. There exist finite sphere points (∞°) in $\sum^*$ at an instant.

6.3.4.2 Cylindrical Curvature and Cylinder Point

According to equation (3.170), a Darboux frame $\{\boldsymbol{R}_P; \boldsymbol{\alpha}, \boldsymbol{n}, \boldsymbol{v}\}$ of Γ_P can be obtained by rotating the Frenet frame $\{\boldsymbol{R}_P; \boldsymbol{\alpha}, \boldsymbol{\beta}, \boldsymbol{\gamma}\}$ about the vector $\boldsymbol{\alpha}$ by a rotation angle, whose value and change rates are determined by equations (3.169), (3.171), and (3.172). The curve Γ_P has fourth-order contact with its ***osculating cylindrical surface***. Substituting equations (6.82) and (6.83) into equations (3.170) and (3.174), we can determine the orientation and position of the axis L and the radius r_0 of the osculating cylindrical surface of Γ_P.

If the curve Γ_P contacts the osculating cylindrical surface at six infinitesimally consecutive positions, equations (3.170) and (3.175) should be satisfied. The normal curvature k_n, geodesic curvature k_g, and geodesic torsion τ_g should meet a contact condition equation, which defines a ***surface of the stationary cylindrical curvature*** in $\sum^*$. There exist infinite points (∞^2) on the surface, referred to as the ***stationary points of cylindrical curvature***. Furthermore, if the curve Γ_P contacts the osculating cylindrical surface at seven infinitesimally consecutive positions, the curvature invariants k_n, k_g, τ_g should meet two contact condition equations. This leads to there being infinite tracing points (∞^1), the ***points of second-order stationary cylindrical curvature***, on a curve in $\sum^*$, which is called the ***curve of second-order stationary cylindrical curvature***. Finally, if the trajectory Γ_P has seventh-order contact with the osculating cylindrical surface, the tracing point P is defined as a ***cylinder point***. At an instant, there would be finite cylinder points (∞°) in the moving body. For higher-order contact of Γ_P and the cylindrical surface, the contact condition equations are so complex that we have to overcome many difficulties in analyzing and solving them.

6.4 Differential Kinematics of Lines

For spatial kinematic geometry, not only the point trajectories but also the line trajectories are discussed. As a ruled surface, the kinematic invariants – three construction parameters – can totally reveal the geometric properties. For spatial linkages, we are interested in the relationship between the geometric properties of the line trajectory and the spatial movement of the rigid body. In this section, the adjoint approach is adopted to describe the line trajectory and reveal its properties, and the characteristics lines, which trace specific trajectories, are readily located.

6.4.1 Frenet Frame

For a moving body $\sum^*$ with $\{\boldsymbol{O}_m;\boldsymbol{i}_m,\boldsymbol{j}_m,\boldsymbol{k}_m\}$ in spatial motion, a point $P(x_{Pm},y_{Pm},z_{Pm})$ of $\sum^*$ traces a spatial point trajectory Γ_P in $\{\boldsymbol{O}_f;\boldsymbol{i}_f,\boldsymbol{j}_f,\boldsymbol{k}_f\}$. A line L of $\sum^*$ passing through P has unit direction vector $\boldsymbol{l}$, as shown in Fig. 6.7. The three components l_{m1}, l_{m2} and l_{m3} of $\boldsymbol{l}$ in $\{\boldsymbol{O}_m;\boldsymbol{i}_m,\boldsymbol{j}_m,\boldsymbol{k}_m\}$ are

$$l_{m1}=\sin\delta_{ml}\cos\theta_{ml},\ l_{m2}=\sin\delta_{ml}\sin\theta_{ml},\ l_{m3}=\cos\delta_{ml} \tag{6.88}$$

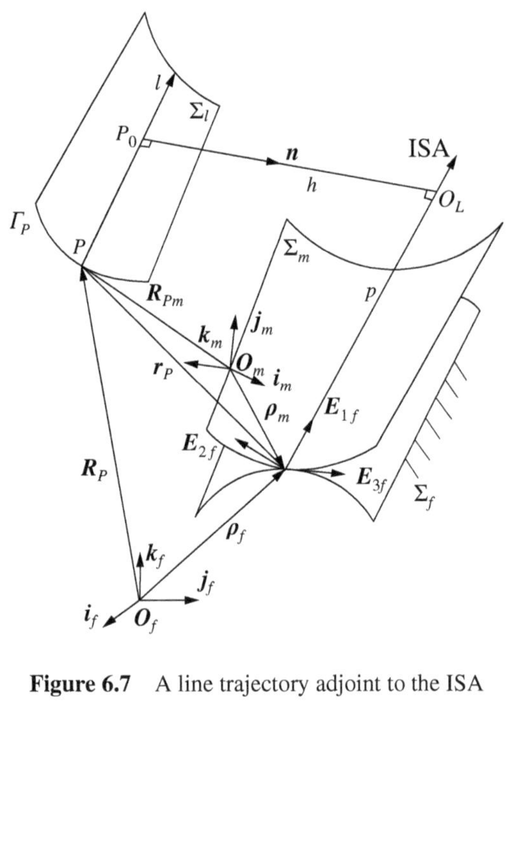

Figure 6.7 A line trajectory adjoint to the ISA

where δ_{ml} is the angle between $\boldsymbol{l}$ and $\boldsymbol{k}_m$, and θ_{ml} is the direction angle of the projection vector of $\boldsymbol{l}$ on the plane $\boldsymbol{O}_m - \boldsymbol{i}_m\boldsymbol{j}_m$. $\boldsymbol{l}$ can also be expressed in the Frenet frame $\{\boldsymbol{\rho}_m; \boldsymbol{E}_{1m}, \boldsymbol{E}_{2m}, \boldsymbol{E}_{3m}\}$ of $\sum_m$ (or $\sum_f$) by equation (6.45), of which three components l_1, l_2 and l_3 are

$$l_1 = \cos\delta_l,\ l_2 = \sin\delta_l\ \cos\theta_l,\ l_3 = \sin\delta_l\ \sin\theta_l \tag{6.89}$$

where δ_l is the angle between $\boldsymbol{l}$ and $\boldsymbol{E}_{1m}$, and θ_l is the direction angle of the projection vector of $\boldsymbol{l}$ on the plane $\boldsymbol{\rho}_m - \boldsymbol{E}_{2m}\boldsymbol{E}_{3m}$.

Now we take $\sum_m$ as an original ruled surface and examine the line L in $\{\boldsymbol{\rho}_m; \boldsymbol{E}_{1m}, \boldsymbol{E}_{2m}, \boldsymbol{E}_{3m}\}$, or the line L adjoint to the ISA at an instant and tracing the trajectory $\sum_l^*$ in $\{\boldsymbol{O}_m; \boldsymbol{i}_m, \boldsymbol{j}_m, \boldsymbol{k}_m\}$. $\sum_l^*$ is an adjoint ruled surface of $\sum_m$, whose vector equation in $\{\boldsymbol{O}_m; \boldsymbol{i}_m, \boldsymbol{j}_m, \boldsymbol{k}_m\}$ is

$$\sum\nolimits_l^* :\ \boldsymbol{R}_l^* = \boldsymbol{R}_P + \mu\boldsymbol{l} = \boldsymbol{\rho}_m + \sum_i u_i\boldsymbol{E}_{im} + \mu\sum_i l_i\boldsymbol{E}_{im} \tag{6.90}$$

where (σ, μ) are the parameters of $\sum_l^*$. Taking the differential of equation (6.90), we have

$$\begin{cases} \mathrm{d}\boldsymbol{R}_l^* = \left(\dfrac{\mathrm{d}\boldsymbol{R}_P}{\mathrm{d}\sigma} + \mu\dfrac{\mathrm{d}\boldsymbol{l}}{\mathrm{d}\sigma}\right)\mathrm{d}\sigma + \boldsymbol{l}\mathrm{d}\mu \\ \dfrac{\mathrm{d}\boldsymbol{R}_P}{\mathrm{d}\sigma} = \left(\alpha_m + \dfrac{\mathrm{d}u_1}{\mathrm{d}\sigma} - u_2\right)\boldsymbol{E}_{1m} + \left(u_1 + \dfrac{\mathrm{d}u_2}{\mathrm{d}\sigma} - \beta_m u_3\right)\boldsymbol{E}_{2m} + \left(\beta_m u_2 + \dfrac{\mathrm{d}u_3}{\mathrm{d}\sigma} + \gamma_m\right)\boldsymbol{E}_{3m} \\ \dfrac{\mathrm{d}\boldsymbol{l}}{\mathrm{d}\sigma} = \left(\dfrac{\mathrm{d}l_1}{\mathrm{d}\sigma} - l_2\right)\boldsymbol{E}_{1m} + \left(l_1 + \dfrac{\mathrm{d}l_2}{\mathrm{d}\sigma} - \beta_m l_3\right)\boldsymbol{E}_{2m} + \left(\beta_m l_2 + \dfrac{\mathrm{d}l_3}{\mathrm{d}\sigma}\right)\boldsymbol{E}_{3m} \\ \boldsymbol{l} = l_1\boldsymbol{E}_{1m} + l_2\boldsymbol{E}_{2m} + l_3\boldsymbol{E}_{3m} \end{cases} \tag{6.91}$$

Since L is a fixed line of $\sum^*$, the coordinates of P and the direction cosines of L do not change in $\{\boldsymbol{O}_m; \boldsymbol{i}_m, \boldsymbol{j}_m, \boldsymbol{k}_m\}$. Based on the absolute fixed line condition equations (3.74), we have

$$\begin{cases} \dfrac{\mathrm{d}u_1}{\mathrm{d}\sigma} = u_2 - \alpha_m,\ \dfrac{\mathrm{d}u_2}{\mathrm{d}\sigma} = \beta_m u_3 - u_1,\ \dfrac{\mathrm{d}u_3}{\mathrm{d}\sigma} = -\beta_m u_2 - \gamma_m \\ \dfrac{\mathrm{d}l_1}{\mathrm{d}\sigma} = l_2,\ \dfrac{\mathrm{d}l_2}{\mathrm{d}\sigma} = \beta_m l_3 - l_1,\ \dfrac{\mathrm{d}l_3}{\mathrm{d}\sigma} = -\beta_m l_2 \end{cases} \tag{6.92}$$

In fact, the trajectory $\sum_l^*$ in $\{\boldsymbol{O}_m; \boldsymbol{i}_m, \boldsymbol{j}_m, \boldsymbol{k}_m\}$ is only a fixed line of $\sum^*$.

On the contrary, we examine the line L of $\sum^*$ in the fixed frame $\{\boldsymbol{O}_f; \boldsymbol{i}_f, \boldsymbol{j}_f, \boldsymbol{k}_f\}$. At an instant, the line L corresponds to the ISA, and for all instants, the ISA generates the fixed axode $\sum_f$ in $\{\boldsymbol{O}_f; \boldsymbol{i}_f, \boldsymbol{j}_f, \boldsymbol{k}_f\}$ while the line L traces a trajectory, a ruled surface $\sum_l$. Consequently, we visualize the fixed axode $\sum_f$ as an original ruled surface, and the ruled surface $\sum_l$ as an adjoint ruled surface of $\sum_f$, which can be expressed by

$$\sum\nolimits_l :\ \boldsymbol{R}_l = \boldsymbol{R}_P + \mu\boldsymbol{l} = \boldsymbol{\rho}_f + \sum_i u_i\boldsymbol{E}_{if} + \mu\sum_i l_i\boldsymbol{E}_{if} \tag{6.93}$$

Similarly, taking the differential of equation (6.93), we have

$$\begin{cases} \mathrm{d}\boldsymbol{R}_l = \left(\dfrac{\mathrm{d}\boldsymbol{R}_P}{\mathrm{d}\sigma} + \mu\dfrac{\mathrm{d}\boldsymbol{l}}{\mathrm{d}\sigma}\right)\mathrm{d}\sigma + \boldsymbol{l}\mathrm{d}\mu \\ \dfrac{\mathrm{d}\boldsymbol{R}_P}{\mathrm{d}\sigma} = \left(\alpha_f + \dfrac{\mathrm{d}u_1}{\mathrm{d}\sigma} - u_2\right)\boldsymbol{E}_{1f} + \left(u_1 + \dfrac{\mathrm{d}u_2}{\mathrm{d}\sigma} - \beta_f u_3\right)\boldsymbol{E}_{2f} + \left(\beta_f u_2 + \dfrac{\mathrm{d}u_3}{\mathrm{d}\sigma} + \gamma_f\right)\boldsymbol{E}_{3f} \\ \dfrac{\mathrm{d}\boldsymbol{l}}{\mathrm{d}\sigma} = \left(\dfrac{\mathrm{d}l_1}{\mathrm{d}\sigma} - l_2\right)\boldsymbol{E}_{1f} + \left(l_1 + \dfrac{\mathrm{d}l_2}{\mathrm{d}\sigma} - \beta_f l_3\right)\boldsymbol{E}_{2f} + \left(\beta_f l_2 + \dfrac{\mathrm{d}l_3}{\mathrm{d}\sigma}\right)\boldsymbol{E}_{3f} \\ \boldsymbol{l} = l_1\boldsymbol{E}_{1f} + l_2\boldsymbol{E}_{2f} + l_3\boldsymbol{E}_{3f} \end{cases} \tag{6.94}$$

Since the Frenet frames $\{\boldsymbol{\rho}_m; \boldsymbol{E}_{1m}, \boldsymbol{E}_{2m}, \boldsymbol{E}_{3m}\}$ of $\sum_m$ and $\{\boldsymbol{\rho}_f; \boldsymbol{E}_{1f}, \boldsymbol{E}_{2f}, \boldsymbol{E}_{3f}\}$ of $\sum_f$ coincide with each other at an instant, the point P has the same coordinates and their change rates in both Frenet frames, and so does the unit vector $\boldsymbol{l}$ of the line L, or u_i, $i = 1, 2, 3$ and l_i, $i = 1, 2, 3$, as well as their derivatives are the same in both equations (6.90) and (6.93). Substituting equation (6.92) into the above equation, we have

$$\begin{cases} \mathrm{d}\boldsymbol{R}_l = \left(\dfrac{\mathrm{d}\boldsymbol{R}_P}{\mathrm{d}\sigma} + \mu\dfrac{\mathrm{d}\boldsymbol{l}}{\mathrm{d}\sigma}\right)\mathrm{d}\sigma + \boldsymbol{l}\mathrm{d}\mu \\ \dfrac{\mathrm{d}\boldsymbol{R}_P}{\mathrm{d}\sigma} = \alpha^*\boldsymbol{E}_{1f} + \beta^*(-u_3\boldsymbol{E}_{2f} + u_2\boldsymbol{E}_{3f}) \\ \dfrac{\mathrm{d}\boldsymbol{l}}{\mathrm{d}\sigma} = \beta^*(-l_3\boldsymbol{E}_{2f} + l_2\boldsymbol{E}_{3f}) \\ \boldsymbol{l} = l_1\boldsymbol{E}_{1f} + l_2\boldsymbol{E}_{2f} + l_3\boldsymbol{E}_{3f} \end{cases} \tag{6.95}$$

From the third expression of the above equation, we derive the arc length σ_l of the spherical image curve of $\boldsymbol{l}$:

$$\mathrm{d}\sigma_l = \left|\frac{\mathrm{d}\boldsymbol{l}}{\mathrm{d}\sigma}\right|\mathrm{d}\sigma = \beta^*\sqrt{1 - l_1^2}\mathrm{d}\sigma \tag{6.96}$$

The second derivative of $\boldsymbol{l}$ with respect to σ is

$$\frac{\mathrm{d}^2\boldsymbol{l}}{\mathrm{d}\sigma^2} = \beta^* l_3\boldsymbol{E}_{1f} - \left(\beta^{*2}l_2 + \frac{\mathrm{d}\beta^*}{\mathrm{d}\sigma}l_3\right)\boldsymbol{E}_{2f} + \left(\frac{\mathrm{d}\beta^*}{\mathrm{d}\sigma}l_2 - \beta^* l_1 - \beta^{*2}l_3\right)\boldsymbol{E}_{3f} \tag{6.97}$$

By equation (6.95), the striction directrix distance of $\sum_l$ is

$$b_l = -\left(\frac{\mathrm{d}\boldsymbol{R}_P}{\mathrm{d}\sigma}\cdot\frac{\mathrm{d}\boldsymbol{l}}{\mathrm{d}\sigma}\right)/\left(\frac{\mathrm{d}\boldsymbol{l}}{\mathrm{d}\sigma}\right)^2 = -(l_2u_2 + l_3u_3)/(1 - l_1^2) \tag{6.98}$$

The vector equation of the striction curve of $\sum_l$ takes the form

$$\boldsymbol{\rho}_l = \boldsymbol{R}_P + b_l\boldsymbol{l} = \boldsymbol{\rho}_f + \sum_i (u_i + b_l l_i)\boldsymbol{E}_{if} \tag{6.99}$$

At an instant, the coordinates of the striction point P_0 of $\sum_l$ in $\{\boldsymbol{\rho}_f; \boldsymbol{E}_{1f}, \boldsymbol{E}_{2f}, \boldsymbol{E}_{3f}\}$ are $(u_1 + b_l l_1, u_2 + b_l l_2, u_3 + b_l l_3)$. The first derivatives of b_l and $\boldsymbol{\rho}_l$ with respect to σ are

$$\begin{cases} \dfrac{\mathrm{d}b_l}{\mathrm{d}\sigma} = \left[\left(l_1 u_2 + l_2 u_1 + l_3 \gamma \right) (1 - l_1^2) - 2 l_1 l_2 (l_2 u_2 + l_3 u_3) \right] / (1 - l_1^2)^2 \\ \dfrac{\mathrm{d}\boldsymbol{\rho}_l}{\mathrm{d}\sigma} = \dfrac{\mathrm{d}\boldsymbol{R}_P}{\mathrm{d}\sigma} + \dfrac{\mathrm{d}b_l}{\mathrm{d}\sigma}\boldsymbol{l} + b_l \dfrac{\mathrm{d}\boldsymbol{l}}{\mathrm{d}\sigma} \end{cases} \tag{6.100}$$

In order to study the geometrical properties of $\sum_l$, the Frenet frame $\{\boldsymbol{\rho}_l; \boldsymbol{E}_1, \boldsymbol{E}_2, \boldsymbol{E}_3\}$ of $\sum_l$ is established as

$$\begin{cases} \boldsymbol{E}_1 = \boldsymbol{l} = \sum\limits_i l_i \boldsymbol{E}_{if} \\ \boldsymbol{E}_2 = \dfrac{\mathrm{d}\boldsymbol{l}}{\mathrm{d}\sigma} \Big/ \left| \dfrac{\mathrm{d}\boldsymbol{l}}{\mathrm{d}\sigma} \right| = -(l_3 \boldsymbol{E}_{2f} - l_2 \boldsymbol{E}_{3f}) / \sqrt{1 - l_1^2} \\ \boldsymbol{E}_3 = \boldsymbol{E}_1 \times \boldsymbol{E}_2 = [(1 - l_1^2)\boldsymbol{E}_{1f} - l_1 l_2 \boldsymbol{E}_{2f} - l_1 l_3 \boldsymbol{E}_{3f}] / \sqrt{1 - l_1^2} \end{cases} \tag{6.101}$$

By the Frenet formulas (3.58) of a ruled surface, the three construction parameters of $\sum_l$ are

$$\begin{cases} \alpha_l = \dfrac{\mathrm{d}\boldsymbol{\rho}_l}{\mathrm{d}\sigma} \cdot \boldsymbol{l} \Big/ \left| \dfrac{\mathrm{d}\boldsymbol{l}}{\mathrm{d}\sigma} \right| = \left[l_1 \alpha^* + \beta^* \left(l_3 u_2 - l_2 u_3 \right) + \dfrac{\mathrm{d}b_l}{\mathrm{d}\sigma} \right] / (\beta^* \sqrt{1 - l_1^2}) \\ \beta_l = \left(\boldsymbol{l}, \dfrac{\mathrm{d}\boldsymbol{l}}{\mathrm{d}\sigma}, \dfrac{\mathrm{d}^2\boldsymbol{l}}{\mathrm{d}\sigma^2} \right) \Big/ \left| \dfrac{\mathrm{d}\boldsymbol{l}}{\mathrm{d}\sigma} \right|^3 = [l_3 + \beta^* l_1 (1 - l_1^2)] / [\beta^* (1 - l_1^2)^{3/2}] \\ \gamma_l = \left(\dfrac{\mathrm{d}\boldsymbol{\rho}_l}{\mathrm{d}\sigma}, \boldsymbol{l}, \dfrac{\mathrm{d}\boldsymbol{l}}{\mathrm{d}\sigma} \right) \Big/ \left| \dfrac{\mathrm{d}\boldsymbol{l}}{\mathrm{d}\sigma} \right|^2 = [\alpha^* (1 - l_1^2) + \beta^* l_1 (l_2 u_3 - l_3 u_2)] / [\beta^* (1 - l_1^2)] \end{cases} \tag{6.102}$$

By equation (3.53), the unit normal vector $\boldsymbol{n}$ of $\sum_l$ at a point (σ_l, μ) is

$$\boldsymbol{n} = (\mu \boldsymbol{E}_3 - \gamma_l \boldsymbol{E}_2) / \sqrt{\mu^2 + \gamma_l^2} \tag{6.103}$$

At the striction point $P_0(\mu = 0)$, the normal vector $\boldsymbol{n}$ of $\sum_l$ is

$$\boldsymbol{n} = -\boldsymbol{E}_2 = (l_3 \boldsymbol{E}_{2f} - l_2 \boldsymbol{E}_{3f}) / \sqrt{1 - l_1^2} \tag{6.104}$$

The unit vector $\boldsymbol{l}$ of the generator is taken as normal vector of a normal plane $\sum_{P0}$ of $\sum_l$ at the striction point P_0, whose vector equation is

$$\boldsymbol{l} \cdot (\boldsymbol{R}_{P0} - \boldsymbol{\rho}_l) = 0 \tag{6.105}$$

where $\boldsymbol{R}_{P0} = \boldsymbol{\rho}_f + x_1 \boldsymbol{E}_{1f} + x_2 \boldsymbol{E}_{2f} + x_3 \boldsymbol{E}_{3f}$ is the vector equation of any point of $\sum_{P0}$. Substituting $\boldsymbol{l} = \sum\limits_i l_i \boldsymbol{E}_{if}$ and equation (6.99) into the above equation, we have the algebraic equation of $\sum_{P0}$ in $\{\boldsymbol{\rho}_m; \boldsymbol{E}_{1m}, \boldsymbol{E}_{2m}, \boldsymbol{E}_{3m}\}$:

$$l_1(x_1 - u_1 - b_l l_1) + l_2(x_2 - u_2 - b_l l_2) + l_3(x_3 - u_3 - b_l l_3) = 0 \tag{6.106}$$

The normal plane $\sum_{P0}$ intersects the ISA ($\boldsymbol{E}_{1f}$) at point O_L, whose coordinates in $\{\boldsymbol{\rho}_m; \boldsymbol{E}_{1m}, \boldsymbol{E}_{2m}, \boldsymbol{E}_{3m}\}$ are

$$x_1 = [u_1 - l_1(l_1u_1 + l_2u_2 + l_3u_3)]/(1 - l_1^2),\ x_2 = 0,\ x_3 = 0 \tag{6.107}$$

We connect $P_0(u_1 + b_l l_1, u_2 + b_l l_2, u_3 + b_l l_3)$ and $O_L([u_1 - l_1(l_1u_1 + l_2u_2 + l_3u_3)]/(1 - l_1^2), 0, 0)$:

$$\overrightarrow{P_0O_L} = \frac{l_2l_3u_3 - l_3^2u_2}{1 - l_1^2}\boldsymbol{E}_{2f} + \frac{l_2l_3u_2 - l_2^2u_3}{1 - l_1^2}\boldsymbol{E}_{3f} = \frac{l_2u_3 - l_3u_2}{\sqrt{1 - l_1^2}}\boldsymbol{n} \tag{6.108}$$

The vector $\overrightarrow{P_0O_L}$ is perpendicular to $\boldsymbol{E}_{1f}$, or $\overrightarrow{P_0O_L} \cdot \boldsymbol{E}_{1f} = 0$. We have the ***striction point characteristics of a line trajectory in spatial movement***:

A line L of a moving body in spatial motion has line trajectory $\sum_l$, whose normal at a striction point is a common normal between the line L and the ISA, or the perpendicular foot of the common normal on L is the striction point of $\sum_l$.

As shown in Fig. 6.7, the distance between L and the ISA along the common normal is designated as h, and the distance between the striction point P_0 of $\sum_l$ and the perpendicular foot of the common normal on the ISA is denoted by p. By equations (6.107) and (6.108), we get

$$\begin{cases} h = (l_2u_3 - l_3u_2)/\sqrt{1 - l_1^2} \\ p = [u_1 - l_1(l_1u_1 + l_2u_2 + l_3u_3)]/(1 - l_1^2) \end{cases} \tag{6.109}$$

where h and p may be positive or negative, depending on the position of the striction point of $\sum_l$ relative to $\sum_f$. It should be noticed that the point P on the line L is a fixed point of $\sum^*$ and it produces a space curve on the line trajectory $\sum_l$, while the striction point of $\sum_l$ corresponds to one point on L at an instant and another point on L at the next instant. So the striction curve may not be a curve traced by a fixed point of the line L.

6.4.2 Striction Curve

In the above discussion, a line L of $\sum^*$ is located by the coordinates (u_1, u_2, u_3) of the point P and the direction cosines (l_1, l_2, l_3) of the unit vector $\boldsymbol{l}$ in the Frenet frame $\{\boldsymbol{\rho}_m; \boldsymbol{E}_{1m}, \boldsymbol{E}_{2m}, \boldsymbol{E}_{3m}\}$ of $\sum_m$ or $\sum_f$. Obviously, the coordinates (u_1, u_2, u_3) have no intrinsic connection with the geometric properties of the line trajectory $\sum_l$, but the position of the striction point of $\sum_l$ does. Based on equation (6.109), the vector equation of the striction curve of $\sum_l$ can be represented as

$$\boldsymbol{\rho}_l = \boldsymbol{\rho}_f + p\boldsymbol{E}_{1f} + h\boldsymbol{E}_{1f} \times \boldsymbol{E}_1/|\boldsymbol{E}_{1f} \times \boldsymbol{E}_1| \tag{6.110}$$

where $|\boldsymbol{E}_{1f} \times \boldsymbol{E}_1| = \sqrt{1 - l_1^2}$. Supposing $H = h/\sqrt{1 - l_1^2}$, the vector equation of $\sum_l$ can be rewritten in $\{\boldsymbol{O}_f; \boldsymbol{i}_f, \boldsymbol{j}_f, \boldsymbol{k}_f\}$ as

$$\sum\nolimits_l : \boldsymbol{R}_l = \boldsymbol{\rho}_l + \mu\boldsymbol{l} = \boldsymbol{\rho}_f + p\boldsymbol{E}_{1f} + H\boldsymbol{E}_{1f} \times \boldsymbol{E}_1 + \mu\boldsymbol{l} \tag{6.111}$$

According to the above equation, $\sum_l$ is dependent of the four parameters l_1, l_2, p, and h. By the geometrical meanings of the above four parameters, we know that the line L will move parallel to the ISA if the value of p varies, and will move toward or away from the ISA if the value of h changes; it rotates around the ISA when l_2 or l_3 changes, and the angle between L and the ISA varies with l_1.

On the contrary, the vector equation of the trajectory $\sum_l^*$ of L in $\{\boldsymbol{O}_m;\boldsymbol{i}_m,\boldsymbol{j}_m,\boldsymbol{k}_m\}$ is

$$\sum\nolimits_l^* : \boldsymbol{R}_l^* = \boldsymbol{\rho}_l^* + \mu\boldsymbol{l}_m = \boldsymbol{\rho}_m + p\boldsymbol{E}_{1m} + H\boldsymbol{E}_{1m}\times\boldsymbol{E}_1 + \mu\boldsymbol{l}_m \tag{6.112}$$

According to the striction point characteristics, the striction point of $\sum_l^*$ is the perpendicular foot on L of the common normal between the ISA and L, and moves on L while L is always collinear with or slides on a fixed line in $\{\boldsymbol{O}_m;\boldsymbol{i}_m,\boldsymbol{j}_m,\boldsymbol{k}_m\}$ or L is a quasi-fixed line. Based on the conditions of a quasi-fixed line, equation (3.75), we have

$$\frac{\mathrm{d}\boldsymbol{l}_m}{\mathrm{d}\sigma} = 0,\ \frac{\mathrm{d}\boldsymbol{\rho}_l^*}{\mathrm{d}\sigma}\times\boldsymbol{l}_m = 0 \tag{6.113}$$

Here, $\frac{\mathrm{d}\boldsymbol{l}_m}{\mathrm{d}\sigma} = 0$ is equivalent to the last three expressions of equation (6.92). Taking the derivative of equation (6.112) with respect to σ and substituting it into $\frac{\mathrm{d}\boldsymbol{\rho}_l^*}{\mathrm{d}\sigma}\times\boldsymbol{l}_m = 0$, we have

$$\begin{cases}\dfrac{\mathrm{d}p}{\mathrm{d}\sigma} = -\alpha_m + \left(l_1l_2p - l_3H + l_1l_3\gamma\right)/(1-l_1^2)\\ \dfrac{\mathrm{d}H}{\mathrm{d}\sigma} = (l_3p + l_1l_2H - l_2\gamma)/(1-l_1^2)\end{cases} \tag{6.114}$$

Based on equation (6.112), the first and second derivatives of the striction directrix distance with respect to σ are

$$\begin{cases}\dfrac{\mathrm{d}b_l}{\mathrm{d}\sigma} = \dfrac{\mathrm{d}\boldsymbol{\rho}_l^*}{\mathrm{d}\sigma}\cdot\boldsymbol{l}_m = \left(l_2p - l_1l_3H + l_3\gamma\right)/(1-l_1^2)\\ \dfrac{\mathrm{d}^2b_l}{\mathrm{d}\sigma^2} = \left[\left(\beta_ml_3 - 2l_1\right)p + (\beta_ml_1 - 2l_3)l_2H + l_3\dfrac{\mathrm{d}\gamma}{\mathrm{d}\sigma} - l_2(\beta_m\gamma + \alpha_m) + 4l_1l_2\dfrac{\mathrm{d}b_l}{\mathrm{d}\sigma}\right]/(1-l_1^2)\end{cases} \tag{6.115}$$

In particular, we are interested in the condition for the striction curve of Σ_l being traced by a fixed point of $\sum^*$, or $\boldsymbol{\rho}_l^*$ is a constant vector, which leads to $\frac{\mathrm{d}\boldsymbol{\rho}_l^*}{\mathrm{d}\sigma}\cdot\boldsymbol{l}_m = 0$ and $\frac{\mathrm{d}\boldsymbol{\rho}_l^*}{\mathrm{d}\sigma}\times\boldsymbol{l}_m = 0$, or

$$\frac{\mathrm{d}b_l}{\mathrm{d}\sigma} = \frac{l_2p - l_1l_3H + l_3\gamma}{1-l_1^2} = 0,\ \frac{\mathrm{d}p}{\mathrm{d}\sigma} = -\alpha_m - l_3H,\ \frac{\mathrm{d}H}{\mathrm{d}\sigma} = \frac{p}{l_3} \tag{6.116}$$

In such a case, the striction curve of Σ_l is traced by a point fixed on L of $\sum^*$, whose position parameters (p, H) should meet equation (6.116).

Differentiating equation (6.110) with respect to σ, combining the Frenet formulas (3.58) with the quasi-fixed line conditions (6.114), we rewrite the construction parameters in terms of (l_1, l_3, p, H) as

$$\begin{cases}\alpha_l = \dfrac{l_1\alpha^* - \beta^*\left(1-{l_1}^2\right)H + \mathrm{d}b_l/\mathrm{d}\sigma}{\beta^*(1-{l_1}^2)^{1/2}}\\ \beta_l = \dfrac{l_3 + \beta^*l_1(1-l_1^2)}{\beta^*(1-l_1^2)^{3/2}}\\ \gamma_l = \dfrac{\alpha^*}{\beta^*} + l_1H\end{cases} \tag{6.117}$$

Based on the above equations, the first derivatives of α_l and γ_l with respect to σ_l are

$$\begin{cases} \dfrac{d\alpha_l}{d\sigma_l} = \left\{ \left[l_2\alpha^* + l_1\dfrac{d\alpha^*}{d\sigma}\left(1 - {l_1}^2\right) + \beta^*\gamma l_2(1 - {l_1}^2) + \dfrac{d\gamma}{d\sigma}l_3 - l_2(\alpha_m + \beta_m\gamma) \right.\right. \\ \left. -\alpha^*\dfrac{d\beta^*}{d\sigma}l_1\left(1 - {l_1}^2\right)/\beta^* \right] + (\beta_m l_1 l_2 - 2l_2l_3)H + [-\beta^* l_3(1 - {l_1}^2) + \beta_m l_3 - 2l_1]p \\ \left. + \left[5l_1l_2 - \dfrac{d\beta^*}{d\sigma}\left(1 - {l_1}^2\right)/\beta^* \right] \dfrac{db_l}{d\sigma} \right\} / [\beta^{*2}(1 - {l_1}^2)^2] \\ \dfrac{d\gamma_l}{d\sigma_l} = \left[\dfrac{d(\alpha^*/\beta^*)}{d\sigma} + (l_1l_3p + l_2H - l_1l_2\gamma)/(1 - l_1^2) \right] / [\beta^*(1 - l_1^2)^{1/2}] \end{cases} \tag{6.118}$$

According to equation (6.117), the condition for a line trajectory Σ_l to be a developable surface ($\gamma_l = 0$) is

$$H = -\frac{\alpha^*}{\beta^* l_1} \quad \text{or}\, h = -\frac{\alpha^*\sqrt{1 - l_1^2}}{\beta^* l_1} \tag{6.119}$$

It shows that the developable condition of a line trajectory Σ_l deals only with parameters l_1, h, and $\frac{\alpha^*}{\beta^*}$, regardless of l_2 (l_3) and p. Especially when Σ_l is a cone, or $\alpha_l = 0$ and $\gamma_l = 0$, the conditions are

$$\begin{cases} H = -\dfrac{\alpha^*}{\beta^* l_1} \\ p = -\left\{ \left[\beta^*\left(1 - {l_1}^2\right)^2/l_1 + l_3 \right] \alpha^*/\beta^* + l_1\alpha^*(1 - {l_1}^2) + l_3\gamma \right\}/l_2 \end{cases} \tag{6.120}$$

Or, the striction points locate on a surface at an instant.

6.4.3 *Spherical Image Curve*

For a rigid body in spatial motion, the moving axode Σ_m has the spherical image curve C_m and the fixed axode Σ_f has the spherical image curve C_f, as shown in Fig. 6.8, which corresponds to the spherical moving and fixed centrodes π_m and π_f in spherical kinematics. The curve C_m rolls on the curve C_f without sliding. A line trajectory Σ_l, or a ruled surface, can be represented by equation (6.111). The spherical image of the unit vector $\boldsymbol{l}$ of the line L of $\sum^*$ is a point P_{sl} attached to C_m, which traces a spherical curve C_l when C_m rolls on C_f. Thus, the spherical curve C_l is the spherical image curve of the generatrices of Σ_l. In other words, the curve C_l is a spherical trajectory of a point in unit spherical movement. The invariant β_l of a line trajectory in spatial movement is equivalent to the geodesic curvature k_g of a point trajectory in spherical movement. When the invariant β_l of Σ_l is zero, according to equations (6.89) and (6.102), we have

$$\sin\theta_l + \beta^* \sin\delta_l \cos\delta_l = 0 \tag{6.121}$$

The above equation defines a curve on a unit sphere, which is defined as the ***spherical image inflection curve***. It is analogous to the spherical inflection curve in spherical kinematics in Chapter 4, but they have different geometrical properties. A line in spatial movement, whose direction parameters (δ_l, θ_l) hold equation (6.121) true, is parallel to a fixed plane at three

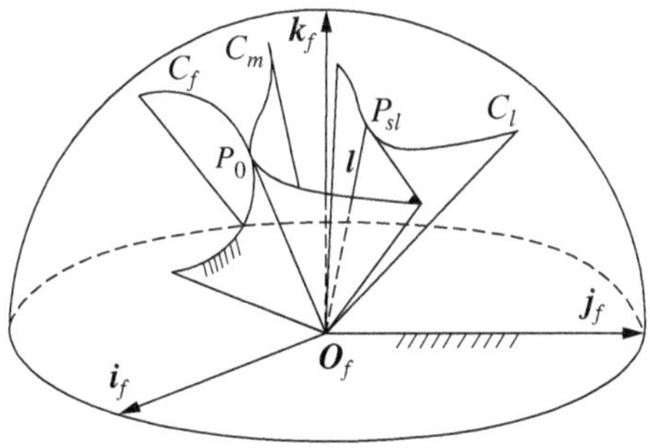

Figure 6.8 The spherical image curve of a line trajectory

infinitesimal positions, but may not be on the plane, while the point on the spherical inflection curve in spherical motion is on a plane passing through the sphere center at three infinitesimal positions at an instant.

According to equation (6.102), we take the derivative of β_l with respect to σ_l:

$$\frac{\mathrm{d}\beta_l}{\mathrm{d}\sigma_l} = \left\{ \beta^* l_2 \left[3 l_1 l_3 - \beta_m \left(1 - l_1^2\right) + \beta^*(1 - l_1^2)\right] - \frac{\mathrm{d}\beta^*}{\mathrm{d}\sigma} l_3(1 - l_1^2) \right\} / [\beta^*(1 - l_1^2)^3] \quad (6.122)$$

Substituting equation (6.89) into the above equation and letting $\frac{\mathrm{d}\beta_l}{\mathrm{d}\sigma_l} = 0$, we have

$$C_{\beta'} : \cot \delta_l = \frac{1}{M \sin \theta_l} + \frac{1}{N \cos \theta_l} \quad (6.123)$$

where M and N are the same as those in equation (4.47), or

$$\frac{1}{M} = \frac{\beta_m - \beta^*}{3}, \quad \frac{1}{N} = \frac{1}{3\beta^*} \frac{\mathrm{d}\beta^*}{\mathrm{d}\sigma}$$

For all parameters (δ_l, θ_l), equation (6.123) determines a spherical curve, called the ***cubic spherical image curve of stationary curvature***. For a line trajectory Σ_l with $\frac{\mathrm{d}\beta_l}{\mathrm{d}\sigma_l} = 0$ at an instant, the generators form a constant inclination angle with respect to a fixed line in a fixed body at four infinitesimal positions.

Furthermore, taking the second derivative of β_l with respect to σ_l and letting it be zero, or $\frac{\mathrm{d}^2\beta_l}{\mathrm{d}\sigma_l^2} = 0$, we get the following equation simplified by equation (6.123):

$$(\tan^2\theta_l + 1) \left[\frac{1}{M^2}(2 - \beta_m M) + \frac{1}{M^2}\left(\frac{3M}{N} + \frac{\mathrm{d}M}{\mathrm{d}\sigma}\right) \tan \theta_l + \frac{1}{N^2}\left(1 + \frac{\mathrm{d}N}{\mathrm{d}\sigma}\right) \tan^2\theta_l \right.$$
$$\left. + \frac{1}{N}\left(\beta_m - \frac{1}{M}\right) \tan^3\theta_l - \frac{1}{N^2}\tan^4\theta_l\right] + \tan^2\theta_l = 0 \quad (6.124)$$

The generators of a line trajectory Σ_l with $\frac{\mathrm{d}\beta_l}{\mathrm{d}\sigma_l} = 0$ and $\frac{\mathrm{d}^2\beta_l}{\mathrm{d}\sigma_l^2} = 0$ form a constant inclination angle with respect to a fixed line at five infinitesimal positions. The directions of these lines are defined as the ***directions of Burmester's line*** in spatial motion at an instant. Since the above equation is a sextic algebraic equation, the number of real roots does not exceed six. Therefore, there exist at most six directions of Burmester's line in the moving body in spatial motion.

6.4.4 Connecting Kinematic Pairs

The construction parameters of a line trajectory Σ_l are derived and expressed in the Frenet frame as equation (6.117), indicating the connections between the kinematics of $\sum^*$ and the geometry of the line trajectory. Based on the discussion in Section 3.2, Σ_l will be a constraint ruled surface Σ_g if they have identical construction parameters. The line L of the moving body $\sum^*$ can be designated as an axis of the kinematic pair, and coincides with the characteristic line L_g of the end of an open kinematic chain, or a binary link, and the fixed line of the constraint ruled surface is the axis of the fixed joint of the binary link. Actually, the trajectory Σ_l of the lines in spatial movement may be close to several kinds of regular constraint ruled surface Σ_g in terms of construction parameters at an instant. It may be a good choice if Σ_l and Σ_g have higher-order contact along a line at an instant. Based on the higher-order contact conditions discussed in Chapter 3, a constraint surface can be constructed by means of the Frenet frame and invariants of Σ_l, and the line of $\sum^*$ is visualized as an axis of the kinematic pair, just like the Burmester point in planar kinematics is designated as the center of the moving joint R of a binary R–R.

Herein, the moving joint of a binary link, connected with $\sum^*$, is first designated as a kinematic pair C since it needs no more additional conditions for Σ_l and Σ_g contacting in higher order. It includes three pairs H, R, and P, corresponding to an identical constraint ruled surface. We may prefer to use the kinematic pair with simpler structure and fewer DOFs if it is good enough both in terms of geometrical constraint and kinematic constraint. According to the definitions of the kinematic pairs C, H, R, and P, we can identify the different relative motions between $\sum^*$ and the binary link along the characteristic line. The condition for a binary link to connect $\sum^*$ with the R-pair is that the relative sliding velocity is zero, with the P-pair is that the relative rotating angular velocity is zero, while with the H-pair is that both the relative sliding velocity and relative angular velocity are in a linear relationship. All of the above conditions are called the ***kinematic conditions for a line of a moving body to be a characteristic line***, or the ***connecting conditions of the kinematic pairs***.

A line trajectory Σ_l has Frenet frame $\{\boldsymbol{\rho}_l; \boldsymbol{E}_1, \boldsymbol{E}_2, \boldsymbol{E}_3\}$ and arc length σ_l of the spherical image curve of $\boldsymbol{l}$, as shown in Fig. 6.9(a), and a constraint ruled surface Σ_g of a binary link has Frenet frame $\{\boldsymbol{\rho}_g; \boldsymbol{E}_{1g}, \boldsymbol{E}_{2g}, \boldsymbol{E}_{3g}\}$ and arc length σ_g of the spherical image curve of $\boldsymbol{l}_g$, as shown in Fig. 6.9(b). When Σ_l contacts Σ_g in second order or higher, they have common generatrices at four infinitesimal positions, or the line L of $\sum^*$ is instantaneously coincident with the line L_g of the binary link. To identify the relative motion between the line L and the line L_g, we first set the reference point B_l fixed on L and point B_g on L_g respectively. The two points coincide with each other at the original instant, and then the relative motion of the two points is adopted to describe the relative translation of the two lines.

Based on the equation of a ruled surface in Chapter 3, the position vector of the reference point B_g on L_g is

$$\boldsymbol{R}_{Bg} = \boldsymbol{R}_P + u_{Bg}\boldsymbol{l}_g = \boldsymbol{\rho}_g + (u_{Bg} - b_{lg})\boldsymbol{l}_g \tag{6.125}$$

where u_{Bg} is the distance from a fixed point P to B_g on the generator (characteristic line L_g) of Σ_g. The velocity of B_g, the derivative of equation (6.125) with respect to the time t, can be derived as

$$\boldsymbol{V}_{Bg} = \frac{\mathrm{d}\boldsymbol{R}_{Bg}}{\mathrm{d}t} = \frac{\mathrm{d}\sigma_g}{\mathrm{d}t}\left[\left(\alpha_g - \frac{\mathrm{d}b_{lg}}{\mathrm{d}\sigma_g}\right)\boldsymbol{E}_{g1} + (u_{Bg} - b_{lg})\boldsymbol{E}_{g2} + \gamma_g \boldsymbol{E}_{g3}\right] \tag{6.126}$$

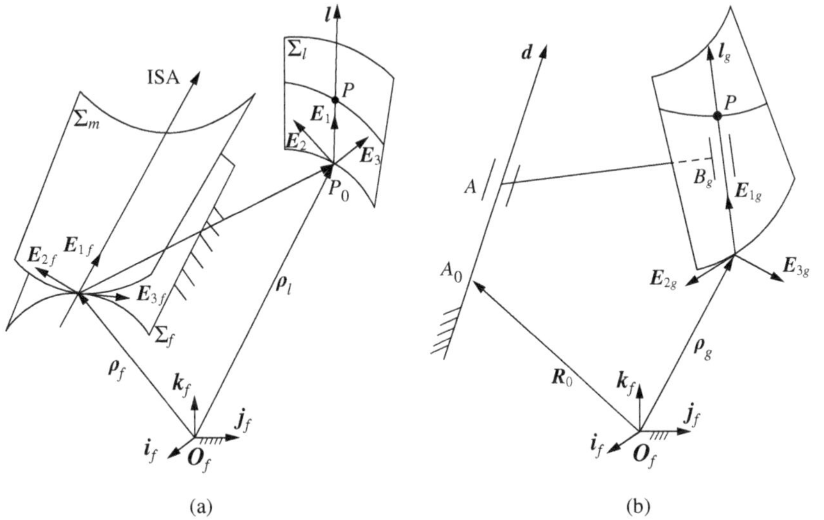

Figure 6.9 Kinematic connecting conditions of a moving joint (a) A line trajectory (b) A constraint ruled surface of a binary link

where α_g and γ_g are the construction parameters of Σ_g. Since points B_g and P are all fixed on L_g, the distance u_{Bg} is constant, or $\frac{du_{Bg}}{d\sigma_g} = 0$. As mentioned in Chapter 3, a binary link rotates around the fixed axis of the fixed joint and the angular velocity vector is $\boldsymbol{\omega}_g = \omega_g \boldsymbol{d}$, where $\boldsymbol{d}$ is the unit vector of the fixed axis and ω_g is the angular velocity, or $\omega_g = \frac{d\varphi}{dt} = \frac{d\sigma_g}{dt \cdot \sin\delta_g}$. The angular velocity vector $\boldsymbol{\omega}_g$ can be expressed in the Frenet frame $\{\boldsymbol{\rho}_g; \boldsymbol{E}_{g1}, \boldsymbol{E}_{g2}, \boldsymbol{E}_{g3}\}$ of Σ_g, or

$$\boldsymbol{\omega}_g = \omega_g \boldsymbol{d} = \frac{d\sigma_g}{dt} \cot\delta_g \cdot \boldsymbol{E}_{g1} + \frac{d\sigma_g}{dt} \boldsymbol{E}_{g3} \tag{6.127}$$

On the contrary, the line L passing through the point P of $\sum^*$ traces a line trajectory Σ_l, whose vector equation is described by equations (6.91) and (6.97), or $\boldsymbol{R}_l = \boldsymbol{R}_P + u_l \boldsymbol{l} = \boldsymbol{\rho}_l + (u_l - b_l)\boldsymbol{l}$. The reference point B_l with distance u_{Bl} to the fixed point P traces a curve Γ_{Bl} on the line trajectory Σ_l, and the vector equation of Γ_{Bl} is

$$\boldsymbol{R}_{Bl} = \boldsymbol{R}_P + u_{Bl} \boldsymbol{l} = \boldsymbol{\rho}_l + (u_{Bl} - b_l)\boldsymbol{l} \tag{6.128}$$

Since points B_l and P are all fixed on L, the distance u_{Bl} is constant, or $\frac{du_{Bl}}{d\sigma_l} = 0$. By the Frenet formulas of a ruled surface, the derivative of equation (6.128) with respect to time t, or the velocity vector of B_l, is

$$\boldsymbol{V}_{Bl} = \frac{d\boldsymbol{R}_{Bl}}{dt} = \frac{d\sigma_l}{dt} \left[\left(\alpha_l - \frac{db_l}{d\sigma_l} \right) \boldsymbol{E}_1 + (u_{Bl} - b_l)\boldsymbol{E}_2 + \gamma_l \boldsymbol{E}_3 \right] \tag{6.129}$$

At an instant, $\sum^*$ makes helical motion about the ISA. The instantaneous angular velocity vector $\boldsymbol{\omega}_l = \omega_l \boldsymbol{E}_{1f}$ can be described by the Frenet frame $\{\boldsymbol{\rho}_l; \boldsymbol{E}_1, \boldsymbol{E}_2, \boldsymbol{E}_3\}$ of Σ_l due to the equation $\boldsymbol{E}_{1f} = l_1\boldsymbol{E}_1 + \sqrt{1 - l_1^2}\boldsymbol{E}_3$. Combining equations (6.62) with (6.96), we have $\omega_l = \beta^* \frac{d\sigma}{dt} = \frac{1}{\sqrt{1-l_1^2}} \frac{d\sigma_l}{dt}$. The angular velocity vector has another form

$$\boldsymbol{\omega}_l = \omega_l \boldsymbol{E}_{1f} = \frac{d\sigma_l}{dt} \frac{l_1}{\sqrt{1 - l_1^2}} \boldsymbol{E}_1 + \frac{d\sigma_l}{dt} \boldsymbol{E}_3 \tag{6.130}$$

According to Theorem 3.8 and equation (3.190), when Σ_l contacts Σ_g to second order or higher, we have $\boldsymbol{E}_{ig} = \boldsymbol{E}_i$, $i = 1, 2, 3$ and $d\sigma_g = d\sigma_l$, $\alpha_g = \alpha_l$, $\beta_g = \beta_l$, $\gamma_g = \gamma_l$. The angular velocity vector of $\sum^*$ relative to the binary link (or the end link of the open chain) can be obtained by equations (6.127) and (6.130):

$$\boldsymbol{\omega} = \boldsymbol{\omega}_l - \boldsymbol{\omega}_g = \frac{d\sigma_l}{dt} \left(\frac{l_1}{\sqrt{1 - l_1^2}} - \beta_l \right) \boldsymbol{E}_1 = \frac{d\sigma}{dt}(l_1 - \beta_l \sqrt{1 - l_1^2})\beta^* \boldsymbol{E}_1 = \Delta_\omega \boldsymbol{E}_1 \tag{6.131}$$

where $\Delta_\omega = \frac{d\sigma_l}{dt} \left(\frac{l_1}{\sqrt{1-l_1^2}} - \beta_l \right) = \frac{d\sigma}{dt}(l_1 - \beta_l \sqrt{1 - l_1^2})\beta^* = -\frac{l_3}{1-l_1^2} \frac{d\sigma}{dt}$.

Letting the reference point B_l of L coincide with B_g of L_g, or $u_{Bl} = u_{Bg}$, $b_l = b_{lg}$, the relative velocity vector of the two points can be obtained by equations (6.126) and (6.129):

$$\boldsymbol{V}_{Bl} - \boldsymbol{V}_{Bg} = \frac{d\sigma_l}{dt} \left(-\frac{db_l}{d\sigma_l} + \frac{db_{lg}}{d\sigma_l} \right) \boldsymbol{E}_1 = \Delta_V \boldsymbol{E}_1 \tag{6.132}$$

where $\Delta_V = \left(-\frac{db_l}{d\sigma_l} + \frac{db_{lg}}{d\sigma_l} \right) \frac{d\sigma_l}{dt}$.

In particular, the striction curve of Σ_g is a point trajectory traced by a point of a binary link, or $\frac{db_{lg}}{d\sigma_l} = 0$ if Σ_g is a constant axis ruled surface.

Therefore, the connecting conditions of the kinematic pairs R, P, H, and C are summarized as follows.

1. R-*pair*. If Σ_l contacts Σ_g to second order, the line L of $\sum^*$ coincides with the line L_g of a binary link at three infinitesimal positions. $\sum^*$ and the binary link (the end link of an open chain) do not have relative motion along the direction vector $\boldsymbol{l}(\boldsymbol{l}_g)$, or $\Delta_V = 0$ and $\frac{d\Delta_V}{dt} = 0$. Then, the ***connecting conditions of the* R-*pair*** are

$$\begin{cases} -\dfrac{db_l}{d\sigma_l} + \dfrac{db_{lg}}{d\sigma_l} = 0 \\ -\dfrac{d^2 b_l}{d\sigma_l^2} + \dfrac{d^2 b_{lg}}{d\sigma_l^2} = 0 \end{cases} \tag{6.133}$$

If the constraint ruled surface Σ_g is a constant axis ruled surface, or $\frac{db_{lg}}{d\sigma_l} = 0$ and $\frac{d^2 b_{lg}}{d\sigma_l^2} = 0$, equation (6.133) can be simplified into $\frac{db_l}{d\sigma_l} = 0$ and $\frac{d^2 b_l}{d\sigma_l^2} = 0$. Combining equation (6.116), we have

$$\begin{cases} \dfrac{\mathrm{d}b_l}{\mathrm{d}\sigma} = (l_2 p - l_1 l_3 H + l_3 \gamma)/(1 - l_1^2) = 0 \\ \dfrac{\mathrm{d}^2 b_l}{\mathrm{d}\sigma^2} = \Big[(\beta_m l_3 - 2l_1)p + (\beta_m l_1 - 2l_3) l_2 H \\ \qquad + l_3 \dfrac{\mathrm{d}\gamma}{\mathrm{d}\sigma} - l_2\left(\beta_m \gamma + \alpha_m\right) + 4 l_1 l_2 \dfrac{\mathrm{d}b_l}{\mathrm{d}\sigma}\Big] / (1 - l_1^2) = 0 \end{cases} \tag{6.134}$$

When Σ_l has third-order contact with Σ_g, the line L coincides with L_g at four infinitesimal positions, and the connecting conditions of the R-pair are $\Delta_V = 0$, $\frac{\mathrm{d}\Delta_V}{\mathrm{d}t} = 0$, and $\frac{\mathrm{d}^2\Delta_V}{\mathrm{d}t^2} = 0$.

2. P-*pair*. If Σ_l contacts Σ_g to second order, $\sum^*$ and the binary link (the end link of an open chain) do not have relative rotation around the direction vector $\boldsymbol{l}(\boldsymbol{l}_g)$, or $\Delta_\omega = 0$ and $\frac{\mathrm{d}\Delta_\omega}{\mathrm{d}t} = 0$. According to equation (6.131), the ***connecting conditions of the* P-*pair*** are

$$l_3 = 0, \ \beta_m = 0 \tag{6.135}$$

In this case, the spatial movement of $\sum^*$ degenerates to a simpler form, so it is not discussed in this book.

3. H-*pair*. If Σ_l contacts Σ_g to second order, $\sum^*$ and the end link have relative rotation around the line $L(L_g)$ and relative motion along the line $L(L_g)$, but the relative angular and linear displacements are linearly related to each other. The ***connecting condition of the* H-*pair*** is $\frac{\Delta_V}{\Delta_\omega} = \frac{1-l_1{}^2}{l_3}\left(\frac{\mathrm{d}b_l}{\mathrm{d}\sigma} - \frac{\mathrm{d}b_{lg}}{\mathrm{d}\sigma}\right) = \text{const.}$ or $\frac{\mathrm{d}}{\mathrm{d}t}\left(\frac{\Delta_V}{\Delta_\omega}\right) = 0$, that is

$$\frac{1 - l_1{}^2}{l_3}\left[\frac{\mathrm{d}^2 b_l}{\mathrm{d}\sigma^2} - \frac{\mathrm{d}^2 b_{lg}}{\mathrm{d}\sigma_l{}^2}\left(\frac{\mathrm{d}\sigma_l}{\mathrm{d}\sigma}\right)^2 - \frac{\mathrm{d}b_{lg}}{\mathrm{d}\sigma_l}\frac{\mathrm{d}^2\sigma_l}{\mathrm{d}\sigma^2}\right]$$
$$- \frac{2l_1 l_2 l_3 - \beta_m l_2 (1 - l_1{}^2)}{l_3{}^2}\left(\frac{\mathrm{d}b_l}{\mathrm{d}\sigma} - \frac{\mathrm{d}b_{lg}}{\mathrm{d}\sigma_l}\frac{\mathrm{d}\sigma_l}{\mathrm{d}\sigma}\right) = 0 \tag{6.136}$$

Particularly, when Σ_g is a constant axis ruled surface, we have $\frac{\mathrm{d}b_{lg}}{\mathrm{d}\sigma_l} = 0$ and $\frac{\mathrm{d}^2 b_{lg}}{\mathrm{d}\sigma_l^2} = 0$. Then, equation (6.136) can be simplified into

$$\frac{1 - l_1{}^2}{l_3}\frac{\mathrm{d}^2 b_l}{\mathrm{d}\sigma^2} - \frac{2l_1 l_2 l_3 - \beta_m l_2 (1 - l_1{}^2)}{l_3{}^2}\frac{\mathrm{d}b_l}{\mathrm{d}\sigma} = 0 \tag{6.137}$$

If Σ_l has third-order contact with Σ_g, the lines L and L_g coincide with each other at four infinitesimal positions, and the connecting conditions of the H-pair are $\frac{\mathrm{d}}{\mathrm{d}t}\left(\frac{\Delta_V}{\Delta_\omega}\right) = 0$ and $\frac{\mathrm{d}^2}{\mathrm{d}t^2}\left(\frac{\Delta_V}{\Delta_\omega}\right) = 0$.

4. C-*pair*. For higher-order contact between Σ_l and Σ_g, if Δ_V and Δ_ω have an arbitrary functional relationship, the C-pair is adopted to connect the moving body and the end link, which is equivalent to a helical pair with arbitrary leads.

6.4.5 Constant Axis Curvature

To reveal the geometrical properties of line trajectories in spatial movement, the generalized curvature theory of a line trajectory will be discussed by analogy to the curvature theory in both planar and spherical kinematics, such as the Euler–Savary equation. In particular, based on the constant inclination curvature and the constant axis curvature introduced in chapter 3, the relationships among the generalized curvature surface, the tracing line and the axodes (ISA) of the moving body are studied, which is a new topic in the spatial kinematic geometry literature.

6.4.5.1 Constant Inclination Curvature

A line trajectory Σ_l contacts a constant inclination curvature surface $\sum_{C'P'C}$ at the common generatrix L in second order. Similar to the curvature circle of a planar curve at three consecutive points, three consecutive generatrices of Σ_l correspond to a constant inclination curvature surface. Based on equation (3.104), the unit direction vector $\boldsymbol{d}$ of the fixed axis L_A of the ***constant inclination curvature surface*** can be represented by the Frenet frame $\{\boldsymbol{\rho}_l; \boldsymbol{E}_1, \boldsymbol{E}_2, \boldsymbol{E}_3\}$ of Σ_l, or

$$\begin{cases} \boldsymbol{d} = \cos\delta \boldsymbol{E}_1 + \sin\delta \boldsymbol{E}_3 \\ \cot\delta = \beta_l = \dfrac{l_3 + \beta^* l_1(1 - l_1^2)}{\beta^*(1 - l_1^2)^{3/2}} = \dfrac{\sin\theta_l + \beta^* \cos\delta_l \sin\delta_l}{\beta^* \sin^2\delta_l} \end{cases} \tag{6.138}$$

The second expression of the above equation can be rewritten as

$$\cot(\delta - \delta_l) + \cot\delta_l = -\frac{\beta^*}{\sin\theta_l} \tag{6.139}$$

The above equation is the ***Euler–Savary analogue of directions of a line trajectory***, or the ***Euler–Savary analogue of the spherical image curve of a line trajectory***, which is the same as that of a point trajectory in spherical movement. It implies that the spherical image of the unit direction vector of a line in spatial movement corresponds to a point in spherical movement.

If Σ_l has third-order contact with the constant inclination curvature surface $\sum_{C'P'C}$, the invariant β_l should meet $\frac{\mathrm{d}\beta_l}{\mathrm{d}\sigma_l} = 0$. Furthermore, if Σ_l has fourth-order contact with $\sum_{C'P'C}$, β_l should meet $\frac{\mathrm{d}\beta_l}{\mathrm{d}\sigma_l} = \frac{\mathrm{d}^2\beta_l}{\mathrm{d}\sigma_l^2} = 0$ simultaneously. In this case, there exist a few directions in the moving body, referred to as the Burmester directions for the spatial movement, whose conditions have been discussed in Section 6.4.3.

The constant inclination curvature surface of Σ_l corresponds to a dyad or an open kinematic chain with three kinematic pairs C′–P′–C, and the line L of $\sum^*$ is the characteristic line $L_{C'P'C}$. In particular, two C-pairs of the open kinematic chain can be replaced by H, R respectively, and the constant inclination curvature surface $\sum_{C'P'C}$ will degenerate to different constraint ruled surfaces, such as $\sum_{RP'C}$, $\sum_{HP'C}$, $\sum_{RP'H}$, $\sum_{RP'R}$, $\sum_{HP'R}$, $\sum_{HP'H}$, etc., while the characteristic line $L_{C'P'C}$ will be L_{RPC}, L_{HPC}, $L_{RP'R}$, $L_{RP'H}$, $L_{HP'R}$, and $L_{HP'H}$. The connecting conditions of the moving joints for these constraint ruled surfaces, as mentioned in Section 6.4.4, will not be discussed in this book.

6.4.5.2 Constant Axis Curvature

The ***constant axis curvature***, or the C–C curvature, as a typical generalized curvature of a line trajectory, will be discussed in this section. If the connecting conditions of the kinematic pairs are considered, the constant axis curvature can also degenerate to several different generalized curvatures including C–R, C–H, R–C, H–C, R–R, R–H, H–R, and H–H curvatures, which will be introduced in the following sections.

The Euler–Savary Analogue of a Line Trajectory

As mentioned in Theorem 3.6 in Section 3.5.2, the constant axis curvature includes β_l = const. and $\alpha_l - \beta_l\gamma_l$ = const., which reveal that the spherical image curve of Σ_l is a circle, and the striction curve of Σ_l is a cylindrical curve.

A line trajectory Σ_l contacts a constant axis curvature surface $\sum_{C'C}$ in second order along the common generatrix L, and their Frenet frames are instantaneously coincident with each other. $\sum_{C'C}$ can be located in the Frenet frame of Σ_l by the contact conditions. According to the striction point characteristics of a line trajectory in Section 6.4.1 and the property of a constant axis ruled surface $\sum_{C'C}$ in Section 3.5.2, the unit direction vector $\boldsymbol{d}$ of the fixed axis L_A of $\sum_{C'C}$ is $\boldsymbol{d} = \cos\delta\boldsymbol{E}_1 + \sin\delta\boldsymbol{E}_3$. The position of $\sum_{C'C}$ can be located by the normal of Σ_l at the striction point B, or the point A on L_A of $\sum_{C'C}$ is located on the common normal between L_A and L, as shown in Fig. 6.10. The position vector of A is

$$\begin{cases} \boldsymbol{R}_A = \boldsymbol{\rho}_l + \eta\boldsymbol{E}_2 \\ \eta = \dfrac{\alpha_l - \beta_l\gamma_l}{1+\beta_l^2} \end{cases} \tag{6.140}$$

where η is the distance between points A and B along the common normal, or the radius of the cylinder located by the striction curve of $\sum_{C'C}$. The positive or negative of the value of η determines whether the fixed axis L_A locates on the forward or inverse direction of the vector $\boldsymbol{E}_2$ of Σ_l at the striction point.

Substituting the construction parameters α_l, β_l, and γ_l of equation (6.117) into equation (6.140), we have

$$\begin{aligned} \eta &= \frac{l_2\beta^* p - [2l_1l_3\beta^* + \beta^{*2}(1-l_1^2)]H - (\alpha^* - \beta^*\gamma)l_3}{\beta^{*2}(1-l_1^2)^{3/2}(1+\beta_l^2)} \\ &= \frac{\beta^* p\cos\theta_l - (\alpha^* - \beta^*\gamma)\sin\theta_l - \beta^* H(2\cos\delta_l\sin\theta_l + \beta^*\sin\delta_l)}{\beta^{*2}\sin^2\delta_l(1+\beta_l^2)} \end{aligned} \tag{6.141}$$

Combining equations (6.138) with (6.141), we get

$$\frac{\eta + h}{\sin^2(\delta - \delta_l)} - \frac{h}{\sin^2\delta_l} = -\frac{\alpha^* - \beta^*\gamma}{\sin\theta_l} + \frac{\cos\theta_l}{\sin^2\theta_l}\beta^* p \tag{6.142}$$

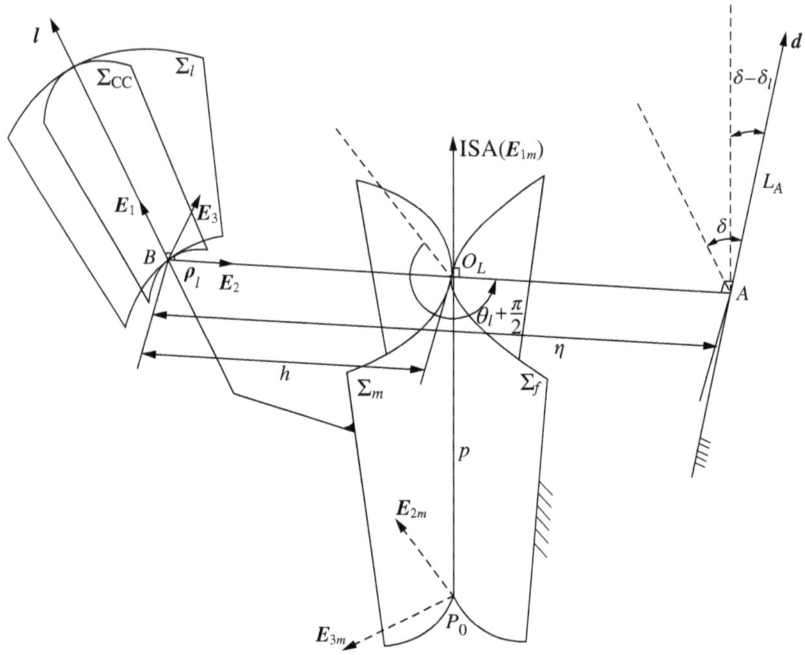

Figure 6.10 Euler–Savary analogue of a line trajectory

Based on the above equation and equation (6.139), both the direction and the position of the constant axis curvature surface $\sum_{C'C}$ of Σ_l can be expressed as

$$\begin{cases} \cot\left(\delta-\delta_l\right)+\cot\delta_l=-\dfrac{\beta^*}{\sin\theta_l} \\ \dfrac{\eta+h}{\sin^2(\delta-\delta_l)}-\dfrac{h}{\sin^2\delta_l}=-\dfrac{\alpha^*-\beta^*\gamma}{\sin\theta_l}+\dfrac{\cos\theta_l}{\sin^2\theta_l}\beta^* p \end{cases} \tag{6.143}$$

At an instant, the above equation is called the ***Euler–Savary analogue of a line trajectory*** in spatial movement, which includes two parts. The first one is the ***Euler–Savary analogue of the line directions*** of Σ_l, and it reveals the relationship among the unit direction $\boldsymbol{l}(\delta_l,\theta_l)$ of the generatrix of Σ_l, the unit direction $\boldsymbol{d}(\delta)$ of the fixed axis of $\sum_{C'C}$, and the induced construction parameter β^* of the axodes. The second one is the ***Euler–Savary analogue of the line positions*** of Σ_l, which represents the relationship among the position of a line L with parameters (h,p), the position of the fixed axis of $\sum_{C'C}$, and the construction parameters $(\alpha^*,\beta^*,\gamma)$ of the axodes.

The Euler–Savary analogue of a line trajectory in spatial movement is similar to the Euler–Savary equation (analogue) of a point trajectory in planar, spherical, and spatial movement not only in connotation but also in expression form with the curvature constant axis surface corresponding to the curvature circle (cone, sphere, cylinder, etc.). Therefore, it exposes an important fact that the properties of a line's trajectory can be described and studied by two curves, a spherical curve and a cylindrical curve, which provides a theoretical basis for the kinematic synthesis of spatial linkages. Although the striction curve of the constant axis curvature surface $\sum_{C'C}$ is a cylindrical curve, the direction of the cylindrical surface's axis depends on the spherical image curve of Σ_l, which is different from the osculating cylindrical surface of a point trajectory.

For some special cases, the Euler–Savary analogue of the line trajectory may degenerate to different expression forms, which are introduced in the following sections.

The Directional Singularity

At an instant, if the line L of $\sum^*$ is parallel to the ISA (the parameter δ_l is 0 or π), the direction cosines of L in $\{\boldsymbol{\rho}_m; \boldsymbol{E}_{1m}, \boldsymbol{E}_{2m}, \boldsymbol{E}_{3m}\}$ are $l_1 = 1, l_2 = l_3 = 0$. In this case, the kinematic invariant β_l tends to be infinite with the spherical image curve of the generatrices being a cusp point at this instant. The three consecutive generatrices of Σ_l are parallel to the ISA and determine an osculating cylindrical surface, whose position and radius can be obtained by the principal curvature of $\sum_l$ in the direction perpendicular to the line L.

As the spatial movement of $\sum^*$, a point P of L traces a spatial curve Γ_P on $\sum_l$, which can be taken as the directrix of $\sum_l$. The vector equation of Γ_P is expressed by equation (6.53) and the differential of arc length σ_P is $\mathrm{d}\sigma_P = \sqrt{\alpha^{*2} + \beta^{*2}(u_2^2 + u_3^2)}\mathrm{d}\sigma$. By equation (3.28), the curvature vector of Γ_P can be written as

$$k_P \boldsymbol{\beta}_P = \frac{\mathrm{d}^2 \boldsymbol{R}_P}{\mathrm{d}\sigma_P^2} = \frac{\mathrm{d}^2 \boldsymbol{R}_P}{\mathrm{d}\sigma^2}\left(\frac{\mathrm{d}\sigma}{\mathrm{d}\sigma_P}\right)^2 + \frac{\mathrm{d}\boldsymbol{R}_P}{\mathrm{d}\sigma}\frac{\mathrm{d}^2\sigma}{\mathrm{d}\sigma_P^2} \tag{6.144}$$

where $\frac{\mathrm{d}\boldsymbol{R}_P}{\mathrm{d}\sigma}$ and $\frac{\mathrm{d}^2\boldsymbol{R}_P}{\mathrm{d}\sigma^2}$ are obtained respectively from equations (6.56) and (6.57). The unit normal vector of $\sum_l$ at any point is

$$\boldsymbol{n} = \left(\frac{\mathrm{d}\boldsymbol{R}_P}{\mathrm{d}\sigma_P} \times \boldsymbol{l}\right) / \left|\frac{\mathrm{d}\boldsymbol{R}_P}{\mathrm{d}\sigma_P} \times \boldsymbol{l}\right| = \frac{u_2 \boldsymbol{E}_{2f} + u_3 \boldsymbol{E}_{3f}}{\sqrt{u_2^2 + u_3^2}} \tag{6.145}$$

Based on equation (3.29), the normal curvature of $\sum_l$ in the tangential direction of Γ_P is

$$\begin{aligned} k_{nP} &= k_P \boldsymbol{\beta}_P \cdot \boldsymbol{n} = \left(\frac{\mathrm{d}\sigma}{\mathrm{d}\sigma_P}\right)^2 \frac{\mathrm{d}^2 \boldsymbol{R}_P}{\mathrm{d}\sigma^2} \cdot \boldsymbol{n} \\ &= \frac{1}{\sqrt{u_2^2 + u_3^2}} \frac{(\alpha^* + \beta^*\gamma)u_2 - \beta^{*2}(u_2^2 + u_3^2) - \beta^* u_1 u_3}{\alpha^{*2} + \beta^{*2}(u_2^2 + u_3^2)} \end{aligned} \tag{6.146}$$

Since $\sum_l$ and the osculating cylindrical surface have second-order contact along the generatrix, the direction along the generatrix is just one of the two principal directions. The angle ϕ between the tangent line of Γ_P and the generatrix of $\sum_l$ is defined by

$$\cos\phi = \boldsymbol{\alpha}_P \cdot \boldsymbol{E}_{1f} = \frac{\mathrm{d}\sigma}{\mathrm{d}\sigma_P}\frac{\mathrm{d}\boldsymbol{R}_P}{\mathrm{d}\sigma} \cdot \boldsymbol{E}_{1f} = \frac{\alpha^*}{\sqrt{\alpha^{*2} + \beta^{*2}(u_2^2 + u_3^2)}} \tag{6.147}$$

According to Euler's formula (3.36), the normal curvature k_{n1} of $\sum_l$ in the direction perpendicular to the direction of the generatrix is

$$k_{n1} = \frac{k_{nP}}{\sin^2\phi} = \frac{(\alpha^* + \beta^*\gamma)u_2 - \beta^{*2}(u_2^2 + u_3^2) - \beta^* u_1 u_3}{\beta^{*2}(u_2^2 + u_3^2)^{3/2}} \tag{6.148}$$

The radius r_0 of the osculating cylindrical surface of $\sum_l$ is determined by the above equation, or $r_0 = \frac{1}{k_{n1}}$. In $\{O_f; i_f, j_f, k_f\}$, the axis L_C of the osculating cylindrical surface passes through the reference point C and is parallel to the generatrix L, as shown in Fig. 6.11, whose position vector is

$$\begin{aligned} R_C &= R_P + r_0 n \\ &= R_P + \frac{\beta^{*2}(u_2^2 + u_3^2)}{(\alpha^* + \beta^*\gamma)u_2 - \beta^{*2}(u_2^2 + u_3^2) - \beta^* u_1 u_3}(u_2 E_{2f} + u_3 E_{3f}) \end{aligned} \tag{6.149}$$

Since the line L is parallel to the ISA instantaneously, the coordinates of P in $\{\rho_m; E_{1m}, E_{2m}, E_{3m}\}$ can be supposed to be

$$u_1 = 0,\ u_2 = r\cos\psi,\ u_3 = r\sin\psi \tag{6.150}$$

Substituting the above equation into equation (6.149), we have

$$R_C = R_P + r_0 n = R_P + r_0 \cos\psi E_{2f} + r_0 \sin\psi E_{3f} \tag{6.151}$$

where the radius r_0 of the osculating cylindrical surface of $\sum_l$ can be derived from equation (6.148) as

$$r_0 = \frac{\beta^{*2} r^2}{(\alpha^* + \beta^*\gamma)\cos\psi - \beta^{*2} r} \tag{6.152}$$

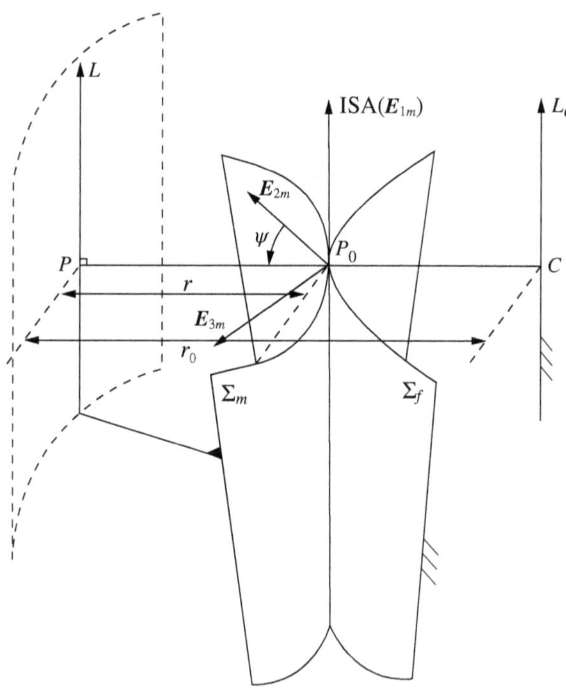

Figure 6.11 Euler–Savary analogue of a line trajectory for the line parallel to the ISA

The above equation takes another form:

$$\frac{1}{r} - \frac{1}{r_0 + r} = \frac{\beta^{*2}}{(\alpha^* + \beta^*\gamma)\cos\psi} \tag{6.153}$$

The above equation can be referred to as the ***Euler–Savary analogue of a cylindrical trajectory*** in the special case where the generatrix is parallel to the ISA instantaneously. When the radius r_0 tends to infinity, the above equation can be written as

$$r - \frac{\alpha^* + \beta^*\gamma}{\beta^{*2}}\cos\psi = 0 \tag{6.154}$$

The equation defines a cylindrical surface, whose any generatrix keeps parallel and locates on a plane at three infinitesimal positions. When the parameter r is zero, the line is just the ISA, and the radius of the osculating cylindrical surface degenerates to zero, which means the ISA keeps fixed at three infinitesimal positions.

The Inflection Line

Equation (6.141) can be rewritten as

$$\eta = -\frac{2\beta^* \cot\delta_l \sin\theta_l + \beta^{*2}}{\sin^2\theta_l/\sin^2\delta_l + 2\beta^* \cot\delta_l \sin\theta_l + \beta^{*2}}\left[h - \frac{\beta^* \cos\theta_l p - (\alpha^* - \beta^*\gamma)\sin\theta_l}{2\beta^* \cot\delta_l \sin\theta_l + \beta^{*2}}\right] \tag{6.155}$$

When η is equal to zero, we have

$$h - \frac{\beta^* \cos\theta_l p - (\alpha^* - \beta^*\gamma)\sin\theta_l}{2\beta^* \cot\delta_l \sin\theta_l + \beta^{*2}} = 0 \tag{6.156}$$

Given the angle δ between the line L and the fixed line L_A, the parameter δ_l for any θ_l can be solved from equation (6.143). Equation (6.156) defines a line on the plane determined by two vectors $\boldsymbol{E}_2$ and $\boldsymbol{F}_{1m}(\boldsymbol{F}_{1f})$. As the parameter θ_l varies from 0 to 2π, a corresponding ruled surface can be obtained, any point of which corresponds to a line L of $\sum^*$ with parameters $(p, h, \theta_l, \delta_l)$, and the line L intersects a fixed line with inclination angle δ at three infinitesimal positions.

In particular, if $\delta = \pi/2$, or $\beta_l = \cot\delta = 0$, through Equation (6.138), we have

$$\sin\theta_l + \beta^* \cos\delta_l \sin\delta_l = 0 \tag{6.157a}$$

or

$$\sin^2\delta_l = \frac{\beta^{*2} \pm \sqrt{\beta^{*4} - 4\beta^{*2}\sin^2\theta_l}}{2\beta^{*2}} \tag{6.157b}$$

For a line L of $\sum^*$, if its line trajectory $\sum_l$ holds the condition equations $\eta = 0$ and $\beta_l = 0$ true, the line L orthogonally intersects with a fixed line at three infinitesimal positions, and is called an ***inflection line***, similar to an inflection point in planar kinematics. At an instant, there exist ∞^2 such lines in $\sum^*$, called the ***inflection line congruence***.

In this case, equation (6.155) can be simplified as

$$\eta = -(1 - \cot^2\delta_l)\left[h - \frac{\beta^* p \cos\theta_l - (\alpha^* - \beta^*\gamma)\sin\theta_l}{\beta^{*2}(\sin^2\delta_l - \cos^2\delta_l)}\right] \tag{6.158}$$

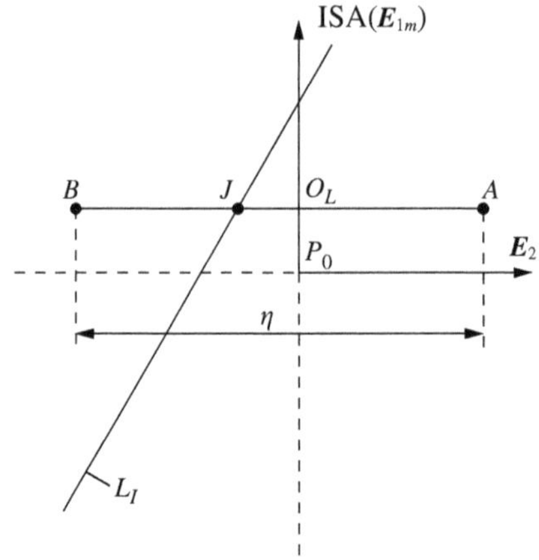

Figure 6.12 Position relationship among the tracing line, fixed axis and the inflection line

As shown in Fig. 6.12, for a line L with parameters (θ_l, δ_l) in the Frenet frame $\{\boldsymbol{\rho}_m; \boldsymbol{E}_{1m}, \boldsymbol{E}_{2m}, \boldsymbol{E}_{3m}\}$ of $\sum_m$ (or $\sum_f$), whose parameter θ_l is given first, another parameter δ_l can be solved by equation (6.157), or the direction of the line L can be determined. On the plane comprised of $\boldsymbol{E}_2$ and $\boldsymbol{E}_{1m}(\boldsymbol{E}_{1f})$, equation (6.156) defines a line L_I. The position point J of an inflection line on L_I, the striction point B on L and the fixed point A on the fixed axis of $\sum_{C'C}$ have the following position relation

$$\overrightarrow{AB} = (1 - \cot^2\delta_l)\overrightarrow{JB} \tag{6.159}$$

The above equation indicates the position relationship among the line L, the fixed axis L_A of $\sum_{C'C}$, and the inflection line.

6.4.5.3 Stationary Line Congruence

A line L in spatial movement, whose trajectory $\sum_l$ with $\frac{d\beta_l}{d\sigma} = 0$ and $\frac{d\eta}{d\sigma} = 0$ contacts $\sum_{C'C}$ in third order, is called a ***stationary line of constant axis curvature***. Combining the conditions $\frac{d\beta_l}{d\sigma} = 0$, $\frac{d\eta}{d\sigma} = 0$ with equation (6.140), we have

$$\frac{d\beta_l}{d\sigma_l} = 0, \ \frac{d\alpha_l}{d\sigma_l} - \beta_l\frac{d\gamma_l}{d\sigma_l} = 0 \tag{6.160}$$

In the above equations, $\frac{d\beta_l}{d\sigma} = 0$ leads to equation (6.123). The direction parameters (θ_l, δ_l) of the line L have been discussed in Section 6.4.3. Equation (6.118) can be rewritten as

$$\begin{cases} \dfrac{d\alpha_l}{d\sigma_l} = \dfrac{a_{11}p + a_{12}H + a_{13}}{\beta^{*2}(1 - l_1^2)^2} \\ \dfrac{d\gamma_l}{d\sigma_l} = \dfrac{a_{21}p + a_{22}H + a_{23}}{\beta^{*}(1 - l_1^2)^{1/2}} \end{cases} \tag{6.161}$$

where the coefficients are

$$\begin{cases} a_{11} = l_3\beta_m - l_3(1-l_1^2)\beta^* - 2l_1 - l_2\dfrac{\mathrm{d}\beta^*/\mathrm{d}\sigma}{\beta^*} + \dfrac{5l_1l_2^2}{1-l_1^2} \\ a_{12} = l_1l_2\beta_m + l_1l_3\dfrac{\mathrm{d}\beta^*/\mathrm{d}\sigma}{\beta^*} - 2l_2l_3 - \dfrac{5l_1^2l_2l_3}{1-l_1^2} \\ a_{13} = l_2(\alpha^* - \alpha_m - \beta_m\gamma) + l_2(1-l_1^2)\beta^*\gamma + l_1(1-l_1^2)\beta^*\dfrac{\mathrm{d}(\alpha^*/\beta^*)}{\mathrm{d}\sigma} + l_3\beta^*\dfrac{\mathrm{d}(\gamma/\beta^*)}{\mathrm{d}\sigma} + \dfrac{5l_1l_2l_3\gamma}{1-l_1^2} \\ a_{21} = l_1l_3/(1-l_1^2) \\ a_{22} = l_2/(1-l_1^2) \\ a_{23} = \dfrac{\mathrm{d}(\alpha^*/\beta^*)}{\mathrm{d}\sigma} - \dfrac{l_1l_2\gamma}{1-l_1^2} \end{cases}$$

Substituting the above equation and equation (6.117) into equation (6.160), we get the conditions for the line trajectory $\sum_l$ to contact $\sum_{C'C}$ in third order, or the equations of the stationary line of constant axis curvature:

$$\begin{cases} \cot\delta_l = \dfrac{1}{M\sin\theta_l} + \dfrac{1}{N\cos\theta_l} \\ a_{31}p + a_{32}H + a_{33} = 0 \end{cases} \tag{6.162}$$

where

$$\begin{cases} a_{31} = a_{11} - [l_3 + \beta^* l_1(1-l_1^2)]a_{21}, a_{32} = a_{12} - [l_3 + \beta^* l_1(1-l_1^2)]a_{22} \\ a_{33} = a_{13} - [l_3 + \beta^* l_1(1-l_1^2)]a_{23} \end{cases}$$

This equation indicates that there exists line congruence in the moving body $\sum^*$, called the ***stationary line congruence of constant axis curvature***. Any line $L_{C'C}^{(3)}$ of the line congruence has a constant inclination angle and the same distance with a fixed line at four infinitesimal positions.

In particular, a line L of $\sum^*$, whose trajectory $\sum_l$ has $\beta_l = 0$, $\eta = 0$, $\frac{\mathrm{d}\beta_l}{\mathrm{d}\sigma} = 0$, and $\frac{\mathrm{d}\eta}{\mathrm{d}\sigma} = 0$ simultaneously is regarded as a ***Ball line***, like the Ball point in the planar kinematics. The Ball line intersects a fixed line orthogonally at four infinitesimal positions, and is constrained by

$$\begin{cases} \sin\theta_l + \beta^*\cos\delta_l\sin\delta_l = 0 \\ \cot\delta_l = \dfrac{1}{M\sin\theta_l} + \dfrac{1}{N\cos\theta_l} \\ a_{31}p + a_{32}H + a_{33} = 0 \\ a_{41}p + a_{42}H + a_{43} = 0 \end{cases} \tag{6.163}$$

where

$$a_{41} = l_2\beta^*,\ a_{42} = -2l_1l_3\beta^* - \beta^{*2}(1 - l_1^2),\ a_{43} = -(\alpha^* - \beta^*\gamma)l_3$$

The first two expressions of equation (6.163) determine the direction of the line L and can be combined and simplified into

$$\sin^2\delta_l(f_1\cos^8\delta_l + f_2\cos^6\delta_l + f_3\cos^4\delta_l + f_4\cos^2\delta_l + f_5) = 0 \tag{6.164}$$

where

$$\begin{cases} f_1 = -\beta^{*4} \\ f_2 = \beta^{*4} - 2\beta^{*3}/M \\ f_3 = 2\beta^{*3}/M - \beta^{*2}/N^2 - \beta^{*2}/M^2 - \beta^{*2} \\ f_4 = \beta^{*2}/N^2 + \beta^{*2}/M^2 - 2\beta^*/M \\ f_5 = -1/M^2 \end{cases}$$

Since $\sin\delta_l \neq 0$, the above equation is the eight-degree algebraic equation of $\cos\delta_l$ and can be solved to get the eight sets of direction angles. Actually, there may exist four directions of lines due to the spherical symmetry. The position of a line or the parameters (p, H) in equation (6.163) correspond to the direction angles (δ_l, θ_l) solved from equation (6.164) and first expression of equation (6.163). A Ball line is an intersection line between the inflection line congruence, equation (6.158), and the stationary line congruence of constant axis curvature, equation (6.160).

6.4.5.4 Constant Axis Line

A line L of $\sum^*$ in spatial movement, whose trajectory $\sum_l$ with $\frac{d\beta_l}{d\sigma} = 0$, $\frac{d\eta}{d\sigma} = 0$, $\frac{d^2\beta_l}{d\sigma^2} = 0$, and $\frac{d^2\eta}{d\sigma^2} = 0$, contacts $\sum_{C'C}$ in fourth order, is called the ***constant axis line***. The kinematic invariants of $\sum_l$ have to satisfy

$$\begin{cases} \dfrac{d\beta_l}{d\sigma_l} = 0,\ \dfrac{d^2\beta_l}{d\sigma_l{}^2} = 0 \\ \dfrac{d\alpha_l}{d\sigma_l} - \beta_l\dfrac{d\gamma_l}{d\sigma_l} = 0,\ \dfrac{d^2\alpha_l}{d\sigma_l{}^2} - \beta_l\dfrac{d\gamma_l{}^2}{d\sigma_l{}^2} = 0 \end{cases} \tag{6.165}$$

Combining equations (6.123), (6.124) and (6.161), we have

$$\begin{cases} \cot\delta_l = \dfrac{1}{M\sin\theta_l} + \dfrac{1}{N\cos\theta_l} \\ (\tan^2\theta_l + 1)\left[\frac{1}{M^2}(2 - \beta_m M) + \frac{1}{M^2}\left(\frac{3M}{N} + \frac{dM}{d\sigma}\right)\tan\theta_l + \frac{1}{N^2}\left(1 + \frac{dN}{d\sigma}\right)\tan^2\theta_l \right. \\ \left. + \frac{1}{N}\left(\beta_m - \frac{1}{M}\right)\tan^3\theta_l - \frac{1}{N^2}\tan^4\theta_l\right] + \tan^2\theta_l = 0 \\ a_{31}p + a_{32}H + a_{33} = 0 \\ a_{51}p + a_{52}H + a_{53} = 0 \end{cases} \tag{6.166}$$

where

$$\begin{cases} a_{51} = \dfrac{da_{31}}{d\sigma} + \dfrac{l_1 l_2 a_{31} + l_3 a_{32}}{1 - l_1^2} \\ a_{52} = \dfrac{da_{32}}{d\sigma} + \dfrac{l_1 l_2 a_{32} - l_3 a_{31}}{1 - l_1^2} \\ a_{53} = \dfrac{da_{33}}{d\sigma} + \dfrac{(l_1 l_3 a_{31} - l_2 a_{32})\gamma}{1 - l_1^2} - a_{31}\alpha_m \end{cases}$$

The first two expressions of equation (6.166) determine the direction of the characteristic line $L^{(4)}_{C'C}$, as discussed in spherical kinematics in Chapter 4. There exist at most six directions of lines in $\sum^*$, and the position of $L^{(4)}_{C'C}$ can be solved by the last two expressions of equation (6.166), or the parameters (p, H) corresponding to the direction angles (δ_l, θ_l) solved from the first two expressions of equation (6.166).

6.4.5.5 Connecting Kinematic Pairs

A line of $\sum^*$ in spatial movement, whose trajectory Σ_l contacts a constant axis surface $\sum_{C'C}$ in second order or higher, can be designated as a characteristic line $L_{C'C}$, the axis of a moving joint C, to connect the moving body $\sum^*$ with a binary link C′–C. As mentioned in Section 6.4.4, the C-pair can be replaced by an R-pair or an H-pair to connect the binary link and $\sum^*$. If $\sum^*$ and the binary link only have relative rotation or make the relative helical motion, the constraint ruled surfaces are still the constant axis surface $\sum_{C'C}$ in geometry, while the characteristic line $L_{C'C}$ will be the characteristic line $L_{C'R}$ or $L_{C'H}$ respectively.

1. *Characteristic line* $L_{C'R}$. If the C-pair of the binary link C′– C is replaced by an R-pair to connect the moving body and the binary link, the characteristic line $L_{C'C}$ becomes the characteristic line $L_{C'R}$. The striction curve of $\sum_{C'R}$, a cylindrical curve, is still traced by a point of the binary link C′–R, and we have $\frac{db_{lg}}{d\sigma_l} = 0$ and $\frac{d^2 b_{lg}}{d\sigma_l^2} = 0$. According to equation (6.134), when the line trajectory Σ_l contacts $\sum_{C'C}$ in second order, the conditions for an R-pair to connect the moving body and the binary link are

$$\frac{db_l}{d\sigma} = 0, \ \frac{d^2 b_l}{d\sigma^2} = 0 \tag{6.167a}$$

Substituting equation (6.115) into the above equations, we have

$$\begin{cases} a_{61}p + a_{62}H + a_{63} = 0 \\ a_{71}p + a_{72}H + a_{73} = 0 \end{cases} \tag{6.167b}$$

where

$$\begin{cases} a_{61} = l_2, \ a_{62} = -l_1 l_3, \ a_{63} = l_3\gamma \\ a_{71} = \beta_m l_3 - 2l_1, \ a_{72} = l_2(\beta_m l_1 - 2l_3) \\ a_{73} = l_3 d\gamma/d\sigma - l_2(\alpha_m + \beta_m\gamma) \end{cases}$$

Based on equation (6.167), we can solve the position parameters (p, H) of the characteristic line $L_{C'R}^{(2)}$:

$$\begin{cases} p = \dfrac{l_1 l_3^2 \mathrm{d}\gamma/\mathrm{d}\sigma - l_1 l_2 l_3(\alpha_m + \beta_m \gamma) + l_3\gamma(\beta_m l_1 l_2 - 2l_2 l_3)}{-\beta_m l_1(1 - l_1^2) + 2l_3(1 - l_3^2)} \\ H = \dfrac{l_2 l_3 \mathrm{d}\gamma/\mathrm{d}\sigma - l_2^2(\alpha_m + \beta_m \gamma) - l_3\gamma(\beta_m l_3 - 2l_1)}{-\beta_m l_1(1 - l_1^2) + 2l_3(1 - l_3^2)} \end{cases} \tag{6.168}$$

Therefore, the direction parameters (δ_l, θ_l) of $L_{C'R}^{(2)}$ can be given randomly, and the direction cosines (l_1,l_2,l_3) of $L_{C'R}^{(2)}$ in $\{\boldsymbol{\rho}_m; \boldsymbol{E}_{1m}, \boldsymbol{E}_{2m}, \boldsymbol{E}_{3m}\}$ can be calculated by equation (6.89). Substituting (l_1,l_2,l_3) into equation (6.168), we can locate the position of $L_{C'R}^{(2)}$ in $\{\boldsymbol{\rho}_m; \boldsymbol{E}_{1m}, \boldsymbol{E}_{2m}, \boldsymbol{E}_{3m}\}$. There exist infinite (∞^2) characteristic lines $L_{C'R}^{(2)}$ in $\sum^*$ at an instant, with each one being designated as the axis of the R-pair of a binary link C′–R at three infinitesimal positions.

Based on equation (6.160), if a line trajectory $\sum_l$ contacts $\sum_{C'C}$ in third order, the conditions for a characteristic line $L_{C'C}$ to be a characteristic line $L_{C'R}^{(3)}$ are $\frac{\mathrm{d}\beta_l}{\mathrm{d}\sigma_l} = 0$, $\frac{\mathrm{d}\alpha_l}{\mathrm{d}\sigma_l} - \beta_l \frac{\mathrm{d}\gamma_l}{\mathrm{d}\sigma_l} = 0$, $\frac{\mathrm{d}b_l}{\mathrm{d}\sigma} = 0$, $\frac{\mathrm{d}^2 b_l}{\mathrm{d}\sigma^2} = 0$, and $\frac{\mathrm{d}^3 b_l}{\mathrm{d}\sigma^3} = 0$. No such $L_{C'R}^{(3)}$ can be located in the moving body at four infinitesimal positions since four parameters of the characteristic line have to satisfy the five constraint equations.

2. *Characteristic line* $L_{C'H}$. When the line trajectory Σ_l contacts $\sum_{C'C}$ in second order, and the C-pair of the binary link C′–C is replaced by an H-pair to connect the moving body and the binary link, the characteristic line $L_{C'C}$ becomes the characteristic line $L_{C'H}$, whose constraint condition is $\frac{\mathrm{d}}{\mathrm{d}\sigma}\left(\frac{\Delta_V}{\Delta_\omega}\right) = 0$. Substituting equation (6.167) into equation (6.137), we have

$$a_{81}p + a_{82}H + a_{83} = 0 \tag{6.169}$$

where

$$\begin{cases} a_{81} = \beta_m\left(1 - l_1^2\right) - \dfrac{2l_1 l_3^3}{1 - l_1^2}, \quad a_{82} = -\dfrac{2l_2 l_3^2}{1 - l_1^2} \\ a_{83} = l_3^2 \dfrac{\mathrm{d}\gamma}{\mathrm{d}\sigma} - l_2 l_3 \alpha_m + \dfrac{2l_1 l_2 l_3^2 \gamma}{1 - l_1^2} \end{cases}$$

Therefore, the direction parameters (δ_l, θ_l) of the characteristic line $L_{C'H}^{(2)}$ can be given arbitrarily. The direction cosines (l_1,l_2,l_3) of $L_{C'H}^{(2)}$ in $\{\boldsymbol{\rho}_m; \boldsymbol{E}_{1m}, \boldsymbol{E}_{2m}, \boldsymbol{E}_{3m}\}$ can be calculated by equation (6.89) and substituted into the above equation, and the values of the coefficients a_{81}, a_{82}, and a_{83} are obtained. There exist infinite (∞^3) characteristic lines $L_{C'H}^{(2)}$ in $\sum^*$ at an instant, with each one being designated as the axis of the H-pair of binary link C′–H at three infinitesimal positions.

When Σ_l contacts $\sum_{C'C}$ in third order, the conditions for a characteristic line $L_{C'C}$ to be a characteristic line $L_{C'H}^{(3)}$ are $\frac{\mathrm{d}\beta_l}{\mathrm{d}\sigma} = 0$, $\frac{\mathrm{d}\alpha_l}{\mathrm{d}\sigma} - \beta_l \frac{\mathrm{d}\gamma_l}{\mathrm{d}\sigma} = 0$, $\frac{\mathrm{d}}{\mathrm{d}\sigma}\left(\frac{\Delta_V}{\Delta_\omega}\right) = 0$, and $\frac{\mathrm{d}^2}{\mathrm{d}\sigma^2}\left(\frac{\Delta_V}{\Delta_\omega}\right) = 0$, or

$$\begin{cases} \cot\delta_l = \dfrac{1}{M\sin\theta_l} + \dfrac{1}{N\cos\theta_l} \\ a_{31}p + a_{32}H + a_{33} = 0 \\ a_{81}p + a_{82}H + a_{83} = 0 \\ a_{91}p + a_{92}H + a_{93} = 0 \end{cases} \tag{6.170}$$

where

$$\begin{cases} a_{91} = \dfrac{\mathrm{d}a_{81}}{\mathrm{d}\sigma} + \dfrac{l_1 l_2 a_{81} + l_3 a_{82}}{1 - l_1^2}, \quad a_{92} = \dfrac{\mathrm{d}a_{82}}{\mathrm{d}\sigma} + \dfrac{l_1 l_2 a_{82} - l_3 a_{81}}{1 - l_1^2} \\ a_{93} = -a_{81}\alpha_m + \dfrac{l_1 l_3 a_{81}\gamma - l_2 a_{82}\gamma}{1 - l_1^2} + \dfrac{\mathrm{d}a_{83}}{\mathrm{d}\sigma} \end{cases}$$

In order that Equation (6.170) has solutions, the determinant of the coefficients matrix of the last three expressions in equation (6.170) should be zero, that is

$$\begin{vmatrix} a_{31} & a_{32} & a_{33} \\ a_{81} & a_{82} & a_{83} \\ a_{91} & a_{92} & a_{93} \end{vmatrix} = 0 \tag{6.171}$$

or

$$a_{31}(a_{82}a_{93} - a_{83}a_{92}) - a_{32}(a_{81}a_{93} - a_{91}a_{83}) + a_{33}(a_{81}a_{92} - a_{82}a_{91}) = 0$$

The direction parameters (δ_l, θ_l) of $L_{\mathrm{C'H}}^{(3)}$ can be solved by both equation (6.171) and the first expression of equation (6.170), and then the position parameters (p, H) can be solved by equation (6.170). There exist finite characteristic lines $L_{\mathrm{C'H}}^{(3)}$ in $\sum^*$ at an instant, with each one being designated as the axis of the H-pair of a binary link C′–H at four infinitesimal positions.

6.4.6 *Constant Parameter Curvature*

When a line trajectory Σ_l contacts a constant axis surface $\sum_{\mathrm{C'C}}$ in second order or higher, the tracing line of $\sum^*$ in spatial movement can be designated as a characteristic line $L_{\mathrm{C'C}}$ or the axis of the moving joint C, while the fixed axis of $\sum_{\mathrm{C'C}}$ is the axis of the fixed joint C, connecting a binary link C′–C and the base link. If the fixed joint C can be replaced by an R-pair or an H-pair, corresponding to the binary links R–C and H–C, the constant axis ruled surface $\sum_{\mathrm{C'C}}$ degenerates to a constant parameter surface in geometrical shape, such as a hyperboloid of one sheet or a helicoid, denoted by $\sum_{\mathrm{RC}}$ and $\sum_{\mathrm{HC}}$ respectively, and the characteristic line L_{CC} becomes L_{RC} and L_{HC}. Even more, if the moving joint C is also replaced by an R-pair or an H-pair to connect the binary link and $\sum^*$, the characteristic lines L_{RC} and L_{HC} will be L_{RR}, L_{RH}, and L_{HR}, L_{HH}, while the constraint ruled surfaces are still a hyperboloid of one sheet or a helicoid in geometrical shape, with additional kinematic constraints. It should be noted that a moving body, connected by a binary link R–C, has relative motion to the base link, and inverted relative motion if connected by a binary link C–R, which is the same for H–C and C–H, H–R and R–H.

6.4.6.1 Characteristic Line L_{RC}

When a line trajectory Σ_l contacts a constant axis surface $\sum_{C'C}$ in second order or higher, and the fixed joint C′ is replaced by an R-pair to connect the binary link C′−C to the base link, the constraint ruled surface is $\sum_{RC}$ with characteristic line L_{RC}, and the construction parameters α_l, β_l, γ_l of $\sum_{RC}$ are all constant and meet equation (1.134). Based on the equation (3.190), the conditions for a line trajectory Σ_l to contact $\sum_{RC}$ in second order are

$$\alpha_l\beta_l + \gamma_l = 0,\ \frac{d\gamma_l}{d\sigma_l} = 0 \tag{6.172}$$

Both the above equations and the constant axis curvature, $\beta_l = \text{const.}$ and $\alpha_l - \beta_l\gamma_l = \text{const.}$, are referred to as the ***hyperbolic curvature*** or **R−C *curvature***. The unit direction vector $\boldsymbol{d}$ of the fixed axis L_A of $\sum_{RC}$ is also determined by equation (6.138). The position vector of the fixed point A of L_A is

$$\begin{cases} \boldsymbol{R}_A = \boldsymbol{\rho}_l + \eta\boldsymbol{E}_2 \\ \eta = -\dfrac{\gamma_l}{\beta_l} = -\dfrac{\left(\alpha^* + \beta^* l_1 H\right)(1-l_1^2)^{3/2}}{l_3 + \beta^* l_1(1-l_1^2)} \end{cases} \tag{6.173}$$

Substituting equations (6.117) and (6.118) into equation (6.172), we have

$$\begin{cases} p = \left\{ -\dfrac{d(\alpha^*/\beta^*)}{d\sigma}[\beta_l l_1 l_3 + l_3(1-l_1^2)^{1/2}] + l_1 l_2 l_3(1-l_1^2)^{-1/2}\gamma - l_1 l_2\beta_l\alpha^* \right. \\ \left. -l_2(1-l_1^2)^{1/2}\alpha^* - l_2 l_3\beta_l\gamma \right\} /[\beta_l(l_2^2 + l_1^2 l_3^2)(1-l_1^2)^{-1} + l_1 l_3^2(1-l_1^2)^{-1/2}] \\ H = \left[-\dfrac{d(\alpha^*/\beta^*)}{d\sigma}\beta_l l_2 + l_1^2 l_3\beta_l\alpha^* + l_1 l_3(1-l_1^2)^{1/2}\alpha^* + l_1\beta_l\gamma \right] \\ \quad /[\beta_l(l_2^2 + l_1^2 l_3^2)(1-l_1^2)^{-1} + l_1 l_3^2(1-l_1^2)^{-1/2}] \end{cases} \tag{6.174}$$

For an arbitrary direction (δ_l, θ_l) of a characteristic line $L_{RC}^{(2)}$, the direction cosines (l_1, l_2, l_3) of $L_{RC}^{(2)}$ in $\{\boldsymbol{\rho}_m; \boldsymbol{E}_{1m}, \boldsymbol{E}_{2m}, \boldsymbol{E}_{3m}\}$ can be obtained through equation (6.89) and the position of the characteristic line $L_{RC}^{(2)}$, or the parameters (p, H), can be located by the above equations. There exist infinite (∞^2) $L_{RC}^{(2)}$ in $\sum^*$ corresponding to infinite $L_{C'R}^{(2)}$ (∞^2) in $\sum^*$.

For a line trajectory $\sum_l$ to contact $\sum_{C'C}$ in third order, the contact conditions can be written as $\alpha_l\beta_l + \gamma_l = 0$, $\frac{d\gamma_l}{d\sigma_l} = 0$, $\frac{d\beta_l}{d\sigma_l} = 0$, $\frac{d^2\gamma_l}{d\sigma_l^2} = 0$, and $\frac{d\alpha_l}{d\sigma_l} = 0$. They are also the constraint equations for the stationary line of R−C curvature, which may not exist generally, since four parameters of a line $L_{RC}^{(3)}$ in $\sum^*$ are constrained by five constraint equations.

6.4.6.2 Characteristic Line L_{RR}

When a line trajectory $\sum_l$ contacts $\sum_{RR}$ in second order, the moving joint C-pair can be replaced by an R-pair to connect the moving body with the binary link R−C. The constraint ruled surface is $\sum_{RR}$ and characteristic line is L_{RR}. Based on equations (6.172) and (6.134), the constraint equations of the characteristic line L_{RR} are

$$\alpha_l\beta_l + \gamma_l = 0, \ \frac{d\gamma_l}{d\sigma_l} = 0, \ \frac{db_l}{d\sigma} = 0, \ \frac{d^2 b_l}{d\sigma^2} = 0 \tag{6.175}$$

By substituting equations (6.117), (6.118), and (6.134) into the above equation, the parameters $(p, H, \theta_l, \delta_l)$ of the characteristic line $L_{RR}^{(2)}$ satisfy

$$\begin{cases} \alpha_l\beta_l + \gamma_l = 0 \\ a_{21}p + a_{22}H + a_{23} = 0 \\ a_{61}p + a_{62}H + a_{63} = 0 \\ a_{71}p + a_{72}H + a_{73} = 0 \end{cases} \tag{6.176}$$

The parameter H can be solved from the first expression in the above equation:

$$H = \frac{l_1 l_3 \alpha^* + \alpha^*\beta^*(1 - l_1^2)}{\beta^*(1 - l_1^2)l_3} = \frac{l_1}{1 - l_1^2}\frac{\alpha^*}{\beta^*} + \frac{\alpha^*}{l_3} \tag{6.177}$$

Substituting H in the above equation into the third expression of equation (6.176), we have the parameter p as

$$p = \frac{l_1^2 l_3}{l_2(1 - l_1^2)}\frac{\alpha^*}{\beta^*} + \frac{l_1\alpha^* - l_3\gamma}{l_2} \tag{6.178}$$

Substituting the parameters (p, H), equations (6.177) and (6.178) into the second and fourth expressions of equation (6.176), we have

$$\begin{cases} \dfrac{l_1^2 l_3^2 + l_2^2}{l_2}\left(\dfrac{l_1}{1 - l_1^2}\dfrac{\alpha^*}{\beta^*} + \dfrac{\alpha^*}{l_3}\right) + (1 - l_1^2)\left[\dfrac{d(\alpha^*/\beta^*)}{d\sigma} - \dfrac{l_1}{l_2}\gamma\right] = 0 \\ \dfrac{l_1^2}{l_2}\dfrac{\alpha^*\beta_m}{\beta^*} - \dfrac{2l_1 l_3(1 - l_3^2)}{l_2(1 - l_1^2)}\dfrac{\alpha^*}{\beta^*} + \dfrac{l_1(1 - l_1^2)}{l_2 l_3}\alpha^*\beta_m - \dfrac{1 - l_1^2}{l_2}\beta_m\gamma - l_2\alpha_m \\ -\dfrac{2(1 - l_3^2)}{l_2}\alpha^* + \dfrac{2l_1 l_3}{l_2}\gamma + l_3\dfrac{d\gamma}{d\sigma} = 0 \end{cases} \tag{6.179}$$

The finite solutions of the direction parameters (δ_l, θ_l) of $L_{RR}^{(2)}$ can be solved by the above transcendental equations. Then through Equations (6.177) and (6.178), the position parameters (p, H) of $L_{RR}^{(2)}$ can be obtained. There exist finite $L_{RR}^{(2)}$ in $\sum^*$ at an instant, which lie on a constraint ruled surface $\sum_{RR}$ at three infinitesimal positions.

6.4.6.3 Characteristic Line L_{RH}

A moving joint H-pair is used to connect $\sum^*$ and a binary link R–H, the constraint ruled surface is $\sum_{RH}$ and the characteristic line of $\sum^*$ is L_{RH}, whose line trajectory $\sum_l$ contacts $\sum_{RH}$ in second order. The constraint equations of the characteristic line $L_{RH}^{(2)}$ of $\sum^*$ are $\alpha_l\beta_l + \gamma_l = 0$,

$\frac{d\gamma_l}{d\sigma_l} = 0$, and $\frac{d}{d\sigma}\left(\frac{\Delta_V}{\Delta_\omega}\right) = 0$. Based on equations (6.174) and (6.169), we have

$$\begin{cases} a_{81}p + a_{82}H + a_{83} = 0 \\ p = \left\{ -\dfrac{d(\alpha^*/\beta^*)}{d\sigma}[\beta_l l_1 l_3 + l_3(1 - l_1^2)^{1/2}] + l_1 l_2 l_3 (1 - l_1^2)^{-1/2}\gamma - l_1 l_2 \beta_l \alpha^* \right. \\ \left. -l_2\left(1 - l_1^2\right)^{1/2}\alpha^* - l_2 l_3 \beta_l \gamma \right\} / [\beta_l(l_2^2 + l_1^2 l_3^2)(1 - l_1^2)^{-1} + l_1 l_3^2 (1 - l_1^2)^{-1/2}] \\ H = \left[-\dfrac{d(\alpha^*/\beta^*)}{d\sigma}\beta_l l_2 + l_1^2 l_3 \beta_l \alpha^* + l_1 l_3 (1 - l_1^2)^{1/2}\alpha^* + l_1 \beta_l \gamma \right] \\ \quad /[\beta_l(l_2^2 + l_1^2 l_3^2)(1 - l_1^2)^{-1} + l_1 l_3^2 (1 - l_1^2)^{-1/2}] \end{cases} \tag{6.180}$$

A line $L_{RH}^{(2)}$ with four parameters $(\delta_l, \theta_l, p, h)$ cannot be determined completely by the above three equations. There exist infinite (∞^1) $L_{RH}^{(2)}$ in $\sum^*$ at any instant, which lie on a constraint ruled surface $\sum_{RH}$ at three infinitesimal positions.

6.4.6.4 Characteristic Line L_{HC}

If the fixed joint pair C′ of the binary link C′–C is replaced by an H-pair, the constraint ruled surface $\sum_{C'C}$ degenerates to $\sum_{HC}$ and the characteristic line is L_{HC}. The constant parameters α_l, β_l, γ_l are all constant. Based on equation (3.190) and the properties of the helicoids, the contact condition for a line trajectory Σ_l to contact $\sum_{HC}$ in second order is

$$\frac{d\gamma_l}{d\sigma_l} = 0 \tag{6.181}$$

According to equation (6.161), the above equation can be rewritten as

$$a_{21}p + a_{22}H + a_{23} = 0 \tag{6.182}$$

A characteristic line $L_{HC}^{(2)}$ has four parameters $(\delta_l, \theta_l, p, H)$, which hold the above equation true. There exist infinite (∞^3) $L_{HC}^{(2)}$ in $\sum^*$ at an instant, with each one lying on a constraint ruled surface $\sum_{HC}$ at three infinitesimal positions, whose unit direction vector $\boldsymbol{d}$ and reference point of the fixed axis L_A are determined by equations (6.138) and (6.140). The helix parameter of $\sum_{HC}$ is

$$p_{HC} = \frac{\alpha_l \beta_l + \gamma_l}{1 + \beta_l^2} \tag{6.183}$$

If the line trajectory $\sum_l$ contacts $\sum_{HC}$ in third order, the contact conditions are $\frac{d\alpha_l}{d\sigma_l} = \frac{d\beta_l}{d\sigma_l} = \frac{d\gamma_l}{d\sigma_l} = 0$ and $\frac{d^2\gamma_l}{d\sigma_l^2} = 0$. Based on equations (6.117) and (6.118), we have

$$\begin{cases} a_{11}p + a_{12}H + a_{13} = 0 \\ a_{21}p + a_{22}H + a_{23} = 0 \\ b_{11}p + b_{12}H + b_{13} = 0 \\ \cot\delta_l = \dfrac{1}{M\sin\theta_l} + \dfrac{1}{N\cos\theta_l} \end{cases} \tag{6.184}$$

where

$$\begin{cases} b_{11} = \dfrac{2l_2l_3\left(1+l_1^2\right)}{1-l_1^2} - \beta_m l_1 l_2,\ b_{12} = \dfrac{2l_1(l_2^2-l_3^2)}{1-l_1^2} + \beta_m l_3 \\ b_{13} = \dfrac{2l_1^2l_3^2-2l_2^2}{1-l_1^2}\gamma + \dfrac{\mathrm{d}^2(\alpha^*/\beta^*)}{\mathrm{d}\sigma^2}(1-l_1^2) - \dfrac{\mathrm{d}\gamma}{\mathrm{d}\sigma}l_1l_2 - (\alpha_m+\beta_m\gamma)l_1l_3 \end{cases}$$

The condition for the above equations to have solutions is the determinant of the coefficient matrix of the first three expressions in equation (6.184) being zero:

$$\begin{vmatrix} a_{11} & a_{12} & a_{13} \\ a_{21} & a_{22} & a_{23} \\ b_{11} & b_{12} & b_{13} \end{vmatrix} = 0 \tag{6.185}$$

or

$$a_{11}(a_{22}b_{13} - b_{12}a_{23}) - a_{12}(a_{21}b_{13} - b_{11}a_{23}) + a_{13}(a_{21}b_{12} - a_{22}b_{11}) = 0$$

The finite solutions of the direction parameters (δ_l, θ_l) of $L_{\mathrm{HC}}^{(3)}$ can be calculated by both the above equation and the fourth expression of equation (6.184), and the corresponding position parameters (p, H) can also be solved by equation (6.184). There exist finite lines $L_{\mathrm{HC}}^{(3)}$ in $\sum^*$ at an instant, which lie on a constraint ruled surface $\sum_{\mathrm{HC}}$ at four infinitesimal positions.

6.4.6.5 Characteristic Line L_{HR}

If the moving joint C-pair is replaced by an R-pair to connect $\sum^*$ and the binary link H–R, we have the corresponding constraint ruled surface $\sum_{\mathrm{HR}}$ and characteristic line L_{HR}. The conditions for a line trajectory $\sum_l$ to contact $\sum_{\mathrm{HR}}$ in second order are $\frac{\mathrm{d}\gamma_l}{\mathrm{d}\sigma_l} = 0$, $\frac{\mathrm{d}b_l}{\mathrm{d}\sigma} = 0$, $\frac{\mathrm{d}^2 b_l}{\mathrm{d}\sigma^2} = 0$, which can be expanded as

$$\begin{cases} a_{21}p + a_{22}H + a_{23} = 0 \\ a_{61}p + a_{62}H + a_{63} = 0 \\ a_{71}p + a_{72}H + a_{73} = 0 \end{cases} \tag{6.186}$$

The condition for the above equations to have solutions is the determinant of the coefficient matrix being zero:

$$\begin{vmatrix} a_{21} & a_{22} & a_{23} \\ a_{61} & a_{62} & a_{63} \\ a_{71} & a_{72} & a_{73} \end{vmatrix} = 0 \tag{6.187}$$

or

$$(l_1^2l_3^2 + l_2^2)\left(l_2\alpha_m - l_3\frac{\mathrm{d}\gamma}{\mathrm{d}\sigma}\right) + l_2(1-l_1^2)^2\beta_m\gamma$$
$$+ \frac{\mathrm{d}(\alpha^*/\beta^*)}{\mathrm{d}\sigma}(1-l_1^2)[\beta_m l_1(1-l_1^2) - 2l_3(1-l_3^2)] = 0$$

There are three equations to determine the characteristic line $L_{\mathrm{HR}}^{(2)}$ with four parameters $(\delta_l, \theta_l, p, h)$, which means that there exist infinite numbers of lines (∞^1) $L_{\mathrm{HR}}^{(2)}$ in $\sum^*$ at an instant, which lie on a constraint ruled surface $\sum_{\mathrm{HR}}$ at three infinitesimal positions.

6.4.6.6 Characteristic Line L_{HH}

If the moving joint H-pair is used to connect a binary link H–H and $\sum^*$, the constraint ruled surface is $\sum_{\mathrm{HH}}$ and the characteristic line is L_{HH}. The conditions for a line trajectory $\sum_l$ to contact $\sum_{\mathrm{HH}}$ in second order are $\frac{\mathrm{d}\gamma_l}{\mathrm{d}\sigma_l} = 0$ and $\frac{\mathrm{d}}{\mathrm{d}\sigma}\left(\frac{\Delta_V}{\Delta_\omega}\right) = 0$, which can be expanded into

$$\begin{cases} a_{21}p + a_{22}H + a_{23} = 0 \\ a_{81}p + a_{82}H + a_{83} = 0 \end{cases} \tag{6.188}$$

The position parameters (p, H) of the characteristic line $L_{\mathrm{HH}}^{(2)}$ can be solved as

$$\begin{cases} p = -\dfrac{l_2 l_3^2 \mathrm{d}\gamma/\mathrm{d}\sigma - l_2^2 l_3 \alpha_m + 2 l_2 l_3^2 \mathrm{d}\left(\alpha^*/\beta^*\right)/\mathrm{d}\sigma}{\beta_m l_2 (1 - l_1^2)} \\ H = \dfrac{l_1 l_3^3 \mathrm{d}\gamma/\mathrm{d}\sigma - l_1 l_2 l_3^2 \alpha_m - [\beta_m (1 - l_1^2)^2 - 2 l_1 l_3^3] \mathrm{d}(\alpha^*/\beta^*)/\mathrm{d}\sigma + l_1 l_2 (1 - l_1^2) \beta_m \gamma}{\beta_m l_2 (1 - l_1^2)} \end{cases} \tag{6.189}$$

The direction parameters (δ_l, θ_l) of the characteristic line $L_{\mathrm{HH}}^{(2)}$ can be given arbitrarily, and there exist infinite (∞^2) numbers of lines $L_{\mathrm{HH}}^{(2)}$ in $\sum^*$ at any instant, which lie on a ruled surface $\sum_{\mathrm{HH}}$ at three infinitesimal positions.

For a moving body $\sum^*$ in spatial movement, as far as we have discussed, the instantaneous characteristic lines of $\sum^*$ can be summarized in Table 6.1.

Table 6.1 The instantaneous characteristic lines

Constraint ruled surfaces	Characteristic lines	Number of infinitesimal positions	Number of characteristic lines
$\sum_{\mathrm{CC}}$	$L_{\mathrm{CC}}^{(2)}$	3	∞^4
	$L_{\mathrm{CC}}^{(3)}$	4	∞^2
	$L_{\mathrm{CC}}^{(4)}$	5	≤ 6
$\sum_{\mathrm{CR}}$	$L_{\mathrm{CR}}^{(2)}$	3	∞^2
$\sum_{\mathrm{CH}}$	$L_{\mathrm{CH}}^{(2)}$	3	∞^3
	$L_{\mathrm{CH}}^{(3)}$	4	∞^0
$\sum_{\mathrm{RC}}$	$L_{\mathrm{RC}}^{(2)}$	3	∞^2
$\sum_{\mathrm{RR}}$	$L_{\mathrm{RR}}^{(2)}$	3	∞^0
$\sum_{\mathrm{RH}}$	$L_{\mathrm{RH}}^{(2)}$	3	∞^1
$\sum_{\mathrm{HC}}$	$L_{\mathrm{HC}}^{(2)}$	3	∞^3
	$L_{\mathrm{HC}}^{(3)}$	4	∞^0
$\sum_{\mathrm{HR}}$	$L_{\mathrm{HR}}^{(2)}$	3	∞^1
$\sum_{\mathrm{HH}}$	$L_{\mathrm{HH}}^{(2)}$	3	∞^2

6.5 Differential Kinematics of Spatial Four-Bar Linkage RCCC

The typical spatial four-bar linkage, RCCC linkage, is analyzed in this section to reveal the approach of the spatial kinematic differential geometry. A hyperboloid of one sheet, the constraint ruled surface of the binary link R–C, is taken as the original ruled surface to examine the trajectories of the points and lines of the coupler by the Frenet frame of the hyperboloid. Then, the analytical equations of the moving axode and fixed axode for the coupler link relative to the base link are derived. Based on the fixed and moving axodes, the generalized curvature theories of both a point trajectory and a line trajectory are discussed by numerical calculation.

6.5.1 Adjoint Expression

As shown in Fig. 6.13, a spatial RCCC linkage with dimension parameters $\alpha_{01} = 30°$, $\alpha_{12} = 55°$, $\alpha_{23} = 45°$, $\alpha_{30} = 60°$, $a_0 = 5$, $a_1 = 2$, $a_2 = 4$, $a_3 = 3$, $h_0 = 0$ is given in Ref. [2]. For convenience of later description, the Cartesian coordinate systems are respectively built up on different links by the Denavit–Hartenberg appointment, such as $\{\boldsymbol{R}_A; \boldsymbol{x}_0, \boldsymbol{y}_0, \boldsymbol{z}_0\}$ of the base link 0, $\{\boldsymbol{R}_B; \boldsymbol{x}_1, \boldsymbol{y}_1, \boldsymbol{z}_1\}$ of link 1, $\{\boldsymbol{R}_D; \boldsymbol{x}_3, \boldsymbol{y}_3, \boldsymbol{z}_3\}$ of link 3, and $\{\boldsymbol{R}_C; \boldsymbol{x}_2, \boldsymbol{y}_2, \boldsymbol{z}_2\}$ of the coupler link 2.

The displacement parameters of the RCCC linkage, θ_1, θ_2, θ_3, θ_0, h_1, h_2, and h_3, are solved completely in Ref. [3] by setting θ_1 as an independent variable, or the input rotation angle. The other displacement parameters, or the dependent variables, and their all-order derivatives with respect to θ_1, are calculated and listed in Appendix A.

In order to describe the spatial motion of the coupler link 2 of the linkage RCCC, the open kinematic chain, comprised of the coupler link 2, the frame link 0, and the input link 1, is taken

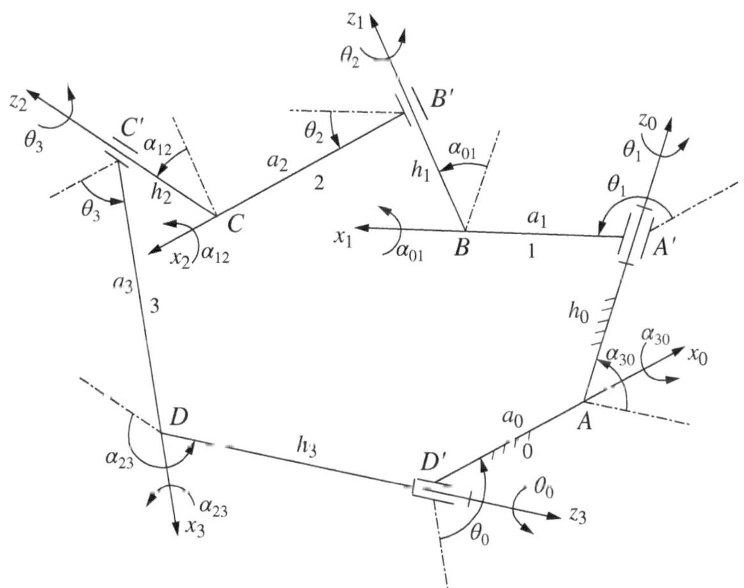

Figure 6.13 A spatial linkage RCCC

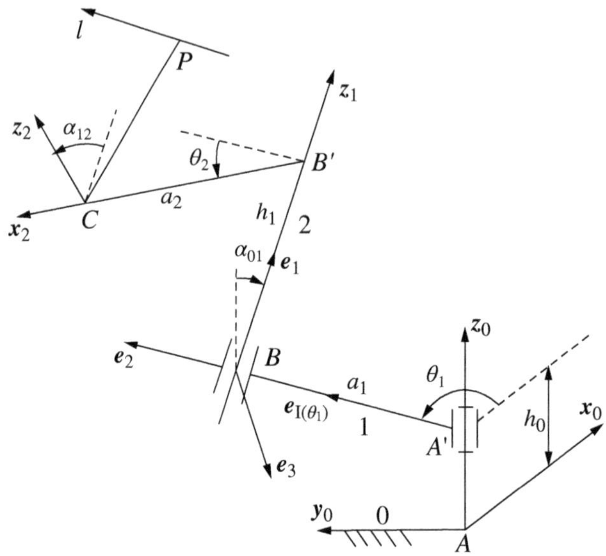

Figure 6.14 The constraint ruled surface $\sum_{\mathrm{RC}}$

as a typical binary link (R–C) as shown in Fig. 6.14. The dimension parameters are h_0, a_1, and α_{01}, and the displacement parameters θ_2 and h_1 are functions of θ_1, which are determined by the displacement solution of the RCCC linkage. The axis z_1 of the C-pair at the point B, the characteristic line, traces the constraint ruled surface $\sum_{\mathrm{RC}}$ in $\{\boldsymbol{R}_A; \boldsymbol{x}_0, \boldsymbol{y}_0, \boldsymbol{z}_0\}$, a hyperboloid of one sheet. A curve Γ_B of $\sum_{\mathrm{RC}}$, a circle, traced by point B of the binary link R–C (not the coupler point), is just the striction curve of $\sum_{\mathrm{RC}}$, whose vector equation in $\{\boldsymbol{R}_A; \boldsymbol{x}_0, \boldsymbol{y}_0, \boldsymbol{z}_0\}$ is

$$\boldsymbol{\rho}_B = h_0 \boldsymbol{z}_0 + a_1 \boldsymbol{e}_{\mathrm{I}(\theta_1)} \tag{6.190}$$

where $\boldsymbol{e}_{\mathrm{I}(\theta_1)}$ is vector function of a unit circle of θ_1 and a_1 is the length of link 1, the distance between the point A' and the point B, the two perpendicular feet of the common normal between the two axes of the R-pair and the C-pair. The unit vector $\boldsymbol{z}_1$ and the differential of the arc length σ of its spherical image curve are represented as

$$\begin{cases} \boldsymbol{z}_1 = \cos\alpha_{01}\boldsymbol{z}_0 - \sin\alpha_{01}\boldsymbol{e}_{\mathrm{II}(\theta_1)} \\ \mathrm{d}\sigma = \left|\dfrac{\mathrm{d}\boldsymbol{z}_1}{\mathrm{d}\theta_1}\right| \mathrm{d}\theta_1 = \sin\alpha_{01}\mathrm{d}\theta_1 \end{cases} \tag{6.191}$$

where α_{01} is the angle between the two axes of the R-pair and the C-pair. The vector equation of the constraint ruled surface $\sum_{\mathrm{RC}}$ is

$$\sum\nolimits_{\mathrm{RC}} : \boldsymbol{R} = \boldsymbol{\rho}_B + \lambda \boldsymbol{z}_1 \tag{6.192}$$

The Frenet frame $\{\boldsymbol{\rho}_B; \boldsymbol{e}_1, \boldsymbol{e}_2, \boldsymbol{e}_3\}$ of $\sum_{\mathrm{RC}}$ is

$$\begin{cases} \boldsymbol{e}_1 = \boldsymbol{z}_1 = \cos\alpha_{01}\boldsymbol{z}_0 - \sin\alpha_{01}\boldsymbol{e}_{\mathrm{II}(\theta_1)} \\ \boldsymbol{e}_2 = \mathrm{d}\boldsymbol{e}_1/\mathrm{d}\sigma = \boldsymbol{e}_{\mathrm{I}(\theta_1)} \\ \boldsymbol{e}_3 = \sin\alpha_{01}\boldsymbol{z}_0 + \cos\alpha_{01}\boldsymbol{e}_{\mathrm{II}(\theta_1)} \end{cases} \tag{6.193}$$

Based on the differential formulas (3.58), the construction parameters of $\sum_{RC}$ are

$$\alpha = -a_1,\ \beta = \cot\alpha_{01},\ \gamma = a_1 \cot\alpha_{01} \tag{6.194}$$

where $\sum_{RC}$ is a hyperboloid of one sheet and $\alpha,\ \beta,\ \gamma$ are all constant.

A coupler point $P(x_{Pm}, y_{Pm}, z_{Pm})$ of link 2 in $\{\boldsymbol{R}_C;\boldsymbol{x}_2,\boldsymbol{y}_2,\boldsymbol{z}_2\}$ has coordinates (u_1, u_2, u_3) in the Frenet frame $\{\boldsymbol{\rho}_B;\boldsymbol{e}_1,\boldsymbol{e}_2,\boldsymbol{e}_3\}$ and coordinates (x_{Pf}, y_{Pf}, z_{Pf}) in $\{\boldsymbol{R}_A;\boldsymbol{x}_0,\boldsymbol{y}_0,\boldsymbol{z}_0\}$, whose relationships are

$$\begin{bmatrix} u_1 \\ u_2 \\ u_3 \\ 1 \end{bmatrix} = \begin{bmatrix} 0 & \sin\alpha_{12} & \cos\alpha_{12} & h_1 \\ \cos\theta_2 & -\cos\alpha_{12}\sin\theta_2 & \sin\alpha_{12}\sin\theta_2 & a_2\cos\theta_2 \\ \sin\theta_2 & \cos\alpha_{12}\cos\theta_2 & -\sin\alpha_{12}\cos\theta_2 & a_2\sin\theta_2 \\ 0 & 0 & 0 & 1 \end{bmatrix} \begin{bmatrix} x_{Pm} \\ y_{Pm} \\ z_{Pm} \\ 1 \end{bmatrix} \tag{6.195}$$

and

$$\begin{bmatrix} x_{Pf} \\ y_{Pf} \\ z_{Pf} \\ 1 \end{bmatrix} = \begin{bmatrix} \sin\alpha_{01}\sin\theta_1 & \cos\theta_1 & -\cos\alpha_{01}\sin\theta_1 & a_1\cos\theta_1 \\ -\sin\alpha_{01}\cos\theta_1 & \sin\theta_1 & \cos\alpha_{01}\cos\theta_1 & a_1\sin\theta_1 \\ \cos\alpha_{01} & 0 & \sin\alpha_{01} & h_0 \\ 0 & 0 & 0 & 1 \end{bmatrix} \begin{bmatrix} u_1 \\ u_2 \\ u_3 \\ 1 \end{bmatrix} \tag{6.196}$$

As for the movement of the linkage RCCC with one degree of freedom, the parameters θ_2, h_1 of the binary link R–C are definite functions of θ_1, and both the position and orientation of link 2 are determined by six parameters (h_0, a_1, α_{01}, h_1, θ_2, θ_1), where h_0, a_1, α_{01} are all constant. A point P of link 2 can be examined in the Frenet frame $\{\boldsymbol{\rho}_B;\boldsymbol{e}_1,\boldsymbol{e}_2,\boldsymbol{e}_3\}$ of $\sum_{RC}$. Its trajectory $\varGamma_P$ in $\{\boldsymbol{R}_A;\boldsymbol{x}_0,\boldsymbol{y}_0,\boldsymbol{z}_0\}$ is the adjoint coupler curve of $\sum_{RC}$, while $\sum_{RC}$ is taken as an original ruled surface. The vector equation of $\varGamma_P$ is

$$\varGamma_P:\ \boldsymbol{R}_P = \boldsymbol{\rho}_B + u_1\boldsymbol{e}_1 + u_2\boldsymbol{e}_2 + u_3\boldsymbol{e}_3 \tag{6.197}$$

where (u_1, u_2, u_3), the coordinates of P in $\{\boldsymbol{\rho}_B;\boldsymbol{e}_1,\boldsymbol{e}_2,\boldsymbol{e}_3\}$, are derived by equation (6.195) as

$$\begin{cases} u_1 = h_1 + y_{Pm}\sin\alpha_{12} + z_{Pm}\cos\alpha_{12} \\ u_2 = a_2\cos\theta_2 + x_{Pm}\cos\theta_2 - y_{Pm}\cos\alpha_{12}\sin\theta_2 + z_{Pm}\sin\alpha_{12}\sin\theta_2 \\ u_3 = a_2\sin\theta_2 + x_{Pm}\sin\theta_2 + y_{Pm}\cos\alpha_{12}\cos\theta_2 - z_{Pm}\sin\alpha_{12}\cos\theta_2 \end{cases} \tag{6.198}$$

Based on equation (6.197) and the constant construction parameters $\alpha,\ \beta,\ \gamma$ of $\sum_{RC}$, the derivatives of $\varGamma_P$ with respect to its arc length σ are

$$\begin{cases} \dfrac{d\boldsymbol{R}_P}{d\sigma} = \left(\alpha + \dfrac{du_1}{d\sigma} - u_2\right)\boldsymbol{e}_1 + \left[u_1 - \left(\beta + \dfrac{d\theta_2}{d\sigma}\right)u_3\right]\boldsymbol{e}_2 + \left[\gamma + \left(\beta + \dfrac{d\theta_2}{d\sigma}\right)u_2\right]\boldsymbol{e}_3 \\ \dfrac{d^2\boldsymbol{R}_P}{d\sigma^2} = \left[\dfrac{d^2u_1}{d\sigma^2} + \left(\beta + 2\dfrac{d\theta_2}{d\sigma}\right)u_3 - u_1\right]\boldsymbol{e}_1 + \left[\alpha - \beta\gamma + 2\dfrac{du_1}{d\sigma} - u_2\right. \\ \left. -\left(\beta + \dfrac{d\theta_2}{d\sigma}\right)^2 u_2 - \dfrac{d^2\theta_2}{d\sigma^2}u_3\right]\boldsymbol{e}_2 + \left[\beta u_1 - \left(\beta + \dfrac{d\theta_2}{d\sigma}\right)^2 u_3 + \dfrac{d^2\theta_2}{d\sigma^2}u_2\right]\boldsymbol{e}_3 \end{cases} \tag{6.199}$$

By equation (6.198), the derivatives of u_1, u_2, and u_3 are

$$\frac{\mathrm{d}u_1}{\mathrm{d}\sigma} = \frac{\mathrm{d}h_1}{\mathrm{d}\sigma}, \ \frac{\mathrm{d}^2u_1}{\mathrm{d}\sigma^2} = \frac{\mathrm{d}^2h_1}{\mathrm{d}\sigma^2}, \ \frac{\mathrm{d}u_2}{\mathrm{d}\sigma} = -\frac{\mathrm{d}\theta_2}{\mathrm{d}\sigma}u_3, \ \frac{\mathrm{d}u_3}{\mathrm{d}\sigma} = \frac{\mathrm{d}\theta_2}{\mathrm{d}\sigma}u_2 \tag{6.200}$$

A line L with unit direction vector $\boldsymbol{l} = (l_{m1}, l_{m2}, l_{m3})$ in $\{\boldsymbol{R}_C; \boldsymbol{x}_2, \boldsymbol{y}_2, \boldsymbol{z}_2\}$, passing through point P of link 2, is examined in the Frenet frame $\{\boldsymbol{\rho}_B; \boldsymbol{e}_1, \boldsymbol{e}_2, \boldsymbol{e}_3\}$ of $\sum_{\mathrm{RC}}$, and its trajectory, a ruled surface $\sum_l$ in $\{\boldsymbol{R}_A; \boldsymbol{x}_0, \boldsymbol{y}_0, \boldsymbol{z}_0\}$, is an adjoint ruled surface of $\sum_{\mathrm{RC}}$, while $\sum_{\mathrm{RC}}$ is taken as an original ruled surface. The vector equation of $\sum_l$ is

$$\begin{aligned}\sum\nolimits_l : \boldsymbol{R} &= \boldsymbol{R}_P + \mu\boldsymbol{l} \\ &= \boldsymbol{\rho}_B + u_1\boldsymbol{e}_1 + u_2\boldsymbol{e}_2 + u_3\boldsymbol{e}_3 + \mu(l_1\boldsymbol{e}_1 + l_2\boldsymbol{e}_2 + l_3\boldsymbol{e}_3)\end{aligned} \tag{6.201}$$

where $\boldsymbol{l} = l_1\boldsymbol{e}_1 + l_2\boldsymbol{e}_2 + l_3\boldsymbol{e}_3$. (l_1, l_2, l_3) are the components (direction cosines) of $\boldsymbol{l}$ in $\{\boldsymbol{\rho}_B; \boldsymbol{e}_1, \boldsymbol{e}_2, \boldsymbol{e}_3\}$ and derived by equation (6.195) as

$$\begin{cases} l_1 = l_{2m}\sin\alpha_{12} + l_{3m}\cos\alpha_{12} \\ l_2 = l_{1m}\cos\theta_2 - l_{2m}\cos\alpha_{12}\sin\theta_2 + l_{3m}\sin\alpha_{12}\sin\theta_2 \\ l_3 = l_{1m}\sin\theta_2 + l_{2m}\cos\alpha_{12}\cos\theta_2 - l_{3m}\sin\alpha_{12}\cos\theta_2 \end{cases} \tag{6.202}$$

The derivatives of equation (6.201) with respect to the arc length σ can be written as

$$\begin{cases} \dfrac{\mathrm{d}\boldsymbol{l}}{\mathrm{d}\sigma} = -l_2\boldsymbol{e}_1 + \left[l_1 - \left(\dfrac{\mathrm{d}\theta_2}{\mathrm{d}\sigma} + \beta\right)l_3\right]\boldsymbol{e}_2 + \left(\dfrac{\mathrm{d}\theta_2}{\mathrm{d}\sigma} + \beta\right)l_2\boldsymbol{e}_3 \\ \dfrac{\mathrm{d}^2\boldsymbol{l}}{\mathrm{d}\sigma^2} = \left[-l_1 + \left(2\dfrac{\mathrm{d}\theta_2}{\mathrm{d}\sigma} + \beta\right)l_3\right]\boldsymbol{e}_1 - \left[l_2 + \dfrac{\mathrm{d}^2\theta_2}{\mathrm{d}\sigma^2}l_3 + \left(\dfrac{\mathrm{d}\theta_2}{\mathrm{d}\sigma} + \beta\right)^2 l_2\right]\boldsymbol{e}_2 \\ \quad + \left[\beta l_1 + \dfrac{\mathrm{d}^2\theta_2}{\mathrm{d}\sigma^2}l_2 - \left(\dfrac{\mathrm{d}\theta_2}{\mathrm{d}\sigma} + \beta\right)^2 l_3\right]\boldsymbol{e}_3 \end{cases} \tag{6.203}$$

According to equation (6.202), the derivatives of l_1, l_2, and l_3 are

$$\frac{\mathrm{d}l_1}{\mathrm{d}\sigma} = 0, \ \frac{\mathrm{d}l_2}{\mathrm{d}\sigma} = -\frac{\mathrm{d}\theta_2}{\mathrm{d}\sigma}l_3, \ \frac{\mathrm{d}l_3}{\mathrm{d}\sigma} = \frac{\mathrm{d}\theta_2}{\mathrm{d}\sigma}l_2 \tag{6.204}$$

Thus, the geometrical properties of both Γ_P and $\sum_l$ can be calculated and analyzed by equations (6.199), (6.200) and equations (6.203) and (6.204).

6.5.2 Axodes

In Section 6.2, the vector equations of the ISA of $\sum^*$ in general spatial movement are derived as equations (6.23) and (6.24) and described by the Frenet frame of the original ruled surface $\sum_k$. For the spatial RCCC four-bar linkage, the coupler link 2 corresponds to the moving body $\sum^*$, and the axis z_1 of the C-pair of link 2 corresponds to the reference line L_k of $\sum^*$. Hence, the vector equations of the ISA of link 2 can be written in the same way, or the reference line L_k of $\sum^*$ is replaced by the axis z_1 of the C-pair of link 2, and then the original ruled surface

$\sum_k$ is replaced by the constraint ruled surface $\sum_{RC}$, a hyperboloid of one sheet. The variables in equations (6.23) and (6.24) can be expressed by the parameters of the binary link R–C as

$$\begin{cases} \alpha_k = -a_1, \ \beta_k = \cot\alpha_{01}, \ \gamma_k = a_1 \ \cot\alpha_{01} \\ d\sigma_k = \sin\alpha_{01} \cdot d\theta_1, \ b_k = -h_1, \ \theta = -\theta_2 \end{cases} \tag{6.205}$$

According to equations (6.23) and (6.24), the ISA of link 2 with reference point Q and unit direction vector $\boldsymbol{s}$ are represented by

$$\begin{cases} \boldsymbol{R}_Q = \boldsymbol{\rho}_B + u_2\boldsymbol{e}_2, \ \boldsymbol{s} = s_1\boldsymbol{e}_1 + s_3\boldsymbol{e}_3 \\ s_1 = \left(\cot\alpha_{01} + \dfrac{d\theta_2}{d\theta_1}/\sin\alpha_{01}\right)/\omega_{s0}, s_3 = 1/\omega_{s0} \\ u_2 = \left[-a_1 + \dfrac{dh_1}{d\theta_1}/\sin\alpha_{01} - a_1 \ \cot\alpha_{01}\left(\cot\alpha_{01} + \dfrac{d\theta_2}{d\theta_1}/\sin\alpha_{01}\right)\right]/\omega_{s0}^2 \\ \omega_{s0} = \left[1 + \left(\cot\alpha_{01} + \dfrac{d\theta_2}{d\theta_1}/\sin\alpha_{01}\right)^2\right]^{1/2} \end{cases} \tag{6.206}$$

θ_2 and h_1 are the functions of θ_1, which are given by equations (A.4b) and (A.6) in Appendix A. We can get the functions of the position vector $\boldsymbol{R}_Q$ and the direction vector $\boldsymbol{s}$ of θ_1 conveniently by symbolic toolbox of Matlab®.

Based on equations (6.34) and (6.44), the relationship between the Frenet frame $\{\boldsymbol{\rho}_m; \boldsymbol{E}_{1m}, \boldsymbol{E}_{2m}, \boldsymbol{E}_{3m}\}$ of $\sum_m$ and the coordinate system $\{\boldsymbol{R}_C; \boldsymbol{x}_2, \boldsymbol{y}_2, \boldsymbol{z}_2\}$ is

$$\begin{cases} \boldsymbol{\rho}_m = (-a_2 + u_2\cos\theta_2 + b_m s_3 \sin\theta_2)\boldsymbol{x}_2 + (-h_1\sin\alpha_{12} + b_m s_1 \sin\alpha_{12} - u_2\cos\alpha_{12}\sin\theta_2 \\ +b_m s_3\cos\alpha_{12}\cos\theta_2)\boldsymbol{y}_2 + (-h_1\cos\alpha_{12} + b_m s_1\cos\alpha_{12} + u_2\sin\alpha_{12}\sin\theta_2 - b_m s_3\sin\alpha_{12}\cos\theta_2)\boldsymbol{z}_2 \\ \boldsymbol{E}_{1m} = s_3\sin\theta_2\boldsymbol{x}_2 + (s_1\sin\alpha_{12} + s_3\cos\alpha_{12}\cos\theta_2)\boldsymbol{y}_2 + (s_1\cos\alpha_{12} - s_3\sin\alpha_{12}\cos\theta_2)\boldsymbol{z}_2 \\ \boldsymbol{E}_{2m} = \dfrac{1}{|d\boldsymbol{s}_m/d\theta_1|}\left[\left(\dfrac{d\theta_2}{d\theta_1}s_3\cos\theta_2 + \dfrac{ds_3}{d\theta_1}\sin\theta_2\right)\boldsymbol{x}_2 + \left(\dfrac{ds_1}{d\theta_1}\sin\alpha_{12} - \dfrac{d\theta_2}{d\theta_1}s_3\cos\alpha_{12}\sin\theta_2\right.\right. \\ \left.\left. + \dfrac{ds_3}{d\theta_1}\cos\alpha_{12}\cos\theta_2\right)\boldsymbol{y}_2 + \left(\dfrac{ds_1}{d\theta_1}\cos\alpha_{12} + \dfrac{d\theta_2}{d\theta_1}s_3\sin\alpha_{12}\sin\theta_2 - \dfrac{ds_3}{d\theta_1}\sin\alpha_{12}\cos\theta_2\right)\boldsymbol{z}_2\right] \\ \boldsymbol{E}_{3m} = \dfrac{1}{|d\boldsymbol{s}_m/d\theta_1|}\left\{\left[\left(\dfrac{ds_1}{d\theta_1}s_3 - s_1\dfrac{ds_3}{d\theta_1}\right)\cos\theta_2 + s_1 s_3\dfrac{d\theta_2}{d\theta_1}\sin\theta_2\right]\boldsymbol{x}_2 + \left[-s_3^2\dfrac{d\theta_2}{d\theta_1}\sin\alpha_{12}\right.\right. \\ \left. -\left(\dfrac{ds_1}{d\theta_1}s_3 - s_1\dfrac{ds_3}{d\theta_1}\right)\cos\alpha_{12}\sin\theta_2 + s_1 s_3\dfrac{d\theta_2}{d\theta_1}\cos\alpha_{12}\cos\theta_2\right]\boldsymbol{y}_2 \\ \left. + \left[-s_3^2\dfrac{d\theta_2}{d\theta_1}\cos\alpha_{12} + \left(\dfrac{ds_1}{d\theta_1}s_3 - s_1\dfrac{ds_3}{d\theta_1}\right)\sin\alpha_{12}\sin\theta_2 - s_1 s_3\dfrac{d\theta_2}{d\theta_1}\sin\alpha_{12}\cos\theta_2\right]\boldsymbol{z}_2\right\} \end{cases} \tag{6.207}$$

The ISA of link 2 produces the fixed axode $\sum_f$ in $\{\boldsymbol{R}_A; \boldsymbol{x}_0, \boldsymbol{y}_0, \boldsymbol{z}_0\}$ by the coordinate transformation equation (6.196) and the construction parameters α_f, β_f, and γ_f of $\sum_f$ are derived by equation (6.35), as shown in Fig. 6.15. For $\theta_1 \in (0, 2\pi)$, the striction curve and the generatrices of $\sum_f$ are shown in Fig. 6.16.

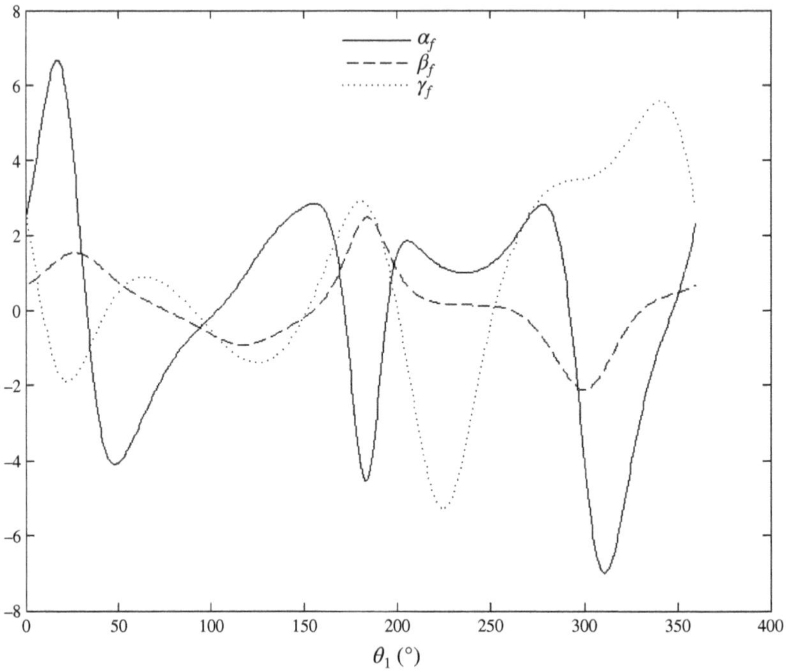

Figure 6.15 Construction parameters of the fixed axode

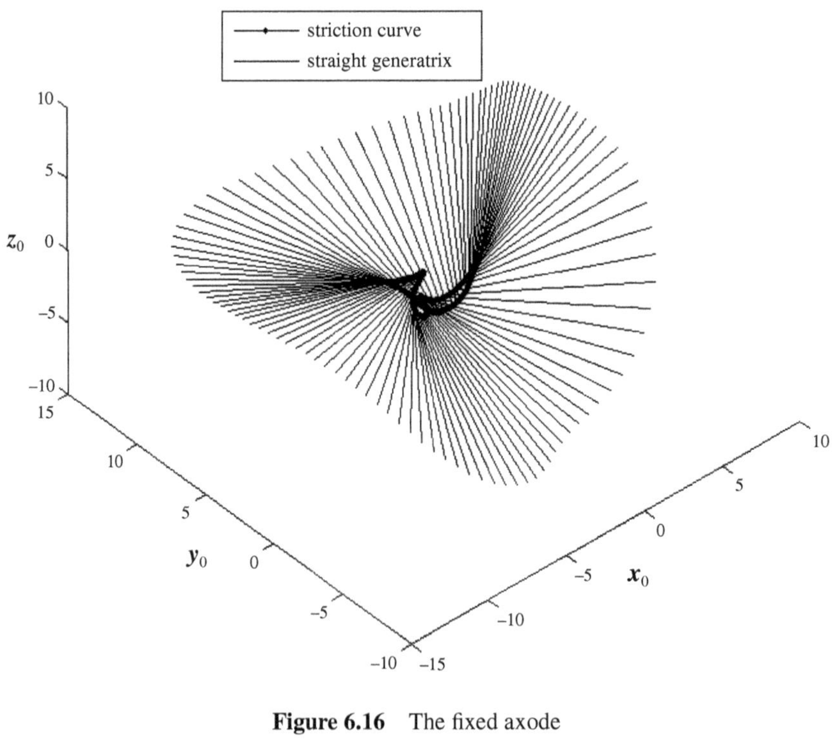

Figure 6.16 The fixed axode

The ISA of link 2 is examined in $\{\boldsymbol{R}_C; \boldsymbol{x}_2, \boldsymbol{y}_2, \boldsymbol{z}_2\}$ as the moving axode $\sum_m$ by coordinate transformation equation (6.195) and the construction parameters α_m, β_m, and γ_m of $\sum_m$ are derived by equation (6.46), as shown in Fig. 6.17. For $\theta_1 \in (0, 2\pi)$, the striction curve and the generatrices of the moving axode $\sum_m$ are shown in Fig. 6.18.

The induced construction parameters α^* and β^* of both $\sum_f$ and $\sum_m$ are $\alpha^* = \alpha_f - \alpha_m$ and $\beta^* = \beta_f - \beta_m$, as shown in Fig. 6.19. $\sum_m$ and $\sum_f$ are partly shown for $\theta_1 \in (0.09, 0.39)$ in Fig. 6.20(a). They contact each other tangentially along the ISA at the instant $\theta_1 = 0.24$. The two spherical image curves corresponding to $\sum_f$ and $\sum_m$ contact tangentially at the spherical image point of the ISA, as shown in Fig. 6.20(b). The construction parameters of $\sum_f$ and $\sum_m$, and the position and orientation parameters of the ISA, are listed in Table 6.2 for $\theta_1 = 0.24$ and $\theta_1 = 1.00$.

6.5.3 *Point Trajectory*

To reveal the curvature of a point trajectory in spatial movement, at the instant $\theta_1 = 0.24$ of the linkage RCCC, the instantaneous kinematics of the coupler points is analyzed, including the geodesic curvature, geodesic inflection points, sphere curvature, cylindrical curvature.

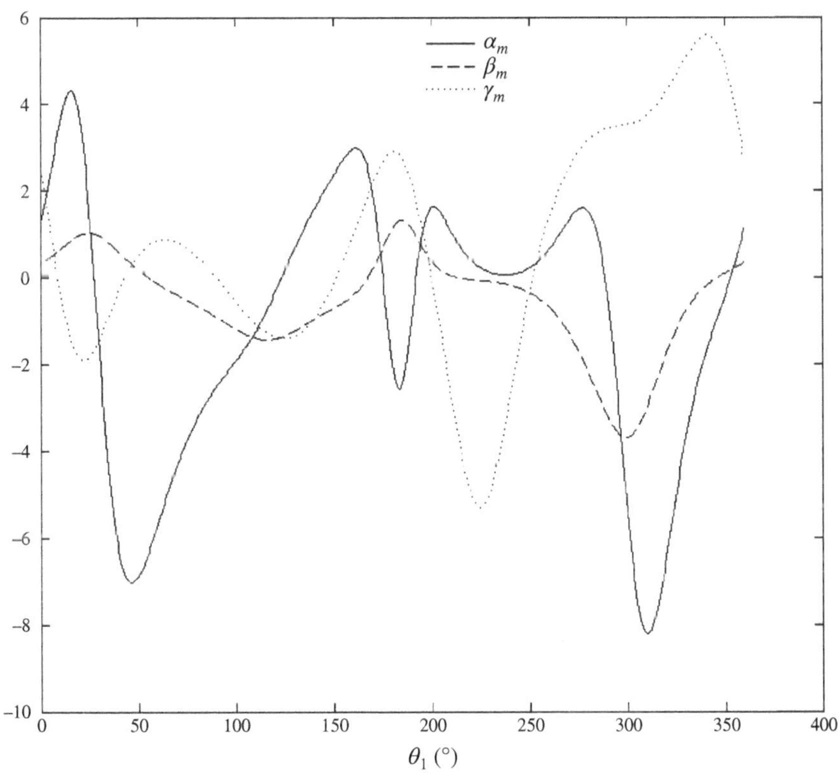

Figure 6.17 Construction parameters of the moving axode

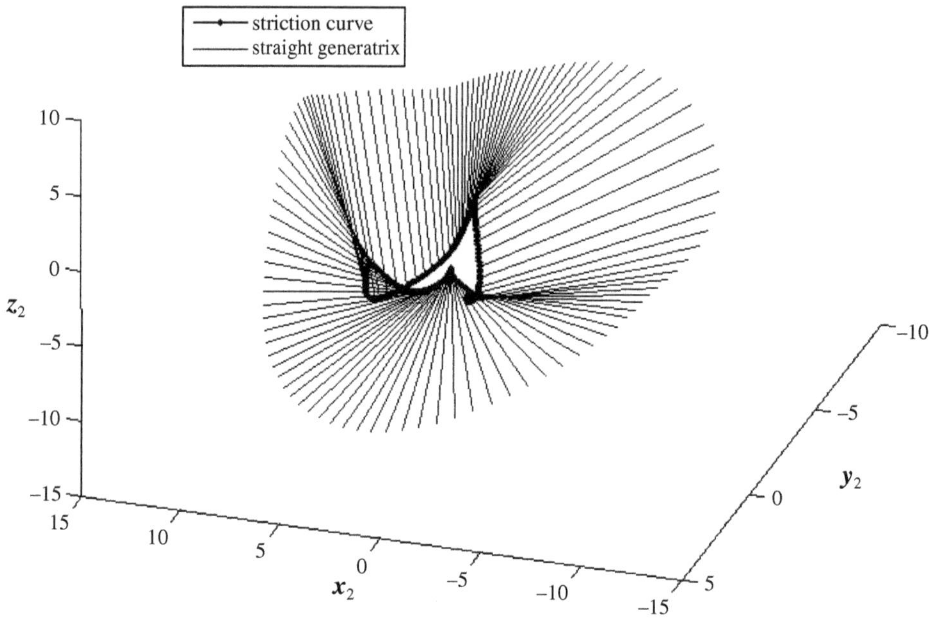

Figure 6.18 Moving axode

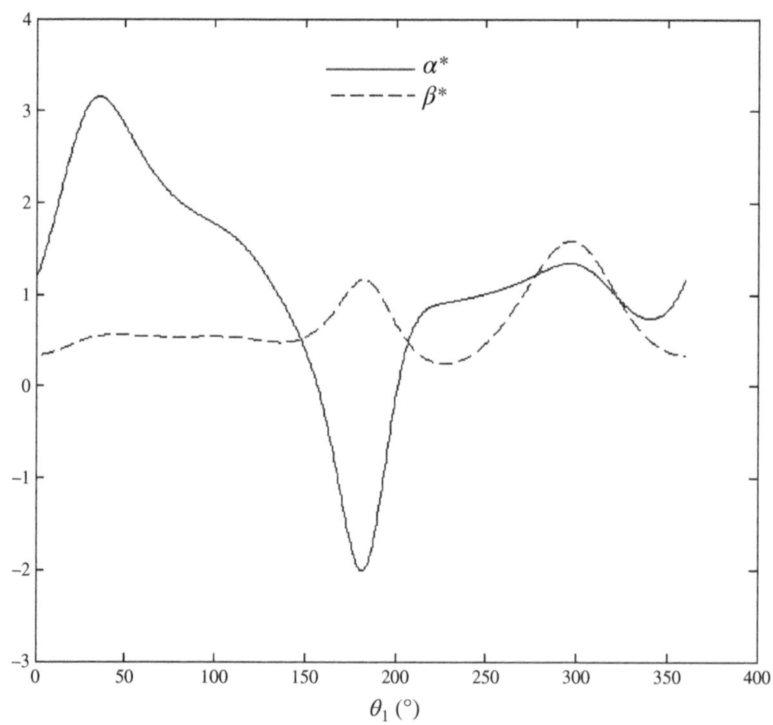

Figure 6.19 Induced construction parameters of the axodes

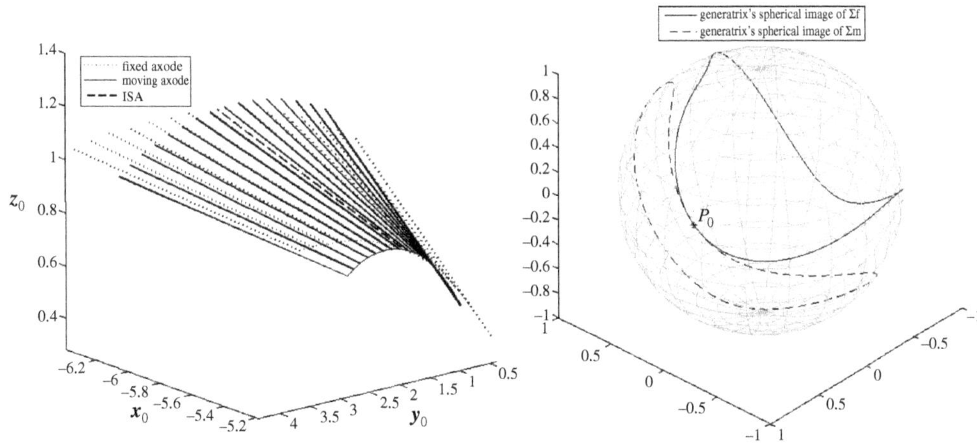

(a) (b)

Figure 6.20 The moving and fixed axodes for the RCCC linkage (a) Axodes (b) Spherical image curves

Table 6.2 Instantaneous parameters of the axodes for the RCCC linkage

	θ_1 =0.24			θ_1 =1.00		
Fixed axode $\sum_f$	α_f 6.3018	β_f 1.1906	γ_f −1.0712	α_f −3.4818	β_f 0.4993	γ_f 0.8023
Moving axode $\sum_m$	α_m 4.1982	β_m 0.7914	γ_m −1.0712	α_m −6.0744	β_m −0.0498	γ_m 0.8023
Induced construction parameters	α^* 2.1036	β^* 0.3991		α^* 2.5926	β^* 0.5490	
Spherical image point of $\sum_f$ (in $\{\boldsymbol{R}_A; x_0, y_0, z_0\}$)	s_{f1} −0.2292	s_{f2} 0.9366	s_{f3} 0.2649	s_{f1} −0.8385	s_{f2} 0.5384	s_{f3} 0.0833
Spherical image point of $\sum_m$ (in $\{\boldsymbol{R}_C; x_2, y_2, z_2\}$)	s_{m1} 0.6619	s_{m2} −0.6118	s_{m3} 0.4331	s_{m1} 0.9046	s_{m2} −0.3452	s_{m3} −0.2500
Striction point of $\sum_f$ (in $\{\boldsymbol{R}_A; x_0, y_0, z_0\}$)	ρ_{f1} −5.3922	ρ_{f2} 1.3628	ρ_{f3} 0.7157	ρ_{f1} −4.7377	ρ_{f2} 2.2454	ρ_{f3} 0.4344
Striction point of $\sum_m$ (in $\{\boldsymbol{R}_C; x_2, y_2, z_2\}$)	ρ_{m1} 2.8311	ρ_{m2} 1.7957	ρ_{m3} −2.1889	ρ_{m1} 0.7004	ρ_{m2} 2.1017	ρ_{m3} −1.8310

6.5.3.1 Geodesic Curvature

At the instant $\theta_1 = 0.24$, if the principal normal of a coupler curve Γ_P passes through the ISA, or the normal curvature k_n is zero, the points lie on the geodesic curvature distribution surface. Substituting the induced construction parameters of the axodes into equation (6.73), the instantaneous distribution surface can be expressed in the Frenet frame $\{\boldsymbol{\rho}_m; \boldsymbol{E}_{1m}, \boldsymbol{E}_{2m}, \boldsymbol{E}_{3m}\}$ of $\sum_m$ as

$$(u_2^2 + u_3^2)(0.6717 + 0.1593u_3) + 3.5257u_3 + 0.8396u_1u_2 = 0 \qquad (6.208)$$

According to the equation, the distribution surface is drawn in Fig. 6.21, for $u_1 \in [-100, 100]$ and $u_2 \in [-100, 100]$.

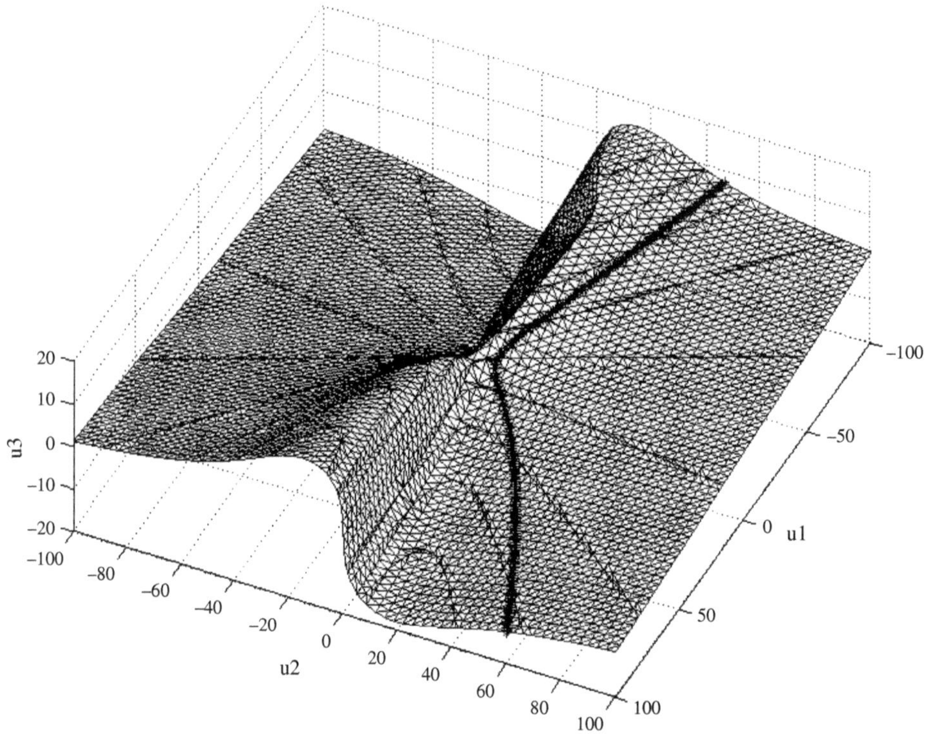

Figure 6.21 Instantaneous geodesic curvature distribution surface

A coupler point P with coordinates $u_1 = 4.5738$, $u_2 = -2$, $u_3 = 1$ in $\{\boldsymbol{\rho}_m; \boldsymbol{E}_{1m}, \boldsymbol{E}_{2m}, \boldsymbol{E}_{3m}\}$ locates on the instantaneous distribution surface, and traces coupler curve Γ_P in $\{\boldsymbol{R}_A; \boldsymbol{x}_0, \boldsymbol{y}_0, \boldsymbol{z}_0\}$ of the linkage RCCC, as shown in Fig. 6.22. Substituting the induced construction parameters of the axodes and $u_1 = 4.5738$ into equations (6.74) and (6.78), we get the diameter $D_{g0} = 15.5557$ of the geodesic inflection circle and coordinates of its center point O, listed in Table 6.3. The principal curvature of Γ_P is equal to the geodesic curvature and the principal normal orthogonally intersects with the ISA at point O'. The geodesic inflection circle just lies on the plane $\{\boldsymbol{O}'; \boldsymbol{E}_{2m}, \boldsymbol{E}_{3m}\}$.

On the plane $\{\boldsymbol{O}'; \boldsymbol{E}_{2m}, \boldsymbol{E}_{3m}\}$, the line $\overline{PO'}$ intersects the geodesic inflection circle at the geodesic inflection point J_P, whose coordinates are listed in Table 6.3. Based on the spatial Euler–Savary equation (6.77c), we have the principal radius of curvature $|\overrightarrow{PO_P}| = 1.9544$ by $|\overrightarrow{PJ_P}| = 16.7720$ and $|\overrightarrow{PO'}| = 2.2361$.

6.5.3.2 Geodesic Inflection Surface

At the instant $\theta_1 = 0.24$ of the linkage RCCC, the geodesic inflection surface in the coupler, defined by equation (6.81), is comprised of a family of geodesic inflection circles lying on the planes perpendicular to the ISA, as shown in Fig. 6.23 for $u_1 \in (-100, 100)$. The geodesic inflection circle denoted by "*" at $u_1 = 50$ has diameter $D_{g0} = 125.7083$, and center's coordinates $x_0 = 50$, $y_0 = 5.2601$, $z_0 = -62.6337$ in $\{\boldsymbol{\rho}_m; \boldsymbol{E}_{1m}, \boldsymbol{E}_{2m}, \boldsymbol{E}_{3m}\}$.

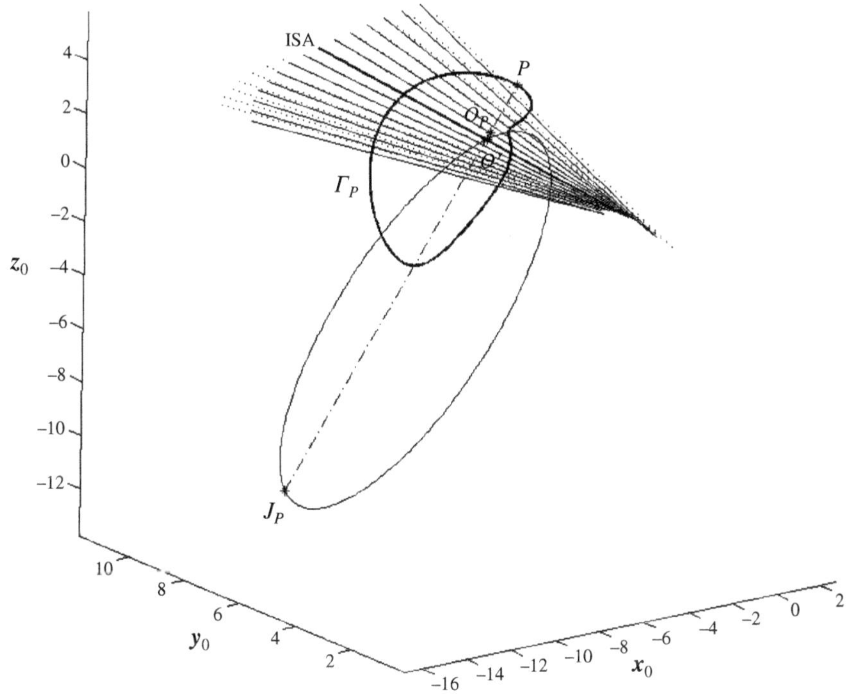

Figure 6.22 Geodesic Euler–Savary equation

Table 6.3 Coordinate parameters for the spatial Euler–Savary equation

Coordinate parameters		P	J_P	O_P	O
Polar coordinates in	r	2.2361	14.5341	0.2817	7.7779
$\{O'; E_{2m}, E_{3m}\}$	θ	2.6779	−0.4636	2.6779	−0.8281
Cartesian coordinates in	x	−5.5109	−12.4836	−6.3234	−7.0219
$\{R_A; x_0, y_0, z_0\}$	y	5.3073	7.8534	5.6040	7.6253
	z	3.9322	−11.1056	2.1798	−5.5724

At this instant, the geodesic inflection surface in Fig. 6.23 intersects the distribution surface in Fig. 6.21, at the inflection curve, as shown by the symbol "*" in Fig. 6.21.

6.5.3.3 Stationary Sphere Curvature

A coupler point $P(x_{Pm}, y_{Pm}, z_{Pm})$ locates at (1, 1, 1) in $\{\boldsymbol{R}_C; \boldsymbol{x}_2, \boldsymbol{y}_2, \boldsymbol{z}_2\}$. At the instant $\theta_1 = 0.24$, the Darboux frame $\{\boldsymbol{R}_P; \boldsymbol{E}_1, \boldsymbol{E}_2, \boldsymbol{E}_3\}$, the curvatures k_n, k_g, τ_g and their derivatives of the coupler curve Γ_P are listed in Table 6.4. The Frenet frame $\{\boldsymbol{R}_P; \boldsymbol{\alpha}, \boldsymbol{\beta}, \boldsymbol{\gamma}\}$ can also be established, and the curvatures k, τ as well as their derivatives are obtained, which are also listed in Table 6.4.

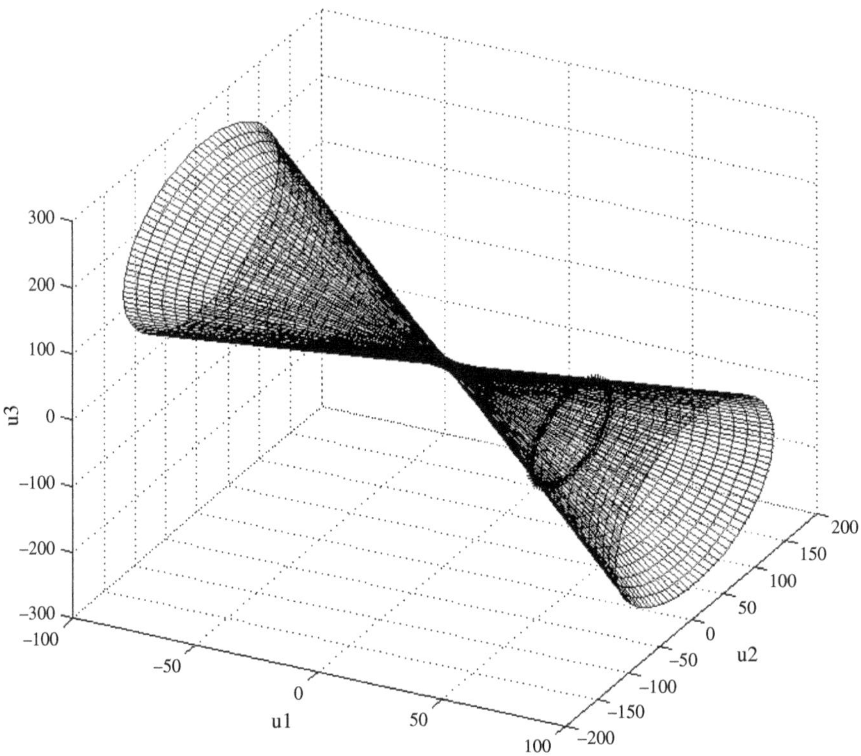

Figure 6.23 Geodesic inflection circle and geodesic inflection surface

Table 6.4 The moving frames and curvatures of the coupler curve

Coupler point P		Coordinates in $\{\rho_m; E_{1m}, E_{2m}, E_{3m}\}$			Coordinates in $\{R_A; x_0, y_0, z_0\}$		
		0.6559	−3.4845	−1.2582	−2.1188	2.4365	2.2279
Darboux frame $\{R_P; E_1, E_2, E_3\}$	E_1	0.8181	0.1953	−0.5409	−0.0118	0.9547	−0.2974
	E_2	−0.5751	0.2779	−0.7695	0.3818	−0.2706	−0.8837
	E_3	0	0.9406	0.3396	−0.9242	−0.1240	−0.3613
Frenet frame $\{R_P; \alpha, \beta, \gamma\}$	α	0.8181	0.1953	−0.5409	−0.0118	0.9547	−0.2974
	β	0.0033	0.9387	0.3439	−0.9263	−0.1225	−0.3563
	γ	0.5749	−0.2831	0.7673	−0.3766	0.2713	0.8858
Curvatures and derivatives	k_n		k_g	τ_g	$dk_n/d\sigma_P$	$dk_g/d\sigma_P$	
	−0.0018		0.3142	−0.1890	−0.1158	−0.1877	
	k		τ	$dk/d\sigma_P$	$d^2k/d\sigma_P^2$	$d\tau/d\sigma_P$	
	0.3143		0.1829	−0.1871	0.4698	−0.2412	

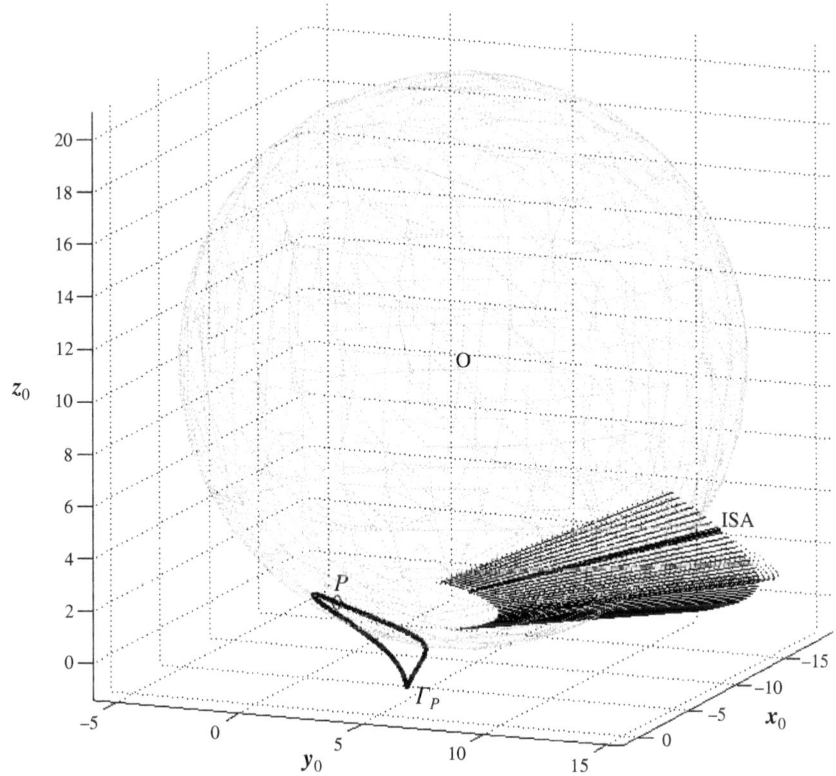

Figure 6.24 Instantaneous curvature sphere of the coupler curve

The curvature sphere of Γ_P in $\{R_A; x_0, y_0, z_0\}$ is located by equation (6.86), whose center point has position vector $\boldsymbol{R}_S = (-8.9668, 4.8573, 10.2688)^T$ and radius is $r_0 = 10.8357$, as shown in Fig. 6.24.

The surface of the stationary sphere curvature can be obtained by substituting the parameters of the axodes, the curvatures of Γ_P, and their derivatives into equation (6.87), as shown in Fig. 6.25.

6.5.3.4 Osculating Cylindrical Surface

At the instant $\theta_1 = 0.24$, by substituting the parameters of the axodes in Table 6.2, the curvatures and torsion of Γ_P, as well as their derivatives into equation (3.172), the sextic algebraic equation about $\sin\theta$ or $\cos\theta$ is obtained as

$$-0.0878 \sin^4\theta \cos^2\theta + 0.0715 \sin^3\theta \cos\theta - 0.3030 \sin^2\theta \cos^2\theta$$
$$+0.0033 \sin^4\theta + 0.0215 \sin\theta \cos^3\theta + 0.0350 \cos^2\theta = 0 \tag{6.209}$$

Four solutions of θ are $\theta_{(1)} = 0.3968$, $\theta_{(2)} = 1.3285$, $\theta_{(3)} = 4.7511$, $\theta_{(4)} = 5.9837$, and four osculating cylindrical surfaces of Γ_P are located by equation (3.174), whose orientation and position of axis and radius are listed in Table 6.5.

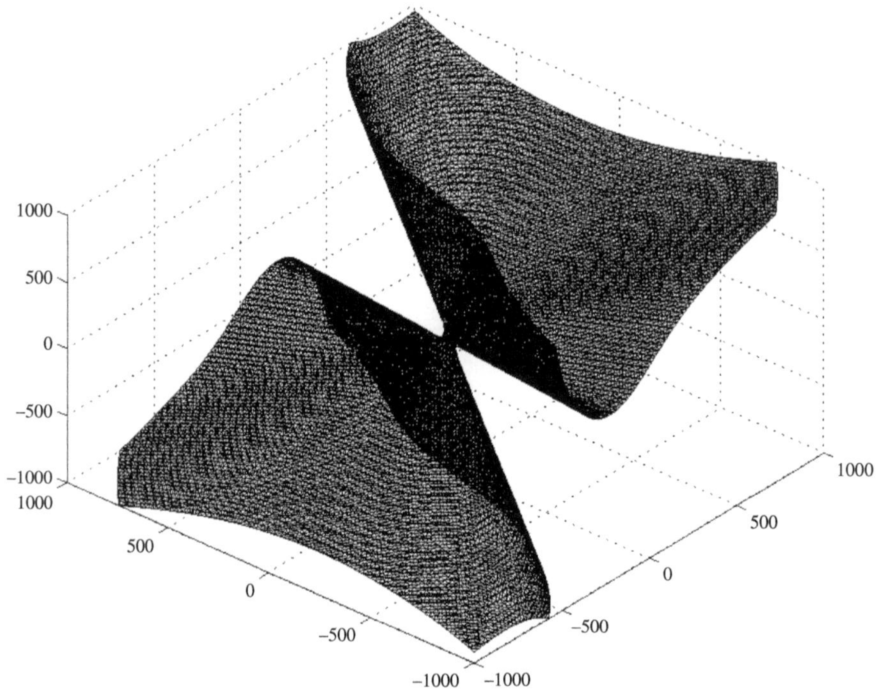

Figure 6.25 The instantaneous surface of the stationary sphere curvature

Table 6.5 Parameters of the osculating cylindrical surface

Osculating cylindrical surface	Position vector $\boldsymbol{R}_A$ of the fixed point A on the axis (in $\{\boldsymbol{R}_A; \boldsymbol{x}_0, \boldsymbol{y}_0, \boldsymbol{z}_0\}$)			Unit direction vector $\boldsymbol{a}$ of the axis (in $\{\boldsymbol{R}_A; \boldsymbol{x}_0, \boldsymbol{y}_0, \boldsymbol{z}_0\}$)			Radius r_0
1	−3.2588	2.4273	2.2436	−0.0035	0.9522	0.3054	1.1402
2	−9.7252	5.4651	12.2498	0.7972	0.3309	0.5050	12.9410
3	−1.2667	1.7460	−0.0219	−0.1757	0.9205	−0.3490	2.5028
4	−2.5159	2.3353	1.9191	−0.2594	0.9656	0.0174	0.5131

One osculating cylindrical surface of Γ_P, determined by the first line parameters in Table 6.5, and the axodes (ISA), are shown in Fig. 6.26.

6.5.4 Line Trajectory

At instant $\theta_1 = 0.24$ of the linkage RCCC, the construction parameters of the moving axode and fixed axode are listed in Table 6.2, and then the coupler curve Γ_P traced by a coupler point $P(1, 1, 1)$ in $\{\boldsymbol{R}_C; \boldsymbol{x}_2, \boldsymbol{y}_2, \boldsymbol{z}_2\}$ has the Frenet frame and the curvature invariants in Table 6.4.

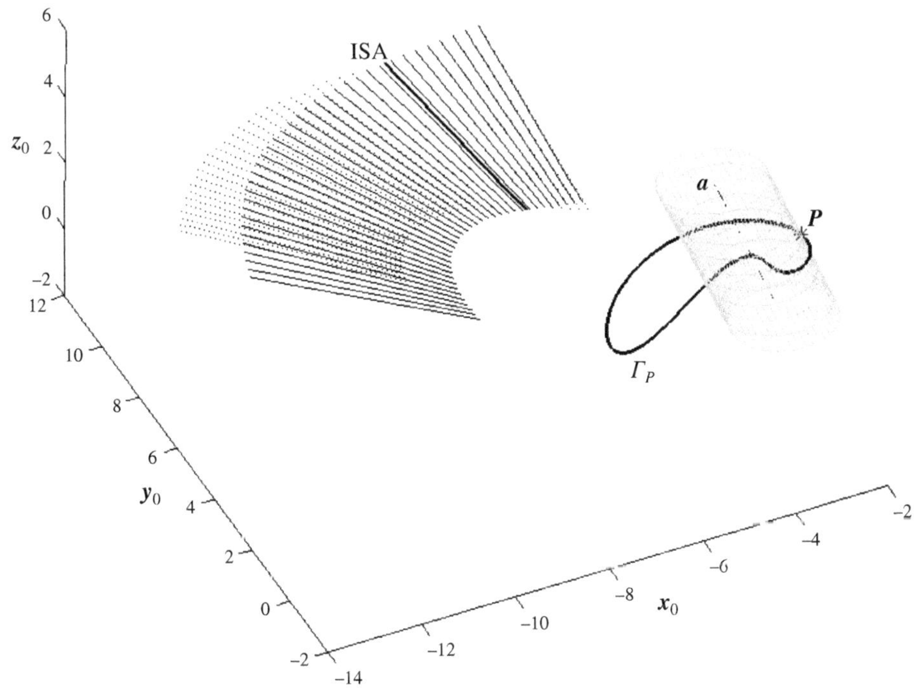

Figure 6.26 Instantaneous osculating cylindrical surface of the coupler curve

Table 6.6 The Frenet frame and the construction parameters of a line trajectory

Line L		Coordinates in $\{\rho_m;E_{1m},E_{2m},E_{3m}\}$			Coordinates in $\{R_A;x_0,y_0,z_0\}$		
		0.6619	0.4278	0.6155	−0.8493	0.4021	0.3421
Frenet frame $\{\rho_l;E_1,E_2,E_3\}$	ρ_l	3.3243	−1.7598	1.2232	−5.5427	4.0575	3.6069
	E_1	0.6619	0.4278	0.6155	−0.8493	0.4021	0.3421
	E_2	0	−0.8211	0.5708	0.2853	−0.1955	0.9383
	E_3	0.7496	−0.3778	−0.5435	0.4442	0.8945	0.0513
Construction parameters	α_l 0.1212			β_l 4.5442		γ_l 7.1625	

A line L with unit direction vector l_m ($l_{m1}=1$, $l_{m2}=0$, $l_{m3}=0$) in $\{R_C;x_2,y_2,z_2\}$ of the coupler passing through the point P traces a ruled surface Σ_l in $\{R_A;x_0,y_0,z_0\}$, as shown in Fig. 6.27, whose Frenet frame $\{\rho_l;E_1,E_2,E_3\}$ and construction parameters α_l, β_l, γ_l are presented in Table 6.6. The change curves of the three construction parameters during the motion cycle of the linkage are depicted in Figure 6.28. The striction directrix distance b_l of Σ_l is 4.0315 by equation (6.98), and the striction point locates at ($h = 2.1430, p = 3.3243$) in $\{\rho_m;E_{1m},E_{2m},E_{3m}\}$ by equation (6.109).

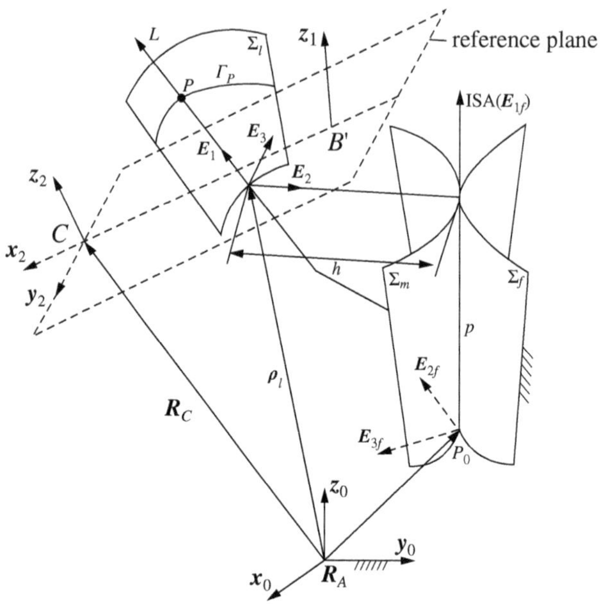

Figure 6.27 The parameters of line trajectories and the reference plane

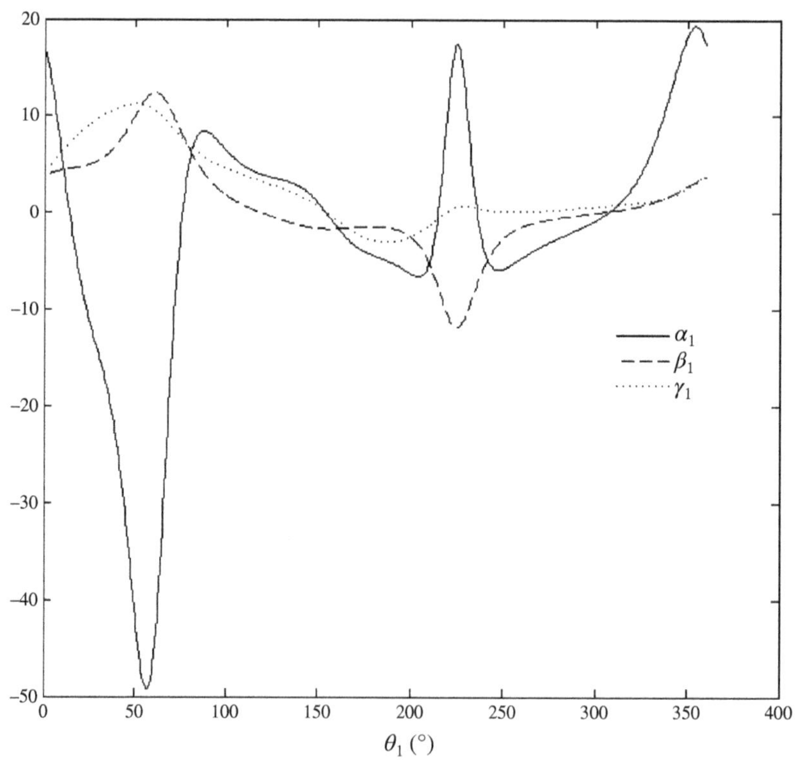

Figure 6.28 Construction parameters of a line trajectory

6.5.4.1 Constant Axis Curvature

The line $L\ (p, h, \delta_l, \theta_l)$ is located at $(3.3243, 2.1430, 0.8474, 0.9634)$ in $\{\boldsymbol{\rho}_m; \boldsymbol{E}_{1m}, \boldsymbol{E}_{2m}, \boldsymbol{E}_{3m}\}$ and traces a ruled surface Σ_l, which corresponds to a constant axis curvature surface. The radius η, inclination angle δ meeting $\cot\delta = \beta_l = 4.5442$, unit direction vector $\boldsymbol{d}$ and reference point P_A of the fixed axis L_{PA} in $\{\boldsymbol{R}_A; \boldsymbol{x}_0, \boldsymbol{y}_0, \boldsymbol{z}_0\}$ are all presented in Table 6.7 and shown in Fig. 6.29.

6.5.4.2 Stationary Line

At instant $\theta_1 = 1.00$ of the linkage RCCC, there exists a cubic spherical image curve $C_{\beta'}$ of stationary curvature for the directions of lines of link 2:

$$C_{\beta'}\ :\ \cot\delta_l = \frac{1}{-5.0098\ \sin\theta_l} + \frac{1}{-26.2120\ \cos\theta_l} \tag{6.210}$$

The curve $C_{\beta'}$ is shown in Fig. 6.30.

A characteristic line $L^{(3)}_{C'C}$ with parameters $(\delta_l, \theta_l, p, H)$ in link 2, whose direction corresponds to a point on the spherical image curve $C_{\beta'}$ of stationary curvature, traces a line trajectory Σ_l, which contacts the constraint ruled surface $\sum_{C'C}$ in third order. Based on equation (6.210), the direction parameters (δ_l, θ_l) of $L^{(3)}_{C'C}$ are given as $\theta_l = 5\pi/4$, $\delta_l = 1.2464$; the position parameters (p, H) of $L^{(3)}_{C'C}$ can be solved by equation (6.162) in $\{\boldsymbol{\rho}_m; \boldsymbol{E}_{1m}, \boldsymbol{E}_{2m}, \boldsymbol{E}_{3m}\}$.

In particular, if the four infinitesimal positions of $L^{(3)}_{C'C}$ intersect orthogonally with a fixed line, or it is a ***Ball line*** $\left(\eta = 0,\ \delta = \frac{\pi}{2}\right)$, the parameters (δ_l, θ_l) can be solved by equations (6.163) and (6.164). The coefficients of equation (6.164) are $f_1 = -0.0909$, $f_2 = 0.1569$, $f_3 = -0.3800$, $f_4 = 0.2316$, and $f_5 = -0.0398$ at the instant $\theta_1 = 1.00$. The position parameters (p,H) of two lines $L^{(3)}_{C'C}$ corresponding to the two sets of (δ_l, θ_l) in $\{\boldsymbol{\rho}_m; \boldsymbol{E}_{1m}, \boldsymbol{E}_{2m}, \boldsymbol{E}_{3m}\}$ are determined by

$$\begin{cases} 1.7375p + 0.7557H + 5.4832 = 0 \\ 0.4290p - 0.0613H + 0.4546 = 0 \end{cases} \quad \text{and} \quad \begin{cases} 1.6306p - 0.7716H - 7.0154 = 0 \\ -0.4157p - 0.0436H + 0.4512 = 0 \end{cases} \tag{6.211}$$

The solutions of the above equation are listed in Table 6.8.

Table 6.7 The constant axis surface of a line trajectory

Fixed included angle (δ)	Radius (η)	Orientation and position of the fixed axis L_{P_A} (in $\{\boldsymbol{R}_A; \boldsymbol{x}_0, \boldsymbol{y}_0, \boldsymbol{z}_0\}$)					
		Unit direction vector $\boldsymbol{d}$			Position vector $\boldsymbol{R}_{P_A}$		
0.2166	−1.4978	−0.7340	0.5849	0.3451	−5.9701	4.3503	2.2015

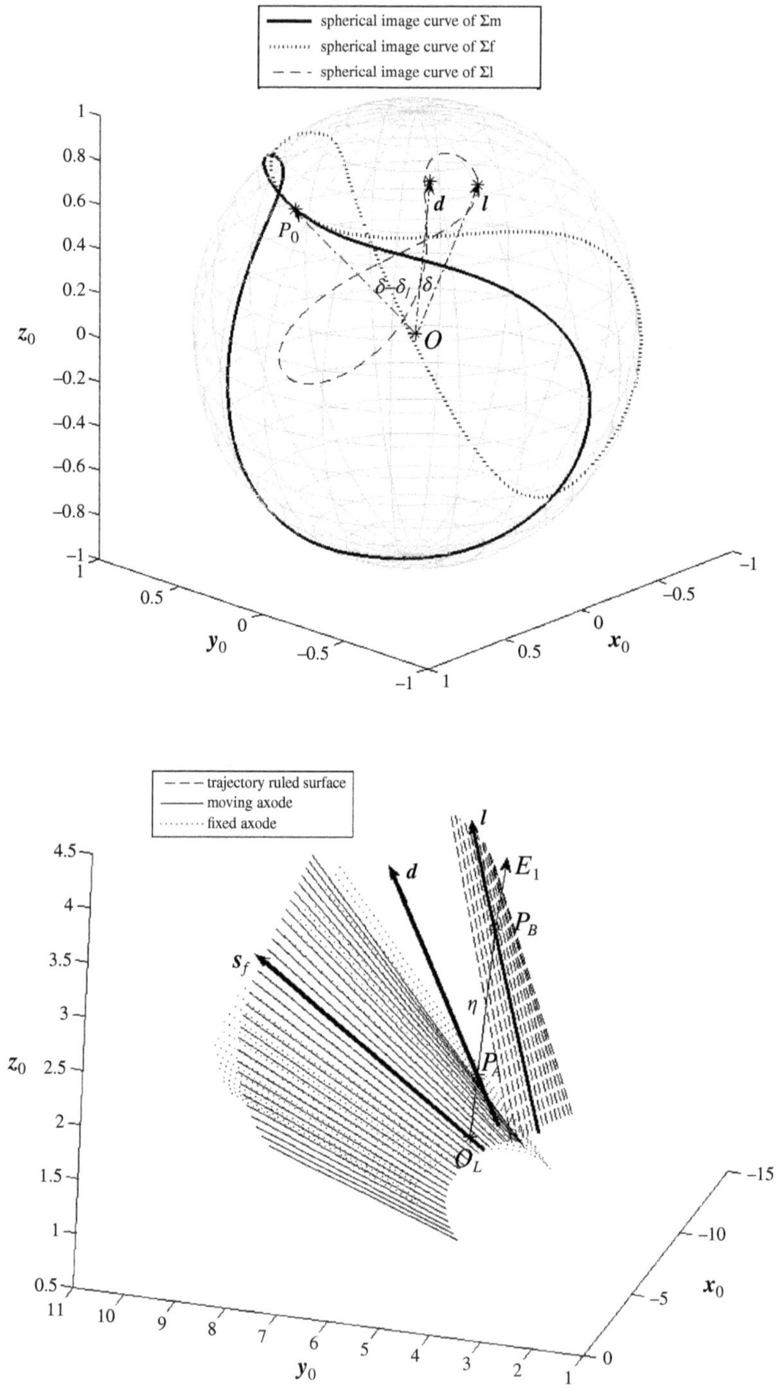

Figure 6.29 (a) The spherical image curve of a line trajectory (b) The Euler–Savary analogue of a line trajectory

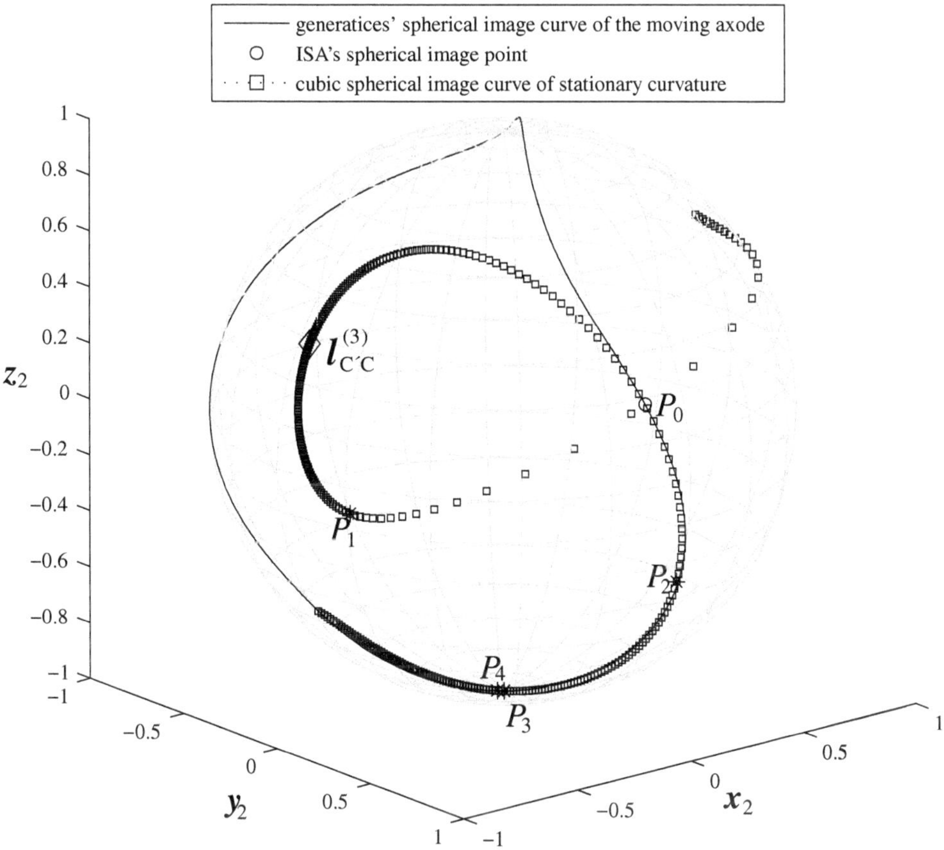

Figure 6.30 Cubic spherical image curve of stationary curvature

6.5.4.3 Constant Axis Line

There exist at most six directions of Burmester's lines at an instant by equations (6.123) and (6.124), or six instantaneous directions of the characteristic lines $L_{C'C}^{(4)}$ in coupler link 2 of the linkage RCCC. The directions of the axes z_1 and z_2 are two of the six directions at any instant, and at the instant $\theta_1 = 1.00$, the directions can be solved by

$$\begin{cases} \cot\delta_l = \dfrac{1}{7.6474\ \sin\theta_l} + \dfrac{1}{3.4307\ \cos\theta_l} \\ -0.0015\tan^6\theta_l - 0.0057\tan^5\theta_l + 0.0047\tan^4\theta_l + 0.5782\tan^3\theta_l \\ +1.0759\tan^2\theta_l + 0.5840\ \tan\theta_l + 0.0697 = 0 \end{cases} \tag{6.212}$$

Four directions of $L_{C'C}^{(4)}$ can be solved through the above equation in $\{\rho_m; \boldsymbol{E}_{1m}, \boldsymbol{E}_{2m}, \boldsymbol{E}_{3m}\}$, whose parameters (δ_l, θ_l) are listed in Table 6.8. In the frame $\{\boldsymbol{R}_C; \boldsymbol{x}_2, \boldsymbol{y}_2, \boldsymbol{z}_2\}$ of the coupler

Table 6.8 The instantaneous characteristic lines of link 2 at the instant $\theta_1 = 1.00$

	Direction cosines in $\{\boldsymbol{\rho}_m; \boldsymbol{E}_{1m}, \boldsymbol{E}_{2m}, \boldsymbol{E}_{3m}\}$			Parameters $(\delta_l, \theta_l, p, H)$ in $\{\boldsymbol{\rho}_m; \boldsymbol{E}_{1m}, \boldsymbol{E}_{2m}, \boldsymbol{E}_{3m}\}$			
$L^{(3)}_{C'C}(\infty^2)$	l_1	l_2	l_3	δ_l	θ_l	p	H
	0.3187	−0.6702	−0.6702	1.2464	$5\pi/4$	$0.3246p - 1.5000H - 8.3867 = 0$	
					(∞^1)	(∞^1)	
$L^{(3)}_{C'C}(\infty^0)$	0.5872	0.7814	−0.2112	0.9432	6.0192	−1.5780	−3.6273
$\left(\eta = 0,\ \delta = \frac{\pi}{2}\right)$	0.6187	−0.7571	−0.2097	0.9037	3.4117	1.6793	−5.6601
$L^{(4)}_{C'C}(\infty^0)$	0.4262	−0.1281	−0.8955	1.1305	4.5703	−5.2179	−2.9518
	0.7642	0.6363	−0.1052	0.7009	6.1193	−4.5109	11.7035
	0.2500	0.7378	−0.6271	1.3182	5.5787	0.0982	−2.2858
	0.2422	0.7255	−0.6441	1.3262	5.5571	13.8430	−7.2734
$L^{(2)}_{RC}(\infty^2)$	0.4262	−0.1281	−0.8955	1.1305	4.5703	−5.2179	−2.9518
				(∞^1)	(∞^1)		
$L^{(2)}_{RR}(\infty^0)$*	0.4190	0.2883	−0.8610	1.1385	5.0355	−3.2186	−0.2993
	0.8377	−0.5242	−0.1532	0.5778	3.4258	−0.0955	−1.8266
$L^{(3)}_{HC}(\infty^0)$	0.4262	−0.1281	−0.8955	1.1305	4.5703	−5.2179	−2.9518
	0.5409	−0.7936	−0.2786	0.9993	3.4792	3.7120	−2.8994
$L^{(2)}_{HR}(\infty^1)$	0.7071	0.4234	0.5663	$\pi/4$	0.9289	1.4553	2.6732
	0.7071	−0.4068	−0.5784	(∞^1)	−2.1837	−2.0959	−0.9500
$L^{(2)}_{HH}(\infty^2)$	0.7071	0.5000	0.5000	$\pi/4$	$\pi/4$	16.5351	−8.1759
				(∞^1)	(∞^1)		

Note: The characteristic line $L^{(2)}_{RR}$ of link 2 is located at the instant.$\theta_1 = 2.00$.

link, the direction cosines of the four directions are

$$\begin{cases} l^{(1)}_{m1} = 0,\ l^{(1)}_{m2} = -0.8192,\ l^{(1)}_{m3} = -0.5736 \\ l^{(2)}_{m1} = 0.6806,\ l^{(2)}_{m2} = 0.0955,\ l^{(2)}_{m3} = -0.7264 \\ l^{(3)}_{m1} = 0,\ l^{(3)}_{m2} = 0,\ l^{(3)}_{m3} = -1 \\ l^{(4)}_{m1} = -0.0149,\ l^{(4)}_{m2} = -0.0167,\ l^{(4)}_{m3} = -0.9997 \end{cases} \tag{6.213}$$

The equation corresponds to four points P_1, P_2, P_3, and P_4 on the cubic spherical image curve of stationary curvature, as shown in Fig. 6.30, and the points P_1 and P_3 are respectively the spherical image points of the axes z_1 and z_2.

The positions (striction points) of the instantaneous characteristic lines $L^{(4)}_{C'C}$ in link 2 can be obtained by substituting the direction cosines of $L^{(4)}_{C'C}$ in Table 6.8 into equation (6.166). Taking the point P_3 with $\delta^{(3)}_l = 1.3182$ and $\theta^{(3)}_l = 5.5787$ for instance, the corresponding position parameters (p,H) can be calculated by

$$\begin{cases} 0.5808p + 1.4804H + 3.3269 = 0 \\ 0.3259p + 0.7364H + 1.6514 = 0 \end{cases} \tag{6.214}$$

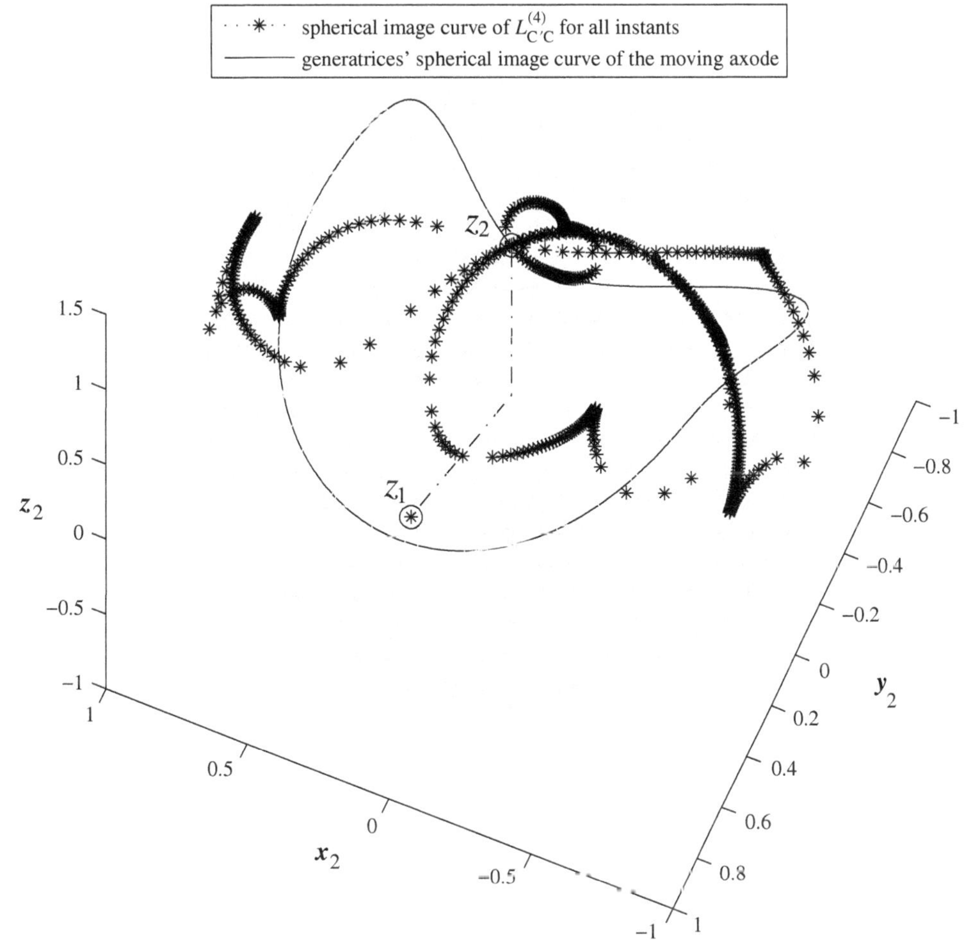

Figure 6.31 Spherical image curve of the lines $L^{(4)}_{C'C}$ for $\theta_1 \in (0, 2\pi)$

The solution $p = 0.0982$, $H = -2.2858$ locates the position of a line $L^{(4)}_{C'C}$ with direction parameters $(\delta_l^{(3)}, \theta_l^{(3)})$, which is the axis z_2 of the moving joint C of link 2 with reference point at $(0, 0, -1)$, and striction point at $(0, 0, -1.8556)$ in $\{\boldsymbol{R}_C; \boldsymbol{x}_2, \boldsymbol{y}_2, z_2\}$. The position parameters (p,H) of striction points of other three lines $L^{(4)}_{C'C}$ corresponding to three spherical image points P_1, P_2, and P_4 are listed in Table 6.8.

For link 2 of the linkage RCCC at an instant θ_1, there are at most six instantaneous directions of the characteristic lines $L^{(4)}_{C'C}$, which correspond the spherical image points. They may form several spherical image curves on a moving unit spherical surface for all instants $\theta_1 \in (0, 2\pi)$, as shown in Fig. 6.31, while the spherical image points P_1 and P_3 always appear at any instant since they correspond to the axes z_1 and z_2 of link 2.

The positions of the instantaneous characteristic lines $L^{(4)}_{C'C}$, or the locations of the striction points of $L^{(4)}_{C'C}$ corresponding to the instantaneous directions, can be calculated by

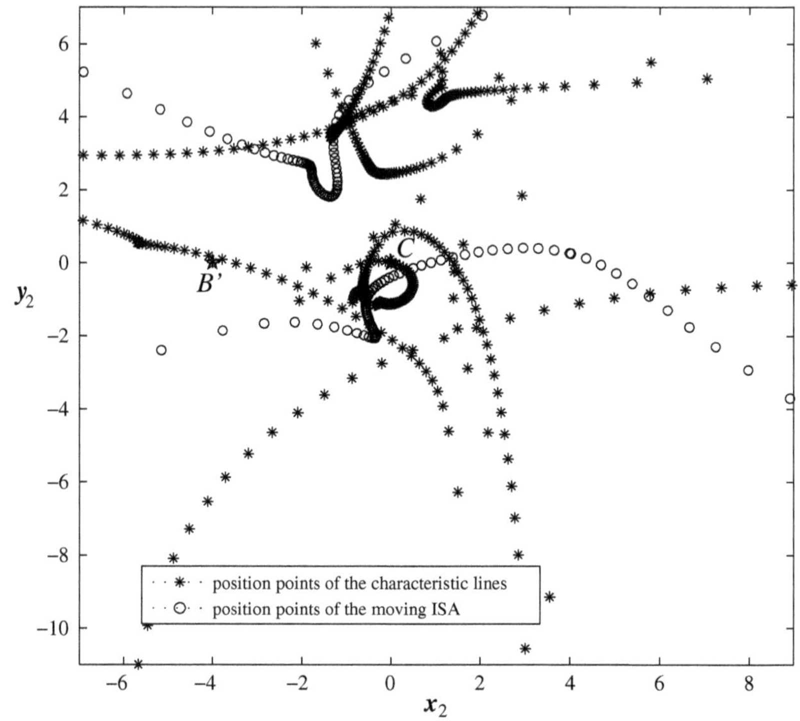

Figure 6.32 The reference points of the lines $L^{(4)}_{C'C}$ on the reference plane

equation (6.166). To reveal the distribution of the positions of $L^{(4)}_{C'C}$ clearly, we prefer to choose the reference point of $L^{(4)}_{C'C}$ on the reference plane $\boldsymbol{R}_C - \boldsymbol{x}_2\boldsymbol{y}_2$, as shown in Fig. 6.27 (rather than the striction points since they locate at different places in link 2). For link 2 of the linkage RCCC at an instant θ_1, there are at most six instantaneous lines $L^{(4)}_{C'C}$, which correspond to their spherical image points and their reference points. For all instants $\theta_1 \in (0, 2\pi)$, the directions of lines $L^{(4)}_{C'C}$ correspond to the spherical image curves on the moving unit spherical surface in Fig. 6.31, and the reference point curves on the reference plane $\boldsymbol{R}_C - \boldsymbol{x}_2\boldsymbol{y}_2$ is shown in Fig. 6.32. Meanwhile, the points B' and C always appear to be the reference points of instantaneous lines $L^{(4)}_{C'C}$ at any instant, like the moving joint of a coupler link for a planar four-bar linkage, corresponding to the reference points of the axes z_1 and z_2 of two C-pairs in link 2.

6.5.4.4 Constant Parameter Characteristic Lines

A line of link 2 of the RCCC linkage at an instant θ_1 may be visualized as the degenerated lines of the characteristic line L_{CC}, such as $L^{(2)}_{RC}$, $L^{(2)}_{RR}$, $L^{(3)}_{HC}$, $L^{(2)}_{HR}$, $L^{(2)}_{HH}$, etc. in some special cases.

Characteristic Line $L^{(2)}_{RC}$

Based on equations (6.172) and (6.174), an instantaneous characteristic line $L^{(2)}_{RC}$ $(\delta_l, \theta_l, p, H)$ in link 2 of the linkage RCCC at an instant θ_1 can be calculated if two direction parameters (δ_l, θ_l)

are given at first. For instance, the direction parameters of a line are designated as $\delta_l = 1.1305$, $\theta_l = 4.5703$ at the instant θ_1=1.00. The striction point parameters of $L_{\rm RC}^{(2)}$ can be calculated by means of equations (6.89) and (6.174), as listed in Table 6.8. The direction cosines of $L_{\rm RC}^{(2)}$ are (0, −0.8192, −0.5736) and the striction point has coordinates (−4.0000, 2.3715, 1.6605) in $\{\boldsymbol{R}_C;\boldsymbol{x}_2,\boldsymbol{y}_2,\boldsymbol{z}_2\}$. Of course, the axis z_1 is one of the lines $L_{\rm RC}^{(2)}$ in link 2.

Characteristic Line $L_{\rm RR}^{(2)}$

Based on equation (6.176), there exist finite instantaneous characteristic lines $L_{\rm RR}^{(2)}$ $(\delta_l, \theta_l, p, H)$ in link 2 at an instant, maybe even none at all. We prefer to choose the instant θ_1=2.00 for the linkage RCCC other than θ_1=1.00. Two sets of direction parameters (δ_l, θ_l) of $L_{\rm RR}^{(2)}$, (1.1385, 5.0355) of line L_1 and (0.5778, 3.4258) of line L_2, are solved by equation (6.179); substituting them into equations (6.177) and (6.178), we have the two instantaneous characteristic lines $L_{\rm RR}^{(2)}$ with striction point parameters (*p*,*H*) as listed in Table 6.8. The unit direction vector (l_{m1},l_{m2},l_{m3}) and the striction point $P(x_{Pm}, y_{Pm}, z_{Pm})$ in $\{\boldsymbol{R}_C;\boldsymbol{x}_2,\boldsymbol{y}_2,\boldsymbol{z}_2\}$ are (−0.0204, −0.7684, −0.6396) and (−2.9052, 1.7860, 2.7803) for line L_1, and (0.8524, −0.0193, −0.5226) and (−1.5413, 3.3520, 0.2247) for line L_2. For all instants $\theta_1 \in (0, 2\pi)$, the instantaneous characteristic lines $L_{\rm RR}^{(2)}$ of link 2 can be located in the same way.

Characteristic Line $L_{\rm HC}^{(3)}$

According to equation (6.184), there exist finite instantaneous characteristic lines $L_{\rm HC}^{(3)}$ $(\delta_l, \theta_l, p, H)$ in link 2 at an instant θ_1. At the instant θ_1=1.00, the two instantaneous characteristic lines $L_{\rm HC}^{(3)}$-1 and $L_{\rm HC}^{(3)}$-2 are solved by equations (6.184) and (6.185), whose direction and striction point parameters $(\delta_l, \theta_l, p, H)$ are listed in Table 6.8. The unit direction vector (l_{m1},l_{m2},l_{m3}) and the striction point $P(x_{Pm}, y_{Pm}, z_{Pm})$ in $\{\boldsymbol{R}_C;\boldsymbol{x}_2,\boldsymbol{y}_2,\boldsymbol{z}_2\}$ are (0, −0.8192, −0.5736) and (−4.000, 2.3715, 1.6605) for $L_{\rm HC}^{(3)}$-1, and (0.3295, −0.9030, 0.2756) and (4.9886, 1.7818, −0.7201) for $L_{\rm HC}^{(3)}$−2. For all instants $\theta_1 \in (0, 2\pi)$, the instantaneous characteristic lines $L_{\rm HC}^{(3)}$ of link 2 can be located in the same way.

Characteristic Line $L_{\rm HR}^{(2)}$

By equation (6.186), there are infinite (∞^1) instantaneous characteristic lines $L_{\rm HR}^{(2)}$ in link 2 at an instant. At the instant θ_1=1.00, the parameter δ_l is designated as $\pi/4$, two values of the parameter θ_l of $L_{\rm HR}^{(2)}$ can be solved by equation (6.187). The striction point parameters (*p*,*H*) of the two lines $L_{\rm HR}^{(2)}$ in $\{\boldsymbol{\rho}_m;\boldsymbol{E}_{1m},\boldsymbol{E}_{2m},\boldsymbol{E}_{3m}\}$ can be calculated by equations (6.89) and (6.187), as listed in Table 6.8. The unit direction vector (l_{m1}, l_{m2}, l_{m3}) and the striction point $P(x_{Pm}, y_{Pm}, z_{Pm})$ in $\{\boldsymbol{R}_C;\boldsymbol{x}_2,\boldsymbol{y}_2,\boldsymbol{z}_2\}$ are (0.9016, 0.4113, −0.1340) and (2.4153, 1.3208, −0.3681) for the first line, while the unit direction vector and the striction point are (0.3755, −0.8962, −0.2393) and (−1.0614, 2.7038, −0.6596) for the second line.

Characteristic Line $L_{\rm HH}^{(2)}$

By equation (6.188), there are infinite (∞^2) instantaneous characteristic lines $L_{\rm HH}^{(2)}$ in link 2 at an instant. At the instant θ_1=1.00, the direction parameters of the line are assumed to be $\delta_l = \theta_l = \pi/4$. The corresponding striction point parameters of $L_{\rm HH}^{(2)}$ can be calculated by equations (6.89) and (6.188), as listed in Table 6.8. The unit direction vector (l_{m1}, l_{m2}, l_{m3})

and the striction point $P(x_{Pm}, y_{Pm}, z_{Pm})$ in $\{\boldsymbol{R}_C; \boldsymbol{x}_2, \boldsymbol{y}_2, \boldsymbol{z}_2\}$ are (0.8776, 0.4194, −0.2321) and (14.1460, −3.5299, −11.5435).

6.6 Discussion

As discussed in Chapters 1 and 4, the instantaneous kinematic geometry of a rigid body in planar and spherical movement has abundant content and expressible form in terms of moving and fixed centrodes, which might be developed to those in spatial movement; a theoretical system of instantaneous kinematic geometry has been expected in the field of kinematics of mechanisms for several decades. However, the kinematic geometry in spatial movement contains much greater complexity than that in planar and spherical movement, not only for point trajectories but also for line trajectories, and it is difficult to compare these with either the constraint curves or the constraint surfaces, other than the easy examples of a line or circle in planar kinematic geometry. Hence, it is the natural starting point for studying instantaneous kinematic geometry to take advantage of the kinematic invariants, the curvatures or generalized curvatures, of both point trajectories and line trajectories.

Müller [1] represented the general higher path curvature in planar kinematics based on the evolute curvature. Freudenstein [4] developed Müller's work and showed how to describe curves to nth order by $n-2$ characteristic numbers. The curvature theory in planar kinematics – like the cubic of stationary curvature and the Burmester point – are presented. Based on the canonical systems presented by Bottema [5, 6], Veldkamp [7] elaborated on Freudenstein's work to develop the same planar higher path curvature theory through instantaneous invariants. Soni, Siddhanty, and Ting [8] presented the characteristic numbers for the line envelope in plane kinematics, and then developed tangent line higher-order envelope curvature theory. Kamphius [9] described the spherical kinematics through the canonical reference system and instantaneous angular velocity vector, obtaining the Ball axes and Burmester axes. Yang and Roth [10] derived the characteristic numbers to define the higher-order curvature of a spherical curve. They applied Kamphius' invariants to obtain the characteristic numbers of the point path in spherical motion, and then established the general higher- order curvature theory for spherical kinematics. Veldkamp [11] extended the canonical system and instantaneous invariants, introduced by Bottema in plane kinematics, to spatial kinematics. In terms of instantaneous invariants of spatial motion, Hsia and Yang [12] derived the analytical expressions for curvature and torsion of the point path, and then obtained equations for the loci of points in the moving body to be straight lines, helical curves, and spherical curves at a particular instant. However, the higher-order curvature theory and the approximation of the point path and cylindrical surface were not discussed. Yang, Kirson, and Roth [13] extended the evolute for a planar curve to the evolute for a spatial ruled surface. Three characteristic scalars, independent of stretch/rotation transformations, are defined to describe the infinitesimal properties of a ruled surface to second order. McCarthy and Roth [14] derived three curvature parameters to define the local shape of a ruled surface by the moving frame and studied all lines of a moving body instantaneously generated for special trajectory ruled surfaces. McCarthy [15] derived the dual formulations of the curvature theory of ruled surfaces and exposed the fundamental curvature functions to characterize the geometry of a ruled surface. Ting and Soni [16, 17] first researched the curvature theory of the plane in spherical and spatial motion based on the canonical systems and instantaneous invariants. The geodesic curvature of the spherical image of the plane envelope, the torsion and curvature of the associated edge of regression are adopted to define the characteristic equations of the plane envelope. It is particularly mentioned that Bottema and Roth [18] systematically studied the theoretical kinematics from planar to spatial motion by the instantaneous invariants, but the geometrical properties are hidden in the algebraic formulations, which need exposing.

The spatial motion of a rigid body can be treated as Chasles' motion [19], where the displacement can be viewed as a rotation about an axis and a translation along the axis, introduced by Mozzi [20] and defined as a screw axis. With the movement of the rigid body, the screw axis forms a moving axode in the moving body and a fixed axode in the fixed frames respectively. Both moving centrode and fixed centrode have revealed their roles in instantaneous planar kinematic geometry, and the axodes might be expected to occupy a similarly important fundamental place in spatial instantaneous kinematic geometry and develop a unified axode-based theory. However, most researches pay more attention to the screw motion or the screw theory. Ball [21] systematically studied the theory of screws. Then, Clifford [22] presented the bi-quaternions. Study [23] came up with the dual angle. Brand [24] gave the systematic operation rules of screws. All these mathematical tools were adopted by researchers [18, 25–30] on mechanisms to study the kinematic analysis and synthesis of mechanisms. Dimentberg [31] introduced the algebra of dual vectors and obtained a displacement equation for a number of spatial four-bar mechanisms. Then, Yang [32, 33] applied the dual number matrices or the dual quaternion algebra in the analysis of spatial linkages. Woo and Freudenstein [34] studied the rigid body motion through the line geometry and gave an algebraic formulation of the mechanical systems. The first full-scale study of spatial motion, a spatial displacement, was mapped into high-dimensional space by Study [23]; the space is named "soma space." Ravani and Roth [35] mapped a spatial displacement into a point of a dual projective three-space. The kinematic properties of spatial motion can be converted into the geometric properties of the image curve in the projective space. McCarthy and Ravani [36] introduced the mapping approach to characterize the corresponding movement of a rigid body by the intrinsic properties of an image curve instantaneously. Lee, Yang, and Ravani [37] presented explicit equations for the instantaneous invariants of spatial kinematics. Similarly, Köse [38, 39] represented the invariants of a line trajectory in spatial motion by the dual curve on the dual unit fixed sphere. The dual curvature theories were set up, and the special points on the dual curve reveal the sets of lines with spatial kinematic meaning.

Although the axodes were referred to by some of these references, such as McCarthy and Ravani [36], the geodesic curvatures of the polhodes of the axodes for spherical motion are derived, which were not connected to those of spatial kinematics. Skreiner [40] studied the higher-order properties of spatial motion based on the local properties of the axodes. Bokelberg, Hunt, and Ridley [41, 42] introduced the concept of a "differential screw" to characterize the difference between two successive ISAs on an axode. The inflections in point trajectory and the acceleration of points were further examined. The dual vector was adopted by Stachel [43] to describe the instantaneous screw, and the spatial Euler–Savary formula for a line in spatial motion was derived, dating back to Disteli [44]. Dooner and Griffis [45] presented the spatial Euler–Savary equations for envelopes of the axodes. The hyperboloids osculating to the axodes were adopted to reflect the properties of the axodes, which can only be applied in some special and simple cases. Schutter [46] defined six scalar functions based on three infinitesimal positions of the ISA, and gave an algorithm to reconstruct a moving frame to represent the movement of a rigid body. The author [47–49] focused on the invariants and invariance of both constraint curves and surfaces, described spatial instantaneous kinematic geometry by the Frenet frame of the axodes, compared the invariants and invariance of the point trajectory or line trajectory with those of constraint curves and surfaces – in particular for a line in spatial movement, where the unit direction vector of a line is mapped onto a unit spherical surface, which is equivalent to a point in spherical movement, and builds a bridge from planar to spatial movement.

So far, the invariants of both the moving axode and the fixed axode reveal the intrinsic quality of the spatial movement of a moving body, which is fundamental in instantaneous spatial kinematic geometry. The point trajectory and line trajectory can be concisely and intuitively expressed by the axodes and their invariants, and compared with the constraint curves and surfaces in terms of invariants to locate the characteristic points or lines.

This book simply introduces the invariants and invariance for a few constraint curves and surfaces corresponding to the binary links – such as the cylinder curvature, sphere curvature, and constant axis curvature – which are just the beginning. The invariance of the constraint curves and surfaces has been studied so little that it only exists in the binary links, even less while the constraint curves and surfaces correspond to both binary links and open kinematic chains such as the binary S–C, open chain R–R–R, C–R–R, etc. including the one DOF and multiple DOFs, which are the theoretical basis of the kinematic synthesis of spatial linkages. Therefore, the spatial kinematic geometry, both in content and approach, has to be developed, even in the classical contents of differential geometry.

References

[1] R. Müller, Üeber die Krümmung der Bahnevoluten bei starren ebenen Systeme. *Zeitschrift für Mathematik und Physik*, **36**, 193–205, 1891.

[2] A.T. Yang and F. Freudenstein, Application of dual-number quaternion algebra to the analysis of spatial mechanisms. *ASME Journal of Applied Mechanics*, **31**(2), 300–308, 1964.

[3] Q.X. Zhang, *Analysis and Synthesis of Spatial Mechanisms*, Vol. **1**. China Machine Press: Beijing, 1984 (in Chinese).

[4] F. Freudenstein, Higher path-curvature analysis in plane kinematics. *ASME Journal of Engineering for Industry*, **87**(2), 184–190, 1965.

[5] O. Bottema, On Cardan positions for the plane motion of a rigid body. *Proceedings Koninklijke Nederlandse Akad. Van Wetenschappen, Series A*, **52**, 643–651, 1949.

[6] O. Bottema, Some remarks on theoretical kinematics, I. On instantaneous invariants. Proceedings of the International Conference for Teachers of Mechanisms, Yale University, New Haven, CT, 15–164, 1961.

[7] G.R. Veldkamp, Some remarks on higher curvature theory. *ASME Journal of Engineering for Industry*, **89**(1), 84–86, 1967.

[8] A.H. Soni, M.N. Siddhanty, and K.L. Ting, Higher order, planar tangent-line envelope curvature theory. *ASME Journal of Mechanical Design*, **101**(4), 563–568, 1979.

[9] H.J. Kamphuis, Application of spherical instantaneous kinematics to the spherical slider-crank mechanism. *Journal of Mechanisms*, **4**(1), 43–56, 1969.

[10] A.T. Yang and B. Roth, Higher-order path curvature in spherical kinematics. *ASME Journal of Engineering for Industry*, **95**(2), 612–616, 1973.

[11] G.R. Veldkamp, Canonical systems and instantaneous invariants in spatial kinematics. *Journal of Mechanisms*, **2**(3), 329–388, 1967.

[12] L.M. Hsia and A.T. Yang, On the intrinsic properties of point trajectory in three-dimensional kinematics. *ASME Journal of Mechanical Design*, **107**(3), 401–405, 1985.

[13] A.T. Yang, Y. Kirson, and B. Roth, On a kinematic curvature theory for ruled surfaces. Proceedings of the Fourth World Congress on the Theory of Machines and Mechanisms, Newcastle-upon-Tyne, pp. 737–742, 1975.

[14] J.M. McCarthy and B. Roth, The curvature theory of line trajectories in spatial kinematics. *ASME Journal of Mechanical Design*, **103**(4), 718–724, 1981.

[15] J.M. McCarthy, On the scalar and dual formulations of the curvature theory of line trajectories. *Journal of Mechanisms, Transmissions, and Automation in Design*, **109**(1), 101–106, 1987.
[16] K.L. Ting and A.H. Soni, Instantaneous kinematics of a plane in space motion. *ASME Journal of Mechanisms, Transmissions, and Automation in Design*, **105**(3), 560–567, 1983.
[17] K.L. Ting and A.H. Soni, Instantaneous kinematics of a plane in spherical motion. *ASME Journal of Mechanisms, Transmissions, and Automation in Design*, **105**(3), 552–559, 1983.
[18] O. Bottema and B. Roth, *Theoretical Kinematics*. North-Holland, New York, 1979.
[19] M. Chasles, Note sur les Proprietes Generales du Systeme de Deux Corps Semblables entr'eux, Bullettin de Sciences Mathematiques, Astronomiques Physiques et Chimiques, Baron de Ferussac, Paris, pp. 321–326, 1830.
[20] G. Mozzi, Discorso matematico sopra il rotamento momentaneo dei corpi, Stamperia di Donato Campo, Napoli, 1763.
[21] R.S. Ball, *A Treatise on the Theory of Screws*. Cambridge University Press, Cambridge, 1900.
[22] W.K. Clifford, Preliminary sketch of bi-quaternion. *Proceedings of London Mathematical Society*, **4**(64/65), 381–395, 1873.
[23] E. Study, *Die Geometrie der Dynamen*. Teubner, Leipzig, 1903.
[24] L. Brand, *Vector and Tensor Analysis*. Wiley, Chichester, 1947.
[25] K.H. Hunt, *Kinematic Geometry of Mechanisms*. Clarendon Press, Oxford, 1978.
[26] J.M. McCarthy, *An Introduction to Theoretical Kinematics*. MIT Press, London, 1990.
[27] J. Duffy, *Statics and Kinematics with Applications to Robotics*. Cambridge University Press, New York, 1996.
[28] J.K. Davidson and K.H. Hunt, *Robots and Screw Theory: Applications of Kinematics and Statics to Robotics* Oxford University Press, New York, 2004.
[29] J.S. Dai, *Screw Algebra and Kinematics Approaches of Mechanisms and Robotics*. Springer, London, 2013.
[30] Z. Huan, Y.S. Zhao, and T.S. Zhao, *Advanced Spatial Mechanisms*. Higher Education Press, Beijing, 2006 (in Chinese).
[31] F.M. Dimentberg, *The Screw Calculus and Its Application to Mechanics*. Nauka, Moscow, 1965 (in Russian).
[32] A.T. Yang, Displacement analysis of spatial five-link mechanisms using 3*3 matrices with dual-number elements. *ASME Journal of Engineering for Industry*, **91**(1), 152–156, 1969.
[33] G.R. Pennock and A.T. Yang, Application of dual-number matrices to the inverse kinematics problem of robot manipulators. *ASME Journal of Mechanisms, Transmissions, and Automation in Design*, **107**(2), 201–208, 1985.
[34] L. Woo and F. Freudenstein, Application of line geometry to theoretical kinematics and the kinematic analysis of mechanical systems. *Journal of Mechanisms*, **5**(3), 417–460, 1970.
[35] B. Ravani and B. Roth, Mappings of spatial kinematics. *ASME Journal of Mechanisms, Transmissions, and Automation in Design*, **106**, 341–347, 1984.
[36] J.M. McCarthy and B. Ravani, Differential kinematics of spherical and spatial motions using kinematic mapping. *ASME Journal of Applied Mechanics*, **53**(1), 15–22, 1986.
[37] C. Lee, A.T. Yang, and B. Ravani, Coordinate system independent form of instantaneous invariants in spatial kinematics. *ASME Journal of Mechanical Design*, **115**(4), 946–952, 1993.
[38] Ö. Köse, Kinematic differential geometry of a rigid body in spatial motion using dual vector calculus: Part I. *Applied Mathematics and Computation*, **183**(1), 17–29, 2006.
[39] Ö. Köse, C.C. Sarıoğlu, B. Karabey, and İ. Karakılıç, Kinematic differential geometry of a rigid body in spatial motion using dual vector calculus: Part II. *Applied Mathematics and Computation*, **182**(1), 333–358, 2006.
[40] M. Skreiner, A study of the geometry and the kinematics of instantaneous spatial motion. *Journal of Mechanisms*, **1**(2), 115–143, 1966.
[41] E.H. Bokelberg, K.H. Hunt, and P.R. Ridley, Spatial motion-I: Points of inflection and the differential geometry of screws. *Mechanism and Machine Theory*, **27**(1), 1–15, 1992.
[42] P.R. Ridley, E.H. Bokelberg, and K.H. Hunt, Spatial motion-II: Acceleration and the differential geometry of screws. *Mechanism and Machine Theory*, **27**(1), 17–35, 1992.
[43] H. Stachel, Instantaneous spatial kinematics and the invariants of the axodes. Proceedings of a Symposium Commemorating the Legacy, Works, and Life of Sir Robert Stawell Ball Upon the 100th Anniversary of A Treatise on the Theory of Screws, pp. 1–14, July 9–11, 2000.
[44] M. Disteli, Über das Analogon der Savaryschen Formel und Konstruktion in der kinematischen Geometrie des Raumes. *Zeitschrift für Mathematic und Physik*, **62**, 261–309, 1914.
[45] D.B. Dooner and M.W. Griffis, On spatial Euler–Savary equations for envelopes. *ASME Journal of Mechanical Design*, **129**(8), 865–875, 2006.
[46] J.D. Schutter, Invariant description of rigid body motion trajectories. *ASME Journal of Mechanisms and Robotics*, **2**(1), 011001, 2009.

[47] D.L. Wang, J. Liu, and D.Z. Xiao, Kinematic differential geometry of a rigid body in spatial motion-I: A new adjoint approach and instantaneous properties of a point trajectory in spatial kinematics. *Mechanism and Machine Theory*, **32**(4), 419–432, 1997.

[48] D.L. Wang, J. Liu, and D.Z. Xiao, Kinematic differential geometry of a rigid body in spatial motion-II: A new adjoint approach and instantaneous properties of a line trajectory in spatial kinematics. *Mechanism and Machine Theory*, **32**(4), 433–444, 1997.

[49] D.L. Wang, J. Liu, and D.Z. Xiao, Kinematic differential geometry of a rigid body in spatial motion-III: Distribution of characteristic lines in the moving body in spatial motion. *Mechanism and Machine Theory*, **32**(4), 445–457, 1997.

7

Discrete Kinematic Geometry and Saddle Synthesis of Spatial Linkages

The kinematic synthesis of spatial linkages, either the open kinematic chain or the closed kinematic chain, for the given spatial movement of a rigid body is to determine the type and dimensions of a spatial linkage, or to search out the characteristic points or lines of the moving body, whose trajectories are the constraint curves or constraint ruled surfaces corresponding to the binary links or open kinematic chains. As mentioned in Chapter 3, several different kinematic pairs may form many kinds of binary link or open kinematic chain, even spatial linkages, leading to much more content and even more complexity for the kinematic synthesis of spatial linkages. Based on the invariants and invariance of the constraint curves and surfaces in Chapter 3, the local geometrical properties of trajectories of both points and lines, or at infinitesimally consecutive positions, are studied in Chapter 6 and the location of instantaneous characteristic points and lines of a moving body are also discussed. In this chapter, we focus on the global properties of the discrete trajectories of the points or lines, or the discrete kinematic geometry at separated positions and the kinematic synthesis of spatial linkages. The characteristic points of a moving body, whose trajectories are approximate to spherical curves or cylindrical curves – the constraint curves of binary links S–S or C′–S, are discussed in Sections 7.2 and 7.3. Then, the constant axis lines of the moving body are introduced in Sections 7.4 and 7.5, whose trajectories are approximate to the constraint ruled surface of the binary link C′–C′ and the degenerate cases, such as the characteristic lines for binary links R–C, R–R, H–C, H–R, H–H, etc are discussed. All of these are referred to as the elements of the discrete kinematic geometry of spatial linkages. As a result, the spatial four-bar linkages are synthesized by locating the characteristic points and characteristic lines in the moving body in Section 7.6.

It is not easy to intuitively image the kinematic synthesis of spatial linkages. The screw algebra is usually adopted to establish closed-loop equations for a binary link or open kinematic chain and the dimension parameters can be solved directly. (which is valid for few locations, but inapplicable for multiple positions of a moving body). Based on the invariants of both

Kinematic Differential Geometry and Saddle Synthesis of Linkages, First Edition. Delun Wang and Wei Wang.

the constraint curves and constraint surfaces in Chapter 3, we develop the saddle point programming for saddle curve fitting in Chapters 2 and 5 to that of space curve fitting and ruled surface fitting in this chapter. The characteristic points or lines of a moving body can be optimally located in terms of two space curves' fitting, spherical curve fitting and cylindrical curve fitting. The approach for kinematic synthesis, from plane, spherical to spatial linkages, is systematically established.

7.1 The Displacement Matrix

To describe the spatial movement of a moving rigid body $\sum^*$ to a fixed rigid body $\sum$, a moving Cartesian coordinate system $\{\boldsymbol{O}_m; \boldsymbol{i}_m, \boldsymbol{j}_m, \boldsymbol{k}_m\}$ with origin point O_m is set up on $\sum^*$ and the fixed system $\{\boldsymbol{O}_f; \boldsymbol{i}_f, \boldsymbol{j}_f, \boldsymbol{k}_f\}$ with origin point O_f is established on the base body $\sum$. The relative rotation angles of the two frames are θ_1, θ_2, θ_3 and the origin point O_m has coordinates $(x_{Omf}, y_{Omf}, z_{Omf})$ in $\{\boldsymbol{O}_f; \boldsymbol{i}_f, \boldsymbol{j}_f, \boldsymbol{k}_f\}$, which are introduced in Chapter 6. The spatial movement of $\sum^*$ can be described by the displacements of three different reference points or a reference line with a reference point in $\{\boldsymbol{O}_f; \boldsymbol{i}_f, \boldsymbol{j}_f, \boldsymbol{k}_f\}$. The displacements may be continuous as discussed in Chapter 6, or discrete (separated positions) as introduced in this chapter. The movement of $\sum^*$ can be represented by the transformation matrix, or the displacement matrix, which is conveniently applied to the kinematic synthesis of linkages.

A point P of $\sum^*$ has Cartesian coordinates (x_{Pm}, y_{Pm}, z_{Pm}) in $\{\boldsymbol{O}_m; \boldsymbol{i}_m, \boldsymbol{j}_m, \boldsymbol{k}_m\}$, whose position vector is

$$\boldsymbol{R}_{Pm} = x_{Pm}\boldsymbol{i}_m + y_{Pm}\boldsymbol{j}_m + z_{Pm}\boldsymbol{k}_m \tag{7.1}$$

The displacement of P can be located by the rectangular coordinates (x_{Pf}, y_{Pf}, z_{Pf}) in $\{\boldsymbol{O}_f; \boldsymbol{i}_f, \boldsymbol{j}_f, \boldsymbol{k}_f\}$, and described by the displacement matrix with six parameters $(x_{Omf}, y_{Omf}, z_{Omf}, \theta_1, \theta_2, \theta_3)$ as

$$\begin{bmatrix} x_{Pf} \\ y_{Pf} \\ z_{Pf} \\ 1 \end{bmatrix} = [\mathbf{M}] \cdot \begin{bmatrix} x_{Pm} \\ y_{Pm} \\ z_{Pm} \\ 1 \end{bmatrix} \tag{7.2}$$

where

$$[\mathbf{M}] = \begin{bmatrix} \mathrm{c}\theta_3\mathrm{s}\theta_2 + \mathrm{c}\theta_1\mathrm{s}\theta_3\mathrm{c}\theta_2 & -\mathrm{s}\theta_3\mathrm{s}\theta_2 + \mathrm{c}\theta_1\mathrm{c}\theta_3\mathrm{c}\theta_2 & \mathrm{s}\theta_1\mathrm{c}\theta_2 & x_{Omf} \\ -\mathrm{c}\theta_3\mathrm{c}\theta_2 + \mathrm{c}\theta_1\mathrm{s}\theta_3\mathrm{s}\theta_2 & \mathrm{s}\theta_3\mathrm{c}\theta_2 + \mathrm{c}\theta_1\mathrm{c}\theta_3\mathrm{s}\theta_2 & \mathrm{s}\theta_1\mathrm{s}\theta_2 & y_{Omf} \\ -\mathrm{s}\theta_1\mathrm{s}\theta_3 & -\mathrm{s}\theta_1\mathrm{c}\theta_3 & \mathrm{c}\theta_1 & z_{Omf} \\ 0 & 0 & 0 & 1 \end{bmatrix}$$

Here, s and c are abbreviations of functions sine and cosine. [**M**] is referred to as the ***spatial displacement matrix***, which reveals the position relationship between $\sum^*$ and $\sum$. There is another way to obtain the six motion parameters of $\sum^*$ in $\{\boldsymbol{O}_f; \boldsymbol{i}_f, \boldsymbol{j}_f, \boldsymbol{k}_f\}$. Three points A, B, and C of $\sum^*$, which are not co-linear, have the displacements (or coordinates) (x_A, y_A, z_A), (x_B, y_B, z_B), and (x_C, y_C, z_C) in $\{\boldsymbol{O}_f; \boldsymbol{i}_f, \boldsymbol{j}_f, \boldsymbol{k}_f\}$. According to Ref. [1], the fourth point D has to be located, whose displacements are

$$\begin{cases} x_D = x_C + N_x/\sqrt{N_x^2 + N_y^2 + N_z^2} \\ y_D = y_C + N_y/\sqrt{N_x^2 + N_y^2 + N_z^2} \\ z_D = z_C + N_z/\sqrt{N_x^2 + N_y^2 + N_z^2} \end{cases} \tag{7.3}$$

where

$$\begin{cases} N_x = (y_A - y_C)(z_B - z_C) - (y_B - y_C)(z_A - z_C) \\ N_y = (z_A - z_C)(x_B - x_C) - (z_B - z_C)(x_A - x_C) \\ N_z = (x_A - x_C)(y_B - y_C) - (x_B - x_C)(y_A - y_C) \end{cases}$$

This leads to the four points A, B, C, and D are not coplanar. They respectively form column matrices (A_{I}), (B_{I}), (C_{I}), (D_{I}) at the position I and (A_{II}), (B_{II}), (C_{II}), (D_{II}) at the position II. Based on all the eight column matrices, we have the displacement matrix $[\mathbf{M}_{\mathrm{I\,II}}]$ of the moving body $\sum^*$ from position I to position II as

$$[\mathbf{M}_{\mathrm{I\,II}}] = \begin{bmatrix} x_{A\mathrm{II}} & x_{B\mathrm{II}} & x_{C\mathrm{II}} & x_{D\mathrm{II}} \\ y_{A\mathrm{II}} & y_{B\mathrm{II}} & y_{C\mathrm{II}} & y_{D\mathrm{II}} \\ z_{A\mathrm{II}} & z_{B\mathrm{II}} & z_{C\mathrm{II}} & z_{D\mathrm{II}} \\ 1 & 1 & 1 & 1 \end{bmatrix} \cdot \begin{bmatrix} x_{A\mathrm{I}} & x_{B\mathrm{I}} & x_{C\mathrm{I}} & x_{D\mathrm{I}} \\ y_{A\mathrm{I}} & y_{B\mathrm{I}} & y_{C\mathrm{I}} & y_{D\mathrm{I}} \\ z_{A\mathrm{I}} & z_{B\mathrm{I}} & z_{C\mathrm{I}} & z_{D\mathrm{I}} \\ 1 & 1 & 1 & 1 \end{bmatrix}^{-1} \tag{7.4}$$

In particular, letting the moving frame $\{\boldsymbol{O}_m; \boldsymbol{i}_m, \boldsymbol{j}_m, \boldsymbol{k}_m\}$ of $\sum^*$ at position I be coincident with the fixed frame $\{\boldsymbol{O}_f; \boldsymbol{i}_f, \boldsymbol{j}_f, \boldsymbol{k}_f\}$ of $\sum$, we can get the displacement matrix of $\sum^*$ relative to $\sum$ at any position from equation (7.4). Hence, through the equivalence of the spatial displacement matrix $[\mathbf{M}_{\mathrm{I\,II}}]$ in equation (7.4) and $[\mathbf{M}]$ in equation (7.2), the parameters $(x_{Omf}, y_{Omf}, z_{Omf}, \theta_1, \theta_2, \theta_3)$ can be solved. Hereinafter, the parameters $(x_{Omf}, y_{Omf}, z_{Omf}, \theta_1, \theta_2, \theta_3)$ are directly given to define the spatial movement of $\sum^*$.

For a straight line L of $\sum^*$ with direction parameters (ϕ, φ) in $\{\boldsymbol{O}_m; \boldsymbol{i}_m, \boldsymbol{j}_m, \boldsymbol{k}_m\}$, the unit direction vector can be written as

$$\boldsymbol{l}_m = \sin\phi\,\cos\varphi\boldsymbol{i}_m + \sin\phi\,\sin\varphi\boldsymbol{j}_m + \cos\phi\boldsymbol{k}_m \tag{7.5}$$

If the line L is described in the fixed frame $\{\boldsymbol{O}_f; \boldsymbol{i}_f, \boldsymbol{j}_f, \boldsymbol{k}_f\}$, the direction cosines (l_x, l_y, l_z) are

$$\begin{bmatrix} l_x \\ l_y \\ l_z \end{bmatrix} = [\mathbf{R}] \cdot \begin{bmatrix} \mathrm{s}\phi\mathrm{c}\varphi \\ \mathrm{s}\phi\mathrm{s}\varphi \\ \mathrm{c}\phi \end{bmatrix} \tag{7.6}$$

where

$$[\mathbf{R}] = \begin{bmatrix} \mathrm{c}\theta_3\mathrm{s}\theta_2 + \mathrm{c}\theta_1\mathrm{s}\theta_3\mathrm{c}\theta_2 & \mathrm{s}\theta_3\mathrm{s}\theta_2 + \mathrm{c}\theta_1\mathrm{c}\theta_3\mathrm{c}\theta_2 & \mathrm{s}\theta_1\mathrm{c}\theta_2 \\ -\mathrm{c}\theta_3\mathrm{c}\theta_2 + \mathrm{c}\theta_1\mathrm{s}\theta_3\mathrm{s}\theta_2 & \mathrm{s}\theta_3\mathrm{c}\theta_2 + \mathrm{c}\theta_1\mathrm{c}\theta_3\mathrm{s}\theta_2 & \mathrm{s}\theta_1\mathrm{s}\theta_2 \\ -\mathrm{s}\theta_1\mathrm{s}\theta_3 & -\mathrm{s}\theta_1\mathrm{c}\theta_3 & \mathrm{c}\theta_1 \end{bmatrix}$$

Here, the rotation matrix $[\mathbf{R}]$ is a sub-matrix of the displacement matrix $[\mathbf{M}]$.

The continuous spatial motion of $\sum^*$ have been introduced in Chapter 6. We now focus on the discrete displacements or the separated positions of $\sum^*$. The motion parameters in equation

(7.2) or (7.6) is added a superscript bracket (i), such as $(x_{Omf}^{(i)}, y_{Omf}^{(i)}, z_{Omf}^{(i)}; \theta_1^{(i)}, \theta_2^{(i)}, \theta_3^{(i)})$, and define a specific position of $\sum^*$. Hence, a point of $\sum^*$ produces discrete displacements in the fixed frame, which can be represented by a discrete point set $\{\boldsymbol{R}_P^{(i)}\}$. Equation (7.2) can be rewritten in discrete form as

$$\boldsymbol{R}_P^{(i)} = \begin{bmatrix} x_{Pf}^{(i)} \\ y_{Pf}^{(i)} \\ z_{Pf}^{(i)} \end{bmatrix} = \begin{bmatrix} x_{Omf}^{(i)} \\ y_{Omf}^{(i)} \\ z_{Omf}^{(i)} \end{bmatrix} + [\mathbf{R}^{(i)}] \cdot \begin{bmatrix} x_{Pm} \\ y_{Pm} \\ z_{Pm} \end{bmatrix} \tag{7.7}$$

where

$$[\mathbf{R}^{(i)}] = \begin{bmatrix} c\theta_3^{(i)} s\theta_2^{(i)} + c\theta_1^{(i)} s\theta_3^{(i)} c\theta_2^{(i)} & -s\theta_3^{(i)} s\theta_2^{(i)} + c\theta_1^{(i)} c\theta_3^{(i)} c\theta_2^{(i)} & s\theta_1^{(i)} c\theta_2^{(i)} \\ -c\theta_3^{(i)} c\theta_2^{(i)} + c\theta_1^{(i)} s\theta_3^{(i)} s\theta_2^{(i)} & s\theta_3^{(i)} c\theta_2^{(i)} + c\theta_1^{(i)} c\theta_3^{(i)} s\theta_2^{(i)} & s\theta_1^{(i)} s\theta_2^{(i)} \\ -s\theta_1^{(i)} s\theta_3^{(i)} & -s\theta_1^{(i)} c\theta_3^{(i)} & c\theta_1^{(i)} \end{bmatrix}$$

A straight line L of $\sum^*$ with direction parameters (ϕ, φ) traces a discrete ruled surface in $\{\boldsymbol{O}_f; \boldsymbol{i}_f, \boldsymbol{j}_f, \boldsymbol{k}_f\}$ or a discrete line set $\{\boldsymbol{R}_L^{(i)}\}$:

$$\boldsymbol{R}_L^{(i)} = \boldsymbol{R}_P^{(i)} + \mu \boldsymbol{l}^{(i)} = \begin{bmatrix} x_{Pf}^{(i)} \\ y_{Pf}^{(i)} \\ z_{Pf}^{(i)} \end{bmatrix} + \mu [\mathbf{R}^{(i)}] \cdot \begin{bmatrix} s\phi c\varphi \\ s\phi s\varphi \\ c\phi \end{bmatrix} \tag{7.8}$$

The space curve or ruled surface traced by a point or line of $\sum^*$ in discrete spatial movement can be represented by a discrete point set $\{\boldsymbol{R}_P^{(i)}\}$ or a discrete line set $\{\boldsymbol{R}_L^{(i)}\}$, which are determined by equation (7.7) or (7.8) with the discrete parameters $(x_{Omf}^{(i)}, y_{Omf}^{(i)}, z_{Omf}^{(i)}, \theta_1^{(i)}, \theta_2^{(i)}, \theta_3^{(i)})$ or the discrete rotation matrix $[\mathbf{R}^{(i)}]$. Thus, the discrete point set $\{\boldsymbol{R}_P^{(i)}\}$ and the discrete line set $\{\boldsymbol{R}_L^{(i)}\}$, or $\{(\boldsymbol{R}_P^{(i)}, \boldsymbol{l}^{(i)})\}$, can be taken as known conditions in this chapter.

7.2 Saddle Sphere Point P_{SS}

The open kinematic chain S–S is a current binary link of spatial linkages, whose constraint curve is a spherical curve. It is naturally one of the basic elements in the discrete kinematic geometry of spatial linkages. The global comparison of the discrete trajectories traced by the points of $\sum^*$ with a spherical curve is discussed intensively.

7.2.1 Spherical Surface Fitting

Assume the discrete spatial movement of $\sum^*$ or multiple separated positions of $\sum^*$ in the fixed frame $\{\boldsymbol{O}_f; \boldsymbol{i}_f, \boldsymbol{j}_f, \boldsymbol{k}_f\}$ are prescribed. A point P of $\sum^*$ with Cartesian coordinates (x_{Pm}, y_{Pm}, z_{Pm})

in the moving frame $\{\boldsymbol{O}_m; \boldsymbol{i}_m, \boldsymbol{j}_m, \boldsymbol{k}_m\}$ traces a spatial discrete trajectory, or a discrete point set $\{\boldsymbol{R}_P^{(i)}\}$ described by equation (7.7). We are interested in the location of the special point of $\sum^*$, whose discrete point set $\{\boldsymbol{R}_P^{(i)}\}$ lies precisely or approximately on a spherical surface corresponding to the binary link S–S. As mentioned in Chapter 3, the normal curvature of a spherical curve is constant and the geodesic torsion is zero, or the distances from the sphere center to all points of the curve are constant. From the geometrical perspective of the invariants of a spherical curve, we can estimate the approximation of a spatial discrete point set and a spherical surface through the normal distances.

Similar to the circle fitting in Chapter 2, a spherical surface is taken to fit the discrete point set $\{\boldsymbol{R}_P^{(i)}\}$ and the fitting errors depend on the radius r and the position coordinates (x_C, y_C, z_C) of the sphere center if they are simultaneously given. Here, the fitting error is not only measured in the normal direction of the spherical surface, but also the maximum value is taken to estimate the approximation of the discrete points and the fitting spherical surface, although the maximum error may exist at several points of $\{\boldsymbol{R}_P^{(i)}\}$. A spherical surface only with a given radius r is used to fit the discrete trajectory $\{\boldsymbol{R}_P^{(i)}\}$, and the position coordinates (x_C, y_C, z_C) of the sphere center are adaptively determined by letting the (maximum) fitting error be minimum, referred to as a ***floating fitting spherical surface***, as shown in Fig. 7.1. Obviously, the fitting error varies with the radius r and there must exist a specific value of the radius r, leading to the (maximum) fitting error being the minimum, or a saddle spherical surface.

Definition 7.1 Based on the given discrete point set $\{\boldsymbol{R}_P^{(i)}\}$, a spherical surface, adaptively determined by letting the maximum fitting error be minimum, is defined as a ***saddle spherical surface***. The maximum fitting error is referred to as the ***saddle spherical surface error***.

As to Definition 7.1, for all spherical surfaces fitting the given discrete point set $\{\boldsymbol{R}_P^{(i)}\}$, the saddle spherical surface's (maximum) fitting error is minimum, or there does not exist any other fitting spherical surface with less (maximum) fitting error. The saddle spherical surface

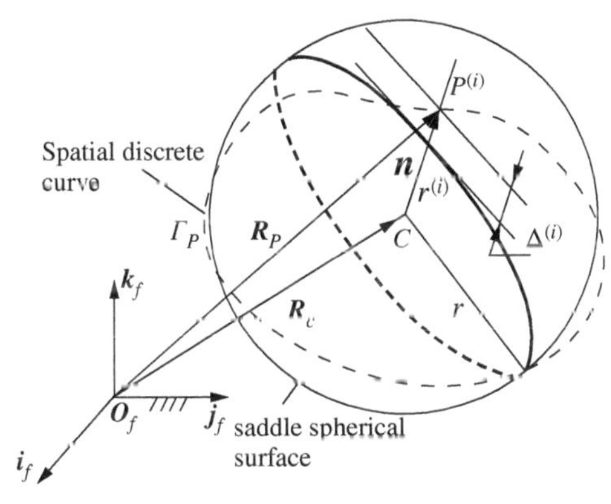

Figure 7.1 The saddle spherical surface fitting of a discrete curve

has optimum parameters since the saddle spherical surface error is maximum for all discrete points in $\{\boldsymbol{R}_P^{(i)}\}$ and minimum for all fitting spherical surfaces with parameters (x_C, y_C, z_C, r). We set up the mathematical model of saddle spherical surface fitting:

$$\begin{cases} \Delta_{SS} = \min\limits_{\boldsymbol{x}} \max\limits_{1\le i\le n} \left\{\Delta^{(i)}(\boldsymbol{x})\right\} \\ \quad = \min\limits_{\boldsymbol{x}} \max\limits_{1\le i\le n} \left\{\left|\sqrt{(x_P^{(i)} - x_C)^2 + (y_P^{(i)} - y_C)^2 + (z_P^{(i)} - z_C)^2} - r\right|\right\} \\ \text{s.t.} \quad r \in (0, +\infty) \\ \boldsymbol{x} = (x_C, y_C, z_C, r)^T \end{cases} \tag{7.9}$$

where the objective function $\{\Delta^{(i)}(\boldsymbol{x})\}$ is the normal fitting error set of the discrete point set $\{\boldsymbol{R}_P^{(i)}\}$ and a fitting spherical surface. The optimization variables $\boldsymbol{x} = (x_C, y_C, z_C, r)^T$ are the sphere center's coordinates (x_C, y_C, z_C) and the radius r of the fitting spherical surface. n is the number of discrete points of $\{\boldsymbol{R}_P^{(i)}\}$, and Δ_{SS} is the saddle spherical surface error for $\{\boldsymbol{R}_P^{(i)}\}$.

The saddle spherical surface, determined by the mathematical model (7.9) for a given discrete point set $\{\boldsymbol{R}_P^{(i)}\}$, has the same properties as the saddle circle in Chapter 2, such as adaptivity, uniqueness, and comparability. For the mathematical model (7.9), the initial values of the optimization variables have to be given first. For convenience of computation, a fitting spherical surface can be roughly determined by the least-square method for $\{\boldsymbol{R}_P^{(i)}\}$, whose position parameters and radius can be taken as the initial values of the optimization variables. The normal deviation e_i between $\{\boldsymbol{R}_P^{(i)}\}$ and the fitting spherical surface can be represented as

$$e_i^2 = (x_P^{(i)} - x_C)^2 + (y_P^{(i)} - y_C)^2 + (z_P^{(i)} - z_C)^2 - r^2, \ i = 1, \ldots, n \tag{7.10}$$

The quadratic sum of the normal deviation e_i is

$$\begin{aligned} F &= \sum_{i=1}^{n} e_i^2 = \sum_{i=1}^{n} [(x_P^{(i)} - x_C)^2 + (y_P^{(i)} - y_C)^2 + (z_P^{(i)} - z_C)^2 - r^2]^2 \\ &= \sum_{i=1}^{n} [(x_P^{(i)})^2 + (y_P^{(i)})^2 + (z_P^{(i)})^2 + A \cdot x_P^{(i)} + B \cdot y_P^{(i)} + C \cdot z_P^{(i)} + D]^2 \end{aligned} \tag{7.11}$$

where $A = -2x_C$, $B = -2y_C$, $C = -2z_C$, $D = x_C^2 + y_C^2 + z_C^2 - r^2$. Differentiating the above equation with respect to A, B, C, and D and letting the derivatives be zero, we have

$$\begin{cases} \dfrac{\partial F}{\partial A} = 2\sum\limits_{i=1}^{n} [(x_P^{(i)})^2 + (y_P^{(i)})^2 + (z_P^{(i)})^2 + A \cdot x_P^{(i)} + B \cdot y_P^{(i)} + C \cdot z_P^{(i)} + D] \cdot x_P^{(i)} = 0 \\ \dfrac{\partial F}{\partial B} = 2\sum\limits_{i=1}^{n} [(x_P^{(i)})^2 + (y_P^{(i)})^2 + (z_P^{(i)})^2 + A \cdot x_P^{(i)} + B \cdot y_P^{(i)} + C \cdot z_P^{(i)} + D] \cdot y_P^{(i)} = 0 \\ \dfrac{\partial F}{\partial C} = 2\sum\limits_{i=1}^{n} [(x_P^{(i)})^2 + (y_P^{(i)})^2 + (z_P^{(i)})^2 + A \cdot x_P^{(i)} + B \cdot y_P^{(i)} + C \cdot z_P^{(i)} + D] \cdot z_P^{(i)} = 0 \\ \dfrac{\partial F}{\partial D} = 2\sum\limits_{i=1}^{n} [(x_P^{(i)})^2 + (y_P^{(i)})^2 + (z_P^{(i)})^2 + A \cdot x_P^{(i)} + B \cdot y_P^{(i)} + C \cdot z_P^{(i)} + D] = 0 \end{cases} \tag{7.12}$$

We get the equation system of the variables A, B, C, and D, that is

$$\begin{cases} a_{11}A + a_{12}B + a_{13}C + a_{14}D = a_{15} \\ a_{21}A + a_{22}B + a_{23}C + a_{24}D = a_{25} \\ a_{31}A + a_{32}B + a_{33}C + a_{34}D = a_{35} \\ a_{41}A + a_{42}B + a_{43}C + a_{44}D = a_{45} \end{cases} \tag{7.13}$$

where

$$a_{11} = \sum_{i=1}^{n} (x_P^{(i)})^2,\ a_{12} = \sum_{i=1}^{n} x_P^{(i)} y_P^{(i)},\ a_{13} = \sum_{i=1}^{n} x_P^{(i)} z_P^{(i)},\ a_{14} = \sum_{i=1}^{n} x_P^{(i)},$$

$$a_{15} = -\sum_{i=1}^{n} [(x_P^{(i)})^3 + x_P^{(i)}(y_P^{(i)})^2 + x_P^{(i)}(z_P^{(i)})^2]$$

$$a_{21} = \sum_{i=1}^{n} x_P^{(i)} y_P^{(i)},\ a_{22} = \sum_{i=1}^{n} (y_P^{(i)})^2,\ a_{23} = \sum_{i=1}^{n} y_P^{(i)} z_P^{(i)},\ a_{24} = \sum_{i-1}^{n} y_P^{(i)},$$

$$a_{25} = -\sum_{i=1}^{n} [(x_P^{(i)})^2 y_P^{(i)} + (y_P^{(i)})^3 + y_P^{(i)}(z_P^{(i)})^2]$$

$$a_{31} = \sum_{i=1}^{n} x_P^{(i)} z_P^{(i)},\ a_{32} = \sum_{i=1}^{n} y_P^{(i)} z_P^{(i)},\ a_{33} = \sum_{i=1}^{n} (z_P^{(i)})^2,\ a_{34} = \sum_{i=1}^{n} z_P^{(i)},$$

$$a_{35} = -\sum_{i=1}^{n} [(x_P^{(i)})^2 z_P^{(i)} + (y_P^{(i)})^2 z_P^{(i)} + (z_P^{(i)})^3]$$

$$a_{41} = \sum_{i=1}^{n} x_P^{(i)},\ a_{42} = \sum_{i=1}^{n} y_P^{(i)},\ a_{43} = \sum_{i=1}^{n} z_P^{(i)},\ a_{44} = n,\ a_{45} = -\sum_{i=1}^{n} [(x_P^{(i)})^2 + (y_P^{(i)})^2 + (z_P^{(i)})^2]$$

From the equation system, the variables A, B, C, and D can be solved. The initial values of the optimization variables or the center point coordinates (x_C, y_C, z_C) and the radius r of the fitting spherical surface can be assigned as

$$\begin{cases} x_C = -\dfrac{A}{2},\ y_C = -\dfrac{B}{2},\ z_C = -\dfrac{C}{2} \\ r = \sqrt{x_C^2 + y_C^2 + z_C^2 - D} \end{cases} \tag{7.14}$$

For a given discrete trajectory $\{\boldsymbol{R}_P^{(i)}\}$ in the fixed frame $\{\boldsymbol{O}_f; \boldsymbol{i}_f, \boldsymbol{j}_f, \boldsymbol{k}_f\}$, the saddle spherical surface can be located by the function **fminimax** in Matlab® based on Definition 7.1 and the mathematic model (7.9). The parameters of **fminimax** are set in the same way as those for saddle circle fitting in Chapter 2, and a ***sub-program ASSF*** is constructed for saddle spherical surface fitting to $\{\boldsymbol{R}_P^{(i)}\}$. The input parameters are the coordinates of the discrete points and the output parameters are the center point's coordinates and the radius of the saddle spherical surface, and the saddle spherical surface error Δ_{SS}.

7.2.2 Saddle Spherical Surface

For convenience in discussing the properties of the saddle spherical surface errors, the given spatial trajectory Γ_P is first assumed to be a continuous curve. The position vector of the saddle spherical surface's center is $\boldsymbol{R}_C$, and the radius is r. As shown in Fig. 7.1, for any point P of Γ_P, the normal fitting error to the saddle spherical surface is Δ, and the position vector $\boldsymbol{R}_P$ of P can be written as

$$\boldsymbol{R}_P = \boldsymbol{R}_C + (r + \Delta)\boldsymbol{n} \tag{7.15}$$

The sign of Δ is positive if point P lies outside the saddle spherical surface and negative if P is inside. Δ can be derived from the above equation:

$$\Delta = (\boldsymbol{R}_P - \boldsymbol{R}_C) \cdot \boldsymbol{n} - r \tag{7.16}$$

For a saddle spherical surface of Γ_P by equation (7.9), the position vector $\boldsymbol{R}_C$ is a constant vector and the radius r is a constant. By differentiating equation (7.16) with respect to the arc length s of Γ_P, we have

$$\frac{\mathrm{d}\Delta}{\mathrm{d}s} = \frac{\mathrm{d}\boldsymbol{R}_P}{\mathrm{d}s} \cdot \boldsymbol{n} + (\boldsymbol{R}_P - \boldsymbol{R}_C) \cdot \frac{\mathrm{d}\boldsymbol{n}}{\mathrm{d}s} \tag{7.17}$$

Substituting vector expression $\boldsymbol{R}_P - \boldsymbol{R}_C = (r + \Delta)\boldsymbol{n}$ into the above equation, we have $\frac{\mathrm{d}\Delta}{\mathrm{d}s} = \frac{\mathrm{d}\boldsymbol{R}_P}{\mathrm{d}s} \cdot \boldsymbol{n}$. The condition for Δ to be an extreme value is $\frac{\mathrm{d}\Delta}{\mathrm{d}s} = 0$ at a point P of Γ_P, or

$$\frac{\mathrm{d}\boldsymbol{R}_P}{\mathrm{d}s} \cdot \boldsymbol{n} = 0 \tag{7.18}$$

This means that when the spatial curve Γ_P and its saddle spherical surface have a common normal, the normal error Δ gets an extreme value. Similar to that of the saddle circle of a planar curve, the extreme value of the fitting errors appears at multiple points on Γ_P but may not be the saddle spherical surface error, such as the points $P^{(k+1)}$ and $P^{(k+2)}$ in Fig. 7.2.

A saddle spherical surface of a given spatial curve Γ_P can be determined totally by five parameters, (x_C, y_C, z_C, r) and the saddle spherical surface error Δ_{SS}, which can be solved by equation (7.15) or (7.16) at five points on Γ_P (or at least five points of Γ_P have to hold equation (7.18) true).

According to the theory of saddle point programming [2], we can estimate the degree of approximation of a given spatial curve and a spherical surface. Spherical surface fitting by letting the maximum fitting error be the minima is equivalent to the given spatial curve being contained by two concentric spherical surfaces. The given spatial curve may have five tangent points at least with the two concentric spherical surfaces, leading to the maximum normal fitting errors being the minima with the same value, as shown in Fig. 7.2. The five tangent points distribute on both sides of the saddle spherical surface in the manner two are inner to the saddle spherical surface and another three are outer, or vice versa.

Therefore, the five tangent points of Γ_P geometrically locate its saddle spherical surface and are defined as the ***saddle spherical surface fitting points*** of Γ_P. The absolute values of the normal fitting errors at the five points are the same and equal to the saddle spherical surface error. The given spatial curve lies on a spherical surface if the saddle spherical surface error is zero.

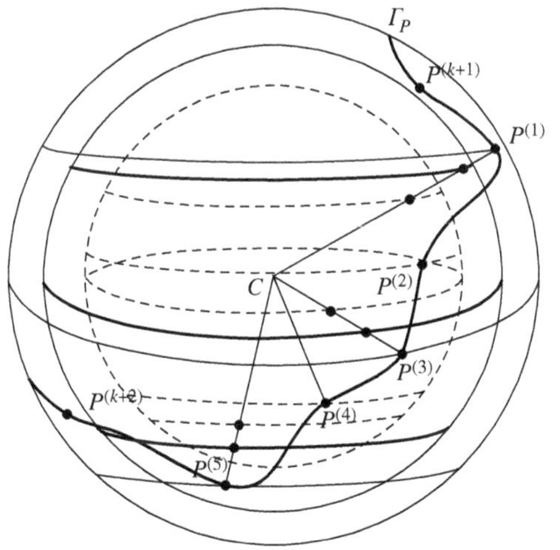

Figure 7.2 Two concentric spheres of a space curve

We now switch to the discrete trajectory $\{\boldsymbol{R}_P^{(i)}\}$ of a point P of $\sum^*$, rather than a continuous curve Γ_P. The five saddle spherical surface fitting points have to be selected from the discrete point set $\{\boldsymbol{R}_P^{(i)}\}$. By equation (7.16), the normal error $\Delta^{(i)}$ of any point in $\{\boldsymbol{R}_P^{(i)}\}$ is

$$\begin{cases} \Delta^{(i)} = |\boldsymbol{R}_P^{(i)} - \boldsymbol{R}_C| - r = \sqrt{(x_P^{(i)} - x_C)^2 + (y_P^{(i)} - y_C)^2 + (z_P^{(i)} - z_C)^2} - r \\ \boldsymbol{R}_P^{(i)} = [\mathbf{M}^{(i)}] \cdot \boldsymbol{R}_{Pm} \end{cases} \tag{7.19}$$

The saddle spherical surface error Δ_{SS} is the maximum absolute value of all $\Delta^{(i)}$:

$$\Delta_{SS} = \max_i\{|\Delta^{(i)}|\} = \max_i\{|\sqrt{(x_P^{(i)} - x_C)^2 + (y_P^{(i)} - y_C)^2 + (z_P^{(i)} - z_C)^2} - r|\} \tag{7.20}$$

It indicates that the saddle spherical surface error of a discrete point set $\{\boldsymbol{R}_P^{(i)}\}$ can be obtained by numerical calculation and comparison among all the normal errors $\Delta^{(i)}$ of $\{\boldsymbol{R}_P^{(i)}\}$ by equation (7.20). The positions of the moving body $\sum^*$ corresponding to the five saddle spherical surface fitting points are referred to as the ***saddle spherical surface fitting positions***. The mathematical model of saddle spherical surface fitting sets up the relationships among the separated positions of $\sum^*$, the discrete point set $\{\boldsymbol{R}_P^{(i)}\}$, and its saddle spherical surface error Δ_{SS}.

7.2.3 Five Positions

Five separated positions of a moving body $\sum^*$ in spatial movement are prescribed in $\{\boldsymbol{O}_f; \boldsymbol{i}_f, \boldsymbol{j}_f, \boldsymbol{k}_f\}$ and numbered 1, 2, 3, 4, 5. A point with Cartesian coordinates (x_{Pm}, y_{Pm}, z_{Pm}) in $\{\boldsymbol{O}_m; \boldsymbol{i}_m, \boldsymbol{j}_m, \boldsymbol{k}_m\}$ traces a spatial trajectory $\{\boldsymbol{R}_P^{(i)}\}$, comprised of five points $P^{(1)}$, $P^{(2)}$, $P^{(3)}$, $P^{(4)}$, and $P^{(5)}$ in $\{\boldsymbol{O}_f; \boldsymbol{i}_f, \boldsymbol{j}_f, \boldsymbol{k}_f\}$. Based on the mathematical model (7.9), we call the sub-program ASSF and obtain the saddle spherical surface of $\{\boldsymbol{R}_P^{(i)}\}$, called the ***five-point***

saddle spherical surface. In this case, the five points $P^{(1)}$, $P^{(2)}$, $P^{(3)}$, $P^{(4)}$, and $P^{(5)}$ are the saddle spherical surface fitting points and the five positions of $\sum^*$ are the saddle spherical surface fitting positions.

A spatial discrete trajectory, traced by a point of $\sum^*$ with five separated positions, corresponds to a five-point saddle spherical surface. According to saddle point programming, the maximum fitting errors achieve the minimum if and only if the fitting errors are identical to an extreme value at the five fitted points $P^{(1)}$, $P^{(2)}$, $P^{(3)}$, $P^{(4)}$, and $P^{(5)}$, which distribute on both sides of the five-point saddle spherical surface. There are many distribution cases for the five points. For convenience in introducing the properties of the saddle spherical surface error, we just discuss one case, $P^{(1)}P^{(2)}-P^{(3)}P^{(4)}P^{(5)}$, of the ten distribution cases. The fitting spherical surface determined by any distribution case is referred to as the ***distributing spherical surface*** of the five points.

For distribution case $P^{(1)}P^{(2)}-P^{(3)}P^{(4)}P^{(5)}$, two points $P^{(1)}$, $P^{(2)}$ are inside and another three points $P^{(3)}$, $P^{(4)}$, $P^{(5)}$ are outside the distributing spherical surface. By equation (7.19), the distributing spherical surface error $\Delta_{12,345}$ meets

$$\begin{cases}(r-\Delta_{12,345})^2=(\boldsymbol{R}_P^{(i)}-\boldsymbol{R}_C)^2, \ i=1,\ 2\\(r+\Delta_{12,345})^2=(\boldsymbol{R}_P^{(i)}-\boldsymbol{R}_C)^2, \ i=3,\ 4,\ 5\end{cases} \tag{7.21}$$

Substituting equation (7.7) into the above equation, we have a system of equations including five equations of eight variables, coordinates (x_{Pm}, y_{Pm}, z_{Pm}) of P in $\{\boldsymbol{O}_m;\boldsymbol{i}_m,\boldsymbol{j}_m,\boldsymbol{k}_m\}$, position and radius parameters (x_C, y_C, z_C, r) of the distributing spherical surface, and the distributing spherical surface error $\Delta_{12,345}$. By eliminating the variables (x_C, y_C, z_C) and r from equation (7.21), a higher-order algebra equation of (x_{Pm}, y_{Pm}, z_{Pm}) and $\Delta_{12,345}$ can be obtained:

$$f(x_{Pm}, y_{Pm}, z_{Pm}, \Delta_{12,345})=0 \tag{7.22}$$

For the trajectory with five points $P^{(1)}$, $P^{(2)}$, $P^{(3)}$, $P^{(4)}$, and $P^{(5)}$, there are another nine distributing spherical surfaces with corresponding errors, such as $\Delta_{123,45}$, $\Delta_{124,35}$, $\Delta_{125,34}$, $\Delta_{134,25}$, $\Delta_{135,24}$, $\Delta_{145,23}$, $\Delta_{234,15}$, $\Delta_{235,14}$, $\Delta_{245,13}$. In a similar way, these distributing spherical surface errors can be obtained. According to Definition 7.1, the five-point saddle spherical surface error Δ_{12345} of $\{\boldsymbol{R}_P^{(i)}\}$, $i=1,\ 2,\ 3,\ 4,\ 5$ is the minimum of the ten distributing spherical surface errors.

For a moving body $\sum^*$ with five separated positions, which are defined by the motion parameters in the first five lines of Table 7.1, the point $P(x_{Pm}, y_{Pm}, z_{Pm})$ of $\sum^*$ in $\{\boldsymbol{O}_m;\boldsymbol{i}_m,\boldsymbol{j}_m,\boldsymbol{k}_m\}$ traces five discrete points in $\{\boldsymbol{O}_f;\boldsymbol{i}_f,\boldsymbol{j}_f,\boldsymbol{k}_f\}$. With the coordinates (x_{Pm}, y_{Pm}, z_{Pm}) taken as the independent variables and the five-point saddle spherical surface error Δ_{12345} as the dependent variable, the error hypersurface Σ_{12345}^{SS} in four-dimensional space can be constructed by equation (7.24). On the intersection planes between the error hypersurface surface Σ_{12345}^{SS} and three coordinate planes, $\boldsymbol{O}_m-\boldsymbol{x}_m\boldsymbol{y}_m$, $\boldsymbol{O}_m-\boldsymbol{x}_m\boldsymbol{z}_m$, and $\boldsymbol{O}_m-\boldsymbol{y}_m\boldsymbol{z}_m$, the points with the same saddle spherical surface errors are connected by contour lines, as shown in Fig. 7.3. In particular, the points of $\sum^*$, whose discrete trajectory lie on a sphere or the saddle spherical surface errors are zero, constitute a quartic surface [3], called the ***sphere point surface***, as shown in Fig. 7.4(a). The intersection curves of the sphere point surface with the plane $z_m=25$ are shown in Fig. 7.4(b) with heavy lines.

Table 7.1 Nine separated spatial positions of the rigid body

	$x_{Omf}^{(i)}$	$y_{Omf}^{(i)}$	$z_{Omf}^{(i)}$	$\theta_1^{(i)}$	$\theta_2^{(i)}$	$\theta_3^{(i)}$
1	0	0	0	0	π/2	0
2	−0.6435	0.5994	0.8447	0.1671	0.0196	1.5959
3	−1.7102	1.2385	2.1826	0.3514	0.2152	1.4284
4	−2.7024	1.5400	3.8675	0.5292	0.3880	1.2530
5	−3.3926	1.5194	5.5368	0.6867	0.5578	1.0740
6	−3.8524	1.3467	6.8812	0.8193	0.7236	0.8910
7	−4.2783	1.1959	7.6914	0.9264	0.8785	0.6970
8	−4.8537	1.2692	7.7800	1.0065	1.0083	0.4779
9	−5.6016	1.9542	6.9018	1.0467	1.0895	0.2103

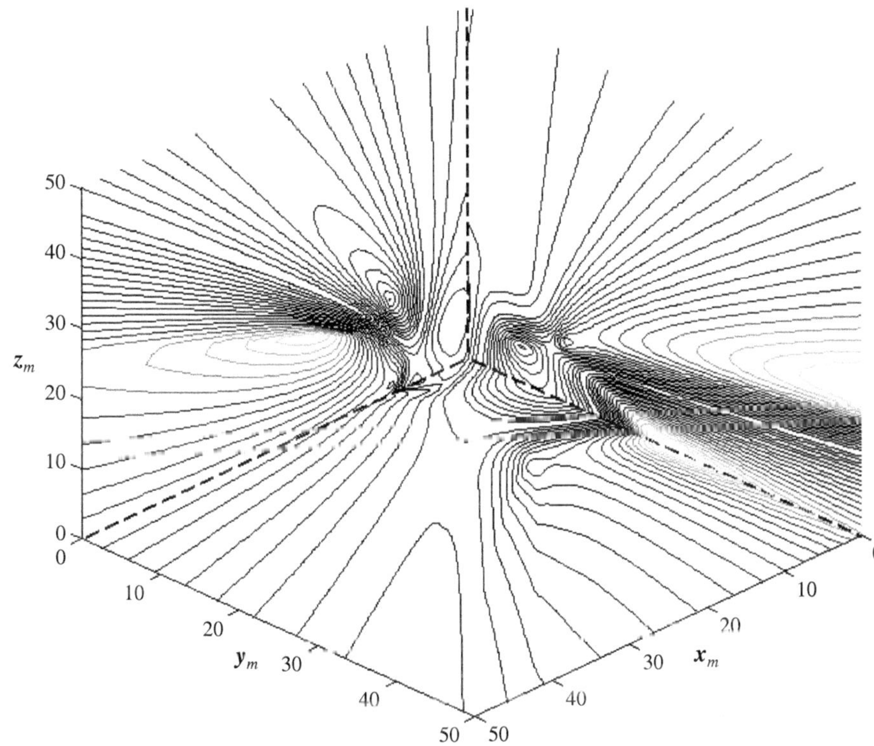

Figure 7.3 The five-position saddle spherical surface error

7.2.4 *Six Positions*

Six separated positions of a moving body $\sum^*$ in $\{\boldsymbol{O}_f; \boldsymbol{i}_f, \boldsymbol{j}_f, \boldsymbol{k}_f\}$ are numbered 1, 2, 3, 4, 5, 6. A point P with coordinates (x_{Pm}, y_{Pm}, z_{Pm}) in $\{\boldsymbol{O}_m; \boldsymbol{i}_m, \boldsymbol{j}_m, \boldsymbol{k}_m\}$ traces a spatial discrete trajectory, comprised of six discrete points $P^{(1)}$, $P^{(2)}$, $P^{(3)}$, $P^{(4)}$, $P^{(5)}$, and $P^{(6)}$ in $\{\boldsymbol{O}_f; \boldsymbol{i}_f, \boldsymbol{j}_f, \boldsymbol{k}_f\}$. Based

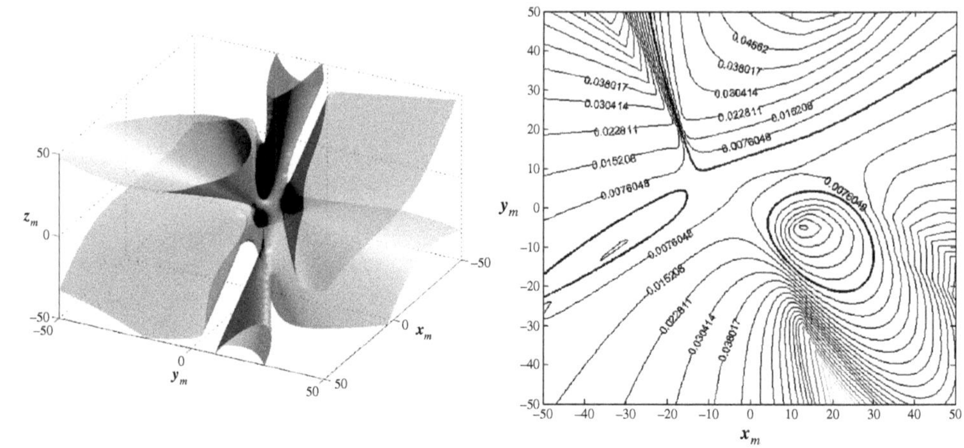

(a) (b)

Figure 7.4 The sphere point surface and intersecting curve for five positions

on the mathematical model (7.9), we call the sub-program ASSF and obtain the saddle spherical surface of $\{\boldsymbol{R}_P^{(i)}\}$, or a ***six-point saddle spherical surface***. Similar to the five positions, with the coordinates (x_{Pm}, y_{Pm}, z_{Pm}) of the points of $\sum^*$ taken as the independent variables and the six-point saddle spherical surface error Δ_{123456} as the dependent variable, the error hypersurface Σ_{123456}^{SS} in four-dimensional space can be constructed. On the three coordinate planes, $\boldsymbol{O}_m\text{–}\boldsymbol{x}_m\boldsymbol{y}_m$, $\boldsymbol{O}_m\text{–}\boldsymbol{x}_m\boldsymbol{z}_m$, and $\boldsymbol{O}_m\text{–}\boldsymbol{y}_m\boldsymbol{z}_m$, the points with the same saddle spherical surface errors are connected by contour lines, as shown in Fig. 7.5(a). The contour plot on the intersection plane z_m=10 is shown in Fig. 7.5(b). The six separated positions of $\sum^*$ are defined by the motion parameters of the first six lines of Table 7.1.

As mentioned in the previous section, a saddle spherical surface can be located by five fitting points of a discrete trajectory $\{\boldsymbol{R}_P^{(i)}\}$, $i = 1, 2, 3, 4, 5, 6$, corresponding to five fitting positions of $\sum^*$. Different points of $\sum^*$ may correspond to different combinations of five fitting positions. There are in total six combinations, or six cases of five fitting positions, such as 12345, 12346, 12356, 12456, 13456, and 23456. For six positions of $\sum^*$, the points of $\sum^*$, whose discrete trajectories correspond to the same combination of saddle spherical surface fitting positions, distribute in specific areas of $\sum^*$, called the ***five-position areas of saddle spherical surface***, such as area I corresponding to positions 12345, area II to 12346, area III to 12356, area IV to 12456, area V to 13456, and area VI to 23456 in Fig. 7.5. Obviously, there are ten distributing spherical surfaces for each five-position combination, whose errors can be solved by equation (7.21). The saddle spherical surface of a discrete trajectory with six points $P^{(1)}$, $P^{(2)}$, $P^{(3)}$, $P^{(4)}$, $P^{(5)}$, and $P^{(6)}$ is only a distributing spherical surface of five points, which makes the maximum fitting error of the six-point saddle spherical surface be minimum.

For $\sum^*$ with six given positions, the error hypersurface Σ_{123456}^{SS} is naturally comprised of segments of the distributing spherical surface error hypersurfaces. The points located on the boundaries of the segments, or the distribution areas, whose coordinates can be solved through equation (7.22), must have the characteristics of the adjacent areas. The discrete trajectory with six points of the boundary point may have two or more five-point distributing spherical

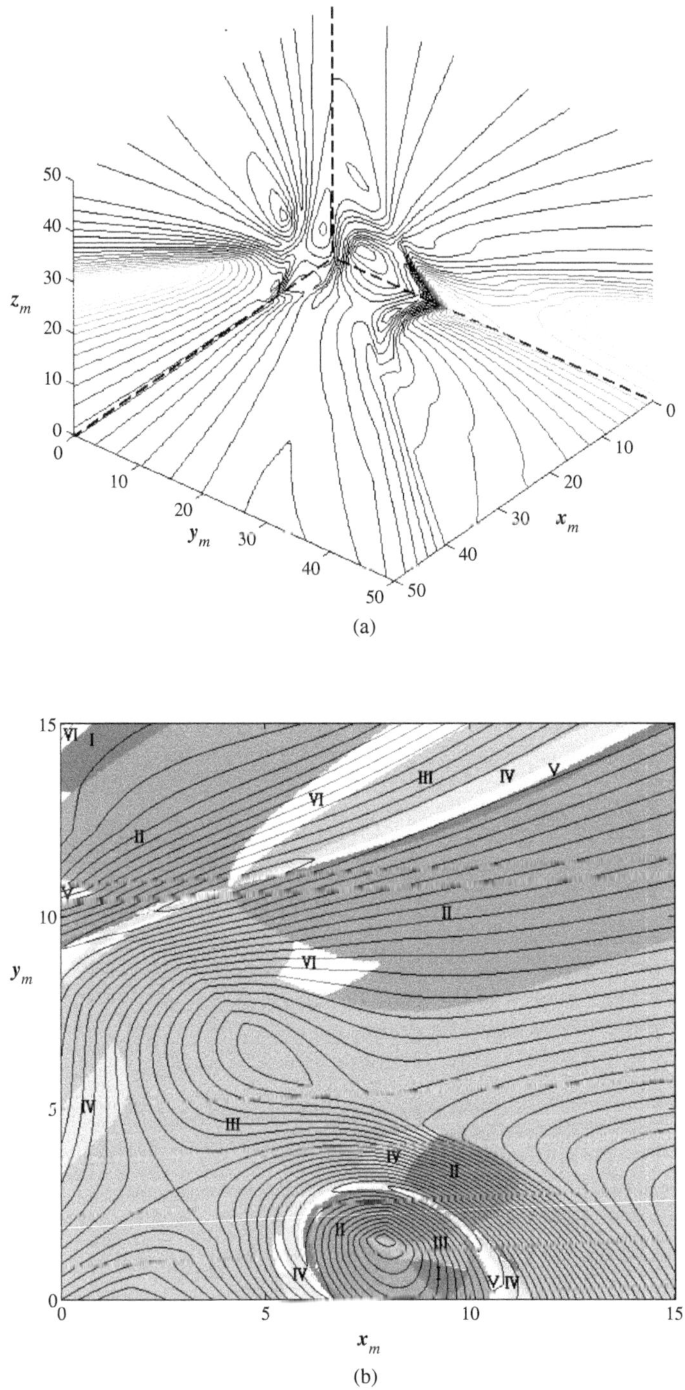

Figure 7.5 The six-position saddle spherical surface errors

surfaces, with the same distributing spherical surface error. If these five-point distributing spherical surfaces are coincident with each other, new characteristic points on the boundary of the error surface would appear, such as the ***sphere points***, which constitute a ***sphere point curve*** on the rigid body $\sum^*$ with six separated positions. Therefore, the error hypersurface Σ^{SS}_{123456} is a piecewise higher-order algebraic surface, and has multiple peaks and valleys. The geometrical properties of the saddle spherical surface errors need to be studied further by mathematical tools such as modern differential geometry.

7.2.5 Multiple Positions

If multiple separated spatial positions of the moving body $\sum^*$ are given, a point of $\sum^*$ traces a discrete trajectory with multiple points. Based on saddle point programming, the saddle spherical surface of the multiple points is still determined totally by the five saddle spherical surface fitting points, no matter how many points the trajectory has. For all points in $\sum^*$, the saddle spherical surface can be located in the same way as in the six-position case. Therefore, no matter how many separated positions of the moving body are prescribed, it can also be regarded as five-position combinations. For n given positions, there are C_n^5 combinations of five positions, and there exist more five-position areas in $\sum^*$.

The more separated positions of the moving body are given, the smaller are the five-position distribution areas. When the multiple separated positions of the rigid body tend to infinity, a discrete trajectory traced by a point of $\sum^*$ would tend to be a continuous curve, whose saddle spherical surface is still determined by the five fitting points. The five fitting points may vary continuously as the tracing point moves continuously on the rigid body.

7.2.6 Saddle Sphere Point

For a given spatial discrete movement of $\sum^*$ with discrete motion parameters $(x^{(i)}_{Omf}, y^{(i)}_{Omf}, z^{(i)}_{Omf};\ \theta^{(i)}_1, \theta^{(i)}_2, \theta^{(i)}_3)$, different points of $\sum^*$ will produce different discrete trajectories $\{\boldsymbol{R}^{(i)}_P\}$, whose geometrical shapes vary gradually and continuously. The saddle spherical surface errors, corresponding to the trajectories, are the nonlinear functions of the tracing points' coordinates, and the error surface of the saddle spherical surface is a higher-order algebraic hypersurface in the multiple-position case. Based on this fact, some points' saddle spherical surface errors may have extreme values, or minimum values, similar to the saddle circle points in Chapter 2.

Definition 7.2 For the moving body in spatial movement with given discrete parameters $(x^{(i)}_{Omf}, y^{(i)}_{Omf}, z^{(i)}_{Omf};\ \theta^{(i)}_1, \theta^{(i)}_2, \theta^{(i)}_3)$, a point of the moving body is defined as a ***saddle sphere point***, whose trajectory corresponds to a saddle spherical surface error which achieves a minimum with respect to other points in its neighborhood. The minimum saddle spherical surface error is called the ***saddle sphere point error***.

In particular, a saddle sphere point is a sphere point if the saddle sphere point error is zero. The saddle point programming needs to be carried out twice to locate the saddle sphere points. The first one is for the saddle spherical surface fitting and the second one is for the location of the saddle sphere point.

The mathematical model of the saddle sphere points can be established as

$$\begin{cases} \delta_{SS} = \min \Delta_{SS}(z) \\ z = (x_{Pm}, y_{Pm}, z_{Pm})^T \end{cases} \tag{7.23}$$

where $\Delta_{SS}(z)$ is the objective function, or the saddle spherical surface error for any point of $\sum^*$, obtained by equation (7.9). $z = (x_{Pm}, y_{Pm}, z_{Pm})^T$ are the optimization variables, or the rectangular coordinates of the point in $\sum^*$. δ_{SS} is the saddle sphere point error.

It is necessary to give the initial values for solving equation (7.23). We do not know how many saddle sphere points exist on $\sum^*$ with a given movement and where they locate on the moving body. It may be a good idea to assign a number of initial points in the different valleys of the saddle spherical surface error hypersurface where the saddle sphere points might be. The suggested area is defined by the following equation:

$$a = \max(\max_{1\le i\le n} x_{Omf}^{(i)} - \min_{1\le i\le n} x_{Omf}^{(i)}, \max_{1\le i\le n} y_{Omf}^{(i)} - \min_{1\le i\le n} y_{Omf}^{(i)}, \max_{1\le i\le n} z_{Omf}^{(i)} - \min_{1\le i\le n} z_{Omf}^{(i)})$$

$$-c_0 a \le x_{Pm}, y_{Pm}, z_{Pm} \le c_0 a \tag{7.24}$$

where c_0 is the coefficient of the searching area.

Based on the mathematical model of the saddle sphere point, equation (7.23) is solved by the function **fmincon** in Matlab with parameters set according to the optimization of the saddle circle points in Chapter 2. A ***sub-program ASSP*** is constructed to locate the saddle sphere points in the moving body with separated positions. The inputs are the motion parameters of the prescribed separated positions and the outputs are the coordinates and errors of the saddle sphere points.

The geometrical shape of a discrete trajectory depends on both the spatial movement of $\sum^*$ and the position of the tracing point in $\sum^*$. The saddle spherical surface errors are nonlinear functions of the coordinates of the point in $\sum^*$, expressed by the higher-order algebraic equations. Therefore, there must be extreme values of the saddle spherical surface errors, or the valleys of the error hypersurface, for all points of $\sum^*$. Based on Definitions 7.1 and 7.2, we have:

Theorem 7.1 *There exist several saddle sphere points in the moving body in non-degenerate spatial motion.*

So far, the properties of both the saddle spherical surface and the saddle sphere point, such as adaptivity, minimum, and comparability, have been described clearly by Definitions 7.1 and 7.2. Theorem 7.1 tells us directly the existence of the saddle sphere points in $\sum^*$. All of these may shed light on the existence of the optimum solutions and the convergence of the effective algorithm in saddle point programming.

There may be a few saddle sphere points in the moving body $\sum^*$ with given discrete spatial movement, which locate on different local areas of $\sum^*$ and have different saddle spherical surface errors. A saddle sphere point has to be searched out for the synthesis of spatial four-bar linkage RRSS. Of course, the saddle sphere point error may depend on the properties of the given spatial movement, which is usually not zero. The saddle sphere points will be the sphere points in $\sum^*$ when the saddle sphere point errors happen to be zero in the case the number of the given separated positions is less than seven. These are just the special cases of the saddle sphere points and can also be located by the above mathematical model, as described in equation (7.23). The sphere points of the moving body will be finite for seven positions. They will be infinite and constitute the sphere point curve for six separated positions and form the sphere point surface for five separated positions. Some numeral examples will be given later. Based on the saddle sphere point errors, the sphere point surface, sphere point curve, and sphere points, as well as the saddle sphere points, corresponding to five, six, seven, and even more separated positions of a moving body can be shown on their respective figures, which may be examined globally to reveal the evolution of the error surfaces with different numbers of prescribed positions. In particular, the theories of the discrete kinematic geometry for both the finite and the multiple separated positions, are connected by the saddle point programming approach for spherical surface fitting, and the spherical curvature of a discrete trajectory may be studied from a new viewing angle.

Example 7.1 The coupler link of a spatial RRSS linkage has 72 discrete positions relative to the base link, whose parameters are given in Table B.1. The coordinate transformation is described by formula (B.3) in Appendix B. Determine: (1) the saddle spherical surface errors in the coupler link, (2) the saddle sphere points in the coupler link.

Solution

1. **The saddle spherical surface errors of discrete point trajectories**
   In the area $x_m \in [0, 1]$, $y_m \in [0, 0.5]$, $z_m \in [0, 0.5]$ of the coupler link, 34 476 points can be selected in steps of 0.02 in three directions of axes $\boldsymbol{x}_2$, $\boldsymbol{y}_2$, and $\boldsymbol{z}_2$. For the discrete trajectories traced by these points, the saddle spherical surface errors Δ_{SS} can be calculated by calling the sub-program ASSF. Hence, with saddle spherical surface error Δ_{SS} and coordinates (x_{Pm}, y_{Pm}, z_{Pm}) of coupler points as the dependent and independent variables, the error hypersurface can be constructed for the 72 positions of the coupler link. Three coordinate planes in cutaway view are taken to reveal the saddle spherical surface errors Δ_{SS} by the contour lines with the same values, as shown in Fig. 7.6. The line $B'C$ is the axis $\boldsymbol{x}_2$, $B'B$ is the axis $\boldsymbol{z}_2$, and the point C is the center point of the pair S.
   To see it clearly, the area $x_m \in [-0.1, 1.2]$, $z_m \in [-0.5, 0.5]$ on the coordinate plane $\boldsymbol{R}_{B'} - \boldsymbol{x}_2\boldsymbol{z}_2$ is amplified, as shown in Fig. 7.7. Of the error surface, there exist two valleys, the long and narrow area along line BB' ($\boldsymbol{z}_2$) of the axis of the R-pair and the vicinity area of the center point C of the S-pair of the coupler link.
2. **The saddle sphere points**
   In the area $x_m \in [-0.5, 1.2]$, $y_m \in [-0.5, 0.5]$, $z_m \in [-0.5, 0.5]$ of the coupler, 50 initial points are randomly generated by the program, and the saddle sphere points are searched by calling the sub-program ASSP based on the optimization model (7.23). Of the optimization results, 21 saddle sphere points locate in the area around the line BB' ($\boldsymbol{z}_2$), and 12 in the vicinity area of the point C. Similarly, if the searching area is put on the coordinate plane $\boldsymbol{R}_{B'} - \boldsymbol{x}_2\boldsymbol{z}_2$ and defined by $x_m \in [-0.1, 1.2]$, $z_m \in [-0.5, 0.5]$, 6 saddle sphere points are in the area around the line BB' ($\boldsymbol{z}_2$) and 3 saddle sphere points are in the vicinity area of the point C when 15 initial points are randomly given. The optimization results show the convergence of the optimization program of the saddle sphere point.

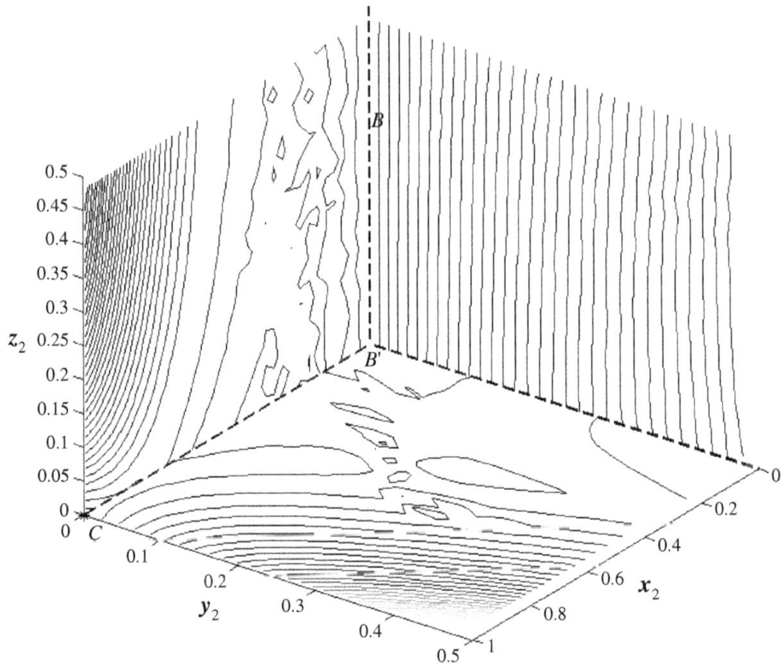

Figure 7.6 The saddle spherical surface errors of the coupler for RRSS

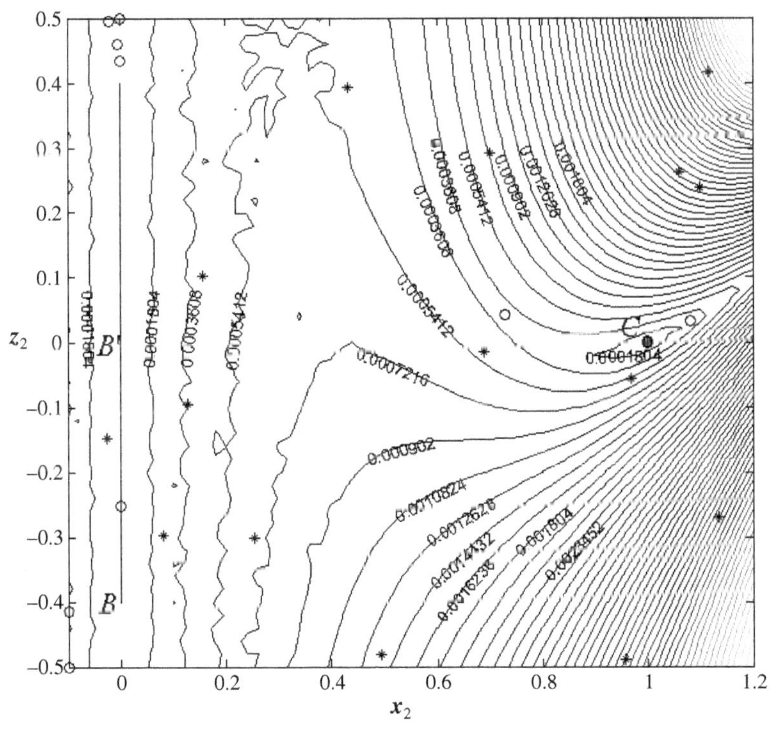

Figure 7.7 The saddle spherical surface errors on the plane $\boldsymbol{R}_{B'} - \boldsymbol{x}_2\boldsymbol{z}_2$ of the coupler

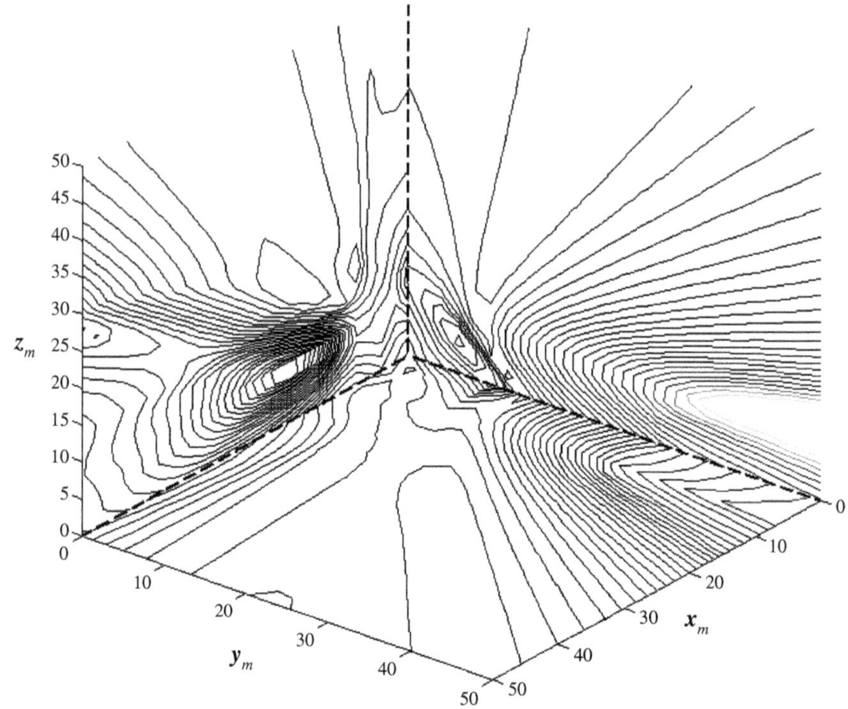

Figure 7.8 The saddle spherical surface errors for nine positions

Example 7.2 The nine separated spatial positions of the moving body are described by parameters $(x_{Omf}^{(i)}, y_{Omf}^{(i)}, z_{Omf}^{(i)}; \theta_1^{(i)}, \theta_2^{(i)}, \theta_3^{(i)})$ listed in Table 7.1. Determine: (1) the saddle spherical surface errors on the moving body, (2) the saddle sphere points of the moving body.

Solution

1. **The saddle spherical surface errors**
   In the area $x_m \in [-50, 50]$, $y_m \in [-50, 50]$, $z_m \in [-50, 50]$ of the moving body, 132 651 points are selected in steps of 2 in every coordinate direction $\boldsymbol{i}_m, \boldsymbol{j}_m, \boldsymbol{k}_m$. These points trace the discrete trajectories in the fixed frame, whose saddle spherical surface errors Δ_{SS} are calculated by calling the sub-program ASSF. Hence, the saddle spherical surface error Δ_{SS} and the coordinates (x_{Pm}, y_{Pm}, z_{Pm}) of the coupler points constitute the error hypersurface in the nine-position case, as shown in Fig. 7.8 with three coordinate planes in cutaway view and contour lines with the same value.
2. **The saddle sphere points**
   50 initial points are randomly selected by the program, and the saddle sphere points are searched by calling the sub-program ASSP. Among the optimization results, seven saddle sphere points (SSP1–SSP7) with less saddle sphere point errors are listed in Table 7.2.

Table 7.2 The saddle sphere points for nine positions

	Saddle sphere point P_{SS}			Center point of the saddle spherical surface			Radius	Fitting error
	x_{Pm}	y_{Pm}	z_{Pm}	x_C	y_C	z_C	r	δ_{SS}
SSP1	2.9313	16.8188	29.4866	6.6412	9.1170	7.5669	23.5296	0.001870
SSP2	1.9601	18.4231	33.6809	6.6577	9.0162	7.6156	28.1083	0.002161
SSP3	3.5004	15.5434	26.7084	6.5810	9.1471	7.5238	20.4553	0.002304
SSP4	1.8896	20.5898	39.1652	6.7055	8.9653	7.7654	33.8294	0.002428
SSP5	−0.3436	21.6252	41.4724	6.6995	8.8562	7.5904	36.8895	0.002588
SSP6	1.1716	22.5815	43.9251	6.7389	8.9071	7.7942	39.0338	0.002657
SSP7	−1.8370	25.1562	49.7953	6.7485	8.7729	7.6347	46.0425	0.002902

Table 7.3 The saddle sphere points for seven positions

	Saddle sphere point P_{SS}			Center point of the saddle spherical surface			Radius	Fitting error
	x_{Pm}	y_{Pm}	z_{Pm}	x_C	y_C	z_C	r	δ_{SS}
1	12.8276	1.6298	14.1929	6.3511	0.7992	5.1977	11.1152	1.871386×10^{-5}
2	3.8051	2.7585	2.6082	1.2174	1.5594	4.2900	3.3106	0.000369
3	−53.1294	40.4888	−31.8402	24.0429	−18.1841	16.9976	108.5511	0.000735
4	39.2618	−14.8993	31.9234	11.8253	−3.5914	10.0086	36.8890	0.001177
5	44.0958	−18.2164	35.5086	12.1976	−3.9959	10.3748	43.0293	0.001224
6	−19.4239	−7.9387	5.5553	12.7610	15.3454	1.5581	39.9262	0.001410
7	−25.6441	−11.3559	7.4397	12.2402	13.9690	1.4061	45.9685	0.001431

Similarly, for a moving body with seven separated positions, described by the motion parameters $(x_{Omf}^{(i)}, y_{Omf}^{(i)}, z_{Omf}^{(i)}; \theta_1^{(i)}, \theta_2^{(i)}, \theta_3^{(i)})$ of the first seven lines in Table 7.1, the saddle sphere points can be located by calling the sub-program ASSP, and seven saddle sphere points with less fitting errors are listed in Table 7.3. The first saddle sphere point is just a sphere point in the moving body, since the saddle sphere point error is near zero.

For the cases of six and five positions, the saddle spherical surface errors Δ_{SS} are shown respectively in Fig. 7.5 in Section 7.2.4 and Fig. 7.3 in Section 7.2.3, which need not be repeated here.

7.3 Saddle Cylinder Point P_{CS}

Based on the generalized curvature in Chapter 6, a cylinder point traces a spatial trajectory, which contacts a cylindrical curve at eight consecutive points. It corresponds to a constraint cylindrical curve and a binary link (or the open chain) C–S. We now turn from the generalized curvature for infinitesimal successive positions to that for the separated positions, and study the global geometrical properties of a discrete point trajectory in terms of the cylindrical curvature, as well as the degenerate cases for the binary links H–S and R–S.

7.3.1 Cylindrical Surface Fitting

For a rigid body $\sum^*$ in discrete spatial movement, its spatial separated positions are described by the parameters $(x_{Omf}^{(i)}, y_{Omf}^{(i)}, z_{Omf}^{(i)}; \theta_1^{(i)}, \theta_2^{(i)}, \theta_3^{(i)})$ in the fixed frame $\{\boldsymbol{O}_f; \boldsymbol{i}_f, \boldsymbol{j}_f, \boldsymbol{k}_f\}$. A point P of $\sum^*$ with coordinates (x_{Pm}, y_{Pm}, z_{Pm}) in $\{\boldsymbol{O}_m; \boldsymbol{i}_m, \boldsymbol{j}_m, \boldsymbol{k}_m\}$ of $\sum^*$ traces a spatial discrete trajectory, or the discrete point set $\{\boldsymbol{R}_P^{(i)}\}$, which can be expressed by equation (7.7). Now we are interested in the approximation of the discrete trajectory $\{\boldsymbol{R}_P^{(i)}\}$ and a cylindrical curve. Based on the invariance of the cylindrical curve in Chapter 3, the global geometrical property of a cylindrical curve can be represented as the distances from all discrete points to the axis of the cylinder being constant. Hence, a cylindrical surface with a given radius r is taken to fit $\{\boldsymbol{R}_P^{(i)}\}$, referred to as a ***floating fitting cylindrical surface***, as shown in Fig. 7.9. The position and orientation of the axis of the floating fitting cylindrical surface are adaptively determined by the saddle point programming, or letting the maximum normal fitting error be minimum, which is measured in the normal direction of the cylindrical surface. The different values of the radius r correspond to different floating fitting cylindrical surfaces for $\{\boldsymbol{R}_P^{(i)}\}$, one of which is the best fitting cylindrical surface with minimum maximum fitting error.

Definition 7.3 For a given discrete trajectory $\{\boldsymbol{R}_P^{(i)}\}$, a cylindrical surface, whose axis and radius are adaptively determined by letting the maximum normal fitting error be minimum, is defined as a ***saddle cylindrical surface*** and the maximum normal fitting error is referred to as the ***saddle cylindrical surface error***.

The saddle cylindrical surface of the discrete trajectory $\{\boldsymbol{R}_P^{(i)}\}$ is determined by the radius r, both the unit direction vector $\boldsymbol{l}_Q(\xi, \eta)$ and the position vector $R_Q = (x_Q, y_Q, z_Q)^T$ of the axis L_Q, as shown in Fig. 7.9. For convenience of calculation, we take the intersection point Q between the projection plane Π and the axis L_Q as the reference point of L_Q, while the projection plane Π is located by taking the direction vector $\boldsymbol{l}_Q$ as its normal vector and passing through the original point $\boldsymbol{O}_f$ of the fixed frame $\{\boldsymbol{O}_f; \boldsymbol{i}_f, \boldsymbol{j}_f, \boldsymbol{k}_f\}$. Hence, the position vector $\boldsymbol{R}_Q$ is the vector from $\boldsymbol{O}_f$ to the reference point Q, which is perpendicular to the axis L_Q, or $\boldsymbol{R}_Q \cdot \boldsymbol{l}_Q = 0$, and the point Q is just the perpendicular foot. The unit vector $\boldsymbol{l}_Q$ is expressed as $(\sin\xi\cos\eta, \sin\xi\sin\eta, \cos\xi)^T$. We have

$$x_Q \sin\xi \cos\eta + y_Q \sin\xi \sin\eta + z_Q \cos\xi = 0 \tag{7.25}$$

This means that the axis L_Q is located by four independent parameters (x_Q, y_Q, ξ, η), and the saddle cylindrical surface is determined by five parameters (x_Q, y_Q, ξ, η, r).

The distance $h^{(i)}$ from a point $P^{(i)}(x_P^{(i)}, y_P^{(i)}, z_P^{(i)})^T$ of the discrete point set $\{\boldsymbol{R}_P^{(i)}\}$ to the projection plane Π can be calculated by

$$h^{(i)} = (\boldsymbol{R}_P^{(i)} - \boldsymbol{R}_Q) \cdot \boldsymbol{l}_Q = x_P^{(i)} \sin\xi \cos\eta + y_P^{(i)} \sin\xi \sin\eta + z_P^{(i)} \cos\xi \tag{7.26}$$

The point $P^{(i)*}$ on the projection plane Π, corresponding to $P^{(i)}$, or the projection point, has the vector equation

$$\boldsymbol{R}_P^{(i)*} = \boldsymbol{R}_P^{(i)} - h^{(i)}\boldsymbol{l}_Q \tag{7.27}$$

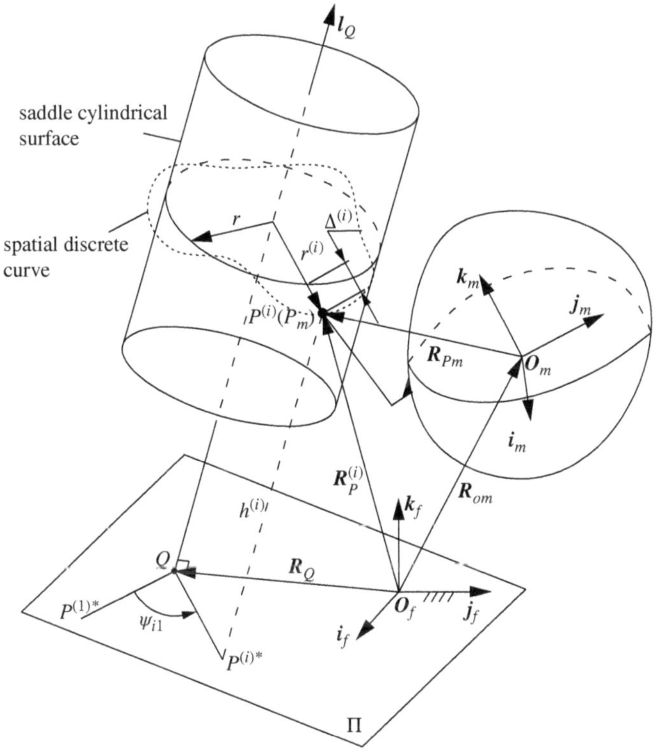

Figure 7.9 The saddle cylindrical surface of a spatial discrete curve

Hence, the distance $r^{(i)}$ from point $P^{(i)}$ of $\{\boldsymbol{R}_P^{(i)}\}$ to the axis L_Q of the saddle cylindrical surface is

$$\begin{aligned} r^{(i)} &= |\boldsymbol{R}_P^{(i)*} - \boldsymbol{R}_Q| \\ &= \sqrt{(x_P^{(i)} - x_Q - h^{(i)} \sin\xi \cos\eta)^2 + (y_P^{(i)} - y_Q - h^{(i)} \sin\xi \sin\eta)^2 + (z_P^{(i)} - z_Q - h^{(i)} \cos\xi)^2} \end{aligned} \tag{7.28}$$

On the projection plane Π, the vector $\overrightarrow{QP^{(i)*}}$ from the reference point Q to the projection point $P^{(i)*}$ forms an included angle ψ_{i1} with the first vector $\overrightarrow{QP^{(1)*}}$ corresponding to the first point $P^{(1)}$:

$$\cos\psi_{i1} = \frac{(\boldsymbol{R}_P^{(i)*} \quad \boldsymbol{R}_Q)}{|\boldsymbol{R}_P^{(i)*} - \boldsymbol{R}_Q|} \cdot \frac{(\boldsymbol{R}_P^{(1)*} - \boldsymbol{R}_Q)}{|\boldsymbol{R}_P^{(1)*} - \boldsymbol{R}_Q|} \tag{7.29}$$

Based on Definition 7.3, the saddle cylindrical surface is the best of all the fitting cylindrical surfaces with different parameters (x_Q, y_Q, ξ, η, r) since the maximum normal fitting error of all discrete points of $\{\boldsymbol{R}_P^{(i)}\}$ achieves the minima. The mathematical model of the saddle cylindrical surface fitting can be established as

$$\begin{cases} \Delta_{\text{CS}} = \min\limits_{\boldsymbol{x}} \max\limits_{1\le i\le n} \left\{\Delta^{(i)}(\boldsymbol{x})\right\} \\ \quad = \min\limits_{\boldsymbol{x}} \max\limits_{1\le i\le n} \{|r^{(i)} - r|\} \\ \text{s.t.} \quad r \in (0, +\infty), \ \xi \in [0, \pi/2], \ \eta \in [0, 2\pi) \\ \boldsymbol{x} = (x_Q, y_Q, \xi, \eta, r)^T \end{cases} \tag{7.30}$$

where Δ_{CS} is the saddle cylindrical surface error of the discrete point set $\{\boldsymbol{R}_P^{(i)}\}$. The objective function $\{\Delta^{(i)}(\boldsymbol{x})\}$ is the normal fitting error set between the discrete point set $\{\boldsymbol{R}_P^{(i)}\}$ and a fitting cylindrical surface, and n is the number of discrete points of $\{\boldsymbol{R}_P^{(i)}\}$. The optimization variables $\boldsymbol{x}$=$(x_Q, y_Q, \xi, \eta, r)^T$ are respectively the position and orientation parameters (x_Q, y_Q, ξ, η) of the axis L_Q and the radius r.

For a discrete point set $\{\boldsymbol{R}_P^{(i)}\}$, the saddle cylindrical surface, determined by the mathematical model of equation (7.30), has similar properties to those of the saddle circle in Chapter 2, such as adaptivity, uniqueness, and comparability. The mathematical model (7.30) can be solved directly by the function **fminimax** in Matlab and the initial values of the optimization variables $\boldsymbol{x} = (x_Q, y_Q, \xi, \eta, r)^T$ have to be given first. Many groups of direction angles (ξ, η) of the axis L_Q might be selected. For any group of direction angles, the point set $\{\boldsymbol{R}_P^{(i)*}\}$ of $\{\boldsymbol{R}_P^{(i)}\}$ on the projection plane Π can be obtained by equations (7.26) and (7.27). Then, through equations (2.12) to (2.15), the fitting circle of $\{\boldsymbol{R}_P^{(i)*}\}$ can be determined by the least-squares method, whose center coordinates and radius provide the initial values of (x_Q, y_Q, r).

For a given discrete trajectory $\{\boldsymbol{R}_P^{(i)}\}$ in the fixed frame $\{\boldsymbol{O}_f; \boldsymbol{i}_f, \boldsymbol{j}_f, \boldsymbol{k}_f\}$, the saddle cylindrical surface can be located by the function **fminimax** in Matlab based on Definition 7.3 and the mathematical model (7.30). The parameters of **fminimax** are set in the same way as the saddle circle fitting in Chapter 2, and a ***sub-program ACSF*** is constructed for saddle cylindrical surface fitting to $\{\boldsymbol{R}_P^{(i)}\}$. The input parameters are the coordinates of the discrete points and the output parameters are the five parameters of the saddle cylindrical surface, and the saddle spherical surface error Δ_{CS}.

7.3.2 Saddle Cylindrical Surface

Based on Definition 7.3, a saddle cylindrical surface is located for a given discrete trajectory Γ_P in $\{\boldsymbol{O}_f; \boldsymbol{i}_f, \boldsymbol{j}_f, \boldsymbol{k}_f\}$, whose saddle cylindrical surface error Δ_{CS} is maximal for all the points on Γ_P, but for all the fitting cylindrical surfaces, the maximal error Δ_{CS} is the smallest.

For convenience in discussing the properties of the saddle cylindrical surface errors, the given spatial trajectory Γ_P is assumed to be a continuous curve at first. The unit direction vector of the axis L_Q of the saddle cylindrical surface is $\boldsymbol{l}_Q$, and the position vector of the point Q on L_Q is $\boldsymbol{R}_Q$. The radius of the saddle cylindrical surface is r. For any point P on Γ_P, the normal fitting error to the saddle cylindrical surface is Δ, and then the position vector $\boldsymbol{R}_P$ of P can be written as

$$\boldsymbol{R}_P = \boldsymbol{R}_Q + (\boldsymbol{R}_P \cdot \boldsymbol{l}_Q) \cdot \boldsymbol{l}_Q + (r + \Delta)\boldsymbol{n} \tag{7.31}$$

The sign of the normal fitting error Δ is positive if point P lies outside the saddle cylindrical surface and negative if P is inside. Then, Δ can be given by the above equation:

$$\Delta = (\boldsymbol{R}_P - \boldsymbol{R}_Q) \cdot \boldsymbol{n} - r \tag{7.32}$$

Since the saddle cylindrical surface is determined by the curve Γ_P, the position vector $\boldsymbol{R}_Q$ and the direction vector $\boldsymbol{l}_Q$ are both constant vectors, and the radius r is a constant. By differentiating equation (7.32) with respect to the arc length s of Γ_P, we have

$$\frac{\mathrm{d}\Delta}{\mathrm{d}s} = \frac{\mathrm{d}\boldsymbol{R}_P}{\mathrm{d}s} \cdot \boldsymbol{n} + (\boldsymbol{R}_P - \boldsymbol{R}_Q) \cdot \frac{\mathrm{d}\boldsymbol{n}}{\mathrm{d}s} \tag{7.33}$$

Substituting vector expression $\boldsymbol{R}_P - \boldsymbol{R}_Q = (\boldsymbol{R}_P \cdot \boldsymbol{l}_Q) \cdot \boldsymbol{l}_Q + (r + \Delta)\boldsymbol{n}$ into the above equation, we have $\frac{\mathrm{d}\Delta}{\mathrm{d}s} = \frac{\mathrm{d}\boldsymbol{R}_P}{\mathrm{d}s} \cdot \boldsymbol{n}$. The condition for Δ to be an extreme value is $\frac{\mathrm{d}\Delta}{\mathrm{d}s} = 0$ at a point P of Γ_P, or

$$\frac{\mathrm{d}\boldsymbol{R}_P}{\mathrm{d}s} \cdot \boldsymbol{n} = 0 \tag{7.34}$$

which means that the normal error Δ gets an extreme value when the spatial curve Γ_P and its saddle cylindrical surface have a common normal. Similarly, the extreme value appears at multiple points on Γ_P but may not be the maximum error, such as the points $P^{(k+1)}$ and $P^{(k+2)}$ in Fig. 7.10.

A saddle cylindrical surface of a given spatial curve Γ_P can be determined by five parameters (x_Q, y_Q, ξ, η, r) and the saddle cylindrical surface error Δ_{CS}, which can be solved simultaneously by equation (7.31) or (7.32) at six points of Γ_P. There are at least six points on Γ_P to hold equation (7.34) true.

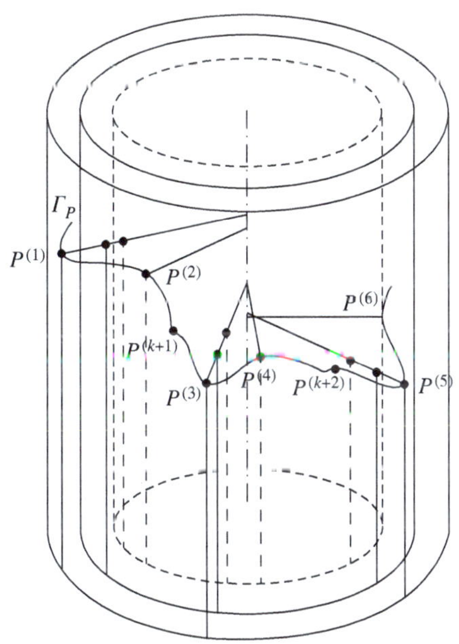

Figure 7.10 Two concentric cylindrical surfaces of a space curve

Based on the saddle point programming for the saddle cylindrical surface fitting [2], we can estimate the degree of approximation of a given spatial curve and a cylindrical surface. The cylindrical surface fitting by letting the maximum fitting error be the minima is equivalent to the given spatial curve being contained by ***two concentric cylindrical surfaces***. The given spatial curve may have at least six tangent points with the two concentric cylindrical surfaces leading to the maximum normal fitting errors being the minima with the same value, as shown in Fig. 7.10. The six tangent points may appear in different distributions. They distribute on both sides of the saddle cylindrical surface, for example, two are inside the saddle cylindrical surface and four are outside, or vice versa, or three are inside and three are outside.

Therefore, the six tangent points of the given curve Γ_P determine its saddle cylindrical surface and are defined as the ***saddle cylindrical surface fitting points*** of Γ_P. At the six points of Γ_P, the absolute values of the normal fitting errors are equal to the saddle cylindrical surface error. The given curve is a cylindrical curve if the saddle cylindrical surface error is zero.

The discrete trajectory $\{\boldsymbol{R}_P^{(i)}\}$, traced by a point of a moving body in discrete movement, rather than a continuous spatial curve, is discussed in the following. If a cylindrical surface is taken to fit the discrete points, by equations (7.31) and (7.32), the normal error $\Delta^{(i)}$ at a point of $\{\boldsymbol{R}_P^{(i)}\}$ can be written in discrete form as

$$\begin{cases} \Delta^{(i)} = |\boldsymbol{R}_P^{(i)} - [(\boldsymbol{R}_P^{(i)} - \boldsymbol{R}_Q) \cdot \boldsymbol{l}_Q]\boldsymbol{l}_Q - \boldsymbol{R}_Q| - r \\ \boldsymbol{R}_P^{(i)} = [\mathbf{M}^{(i)}] \cdot \boldsymbol{R}_{Pm} \end{cases} \tag{7.35}$$

The saddle cylindrical surface error Δ_{CS} is the maximum absolute value of all $\Delta^{(i)}$:

$$\Delta_{CS} = \max\{|\Delta^{(i)}|\} = \max\{||\boldsymbol{R}_P^{(i)} - [(\boldsymbol{R}_P^{(i)} - \boldsymbol{R}_Q) \cdot \boldsymbol{l}_Q]\boldsymbol{l}_Q - \boldsymbol{R}_Q| - r|\},\ i = 1, \ldots, n \tag{7.36}$$

This means that the saddle cylindrical surface error of a discrete point set $\{\boldsymbol{R}_P^{(i)}\}$ can be obtained by numerical calculation and comparison among all the fitting errors $\Delta^{(i)}$ of $\{\boldsymbol{R}_P^{(i)}\}$ by equation (7.36). The positions of the moving body $\sum^*$ corresponding to the six saddle cylindrical surface fitting points are referred to as the ***saddle cylindrical surface fitting positions***. Hence, the mathematical model of saddle cylindrical surface fitting sets up the relationships among the separated positions of $\sum^*$, the discrete point set $\{\boldsymbol{R}_P^{(i)}\}$, and its saddle cylindrical surface error Δ_{CS}.

7.3.3 Six Positions

Six separated spatial positions of a moving body $\sum^*$ are given in $\{\boldsymbol{O}_f; \boldsymbol{i}_f, \boldsymbol{j}_f, \boldsymbol{k}_f\}$ and numbered 1, 2, 3, 4, 5, 6. A point of $\sum^*$ with Cartesian coordinates (x_{Pm}, y_{Pm}, z_{Pm}) in $\{\boldsymbol{O}_m; \boldsymbol{i}_m, \boldsymbol{j}_m, \boldsymbol{k}_m\}$ traces a spatial discrete trajectory, comprised of six discrete points $P^{(1)}$, $P^{(2)}$, $P^{(3)}$, $P^{(4)}$, $P^{(5)}$, and $P^{(6)}$ in $\{\boldsymbol{O}_f; \boldsymbol{i}_f, \boldsymbol{j}_f, \boldsymbol{k}_f\}$. Based on the mathematical model (7.30), we call the sub-program ACSF and obtain the saddle cylindrical surface of $\{\boldsymbol{R}_P^{(i)}\}$, referred to as the ***six-point saddle cylindrical surface***. In this case, the six points $P^{(1)}$, $P^{(2)}$, $P^{(3)}$, $P^{(4)}$, $P^{(5)}$, and $P^{(6)}$ are just the saddle cylindrical surface fitting points and the corresponding six positions are the saddle cylindrical surface fitting positions.

According to the saddle point programming, the six fitting points $P^{(1)}$, $P^{(2)}$, $P^{(3)}$, $P^{(4)}$, $P^{(5)}$, and $P^{(6)}$ distribute on both sides of the six-point saddle cylindrical surface. There are many cases for the six fitting points to distribute on both sides of the saddle cylindrical surface. For convenience in introducing the properties of the saddle cylindrical surface error, we discuss one distribution type, three points inside and three points outside, which corresponds to 10 distribution cases of the six points and 10 fitting spherical surface, referred to as ***distributing cylindrical surfaces*** of the six points.

As to the distribution case $P^{(1)}P^{(3)}P^{(5)}-P^{(2)}P^{(4)}P^{(6)}$, three points $P^{(1)}$, $P^{(3)}$, $P^{(5)}$ are inside and another three points $P^{(2)}$, $P^{(4)}$, $P^{(6)}$ are outside the distributing cylindrical surface. By equation (7.35), the equations of the distributing cylindrical surface error $\Delta_{135,246}$ are

$$\begin{cases} r - \Delta_{135,246} = \left| \boldsymbol{R}_P^{(i)} - [(\boldsymbol{R}_P^{(i)} - \boldsymbol{R}_Q) \cdot \boldsymbol{l}_Q]\boldsymbol{l}_Q - \boldsymbol{R}_Q \right|, & i = 1,\ 3,\ 5 \\ r + \Delta_{135,246} = \left| \boldsymbol{R}_P^{(i)} - [(\boldsymbol{R}_P^{(i)} - \boldsymbol{R}_Q) \cdot \boldsymbol{l}_Q]\boldsymbol{l}_Q - \boldsymbol{R}_Q \right|, & i = 2,\ 4,\ 6 \end{cases} \tag{7.37}$$

Substituting equation (7.7) into the above equations, we have a system of equations including six equations of nine variables, which are the coordinates (x_{Pm}, y_{Pm}, z_{Pm}) of P in $\{\boldsymbol{O}_m; \boldsymbol{i}_m, \boldsymbol{j}_m, \boldsymbol{k}_m\}$, the parameters $(x_Q,\ y_Q,\ \xi,\ \eta,\ r)$ of the distributing cylindrical surface, and the distributing cylindrical surface error $\Delta_{135,246}$. By eliminating the variables $(x_Q,\ y_Q,\ \xi,\ \eta,\ r)$ from the six equations, a higher-order algebraic equation of (x_{Pm}, y_{Pm}, z_{Pm}) and $\Delta_{135,246}$ can be obtained:

$$f(x_{Pm}, y_{Pm}, z_{Pm}, \Delta_{135,246}) = 0 \tag{7.38}$$

For the six points $P^{(1)}$, $P^{(2)}$, $P^{(3)}$, $P^{(4)}$, $P^{(5)}$, and $P^{(6)}$, there are other distribution cases and corresponding distributing cylindrical surface errors. According to Definition 7.3, the six-point saddle spherical surface error Δ_{123456} of $\{\boldsymbol{R}_P^{(i)}\}$, $i = 1,\ 2,\ 3,\ 4,\ 5,\ 6$ is the minimum of all the distributing cylindrical surface errors.

A moving body $\sum^*$ with six separated positions described by parameters of the first six lines of Table 7.1, the point $P(x_{Pm}, y_{Pm}, z_{Pm})$ of $\sum^*$ in $\{\boldsymbol{O}_m; \boldsymbol{i}_m, \boldsymbol{j}_m, \boldsymbol{k}_m\}$ traces six discrete points in $\{\boldsymbol{O}_f; \boldsymbol{i}_f, \boldsymbol{j}_f, \boldsymbol{k}_f\}$. We take (x_{Pm}, y_{Pm}, z_{Pm}) as the independent variables and the six-point saddle cylindrical surface error Δ_{123456} as the dependent variable. The error hypersurface $\Sigma_{123456}^{\mathrm{CS}}$ in four-dimensional space can be constructed. On the intersection planes between the error hypersurface surface $\Sigma_{123456}^{\mathrm{CS}}$ and three coordinate planes, $\boldsymbol{O}_m\text{–}\boldsymbol{x}_m\boldsymbol{y}_m$, $\boldsymbol{O}_m\text{–}\boldsymbol{x}_m\boldsymbol{z}_m$, and $\boldsymbol{O}_m\text{–}\boldsymbol{y}_m\boldsymbol{z}_m$, the points with the same saddle spherical surface errors are connected by contour lines, as shown in Fig. 7.11.

Because the distributing conditions of the fitting points haven't been fully investigated and the multiple extreme values for the saddle cylindrical surface fitting cannot be identified clearly, the contour lines in Fig. 7.11 are not smooth. The algebraic properties of the saddle cylindrical surface error need studying further.

7.3.4 Seven Positions

Seven separated spatial positions of a moving body $\sum^*$ are given in $\{\boldsymbol{O}_f; \boldsymbol{i}_f, \boldsymbol{j}_f, \boldsymbol{k}_f\}$ and numbered 1, 2, 3, 4, 5, 6, 7. A point $P(x_{Pm}, y_{Pm}, z_{Pm})$ of $\sum^*$ in $\{\boldsymbol{O}_m; \boldsymbol{i}_m, \boldsymbol{j}_m, \boldsymbol{k}_m\}$ traces a spatial discrete trajectory $\{\boldsymbol{R}_P^{(i)}\}$ in $\{\boldsymbol{O}_f; \boldsymbol{i}_f, \boldsymbol{j}_f, \boldsymbol{k}_f\}$, comprised of seven discrete points $P^{(1)}\text{–}P^{(7)}$, whose saddle cylindrical surface, or a ***seven-point saddle cylindrical surface***, can be obtained by calling the sub-program ACSF. Similar to the case of six positions, the coordinates (x_{Pm}, y_{Pm}, z_{Pm}) of the points of $\sum^*$ are taken as the independent variables and the seven-point

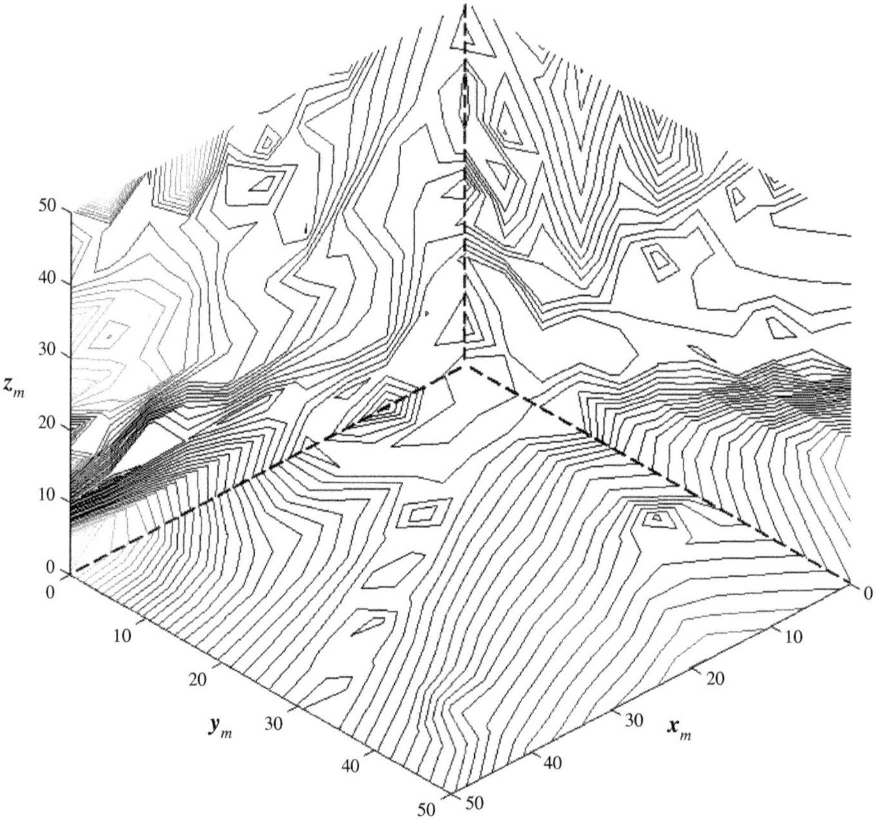

Figure 7.11 The saddle cylindrical surface error for six positions

saddle cylindrical surface error $\Delta_{1234567}$ as the dependent variable. The error hypersurface $\Sigma^{CS}_{1234567}$ in four-dimensional space can be constructed. On the intersection planes between the error hypersurface surface $\Sigma^{CS}_{1234567}$ and the three coordinate planes, $\boldsymbol{O}_m-\boldsymbol{x}_m\boldsymbol{y}_m$, $\boldsymbol{O}_m-\boldsymbol{x}_m\boldsymbol{z}_m$, and $\boldsymbol{O}_m-\boldsymbol{y}_m\boldsymbol{z}_m$, the points with the same saddle cylindrical surface errors are connected by contour lines, as shown in Fig. 7.12(a). The contour plot on the plane z_m=0 is shown in Fig. 7.12(b). The seven separated position of $\sum^*$ are defined by the motion parameters of the first seven lines of Table 7.1.

As mentioned in the previous section, a saddle cylindrical surface can be located by six fitting points of a discrete trajectory $\{\boldsymbol{R}_P^{(i)}\}$, $i = 1, \ldots, 7$, corresponding to six fitting positions of $\sum^*$. Different points of $\sum^*$ may correspond to different combinations of six fitting positions. There are in total seven combinations, or seven fitting-position cases, such as 123456, 123457, 123467, 123567, 124567, 134567, and 234567. For a rigid body $\sum^*$ with seven given positions, the points of $\sum^*$, whose discrete trajectories correspond to the same six saddle cylindrical surface fitting positions, distribute in specific areas of $\sum^*$, called the ***six-position areas of saddle cylindrical surface***, such as area I corresponding to positions 123456, area II to 123457, area III to 123467, area IV to 123567, area V to 124567, area VI to 134567, and area VII to 234567 in Fig. 7.12. Obviously, each six-position combination has many distributing

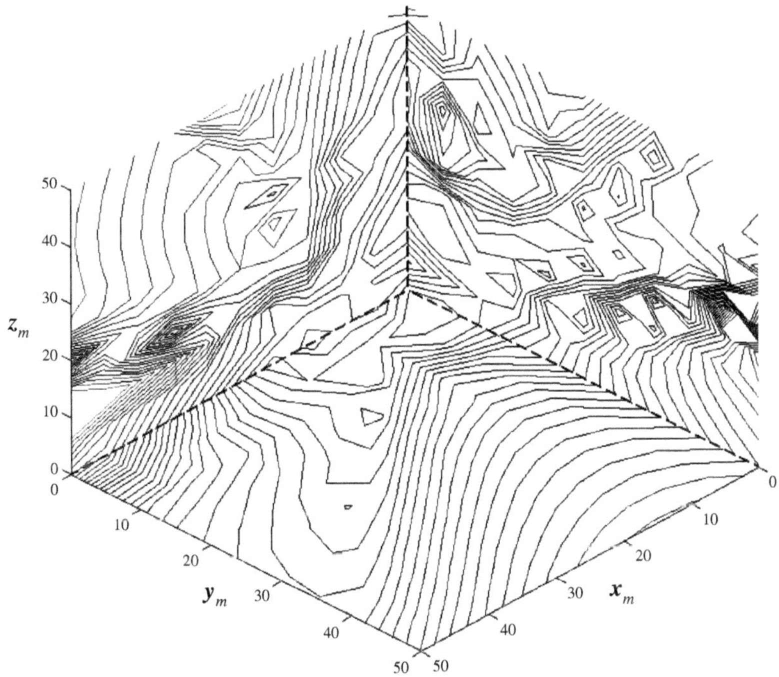

(a)

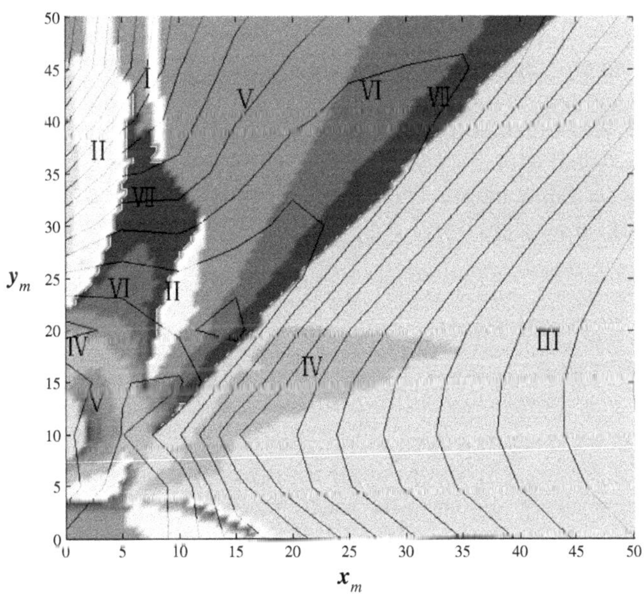

(b)

Figure 7.12 The saddle cylindrical surface error for seven positions

cylindrical surfaces and errors, which can be solved through equation (7.37). The saddle cylindrical surface of the discrete trajectory with seven points $P^{(1)}$, $P^{(2)}$, $P^{(3)}$, $P^{(4)}$, $P^{(5)}$, $P^{(6)}$, and $P^{(7)}$ is only a distributing cylindrical surface of six points, which makes the maximum fitting error of the seven-point saddle cylindrical surface be minimum.

For $\sum^*$ with seven prescribed positions, the error hypersurface $\Sigma_{1234567}^{CS}$ is naturally comprised of segments of the distributing cylindrical surface error hypersurfaces. The points located on the boundaries of the segments or the distribution areas, whose coordinates can be solved through equation (7.38), have the characteristics of the adjacent areas. The discrete trajectory with seven points of the boundary point may have two or more six-point distributing cylindrical surfaces with the same distributing cylindrical surface error. If these six-point distributing cylindrical surfaces coincide with each other, the boundaries of the error surface would have new characteristics to be explored. Therefore, the error hypersurface $\Sigma_{1234567}^{CS}$ is a piecewise higher-order algebraic surface and has multiple peaks and valleys. The geometrical properties of the saddle cylindrical surface errors also need studying further.

7.3.5 Multiple Positions

If multiple separated spatial positions of the moving body $\sum^*$ are given, a point of $\sum^*$ traces a discrete trajectory with multiple points. Based on the saddle point programming, the multiple-point saddle cylindrical surface is still determined totally by the six saddle cylindrical surface fitting points, no matter how many points the trajectory has. For all the points on $\sum^*$, the saddle cylindrical surface can be located in the same way as in the seven-position case. Therefore, no matter how many separated positions of the moving body are given, it can also be regarded as combinations of six positions. For n given positions, there are C_n^6 combinations of six positions, and there exist more six-position areas in $\sum^*$.

The more separated positions of the moving body are given, the smaller are the six-position distribution areas. When the multiple separated positions of the rigid body tend to infinity, a discrete trajectory would tend to a continuous curve, whose saddle cylindrical surface is still determined by the six fitting points. The six fitting points may vary continuously as the point varies continuously on the rigid body.

7.3.6 Saddle Cylinder Point

For a rigid body $\sum^*$ with parameters $(x_{Omf}^{(i)}, y_{Omf}^{(i)}, z_{Omf}^{(i)}; \theta_1^{(i)}, \theta_2^{(i)}, \theta_3^{(i)})$, different points of $\sum^*$ will trace trajectories with different geometrical shapes, which vary gradually and continuously from one tracing point to another. The saddle cylindrical surface errors, corresponding to the trajectories, are the nonlinear functions of the tracing points' coordinates, and have extreme values, or minimum values, similar to the saddle circle errors of the saddle circle points in Chapter 2.

Definition 7.4 For a moving body $\sum^*$ in spatial movement, a point of $\sum^*$ is defined as a ***saddle cylinder point***, whose trajectory corresponds to a saddle cylindrical surface with the minimum saddle cylindrical surface error compared to the trajectories traced by other points in its neighborhood. The minimum error is called the ***saddle cylinder point error***.

In particular, the saddle cylinder point error is zero for a ***cylinder point***. The above definition implies that the saddle point programming should also be conducted twice to locate the saddle cylinder points. The first one is for the saddle cylindrical surface fitting. The second one is for the location of the saddle cylinder point. The mathematical model of saddle cylinder points can be established as

$$\begin{cases} \delta_{\mathrm{CS}} = \min \Delta_{\mathrm{CS}}(z) \\ z = (x_{Pm}, y_{Pm}, z_{Pm})^T \end{cases} \tag{7.39}$$

where δ_{CS} is the saddle cylinder point error, and $\Delta_{\mathrm{CS}}(z)$ is the objective function, or the saddle cylindrical surface errors for all points of $\sum^*$, which can be obtained by equation (7.30). $z = (x_{Pm}, y_{Pm}, z_{Pm})^T$ are the optimization variables, or the rectangular coordinates of the point in $\sum^*$. Many initial values of the optimization variables are randomly generated in the area defined by equation (7.24) for the optimizing search of the saddle cylinder points.

The mathematical model of the saddle cylinder point, as described by equation (7.39), can be solved by the function **fmincon** in Matlab, with parameters set according to the optimization of the saddle circle points in Chapter 2. A ***sub-program ACSP*** is constructed for the location of saddle cylinder points in the moving body with discrete positions. The inputs are the motion parameters of the given positions and the outputs are the coordinates and errors of the saddle cylinder points.

In general, the saddle cylinder point error is the nonlinear function of the tracing point's coordinates, and there exist minimum values at some points of $\sum^*$. Based on Definitions 7.3 and 7.4:

Theorem 7.2 *There exist several saddle cylinder points in the moving body in non-degenerate spatial motion.*

So far, the properties of a saddle cylindrical surface fitting and the saddle cylinder point, such as adaptivity, minimum, comparability, and double saddle point property, have been clearly described by Definitions 7.3 and 7.4. Theorem 7.2 directly points out the existence of the saddle cylinder points in $\sum^*$. Thus, there may be a few saddle cylinder points which locate in different local areas of $\sum^*$ and have different saddle cylinder errors. A saddle cylinder point has to be searched out for the synthesis of a spatial four-bar linkage RRSC. Of course, the saddle cylinder point errors may depend on the properties of the given spatial movement, which is usually not zero. The saddle cylinder points will be the cylinder points when the saddle cylinder point errors happen to be zero in the case the number of given discrete positions is less than eight. They are just the special cases of the saddle cylinder points and can also be located by the above mathematical model, equation (7.31). The cylinder points of the moving body will be finite for eight positions. They will be infinite and form a curve called a ***cylinder point curve*** for seven positions, and a surface called a ***cylinder point surface*** for six positions. Some numeral examples will be given in the following. Based on the saddle cylinder point errors, the cylinder point surface, cylinder point curve, and cylinder points, as well as saddle cylinder points in $\sum^*$, corresponding to six, seven, eight, and even multiple separated positions of $\sum^*$, can respectively be shown on their respective figures, which may be globally examined to reveal the evolution of the error surface with different numbers of given positions. In particular, the theories of the discrete differential geometry for both the finite and the multiple separated positions, are connected by the saddle point programming approach for the cylindrical surface fitting, and the cylindrical curvature of a discrete trajectory may be studied from a new viewing angle.

7.3.7 The Degeneration of the Saddle Cylinder Point (R–S, H–S)

The binary link C–S will degenerate to a binary link R–S or H–S if the kinematic pair C is replaced by the kinematic pair R or H, whose constraint curve, a cylindrical curve, will be a circle or a helical curve.

7.3.7.1 The Saddle Space Circle Point and the Binary Link R–S

As a special case, the constraint curve on the cylindrical surface is degenerated to a circle, and the binary link C–S can be replaced by a binary link R–S. To estimate the approximation between a given spatial discrete trajectory $\{\boldsymbol{R}_P^{(i)}\}$ and a circle, we take a circle to fit $\{\boldsymbol{R}_P^{(i)}\}$, which is referred to as ***saddle space circle fitting***. Based on equation (7.30), the mathematical model of the saddle space circle fitting is rewritten as

$$\begin{cases} \Delta_{\mathrm{RS}} = \min\limits_{\boldsymbol{x}} \max\limits_{1\le i\le n} \left\{\Delta^{(i)}(\boldsymbol{x})\right\} \\ \quad\;\; = \min\limits_{\boldsymbol{x}} \max\limits_{1\le i\le n} \left\{\sqrt{(r^{(i)}-r)^2 + (h^{(i)}-h)^2}\right\} \\ \text{s.t.}\;\; r \in (0,+\infty),\; \xi \in [0,\pi/2],\; \eta \in [0,2\pi) \\ \boldsymbol{x} = (x_Q, y_Q, \xi, \eta, r, h)^T \end{cases} \tag{7.40}$$

where Δ_{RS} is the saddle space circle error of $\{\boldsymbol{R}_P^{(i)}\}$, and the objective function $\{\Delta^{(i)}(\boldsymbol{x})\}$ is the normal fitting error set of the discrete point set $\{\boldsymbol{R}_P^{(i)}\}$ and a fitting circle. The optimization variables $\boldsymbol{x} = (x_Q, y_Q, \xi, \eta, r, h)^T$ are the same as those of equation (7.30) except for the variable h, or the distance between the plane the fitting circle lies on and the reference plane Π. n is the number of discrete points of $\{\boldsymbol{R}_P^{(i)}\}$. $h^{(i)}$ and $r^{(i)}$ can be calculated by equations (7.26) and (7.28) respectively.

Equation (7.40) can be solved by the function **fminimax** in the same way as equation (7.30). As to the initial values of the optimization variables, the fitting plane to $\{\boldsymbol{R}_P^{(i)}\}$ should first be determined through equations (5.8) to (5.12) by the least-squares method, and the direction angles of the normal vector of the plane are the initial values of (ξ, η). By equations (7.26) and (7.27), the projection points $\{\boldsymbol{R}_P^{(i)*}\}$ and the distances $h^{(i)}$ can be obtained by projecting the discrete points $\{\boldsymbol{R}_P^{(i)}\}$ onto the reference plane Π. The initial values of (x_Q, y_Q, r) can also be derived by the least-squares method, or equations (2.12) to (2.15). The initial value of h can be calculated by the equation $h = \frac{1}{n}\sum h^{(i)}$.

The point of $\sum^*$ is defined as a ***saddle space circle point*** P_{RS}, if the saddle spatial circle error of its trajectory achieves the minimum value with respect to other points in its neighborhood. The mathematical model of the saddle space circle points can be written as

$$\begin{cases} \delta_{\mathrm{RS}} = \min \Delta_{\mathrm{RS}}(z) \\ z = (x_{Pm}, y_{Pm}, z_{Pm})^T \end{cases} \tag{7.41}$$

where δ_{RS} is the saddle space circle point error. The objective function $\Delta_{\mathrm{RS}}(z)$ is the saddle space circle error for any point in the moving body $\sum^*$, which can be obtained by equation (7.40). The optimization variables $z = (x_{Pm}, y_{Pm}, z_{Pm})^T$ are the rectangular coordinates of the points of $\sum^*$.

7.3.7.2 The Saddle Helix Point and the Binary Link H–S

As another special case, the constraint curve on the cylindrical surface is degenerated into a helical curve, and the binary link C–S can be replaced by binary H–S. In order to estimate the approximation between a given discrete trajectory $\{\boldsymbol{R}_P^{(i)}\}$ and a helical curve, we take a helical curve to fit $\{\boldsymbol{R}_P^{(i)}\}$, which is referred to as ***saddle helical curve fitting***. Based on equation (7.30), the mathematical model of the saddle helical curve fitting is written as

$$\begin{cases} \Delta_{\text{HS}} = \min\limits_{\boldsymbol{x}} \max\limits_{1\le i\le n} \left\{\Delta^{(i)}(\boldsymbol{x})\right\} \\ \quad\;\; = \min\limits_{\boldsymbol{x}} \max\limits_{1\le i\le n} \left\{\sqrt{(r^{(i)}-r)^2+(h^{(i)}-h^{(1)}-p\psi_{i1})^2}\right\} \\ \text{s.t.} \quad r\in(0,+\infty) \\ \boldsymbol{x}=(x_Q, y_Q, \xi, \eta, r, p)^T \end{cases} \tag{7.42}$$

where the parameters $h^{(i)}$, $r^{(i)}$, and ψ_{i1} can respectively be calculated by equations (7.26), (7.28), and (7.29). Δ_{HS} is the saddle helical curve error of $\{\boldsymbol{R}_P^{(i)}\}$, and the objective function $\{\Delta^{(i)}(\boldsymbol{x})\}$ is the normal fitting error set of all points of $\{\boldsymbol{R}_P^{(i)}\}$ and a fitting helical curve. The optimization variables $\boldsymbol{x}=(x_Q, y_Q, \xi, \eta, r, p)^T$ are the same as those of the mathematical model (7.30), except for p or the helical pitch of the fitting helical curve. n is the number of discrete points of $\{\boldsymbol{R}_P^{(i)}\}$.

Equation (7.42) can be solved by the function **fminimax** in the same way as equation (7.30). Since a saddle helical curve fitting is a special case of a saddle cylindrical surface fitting, the initial values of the optimization variables $\boldsymbol{x}=(x_Q, y_Q, \xi, \eta, r, p)^T$ can refer to those of the saddle cylindrical surface fitting. After the variables (x_Q, y_Q, ξ, η, r) are determined, the relative displacement $h^{(i)}$ and rotation angles ψ_{i1} of the discrete point in $\{\boldsymbol{R}_P^{(i)}\}$ about the axis of the cylindrical surface can be calculated through equations (7.26) and (7.29). The initial value of the helical pitch p is given by

$$p^{(i)} = \frac{h^{(i)}-h^{(1)}}{\psi_{i1}}, \quad p = \frac{\sum\limits_{i=2}^{n} p^{(i)}}{n-1} \tag{7.43}$$

The point of $\sum^*$ is defined as a ***saddle helix point*** P_{HS}, if the saddle helical curve error of its trajectory achieves the minimum value with respect to other points in its neighborhood. The mathematical model for the saddle helix point can be written as

$$\begin{cases} \delta_{\text{HS}} = \min \Delta_{\text{HS}}(\boldsymbol{z}) \\ \boldsymbol{z}=(x_{Pm}, y_{Pm}, z_{Pm})^T \end{cases} \tag{7.44}$$

where δ_{HS} is the saddle helix point error. The objective function $\Delta_{\text{HS}}(\boldsymbol{z})$ is the saddle helical curve error for any point in the moving body $\sum^*$, which can be obtained through equation (7.42). The optimization variables $\boldsymbol{z}=(x_{Pm}, y_{Pm}, z_{Pm})^T$ are the rectangular coordinates of points in $\sum^*$.

Example 7.3 The coupler link of a spatial RCCC linkage has 72 separated positions in the frame of the base link, which are described by the motion parameters listed in Table A.1 and the coordinate transformation formula (A.7) in Appendix A. Determine: (1) the saddle cylindrical surface errors of coupler points; (2) the saddle cylinder points in the coupler link.

Solution

1. **The saddle cylindrical surface errors**
   In the area $x_m \in [-5,1]$, $y_m \in [-2,2]$, $z_m \in [-2,2]$ of the coupler link, 102 541 points can be selected in total in steps of 0.1 in each axis $\boldsymbol{x}_2, \boldsymbol{y}_2$, and z_2. For the discrete trajectories traced by these points, the saddle cylindrical surface errors Δ_{CS} can be calculated by calling the sub-program ACSF. Hence, with the saddle cylindrical surface errors Δ_{CS} and the coordinates (x_{Pm}, y_{Pm}, z_{Pm}) of the coupler point as the dependent and independent variables, the error hypersurface can be constructed, and three coordinate planes in a cutaway view are taken to reveal the saddle cylindrical surface errors Δ_{CS} by the contour lines with the same values, as shown in Fig. 7.13.
   For the coupler points in the area $x_m \in [-6,6]$, $y_m \in [-3,3]$ on the coordinate plane $\boldsymbol{R}_C - \boldsymbol{x}_2\boldsymbol{y}_2$, the saddle cylindrical surface error surface and its contour plot are drawn in Fig. 7.14, on which there exist four valleys and four corresponding bottom points $P_1(-0.0143,-0.6249)$, $P_2(-3.2793,0.2401)$, $P_3(-2.5819,2.2864)$, $P_4(-3.2647,3)$.
2. **The saddle cylinder points**
   In the area $x_m \in [-6,6]$, $y_m \in [-3,3]$, $z_m \in [-3,3]$ of the coupler link, 50 initial points are randomly generated by the program, and the saddle cylinder points are searched by calling the sub-program ACSP based on the optimization model (7.39). Among the optimization results, five saddle cylinder points with less fitting errors are listed in Table 7.4.

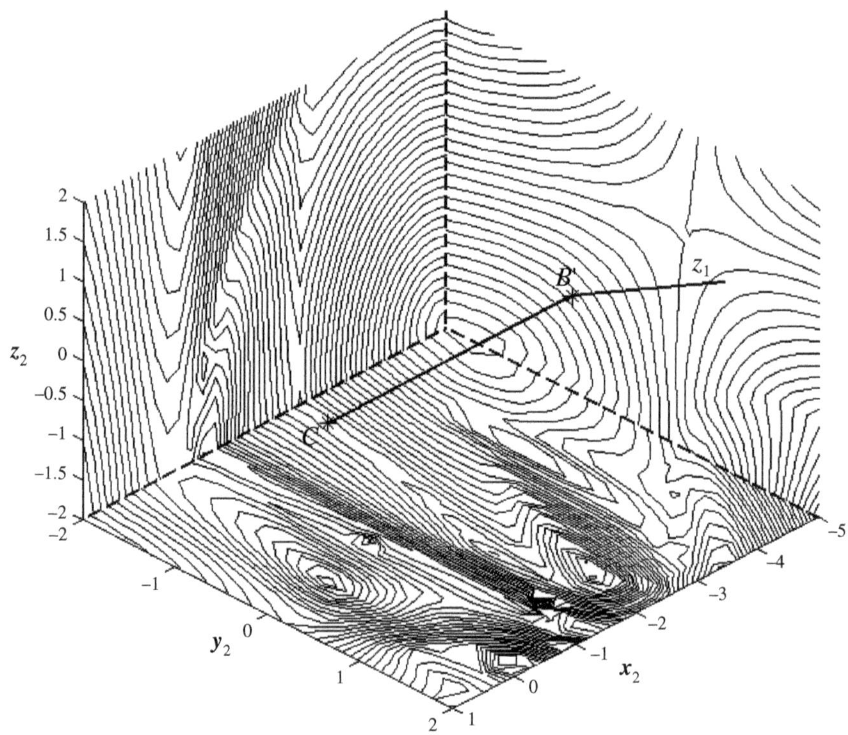

Figure 7.13 The saddle cylindrical surface errors of the coupler for the RCCC linkage

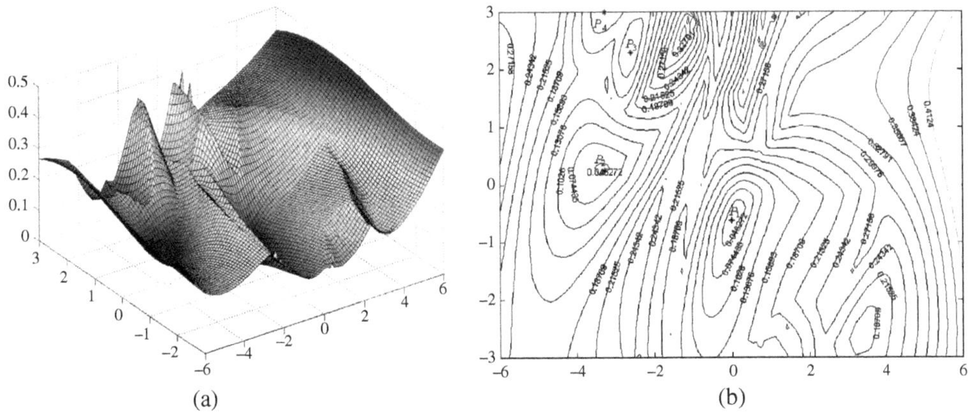

Figure 7.14 The saddle cylindrical surface errors of the coupler plane $\boldsymbol{R}_C - \boldsymbol{x}_2\boldsymbol{y}_2$

Table 7.4 The saddle cylinder points in the coupler link of the RCCC linkage

	Saddle cylinder point			Axis of the saddle cylindrical surface				Radius	Saddle cylinder point error
				Reference point Q		Direction angles (ξ, η)			
	x_{Pm}	y_{Pm}	z_{Pm}	x_Q	y_Q	ξ	η	r	δ_{CS}
1	−0.2296	0.2187	−3	−4.9343	−0.0093	1.1240	1.5382	3.1687	0.006814
2	−0.0154	−0.1100	−1.7348	−5.0430	0.0248	1.0743	1.5417	3.2310	0.012876
3	0.0215	−0.4598	−0.5958	−5.0547	0.0133	1.0595	1.5544	3.5724	0.016086
4	−0.0025	−0.5537	−0.2577	−5.0953	0.0247	1.0635	1.5460	3.7527	0.016334
5	−0.0508	−0.7611	0.5189	−5.0815	0.0201	1.0558	1.5539	4.0943	0.017949

Example 7.4 The nine separated spatial positions of the moving body are described by parameters $(x_{Omf}^{(i)}, y_{Omf}^{(i)}, z_{Omf}^{(i)}; \theta_1^{(i)}, \theta_2^{(i)}, \theta_3^{(i)})$ listed in Table 7.1. Determine: (1) the saddle cylindrical surface errors; (2) the saddle cylinder points.

Solution

1. **The saddle cylindrical surface errors**

   In the area $x_m \in [-50, 50]$, $y_m \in [-50, 50]$, $z_m \in [-50, 50]$ of the moving body, 132 651 points are selected in total in steps of 2 in coordinate directions $\boldsymbol{i}_m, \boldsymbol{j}_m, \boldsymbol{k}_m$ and trace the discrete trajectories in the fixed frame, whose saddle cylindrical surface errors Δ_{CS} can be calculated by calling the sub-program ACSF. Hence, the saddle cylindrical surface error Δ_{CS} and the tracing point coordinates (x_{Pm}, y_{Pm}, z_{Pm}) constitute the error hypersurface for the nine-position case, and three coordinate planes in a cutaway view are taken to reveal the saddle cylindrical surface errors Δ_{CS} by the contour lines with the same values, as shown in Fig. 7.15.

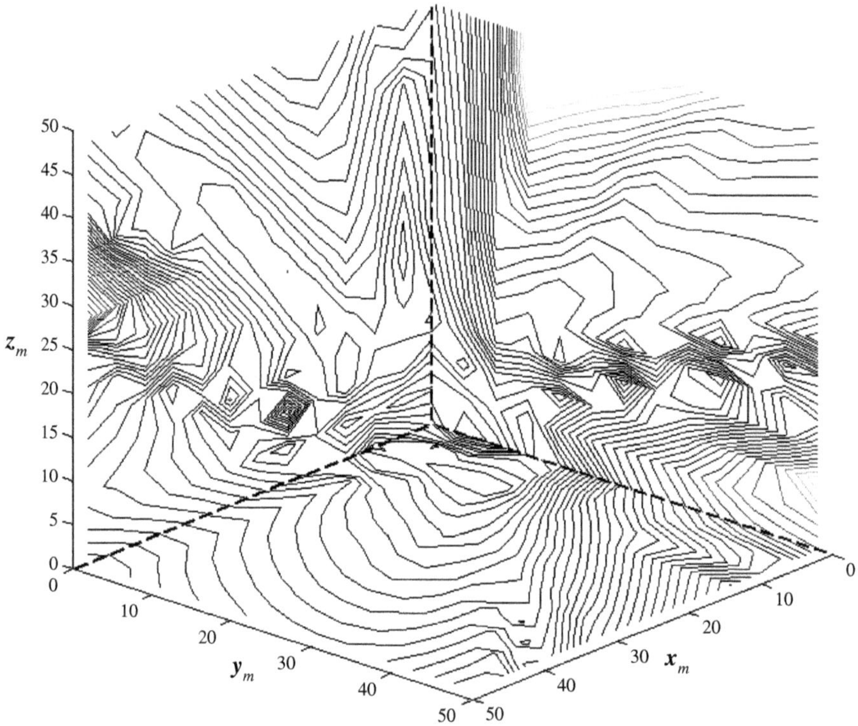

Figure 7.15 The saddle cylindrical surface errors for nine positions

Table 7.5 Seven saddle cylinder points for nine positions

	Saddle cylinder point			The axis of the saddle cylindrical surface				Radius	Saddle cylinder point error
				Reference point Q		Direction angles (ξ, η)			
	x_{Pm}	y_{Pm}	z_{Pm}	x_Q	y_Q	ξ	η	r	δ_{CS}
SCP1	−9.6354	16.1722	−1.7120	−23.8730	46.7247	1.2820	0.5316	34.8225	0.003565
SCP2	−16.8573	20.7756	−6.9014	29.4871	−41.5932	0.3731	2.7076	81.3531	0.003818
SCP3	−20.2257	19.2444	−7.6309	13.0107	−27.8737	0.2737	1.9149	59.9031	0.004498
SCP4	6.4366	21.8357	−15.9559	−6.3547	30.6532	1.4801	0.5547	102.2371	0.004617
SCP5	−0.1368	1.6592	−9.1681	2.3611	−0.7933	1.1883	1.6366	11.8991	0.004652
SCP6	−10.5473	3.1821	−0.1288	−12.5361	5.1831	0.2378	1.4412	2.8929	0.004720
SCP7	−7.7023	1.1333	−14.8945	−0.1893	−2.5651	0.9195	1.5554	18.3279	0.005325

2. **The saddle cylinder points**

100 initial points are randomly selected by the program and the saddle cylinder points are searched by calling the sub-program ACSP. Among the optimization results, seven saddle cylinder points (SCP1–SCP7) with less saddle cylinder point errors are listed in Table 7.5.

Table 7.6 Seven saddle cylinder points for eight positions

	Saddle cylinder point			The axis of the saddle cylindrical surface				Radius	Saddle cylinder point error
				Reference point Q		Direction angles (ξ, η)			
	x_{Pm}	y_{Pm}	z_{Pm}	x_Q	y_Q	ξ	η	r	δ_{CS}
1	−7.7014	6.7076	5.0745	−11.2478	6.7965	0.7125	0.7934	2.6018	0.001031
2	−10.6352	−16.8060	−17.7404	13.4585	15.8983	1.4626	2.4876	48.6626	0.001214
3	9.7953	12.3917	−38.6472	−1.4450	7.8452	0.6325	1.9010	23.3406	0.001360
4	−17.9988	20.0669	−6.9437	14.3883	−26.4300	0.3077	2.5082	58.7560	0.001908
5	6.6012	1.5889	1.9863	−1.3085	−4.5437	0.7726	3.7115	10.9769	0.001997
6	11.4723	17.1630	−47.1102	−3.1042	9.3607	0.5542	1.9826	28.0601	0.002221
7	8.7507	18.9870	−8.7629	6.2427	23.4944	1.3064	0.5147	52.4150	0.002566

Similarly, for a moving body with eight separated positions, described by the motion parameters of the first eight lines of Table 7.1, the saddle cylinder points can be located by calling the sub-program ACSP, and seven saddle cylinder points with less fitting errors are listed in Table 7.6.

For the cases of seven and six positions, the saddle cylindrical surface errors Δ_{CS} have been shown respectively in Figs 7.12 and 7.11 in Section 7.3, and need not be introduced again here.

7.4 Saddle Constant Axis Line L_{CC}

We move on to discuss the global properties of discrete line trajectory in spatial movement by comparing them with the constraint ruled surfaces. It is an interesting but difficult topic, not only with respect to the global properties but also the complexity; in particular, there are many constraint surfaces, corresponding to different variety of binary links or open chains, which are new in the literature. We only deal with the constant axis line and its degenerate cases in the book, based on the theoretical bases provided by Chapters 3 and 6.

7.4.1 Ruled Surface Fitting

A moving body $\sum^*$ with the moving coordinate system $\{\boldsymbol{O}_m; \boldsymbol{i}_m, \boldsymbol{j}_m, \boldsymbol{k}_m\}$ occupies separated positions in the fixed frame $\{\boldsymbol{O}_f, \boldsymbol{i}_f, \boldsymbol{j}_f, \boldsymbol{k}_f\}$, which is described by the given discrete motion parameters $(x_{Omf}^{(i)}, y_{Omf}^{(i)}, z_{Omf}^{(i)}; \theta_1^{(i)}, \theta_2^{(i)}, \theta_3^{(i)})$. A straight line L of $\sum^*$ with direction angles (ϕ, φ) passes through $P\,(x_{Pm}, y_{Pm}, z_{Pm})$ in $\sum^*$ and traces a discrete ruled surface, or a discrete line set $\{\boldsymbol{R}_L^{(i)}\}$ in $\{\boldsymbol{O}_f; \boldsymbol{i}_f, \boldsymbol{j}_f, \boldsymbol{k}_f\}$, which is expressed by equation (7.8). We expect to develop the curve fitting approach to that of ruled surface fitting, such as the constant axis ruled surface $\sum_{C'C}$ corresponding to the binary link C–C. Needless to say, there exist enormous difficulties in estimating the degree of approximation between two ruled surfaces if they are compared directly by the rulings. Fortunately, Theorems 3.5 and 3.6 in Chapter 3 have provided us with the necessary and sufficient conditions for a ruled surface to be a constant axis ruled surface: the image spherical curve of the unit direction vector of the generator is a circle and the striction curve is a cylindrical curve. Hence, the fitting error for a constant axis ruled surface to fit the discrete line set $\{\boldsymbol{R}_L^{(i)}\}$ can be divided into two parts: the spherical image curve fitting error and the striction curve fitting error, which correspond to the direction and position of L in $\sum^*$ respectively.

7.4.2 Saddle Spherical Image Circle Point

As mentioned in Chapter 6, the spatial movement of $\sum^*$ can be regarded as the moving axode Σ_m rolls and slides on the fixed axode Σ_f along the ISA, while the spherical image curve C_m of Σ_m rolls on the spherical image curve C_f of Σ_f without sliding. For a straight line L of $\sum^*$, the unit direction vector $\boldsymbol{l}_m$ in $\{\boldsymbol{O}_m; \boldsymbol{i}_m, \boldsymbol{j}_m, \boldsymbol{k}_m\}$ has a corresponding spherical image point P_{lm}, which is attached to C_m since they belong to $\sum^*$ together. The straight line L traces a trajectory ruled surface Σ_l in $\{\boldsymbol{O}_f; \boldsymbol{i}_f, \boldsymbol{j}_f, \boldsymbol{k}_f\}$, whose spherical image curve C_l on a unit spherical surface is traced by the point P_{lm} as C_m rolls on C_f. Hence, all parallel lines of $\sum^*$ with the same direction but different positions may be mapped onto the same spherical image point attached to the spherical image curve C_m and produce line trajectories, or ruled surfaces, with the identical spherical image curve.

For a given spatial discrete movement of a rigid body $\sum^*$ described by the parameters $(x_{Omf}^{(i)}, y_{Omf}^{(i)}, z_{Omf}^{(i)}; \theta_1^{(i)}, \theta_2^{(i)}, \theta_3^{(i)})$, a line $L(\phi, \varphi)$ of $\sum^*$ traces a discrete line trajectory or a discrete line set $\{\boldsymbol{R}_L^{(i)}\}$ in $\{\boldsymbol{O}_f; \boldsymbol{i}_f, \boldsymbol{j}_f, \boldsymbol{k}_f\}$. The spherical image curve of $\{\boldsymbol{R}_L^{(i)}\}$ is comprised of some discrete points on the fixed unit spherical surface, which are called the ***discrete spherical image point set*** $\{\boldsymbol{l}^{(i)}\} = \{(l_x^{(i)}, l_y^{(i)}, l_z^{(i)})^T\}$ and expressed by the discrete form of equation (7.6). We can estimate the degree of approximation between the discrete spherical image point set $\{\boldsymbol{l}^{(i)}\}$ and a spherical circle or take a spherical circle to fit the discrete spherical image points $\{\boldsymbol{l}^{(i)}\}$ adaptively, which is similar to that in the discrete kinematic geometry of spherical movement in Chapter 5, as shown in Fig. 7.16. The mathematical model of the saddle spherical image circle fitting is

$$\begin{cases} \Delta_{\mathrm{Rr}} = \min\limits_{\boldsymbol{x}} \max\limits_{1\le i\le n} \left\{\Delta^{(i)}(\boldsymbol{x})\right\} \\ = \min\limits_{\boldsymbol{x}} \max\limits_{1\le i\le n} \{\,|\arccos(\sin\xi\ \cos\eta \cdot l_x^{(i)} + \sin\xi\ \sin\eta \cdot l_y^{(i)} + \cos\xi \cdot l_z^{(i)}) - \delta|\} \\ \text{s.t. } \ \xi \in [0, \pi),\ \eta \in [0, 2\pi),\ \delta \in [0, \pi/2] \\ \boldsymbol{x} = (\xi, \eta, \delta)^T \end{cases} \tag{7.45}$$

where Δ_{Rr} is the saddle spherical image circle error of $\{\boldsymbol{l}^{(i)}\}$ and the objective function is the normal fitting error set of $\{\boldsymbol{l}^{(i)}\}$ and a fitting spherical circle. The optimization variables $\boldsymbol{x} = (\xi, \eta, \delta)^T$ are respectively the half-cone angle δ and the direction angles (ξ, η) of the axis of the cone constituted by the fitting spherical circle and the center point of the fixed unit spherical surface. n is the number of discrete spherical image points of $\{\boldsymbol{l}^{(i)}\}$.

In fact, the mathematical model (7.45) of the saddle spherical image circle fitting is identical to the mathematical model (5.7) of the saddle spherical circle fitting in Chapter 5. Hence, the solving process and the initial values of the optimization variables for equation (7.45) can refer to those of equation (5.7) or the sub-program ArRF can be called directly to solve equation (7.45). To identify the saddle spherical image circle fitting from the saddle spherical circle fitting, we stress the discrete spherical image point set $\{\boldsymbol{l}^{(i)}\}$ and give the following definition.

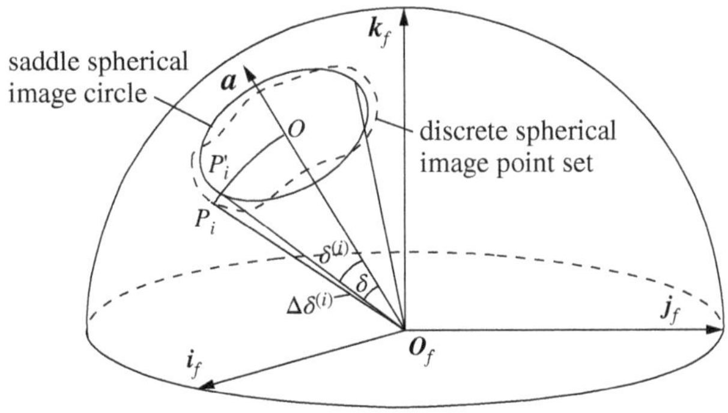

Figure 7.16 The saddle spherical image circle fitting of a discrete line trajectory

Definition 7.5 Based on the discrete spherical image point set $\{\boldsymbol{l}^{(i)}\}$ of a given discrete line set $\{\boldsymbol{R}_L^{(i)}\}$, a spherical image circle, adaptively located by letting the (maximum) normal fitting error be minimum, is defined as a ***saddle spherical image circle***, and the fitting error is referred to as the ***saddle spherical image circle error***.

As mentioned above, the spherical image point corresponding to the direction of parallel lines of $\sum^*$ in spatial movement is equivalent to a point of $\sum^*$ in unit spherical movement. The spherical image curve is viewed as the spherical trajectory traced by the point of $\sum^*$ in unit spherical movement. All properties about the saddle spherical circles in discrete kinematic geometry of spherical movement in Chapter 5 are applicable to the saddle spherical image circle, such as the adaptivity, uniqueness, and comparability, as well as the properties of the error surfaces in discrete positions, both finite and multiple separated. As for the spatial movement of $\sum^*$, the lines of $\sum^*$ have different directions and trace different ruled surfaces, which correspond to different spherical image curves. There exist some lines of $\sum^*$ with specific directions, whose line trajectories correspond to the spherical image curves closest to circles with respect to other lines in its neighborhood, or the saddle spherical image circle errors achieve the minimum values.

Definition 7.6 The direction of a line of the moving body with respect to other directions at its neighborhood, whose corresponding saddle spherical image circle error achieves the minimum, is defined as a ***saddle spherical image circle point***, and its corresponding error is referred to as a ***saddle spherical image circle point error***.

Especially if the saddle spherical image circle point error is zero, the direction of a line on the moving body $\sum^*$ corresponds to a spherical image circle point, or the line trajectory is a constant inclination ruled surface. The saddle spherical image circle point is analogous to the saddle spherical circle point discussed in Chapter 5. We write the mathematical model of the saddle spherical image circle point as

$$\begin{cases} \gamma_{\mathrm{Rr}} = \min \Delta_{\mathrm{Rr}}(z) \\ \text{s.t. } \phi \in [0, \pi/2],\ \varphi \in [0, 2\pi] \\ z = (\phi, \varphi)^T \end{cases} \tag{7.46}$$

where γ_{Rr} is the saddle spherical image circle point error and the objective function $\Delta_{\mathrm{Rr}}(z)$ is the saddle spherical image circle error for any direction of the lines in $\sum^*$, which can be obtained through equation (7.45). The optimization variables $z = (\phi, \varphi)^T$ are the direction angles of any direction in $\sum^*$. Equation (7.46) can be solved directly by the function **fmincon** in Matlab, or by calling the sub-program ArRP.

Based on the theories of the discrete kinematic geometry of spherical movement in Chapter 5, for all lines on $\sum^*$ in spatial movement, the saddle spherical image circle point error is certainly the function of the direction parameters (ϕ, φ) of the lines, or there exist minimum values of the errors.

Theorem 7.3 *There exist several directions of lines in the moving body in a non-degenerate spatial movement, which correspond to the saddle spherical image circle points.*

The spherical circle points and the saddle spherical circle points of $\sum^*$ in discrete spherical movement have been discussed thoroughly in Chapter 5, including the cases of four positions, five positions, and multiple positions. The spherical movement reveals exactly its important role of a bridge, not only from point trajectory to line trajectory, but also from plane to spatial movement. We can easily extend the properties of a point path in plane movement to those in spherical movement, and develop them into spatial movement. And we can also apply the results of point trajectories in spherical movement to line trajectories, especially the directions of the lines of $\sum^*$ in spatial movement.

7.4.3 Saddle Striction Cylinder Point

Now we switch to the second part, or the striction curve fitting for a constant axis ruled surface to fit the discrete line set $\{\boldsymbol{R}_L^{(i)}\}$ in $\{\boldsymbol{O}_f; \boldsymbol{i}_f, \boldsymbol{j}_f, \boldsymbol{k}_f\}$. Based on the first part, or the spherical image circle fitting to the discrete spherical image point set $\{\boldsymbol{l}^{(i)}\}$ of $\{\boldsymbol{R}_L^{(i)}\}$, the spherical image circle (cone) provides the direction angles (ξ, η) of the fixed axis of the constant axis ruled surface $\sum_{\mathrm{C'C}}$ in $\{\boldsymbol{O}_f; \boldsymbol{i}_f, \boldsymbol{j}_f, \boldsymbol{k}_f\}$. In Chapter 3, Theorems 3.5 and 3.6 declare that the striction curve of a constant axis ruled surface $\sum_{\mathrm{C'C}}$ is a cylindrical curve, and the striction point on the generatrix is just one of the perpendicular feet of the common normal between the generatrix and the fixed axis at any instant, which is the theoretical basis of the striction curve fitting in the case of separated positions. Hence, a cylindrical surface or a ***striction cylindrical surface*** is logically taken to fit the striction points of the discrete line set $\{\boldsymbol{R}_L^{(i)}\}$ for location of the fixed axis of $\sum_{\mathrm{C'C}}$. Of course, the direction angles (ξ, η) of the axis of the striction cylindrical surface are determined by the spherical image circle (cone) fitting before the striction cylindrical surface fitting to the striction points, which is different from the saddle cylindrical surface fitting to the discrete point set $\{\boldsymbol{R}_P^{(i)}\}$ in Section 7.2 in that the parameters of the saddle cylindrical surface, including the direction and the position of its axis, are determined totally by $\{\boldsymbol{R}_P^{(i)}\}$.

For the discrete trajectory or discrete line set $\{\boldsymbol{R}_L^{(i)}\}$ in $\{\boldsymbol{O}_f; \boldsymbol{i}_f, \boldsymbol{j}_f, \boldsymbol{k}_f\}$ traced by a line L of $\sum^*$ in spatial movement, a constant axis ruled surface $\sum_{\mathrm{C'C}}$ is taken to fit $\{\boldsymbol{R}_L^{(i)}\}$ and the direction of the fixed axis L_Q of $\sum_{\mathrm{C'C}}$ is first determined by the saddle spherical image circle of the discrete spherical image point set $\{\boldsymbol{l}^{(i)}\}$ of $\{\boldsymbol{R}_L^{(i)}\}$. By equation (7.45), the unit direction vector of L_Q is $\boldsymbol{l}_Q = (\sin\xi\ \cos\eta, \sin\xi\ \sin\eta, \cos\xi)^T$. Similar to the saddle cylindrical surface fitting to the discrete point set $\{\boldsymbol{R}_P^{(i)}\}$ in Section 7.2, the projection plane Π, with $\boldsymbol{l}_Q$ as its normal vector and passing through the original point O_f of $\{\boldsymbol{O}_f; \boldsymbol{i}_f, \boldsymbol{j}_f, \boldsymbol{k}_f\}$ is chosen. The intersection point Q between the projection plane Π and the axis L_Q is viewed as the reference point of L_Q, whose position vector is $\boldsymbol{R}_Q=(x_Q, y_Q, z_Q)^T$ in $\{\boldsymbol{O}_f; \boldsymbol{i}_f, \boldsymbol{j}_f, \boldsymbol{k}_f\}$. The axis L_Q can be expressed as

$$\boldsymbol{R}_{l_Q} = \boldsymbol{R}_Q + \mu \boldsymbol{l}_Q \tag{7.47}$$

At the ith position, according to Theorems 3.5 and 3.6, the line $L^{(i)}$ of $\{\boldsymbol{R}_L^{(i)}\}$ and the axis L_Q have a common normal line $\overline{A^{(i)}B^{(i)}}$, where $B^{(i)}$ is the perpendicular foot on $L^{(i)}$, or a striction point, $A^{(i)}$ is the perpendicular foot on the axis L_Q, as shown in Fig. 7.17. By the vector equations (7.8) of $L^{(i)}$ and (7.47) of L_Q, the position vectors of points $A^{(i)}$ and $B^{(i)}$ are

$$\begin{cases} \boldsymbol{R}_A^{(i)} = \boldsymbol{R}_Q + u^{(i)} \boldsymbol{l}_Q \\ \boldsymbol{R}_B^{(i)} = \boldsymbol{R}_P^{(i)} + v^{(i)} \boldsymbol{l}^{(i)} \end{cases} \tag{7.48}$$

where $u^{(i)}$ and $v^{(i)}$ are the parameters of the two lines respectively. $u^{(i)}$ denotes the displacement of $A^{(i)}$ relative to the reference point Q on L_Q and $v^{(i)}$ reveals the displacement of the striction

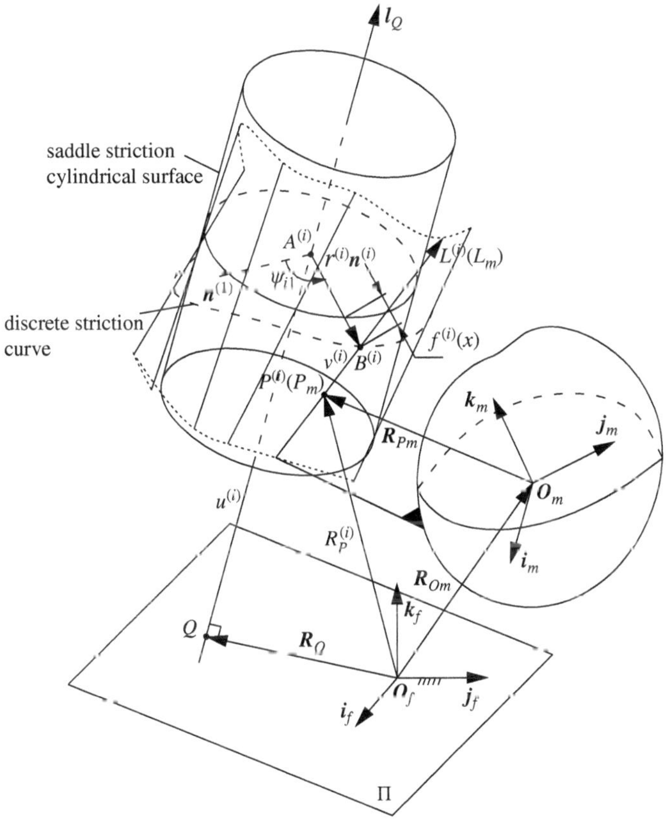

Figure 7.17 The saddle striction cylindrical surface fitting of the discrete striction curve

point $B^{(i)}$ relative to the reference point P on $L^{(i)}$. At the striction point $B^{(i)}$, the unit direction vector $\boldsymbol{n}^{(i)}$ of the common normal line $\overline{A^{(i)}B^{(i)}}$ is

$$\boldsymbol{n}^{(i)} = \frac{\boldsymbol{l}^{(i)} \times \boldsymbol{l}_Q}{|\boldsymbol{l}^{(i)} \times \boldsymbol{l}_Q|} \tag{7.49}$$

The common normal line $\overline{A^{(i)}B^{(i)}}$ represented by the vector $\boldsymbol{R}_B^{(i)} - \boldsymbol{R}_A^{(i)}$ is perpendicular to the lines $L^{(i)}$ and L_Q, or

$$\begin{cases} (\boldsymbol{R}_A^{(i)} - \boldsymbol{R}_B^{(i)}) \cdot \boldsymbol{l}_Q = 0 \\ (\boldsymbol{R}_A^{(i)} - \boldsymbol{R}_B^{(i)}) \cdot \boldsymbol{l}^{(i)} = 0 \end{cases} \tag{7.50}$$

Substituting equation (7.48) into equation (7.50), we derive the parameters $u^{(i)}$ and $v^{(i)}$ as

$$\begin{cases} u^{(i)} = \dfrac{(\boldsymbol{R}_Q - \boldsymbol{R}_P^{(i)}) \cdot [\boldsymbol{l}^{(i)}(\boldsymbol{l}^{(i)} \cdot \boldsymbol{l}_Q) - \boldsymbol{l}_Q]}{1 - (\boldsymbol{l}^{(i)} \cdot \boldsymbol{l}_Q)^2} \\ v^{(i)} = \dfrac{(\boldsymbol{R}_Q - \boldsymbol{R}_P^{(i)}) \cdot [\boldsymbol{l}^{(i)} - (\boldsymbol{l}^{(i)} \cdot \boldsymbol{l}_Q)\boldsymbol{l}_Q]}{1 - (\boldsymbol{l}^{(i)} \cdot \boldsymbol{l}_Q)^2} \end{cases} \tag{7.51}$$

In the case of $i \geq 2$, the unit vector $\boldsymbol{n}^{(i)}$ of the common normal line between the fixed axis L_Q and the line $L^{(i)}$ at the ith position forms an angle ψ_{i1} with $\boldsymbol{n}^{(1)}$ at the first position, which is

$$\cos\psi_{i1} = \boldsymbol{n}^{(i)} \cdot \boldsymbol{n}^{(1)} \tag{7.52}$$

The normal distance $r^{(i)}$ between the line $L^{(i)}$ and the line L_Q, or the length of the common normal line $\overline{A^{(i)}B^{(i)}}$, can be derived by equation (7.49):

$$r^{(i)} = (\boldsymbol{R}_P^{(i)} - \boldsymbol{R}_Q) \cdot \frac{\boldsymbol{l}^{(i)} \times \boldsymbol{l}_Q}{|\boldsymbol{l}^{(i)} \times \boldsymbol{l}_Q|} \tag{7.53}$$

Hence, for a given discrete line set $\{\boldsymbol{R}_L^{(i)}\}$, the direction angles (ξ, η) of the fixed axis L_Q and the half-cone angle δ of the fitting constant axis ruled surface $\sum_{C'C}$ can be determined by the saddle spherical image circle fitting, equation (7.45), and the discrete striction points $\boldsymbol{R}_B^{(i)}$ of $\{\boldsymbol{R}_L^{(i)}\}$ and the normal distance $r^{(i)}$ can be calculated by equations (7.48) to (7.53), which provide the conditions for ***striction cylindrical surface fitting***, or locating the position of the fixed axis L_Q and the radius r of the striction cylindrical surface of the constant axis ruled surface. By referring to the saddle cylindrical surface fitting to the discrete point set $\{\boldsymbol{R}_P^{(i)}\}$ in Section 7.2:

Definition 7.7 For a given discrete line set $\{\boldsymbol{R}_L^{(i)}\}$ with its fixed axis direction, a cylindrical surface is defined as a ***saddle striction cylindrical surface*** whose axis position is located by letting the maximum variation of the normal distance $r^{(i)}$ between discrete lines $L^{(i)}$ of $\{\boldsymbol{R}_L^{(i)}\}$ and the fixed axis line L_Q be minimum, while the maximum variation is referred to as the ***saddle striction cylindrical surface error***.

According to Definition 7.7, the mathematical model of the saddle striction cylindrical surface can be set up as

$$\begin{cases} \Delta_{CC} = \min_{x} \max_{1\leq i\leq n} \left\{ \Delta^{(i)}(\boldsymbol{x}) \right\} \\ \quad = \min_{x} \max_{1\leq i\leq n} \{ |r^{(i)} - r| \} \\ \text{s.t.} \quad r \in (0, +\infty) \\ \boldsymbol{x} = (x_Q, y_Q, r)^T \end{cases} \tag{7.54}$$

where Δ_{CC} is the saddle striction cylindrical surface error of $\{\boldsymbol{R}_L^{(i)}\}$ and the objective function $\Delta^{(i)}(\boldsymbol{x})$ is the deviation between the radius r and the normal distance $r^{(i)}$ at the discrete position. The optimization variables $\boldsymbol{x} = (x_Q, y_Q, r)^T$ are respectively the coordinates (x_Q, y_Q) of the reference point Q of the axis L_Q and the radius r of the fitting cylindrical surface.

For the given discrete line set $\{\boldsymbol{R}_L^{(i)}\}$, the unit direction vector $\boldsymbol{l}_Q$ is determined and the reference plane Π can be constructed after the saddle spherical image circle fitting, whose vector equation is

$$(\boldsymbol{R} - \boldsymbol{R}_Q) \cdot \boldsymbol{l}_Q = 0 \tag{7.55}$$

All striction points of the lines $L^{(i)}$ of $\{\boldsymbol{R}_L^{(i)}\}$ can be projected onto the plane Π to be projection striction points; in fact, the saddle striction cylindrical surface fitting of the discrete striction points is equivalent to the saddle circle fitting to the projection striction points on the plane Π.

For convenience in carrying out the saddle cylindrical surface fitting, the initial values of the optimization variables can simply be provided by the intersection points between the lines $L^{(i)}$ of $\{\boldsymbol{R}_L^{(i)}\}$ and the projection plane Π, rather than the projection striction points, while the intersection points can be derived as follows. The line $L^{(i)}$ of $\{\boldsymbol{R}_L^{(i)}\}$ can be expressed by $\boldsymbol{R} = \boldsymbol{R}_P^{(i)} + \mu^{(i)}\boldsymbol{l}^{(i)}$. We substitute it into equation (7.55) and get $\mu^{(i)} = -\boldsymbol{R}_P^{(i)} \cdot \boldsymbol{l}_Q/(\boldsymbol{l}_Q \cdot \boldsymbol{l}^{(i)})$. The position vector $\boldsymbol{R}_P^{(i)*}$ of the intersection point on the projection plane is

$$\boldsymbol{R}_P^{(i)*} = \boldsymbol{R}_P^{(i)} - \frac{\boldsymbol{R}_P^{(i)} \cdot \boldsymbol{l}_Q}{\boldsymbol{l}_Q \cdot \boldsymbol{l}^{(i)}} \boldsymbol{l}^{(i)} \tag{7.56}$$

The discrete intersection points $\{\boldsymbol{R}_P^{(i)*}\} = \{(x_P^{(i)*}, y_P^{(i)*}, z_P^{(i)*})^T\}$, $i = 1, \ldots, n$ on the projection plane Π are obtained for $\{\boldsymbol{R}_L^{(i)}\}$, and are fitted by a circle under the least-squares method, such as equations (2.12) to (2.15).

As to the optimization model of the saddle striction cylindrical surface, the function **fminimax** in Matlab is applied directly to solve equation (7.54). The ***sub-program ACCF*** is similar to the sub-program ACSF for a saddle cylindrical surface.

A line of $\sum^*$ in spatial movement traces a discrete line set $\{\boldsymbol{R}_L^{(i)}\}$, whose saddle spherical image circle can be determined by equation (7.45) and saddle striction cylindrical surface can be located by equation (7.54). Similarly, all lines of $\sum^*$, which are parallel to this line with the same direction, pass through different reference points of $\sum^*$ and trace different discrete line trajectories $\{\boldsymbol{R}_L^{(i)}\}$ in $\{\boldsymbol{O}_f; \boldsymbol{i}_f, \boldsymbol{j}_f, \boldsymbol{k}_f\}$. Their discrete striction curves will appear of different shapes and magnitudes, leading to different saddle striction cylindrical surfaces with different saddle striction cylindrical surface errors (although they correspond to the identical saddle spherical image circle). In other words, the parameters of the saddle striction cylindrical surface of $\{\boldsymbol{R}_L^{(i)}\}$, including the position parameters of the axis L_Q, the radius r, and the fitting error Δ_{CC}, are functions of the reference point's coordinates (x_{Pm}, y_{Pm}, z_{Pm}) of the tracing line in $\sum^*$, except for the direction parameter (ϕ, φ). Hence, the saddle striction cylindrical surface errors of line trajectories, traced by any group of parallel lines of $\sum^*$, may have extreme values for some reference points of $\sum^*$ with respect to other points at their neighborhoods.

Definition 7.8 The reference points of the moving body $\sum^*$ are defined as the ***saddle striction cylinder points*** if the parallel lines of $\sum^*$ passing these points with a given direction vector $\boldsymbol{l}_m$ trace the trajectories in a fixed frame, which leads to the saddle striction cylindrical surface errors being minimum with respect to other reference points at their neighborhood. The error is referred to as the ***saddle striction cylinder point error***.

Based on Definition 7.8, the mathematical model of the saddle striction cylinder point for the parallel lines of $\sum^*$ with given direction vectors $\boldsymbol{l}_m(\phi, \varphi)$ is set up as

$$\begin{cases} \chi_{CC} = \min \Delta_{CC}(z) \\ z = (x_{Pm}, y_{Pm})^T \end{cases} \tag{7.57}$$

where χ_{CC} is the saddle striction cylinder point error, and the objective function $\Delta_{CC}(z)$ is the saddle striction cylindrical surface error for any of the parallel lines. The optimization variables $z = (x_{Pm}, y_{Pm})^T$ are the coordinates of the reference points on the reference plane $\boldsymbol{O}_m - \boldsymbol{i}_m\boldsymbol{j}_m$ in $\{\boldsymbol{O}_m; \boldsymbol{i}_m, \boldsymbol{j}_m, \boldsymbol{k}_m\}$.

For the mathematical model of the saddle striction cylinder point, equation (7.57) can be solved directly by the function **fmincon** of Matlab, whose parameters are set in the same way as those of the saddle cylinder point. A ***sub-program ACCP*** is constructed for the location of saddle striction cylinder points in the moving body with multiple separated positions. The inputs are the motion parameters of the separated positions and the outputs are the coordinates and errors of the saddle striction cylinder points.

In particular, a saddle striction cylinder point, or the reference point of the line of $\sum^*$, is a striction cylinder point if the saddle striction cylinder point error is zero. This means that the striction points of the discrete line set $\{\boldsymbol{R}_L^{(i)}\}$ locate exactly on a cylindrical surface. It needs to be pointed out that the saddle striction cylinder point is just a reference point on the coordinate plane $\boldsymbol{O}_m - \boldsymbol{i}_m\boldsymbol{j}_m$, where the line of $\sum^*$ passes through, other than a striction point of the line $L^{(i)}$ of $\{\boldsymbol{R}_L^{(i)}\}$. In fact, the striction point of a line $L^{(i)}$ corresponds to a point of the line L, which can be designated as a reference point of $\sum^*$, but it is not a fixed point of $\sum^*$ and changes its position as $\sum^*$ moves, which leads to inconvenience in calculation. In Chapter 6, the striction point at an instant (position) is expressed by the coordinates (h, p) in the Frenet frame of the axodes, which reflects both the kinematic and geometrical meanings.

Equation (7.54) reveals that the saddle striction cylinder errors depend on the kinematic properties of $\sum^*$ and the positions of the lines of $\sum^*$, or are a function of the reference points' coordinates. For a given direction, all reference points of the reference plane $\boldsymbol{O}_m - \boldsymbol{i}_m\boldsymbol{j}_m$ determine parallel straight lines and correspond to different saddle striction cylinder errors.

Theorem 7.4 *There exist a few saddle striction cylinder points in a moving body with a non-degenerate spatial movement, whose corresponding saddle striction cylinder errors achieve the minimum.*

The saddle cylindrical surface fitting for a discrete point trajectory $\{\boldsymbol{R}_P^{(i)}\}$ is discussed in Section 7.3, and the saddle striction cylindrical surface fitting for discrete striction points of a discrete line set $\{\boldsymbol{R}_L^{(i)}\}$, also a discrete space curve, is studied in this section. There could be expected to exist a relationship between them, just like the saddle spherical circle fitting in Chapter 5 and the saddle spherical image circle fitting in this section, but there isn't so far. The kinematic differential geometry of spatial movement at infinitesimally successive positions in Chapter 6 might provide some useful results to this research, such as the geometrical properties of the striction curve of a line trajectory.

7.4.4 Saddle Constant Axis Line

For a given spatial movement of $\sum^*$ with parameters $(x_{Omf}^{(i)}, y_{Omf}^{(i)}, z_{Omf}^{(i)}; \theta_1^{(i)}, \theta_2^{(i)}, \theta_3^{(i)})$, a line of $\sum^*$ in spatial movement traces a discrete line set $\{\boldsymbol{R}_L^{(i)}\}$, whose saddle spherical image circle can be determined by equation (7.45) and saddle striction cylindrical surface can be located by equation (7.54). This reveals that the global geometrical properties of the discrete line trajectory $\{\boldsymbol{R}_L^{(i)}\}$ can be revealed by comparing it with a constant axis ruled surface by the saddle ruled surface fitting. All directions of lines of $\sum^*$ corresponding to the saddle spherical image circle points can be determined by equation (7.46), and the positions of such lines, corresponding to the saddle striction cylinder points, are located by equation (7.57), which are the characteristic lines of $\sum^*$ in discrete kinematic geometry. We can successively locate a characteristic line of $\sum^*$, both the direction and the position, whose line trajectory or discrete line set $\{\boldsymbol{R}_L^{(i)}\}$ in the fixed frame $\{\boldsymbol{O}_f; \boldsymbol{i}_f, \boldsymbol{j}_f, \boldsymbol{k}_f\}$ is approximately on a constant axis ruled surface and corresponds to a binary link C–C. Combining Definitions 7.5, 7.6, 7.7 with 7.8:

Definition 7.9 A line of a moving body in spatial movement, whose direction is defined by a saddle spherical image circle point and position is located by the saddle striction cylinder point, is defined as a ***saddle constant axis line***.

Combining equations (7.46) with (7.57), we can set up the mathematical model of the saddle constant axis line as

$$\begin{cases} \delta_{\mathrm{CC}} = \min\left[\lambda \Delta_{\mathrm{Rr}}(z_1) + (1-\lambda)\Delta_{\mathrm{CC}}(z_2)\right] \\ \text{s.t. } \phi \in [0, \pi/2],\ \varphi \in [0, 2\pi] \\ z = (z_1, z_2)^T \end{cases} \tag{7.58}$$

where the composite objective function is the weighted sum of the saddle spherical image circle error $\Delta_{\mathrm{Rr}}(z_1)$ and the saddle striction cylinder error $\Delta_{\mathrm{CC}}(z_2)$, and λ is the weighting coefficient for balancing $\Delta_{\mathrm{Rr}}(z_1)$ and $\Delta_{\mathrm{CC}}(z_2)$. The optimization variables $z_1 = (\phi, \varphi)^T$ are the direction angles and $z_2 = (x_{Pm}, y_{Pm})^T$ are the coordinates of the reference point of any line. δ_{CC} is the ruled surface fitting error between the discrete line set $\{\boldsymbol{R}_L^{(i)}\}$ and a constant axis ruled surface $\Sigma_{C'C}$, referred to as a ***saddle constant axis line error***.

For a discrete line set $\{\boldsymbol{R}_L^{(i)}\}$ in $\{\boldsymbol{O}_f; \boldsymbol{i}_f, \boldsymbol{j}_f, \boldsymbol{k}_f\}$, according to Theorems 7.3 and 7.4, the saddle spherical image circle points and the corresponding saddle striction cylinder points exists in $\sum^*$. Then, based on Definition 7.9 of a saddle constant axis line, we have

Theorem 7.5 *There certainly exist a few saddle constant axis lines in a moving body in non-degenerate spatial movement.*

Therefore, for a line of $\sum^*$ in spatial movement, the ruling directions of its discrete line trajectory can be described by a point in spherical movement, and the striction points of its discrete line trajectory can be expressed by the reference points in spatial movement, whose properties at the separated spatial positions correspond to the kinematic differential geometry of lines at the infinitesimally consecutive spatial positions in Chapter 6. It would be an interesting field to research, even more attractive and broader than in the previous literature.

The saddle constant axis lines of a moving body are expressed by both the saddle spherical image circle points for the directions and the saddle striction cylinder points for the positions of the lines. The saddle spherical circle points have been discussed in Chapter 5, whose number depends on the properties of the spherical movement. In the case of five separated positions, there are six spherical circle points of the moving body at most. The saddle spherical circle points have been discussed for multiple positions, as well as the saddle spherical image circle points. The striction curves of line trajectories, especially the constraint ruled surfaces, are discussed in Chapter 6. This reveals that there is a function between the coordinates of the reference points and the directions of the lines for generalized curvature, or contact conditions. Similarly, there are a few saddle constant axis lines on the moving body, depending on the properties of spatial movement, particularly on the number of saddle spherical image circle points, while the striction points may correspond to them.

7.5 Degenerate Constant Axis Lines L_{RC} and L_{HC}

A binary link C–C and the constraint constant axis ruled surface $\sum_{C'C}$ correspond to the characteristic line L_{CC} of a moving body, as mentioned in Chapter 3. If the kinematic pair C connecting the binary link to the base body is replaced by a pair R or H, the binary link C–C degenerates to a binary link R–C or H–C, and the constraint ruled surface $\sum_{C'C}$ changes to a hyperboloid of one sheet $\sum_{RC}$ or a helical surface $\sum_{HC}$. The characteristic line of $\sum^*$ will be L_{RC} or L_{HC}, which will be discussed in this section for the discrete movement of the moving body.

7.5.1 Saddle Characteristic Line L_{RC} (R–C, R–R)

The binary link R–C corresponds to the characteristic line L_{RC} and the constraint surface Σ_{RC}, a hyperboloid of one sheet, whose spherical image curve and striction curve are both circles. For the given discrete line set $\{\boldsymbol{R}_L^{(i)}\}$, the degree of the approximation between a hyperboloid of one sheet and $\{\boldsymbol{R}_L^{(i)}\}$ can be estimated by the saddle point programming approach of the ruled surface fitting, including both the saddle spherical image circle fitting and the saddle striction cylindrical surface fitting. The mathematical model of the saddle spherical image circle fitting has been established as equation (7.45). Because the striction curve of Σ_{RC} is a circle, all the striction points $B^{(i)}$ on the discrete lines $L^{(i)}$ of $\{\boldsymbol{R}_L^{(i)}\}$ and the perpendicular foot points $A^{(i)}$ on the fixed axis L_Q are approximately on a plane, that is, the parameters $u^{(i)}$ of equation (7.51) are close to a constant. The saddle striction cylindrical surface fitting is actually a ***saddle striction circle fitting***. Based on equation (7.54), the mathematical model of the saddle striction circle fitting for a hyperboloid of one sheet can be written as

$$\begin{cases} \Delta_{\mathrm{RC}} = \min\limits_{x} \max\limits_{1\leq i\leq n} \left\{\Delta^{(i)}(\boldsymbol{x})\right\} \\ \quad = \min\limits_{x} \max\limits_{1\leq i\leq n} \left\{ \sqrt{(r^{(i)} - r)^2 + (u^{(i)} - u)^2} \right\} \\ \text{s.t.} \quad r \in (0, +\infty) \\ \boldsymbol{x} = (x_Q, y_Q, r, u)^T \end{cases} \tag{7.59}$$

where the objective function $\{\Delta^{(i)}(\boldsymbol{x})\}$ is the set of normal distances of the discrete striction points of $\{\boldsymbol{R}_L^{(i)}\}$ and a fitting circle, and the variables $u^{(i)}$ and $r^{(i)}$ can be calculated by equations (7.51) and (7.53). The optimization variables $\boldsymbol{x} = (x_Q, y_Q, r, u)^T$ contain the coordinates (x_Q, y_Q) of the reference point Q on the fixed axis L_Q in the fixed frame, the radius r of the fitting circle, and the distance u from the center point of the fitting circle to the reference plane Π. Δ_{RC} is the maximum fitting error for a circle to fit the discrete striction points of $\{\boldsymbol{R}_L^{(i)}\}$ adaptively by saddle point programming, or the ***saddle striction circle error***.

Similar to the saddle constant axis lines, there exists a group of parallel lines in the moving body in discrete spatial movement whose the same direction is defined by the saddle spherical image circle point. The parallel lines will trace different discrete line sets $\{\boldsymbol{R}_L^{(i)}\}$ in the fixed frame since they pass through different reference points, whose discrete striction curves will correspond to different saddle striction circles. The saddle striction circle errors of the discrete striction curves vary with the coordinates of the reference points on the lines, and the reference points, corresponding to the minimum values of the saddle striction circle errors, are referred to as the ***saddle striction circle points***. The mathematical model of the saddle striction circle point is built up as

$$\begin{cases} \chi_{\mathrm{RC}} = \min \Delta_{\mathrm{RC}}(z) \\ z = (x_{Pm}, y_{Pm})^T \end{cases} \tag{7.60}$$

where χ_{RC} is the saddle striction circle point error, and the objective function $\Delta_{\mathrm{RC}}(z)$ is the saddle striction circle error for any of the parallel lines. The optimization variables $z = (x_{Pm}, y_{Pm})^T$ are the coordinates of the reference points on the reference plane $\boldsymbol{O}_m - \boldsymbol{i}_m\boldsymbol{j}_m$ in $\{\boldsymbol{O}_m; \boldsymbol{i}_m, \boldsymbol{j}_m, \boldsymbol{k}_m\}$.

Combining equations (7.46), (7.59) with (7.60), the mathematical model of the saddle characteristic line L_{RC} is

$$\begin{cases} \delta_{\mathrm{RC}} = \min[\lambda\Delta_{\mathrm{Rr}}(z_1) + (1-\lambda)\Delta_{\mathrm{RC}}(z_2)] \\ \text{s.t. } \phi \in [0, \pi/2], \ \varphi \in [0, 2\pi] \\ z = (z_1, z_2)^T \end{cases} \tag{7.61}$$

where the composite objective function is the weighted sum of the saddle spherical image circle error $\Delta_{\mathrm{Rr}}(z_1)$ and the saddle striction circle error $\Delta_{\mathrm{RC}}(z_2)$, and λ is the weighting coefficient for balancing $\Delta_{\mathrm{Rr}}(z_1)$ and $\Delta_{\mathrm{RC}}(z_2)$. The optimization variables $z_1 = (\phi, \varphi)^T$ are the direction angles and $z_2 = (x_{Pm}, y_{Pm})^T$ are the coordinates of the reference point of any line. δ_{RC} is the fitting error of the discrete line set $\{\boldsymbol{R}_L^{(i)}\}$ and a hyperboloid of one sheet Σ_{RC}. The above mathematical model can be solved separately in the non-constrained case. The results include the saddle spherical image circle points and the saddle striction circle points, which correspond to the directions and positions of the saddle characteristic lines L_{RC} in the moving body.

Generally, the moving body connected to the fixed body by a binary link R–C, has relative motion to the binary link by rotating around and sliding along the axis of the moving joint C, or the characteristic line L_{RC}. Particularly, if the kinematic pair C which connects the binary link R–C to the moving body is replaced by the R-pair, the binary link R–C degenerates to another binary link R–R and the constraint ruled surface is still a hyperboloid of one sheet, but the relative motion between the moving body and the binary link R–R is just the rotation around the characteristic line without sliding. The characteristic line L_{RC} then degenerates to L_{RR}. Based on equation (7.61) of the saddle characteristic line L_{RC}, the mathematical model for the saddle characteristic line L_{RR} has an additional condition, that the distances $v^{(i)}$ from the reference point P of the line L to the striction point $B^{(i)}$ on each line of the discrete line set $\{\boldsymbol{R}_L^{(i)}\}$ should remain approximately a constant, which is so strict for a line of $\sum^*$ to be a characteristic line L_{RR} or a saddle characteristic line L_{RR} that the spatial movement of $\sum^*$ has to be degenerate. Hence, the saddle characteristic line L_{RC} may be solved at first for a given spatial movement of a moving body by equation (7.61), and then the distances $v^{(i)}$ are estimated to be approximately a constant or not, leading to the characteristic line L_{RC} being regarded as a characteristic line L_{RR} or not.

7.5.2 *Saddle Characteristic Line L_{HC} (H–C, H–R, H–H)*

The binary link H–C corresponds to the characteristic line $L_{HC}$and the constraint surface Σ_{HC}, a helical surface, whose spherical image curve is a circle and striction curve is a cylindrical helix. For the given discrete line set $\{\boldsymbol{R}_L^{(i)}\}$, the degree of the approximation between a helical surface and $\{\boldsymbol{R}_L^{(i)}\}$ can be estimated by the saddle approach of the ruled surface fitting, including both the saddle spherical image circle fitting and the saddle striction cylindrical surface fitting. The mathematical model of the saddle spherical image circle fitting has been established as equation (7.45). The saddle striction cylindrical surface fitting of a helical surface becomes ***saddle striction cylindrical helix fitting***, or all striction points $B^{(i)}$ of the discrete lines $L^{(i)}$ of $\{\boldsymbol{R}_L^{(i)}\}$ locate not only on a cylindrical surface but also on a cylindrical helix, which means that there exists an approximate linear function between the sliding displacement $u^{(i)}$ in equation (7.51) and the rotation displacement ψ_{i1} in equation (7.52). Based on equation (7.54), the mathematical model of the saddle striction cylindrical helix fitting of a helical surface can be written as

$$\begin{cases} \Delta_{\mathrm{HC}} = \min\limits_{\boldsymbol{x}} \max\limits_{1\le i\le n} \left\{\Delta^{(i)}(\boldsymbol{x})\right\} \\ \quad = \min\limits_{\boldsymbol{x}} \max\limits_{1\le i\le n} \left\{ \sqrt{(r^{(i)} - r)^2 + (u^{(i)} - u^{(1)} - p\psi_{i1})^2} \right\} \\ \text{s.t.} \quad r \in (0, +\infty) \\ \boldsymbol{x} = (x_Q, y_Q, r, p)^T \end{cases} \tag{7.62}$$

where the objective function $\{\Delta^{(i)}(\boldsymbol{x})\}$ is the fitting error set of the discrete striction points of $\{\boldsymbol{R}_L^{(i)}\}$ and a fitting cylindrical helix, and the variables $u^{(i)}$ and $r^{(i)}$ can be calculated by equations (7.51) and (7.53). The optimization variables $\boldsymbol{x} = (x_Q, y_Q, r, p)^T$ contain the coordinates (x_Q, y_Q) of the reference point Q on the fixed axis L_Q in the fixed frame, the radius r, and the helical pitch p of the fitting cylindrical helix. Δ_{HC} is the maximum fitting error for a cylindrical helix to fit the discrete striction points of $\{\boldsymbol{R}_L^{(i)}\}$ adaptively by saddle point programming, or the ***saddle striction cylindrical helix error***.

Similarly, for a rigid body in discrete spatial movement, the saddle spherical image circle point defines a direction of parallel lines in the moving body, which trace different discrete line sets $\{\boldsymbol{R}_L^{(i)}\}$ in the fixed frame and correspond to different saddle striction cylindrical helixes. The saddle striction cylindrical helix errors vary with the coordinates of the reference points of the lines, and the reference points, corresponding to the minimum values of the saddle striction cylindrical helix errors, are referred to as the ***saddle striction cylindrical helix points***. The mathematical model of the saddle striction cylindrical helix point is built up as

$$\begin{cases} \chi_{\mathrm{HC}} = \min \Delta_{\mathrm{HC}}(z) \\ z = (x_{Pm}, y_{Pm})^T \end{cases} \tag{7.63}$$

where χ_{HC} is the saddle striction cylindrical helix point error, and the objective function $\Delta_{\mathrm{HC}}(z)$ is the saddle striction cylindrical helix error for any of the parallel lines of $\sum^*$. The optimization variables $z = (x_{Pm}, y_{Pm})^T$ are the coordinates of the reference points on the reference plane $\boldsymbol{O}_m - \boldsymbol{i}_m\boldsymbol{j}_m$ in $\{\boldsymbol{O}_m; \boldsymbol{i}_m, \boldsymbol{j}_m, \boldsymbol{k}_m\}$.

Combining equations (7.46), (7.62) with (7.63), we have the mathematical model of the characteristic line L_{HC} for a helical surface as

$$\begin{cases} \delta_{\mathrm{HC}} = \min[\lambda\Delta_{\mathrm{Rr}}(z_1) + (1-\lambda)\Delta_{\mathrm{HC}}(z_2)] \\ \text{s.t. } \phi \in [0, \pi/2],\ \varphi \in [0, 2\pi] \\ z = (z_1, z_2)^T \end{cases} \tag{7.64}$$

where the composite objective function is the weighted sum of the saddle spherical image circle error $\Delta_{\mathrm{Rr}}(z_1)$ and the saddle striction cylindrical helix error $\Delta_{\mathrm{HC}}(z_2)$, and λ is the weighting coefficient for balancing $\Delta_{\mathrm{Rr}}(z_1)$ and $\Delta_{\mathrm{HC}}(z_2)$. The optimization variables $z_1 = (\phi, \psi)^T$ are the direction angles and $z_2 = (x_{Pm}, y_{Pm})^T$ are the coordinates of the reference point of any line. δ_{HC} is the fitting error between the discrete line set $\{\boldsymbol{R}_L^{(i)}\}$ and a helical surface Σ_{HC}. The above mathematical model can be solved separately in the non-constrained case. The results include the saddle spherical image circle points and the saddle striction cylindrical helix points, which correspond to the directions and positions of the saddle characteristic lines L_{HC} in the moving body.

Generally, the moving body is connected to the fixed body by a binary link H–C, has relative motion to the binary link by rotating around and sliding along the axis of the moving joint C, or the characteristic line L_{HC}. Particularly, if the kinematic pair C which connects the binary link H–C to the moving body is replaced by the R-pair, it degenerates to another binary link H–R and the constraint ruled surface is still a helical surface, but the relative motion between the moving body and the binary link H–R is just the rotation around the characteristic line without sliding. The characteristic line L_{HC} degenerates to L_{HR}. Based on equation (7.64) of the saddle characteristic line L_{HC}, the mathematical model of the saddle characteristic line L_{HR} has an additional condition, that the distances $v^{(i)}$ should remain approximately constant. Hence, the saddle characteristic line L_{HC} may be solved at first for a given spatial movement of $\sum^*$ by equation (7.64), and then the distances $v^{(i)}$ are estimated to be approximately a constant or not, which leads to the line L_{HC} being regarded as a characteristic line L_{HR} or not.

If the relative rotation and sliding motion of the moving body $\sum^*$ and the binary link H–C along the characteristic line are linear dependent, we can replace the C-pair by an H-pair, and have the binary link H–H. The characteristic line L_{HC} then degenerates to L_{HH}. Based on equation (7.64) of the saddle characteristic line L_{HC}, the mathematical model for the saddle characteristic line L_{HH} has an additional condition, that the sliding displacement $v^{(i)}$ should be a linear function of the rotation angle ς_{i1} of $\sum^*$ relative to the binary link, or $p^{(i)} = \frac{v^{(i)}-v^{(1)}}{\varsigma_{i1}}$ is approximately a constant, where ς_{i1} can also be calculated by equation (7.52), or $\cos\varsigma_{i1} = \boldsymbol{n}_m^{(i)} \cdot \boldsymbol{n}_m^{(1)}$ only with the two vectors described in the moving Cartesian frame $\{O_m; i_m, j_m, k_m\}$.

Taking into account the binary link S–C and its degenerate cases such as S–R, and S–H, the constraint surfaces are the constant distance ruled surfaces mentioned in Chapter 3. We can also take the constant distance ruled surfaces to fit a line trajectory and do the same as mentioned above. However, this is not necessary because the binary link S–C is an inverted binary link C–S, discussed in Section 7.2.

Example 7.5 The coupler link of a spatial RCCC linkage has 72 separated positions in the base link, which are described by parameters in Table A.1 and the coordinate transformation formulas (A.7) and (A.8) in Appendix A. Determine: (1) the saddle spherical image circle points; (2) the saddle constant axis lines L_{CC} of the coupler; (3) the saddle characteristic lines L_{RC} of the coupler.

Solution

1. **The saddle spherical image circle points**
   The directions of the lines of the coupler link are selected in steps of 0.05 in the two spherical coordinate directions and mapped onto a unit hemispherical surface $\phi \in [0, \pi/2]$, $\varphi \in [0, 2\pi]$ as 4032 spherical image points. For their discrete spherical image point sets, the saddle spherical image circle errors can be obtained by calling the sub-program ArRF. The error surface is constructed and projected onto the coordinate plane $\boldsymbol{R}_C - \boldsymbol{x}_2\boldsymbol{y}_2$ to be a contour plot, as shown in Fig. 7.18, of which there are four valleys and four corresponding bottom points P_1, P_2, P_3, and P_4, listed in Table 7.7. P_1 and P_2 are the spherical image points of the axes z_1 and z_2 of the two cylindrical pairs, whose saddle spherical circle errors are zero.
   50 initial points are generated by Matlab and the corresponding optimization results are obtained by calling the sub-program ArRP, of which 34 converge to the point P_2, five to the point P_1, two to the point P_3, and two to the point P_4.
2. **The saddle constant axis lines L_{CC}**
   The direction of a line of the coupler link, corresponding to spherical image circle point P_1, is taken as an example and 961 reference points in the area $x_{Pm} \in [-7, -1]$, $y_{Pm} \in [-3, 3]$ on the coordinate plane $\boldsymbol{R}_C - \boldsymbol{x}_2\boldsymbol{y}_2$ are selected in steps of 0.2 as reference points, and 961 lines of the coupler link parallel to the axis z_1 are located. Based on the optimization model (7.54), the saddle striction cylindrical surface error of each line trajectory is obtained by calling the sub-program ACCF, which is viewed as the dependent function while the reference point coordinates (x_{Pm}, y_{Pm}) on the plane $\boldsymbol{R}_C - \boldsymbol{x}_2\boldsymbol{y}_2$ are taken as the independent variables. The error surface of the saddle striction cylindrical surface error can be constructed, and mapped onto the plane as the contour lines shown in Fig. 7.19. There is only one valley in the error surface, and its bottom point (x_{PmCC}, y_{PmCC}) locates at (–4,0).

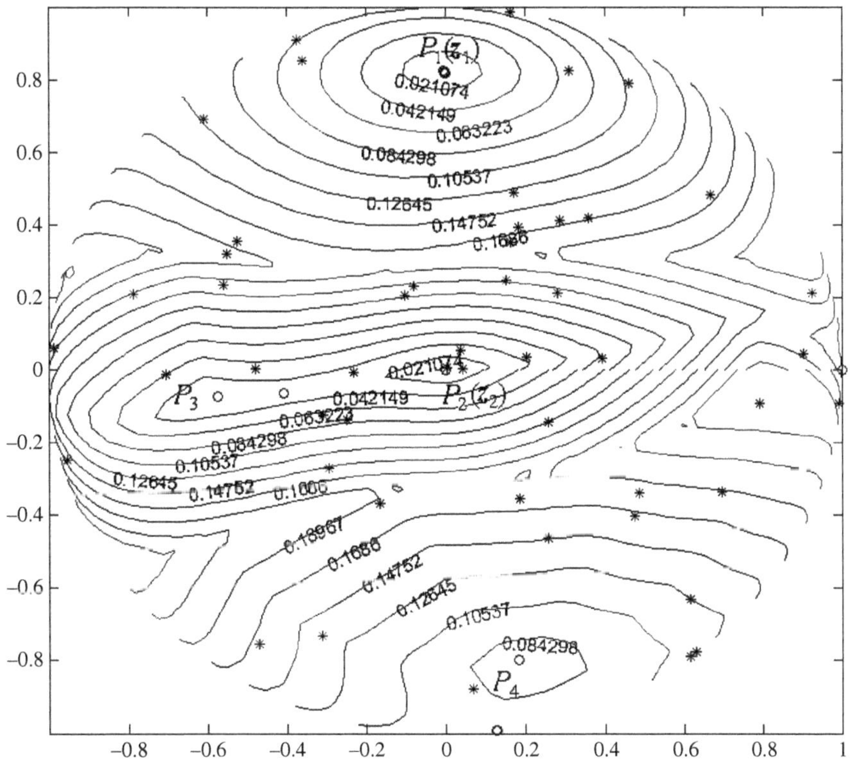

Figure 7.18 The saddle spherical image circle errors for RCCC linkage

Table 7.7 The saddle spherical image circle points

Saddle spherical image circle point	Direction angles		Saddle spherical image circle (cone)		
			Direction angles of the axis		Half-cone angle
	ϕ	φ	ξ	η	δ
P_1	0.9599	π/2	0	π/2	0.5236
P_2	0	π/2	1.0472	π/2	0.7854
P_3	0.6151	3.2699	1.7043	5.4176	1.1176
P_4	0.9587	4.9361	1.8662	0.8539	0.5346

It is just the reference point of the axis z_1 in the coupler link, whose discrete line trajectory and saddle striction cylinder are shown in Fig. 7.20(a).

A line parallel to the axis z_1, whose reference point locates at (−1,−1) on the plane $\boldsymbol{R}_C - \boldsymbol{x}_2\boldsymbol{y}_2$, traces the discrete trajectory with the saddle striction cylindrical surface as shown in Fig. 7.20(b). Figures 7.20(c) and 7.20(d) are respectively the projections of Figs 7.20(a) and 7.20(b) on the plane $\boldsymbol{R}_C - \boldsymbol{x}_2\boldsymbol{y}_2$.

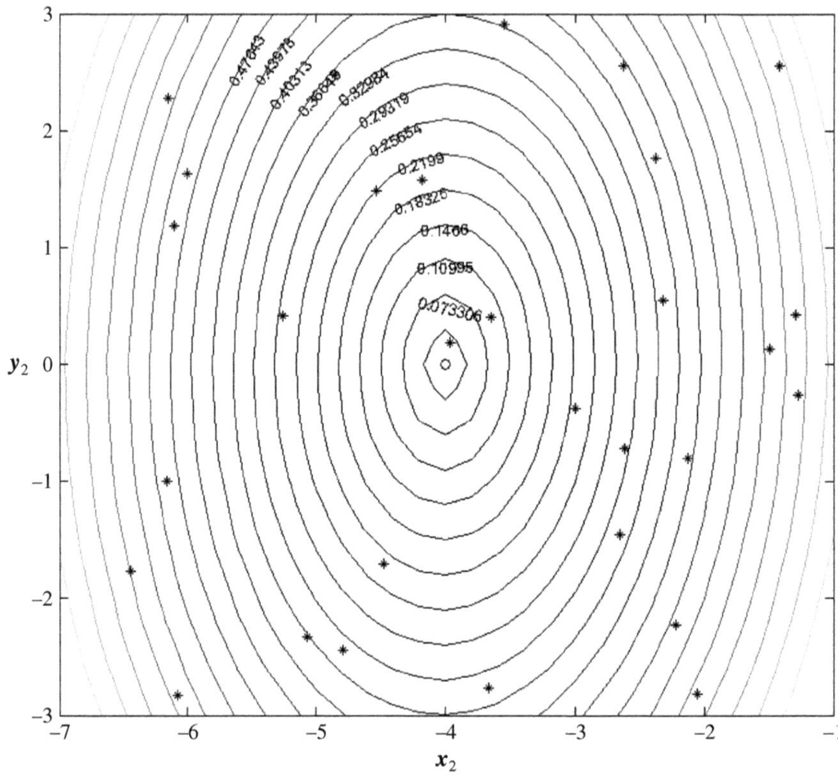

Figure 7.19 The saddle striction cylindrical surface errors for the lines parallel to z_1

Similarly, the lines in the coupler link with direction corresponding to the spherical image circle point P_2, or parallel to the axis z_2, whose reference points locate in the area $x_{Pm} \in [-3, 3]$, $y_{Pm} \in [-3, 3]$ on the plane $\boldsymbol{R}_C - \boldsymbol{x}_2\boldsymbol{y}_2$, trace the line trajectories; their saddle striction cylindrical surface error surface is shown in Fig. 7.21 by contour plot. There is also only a bottom point with coordinates $x_{Pm\text{CC}} = 0,\ y_{Pm\text{CC}} = 0$, which is just the reference point of the axis z_2.

For the lines parallel to the axes z_1 and z_2, the saddle striction cylinder points and the corresponding saddle striction cylindrical surfaces are obtained by calling the sub-program ACCP, listed in Table 7.8, and the saddle constant axis lines in the coupler link can be located by the saddle spherical image circle points and the saddle striction cylinder points.

3. **The saddle characteristic line L_{RC}**
961 lines in the coupler link, parallel to the axis z_1, or corresponding to the spherical image circle point P_1, pass though their reference points in the area $x_{Pm} \in [-7, -1]$, $y_{Pm} \in [-3, 3]$ on the plane $\boldsymbol{R}_C - \boldsymbol{x}_2\boldsymbol{y}_2$ in steps of 0.2 in the two coordinate directions and trace the line trajectories. The contour plot of the error surface of the saddle striction circle errors is shown in Fig. 7.22. In the optimizing search of the striction circle points, 30 initial points are randomly generated, and all optimization results converge to the only saddle striction circle point with coordinates $x_{Pm\text{RC}} = -4,\ y_{Pm\text{RC}} = 0$.

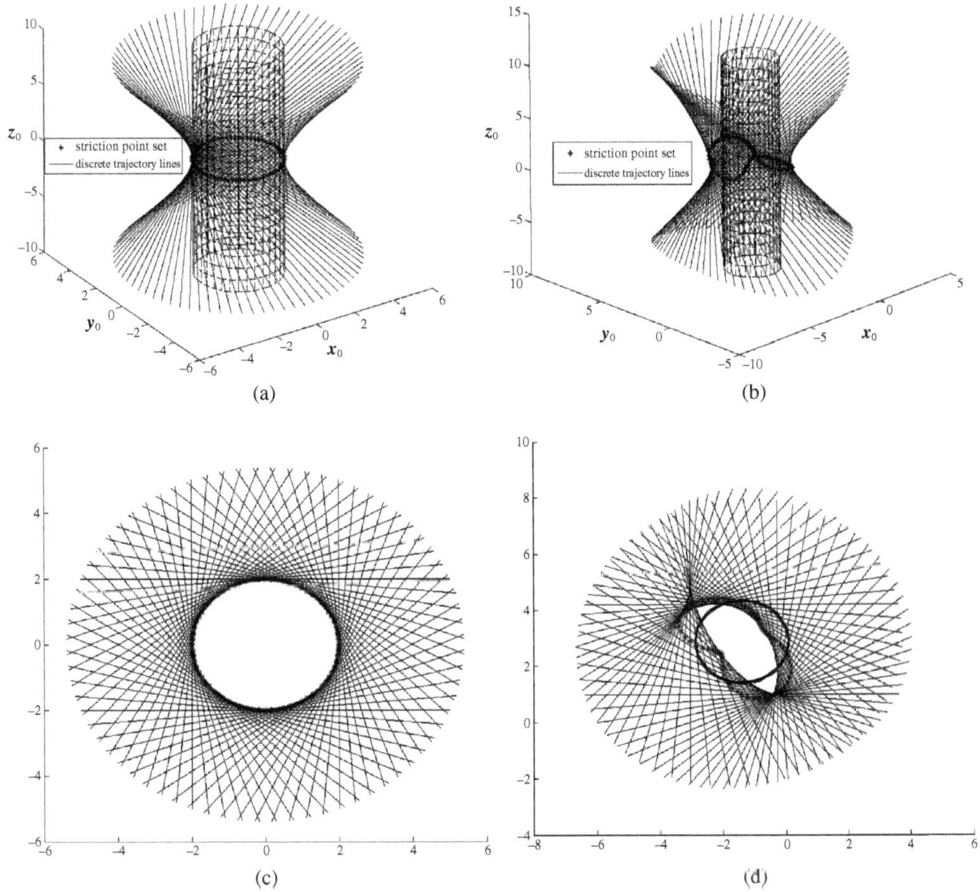

Figure 7.20 The saddle striction cylindrical surfaces of both the axis z_1 and its parallel line

Table 7.8 The saddle striction cylinder (circle) points

Spherical image circle points	Characteristic lines	Saddle striction cylinder (circle) point		Saddle striction cylindrical surface (circle)			Fitting error
				Position of the axis		Radius	
		x_{Pm}	y_{Pm}	x_Q	y_Q	r	χ
P_1	L_{CC}	−4	0	0	0	2	0
	L_{RC}	−4	0	0	0	2	0
P_2	L_{CC}	0	0	−5	0	3	0
	L_{RC}	−0.3292	0.5404	−5.4811	−0.4316	3.3798	0.930542

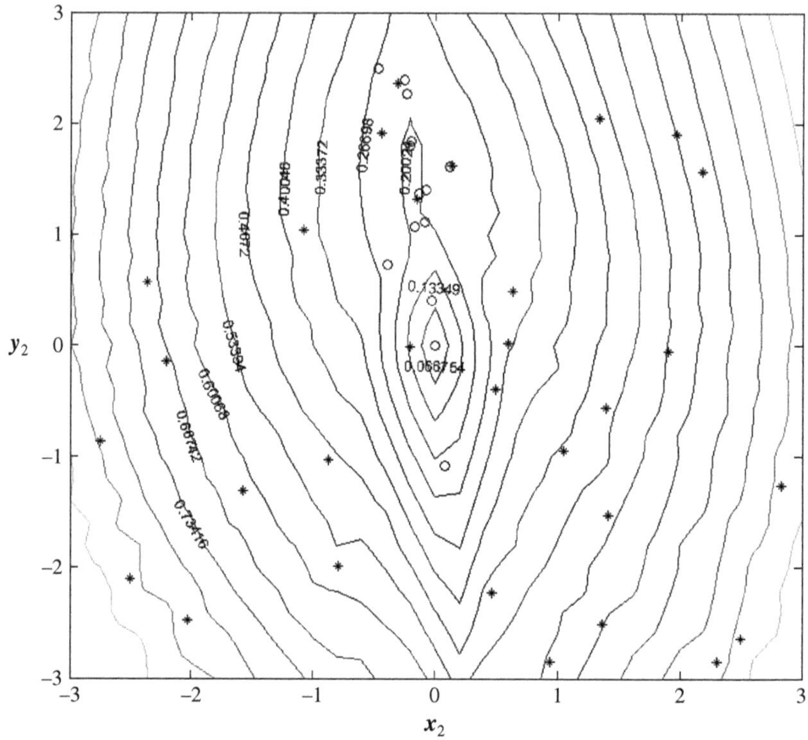

Figure 7.21 The saddle striction cylindrical surface errors for the line parallel to z_2

Similarly, 961 lines in the coupler link, parallel to the axis z_2, or corresponding to the spherical image circle point P_2, pass through their reference points in the area $x_{Pm} \in [-3, 3]$, $y_{Pm} \in [-3, 3]$ on the plane $\boldsymbol{R}_C - \boldsymbol{x}_2\boldsymbol{y}_2$. The contour plot of the error surface of the saddle striction circle errors is shown in Fig. 7.23. Only one saddle striction circle point (x_{PmRC}, y_{PmRC}) locates at $(-0.3292, 0.5404)$. From 30 random initial points, all optimization results converge to the only saddle striction circle point, listed in Table 7.8. The saddle characteristic line L_{RC} in the coupler link is located by the saddle spherical image circle point and the relevant saddle striction circle point.

Example 7.6 The nine separated spatial positions of the moving body are described by the parameters $(x_{Omf}^{(i)}, y_{Omf}^{(i)}, z_{Omf}^{(i)}; \theta_1^{(i)}, \theta_2^{(i)}, \theta_3^{(i)})$ listed in Table 7.1. Determine: (1) the saddle spherical image circle points; (2) the saddle constant axis lines L_{CC}; (3) the saddle characteristic lines L_{RC}, L_{HC}, and L_{RR}.

Solution

1. **The saddle spherical image circle points**
   The saddle spherical image circle errors are calculated by the sub-program ARrP for all line directions of the moving body, and constitute the error surface, which is projected onto the coordinate plane $\boldsymbol{O}_m - \boldsymbol{i}_m\boldsymbol{j}_m$ as a contour plot, shown in Fig. 7.24.

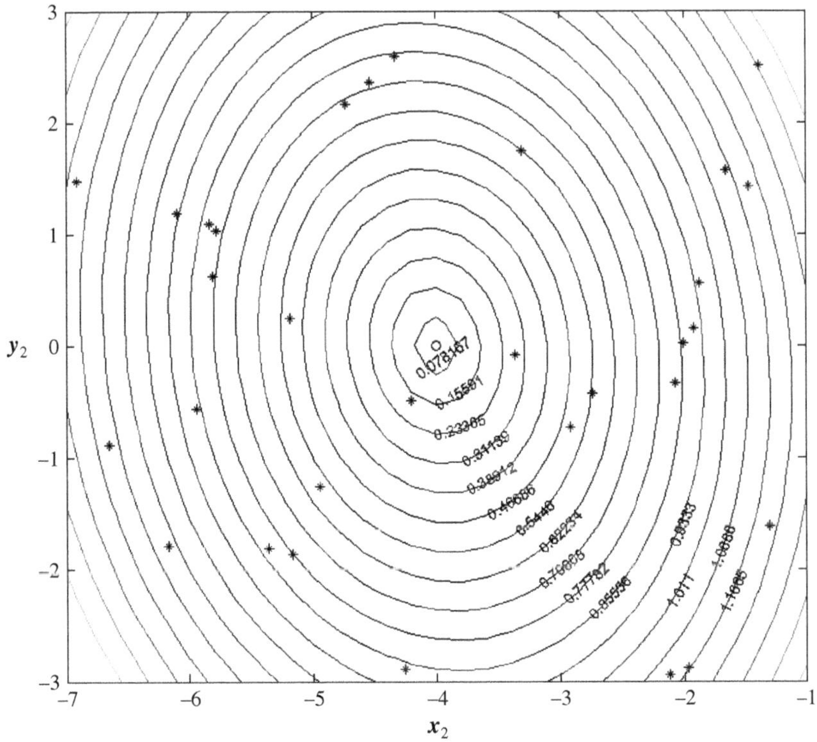

Figure 7.22 The saddle striction circle errors of the lines parallel to z_1

The saddle spherical image circle points in the moving body are obtained by calling the sub-program ArRP, in which the five points (SRrP1–SRrP5) with less saddle spherical image circle errors are listed in Table 7.9.

2. The saddle constant axis line L_{CC}

Any saddle spherical image circle point in Table 7.9, such as SRrP1, provides a line direction corresponding to a group of parallel lines. The saddle striction cylindrical surface errors for these parallel lines can be obtained by calling the sub-program ACCF. The error surface is shown in Fig. 7.25. Based on equation (7.57), the optimizing search of the saddle striction cylinder point can be conducted. From 15 random initial points, the optimized result is the saddle striction cylinder point SRrP1-CCP1, listed in Table 7.10.

Similarly, four other saddle spherical image points in Table 7.9 correspond to their saddle striction cylinder points, listed in Table 7.10. Combining the saddle spherical image circle points in Table 7.9 with the corresponding saddle striction cylinder points in Table 7.10, five saddle constant axis lines L_{CC} of the moving body are located.

3. The saddle characteristic line L_{RC}

For the parallel lines corresponding to a saddle spherical image circle point in Table 7.9, such as SRrP2, the error surface of the saddle striction circle errors is constructed as shown in Fig. 7.26. Based on equation (7.60), the optimizing search of the saddle striction circle point can be conducted. From 15 random initial points, the saddle striction circle point is obtained, SRrP2-RCP2, as listed in Table 7.11.

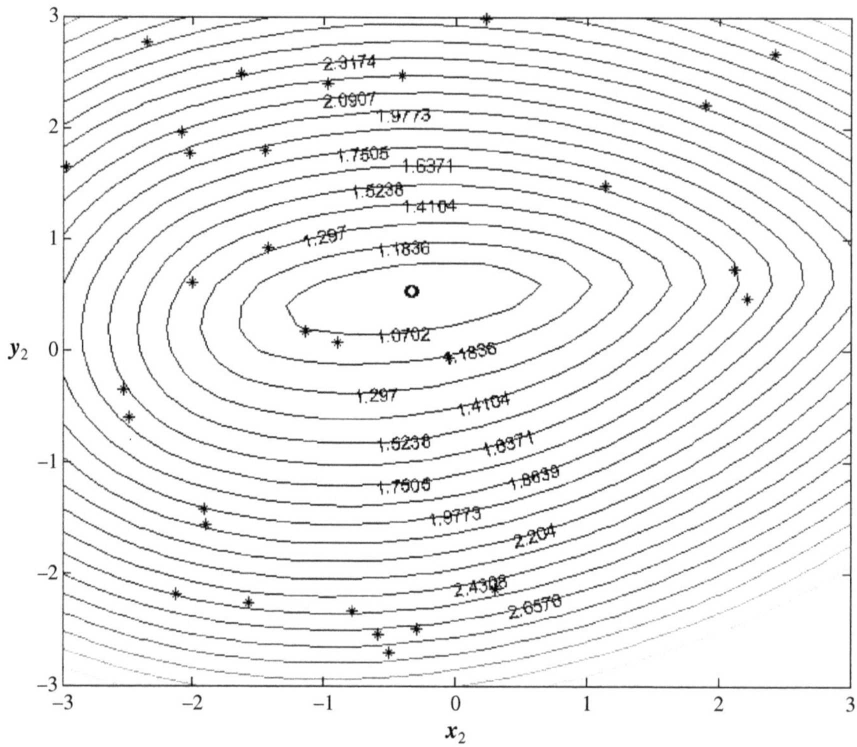

Figure 7.23 The saddle striction circle errors for the lines parallel to z_2

Similarly, another four saddle spherical image points in Table 7.9 correspond to four saddle striction circle points, listed in Table 7.11. By combining the saddle spherical image circle points in Table 7.9 with the corresponding saddle striction circle points in Table 7.11, the saddle characteristic lines L_{RC} of the moving body are located.

4. **The saddle characteristic line L_{RR}**
   As mentioned in Section 7.5.1, a saddle characteristic line L_{RC} of the moving body will be a saddle characteristic line L_{RR} if the distance $v^{(i)}$ between a fixed point on the tracing straight line and the striction point on the discrete line is approximately a constant. By balancing the fitting error χ_{RC} of the saddle characteristic line L_{RC} and the deviation of $v^{(i)}$, five saddle characteristic lines L_{RR} are located, as listed in Table 7.12.
5. **The saddle characteristic line L_{HC}**
   For the parallel lines corresponding to a saddle spherical image circle point in Table 7.9, such as SRrP1, the error surface of the saddle striction cylindrical helix error is constructed as shown in Fig. 7.27. Based on equation (7.63), the optimizing search of the saddle striction cylindrical helix point can be conducted. From 15 initial points, the saddle striction cylindrical helix point SRrP1-HCP1 is obtained, as listed in Table 7.13.

   Similarly, the saddle striction cylindrical helix points can be optimized and listed in Table 7.13, corresponding to another four saddle spherical image points in Table 7.9. By combining the saddle spherical image circle points in Table 7.9 with the corresponding saddle striction cylindrical helix points in Table 7.13, five saddle characteristic lines L_{HC} of the moving body are located.

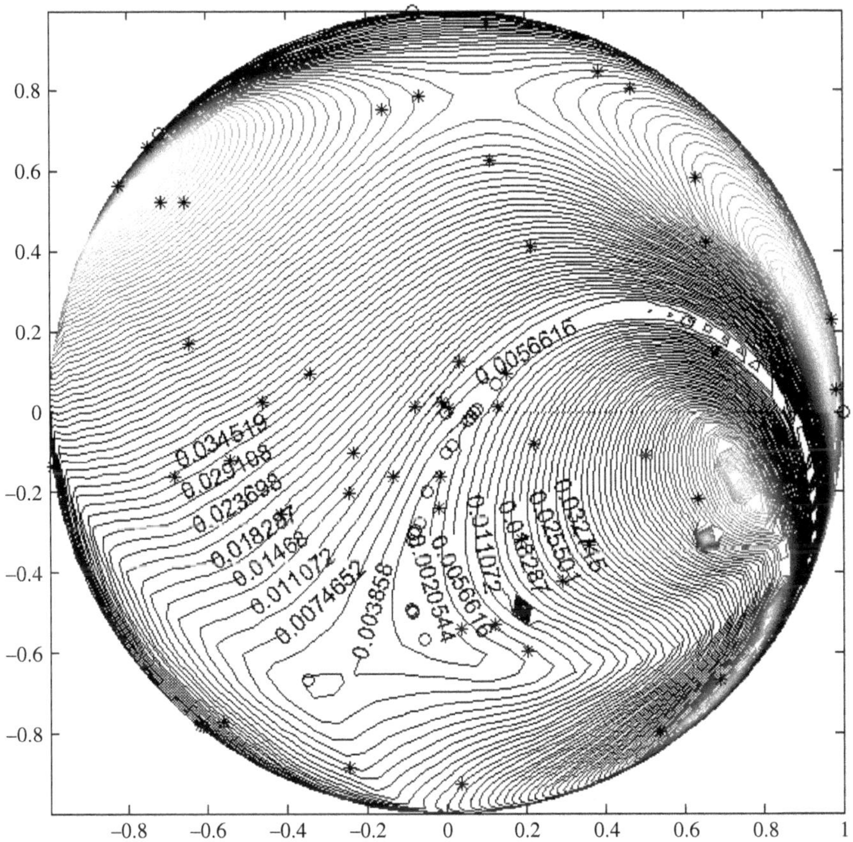

Figure 7.24 The saddle spherical image circle errors for nine positions

Table 7.9 The five saddle spherical circle points for nine positions

No.	Direction angles		Saddle spherical image circle (cone)			Error
			Direction of the axis		Half-cone angle	
	ϕ	φ	ξ	η	δ	γ_{Rr}
SRrP1	0.5296	4.5406	0.0061	4.3882	0.5237	0.000066
SRrP2	0.4996	4.5071	0.0326	1.9795	0.5269	0.000276
SRrP3	0.0724	5.0417	0.4885	1.4313	0.5568	0.002774
SRrP4	0.0634	5.2426	0.5050	1.4268	0.5588	0.002930
SRrP5	0.8513	4.2289	0.4005	3.6787	0.5449	0.003171

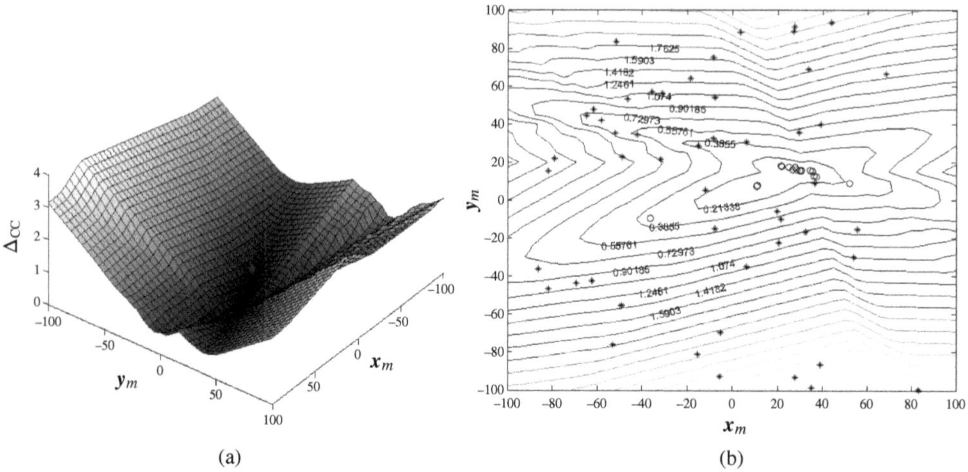

Figure 7.25 The saddle striction cylindrical surface errors for nine positions

Table 7.10 The saddle striction cylinder points for nine positions

	Saddle striction cylinder point		Saddle striction cylindrical surface			Error
			Position of the axis		Radius	
	x_{PmCC}	y_{PmCC}	x_Q	y_Q	r	χ_{CC}
SRrP1-CCP1	11.2604	7.7437	8.8148	4.7889	1.9225	0.012696
SRrP2-CCP2	12.2233	6.9505	9.7533	4.0869	1.9511	0.009349
SRrP3-CCP3	4.4213	5.7202	1.6517	1.8734	2.4233	0.022094
SRrP4-CCP4	4.4969	5.6257	1.7425	1.7309	2.4219	0.021556
SRrP5-CCP5	8.2766	15.3711	3.6662	9.6138	2.1152	0.046447

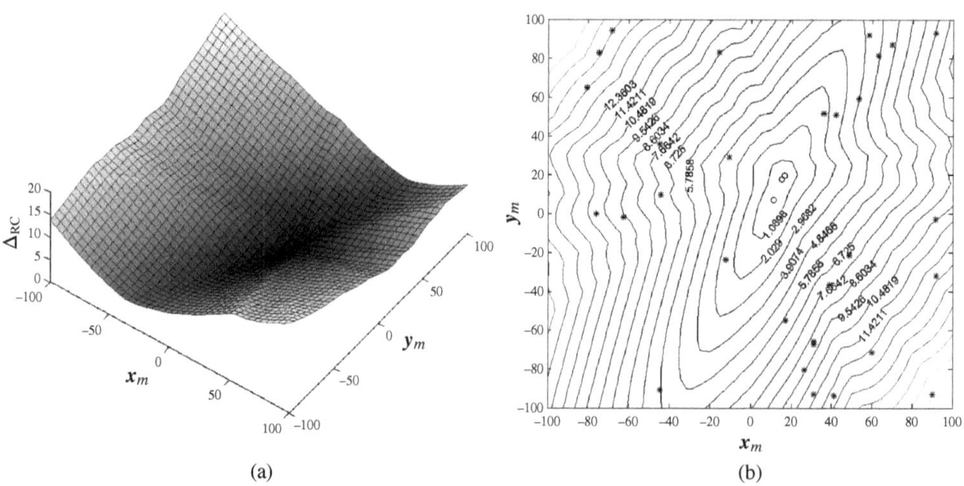

Figure 7.26 The saddle striction circle errors for nine positions

Table 7.11 The saddle striction circle points for nine positions

	Saddle striction circle point		Saddle striction circle			Distance parameters	Error
			Position of axis		Radius		
	x_{PmRC}	y_{PmRC}	x_Q	y_Q	r	u_0	χ_{RC}
SRrP1-RCP1	11.3969	8.3403	9.0264	5.3546	1.8323	5.6285	0.092573
SRrP2-RCP2	11.1327	6.8966	8.5615	3.9857	2.0698	5.6564	0.102669
SRrP3-RCP3	5.0734	1.9250	2.1335	−1.6497	2.8520	7.5150	0.421271
SRrP4-RCP4	4.9446	1.8845	2.0268	−1.7432	2.8620	7.5755	0.423044
SRrP5-RCP5	13.9496	9.9211	8.8491	4.0678	2.0651	1.6212	0.502353

Table 7.12 The positions of L_{RR} for nine positions

	Reference points		Saddle striction circle			Distance parameters		Error
			Position of axis		Radius			
	x_{PmRR}	y_{PmRR}	x_Q	y_Q	r	u_0	v_0	χ_{RR}
SRrP1-RRP1	8.5751	6.6247	5.3485	4.1743	2.1709	4.9324	6.5324	0.610735
SRrP2-RRP2	8.4430	6.4709	5.2162	4.0745	2.1651	5.0288	6.4027	0.606864
SRrP3-RRP3	7.2159	3.5435	4.2057	0.2836	2.5221	7.2536	5.6493	0.573985
SRrP4-RRP4	6.9406	3.5120	3.9241	0.1904	2.5316	7.3069	5.5485	0.577589
SRrP5-RRP5	9.5146	10.9977	4.5964	6.2831	2.0228	1.8803	9.1990	0.745285

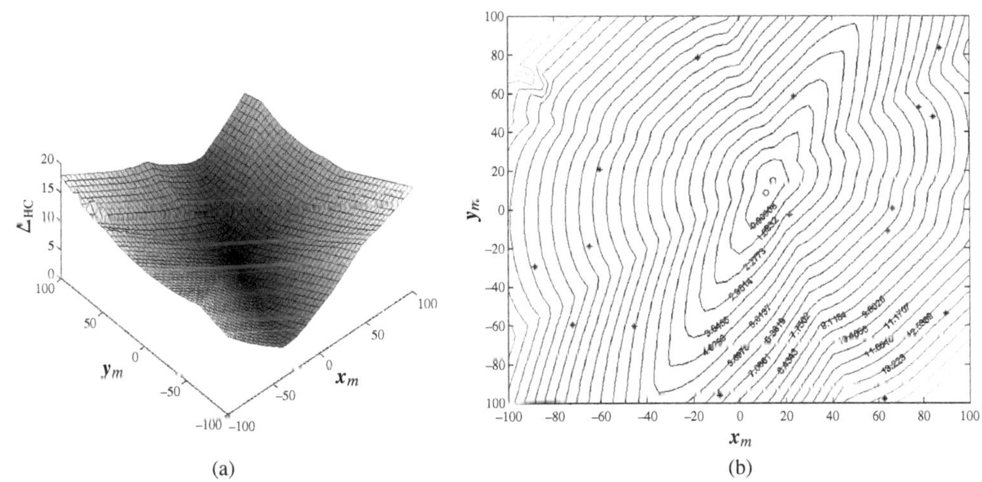

Figure 7.27 The saddle striction cylindrical helix errors for nine positions

Table 7.13 The saddle striction cylindrical helix points for nine positions

	Saddle striction cylindrical helix point		Saddle striction of cylindrical helix				Error
			Position of axis		Helix parameter	Radius	
	x_{PmHC}	y_{PmHC}	x_Q	y_Q	p	r	χ_{HC}
SRrP1-HCP1	11.3405	8.6934	8.9634	5.5992	−0.1879	1.8521	0.115291
SRrP2-HCP2	11.2733	7.4148	8.7522	4.4225	−0.1336	2.0280	0.126913
SRrP3-HCP3	7.0969	2.0330	4.8838	−0.1123	2.3003	1.6747	0.339553
SRrP4-HCP4	6.9631	1.9858	4.7798	−0.1814	2.3648	1.6512	0.344411
SRrP5-HCP5	15.3706	17.8283	9.3098	8.8286	−2.0290	2.4871	0.439190

Example 7.7 Five separated spatial positions of the moving body are described by the parameters $(x_{Omf}^{(i)}, y_{Omf}^{(i)}, z_{Omf}^{(i)}; \theta_1^{(i)}, \theta_2^{(i)}, \theta_3^{(i)})$ of the first five lines of Table 7.1. Determine: (1) the saddle constant axis lines L_{CC} for the five positions; (2) the saddle characteristic lines L_{CC}, L_{RC}, and L_{HC} for the first four positions.

Solution

1. **The characteristic line L_{CC} for five positions**
Five saddle spherical image circle points are optimized and listed in Table 7.14, in which SRrp1 and SRrp2 are spherical image circle points.
Corresponding to the five spherical image circle points in Table 7.14, five saddle striction cylinder points can be optimized and listed in Table 7.15. For the spherical image circle point SRrp1, the error surface of the saddle striction cylindrical surface error is shown in Fig. 7.28.
Similar to Example 7.6, the saddle constant axis line L_{CC} in the moving body can be optimally located with the saddle spherical image circle points and their corresponding saddle striction cylinder points for the five separated positions. By combining the saddle spherical image circle points in Table 7.14 with the corresponding saddle striction cylinder points in Table 7.15. Five saddle constant axis lines L_{CC} of the moving body are located for the five given positions, in which SRrp1-CCP1 and SRrp2-CCP2 are the constant axis lines since their saddle spherical image circle errors and saddle striction cylindrical surface errors are zero.
2. **The characteristic line L_{CC} for four positions**
As mentioned in Chapter 5, there is a cubic spherical image circle point curve on the unit spherical surface in the moving body with four given positions. Five points are selected from the curve and listed in Table 7.16.
Any spherical image circle point in Table 7.16, such as sRrp1, corresponds to the parallel lines in the moving body, and their saddle striction cylindrical surface error surface is constructed as shown in Fig. 7.29. There is a long and narrow bottom on the error surface. The saddle striction cylinder points can easily be calculated and locate on the narrow bottom. Six striction cylinder points are listed in Table 7.17.

Table 7.14 The saddle spherical image circle points for five positions

	Direction angles		Saddle spherical image circle (cone)			Error
			Direction of axis		Half-cone angle	
	ϕ	φ	ξ	η	δ	γ_{Rr}
SRrp1	0.5474	4.5590	0.0282	4.8872	0.5208	3.507691×10^{-7}
SRrp2	1.3722	1.7582	1.4862	2.1912	0.4433	1.819791×10^{-7}
SRrp3	0.5209	4.5062	0.0190	2.7554	0.5247	0.000126
SRrp4	0.6138	3.3364	0.5491	2.5119	0.4476	0.000215
SRrp5	0.6317	3.0514	0.6343	2.3744	0.3952	0.000284

Table 7.15 The saddle striction cylinder points for five positions

	Saddle striction cylinder point		Saddle striction cylindrical surface			Error
			Position of axis		Radius	
	x_{PmCC}	y_{PmCC}	x_Q	y_Q	r	χ_{CC}
SRrp1-CCP1	24.3969	8.0805	23.7723	4.8746	0.0767	4.862255×10^{-9}
SRrp2-CCP2	11.7234	−42.4125	5.8840	3.1900	0.1350	7.094598×10^{-8}
SRrp3-CCP3	31.4924	10.9979	30.8368	7.0247	0.0507	8.255815×10^{-8}
SRrp4-CCP4	−1.6729	7.0351	−0.8534	5.2724	0.0303	2.378511×10^{-8}
SRrp5-CCP5	−0.5002	14.5919	1.1949	10.7093	0.1750	5.785051×10^{-8}

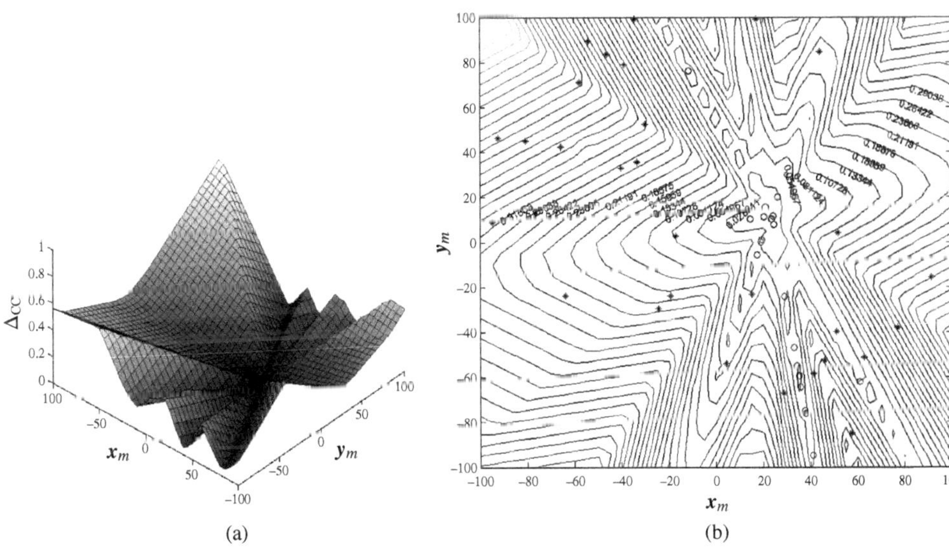

Figure 7.28 The saddle striction cylindrical surface errors for five positions

Table 7.16 Five spherical image circle points for four positions

	Direction angles		Spherical image circle (cone)		
			Direction of axis		Half-cone angle
	ϕ	φ	ξ	η	δ
sRrp 1	0.5008	5.1678	0.3180	0.4819	0.5930
sRrp 2	0.5719	3.1453	0.5764	2.3641	0.4166
sRrp3	1.3828	3.8988	1.0352	3.2681	0.6799
sRrp4	0.9424	5.6508	0.8593	0.7298	1.0347
sRrp5	1.1981	5.5582	1.0449	0.9507	1.4721

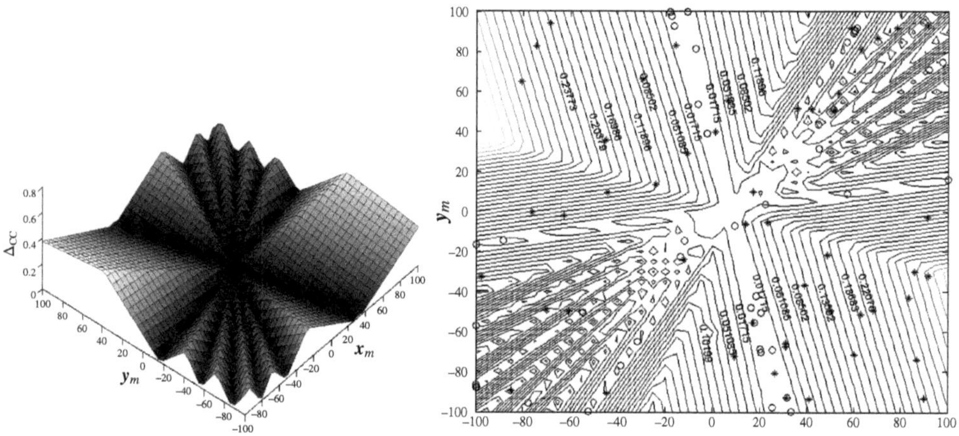

(a) (b)

Figure 7.29 The saddle striction cylindrical surface errors for four positions

Table 7.17 The striction cylinder points for four positions

No.	Striction cylinder point		Striction cylindrical surface			Error
			Position of axis		Radius	
	x_{PmCC}	y_{PmCC}	x_Q	y_Q	r	χ_{CC}
sRrp1-CCP1	31.5112	−92.7641	34.4482	−86.3406	1.4939	2.594458×10^{-7}
sRrp1-CCP2	−2.4604	39.1470	−4.5706	34.1451	0.6061	4.350469×10^{-8}
sRrp1-CCP3	20.4014	−70.2696	43.9920	−57.2926	22.7668	4.413009×10^{-8}
sRrp1-CCP4	16.3110	−47.9193	32.3067	−39.5415	15.6347	4.262750×10^{-8}
sRrp1-CCP5	−6.1763	53.5755	−8.8385	47.3240	0.8358	2.514353×10^{-8}
sRrp1-CCP6	9.4826	−7.2274	9.1468	−8.2126	0.1322	4.542056×10^{-8}

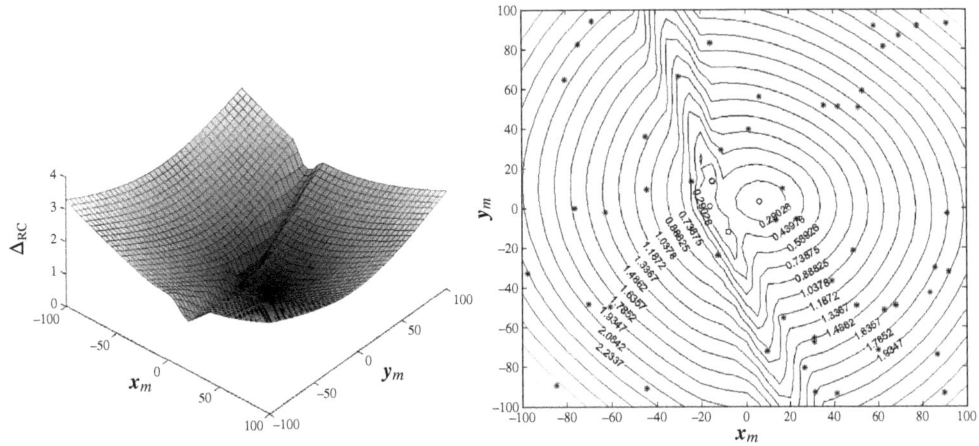

(a) (b)

Figure 7.30 The saddle striction circle errors for four positions

Table 7.18 The saddle striction circle points for four positions

	Saddle striction circle point		Saddle striction circle			Distance parameter	Error
			Position of axis		Radius		
	x_{PmRC}	y_{PmRC}	x_Q	y_Q	r	u_0	χ_{RC}
sRrp1-RCP1	5.8565	7.3001	2.2861	3.9412	2.8898	7.4263	0.072567
sRrp2-RCP2	−15.7457	0.8672	−12.8475	−1.7156	0.4953	6.4804	0.115686
sRrp3-RCP3	23.7275	25.7533	0.2107	7.3055	4.6588	−0.1284	0.092912
sRrp4-RCP4	−2.6034	11.9035	−7.0789	0.9265	8.1600	9.9104	0.041779
sRrp5-RCP5	15.8796	−17.3235	16.4271	−12.1377	0.7725	−4.4632	0.070524

By combining the spherical image circle point sRrp1 with any striction cylinder point in Table 7.17, the constant axis line L_{CC} is located in the moving body with four positions. Similarly, we can do the same for the other four spherical image circle points.

3. **The characteristic line L_{RC} for four positions**

A spherical image circle point in Table 7.16, such as sRrp2, is taken as the direction of the saddle characteristic line L_{RC}, the saddle striction circle error surface is constructed as shown in Fig. 7.30 and the saddle striction circle point can be optimized.

Similarly, the five saddle striction circle points corresponding to the five spherical image circle point in Table 7.16 are listed in Table 7.18.

By combining the spherical image circle points in Table 7.16 with the corresponding saddle striction circle points in Table 7.18, the saddle characteristic lines L_{RC} of the moving body are located.

Table 7.19 Directions of the characteristic line L_{HC} for four positions

No.	Direction angles		Saddle spherical image circle (cone)			Error
			Direction of axis		Half-cone angle	
	ϕ	φ	ξ	η	δ	δ_{Rr}
sRrp1-HCP1	0.8177	3.9937	0.4560	3.3350	0.5196	9.970452×10^{-5}
sRrp2-HCP2	0.5732	3.5388	0.4613	2.5699	0.4756	1.166468×10^{-5}
sRrp3-HCP3	0.5434	3.5036	0.4575	2.5043	0.4704	9.966307×10^{-5}
sRrp4-HCP4	0.6028	3.8970	0.3377	2.8657	0.5081	1.171355×10^{-4}
sRrp5-HCP5	0.5299	3.3845	0.4881	2.4265	0.4548	1.387613×10^{-4}

Table 7.20 Positions of the characteristic lines L_{HC} for four positions

No.	Reference points		Saddle striction cylindrical helix				Error
			Position of axis		Pitch	Radius	
	x_{PmHC}	y_{PmHC}	x_Q	y_Q	p	r	χ_{HC}
sRrp1-HCP1	50.0126	70.5760	33.3708	46.2842	−6.2614	0.1435	4.515468×10^{-5}
sRrp2-HCP2	−10.0970	−58.8456	−11.8548	−49.2300	3.1894	0.6998	7.756429×10^{-5}
sRrp3-HCP3	−32.4102	−11.4616	−20.9904	−8.4653	16.1764	6.4816	2.424753×10^{-4}
sRrp4-HCP4	−72.2316	−27.6843	−61.6316	−16.4657	15.9758	0.3084	1.166468×10^{-5}
sRrp5-HCP5	−25.3769	−8.6345	−15.6901	−6.9589	14.3857	6.0257	1.370187×10^{-4}

4. **The saddle characteristic line L_{HC} for four positions**
   According to the kinematic differential geometry of spatial movement in Chapter 6, there are finite characteristic lines L_{HC} in the moving body at four infinitesimally separated positions. There should be finite characteristic lines L_{HC} in the moving body with four given separated positions. Based on the optimization model (7.64), five saddle characteristic lines L_{HC} in the moving body are located, whose directions and positions are determined respectively by the saddle spherical image points and the saddle striction cylindrical helix points, listed in Tables 7.19 and 7.20.

7.6 The Saddle Kinematic Synthesis of Spatial Four-Bar Linkages

The binary links, or open kinematic chains, comprised of the links and kinematic pairs are of more than 10 kinds, such as S–S, C–S, R–S, H–S, C–C, R–C, H–C, R–R, H–H, H–R mentioned in the above sections, since the kinematic pairs can be combined differently. The binary links or open kinematic chains may constitute many types of spatial linkage, including the closed-loop spatial linkages or open spatial linkages, with single DOF or multiple

DOFs. We just discuss the saddle kinematic synthesis for some of the spatial four-bar linkages, the closed-loop linkages with single DOF. This is because of not only our appreciation both the simpler and the wider application cases of the spatial four-bar linkages, but also the less theoretical basis of the spatial five-bar linkages and more. We expect to develop the saddle approach to be a systematic one for kinematic synthesis of linkages, from planar, spherical to spatial linkages, including both four-bar and multi-bar linkages, even the open kinematic chains, or manipulators.

7.6.1 A Brief Introduction

A space movement of a rigid body may be continuous (even in high order) or discrete (finite positions and multiple positions), corresponding to three types of kinematic synthesis of spatial linkages: continuous synthesis, precise synthesis, and approximate synthesis, like those of planar linkages. On the other hand, the space movement of a rigid body may be given as the following three cases: a moving body to occupy the given positions(currently referred to as the position synthesis, or rigid body guidance), a point/line of the moving body to reproduce a given trajectory (referred to as the trajectory synthesis) and one body moves relative to another with a displacement function(called as the function generator), which can be converted into each other. The position synthesis is taken as a typical example to reveal the descriptions of spatial movement and transformation among the three kinds of kinematic synthesis of spatial linkages.

7.6.1.1 The Continuous Position Synthesis

The continuous synthesis of linkages for the given continuous movement of a rigid body is to find out the characteristic points/lines of the moving body, whose trajectories are the constraint curves or constraint surfaces. A point P of a moving body has a continuous displacement or a trajectory Γ_P in the fixed frame, which can be expressed by equation (7.2). By taking the derivative of equation (7.2) with respect to the arc length of Γ_P and substituting it into equation (3.12), we obtain the invariants of Γ_P, such as the curvature k_P and the torsion τ_P. Then we can judge whether P is a sphere point or cylinder point or not by k_P and τ_P according to Theorems 3.3 and 3.4. A line L of a moving body has a continuous displacement, or a ruled surface Σ_l in the fixed frame, which can be described by equations (7.2) and (7.6). By taking the derivatives of the two equations with respect to the arc length of the spherical image curve of the generatrices of Σ_l and substituting them into equations (3.60) to (3.62), we can obtain the invariants, that is, the construction parameters α_l, β_l, γ_l of Σ_l. Hence, we estimate whether the line of a moving body is a characteristic line or not by α_l, β_l, and γ_l according to Theorems 3.5 and 3.6, such as the characteristic lines L_{CC}, L_{RC}, and L_{HC} corresponding to the binary links C′−C, R−C, and H−C.

In fact, Chapter 6 has discussed these intensively and presented us with the results. We can only get instantaneous solutions for the location of the characteristic points and characteristic lines in the moving body, in other words, the point trajectories or line trajectories contact the constraint curves/ruled surfaces to higher order but can't globally be the constraint curves/ruled surfaces unless the spatial movement is degenerate.

7.6.1.2 The Discrete Position Synthesis

The spatial movement of a rigid body is given as distinct separated positions in the fixed frame, and a spatial linkage is to be designed to guide the couple to occupy the prescribed positions, which is referred to as the discrete position synthesis of linkages in this book. Traditionally, the design for the moving body to occupy a few successive distinct separated locations exactly is named as the precise synthesis, while that for the moving body to pass through more distinct separated locations approximately is called the approximate synthesis, or optimum synthesis. The results of precise synthesis are viewed as the special cases, or the precise results of the discrete position synthesis in this book.

For the precise synthesis of spatial linkages, the characteristic points or lines of the moving body have to be determined, whose trajectories are on the constraint surfaces, such as the spherical surface, cylindrical surface, constant axis surface, and constant parameter surface, etc., or the loop equations can be set up according to Theorems 3.3 to 3.6. The position of the characteristic points or lines can be located by solving the loop equations. Hence, the different constraint surfaces have to be treated as different cases and different loop equations are established by vector algebra or matrices.

For the approximate synthesis of spatial linkages with multiple positions, there exists enormous difficulty in solving the loop equations because of the complexity of the constraint surfaces. We might expect to treat both the points and lines in spatial movement as the points in planar movement. It is a key fact to take advantage of the global geometrical properties or the invariants of the constraint surfaces, especially the ruled surfaces, like the curvature radius of plane curves. Based on the invariants of the constraint surfaces in Chapter 3, we develop the saddle approach of the adaptive curve fitting to that of the space curve fitting and ruled surface fitting and convert the location of the characteristic points and lines on the moving body to spherical curve fitting and cylindrical curve fitting optimally for both the finite and multiple separated positions in spatial movement.

7.6.1.3 The Function Generator

A spatial linkage is to be determined whose input and output links (two binary links) make relative movement described by a given function, which is referred to as the function generator of spatial linkages. The function generator is similar to the position synthesis in content, since the movements of the two binary links may be either translation displacement or rotation angular displacement, or even both, but is more complex in expression forms because the two binary links move simultaneously relative to the base link, as shown in Fig. 7.31. Link 1 and link 3 have angular displacements with respect to link 0, or the base link.

In general, the position synthesis is a typical case for kinematic synthesis of spatial linkages and the function generator is usually converted into the position synthesis. Hence, we choose two kinematic pairs or joints to connect the binary link and the base link, which leads to several possibilities since there exist several combinations of spatial kinematic pairs other than just one for spherical linkages and two for plane linkages. Two fixed joints, kinematic pairs R and C, are taken as a typical example to illustrate how to convert the function generator into the position synthesis.

The functions between input and output links (two binary links) are given as $\varphi_3 = \varphi_3(\varphi_1)$ and $h_3 = h_3(\varphi_1)$, and a spatial linkage is to be synthesized whose base link has two kinematic pairs R and C. First, we establish a fixed Cartesian coordinate system $\{\boldsymbol{R}_O;\boldsymbol{x}_0,\boldsymbol{y}_0,\boldsymbol{z}_0\}$ in the fixed frame 0 and two moving coordinate systems $\{\boldsymbol{R}_{O_1};\boldsymbol{x}_1,\boldsymbol{y}_1,\boldsymbol{z}_1\}$ and $\{\boldsymbol{R}_{O_3};\boldsymbol{x}_3,\boldsymbol{y}_3,\boldsymbol{z}_3\}$ in

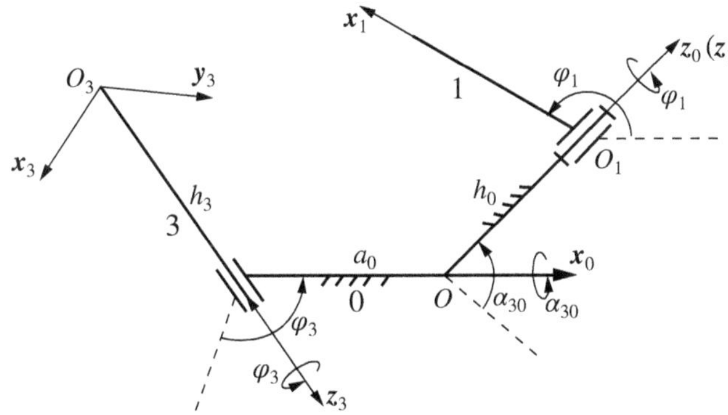

Figure 7.31 The function between the input and output of spatial linkages

links 1 and 3 respectively. For a point P of link 1 with rectangular coordinates (x_1, y_1, z_1) in $\{\boldsymbol{R}_{O_1}; \boldsymbol{x}_1, \boldsymbol{y}_1, \boldsymbol{z}_1\}$, it can be referred to as (x_0, y_0, z_0) in $\{\boldsymbol{R}_O; \boldsymbol{x}_0, \boldsymbol{y}_0, \boldsymbol{z}_0\}$. The coordinate transformation from link 1 to frame 0 is

$$\begin{bmatrix} x_0 \\ y_0 \\ z_0 \\ 1 \end{bmatrix} = \begin{bmatrix} c\varphi_1 & -s\varphi_1 & 0 & 0 \\ s\varphi_1 & c\varphi_1 & 0 & 0 \\ 0 & 0 & 1 & h_0 \\ 0 & 0 & 0 & 1 \end{bmatrix} \begin{bmatrix} x_1 \\ y_1 \\ z_1 \\ 1 \end{bmatrix} \tag{7.65}$$

The coordinate transformation from link 3 to frame 0 is

$$\begin{bmatrix} x_0 \\ y_0 \\ z_0 \\ 1 \end{bmatrix} = \begin{bmatrix} c\varphi_3 & -s\varphi_3 & 0 & -a_0 \\ s\varphi_3 c\alpha_{30} & c\varphi_3 c\alpha_{30} & -s\alpha_{30} & h_3 s\alpha_{30} \\ s\varphi_3 s\alpha_{30} & c\varphi_3 s\alpha_{30} & c\alpha_{30} & -h_3 c\alpha_{30} \\ 0 & 0 & 0 & 1 \end{bmatrix} \begin{bmatrix} x_3 \\ y_3 \\ z_3 \\ 1 \end{bmatrix} \tag{7.66}$$

Hence, the coordinate transformation from link 3 to link 1 is

$$\begin{bmatrix} x_1 \\ y_1 \\ z_1 \\ 1 \end{bmatrix} = [\mathbf{M}_{31}] \begin{bmatrix} x_3 \\ y_3 \\ z_3 \\ 1 \end{bmatrix} = [\mathbf{R}_{31}][\mathbf{T}_{31}] \begin{bmatrix} x_3 \\ y_3 \\ z_3 \\ 1 \end{bmatrix} \tag{7.67}$$

where

$$[\mathbf{R}_{31}] = \begin{bmatrix} c\varphi_3 c\varphi_1 + s\varphi_3 c\alpha_{30} s\varphi_1 & -s\varphi_3 c\varphi_1 + c\varphi_3 c\alpha_{30} s\varphi_1 & -s\alpha_{30} s\varphi_1 & 0 \\ -c\varphi_3 s\varphi_1 + s\varphi_3 c\alpha_{30} c\varphi_1 & s\varphi_3 s\varphi_1 + c\varphi_3 c\alpha_{30} c\varphi_1 & -s\alpha_{30} c\varphi_1 & 0 \\ s\varphi_3 s\alpha_{30} & c\varphi_3 s\alpha_{30} & c\alpha_{30} & 0 \\ 0 & 0 & 0 & 1 \end{bmatrix}$$

$$[\mathbf{T}_{31}] = \begin{bmatrix} 1 & 0 & 0 & -a_0 c\varphi_1 + h_3 s\alpha_{30} s\varphi_1 \\ 0 & 1 & 0 & -a_0 s\varphi_1 + h_3 s\alpha_{30} c\varphi_1 \\ 0 & 0 & 1 & -h_0 - h_3 c\alpha_{30} \\ 0 & 0 & 0 & 1 \end{bmatrix}$$

Comparing the above equation with equation (7.2), or $[\mathbf{M}_{31}]$ with $[\mathbf{M}]$ and $[\mathbf{R}_{31}]$ with $[\mathbf{R}]$, we have the motion parameters of link 3 relative to link 1:

$$\begin{cases} x_{Omf} = -a_0 c\varphi_1 + h_3 s\alpha_{30} s\varphi_1, y_{Omf} = -a_0 s\varphi_1 + h_3 s\alpha_{30} c\varphi_1 \\ z_{Omf} = -h_0 - h_3 c\alpha_{30}, \theta_1 = \alpha_{30}, \theta_2 = 3\pi/2 - \varphi_1, \theta_3 = \varphi_3 + \pi \end{cases} \tag{7.68}$$

For discrete synthesis, the above equation can be discretized and substituted into equations (7.7) and (7.8), and the points and lines of link 3 will trace the discrete point set $\{\boldsymbol{R}_P^{(i)}\}$ and the discrete line set $\{\boldsymbol{R}_L^{(i)}\}$ in $\{\boldsymbol{R}_{O_1}; \boldsymbol{x}_1, \boldsymbol{y}_1, \boldsymbol{z}_1\}$. Hence, the function generator has been converted into the position synthesis of spatial linkages.

7.6.1.4 The Trajectory Synthesis

For a required trajectory, either point trajectory or line trajectory in the fixed frame $\{\boldsymbol{O}_f; \boldsymbol{x}_f, \boldsymbol{y}_f, \boldsymbol{z}_f\}$, a spatial linkage needs to be synthesized, whose coupler point or coupler line reproduces the required trajectory. Both precise and approximate synthesis for spatial linkages is more difficult than that for plane and spherical linkages, since the space trajectories are much more complex in both parameters and geometrical shapes, revealing a wider field to be studied. On the other hand, when a point or line of a moving body moves along a given trajectory, the moving body may move with additional degrees of freedom, which means that theoretically there exist multiple solutions for trajectory synthesis of spatial linkages. For convenience of computation, the trajectory synthesis is generally converted into position synthesis.

A point trajectory $\Gamma : x = x(t),\ y = y(t),\ z = z(t)$ is given in the fixed frame $\{\boldsymbol{O}_f; \boldsymbol{x}_f, \boldsymbol{y}_f, \boldsymbol{z}_f\}$. A dyad, or a three-link chain (a fixed link, a binary link, and the end link), is usually designated to guide a third link moving with three DOFs and a point of the third link is required to move along the given trajectory Γ, which leads to the fact that the third link, or the floating link, moves in one DOF just like for position synthesis. So here, the trajectory synthesis has been converted into position synthesis.

In order to illustrate the conversion process, a dyad with R–C is taken as a typical example, shown in Fig. 7.32, and the point P_m of the third link BB' moves along the given trajectory Γ. The coordinate systems $\{\boldsymbol{R}_{A'}; \boldsymbol{x}_0, \boldsymbol{y}_0, \boldsymbol{z}_0\}$ and $\{\boldsymbol{R}_{B'}; \boldsymbol{x}_2, \boldsymbol{y}_2, \boldsymbol{z}_2\}$ are established in the base link AA' and the third link BB' respectively, as shown in Fig. 7.32. The parameters of a dyad are designated as:

x_A, y_A – the coordinates of the reference point A on the axis L_A of an R-pair in $\{\boldsymbol{O}_f; \boldsymbol{x}_f, \boldsymbol{y}_f, \boldsymbol{z}_f\}$, which connects the binary link $A'B$ to the base link AA'.

ξ_A, η_A – the direction angles of the axis L_A in $\{\boldsymbol{O}_f; \boldsymbol{x}_f, \boldsymbol{y}_f, \boldsymbol{z}_f\}$.

h_0 – the distance between the reference point A of the axis L_A and the reference point A' of an R-pair.

a_1 – the length of the binary link $A'B$, or the length along the common normal between the axis L_A and the axis L_B of a C-pair.

α_{01} – the angle between the axis L_A and the axis L_B.

r_P, φ_P, z_P – the cylindrical coordinates of the point P_m of the third link BB' in $\{\boldsymbol{R}_{B'}; \boldsymbol{x}_2, \boldsymbol{y}_2, \boldsymbol{z}_2\}$.

Thus, a dyad with seven parameters, x_A, y_A, ξ_A, η_A, h_0, a_1, α_{01}, has to be constructed first if it is used to reproduce a required trajectory Γ, while the tracing point $P_m(r_P, \varphi_P, z_P)$ should

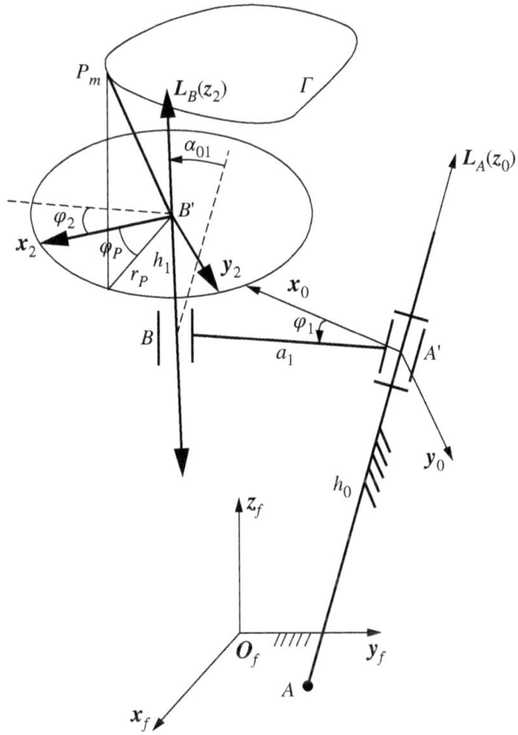

Figure 7.32 A dyad for the trajectory synthesis of spatial linkages

be properly chosen to reach each point of Γ. A point of the third link BB' with coordinates (x_2, y_2, z_2) in $\{\boldsymbol{R}_{B'}; \boldsymbol{x}_2, \boldsymbol{y}_2, \boldsymbol{z}_2\}$ will be examined in $\{\boldsymbol{R}_{A'}, \boldsymbol{x}_0, \boldsymbol{y}_0, \boldsymbol{z}_0\}$ as (x_0, y_0, z_0), or

$$\begin{bmatrix} x_0 \\ y_0 \\ z_0 \\ 1 \end{bmatrix} = [\mathbf{M}_{20}] \cdot \begin{bmatrix} x_2 \\ y_2 \\ z_2 \\ 1 \end{bmatrix} = [\mathbf{R}_{20}] \cdot [\mathbf{T}_{20}] \cdot \begin{bmatrix} x_2 \\ y_2 \\ z_2 \\ 1 \end{bmatrix} \tag{7.69}$$

where

$$[\mathbf{R}_{20}] = \begin{bmatrix} c\varphi_2 c\varphi_1 - c\alpha_{01} s\varphi_1 s\varphi_2 & -s\varphi_2 c\varphi_1 - c\alpha_{01} s\varphi_1 c\varphi_2 & s\alpha_{01} s\varphi_1 & 0 \\ c\varphi_2 s\varphi_1 + c\alpha_{01} c\varphi_1 s\varphi_2 & -s\varphi_2 s\varphi_1 + c\alpha_{01} c\varphi_1 c\varphi_2 & -s\alpha_{01} c\varphi_1 & 0 \\ s\alpha_{01} s\varphi_2 & s\alpha_{01} c\varphi_2 & c\alpha_{01} & 0 \\ 0 & 0 & 0 & 1 \end{bmatrix}$$

$$[\mathbf{T}_{20}] = \begin{bmatrix} 1 & 0 & 0 & a_1 c\varphi_1 + h_1 s\alpha_{01} s\varphi_1 \\ 0 & 1 & 0 & a_1 s\varphi_1 - h_1 s\alpha_{01} c\varphi_1 \\ 0 & 0 & 1 & h_1 c\alpha_{01} \\ 0 & 0 & 0 & 1 \end{bmatrix}$$

This is a coordinate transformation in different coordinate systems. Hence, the parameters of the third link BB' relative to the base link AA' can be expressed as

$$\begin{cases} x_{Omf} = a_1 c\varphi_1 + h_1 s\alpha_{01} s\varphi_1, \; y_{Omf} = a_1 s\varphi_1 - h_1 s\alpha_{01} c\varphi_1 \\ z_{Omf} = h_1 c\alpha_{01}, \; \theta_1 = \alpha_{01}, \; \theta_2 = 3\pi/2 + \varphi_1, \; \theta_3 = \varphi_2 + \pi \end{cases} \tag{7.70}$$

Each position of the third link BB' can be located by the parameters $(\varphi_1, \varphi_2, h_1)$, which are solved from equation (7.69) by matching to each point of Γ, as the point P_m moves along Γ. Substituting the parameters $(x_{Omf}, y_{Omf}, z_{Omf}; \theta_1, \theta_2, \theta_3)$ of equation (7.70) into equations (7.7) and (7.8) for a point or a line of the moving link BB', we have the discrete point set $\{\boldsymbol{R}_P^{(i)}\}$ or discrete line set $\{\boldsymbol{R}_L^{(i)}\}$ in the base link. Hence, the trajectory synthesis has been converted into position synthesis of spatial linkages.

Of course, there are various kinds of kinematic pairs for spatial linkages. A dyad can be formed by different kinematic pairs, corresponding to different open chains. In particular, for line trajectory synthesis, this may lead to different results from those of point trajectory synthesis, or different spatial linkages appear. It is one of the causes for multiple solutions of the trajectory synthesis of spatial linkages, which may provide more choices for designers in practical engineering. It is important for trajectory synthesis to take the seven reasonable construction parameters x_A, y_A, ξ_A, η_A, h_0, a_1, α_{01} and the position of the tracing point $P_m(r_P, \varphi_P, z_P)$ properly; which can be assigned as optimization variables, but good initial values have to be set.

7.6.2 The Spatial Linkage RCCC

A spatial RCCC linkage contains an R-pair and three C-pairs, or two binary links (R–C and C–C). The binary link R–C is usually chosen as input link while the binary link C–C is the output link. It is significant in both theoretical research and practical application to study the spatial RCCC linkage, since it can degenerate to different linkages in different cases, such as a planar four-bar linkage, a spherical four-bar linkage and different spatial four-bar linkages without S-pair.

The synthesis of a spatial four-bar linkage includes the selecting of the type of kinematic pairs and the determining of the dimensions of links, under the constraints lead by the requirements in geometry, kinematics, and dynamics. Unfortunately, the constraint conditions of spatial four-bar linkages are not explicit, compared with those of planar four-bar linkages or spherical four-bar linkages; even the existence conditions of a crank, or Grashof kinematic chain, have not yet been fully revealed in the literature. We would like to apply the constraint equations in the kinematic synthesis of planar linkages to that of spatial linkages, but we cannot do this except for the geometrical error. Hence, the geometrical error, including point trajectory error and line trajectory error, is measured as a normal error and estimated by the saddle approach. In particular, a line trajectory can be converted into a spherical image curve and a striction curve, which may provide a new view on the kinematic synthesis of spatial linkages.

7.6.2.1 The Mathematical Model

A spatial linkage RCCC has two binary links, R–C and C–C, which correspond to the characteristic lines L_{CC} and L_{RC} of the coupler. The constraint surfaces are a constant axis ruled surface and a hyperboloid of one sheet respectively, as shown in Fig. 7.33.

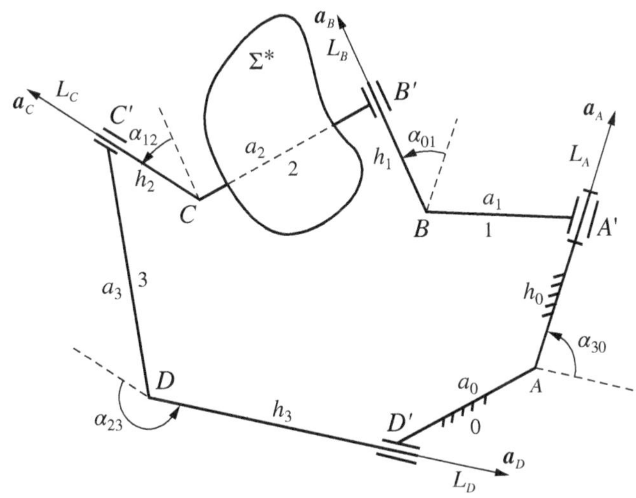

Figure 7.33 The spatial four-bar linkage RCCC

Hence, the saddle kinematic synthesis of the linkage RCCC is to locate the saddle characteristic lines L_{CC} and L_{RC} on a moving body Σ^* with the given movement. For a given discrete movement of Σ^*, two arbitrary lines of Σ^* will trace two discrete line sets $\{\boldsymbol{R}_{L1}^{(i)}\}$ and $\{\boldsymbol{R}_{L2}^{(i)}\}$. We take a constant axis ruled surface and a hyperboloid of one sheet to fit the two discrete line sets $\{\boldsymbol{R}_{L1}^{(i)}\}$ and $\{\boldsymbol{R}_{L2}^{(i)}\}$ respectively, and let the weighted sum of their fitting errors be minimum. The mathematical model for the saddle kinematic synthesis of a spatial linkage RCCC is

$$\begin{cases} \min F(\mathbf{Z}) = \min[\lambda\delta_{CC}(\mathbf{Z}_1) + (1-\lambda)\delta_{RC}(\mathbf{Z}_2)] \\ \text{s.t. } g_j(\mathbf{Z}) < 0,\ j = 1, \ldots, k \end{cases} \tag{7.71}$$

where λ is the weighting coefficient. The above equation can be explained in detail as follows:

A. **The objective function and the optimization variables**

The objective function

The weighted sum of both the error $\delta_{CC}(\mathbf{Z}_1)$ of a saddle constant axis line L_{CC} and the error $\delta_{RC}(\mathbf{Z}_2)$ of a saddle characteristic line L_{RC} is taken as the objective function, whose definitions and solutions are introduced in Sections 7.4 and 7.5.

The optimization variables

The optimization variables $\mathbf{Z}_1 = (\phi_1, \varphi_1, x_{Pm1}, y_{Pm1})^T$ and $\mathbf{Z}_2 = (\phi_2, \varphi_2, x_{Pm2}, y_{Pm2})^T$ are respectively the parameters of two lines on the rigid body, or the coordinates of the reference point and the direction of the line. The domains of the optimization variables are $x_{Pmi}, y_{Pmi} \in (-\infty, +\infty)$, $\phi_i \in [0, \pi)$, $\varphi_i \in [0, 2\pi)$, i=1, 2.

B. **The constraint equations**

As mentioned above, the saddle characteristic line L_{RC}, corresponding to a binary link R–C, or link 1, is assigned as the axis of a moving joint C-pair in Σ^* and the fixed axis of the hyperboloid of one sheet is the axis of a fixed joint R-pair. The saddle characteristic line L_{CC}, corresponding to a binary link C–C, or link 3, is designated as another moving joint C-pair in Σ^*, and the fixed axis of the constant axis ruled surface is the axis of a fixed joint C-pair. The dimensions of the synthesized RCCC linkage are

$$\begin{cases} \alpha_{01} = \delta_{\mathrm{RC}},\ \alpha_{12} = \cos^{-1}\left(\boldsymbol{a}_B \cdot \boldsymbol{a}_C\right),\ \alpha_{23} = \delta_{\mathrm{CC}},\ \alpha_{30} = \cos^{-1}(\boldsymbol{a}_A \cdot \boldsymbol{a}_D) \\ a_1 = r_{\mathrm{RC}},\ a_2 = \left| \dfrac{\left(x_{Pm1} - x_{Pm2}\right)\Delta_1 + (y_{Pm1} - y_{Pm2})\Delta_2 + (z_{Pm1} - z_{Pm2})\Delta_3}{\sqrt{\Delta_1^2 + \Delta_2^2 + \Delta_3^2}} \right|,\ a_3 = r_{\mathrm{CC}} \\ a_0 = \left| \dfrac{\left(x_{QA} - x_{QD}\right)\nabla_1 + (y_{QA} - y_{QD})\nabla_2 + (z_{QA} - z_{QD})\nabla_3}{\sqrt{\nabla_1^2 + \nabla_2^2 + \nabla_3^2}} \right| \\ h_0 = \dfrac{\begin{array}{c}(a_{A1} - a_{D1}\cos\alpha_{30})(x_{A'} - x_{QD}) + (a_{A2} - a_{D2}\cos\alpha_{30})(y_{A'} - y_{QD}) \\ + (a_{A3} - a_{D3}\cos\alpha_{30})(z_{A'} - z_{QD})\end{array}}{\sin^2\alpha_{30}} \end{cases} \tag{7.72}$$

where

$$\begin{cases} \boldsymbol{a}_A = \left(a_{A1}, a_{A2}, a_{A3}\right)^T = (\sin\xi_A\cos\eta_A, \sin\xi_A\sin\eta_A, \cos\xi_A)^T \\ \boldsymbol{a}_B = (a_{B1}, a_{B2}, a_{B3})^T = (\sin\phi_B\cos\varphi_B, \sin\phi_B\sin\varphi_B, \cos\phi_B)^T \\ \boldsymbol{a}_C = (a_{C1}, a_{C2}, a_{C3})^T = (\sin\phi_C\cos\varphi_C, \sin\phi_C\sin\varphi_C, \cos\phi_C)^T \\ \boldsymbol{a}_D = (a_{D1}, a_{D2}, a_{D3})^T = (\sin\xi_D\cos\eta_D, \sin\xi_D\sin\eta_D, \cos\xi_D)^T \\ \Delta_1 = a_{B2}a_{C3} - a_{B3}a_{C2},\ \Delta_2 = a_{B3}a_{C1} - a_{B1}a_{C3},\ \Delta_3 = a_{B1}a_{C2} - a_{B2}a_{C1} \\ \nabla_1 = a_{A2}a_{D3} - a_{A3}a_{D2},\ \nabla_2 = a_{A3}a_{D1} - a_{A1}a_{D3},\ \nabla_3 = a_{A1}a_{D2} - a_{A2}a_{D1} \\ x_{A'} = x_{QA} + u_{\mathrm{RC}}a_{A1},\ y_{A'} = y_{QA} + u_{\mathrm{RC}}a_{A2},\ z_{A'} = z_{QA} + u_{\mathrm{RC}}a_{A3} \end{cases}$$

Considering the requirements of geometrical dimension performance, we may list some constraint conditions as follows.

The length of each link for a linkage has to be larger than a designated positive value first, in order to be easily manufactured:

$$g_1(\mathbf{Z}) = \varepsilon - \min(a_1, a_2, a_3, a_0) \le 0 \tag{7.73}$$

where ε is the permitted minimum positive value for the restriction of the link length.

The condition of the Grashof kinematic chain for the spherical image four-bar linkage of an RCCC linkage is

$$g_2(\mathbf{Z}) = 2\{\max(\alpha_{01}, \alpha_{12}, \alpha_{23}, \alpha_{30}) + \min(\alpha_{01}, \alpha_{12}, \alpha_{23}, \alpha_{30})\} - \sum_{i=0}^{3} \alpha_{i,i+1} \le 0 \tag{7.74}$$

If a crank is possessed by the spherical image four-bar linkage, the link lengths will meet

$$\begin{cases} g_3(\mathbf{Z}) = \min(\alpha_{01}, \alpha_{30}) - \alpha_{12} \le 0 \\ g_4(\mathbf{Z}) = \min(\alpha_{01}, \alpha_{30}) - \alpha_{23} \le 0 \end{cases} \tag{7.75}$$

C. **The optimization algorithm**
For the saddle kinematic synthesis of spatial linkages RCCC under constraint conditions, the optimization model is established as equation (7.71) and the function **fmincon** in Matlab can be adopted directly to solve the equation.

D. **Initial values of the optimization variables**
In the saddle kinematic synthesis of spatial linkages RCCC, the initial values of the optimization variables, or the parameters of two lines of the moving body $\mathbf{Z}_1 = (\phi_1, \varphi_1, x_{Pm1}, y_{Pm1})^T$ and $\mathbf{Z}_2 = (\phi_2, \varphi_2, x_{Pm2}, y_{Pm2})^T$ have to be given first to run the optimization program. The parameters of one saddle characteristic line L_{RC} and one saddle constant axis line L_{CC} are chosen as a group of initial values.

E. **The stop conditions**
Similarly, the stop conditions are set by the function **optimset** in Matlab and the reason for the termination can be identified by the integer value of the output function **exitflag**.

The parameters of the function **fmincon** are set in the same way as the saddle kinematic synthesis of planar four-bar linkages and that of spherical four-bar linkages. A **program for saddle kinematic synthesis of a spatial linkage RCCC with discrete positions**, or **KS-RCCC** for short, is constructed. The input parameters are the separated spatial positions of a moving body and the output parameters are a linkage RCCC with dimensions and performance.

7.6.2.2 The Saddle Kinematic Synthesis for Discrete Positions

Both the saddle characteristic line L_{RC} and the saddle constant axis line L_{CC} are taken as the axes of the two moving joints, C-pair, and the two fixed axes of the corresponding constraint ruled surfaces as those of the two fixed joints, C-pair and R-pair, of the spatial four-bar linkage RCCC. The dimensions can be calculated by equation (7.72). In general, both the fitting errors of the two saddle characteristic lines are usually not zero, depending on the movement properties of the moving body if it is not degenerate. The process of the saddle kinematic synthesis of spatial linkages RCCC is illustrated by the following numerical example.

A spatial linkage RCCC is synthesized, which guides the coupler link to occupy approximately multiple required positions. No matter how many separated positions are required, the mathematical model as described in equation (7.71) and its solving process are the same. The results include the approximate solutions and the precision solutions if it has them, for either multiple separated positions or less than five separated positions.

Example 7.8 The nine separated positions of the moving body in spatial movement are described by the parameters $(x_{Omf}^{(i)}, y_{Omf}^{(i)}, z_{Omf}^{(i)}; \theta_1^{(i)}, \theta_2^{(i)}, \theta_3^{(i)})$ listed in Table 7.1. Synthesize a spatial four-bar linkage RCCC for guiding the coupler link to occupy the given positions.

Solution

According to the mathematical model of the saddle kinematic synthesis of spatial four-bar linkages RCCC, the process of solving equation (7.71), like that for the saddle kinematic synthesis of planar, spherical four-bar linkages, has three parts: preparation of initial values, program setting, and program operation.

Table 7.21 The five initial spatial linkages RCCC

L_{CC}	L_{RC}	Dimension of the RCCC linkage								
		α_{01}	α_{12}	α_{23}	α_{30}	a_1	a_2	a_3	a_0	h_0
SRrP3-CCP3	SRrP1-RCP1	0.5237	0.4673	0.5568	0.4944	1.8323	6.1548	2.4233	6.8108	−1.5850
SRrP4-CCP4	SRrP1-RCP1	0.5237	0.4828	0.5588	0.5110	1.8323	6.0168	2.4219	6.6807	−1.5692
SRrP5-CCP5	SRrP1-RCP1	0.5237	0.3753	0.5449	0.3959	1.8323	6.7958	2.1152	6.3923	8.1865
SRrP1-CCP1	SRrP3-RCP3	0.5568	0.4673	0.5237	0.4944	2.8520	4.6341	1.9225	5.7052	−7.8953
SRrP5-CCP5	SRrP3-RCP3	0.5568	0.8028	0.5449	0.7974	2.8520	4.0506	2.1152	3.7689	−7.9382

1. **The preparation of initial values**
   At first, based on the given nine positions of the moving body, the saddle spherical image circle points can be searched out by the sub-program ArRP. Five optimized saddle spherical image circle points, SRrP1–SRrP5, are listed in Table 7.9. Five saddle striction cylinder points and five saddle striction circle points on the reference plane are obtained, listed respectively in Tables 7.10 and 7.11. Hence, five saddle constant axis lines L_{CC} and five saddle characteristic lines L_{RC} can be located in the moving body. Combining L_{RC} with L_{CC}, we have 20 groups of initial values and 20 spatial RCCC linkages, of which five are listed in Table 7.21.
2. **The program setting**
   For the kinematic synthesis program **KS-RCCC**, the parameter settings of the function **fmincon** can refer totally to the previously introduced contents, including the optimization algorithm, termination conditions, and constraint condition setting. The input nine positions of the moving body are described by the parameters $(x_{Omf}^{(i)}, y_{Omf}^{(i)}, z_{Omf}^{(i)}, \theta_1{}^{(i)}, \theta_2{}^{(i)}, \theta_3{}^{(i)})$, $i = 1, \ldots, 9$ in Table 7.1. The linear constraints **A**, **b**, **Aeq**, **beq** and the nonlinear constraints **nonlcon** are set to null. The lower bound is $\text{lb} = [0, 0, -\infty, -\infty, 0, 0, -\infty, -\infty]$ and the upper bound is $\text{ub} = [\pi, 2\pi, +\infty, +\infty, \pi, 2\pi, +\infty, +\infty]$.
3. **The program operation**
   Based on the initial values of the combination of saddle constant axis lines L_{CC} with saddle characteristic lines L_{RC} listed in Table 7.21, we run the program **KS-RCCC** of the saddle kinematic synthesis of the spatial linkage RCCC for nine discrete positions. The optimized solutions naturally correspond to their initial values for non-constraint optimization programming, since the initial values are the saddle points in geometry. The dimensions of the five linkages RCCC synthesized are also shown in Table 7.21.

7.6.3 The Spatial Linkage RRSS

A spatial linkage RRSS contains two R-pairs and two S-pairs, or two binary links, R–R and S–S. The binary link R–R is usually chosen as an input link, corresponding to the characteristic lines L_{RR} of the coupler, and its constraint surface in the base frame is a hyperboloid of one sheet. The binary link S–S is the output link, corresponding to the sphere point P_{SS} of the coupler, whose constraint surface in the base frame is a spherical surface as shown in Fig. 7.34.

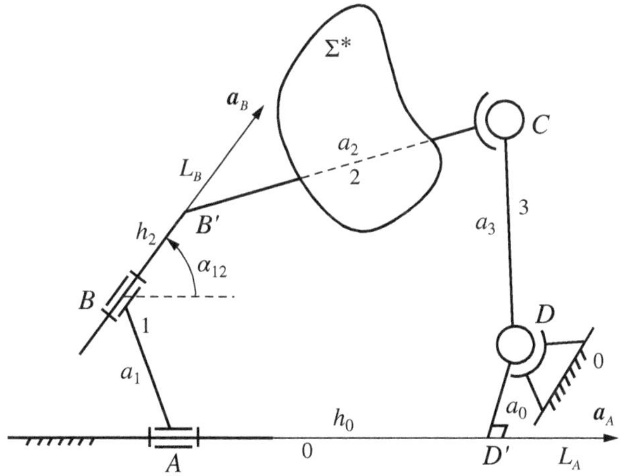

Figure 7.34 A spatial linkage RRSS

7.6.3.1 The Mathematical Model

The saddle kinematic synthesis of a spatial four-bar linkage RRSS is to locate a saddle characteristic line L_{RR} and a saddle sphere point P_{SS} on a moving body. For a given discrete movement of Σ^*, a line and a point of Σ^* will trace a discrete line set $\{\boldsymbol{R}_{L1}^{(i)}\}$ and a discrete point set $\{\boldsymbol{R}_{P1}^{(i)}\}$. We take a hyperboloid of one sheet to fit the discrete line set $\{\boldsymbol{R}_{L1}^{(i)}\}$ and a spherical surface to fit the discrete point set $\{\boldsymbol{R}_{P1}^{(i)}\}$ respectively, letting the weighted sum of the two fitting errors be minimum. The mathematical model for the saddle kinematic synthesis of a spatial four-bar linkage RRSS is

$$\begin{cases} \min F(\mathbf{Z}) = \min(\lambda\Delta_{SS}(\mathbf{Z}_1) + (1-\lambda)\Delta_{RR}(\mathbf{Z}_2)) \\ \text{s.t. } g_j(\mathbf{Z}) \le 0,\ j = 1,\ \dots,k \end{cases} \tag{7.76}$$

where λ is the weighting coefficient. The above equation can be explained in detail as follows.

A. **The objective function and the optimization variables**

The objective function

The weighted sum of both the fitting error $\Delta_{SS}(\boldsymbol{Z}_1)$ of a saddle sphere point P_{SS} and the error $\Delta_{RR}(\boldsymbol{Z}_2)$ of a saddle characteristic line L_{RR} is taken as the objective function, whose definitions and solutions are introduced in Sections 7.2 and 7.5.

The optimization variables

The optimization variables of the optimization model are $\boldsymbol{Z} = (z_1, z_2)^T$, where $\boldsymbol{Z}_1 = (x_{Pm1}, y_{Pm1}, z_{Pm1})^T$ are the coordinates of a point and $\boldsymbol{Z}_2 = (\phi_2, \varphi_2, x_{Pm2}, y_{Pm2})^T$ are the parameters of a line of the moving body. The domain of the optimization variables is $x_{Pmi}, y_{Pmi}, z_{Pm1} \in (-\infty, +\infty)$, $\phi_2 \in [0, \pi)$, $\varphi_2 \in [0, 2\pi)$, i=1, 2.

B. **The constraint equations**

As mentioned above, the saddle characteristic line L_{RR} corresponding to a binary link R–R, or link 1, is assigned as the axis of a moving joint R-pair in Σ^* and the fixed axis of the hyperboloid of one sheet is the axis of a fixed joint R-pair. The saddle sphere point P_{SS} corresponding to a binary link S–S, or link 3, is designated as the moving joint S-pair in Σ^* and the center point of the spherical surface is the center of the fixed joint S-pair. The dimensions of the synthesized RRSS linkage are

$$\begin{cases} \alpha_{12} = \delta_{\text{RR}},\ a_1 = r_{\text{RR}},\ a_3 = r_{\text{SS}} \\ h_2 = a_{B1}(x_{Pm1} - x_{Bm}) + a_{B2}(y_{Pm1} - y_{Bm}) + a_{B3}(z_{Pm1} - z_{Bm}) \\ a_2 = \sqrt{(x_{Pm1} - x_{Bm})^2 + (y_{Pm1} - y_{Bm})^2 + (z_{Pm1} - z_{Bm})^2 - h_2^2} \\ h_0 = a_{A1}(x_D - x_A) + a_{A2}(y_D - y_A) + a_{A3}(z_D - z_A) \\ a_0 = \sqrt{(x_D - x_A)^2 + (y_D - y_A)^2 + (z_D - z_A)^2 - h_0^2} \end{cases} \tag{7.77}$$

where

$$\begin{cases} \boldsymbol{a}_B = (a_{B1}, a_{B2}, a_{B3})^T = (\sin\phi_B \cos\varphi_B, \sin\phi_B \sin\varphi_B, \cos\phi_B)^T \\ \boldsymbol{a}_A = (a_{A1}, a_{A2}, a_{A3})^T = (\sin\xi_A \cos\eta_A, \sin\xi_A \sin\eta_A, \cos\xi_A)^T \\ x_A = x_{QA} + u_{\text{RR}} a_{A1}, y_A = y_{QA} + u_{\text{RR}} a_{A2}, z_A = z_{QA} + u_{\text{RR}} a_{A3} \\ x_{Bm} = x_{Pm2} + v_{\text{RR}} a_{B1},\ y_{Bm} = y_{Pm2} + v_{\text{RR}} a_{B2},\ z_{Bm} = z_{Pm2} + v_{\text{RR}} a_{B3} \end{cases}$$

Considering the requirements of geometrical dimensions and performance in application cases, we may add some constraint conditions.

C. **The optimization algorithm**

For the saddle kinematic synthesis of spatial linkages RRSS under constraint conditions, the optimization model is established as equation (7.76) and the function **fmincon** in Matlab can be adopted directly to solve the equation.

D. **Initial values of the optimization variables**

In the saddle kinematic synthesis of spatial linkages RRSS, the initial values of the optimization variables, or the coordinates $\mathbf{Z}_1 = (x_{Pm1}, y_{Pm1}, z_{Pm1})^T$ of a point and the parameters $\mathbf{Z}_2 = (\phi_2, \varphi_2, x_{Pm2}, y_{Pm2})^T$ of a line of the moving body have to be given first to run the optimization program. The parameters of one saddle characteristic line L_{RR} and saddle sphere point P_{SS} are chosen as a group of initial values.

E. **The stop conditions**

Similarly, the stop conditions are set by the function **optimset** in Matlab and the reason for the termination can be identified by the integer value of the output function **exitflag**.

7.6.3.2 The Saddle Kinematic Synthesis for Discrete Positions

The saddle characteristic line L_{RR} is taken as the axis of the moving joint, R-pair, and the saddle sphere point P_{SS} as the center of the moving joint, S-pair, while the fixed axis of the constraint ruled surface is taken as the fixed axis of the fixed joint, R-pair, and the fixed center point of the spherical surface as the fixed center of the fixed joint, S-pair. The dimensions of the spatial four-bar linkage RRSS can be calculated by equation (7.77). In general, both the saddle characteristic line error and the saddle sphere point error are usually not zero, depending on the movement properties of the moving body if it is not degenerate. The process of saddle kinematic synthesis of spatial linkages RRSS is indicated by the following numerical example.

A spatial linkage RRSS is synthesized, which guides the coupler link to occupy approximately multiple required positions. The mathematical model as described in equation (7.76) and its solving process are independent of the number of given discrete positions. The results include the approximate solutions and the precision solutions if it has them, for either multiple separated positions or less than three separated positions.

Example 7.9 The nine separated spatial positions of the moving body are described by the parameters $(x_{Omf}^{(i)}, y_{Omf}^{(i)}, z_{Omf}^{(i)}; \theta_1^{(i)}, \theta_2^{(i)}, \theta_3^{(i)})$ listed in Table 7.1. Synthesize a spatial linkage RRSS for guiding the coupler link to occupy the given positions.

Solution

By the mathematical model of equation (7.76), the process for the saddle kinematic synthesis of a spatial RRSS linkage is the same as that for the saddle kinematic synthesis of a spatial RCCC linkage. There are three parts: preparation of initial values, program settings, and program operation.

1. **The preparation of initial values**
   At first, based on the given nine positions of the moving body, the saddle spherical image circle points can be searched out by the sub-program ArRP. Five optimized saddle spherical image circle points, SRrP1–SRrP5, are listed in Table 7.9. Five saddle characteristic lines L_{RR} are located in the moving body, listed in Table 7.12. Meanwhile, seven saddle sphere points P_{SS} (SSP1–SSP7) are obtained by calling the sub-program ASSP and listed in Table 7.2. Combining L_{RR} with P_{SS}, 35 groups of initial values, or spatial RRSS linkages, are provided, of which five are listed in Table 7.22.
2. **The program setting**
   For the kinematic synthesis program, the parameter settings include the optimization algorithm, termination conditions, and constraint condition setting, which are the same as those of the sub-program **KS-RCCC**. The inputs, nine positions of the moving body, are described by the parameters in Table 7.1. The linear constraints **A**, **b**, **Aeq**, **beq** and the nonlinear constraints **nonlcon** are set to null. The lower bound is $\mathrm{lb} = [-\infty, -\infty, -\infty, 0, 0, -\infty, -\infty]$ and the upper bound is $\mathrm{ub} = [+\infty, +\infty, +\infty, \pi, 2\pi, +\infty, +\infty]$.
3. **The program operation**
   Based on the initial values, the combinations of saddle sphere points P_{SS} with saddle characteristic lines L_{RR} listed in Table 7.22, we run the program for the saddle kinematic synthesis of spatial four-bar linkages RRSS for nine discrete positions. The optimized solutions correspond geometrically to the five initial values for non-constraint optimization programming. The dimensions of the five synthesized RRSS linkages are also shown in Table 7.22.

Table 7.22 The five initial spatial linkages RRSS

P_{SS}	L_{RR}	Dimensions of the RRSS linkage						
		α_{12}	a_1	a_2	a_3	a_0	h_0	h_2
SSP1	SRrP1-RRP1	0.5237	2.1709	23.8768	23.5296	5.1543	2.5693	14.3274
SSP1	SRrP2-RRP2	0.5269	2.1651	23.2768	23.5296	5.0480	2.7211	15.1637
SSP1	SRrP3-RRP3	0.5568	2.5221	16.0467	23.5296	4.4179	4.0988	22.7506
SSP1	SRrP4-RRP4	0.5588	2.5316	15.6933	23.5296	4.4400	4.1418	23.0230
SSP1	SRrP5-RRP5	0.5449	2.0228	25.0557	23.5296	4.9208	1.0445	8.6586

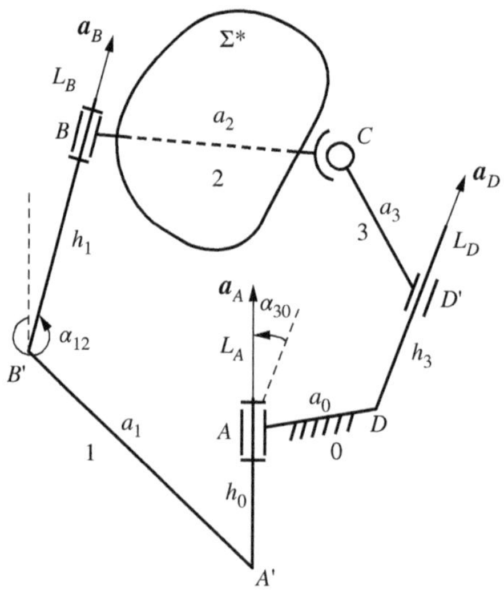

Figure 7.35 The spatial linkage RRSC

7.6.4 *The Spatial Linkage RRSC*

A spatial linkage RRSC contains two R-pairs, one S-pair, and one C-pair or two binary links R–R and C–S. The binary link R–R is usually chosen as an input link, corresponding to the characteristic lines L_{RR} of the coupler, and its constraint surface in the base frame is a hyperboloid of one sheet. The binary link C–S is the output link, corresponding to the characteristic point P_{CS} of the coupler, whose constraint surface in the base frame is a cylindrical surface as shown in Fig. 7.35.

7.6.4.1 The Mathematical Model

The saddle kinematic synthesis of spatial four-bar linkages RRSC is to locate a saddle characteristic line L_{RR} and a saddle cylinder point P_{SS} on a moving body with the given movement. For a given discrete movement of Σ^*, both a line and a point of Σ^* will trace a discrete line set $\{\boldsymbol{R}_{L1}^{(i)}\}$ and a discrete point set $\{\boldsymbol{R}_{P1}^{(i)}\}$ respectively. We take a hyperboloid of one sheet to fit the discrete line set $\{\boldsymbol{R}_{L1}^{(i)}\}$ and a cylindrical surface to fit the discrete point set $\{\boldsymbol{R}_{P1}^{(i)}\}$ respectively by letting the weighted sum of the two fitting errors be minimum. The mathematical model for the saddle kinematic synthesis of a spatial four-bar linkage RRSC is

$$\begin{cases} \min F(\mathbf{Z}) = \min(\lambda\Delta_{\mathrm{CS}}(\mathbf{Z}_1) + (1-\lambda)\Delta_{\mathrm{RR}}(\mathbf{Z}_2)) \\ \text{s.t. } g_j(\mathbf{Z}) \le 0,\ j = 1, \ldots, k \end{cases} \tag{7.78}$$

where λ is the weighted coefficient. The above equation can be explained in detail as follows.

A. **The objective function and the optimization variables**

The objective function

The weighted sum of both the fitting error $\Delta_{\mathrm{CS}}(\mathbf{Z}_1)$ of a saddle cylinder point P_{CS} and the error $\Delta_{\mathrm{RR}}(\mathbf{Z}_2)$ of a saddle characteristic line L_{RR} is taken as the objective function, whose definitions and solutions are introduced in Sections 7.3 and 7.5.

The optimization variables

The optimization variables of the optimization model are $\mathbf{Z} = (z_1, z_2)^T$, where $\mathbf{Z}_1 = (x_{Pm1}, y_{Pm1}, z_{Pm1})^T$ are the coordinates of a point and $\mathbf{Z}_2 = (\phi_2, \varphi_2, x_{Pm2}, y_{Pm2})^T$ are the parameters of a line on the moving body. The domain of the optimization variables is $x_{Pmi}, y_{Pmi}, z_{Pm1} \in (-\infty, +\infty)$, $\phi_2 \in [0, \pi)$, $\varphi_2 \in [0, 2\pi)$, i=1, 2.

B. **The constraint equations**

As mentioned above, the saddle characteristic line L_{RR}, corresponding to a binary link R–R, or link 1, is assigned as the axis of the moving joint R-pair in Σ^* and the fixed axis of the hyperboloid of one sheet is the axis of the fixed joint R-pair. The saddle cylinder point P_{CS} corresponding to a binary link C–S, or link 3, is designated as the moving joint S-pair in Σ^* and the axis of the cylindrical surface is that of the fixed joint C-pair. The dimensions of the spatial four-bar linkage RRSC synthesized are

$$\begin{cases} a_3 = r_{\mathrm{CS}},\ a_1 = r_{\mathrm{RR}},\ \alpha_{12} = \delta_{\mathrm{RR}} \\ \alpha_{30} = \cos^{-1}(\boldsymbol{a}_A \cdot \boldsymbol{a}_D) \\ h_0 = q - u_{\mathrm{RR}},\ h_1 = w - v_{\mathrm{RR}} \\ a_2 = \sqrt{(x_{Pm1} - x_{Bm})^2 + (y_{Pm1} - y_{Bm})^2 + (z_{Pm1} - z_{Bm})^2} \\ a_0 = \left| \dfrac{(x_{QA} - x_{QD})\,\Delta_1 + (y_{QA} - y_{QD})\Delta_2 + (z_{QA} - z_{QD})\Delta_3}{\sqrt{\Delta_1^2 + \Delta_2^2 + \Delta_3^2}} \right| \end{cases} \tag{7.79}$$

where

$$\begin{cases} \boldsymbol{a}_A = (a_{A1}, a_{A2}, a_{A3})^T = (\sin\xi_A \cos\eta_A, \sin\xi_A \sin\eta_A, \cos\xi_A)^T \\ \boldsymbol{a}_B = (a_{B1}, a_{B2}, a_{B3})^T = (\sin\phi_B \cos\varphi_B, \sin\phi_B \sin\varphi_B, \cos\phi_B)^T \\ \boldsymbol{a}_D = (a_{D1}, a_{D2}, a_{D3})^T = (\sin\xi_D \cos\eta_D, \sin\xi_D \sin\eta_D, \cos\xi_D)^T \\ \Delta_1 = a_{A2}a_{D3} - a_{A3}a_{D2},\ \Delta_2 = a_{A3}a_{D1} - a_{A1}a_{D3},\ \Delta_3 = a_{A1}a_{D2} - a_{A2}a_{D1} \\ w = a_{B1}(x_{Pm1} - x_{Pm2}) + a_{B2}(y_{Pm1} - y_{Pm2}) + a_{B3}(z_{Pm1} - z_{Pm2}) \\ q = \dfrac{\begin{array}{c}(x_{QA} - x_{QD})(a_{D1}\cos\alpha_{30} - a_{A1}) + (y_{QA} - y_{QD})(a_{D2}\cos\alpha_{30} - a_{A2}) \\ +(z_{QA} - z_{QD})(a_{D3}\cos\alpha_{30} - a_{A3})\end{array}}{\sin^2\alpha_{30}} \\ x_{Bm} = x_{Pm2} + wa_{B1},\ y_{Bm} = y_{Pm2} + wa_{B2},\ z_{Bm} = z_{Pm2} + wa_{B3} \end{cases}$$

C. **The optimization algorithm**
For the saddle kinematic synthesis of spatial four-bar linkages RRSC under constraint conditions, the optimization model is established as equation (7.78) and the function **fmincon** in Matlab can be adopted directly to solve the equation.

D. **Initial values of the optimization variables**
In the saddle kinematic synthesis of spatial linkages RRSC, the initial values of the optimization variables, or the coordinates $\mathbf{Z}_1 = (x_{Pm1}, y_{Pm1}, z_{Pm1})^T$ of a point and the parameters $\mathbf{Z}_2 = (\phi_2, \varphi_2, x_{Pm2}, y_{Pm2})^T$ of a line of the moving body, have to be given first to run the optimization program. The parameters of one saddle characteristic line L_{RR} and a saddle cylinder point P_{CS} are chosen as a group of initial values.

E. **The stop conditions**
Similarly, the stop conditions are set by the function **optimset** in Matlab and the reason for the termination can be identified by the integer value of the output function **exitflag**.

7.6.4.2 The Saddle Kinematic Synthesis for Discrete Positions

The saddle characteristic line L_{RR} is taken as the axis of the moving joint, R-pair, and the saddle cylinder point P_{CS} as the center of the moving joint, S-pair, on the moving body. The fixed axes of the hyperboloid of one sheet are taken as the axis of the fixed joint, R-pair, and the fixed axis of the cylinder as the axis of the fixed joint, C-pair, to form the spatial four-bar linkage RRSC. Its dimensions can be calculated by equation (7.79). In general, both the saddle characteristic line error and the saddle cylinder point error are usually not zero, depending on the movement properties of the moving body if it is not degenerate. The process of the saddle kinematic synthesis of spatial linkages RRSC is shown by the following numerical example.

A spatial four-bar linkage RRSC is synthesized, which guides the coupler link to occupy approximately the multiple required positions. The solving process deals with the mathematical model as described in equation (7.78) and is independent of the number of discrete positions, no matter how many separated positions are required. The results include the approximate solutions and the precise solutions if it has them, for either multiple separated positions or fewer than three separated positions.

Example 7.10 The nine separated spatial positions of the moving body are described by the parameters $(x_{Omf}^{(i)}, y_{Omf}^{(i)}, z_{Omf}^{(i)}; \theta_1^{(i)}, \theta_2^{(i)}, \theta_3^{(i)})$ listed in Table 7.1. Synthesize a spatial four-bar linkage RRSC for guiding the coupler link to occupy the given positions.

Solution

Based on the mathematical model of equation (7.78), the process for the saddle kinematic synthesis of a spatial four-bar RRSC linkage, the same as that for the saddle kinematic synthesis of a spatial RCCC linkage, has three parts: preparation of initial values, program settings, and program operation.

1. **The preparation of initial values**
At first, based on the given nine positions of the moving body, five saddle spherical image circle points (SRrP1–SRrP5) are listed in Table 7.9, and five saddle characteristic lines L_{RR} in the moving body are located by the reference points, listed in Table 7.12. Meanwhile,

Table 7.23 The five initial spatial linkages RRSC

P_{CS}	L_{RR}	Dimensions of the spatial RRSC linkage							
		α_{12}	α_{30}	a_1	a_2	a_3	a_0	h_0	h_1
CSP6	SRrP1-RRP1	0.5237	0.2437	2.1709	19.1559	2.8929	17.8596	−9.8070	−3.2789
CSP6	SRrP2-RRP2	0.5269	0.2104	2.1651	18.9915	2.8929	17.6116	−10.8543	−3.1188
CSP6	SRrP3-RRP3	0.5568	0.2508	2.5221	17.7597	2.8929	17.2825	11.5421	−6.1687
CSP6	SRrP4-RRP4	0.5588	0.2673	2.5316	17.4786	2.8929	17.0365	10.6488	−6.2194
CSP6	SRrP5-RRP5	0.5449	0.5769	2.0228	17.7859	2.8929	12.6469	1.3150	2.9352

seven saddle cylinder points P_{CS} (SCP1−SCP7) are obtained by calling the sub-program ACSP and listed in Table 7.5. By combining L_{RR} with P_{CS}, 35 groups of initial values, or 35 spatial RRSC linkages can be formed, of which five are listed in Table 7.23.

2. **The program setting**
   For the kinematic synthesis program, the parameter settings include the optimization algorithm, termination conditions, and constraint condition setting, which are the same as those of the sub-program **KS-RCCC**. The inputs, nine positions of the moving body, are described by the parameters in Table 7.1. The linear constraints **A**, **b**, **Aeq**, **beq** and the nonlinear constraints **nonlcon** are set to null. The lower bound is $lb = [-\infty, -\infty, -\infty, 0, 0, -\infty, -\infty]$ and the upper bound is $ub = [+\infty, +\infty, +\infty, \pi, 2\pi, +\infty, +\infty]$.
3. **The program operation**
   Based on the initial values, the combinations of saddle cylinder points P_{CS} with saddle characteristic lines L_{RR}, listed in Table 7.23, we run the program for the saddle kinematic synthesis of spatial four-bar linkages RRSC for nine discrete positions. The optimized solutions logically correspond to the five initial values for the non constraint optimization programming. The dimensions of the five spatial four-bar linkages RRSC are also shown in Table 7.23.

7.7 Discussion

The spatial linkages, compared with the planar linkages and spherical linkages, have many merits such as compact structure, diverse output, and reliable work, and can be applied to many cases where planar linkages are impossible or complicated. Because of the complexity of spatial movement of the spatial linkages, the design process can't be imagined and conceived, or is difficult to adopt via the intuitive trial-and-error method. The kinematic geometry, the theoretical basis of the kinematic synthesis of spatial linkages, is extremely important.

Although Bagci [4] studied the kinematic synthesis of the spherical four-bar linkage at finitely separated positions through the geometry graph, the solving procedure is sophisticated and of low precision. If the geometry graph is extended to the kinematic synthesis of spatial linkages, the graphic process will be very complex and unrealistic. So, the graphic method must be confined to very simple cases. Like the traditional atlas method, the coupler curve of the spatial linkage is so intricate that configurations of the projection lines of the coupler curve on the different projection planes vary dramatically. The atlas of the coupler curves of spatial linkages can be established for path synthesis. Chu and Sun [5, 6] applied the Fourier series theory to the kinematic synthesis of spatial linkages. The relationship between the harmonic component of the output and the basic dimensional types of the linkage mechanism was discovered. Based on this relationship, or numerical atlas database, a new approach for the kinematic synthesis of spatial linkages is presented.

With the development of computer technology and its wide applications, the analytic method and the optimization method are gradually being adopted to accomplish the kinematic synthesis of spatial linkages. Hartenberg and Denavit [7] introduced the symbolic notation to describe the spatial linkages, in which the method of matrix analysis was adopted to obtain the displacement equation of the spatial linkages. The function synthesis of spatial linkages boils down to the solving of a system of nonlinear equations. Roth [8, 9] described the spatial finitely separated positions of a moving body in terms of the screw motion and linear transformations, and derived the equations of the points which lie on a special locus – a sphere, circle, plane, line, or cylinder. Then, the results are applied to the synthesis of spatial linkages. Based on the screw-axis geometry, Roth [10–12] discussed the synthesis of cranks with helical, cylindrical, spherical, revolute, and prismatic joints for both finitely and infinitesimally separated position problems. An "equivalent screw triangle" is introduced to get the design equations, which can be applied to the design of binary links for rigid body guidance, function generation, and similar problems. Chen and Roth [13, 14], James and Roth [15] extended the theory for finite kinematic synthesis to the infinitesimal and hybrid kinematic synthesis. Sandor [16] presented the position equations and constraint equations of spatial linkages by means of vectors and quaternion operators. Suh and Radcliffe [17, 18] presented the displacement, velocity, and acceleration constraint equations for several binary links, such as S–S, R–S, R–R, R–C, and C–C; the screw displacement matrices are adopted to generate the design equations. McCarthy [19] applied the standard form equation presented by Sandor and Erdman [20] for the design of a planar RR chain to the spatial CC chain, and identified the equation of the spatial triangle as the generalization of the standard form equation for the open chain. Innocenti [21] presented the constraint equations for a binary link SS, and proposed an original elimination method; a set of six second-order algebraic equations in six unknowns can be converted into a final univariate polynomial equation of 20th order. Then, a 5-SS Stewart platform can be synthesized to guide the body through seven arbitrarily chosen positions. Lee and Mavroidis [22] applied the Denavit and Hartenberg parameters and 4×4 homogeneous matrices to obtain the constraint equations of a spatial 3R serial chain, which were solved using polynomial homotopy continuation. Perez and McCarthy [23] formulated the constraint equations of spatial serial chains through the screw displacement and the dual quaternion algebra. Bai and Angules [24] adopted the dual algebra to reformulate the constraint equations of CC, RC, and RR dyads. A semigraphical approach was introduced to solve the constraint equations, which can filter out the complex solutions and ease the synthesis process. The linkages CCCC and RCCC for rigid-body guidance at four and five given locations can be completed. Hence, the current kinematic synthesis of spatial linkages, or the precise synthesis, focuses mainly on the constraint equations of the binary links or the open kinematic chains; solving the constraint equations by the analytical method, including the matrix method, dual algebraic and dual quaternion method, etc., and their solutions, can be achieved through algebraic methods [25, 26] or polynomial continuation methods [27], even improved methods.

In Ref. [14], Suh and Radcliffe studied the path generation of a spatial RRSS linkage by the least-squares method and directly searched the solutions. The objective function was the composition of the structural error and the constraint equations. Based on the screw theory,

Angeles [28] derived the constraint equations for the spatial linkage and discussed the approximate function generation of a spatial RSSR linkage, in which the objective function is the least quadratic sum of the constraint equations of the binary link RR. Jamalov, Litvin, and Roth [29] regarded the spatial RCCC linkage as the combination of two open-loop chains. Two special types of 4×4 matrices were used to analyze the RCCC linkage. Based on the geometric characteristics of the constraint surface for the RC binary link of the spatial RSCR linkage, Ananthasuresh and Kramer [30] presented a closed-form solution of the analysis of the RSCR spatial linkage. However, for the optimal synthesis of the linkage for function, path, and motion generation problems, the structural error is still taken as the objective function and the dimension parameters are set as the optimization variables. Jiménez [31] gave the geometrical constraints and the functional constraints of linkages based on the natural coordinates. The objective function is still the quadratic sum of the constraint equations for the optimization synthesis of the linkages. Krovi, Ananthasuresh, and Kumar [32] presented the dimensional synthesis of a spatial dyad with two revolute joints for path-following tasks. The rotation matrix approach was adopted to obtain the precision point synthesis equations. With the least-squares error between corresponding points as the objective function, the five free-choice variables were optimized under the condition that three points were chosen as the design positions. Parikian [33] studied how a coupler curve can be regenerated by several spatial linkages with one DOF, whose condition is that a spatial linkage has to be synthesized for the prescribed curve.

As we know so far for the topics of kinematic synthesis of spatial linkages, most prefer the relatively simple spatial four-bar linkages RSSR [34–38], with different approaches depending on the application situations, and RCCC, mainly with the analytical method for finite precision positions. The kinematic synthesis of spatial four-bar linkages, in particular the optimization approaches, or the approximate synthesis, is rarely discussed, even though the construction errors are usually designated as the objective functions, as well as the constraint equation, and their combination. The optimum solutions depend on the initial values of the design variables; furthermore, some of them are valid only in special cases. The theoretical basis of the kinematic synthesis of spatial linkages, the spatial discrete kinematic geometry, is the stage dealing with finite separated positions by an "equivalent screw triangle," which is the new field to study for computing kinematics.

The spatial discrete kinematic geometry studies discrete movement and compares both the point trajectories and line trajectories with the constraint surfaces globally, which is a difficult issue, and also challenges scholars in the research field of mechanisms and robotics. The authors take the invariants and invariance of the constraint curves and surfaces of simpler linkages as the basic elements. They define the normal error between the trajectories and the constraint curves or surfaces as the fitting error (from the viewpoint of the saddle approach) or the maximum fitting error of the saddle curve (surface) to be the minima, and discuss the distribution of the characteristic points and lines on the moving body. For example, saddle sphere points, saddle cylinder points, saddle constant axis lines, and their degenerate lines, whose trajectories are approximately the constraint surfaces corresponding to the binary links S–S, C–S, C–C, H–C, R–C. They also set up [39, 40] the saddle approach for discrete kinematic synthesis of spatial linkages, corresponding to that of planar and spherical linkages, a unified approach of discrete kinematic synthesis of linkages.

The discrete kinematic geometry, the theoretical basis of the kinematic synthesis of linkages, is in fact to study the invariants and invariance by means of the best uniform approximation between the relative discrete points (lines) and the constraint surfaces in multi-dimensional space, which identifies itself from the current theory of an "equivalent screw triangle." For instance, a constant axis ruled surface approximate to a line trajectory can be converted into a saddle spherical image circle and a saddle striction cylindrical surface. However, the discrete kinematic geometry, including planar, spherical, and spatial, is just a starting point in terms of the approach of saddle point programming. Whatever the finite and multiple separated positions of the moving body, it is neither enough nor thorough enough to study both the geometrical and algebraic properties of the best uniform approximation between the discrete trajectories and the constraint surfaces, such as the saddle sphere points, distributing spherical surfaces, saddle spherical surface and its error, in particular the error surface and its boundaries, whose valleys (including positions and numbers) are relevant for the existence of the solutions of kinematic synthesis. We expect to develop the classical finite separated kinematic geometry, in terms of the roles and equivalent screw triangle, to the discrete kinematic geometry by saddle point programming, discussing the relationship among the tracing points/lines, discrete positions, and discrete trajectories, and revealing the distributions of the characteristic points (lines) on the moving body corresponding to those of the kinematic differential geometry in Chapter 6. The connections among a few relevant separated positions of the moving body are set up by the approach of saddle point programming, particularly the fitting positions of constraint surfaces, which might highlight the existence of discrete invariants of the moving body at multiple positions, like that of the rotation poles at two positions and the screw axis at three positions, corresponding to the axodes at infinitesimal successive positions. Then, the discrete invariants might support the discrete generalized curvature theory for discrete positions or multiple separated positions of the moving body, like the Euler–Savary formula at infinitesimal successive positions. The theoretical system of discrete kinematic geometry, from planar, spherical, to spatial movement, would be set up. We might expect modern differential geometry, such as the differential manifold, fiber bundle, and topology (the moving frame and its differential formula in the classical differential geometry), to play a solid ground role.

References

[1] Q.X. Zhang, *Analysis and Synthesis of Spatial Mechanisms*, Vol. **1**. China Machine Press, Beijing, 1984 (in Chinese).

[2] J. Liu and X.M. Wang, *Saddle Point Programming and Geometric Error Evaluation*. Dalian University of Technology Press, Dalian, 1996 (in Chinese).

[3] K.H. Hunt, *Kinematic Geometry of Mechanism*. Oxford University Press, Oxford, 1978.

[4] C. Bagci, Geometric methods for the synthesis of spherical mechanisms for the generation of functions, paths and rigid-body positions using conformal projections. *Mechanism and Machine Theory*, **19**(1), 113–127, 1984.

[5] J.K. Chu and J.W. Sun, Synthesis of coupler curve of linkage mechanism through Fourier series method. ASME/IFToMM International Conference on Reconfigurable Mechanisms and Robots, London, 2009.

[6] J.W. Sun and J.K. Chu, Fourier method to function synthesis of RCCC mechanism. *IMechE Part C: J. Mechanical Engineering Science*, **223**(2), 503–513, 2008.

[7] R.S. Hartenberg and J. Denavit, *Kinematic Synthesis of Linkages*. McGraw-Hill, New York, 1964.

[8] B. Roth, The kinematics of motion through finitely separated positions. *ASME Journal of Applied Mechanics*, **34**(3), 591–598, 1967.

[9] B. Roth, Finite-position theory applied to mechanism synthesis. *ASME Journal of Applied Mechanics*, **34**(3), 599–605, 1967.

[10] B. Roth, On the screw axes and other special lines associated with spatial displacements of a rigid body. *ASME Journal of Engineering for Industry*, **89**(1), 102–110, 1967.

[11] B. Roth, The design of binary cranks with revolute, cylindric, and prismatic joints. *Journal of Mechanisms*, **3**(2), 61–72, 1968.

[12] L.W. Tsai and B. Roth, Design of dyads with helical, cylindrical, spherical, revolute and prismatic joints. *Mechanism and Machine Theory*, **7**(1), 85–102, 1972.

[13] P. Chen and B. Roth, A unified theory for the finitely and infinitesimally separated position problems of kinematic synthesis. *ASME Journal of Engineering for Industry*, **91**(1), 203–208, 1969.

[14] P. Chen and B. Roth, Design equations for the finitely and infinitesimally separated position synthesis of binary links and combined link chains. *ASME Journal of Engineering for Industry*, **91**(1), 209–219, 1969.

[15] P.A. James and B. Roth, A unified theory for kinematic synthesis. *ASME Journal of Mechanical Design*, **116**(1), 144–154, 1992.

[16] G.N. Sandor, Principles of a general quaternion-operator method of spatial kinematic synthesis. *ASME Journal of Applied Mechanics*, **35**(1), 40–46, 1968.

[17] C.H. Suh, Design of space mechanisms for rigid body guidance. *ASME Journal of Engineering for Industry*, **90**(3), 499–506, 1968.

[18] C.H. Suh and C.W. Radcliffe, *Kinematics and Mechanisms Design*. Wiley, Chichester, 1978.

[19] J.M. McCarthy, The synthesis of planar RR and spatial CC chains and the equation of a triangle. *ASME Journal of Mechanical Design*, **117**(B), 101–106, 1994.

[20] G.N. Sandor and A.G. Erdman, *Advanced Mechanisms Design: Analysis and Synthesis*, Vol. **2**. Prentice-Hall, Englewoods Cliffs, NJ, 1984.

[21] C. Innocenti, Polynomial solution of the spatial Burmester problem. *ASME Journal of Mechanical Design*, **117**(1), 64–68, 1995.

[22] E. Lee and C. Mavroidis, Solving the geometric design problems of spatial 3R robot manipulators using polynomial homotopy continuation. *ASME Journal of Mechanical Design*, **124**(4), 652–661, 2002.

[23] A. Perez and J.M. McCarthy, Dual quaternion synthesis of constrained robotic systems. *ASME Journal of Mechanical Design*, **126**(3), 425–435, 2003.

[24] S. Bai and J. Angeles, A robust solution of the spatial Burmester problem. *ASME Journal of Mechanisms and Robotics*, **4**(3), 031003, 2012.

[25] M. Raghavan and B. Roth, Solving polynomial systems for the kinematic analysis and synthesis of mechanisms and robot manipulators. *ASME Journal of Mechanical Design*, **117**(B), 71–79, 1995.

[26] C. Huang and Y.-J. Chang, Polynomial solution to the five-position synthesis of spatial CC dyads via dialytic elimination. Proceedings of the ASME Design Technical Conference, September 10–13, 2000, Baltimore MD. Paper No. DETC 2000/MECH-14102.

[27] C.W. Wampler, A.J. Sommese, and A.P. Morgan, Numerical continuation methods for solving polynomial systems arising in kinematics. *ASME Journal of Mechanical Design*, **112**(1), 59–68, 1990.

[28] J. Angeles, *Spatial Kinematic Chains: Analysis, Synthesis, Optimization*. Springer, Berlin, 1982.

[29] R.I. Jamalov, F.L. Litvin, and B. Roth, Analysis and design of RCCC linkages. *Mechanism and Machine Theory*, **19**(4/5), 397–407, 1984.

[30] G.K. Ananthasuresh and S.N. Kramer, Analysis and optimal synthesis of the RSCR spatial mechanism. *ASME Journal of Mechanical Design*, **116**(1), 174–181, 1994.

[31] J.M. Jiménez, G. Álvarez, J. Cardenal, and J. Cuadrado, A simple and general method for kinematic synthesis of spatial mechanisms. *Mechanisms and Machine Theory*, **32**(3), 323–341, 1997.

[32] V. Krovi, G.K. Ananthasuresh, and V. Kumar, Kinematic synthesis of spatial R–R dyads for path following with applications to coupled serial chain mechanisms. *ASME Journal of Mechanical Design*, **123**(3), 359–366, 1999.

[33] T.F. Parikian, Multi-generation of coupler curves of spatial linkages. *Mechanism and Machine Theory*, **32**(1), 103–110, 1997.

[34] R. Sancibrian, A. De-Juan, P. Garcia, A. Fernandez, and F. Viadero, Optimal synthesis of function generating spherical RSSR mechanisms. 12th IFToMM World Congress, Besancon, France, June 18–21, 2007.

[35] K.C. Gupta and S.M.K. Kazerounian, Synthesis of fully rotatable R–S–S–R linkages. *Mechanism and Machine Theory*, **18**(3), 19–205, 1983.

[36] L.V.B. Rao and K. Lakshminarayana, Optimal designs of the RSSR crank-rocker mechanism-I. General time ratio. *Mechanism and Machine Theory*, **19**(4/5), 431–441, 1984.

[37] K. Lakshminarayana and L.V.B. Rao, Optimal designs of the RSSR crank-rocker mechanism-II. *Unit time ratio and limits of capability. Mechanism and Machine Theory*, **19**(4/5), 443–448, 1984.

[38] S. Dhall, Computer-aided design of the RSSR function generating spatial mechanism using the selective precision synthesis method. *ASME Journal of Mechanisms, Transmissions, and Automation in Design*, **110**, 378–382, 1988.

[39] D. Wang and S. Wang, New approach for spatial mechanism synthesis with the C–C binary crank by adaptive saddle-fitting. *Chinese Journal of Mechanical Engineering*, **40**(12), 25–30, 2004 (in Chinese).

[40] D. Wang and S. Wang. A unified approach to kinematic synthesis of mechanism by adaptive curve fitting. *Science in China (Series E: Technological Sciences)*, **47**(1), 85–96, 2004.

Appendix A

Displacement Solutions of Spatial Linkages RCCC

A spatial RCCC linkage with an R-pair and three C-pairs, as shown in Fig. A.1 (a typical spatial four-bar linkage), can degenerate to a planar and spherical four-bar linkage and other spatial four-bar linkage without S-pair, which not only has theoretical value but also practical significance. The displacement equations of spatial linkages RCCC are set by a unified model in Ref. [1].

The relationship between the rotation angle θ_1 of the side link 1 relative to the frame link 0 and the rotation angle θ_0 of the side link 3 to the frame link 0 can be described by the following equation:

$$A_{01}\sin\theta_0 + B_{01}\cos\theta_0 + C_{01} = 0 \tag{A.1a}$$

where

$$\begin{cases} A_{01} = \sin\alpha_{23}\sin\alpha_{01}\sin\theta_1 \\ B_{01} = -\sin\alpha_{01}\cos\alpha_{30}\sin\alpha_{23}\cos\theta_1 - \sin\alpha_{23}\cos\alpha_{01}\sin\alpha_{30} \\ C_{01} = \cos\alpha_{23}\cos\alpha_{01}\cos\alpha_{30} - \cos\alpha_{23}\sin\alpha_{01}\sin\alpha_{30}\cos\theta_1 - \cos\alpha_{12} \end{cases} \tag{A.1b}$$

The rotation angle θ_2 of the coupler link 2 relative to the side link 1 has a relationship with the input angle θ_1:

$$A_{12}\sin\theta_2 + B_{12}\cos\theta_2 + C_{12} = 0 \tag{A.2a}$$

where

$$\begin{cases} A_{12} = \sin\alpha_{30}\sin\alpha_{12}\sin\theta_1 \\ B_{12} = -\sin\alpha_{12}\cos\alpha_{01}\sin\alpha_{30}\cos\theta_1 - \cos\alpha_{30}\sin\alpha_{12}\sin\alpha_{01} \\ C_{12} = -\sin\alpha_{30}\cos\alpha_{12}\sin\alpha_{01}\cos\theta_1 + \cos\alpha_{30}\cos\alpha_{12}\cos\alpha_{01} - \cos\alpha_{23} \end{cases} \tag{A.2b}$$

The rotation angle θ_3 of the coupler link 2 relative to the side link 3 has a relationship with the input angle θ_1:

$$A_{31}\sin\theta_3 + B_{31}\cos\theta_3 + C_{31} = 0 \tag{A.3a}$$

Kinematic Differential Geometry and Saddle Synthesis of Linkages, First Edition. Delun Wang and Wei Wang.

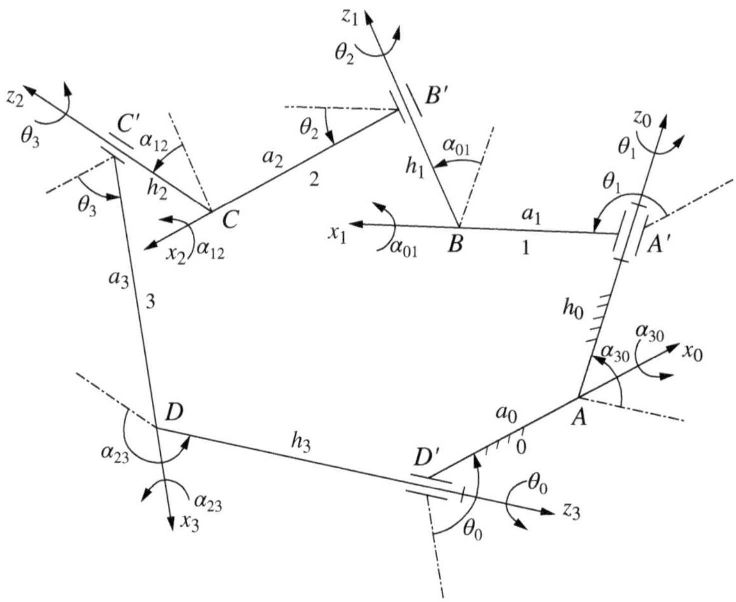

Figure A.1 Spatial RCCC linkage

where

$$\begin{cases} A_{31} = 0 \\ B_{31} = \sin\alpha_{23}\sin\alpha_{12} \\ C_{31} = -\sin\alpha_{01}\sin\alpha_{30}\cos\theta_1 + \cos\alpha_{01}\cos\alpha_{30} - \cos\alpha_{23}\cos\alpha_{12} \end{cases} \tag{A.3b}$$

From the above equations, there is a unified relationship among the dependent variables θ_0, θ_2, θ_3 and the independent variable θ_1, which can be written in the following standard form:

$$A\sin\theta + B\cos\theta + C = 0 \tag{A.4a}$$

where the coefficients A, B, and C are all functions of the input angle θ_1. The solution of the standard equation (A.4a) is

$$\theta = 2\arctan\frac{A \pm \sqrt{A^2 + B^2 - C^2}}{B - C} \tag{A.4b}$$

The rotation angles θ_0, θ_2, θ_3 can be solved by giving the coefficients A, B, and C the corresponding value. To solve the rotation angle θ_0, the coefficients A, B, and C are replaced by A_{01}, B_{01}, and C_{01}. To solve the rotation angles θ_2 and θ_3, the coefficients A, B, C are respectively replaced by A_{12}, B_{12}, C_{12} and A_{31}, B_{31}, C_{31}.

For a given input angle θ_1, the dependent variables θ_0, θ_2, θ_3 may have two solutions by equations (A.4), which have to be judged as final solutions by the following equations:

$$\begin{cases} \sin\theta_2\sin\alpha_{12} = \sin\alpha_{23}\left(\sin\theta_0\cos\theta_1 + \cos\theta_0\sin\theta_1\cos\alpha_{30}\right) + \sin\theta_1\cos\alpha_{23}\sin\alpha_{30} \\ \sin\theta_3\sin\alpha_{23} = \sin\alpha_{30}(\sin\theta_1\cos\theta_2 + \cos\theta_1\sin\theta_2\cos\alpha_{01}) + \sin\theta_2\cos\alpha_{30}\sin\alpha_{01} \end{cases} \tag{A.5}$$

The displacement h_1 on the axis z_1 of the coupler link 2 relative to the side link 1 has the following relation with the input angle θ_1:

$$h_1 = -\frac{a_0 \cos\theta_0 + h_0 \sin\alpha_{30} \sin\theta_0 + a_1(\cos\theta_3 \cos\theta_2 - \cos\alpha_{12} \sin\theta_2 \sin\theta_3) + a_2 \cos\theta_3 + a_3}{\sin\theta_3 \sin\alpha_{12}} \tag{A.6}$$

The higher derivatives of the variables θ_0, θ_2, θ_3, h_1 can be obtained by differentiating equations (A.4) and (A.6) with respect to θ_1; the derivative can be taken by the symbolic operation of Matlab. Substituting the derivatives of θ_2 and h_1 with respect to θ_1, such as $\frac{d\theta_2}{d\theta_1}$, $\frac{d^2\theta_2}{d{\theta_1}^2}$, and $\frac{dh_1}{d\theta_1}$ into the kinematic equations of the RCCC linkage in Chapter 6, we can obtain the kinematic properties of coupler 2, such as the Frenet frames of the axodes and their kinematic invariants.

The dimensions of the spatial RCCC linkage are designated as $\alpha_{01} = 30^\circ$, $\alpha_{12} = 55^\circ$, $\alpha_{23} = 45^\circ$, $\alpha_{30} = 60^\circ$, $a_0 = 5$, $a_1 = 2$, $a_2 = 4$, $a_3 = 3$, $h_0 = 0$. Based on the above displacement equations, the rotation angle θ_2 and the displacement h_1 can be solved as functions of the given input angle θ_1. Their derivatives $\theta_2' = \frac{d\theta_2}{d\theta_1}$, $\theta_2'' = \frac{d^2\theta_2}{d{\theta_1}^2}$, $\theta_2''' = \frac{d^3\theta_2}{d{\theta_1}^3}$ and $h_1' = \frac{dh_1}{d\theta_1}$, $h_1'' = \frac{d^2h_1}{d{\theta_1}^2}$, $h_1''' = \frac{d^3h_1}{d{\theta_1}^3}$ are also obtained and calculated, as shown in Figs A.2 and A.3.

A point $P(x_{Pm}, y_{Pm}, z_{Pm})$ in $\{\boldsymbol{R}_C; \boldsymbol{x}_2, \boldsymbol{y}_2, \boldsymbol{z}_2\}$ of the coupler link 2 traces a coupler curve Γ_P: $\boldsymbol{R}_P = (x_{Pf}, y_{Pf}, z_{Pf})^T$ in the fixed frame $\{\boldsymbol{R}_A; \boldsymbol{x}_0, \boldsymbol{y}_0, \boldsymbol{z}_0\}$, and the relationships between

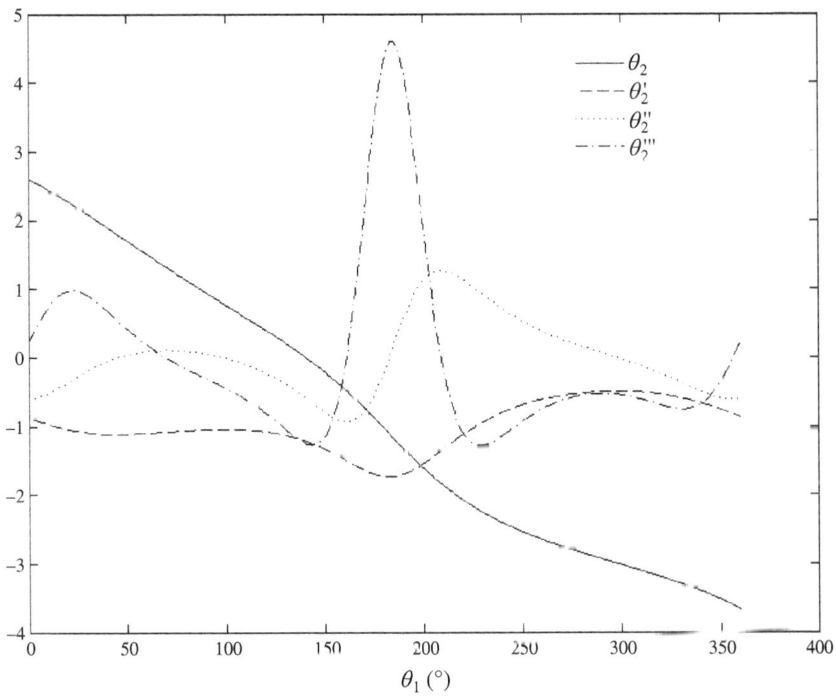

Figure A.2 The relations between the rotation angle θ_2 and the input angle θ_1

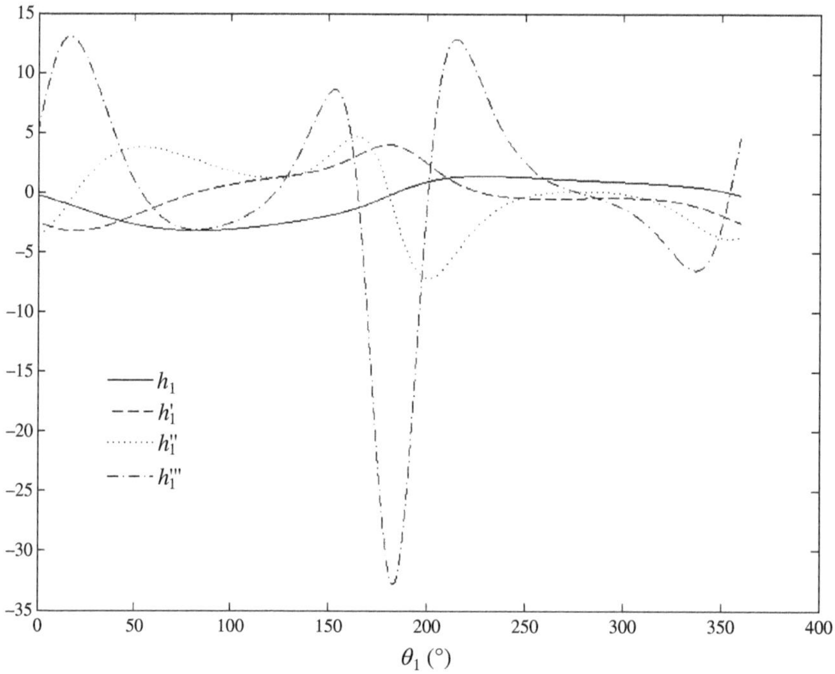

Figure A.3 The relations between the displacement h_1 and the input angle θ_1

(x_{Pf}, y_{Pf}, z_{Pf}) and (x_{Pm}, y_{Pm}, z_{Pm}) are

$$\begin{bmatrix} x_{Pf} \\ y_{Pf} \\ z_{Pf} \\ 1 \end{bmatrix} = [\mathbf{M}_2] \cdot [\mathbf{M}_1] \cdot \begin{bmatrix} x_{Pm} \\ y_{Pm} \\ z_{Pm} \\ 1 \end{bmatrix} \tag{A.7}$$

where

$$[\mathbf{M}_1] = \begin{bmatrix} 0 & \sin\alpha_{12} & \cos\alpha_{12} & h_1 \\ \cos\theta_2 & -\cos\alpha_{12}\sin\theta_2 & \sin\alpha_{12}\sin\theta_2 & a_2\cos\theta_2 \\ \sin\theta_2 & \cos\alpha_{12}\cos\theta_2 & -\sin\alpha_{12}\cos\theta_2 & a_2\sin\theta_2 \\ 0 & 0 & 0 & 1 \end{bmatrix}$$

$$[\mathbf{M}_2] = \begin{bmatrix} \sin\alpha_{01}\sin\theta_1 & \cos\theta_1 & -\cos\alpha_{01}\sin\theta_1 & a_1\cos\theta_1 \\ -\sin\alpha_{01}\cos\theta_1 & \sin\theta_1 & \cos\alpha_{01}\cos\theta_1 & a_1\sin\theta_1 \\ \cos\alpha_{01} & 0 & \sin\alpha_{01} & h_0 \\ 0 & 0 & 0 & 1 \end{bmatrix}$$

A straight line L of the coupler link 2, passing through the point P and having direction $\boldsymbol{l}_m(l_{1m}, l_{2m}, l_{3m})$ in $\{\boldsymbol{R}_C; \boldsymbol{x}_2, \boldsymbol{y}_2, \boldsymbol{z}_2\}$, traces a ruled surface Σ_l: $\{\boldsymbol{R}_L\} = \{\boldsymbol{R}_P + \mu \boldsymbol{l}_f\}$ in

Table A.1 The discrete parameters of the linkage RCCC

No.	$\theta_1^{(i)}$	$\theta_2^{(i)}$	$h_1^{(i)}$	No.	$\theta_1^{(i)}$	$\theta_2^{(i)}$	$h_1^{(i)}$
1	0	2.6124	−0.2098	37	3.1416	−1.0314	−0.3008
2	0.0873	2.5346	−0.4413	38	3.2289	−1.1828	0.0465
3	0.1745	2.4524	−0.6961	39	3.3161	−1.3326	0.3761
4	0.2618	2.3664	−0.9673	40	3.4034	−1.4780	0.6690
5	0.3491	2.2770	−1.2469	41	3.4907	−1.6165	0.9125
6	0.4363	2.1848	−1.5262	42	3.5779	−1.7465	1.1018
7	0.5236	2.0906	−1.7971	43	3.6652	−1.8672	1.2390
8	0.6109	1.9948	−2.0528	44	3.7525	−1.9783	1.3303
9	0.6981	1.8982	−2.2875	45	3.8397	−2.0802	1.3839
10	0.7854	1.8012	−2.4975	46	3.9270	−2.1734	1.4080
11	0.8727	1.7042	−2.6800	47	4.0143	−2.2588	1.4101
12	0.9599	1.6076	−2.8340	48	4.1015	−2.3372	1.3962
13	1.0472	1.5115	−2.9592	49	4.1888	−2.4094	1.3712
14	1.1345	1.4161	−3.0561	50	4.2761	−2.4762	1.3386
15	1.2217	1.3216	−3.1261	51	4.3633	−2.5383	1.3013
16	1.3090	1.2279	−3.1706	52	4.4506	−2.5965	1.2612
17	1.3963	1.1350	−3.1916	53	4.5739	−2.6511	1.2197
18	1.4835	1.0429	−3.1912	54	4.6251	−2.7029	1.1779
19	1.5708	0.9514	−3.1714	55	4.7124	−2.7522	1.1363
20	1.6581	0.8603	−3.1341	56	4.7997	−2.7996	1.0955
21	1.7453	0.7695	−3.0815	57	4.8869	−2.8453	1.0555
22	1.8326	0.6785	−3.0151	58	4.9742	−2.8898	1.0164
23	1.9199	0.5872	−2.9367	59	5.0615	−2.9335	0.9781
24	2.0071	0.4952	−2.8474	60	5.1487	−2.9766	0.9403
25	2.0944	0.4019	−2.7483	61	5.2360	−3.0196	0.9024
26	2.1817	0.3070	−2.6399	62	5.3233	−3.0628	0.8638
27	2.2689	0.2097	−2.5223	63	4.4105	−3.1065	0.8235
28	2.3562	0.1096	2.3947	64	5.4978	3.1320	0.7801
29	2.4435	0.0058	−2.2557	65	5.5851	3.0859	0.7318
30	2.5307	−0.1024	−2.1025	66	5.6723	3.0380	0.6763
31	2.6180	−0.2160	−1.9314	67	5.7596	2.9878	0.6105
32	2.7053	−0.3356	−1.7374	68	5.8469	2.9349	0.5309
33	2.7925	−0.4619	−1.5149	69	5.9341	2.8786	0.4334
34	2.8798	−0.5953	−1.2590	70	6.0214	2.8186	0.3138
35	2.9671	−0.7354	0.9678	71	6.1087	2.7543	0.1683
36	3.0543	−0.8814	−0.6448	72	6.1959	2.6857	−0.0060

$\{\boldsymbol{R}_A; x_0, y_0, z_0\}$. The relationships between $\boldsymbol{l}_m = (l_{1m}, l_{2m}, l_{3m})^T$ and $\boldsymbol{l}_f = (l_{1f}, l_{2f}, l_{3f})^T$ are

$$\begin{bmatrix} l_{1f} \\ l_{2f} \\ l_{3f} \\ 0 \end{bmatrix} = [\mathbf{M}_2] \cdot [\mathbf{M}_1] \cdot \begin{bmatrix} l_{1m} \\ l_{2m} \\ l_{3m} \\ 0 \end{bmatrix} \tag{A.8}$$

If the input angle $\theta_1 \in [0,2\pi]$ is discretized as $\theta_1^{(i)}$, $i = 1, \ldots, 72$ by an interval of 5°, the discrete values of the rotation angle $\theta_2^{(i)}$ and the displacement $h_1^{(i)}$ can be solved by the displacement equations of the RCCC linkage, and are listed in Table A.1.

By substituting $\theta_1^{(i)}, \theta_2^{(i)}, h_1^{(i)}, i = 1, \ldots, 72$ in Table A.1 into equations (A.7) and (A.8), the discrete coupler curve Γ_P, or the discrete point set $\{\boldsymbol{R}_P^{(i)}\}$, $i = 1, \ldots, 72$, and the discrete line trajectory Σ_l, or discrete line set $\{\boldsymbol{R}_L^{(i)}\} = \{\boldsymbol{R}_P^{(i)} + \mu \boldsymbol{l}^{(i)}\}$, $i = 1, \ldots, 72$, are obtained in the base frame $\{\boldsymbol{R}_A; \boldsymbol{x}_0, \boldsymbol{y}_0, \boldsymbol{z}_0\}$.

Reference

[1] Q.X. Zhang, *Analysis and Synthesis of Spatial Mechanisms*, Vol. **1**. China Machine Press, Beijing, 1984 (in Chinese).

Appendix B

Displacement Solutions of the Spatial RRSS Linkage

For the spatial RRSS linkage, the displacements can also be solved according to the method introduced in Ref. [1]. The displacement equation is

$$A \sin \theta_2 + B \cos \theta_2 + C = 0 \tag{B.1a}$$

where

$$\begin{cases} A = 2a_0 a_2 \cos \alpha_{12} \sin \theta_1 - 2h_0 a_2 \sin \alpha_{12} \\ B = 2a_1 a_2 - 2a_0 a_2 \cos \theta_1 \\ C = a_1^2 + h_2^2 + a_2^2 + a_0^2 + h_0^2 - a_3^2 - 2a_0 a_1 \cos \theta_1 - 2a_0 h_2 \sin \alpha_{12} \sin \theta_1 - 2h_0 h_2 \cos \alpha_{12} \end{cases} \tag{B.1b}$$

For the given input rotation angle θ_1, the motion parameter θ_2 of the coupler link 2 can be solved through the above equation:

$$\theta_2 = 2 \arctan \frac{A \pm \sqrt{A^2 + B^2 - C^2}}{B - C} \tag{B.2}$$

Similarly, a point $P(x_{Pm}, y_{Pm}, z_{Pm})$ in $\{\boldsymbol{R}_{B'}, \boldsymbol{x}_2, \boldsymbol{y}_2, \boldsymbol{z}_2\}$ of the coupler link 2 traces a coupler curve Γ_P: $\boldsymbol{R}_P = (x_{Pf}, y_{Pf}, z_{Pf})$ in the fixed frame $\{\boldsymbol{R}_{D'}, \boldsymbol{x}_0, \boldsymbol{y}_0, \boldsymbol{z}_0\}$, and the relationships between (x_{Pf}, y_{Pf}, z_{Pf}) and (x_{Pm}, y_{Pm}, z_{Pm}) are

$$\begin{bmatrix} x_{Pf} \\ y_{Pf} \\ z_{Pf} \\ 1 \end{bmatrix} = [\mathbf{M}_2] \cdot [\mathbf{M}_1] \cdot \begin{bmatrix} x_{Pm} \\ y_{Pm} \\ z_{Pm} \\ 1 \end{bmatrix} \tag{B.3}$$

Kinematic Differential Geometry and Saddle Synthesis of Linkages, First Edition. Delun Wang and Wei Wang.

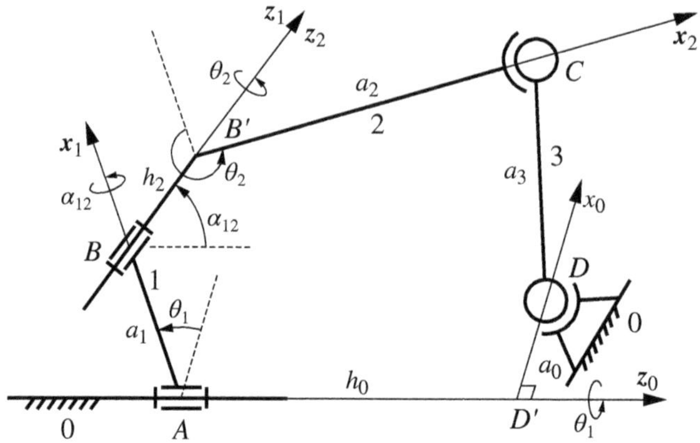

Figure B.1 Spatial RRSS linkage

Table B.1 The discrete motion parameters of the RRSS linkage

No.	$\theta_1^{(i)}$	$\theta_2^{(i)}$	No.	$\theta_1^{(i)}$	$\theta_2^{(i)}$	No.	$\theta_1^{(i)}$	$\theta_2^{(i)}$
1	0	2.1459	25	2.0944	2.4464	49	4.1888	2.3745
2	0.0873	2.1522	26	2.1817	2.4561	50	4.2761	2.3586
3	0.1745	2.1598	27	2.2689	2.4648	51	4.3633	2.3424
4	0.2618	2.1687	28	2.3562	2.4727	52	4.4506	2.3259
5	0.3491	2.1788	29	2.4435	2.4797	53	4.5739	2.3092
6	0.4363	2.1899	30	2.5307	2.4856	54	4.6251	2.2926
7	0.5236	2.2019	31	2.6180	2.4906	55	4.7124	2.2762
8	0.6109	2.2147	32	2.7053	2.4944	56	4.7997	2.2601
9	0.6981	2.2282	33	2.7925	2.4971	57	4.8869	2.2445
10	0.7854	2.2423	34	2.8798	2.4986	58	4.9742	2.2294
11	0.8727	2.2567	35	2.9671	2.4988	59	5.0615	2.2151
12	0.9599	2.2715	36	3.0543	2.4979	60	5.1487	2.2016
13	1.0472	2.2864	37	3.1416	2.4956	61	5.2360	2.1891
14	1.1345	2.3014	38	3.2289	2.4920	62	5.3233	2.1776
15	1.2217	2.3164	39	3.3161	2.4871	63	4.4105	2.1674
16	1.3090	2.3312	40	3.4034	2.4810	64	5.4978	2.1585
17	1.3963	2.3458	41	3.4907	2.4735	65	5.5851	2.1509
18	1.4835	2.3601	42	3.5779	2.4648	66	5.6723	2.1449
19	1.5708	2.3741	43	3.6652	2.4548	67	5.7596	2.1403
20	1.6581	2.3876	44	3.7525	2.4438	68	5.8469	2.1373
21	1.7453	2.4006	45	3.8397	2.4317	69	5.9341	2.1359
22	1.8326	2.4131	46	3.9270	2.4186	70	6.0214	2.1360
23	1.9199	2.4249	47	4.0143	2.4046	71	6.1087	2.1378
24	2.0071	2.4360	48	4.1015	2.3899	72	6.1959	2.1411

where

$$[\mathbf{M}_1] = \begin{bmatrix} \cos\theta_2 & -\sin\theta_2 & 0 & 0 \\ \sin\theta_2 & \cos\theta_2 & 0 & 0 \\ 0 & 0 & 1 & h_2 \\ 0 & 0 & 0 & 1 \end{bmatrix}$$

$$[\mathbf{M}_2] = \begin{bmatrix} \cos\theta_1 & -\cos\alpha_{12}\sin\theta_1 & \sin\alpha_{12}\sin\theta_1 & a_1\cos\theta_1 \\ \sin\theta_1 & \cos\alpha_{12}\cos\theta_1 & -\sin\alpha_{12}\cos\theta_1 & a_1\sin\theta_1 \\ 0 & \sin\alpha_{12} & \cos\alpha_{12} & -h_0 \\ 0 & 0 & 0 & 1 \end{bmatrix}$$

The dimensions of the spatial RCCC linkage are designated as $a_0 = 0.2$, $h_0 = 0.3$, $a_1 = 1.5$, $h_2 = 0.4$, $a_2 = 1$, $a_3 = 1$, $\alpha_{12} = 45°$. We have a group of discrete angles $\theta_1^{(i)}$, $i = 1, \ldots, 72$ by taking 5° as an interval of θ_1 in $[0,2\pi]$ and then solve the rotation angles $\theta_2^{(i)}$ of the coupler link 2 relative to the binary link 1 based on equation (B.2) of the RRSS linkage. The discrete values of the angles $\theta_1^{(i)}$ and $\theta_2^{(i)}$ are listed in Table B.1.

By substituting $\theta_1^{(i)}$, $\theta_2^{(i)}$, $i = 1, \ldots, 72$ in Table B.1 into equation (B.3), the discrete coupler curve Γ_P, or the discrete point set $\{\boldsymbol{R}_P^{(i)}\}$, $i = 1, \ldots, 72$, is obtained in the base frame $\{\boldsymbol{R}_{D'}, x_0, y_0, z_0\}$.

Reference

[1] Q.X. Zhang, *Analysis and Synthesis of Spatial Mechanisms*, Vol. **1**. China Machine Press, Beijing, 1984 (in Chinese).

Index

Kinematic Differential Geometry and Saddle Synthesis of Linkages, First Edition. Delun Wang and Wei Wang.

www.ingramcontent.com/pod-product-compliance
Ingram Content Group UK Ltd.
Pitfield, Milton Keynes, MK11 3LW, UK
UKHW050418051026
15470UKWH00022B/1596

9 781118 255049